Photonics

Principles and Practices

OPTICAL SCIENCE AND ENGINEERING

Founding Editor
Brian J. Thompson
University of Rochester
Rochester, New York

Photonics
Principles and Practices

Abdul Al-Azzawi

CRC Press
Taylor & Francis Group
Boca Raton London New York

CRC Press is an imprint of the
Taylor & Francis Group, an informa business

CRC Press
Taylor & Francis Group
6000 Broken Sound Parkway NW, Suite 300
Boca Raton, FL 33487-2742

© 2007 by Taylor & Francis Group, LLC
CRC Press is an imprint of Taylor & Francis Group, an Informa business

No claim to original U.S. Government works
Printed in the United States of America on acid-free paper
10 9 8 7 6 5 4 3 2 1

International Standard Book Number-10: 0-8493-8290-4 (Hardcover)
International Standard Book Number-13: 978-0-8493-8290-1 (Hardcover)

Library of Congress Cataloging-in-Publication Data

Al-Azzawi, Abdul.
    Photonics : principles and practices / Abdul Al-Azzawi.
        p. cm. -- (Optical science and engineering series ; 123)
    Includes bibliographical references and index.
    ISBN 0-8493-8290-4
    1. Photonics. 2. Optoelectronic devices. I. Title.

TA1520.A43 2006
621.36--dc22   2006044675

Visit the Taylor & Francis Web site at
http://www.taylorandfrancis.com

and the CRC Press Web site at
http://www.crcpress.com

Preface

The world is bathed in light. Light is one of the most familiar and essential things to the world's existence. For many thousands of years, the Sun was its only source of light. Eventually, the ability to create fire, and its by-product, light, led to a profound change in the way humans managed their time. Today, there are many options for creating light. Researchers' understanding of light has spawned many applications of light. Light can be used in fibre communications, and early applications included ship-to-ship communications using Morse code. Infrared remote controls for televisions demonstrated free-space optical communications, using many of the same principles. Optical fibre has revolutionized the way the world interacts. Fibre optics, as a waveguide technology, now provides an essential back-bone for much of the world's high-speed communication networks. Photo-dynamic therapies use light to treat cancers. Light is used to treat those with seasonal disorders. Lasers are now used in medical applications such as re-shaping corneas, cauterizing blood vessels, and removing tattoos. Lasers are also used in industrial applications such as cutting metal, welding, and sensing. New imaging technology permits the creation of flat-panel displays, night vision devices, and autonomous product inspection systems. With so many applications of light, the need for photonics technology and innovation will most certainly grow in the future as new applications emerge to light.

A unique approach is taken in this book to present light and its applications in photonics technology. This book covers the basic theoretical principles and industrial applications of photonics technology, and they are suitable for students, professionals, and professors. Each chapter is presented in two parts: theoretical and practical. The theoretical part has adequate material to cover the whole aspect of the subject. In the practical part, students will apply the learned theoretical concepts in simple and advanced experimental works. Students will learn and gain practical hands-on experience in the photonics subjects. This will assist the students in applying theoretical knowledge to real-world applications. The step-by-step approach and technical illustrations in this book will guide students through each experiment. The experimental work has more than one case in most of the chapters, and sometimes it has sub-cases.

This book is written using simplistic language, and it gives adequate information and instruction to enable students to achieve maximum comprehension. An effort has been made to use the international system of units (SI) throughout the book. The organization of the chapters is designed to provide a solid foundation for today's photonics' students and to upgrade their knowledge. Universal tools, devices, and equipment that are used throughout the experiments are available in any photonics, physics, and material labs. This book provides theoretical and practical aids and is an effective teaching tool, helpful to both professors and students. Simple and advanced subjects are presented by an expert author, and some new photonics subjects appear for the first time in this book.

Care has been taken to label parts clearly and to use colours in diagrams wherever it will aid understanding. An insert containing all colour figures is also included with this book. Some figures are drawn in three-dimensions, where applicable, for easy understanding of the concepts. Colour pictures are used to clearly show parts in a device, system, and experimental set-up.

The book is structured in seven sections that are comprised of thirty-six chapters. Section One through Section Six are divided into many chapters. The sections are listed below.

- **Section One**, *Light* Chapter 1 through Chapter 7 covers light, light and shadow, thermal radiation, light production, light intensity, light and colour, and the laws of light.
- **Section Two**, *Optics* Chapter 8 through Chapter 16 covers plane mirrors, spherical mirrors, lenses, prisms, beamsplitters, light's passing through optical components, optical instruments for viewing applications, polarization of light, and optical materials.

- **Section Three**, *Waves and Diffraction* Chapter 17 through Chapter 21 covers waves, diffraction and interference, diffraction grating, interferometers, and spectrometers.
- **Section Four**, *Optical Fibres* (cables, components, devices, instruments, and systems) Chapter 22 through Chapter 34 covers fibre optic cables, advanced fibre optic cables, light attenuation in optical components, fibre optic cable types and installations, fibre optic connectors, passive fibre optic devices, wavelength division multiplexer, optical amplifiers, optical receivers, lasers, opto-mechanical switches, optical fibre communications, and fibre optic lighting.
- **Section Five**, *Testing* Chapter 35 covers fibre optics testing.
- **Section Six**, *Safety* Chapter 36 covers laboratory safety.
- **Section Seven**, *Miscellaneous* includes a complete glossary of terms, appendices section, references, and index section.

The book includes 858 figures, 118 tables, and 115 experimental cases. The book was developed with generous input from members of the photonics industry, research scientists, and members from academia.

Abdul Al-Azzawi
Algonquin College
Canada

Acknowledgments

This book would not have been possible without the enthusiasm and teamwork of my colleagues' and family's support. In particular, the author would like to thank Mietek Slocinski for his support, time, and energy in working long hours to set-up the labs, take pictures, and fruitful discussion during the years to complete the book.

The author would like to thank Steve Finnegan and Kathy Deugo, and Nicole McGahey for their support in solving various difficulties. Thanks to my daughter, Abeer, and son, Abaida, for their help in reviewing the chapters and making drawings and figures.

The author also extends his thanks to Eng. Monica Havelock for her contribution in working long hours in editing and reviewing the Scientific, materials proofreading, and support.

The author wishes to express his gratitude to colleagues Prof. Devon Galway and Prof. Rao Kollipara for their comments and feedback in reviewing some materials in this book.

The author wishes to thank Gergely Horvath for the hard work and professional manner in editing and proofreading a major portion of this book on a tight schedule. The author likes to thank Madeleine Camm, Andrew Lynch, and Nicolas Lea for reviewing a few chapters in this book.

The author would like to extend his thank to Dr. Robert Weeks, Dr. Charley Bamber, Dr. Imad Hasan, Dr. Wahab Almuhtadi, Dr. Mostefa Mohamed, Dr. Govindanunny Thekkadath, Eng. Nazar Rida, Eng. Mohammed Mohammed, and Peter Casey for reviewing and proposing the materials in this book.

The following list shows the names of the professors who participated in writing a part or all of a chapter:

Chapter 16 *Optical Materials* is written by Dr. Charley Bamber
Chapter 30 *Optical Receivers* is written by Dr. Imad Hasan
Chapter 31 *Lasers* is written by Dr. Abdul Al-Azzawi and Dr. Robert Weeks
Chapter 33 *Optical Fibre Communications* is written by Dr. Wahab Almuhtadi
Chapter 35 *Fibre Optics Testing* is written by Eng. Valerie Dube

Author

Abdul Al-Azzawi, PhD, graduated from the University of Strathclyde in Glasgow, Scotland, UK. He has worked in the photonics manufacturing industry, research (NRC/Canmet), and teaching at Algonquin College, Ontario, Canada. While employed at NRC, he participated in studying energy saving in a residential building and developing the green building assessment programme. As a photonics engineer, he designed new production lines, modified products, developed manufacturing process, and designed new jigs.

At Algonquin College, he has taught mechanical and photonics courses in the mechanical and photonics engineering programmes. He was a member of the founding team of the Photonics Engineering Programmes. He has published three books and many papers, and he has participated in many workshops and conferences around the world. He is the author of the book, *Fibre Optics— Principles and Practices*.

He is the coordinator of the photonics engineering programme at Algonquin College, Ottawa, Ontario, Canada. His special area of interest is optic and optical fibre devices, fibre optic lighting, and fibre optic sensors. He is a member of the professional photonics societies in Canada. He is the recipient of the NISOD Excellence award from the University of Texas at Austin in 2005.

Table of Contents

Chapter 7

Chapter 9

Chapter 14

Chapter 15

Chapter 16

Chapter 23

Chapter 31

Section I

Light

1 The Nature of Light

1.1 INTRODUCTION

Throughout history, humankind has been fascinated with the properties and behaviour of light. Light from the sun served as a catalyst in the formation of life on Earth. Solar and lunar light provided humanity with the celestial timepieces required to measure time.

We live in a world bathed in light. Light is one of the most familiar things in our lives. We see things with eyes that sense the intensity (brightness) and wavelength (colour) of light. We experience light in a variety of other ways as well. For example, we sense radiant heat when our skin is near a warm object. This is due to our skin's reaction to infrared radiation.

We learn almost all of what we know about the world around us from the interaction of materials with electromagnetic waves. Often, the word light is used a little more broadly to include electromagnetic waves, such as in the ultraviolet and infrared waves that are just outside the visible range.

Much of what we know about light has been discovered during the past five centuries. Initially, light was understood to be a particle. Light is now widely understood to be one part of a much larger electromagnetic spectrum. Photons, the smallest resolvable quanta of light, were initially described using particle theory; however, a wave model has since been widely adopted. In the context of the wave model, photons are energy packets moving through space and time.

1.2 THE EVOLUTION OF LIGHT THEORY

In the 17th century, light was considered to be stream of particles that were emitted by a light source. These particles stimulate the sense of sight when entering the eye. The English physicist and mathematician Isaac Newton (1642–1727) was the inventor of the particle theory of light. Newton regarded rays of light as streams of very small particles emitted from a source of light and travelling in straight lines. Newton was able to provide a simple explanation for some known experimental facts concerning the nature of light, specifically, the laws of reflection and refraction.

Most scientists accepted Newton's particle theory of light. However, during Newton's lifetime, another theory was proposed by the Dutch physicist and astronomer Christian Huygens (1629–1695). In 1678, Huygens presented his theory, in which light might be some sort of wave motion. His experiment demonstrated that when the two beams of light intersected, they emerged unchanged, just as in the case of two water or sound waves. Huygens was able to adopt a wave theory of light to derive the laws of reflection and refraction, and to explain double refraction in calcite. This wave theory did not receive immediate acceptance from the scientific community for several reasons. The only waves known at that time were sound and water. It was known that these waves travelled through some sort of medium. On the other hand, light could travel to us from the sun through the vacuum of space. It was agreed that if light was some form of wave motion, the waves would be able to bend around obstacles and corners. This bending is easily observed with both water and sound waves. In this case, it would be easy to see the light around corners. It is now known that light does actually bend around the edges of objects; this phenomenon is known as the diffraction of light, which will be discussed later in this book.

In 1660, experimental evidence for the diffraction of light was discovered by Francesco Grimaldi (1618–1663). Most scientists still rejected the wave theory and continued to adhere to Newton's particle theory for more than a century.

In 1801, the first clear demonstration of the wave theory of light was provided by the English physician, Thomas Young (1773–1829). He performed a significant experiment, which showed that light exhibits interference behaviour. There are two types of interference: constructive and destructive. When two waves are moving in the same direction, the vertical displacement (amplitude) of the combined waveform is greater than that of either wave; this situation is referred to as constructive interference. Conversely, if one wave has a negative displacement, the two waves work to cancel each other when they overlap, and the amplitude of the combined waveform is smaller than that of either wave. This is referred to as destructive interference. Light interference behaviour will be explained later in this book. Most scientists accepted the wave theory of light and more theoretical and experimental work was conducted to further explore it.

In 1821, the French physicist Augustin Fresnel (1788–1827) published the results and analysis of a number of detailed experiments, which dealt with interference of polarized light and diffraction phenomena. He obtained circularly polarized light by means of a special glass prism now known as a Fresnel rhomb. For each of the two components of the polarized light, Fresnel developed the Fresnel Equations, which give the amplitude of light reflected and transmitted at a plane interface separating two optical media.

In 1850, Jean Foucault (1791–1868) provided further evidence of the inadequacy of the particle theory by showing that the speed of light in liquids is less than that in air. According to the particle model of light, the speed of light would be higher in a glass and liquid than in air. Further experimental and theoretical developments during the 19th century led to the general acceptance of the wave theory of light.

In 1873, the most important development concerning the theory of light was the work of a Scottish physicist, James C. Maxwell (1831–1879). Maxwell asserted that light was a form of high-frequency electromagnetic wave. Working in the field of electricity and magnetism, Maxwell created known principles in his set of four Maxwell Equations. These equations predict the speed of an electromagnetic wave in the ether; this turned out to be the true measured speed of light. His theory predicted that these waves should have a speed of about 3×10^8 m/s. Within experimental error, his predicted value is nearly equal to the speed of light measured by sophisticated instruments today. From then on, light was viewed as a particular region of the electromagnetic spectrum of radiation.

In 1887, Heinrich Hertz (1857–1894), a German physicist and pioneering investigator of electromagnetic waves, provided experimental confirmation of Maxwell's theory by producing and detecting electromagnetic waves. Hertz also defined the frequency. Furthermore, Hertz and other scientists and investigators showed that these waves exhibited reflection, refraction, and all the other characteristic properties of waves.

Although the classical theory of electricity and magnetism was able to explain most known properties of light, some subsequent experiments could not be explained by assuming that light is a wave. The most striking discovery of the experiments is the photoelectric effect, which was discovered by Hertz. The photoelectric effect is the ejection of electrons from a metal when its surface is exposed to light. As one example of the difficulties that arose, experiments showed that the kinetic energy of an ejected electron is independent of the light intensity. This was in contradiction to the wave theory, which held that a more intense beam of light should add more energy to the electron. In 1905, an experiment demonstrating of this phenomenon was proposed by Albert Einstein (1879–1955), a German-Swiss physicist. In 1900, Einstein's theory used the concept of the quantum theory developed by Max Planck (1858–1947), a German theoretical physicist. The quantization model assumes that the energy of a light wave is present in bundles of energy called photons. Therefore, the energy is said to be quantized. According to Einstein's theory, the energy of a photon is proportional to the frequency of the electromagnetic wave. The energy of a photon can be defined by Equation (1.1):

$$E = n(hf) \quad n = 0,1,2,3,\ldots \tag{1.1}$$

where

n is a positive integer number
f is the frequency of light (Hz)
h is a constant known as Planck's constant and has a value of 6.6261×10^{-34} (J s).

This idea suggests that light can behave as discrete quanta or "particles" of energy, rather than as waves. Equation (1.1) is thus the mathematical "connection" between the wave nature of light (a wave of frequency f) and the particle nature (photons each with an energy E). Given light of a certain frequency (or wavelength), this equation can be used to calculate the amount of energy in each photon, or vice versa. It is important to note that this theory retains some features of both the wave theory and the particle theory of light.

Einstein used the quantum concept to explain the photoelectric effect, for which the classical physical description was inadequate. Certain metallic materials are photosensitive. That is, when light strikes their surface, electrons can be emitted. The radiant energy supplies the work necessary to free the electrons from the material's surface. The electron interacts with one photon of light as if the electron had been struck by a particle. Yet the photon has wave-like characteristics as implied by the fact that light exhibits interference phenomena of light.

In 1913, the Danish physicist Niels Bohr (1885–1962) incorporated the quantum behaviour of radiation in his explanation of the emission and absorption processes of the hydrogen atom; this provided a physical basis for understanding the hydrogen spectrum. In 1922, the photon model of light came again to the rescue for the American scientist Arthur Compton (1892–1962), who explained the scattering of x-rays from electrons as particle collisions between photons and electrons in which both energy and momentum were conserved.

All such victories for the photon or particle model of light indicated that light could be treated as a particular kind of matter, possessing both energy and momentum. It was the French scientist Luis de Broglie (1892–1987) who saw the other side of the picture. In 1924, Broglie published his theory that subatomic particles exhibit wave properties. He suggested that a particle with momentum M had an associated wavelength λ of

$$M = \frac{h}{\lambda} \tag{1.2}$$

where h is Planck's constant.

Broglie suggested that because electromagnetic waves show particle characteristics, particles should, in some cases, also exhibit wave properties. This prediction was verified experimentally within a few years by the American physicists Clinton Davisson (1881–1958) and Lester Germer (1896–1971) and the British physicist George Thomson (1892–1975). They showed that a beam of electrons scattered by a crystal produces a diffraction pattern characteristic of a wave. The wave concept of a particle led the Austrian physicist Erwin Schrodinger (1887–1961) to develop a so-called wave equation to describe the wave properties of a particle and, more specifically, the wave behaviour of the electron in the hydrogen atom.

Thus the wave-particle duality came full circle. Light behaved like waves in its propagation and in the phenomena of interference and diffraction. Light could, however, also behave as particles in its interaction with matter, as in the photoelectric effect. On the other hand, electrons usually behaved like particles, as observed in the point-like scintillations of a phosphor exposed to a beam of electrons. In other situations, electrons were found to behave like waves, as in the diffraction produced by an electron microscope. Photons and electrons that behaved both as particles and as waves seemed at first an impossible contradiction, since particles and waves are very different entities indeed. Gradually it became clear, to a large extent through the reflections of Niels Bohr and especially in his principle of complementarily, that photons and electrons were neither waves nor particles, but something more complex than either.

In attempting to explain physical phenomena, it is natural use physical models, such as waves and particles. However, it turns out that the behaviour of a photon or an electron is not fully explained by either model. In certain situations, wavelike behaviour may predominate; in other situations, particle-like behaviours may stand out. It is noticeable that there is no singular physical model that adequately handles all cases.

Quantum mechanics, or wave mechanics, as it is often called, deals with all particles, which are localized in space, and so describes both light and matter. Combined with relativity equations, momentum M, wavelength λ, and speed v for both material particles and photons are given by the same general equations:

$$M = \frac{\sqrt{E^2 - m^2 c^4}}{c} \tag{1.3}$$

$$\lambda = \frac{h}{M} = \frac{hc}{\sqrt{E^2 - m^2 c^4}} \tag{1.4}$$

$$v = \frac{Mc^2}{E} = c\sqrt{1 - \frac{m^2 c^4}{E^2}} \tag{1.5}$$

where

m is the rest-mass,
E is the total energy,
mc^2 is the sum of rest-mass energy
v is the speed of light in an optical medium, and
kinetic energy provides the work done to accelerate the particle from rest to its measured speed.

The relativistic mass is given by γm, where γ is the ratio $1/\sqrt{1-(v/c)^2}$. The proper expression for kinetic energy is no longer simply $(1/2)mv^2$, but $mv^2(\gamma-1)$. The relativistic expression for kinetic energy approaches $(1/2)mv^2$ for $v \ll c$.

A crucial difference between particles such as electrons and neutrons and particles like photons is that the photons have zero rest mass. Equation (1.3) through Equation (1.5) then take the simpler form for photons:

$$M = \frac{E}{c} \tag{1.6}$$

$$\lambda = \frac{h}{M} = \frac{hc}{E} \tag{1.7}$$

$$v = \frac{Mc^2}{E} = c \tag{1.8}$$

Thus, while nonzero rest-mass particles like electrons have a limiting speed of c, Equation (1.8) shows that zero rest-mass particles like photons must travel with the constant speed c in a vacuum. The energy of a photon is not a function of its speed but of its frequency, as expressed in Equation (1.1) or in Equation (1.6) and Equation (1.7) taken together. Notice that since a photon has a zero rest mass, there is no distinction between its total energy and its kinetic energy.

In summary, from the above view of these developments, one must regard light as having a dual nature. That is, in some cases light acts like a wave and in others it acts like a particle.

1.3 MEASUREMENTS OF THE SPEED OF LIGHT

Light travels at a high speed of about 3.00×10^8 m/s. Early experimental attempts to measure its speed were unsuccessful. It was believed that light travelled with infinite speed. Many attempts were made to obtain an exact value of the speed of light. The following shows the experimental attempts and theoretical methods that were tried in the last few centuries.

1.3.1 GALILEO'S ATTEMPTS

The first scientific attempt to measure the speed of light was made by the Italian physicist Galileo Galilei (1564–1642), but without success. He used two shuttered lanterns for his experiment. Galileo stationed himself on one hilltop with one lantern and an assistant on another hilltop with a similar lantern. The distance between the two hilltops was measured. Galileo would first open his lantern for an instant, sending a short flash of light to the assistant. As soon as the assistant saw this light he opened his own lantern, sending a flash back to Galileo, who measured the total time elapsed. After numerous repetitions of this experiment at greater distances between observers, Galileo came to the conclusion that they could not open their lanterns fast enough and that light probably travels with an infinite speed.

1.3.2 ROEMER'S METHOD

In 1675, the first successful estimate of the speed of light was made by the Danish astronomer Ole Roemer (1644–1710). His method involved astronomical observations of one of the moons of Jupiter, called Io. At that time, four of Jupiter's 14 moons had been discovered, and the periods of their orbits were known. Io, the innermost moon, has a period of about 42.5 h. This was measured

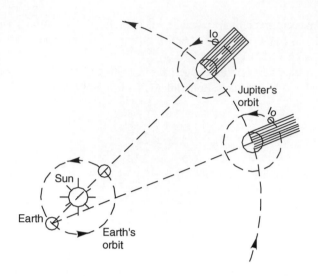

FIGURE 1.1 Schematic diagram of Roemer's method.

by observing the eclipse of Io as it passed behind Jupiter, as shown in Figure 1.1. The period of Jupiter is about 12 years, so as the Earth moves through 180° around the sun, Jupiter revolves through only 15°.

Using the orbital motion of Io as a clock, one would expect a constant period in its orbit over long time intervals. However, Roemer observed a systematic variation in Io's period during each Earth orbit. He found that the periods were larger than average when the Earth receded from Jupiter and smaller than average when approaching Jupiter. However, when Roemer checked to see if the second eclipse did occur at the predicted time, he found that if the Earth was receding from Jupiter, the eclipse was late. In fact, if the interval between observations was three months, the delay was approximately 600 s. Roemer attributed this variation in period to the fact that the distance between the Earth and Jupiter was changing between the observations. In three months (one quarter of the period of the Earth orbit), the light from Jupiter has to travel an additional distance equal to the radius of the Earth's orbit.

Roemer noted that when the Earth is closest to Jupiter, Io's reappearance was 11 min early, and when further from Jupiter, it was 11 min late. Roemer concluded that this total discrepancy of about 22 min is equal to the time required for the light from Jupiter's moon to cross the diameter of the Earth's orbit, as shown in Figure 1.1. Such measurements required the development of a reasonably sophisticated timekeeping device, so that a discrepancy of 22 min could be observed over a 6-month interval. The diameter of the Earth's orbit around the sun (in Roemer's day) was thought to be 2.9010^{11} m. This gives a speed of light c:

$$c = \frac{d}{t} \approx \frac{2.90 \times 10^{11}\,\text{m}}{22\,\text{min} \times \frac{60\,\text{s}}{\text{min}}} \approx 2.20 \times 10^8\,\text{m/s} \qquad (1.9)$$

where

  d is the distance (m),
  t is time (s).

In 1676, Roemer announced a value for the speed of light of 2.25×10^8 m/s. This experiment is important historically because it demonstrated that the speed of light does have a finite numerical value.

1.3.3 Fizeau's Method

In 1849, the first successful laboratory measurement of the speed of light was performed by the French scientist Armand Fizeau (1819–1896). The basic elements of his experimental arrangement are shown in Figure 1.2. The basic idea is to measure the total time it takes light to travel from an intense source to a distant mirror and back. To measure the transit time, Fizeau used a rotating toothed wheel. Light passing through one notch travels to a mirror a considerable distance away, is reflected back, and then, if the rotational speed of the toothed wheel is adjusted properly, passed through the next notch in the wheel. By measuring the rotational speed of the wheel and the distance from the wheel to the mirror, Fizeau was able to obtain a value of 3.13×10^8 m/s for the speed of light using Equation (1.10) and Equation (1.11).

$$t = \frac{\theta}{\varpi} \qquad (1.10)$$

$$c = 2\frac{d}{t} \qquad (1.11)$$

where

θ is the number of revolutions (rev)
ω is the angular speed of the wheel (rev/s)
d is the distance between the light source and the mirror (m)
t is the transit time for one round trip (s).

In 1862, the French scientist Jean Foucault (1819–1868), who worked with Fizeau, greatly improved the accuracy of the experiment by substituting a rotating mirror for the wheel. He was able to obtain a value of 2.977×10^8 m/s for the speed of light in air.

Figure 1.2 demonstrates Fizeau's method, in which the wheel has 450 notches and rotates with a speed of 35 rev/s. A mirror is located at a distance of 9500 m from the notched wheel. Light passing through one notch travels to the mirror and back just in time to pass through the next notch. To calculate the speed of light, the distance travelled $2d$ is divided by the time Δt. To find the time, the wheel rotates from one notch to the next during this time. The time is calculated by the wheel rotation through an angle $\Delta\theta = (1/450)$ rev. Knowing the rotational speed ω of the wheel, it is possible to find the time using the relation $\Delta\theta = \omega\Delta t$. Therefore:

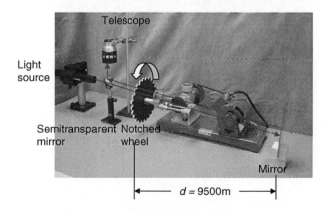

FIGURE 1.2 Fizeau's method to measure the speed of light.

The time required for the wheel to rotate from one notch to the next:

$$\Delta t = \frac{\Delta \theta}{\omega} = \frac{(1/450 \text{ rev})}{35 \text{ rev/s}} = 6.3 \times 10^{-5} \text{ s}$$

Calculate the speed of light by dividing the distance into the time:

$$c = \frac{2d}{\Delta t} = \frac{2(9500 \text{ m})}{6.3 \times 10^{-5} \text{ s}} \approx 3.0 \times 10^{8} \text{ m/s}$$

1.3.4 MICHELSON'S MEASUREMENTS

In the years that followed these earliest experiments, several investigators improved upon Fizeau's apparatus and methods of observation, and obtained more accurate values for the speed of light. The German American physicist Albert A. Michelson (1852–1931) was celebrated for the invention and development of the interferometer, an optical instrument, now named the Michelson stellar interferometer in his honour. In 1877, Michelson used his own improvement of Foucault's rotating mirror method, replacing the toothed wheel by a small eight-sided mirror, as shown in Figure 1.3. If the angular speed of the rotating eight-sided mirror is adjusted correctly, light reflected from one side travels to the fixed mirror, reflects, and can be detected after reflecting from another side that has rotated into place at just the right time. The minimum angular speed must be such that one side of the mirror rotates one-eighth of a revolution during the time it takes for the light to make the round trip between the mirrors. For one of his experiments, Michelson placed mirrors on Mt. San Antonio and Mt. Wilson in California, a distance of 35 km apart. From the value of the minimum angular speed in such experiments, he obtained the value of the speed of light of $c = (2.997\ 96 \pm 0.00004) \times 10^{8}$ m/s in 1926.

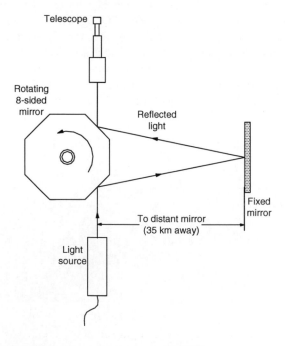

FIGURE 1.3 Michelson's experiment to measure the speed of light.

1.3.5 MAXWELL'S ELECTROMAGNETIC WAVES METHOD

In 1865, Maxwell's theoretical description of electromagnetic waves allowed him to obtain a simple expression for c in terms known as physical quantities. Figure 1.4 shows the propagation of an electromagnetic wave in free space. The wave speed should be a function of the properties of the optical medium through which it propagates. The propagation velocity of an electromagnetic wave is given in Equation (1.12). Maxwell determined theoretically that electromagnetic waves propagate through a vacuum at a speed given by:

$$c = \frac{1}{\sqrt{\varepsilon_0 \mu_0}} \tag{1.12}$$

where the (electric) permittivity of free space is $\varepsilon_0 = 8.85 \times 10^{-12}$ (C^2/N m^2) and the (magnetic) permeability of free space is $\mu_0 = 4\pi \times 10^{-7}$ (T m/A).

$$c = \frac{1}{\sqrt{(8.85 \times 10^{-12} \text{ C}^2/\text{Nm}^2) \times (4\pi \times 10^{-7} \text{ Tm/A})}} = 3.00 \times 10^8 \text{ m/s}$$

The experimental and theoretical values for c agree. Maxwell's success in predicting c provided a basis for inferring that light behaves as a wave consisting of oscillating electric and magnetic fields.

Today the speed of light has been determined with such high accuracy that it is used to define the metre. Now the speed of light in air is defined as:

$$c = 299, 796, 458 \text{ m/s}$$

Although a value of the speed of light in air of $c = 3.00 \times 10^8$ m/s is adequate for most calculations, the speed of light in a vacuum is $c = 299, 792, 458$ m/s. The speed of light has a different value for each medium, such as water, oil, or glass.

Each material has a refractive index. The refractive index or the index of refraction (n) is defined as the ratio between the speed of light in vacuum and the speed of light in a medium (v). For example, the index of refraction for air, water, and crown glass are 1.0003, 1.33, and 1.52, respectively. The refractive index of materials is discussed in detail in this book.

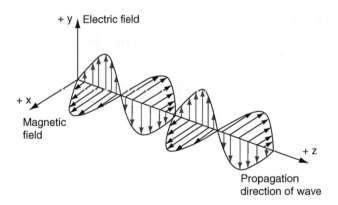

FIGURE 1.4 Propagation of an electromagnetic wave in free space.

1.4 LIGHT SOURCES

Figure 1.5 briefly describes the classification of light sources, which have been built recently and are used by industry for many applications.

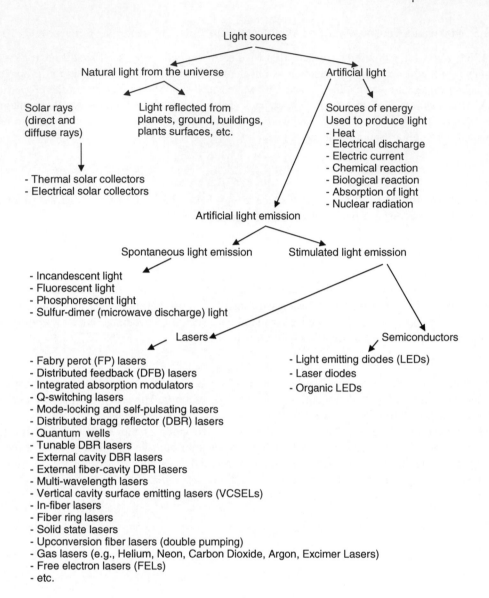

FIGURE 1.5 Classifications of light sources.

1.5 THE ELECTROMAGNETIC SPECTRUM

A few decades after the publication of Maxwell's theory, Hertz showed that electromagnetic waves could indeed be produced by an oscillating magnetic spark. In further experiments, he showed that electromagnetic waves undergo reflection, refraction, diffraction, and interference. Electromagnetic waves behave exactly like light, except that their wavelengths are much greater. The work of Hertz and others laid the foundations for the use of electromagnetic waves in radio communication.

Electromagnetic waves can exist at virtually any frequency (f) or wavelength (λ). The frequency and wavelength are related to the speed (v) of the wave, defined by:

$$v = f\lambda \tag{1.13}$$

For electromagnetic waves travelling through a vacuum or, to a good approximation, through air, the speed is $v = c$, so $c = f\lambda$. Equation (1.13) shows that the frequency and wavelength are inversely related by the travelling wave relationship $\lambda = c/f$. Therefore, the greater the frequency, the shorter the wavelength.

Figure 1.6 shows that electromagnetic waves exist with an enormous range of frequencies, from values less than 10^4 Hz to greater than 10^{22} Hz. It also shows the whole range of electromagnetic

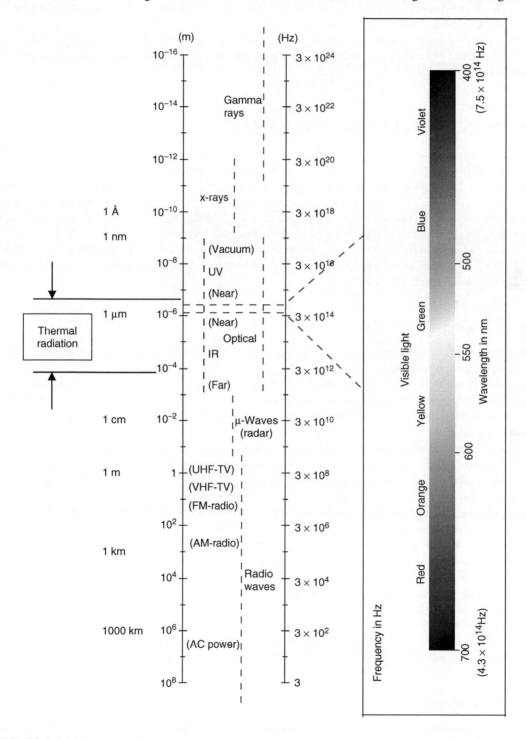

FIGURE 1.6 Electromagnetic wave spectrum.

waves, in order of increasing frequency or decreasing wavelength. Since all these waves travel through a vacuum at the same speed of light $c = 3.00 \times 10^8$ m/s, Equation (1.13) can be used to find the correspondingly wide range of wavelengths. Each range of wavelengths is referred to as a band. A small group of wavelengths within a band is called a channel.

The ordered series of electromagnetic wave frequencies or wavelengths shown in Figure 1.6 is called the electromagnetic spectrum. Historically, regions of the spectrum have been given names, such as visible waves, radio waves, and infrared waves. Although the boundary between adjacent regions is shown as a sharp line in Figure 1.6, the boundary is not so well defined in practice, and the regions often overlap.

1.6 TYPES OF ELECTROMAGNETIC WAVES

In 1867, when Maxwell published the first extensive account of his Electromagnetic Theory, the frequency band was only known to extend from the infrared, across the visible light, to the ultraviolet. Although this region is of major concern in optics, it is a small range of the electromagnetic spectrum. This section lists the main categories into which the electromagnetic spectrum is usually classified, by ranges of frequencies or wavelengths. Table 1.1 lists the frequency and wavelength ranges for the general types of electromagnetic waves.

Somewhere, not far from the centre of the spectrum, there is visible light. Notice that the visible part of the electromagnetic spectrum, so important to life on Earth, is actually the smallest of the frequency bands named above.

1.6.1 RADIOFREQUENCY WAVES

The lowest-frequency electromagnetic waves of particular importance are radio and television waves, in the frequency range of roughly 10^6 to 10^9 Hz. Waves in this frequency range are produced in a variety of ways. The low-frequency (LF), medium-frequency (MF), and high-frequency (HF) bands are reflected by the upper layers of the ionosphere. The short-wave radio frequencies in the HF band are reflected off the E and F layers of the ionosphere at a considerable altitude and can travel great distances.

The straight-line radio waves are normally transmitted around the curvature of the Earth. This is accomplished by reflection off ionic layers in the upper atmosphere. Energetic particles from the sun ionize gas molecules, giving rise to several ion layers. Certain layers reflect radio waves below a specific frequency. By "bouncing" radio waves off these layers, it is easy to send radio transmissions beyond the horizon, to any region of the Earth. Such reflection of radio waves requires the ionic layers to have uniform density. When, from time to time, a solar disturbance produces a shower of energetic particles that upsets this uniformity, a communication blackout can occur, as the radio waves are scattered in many directions rather than reflected in straight lines. To avoid such disruptions, global communications have, in the past, relied largely on transoceanic cables. Now, communications satellites can provide line-of-sight transmission to any point on the globe.

The ionosphere is essentially transparent to the VHF, UHF, and SHF portions of the electromagnetic spectrum. Because transmission from the radiating antenna to the receiver must be line-of-sight in these bands, television and FM radio signals have a limited range.

1.6.2 MICROWAVES

Microwaves with frequencies from 10^9 to about 10^{12} Hz are produced by special vacuum electron tubes called klystrons and magnetrons. Nowadays microwaves are used in communications,

TABLE 1.1
Classifications of Electromagnetic Waves

Types of Waves		Approximate Frequency Range (f) (Hz)	Approximate Wavelength Range (λ) (m)	Source
Power Waves		60	5×10^6	Electric Currents
Radio Waves	AM	$(0.53 \times 10^6) - (1.70 \times 10^6)$	$570 - 186$	Electric Circuits
	FM	$(88.00 \times 10^6) - (108.00 \times 10^6)$	$3.40 - 2.80$	
	TV	$(54.00 \times 10^6) - (890.00 \times 10^6)$	$5.60 - 0.34$	
Microwaves		$10^9 - 10^{11}$	$10^{-1} - 10^{-3}$	Special Vacuum Tubes
Infrared Radiation		$10^{11} - 10^{14}$	$10^{-3} - 10^{-7}$	Warm and Hot Bodies
Visible Light		$(4.00 \times 10^{14}) - (7.00 \times 10^{14})$	10^{-7}	Sun and Lamps Light
Ultraviolet Radiation		$10^{14} - 10^{17}$	$10^{-7} - 10^{-10}$	Very Hot Bodies and Special Lamps
X-rays		$10^{17} - 10^{19}$	$10^{-10} - 10^{-12}$	High-Speed Electron Collisions and Atomic Processes
Gamma Rays		Above 10^{19}	Below 10^{-12}	Nuclear Reactions, Processes in Particle Accelerators, and Natural radioactivity

inter-station television, microwave ovens, and radar applications. They are used in studying the origin of the universe, opening garage doors, guiding planes, and viewing the surface of the planet. They are also quite useful for studying physical optics with experimental arrangements that are scaled up to convenient dimensions.

1.6.3 Infrared Waves

The infrared region of the electromagnetic spectrum lies just beneath red light of the visible light range. Infrared waves have frequencies from about 10^{12} to 4.3×10^{14} Hz. They were first detected by the renowned astronomer Sir William Herschel (1738–1822) in 1800. The infrared radiation, or IR, is often subdivided into four regions: the near IR, near the visible (780–3000 nm); the intermediate IR (3000–6000 nm); the far IR (6000–15000 nm); and the extreme IR (15000 nm–1.0 mm).

Infrared rays are often generated by the rotations and vibrations of molecules. In turn, when infrared rays are absorbed by an object its molecules rotate and vibrate more vigorously, resulting in an increase in the object's temperature. The molecules of any object at a temperature above absolute zero ($-273°C$) will radiate IR, even if only weakly. On the other hand, infrared is emitted in a continuous spectrum from hot bodies, such as electric heaters, glowing coals, and ordinary house radiators. Roughly half the electromagnetic energy from the sun is IR, and the common light bulb actually radiates far more IR than light. The human body radiates IR quite weakly, starting at around 3000 nm. IR is also associated with maintaining the Earth's warmth or average temperature through the greenhouse effect. Incoming visible light (which passes relatively easily through the atmosphere) is absorbed by the Earth's surface and re-radiated as infrared (longer-wavelength) radiation, which is trapped by greenhouse gases, such as carbon dioxide and water vapour.

IR energy is generally measured with a device that responds to the heat generated upon absorption of IR by a blackened surface. There are many types of IR detectors used in different fields, for example, thermocouple, pneumatic, pyroelectric, and bolometer detectors. These detectors rely on temperature-dependent variations in induced voltage, gas volume, permanent electric polarization and resistance, respectively.

The detector can be coupled to a cathode ray tube by way of a scanning system to produce an instantaneous television-like IR picture, known as a thermograph. There are IR spy satellites that look out for rocket launchings, IR resource satellites that look out for crop diseases, and IR astronomical satellites that look out into space. There are "heat-seeking" missiles guided by IR, and IR lasers and telescopes looking out into universe.

Most remote controls operate on a beam of infrared light with a wavelength of about 1000 nm. This infrared light is so close to the visible spectrum and so low in intensity that it cannot be felt as heat.

1.6.4 Visible Light

The visible light region occupies only a very small portion of the total electromagnetic spectrum. The visible spectrum covers a range of less than 4% of the width of the electromagnetic spectrum, as shown in Figure 1.6. It runs from 4×10^{14} to about 7×10^{14} Hz, or a wavelength range of 780–400 nm, respectively. Only the radiation in this region can activate the receptors in our eyes. Visible light emitted or reflected from the objects around us provides visual information about our world. Visible light can be emitted by many sources, such as an incandescent bulb or a hot metal filament. The resulting broad emission spectrum is referred to as thermal radiation, and it is one of the major sources of light. Light is also generated by passing an electric discharge through a gas-filled tube. The atoms of the gas will become excited and emit a stream of photons as visible light. This occurs in an ordinary florescent tube or a gas laser device, such as a Helium Neon laser tube.

Newton was the first to recognize that white light is actually a mixture of all the colours of the visible spectrum. Approximate frequency and wavelength ranges for the various colours of white light are given in Table 1.2. When white light passes through a prism, the white light disperses into its various spectral components from the red to violet colour, like a rainbow. The sun emits white light, as do incandescent light bulbs. Most sources of illumination are white light sources.

TABLE 1.2
Frequency and Wavelength Ranges for the
Various Colours in White Light

Colour	Wavelength (λ) (nm)	Frequency (f) (Hz)
Red	780 – 622	$(384 - 482) \times 10^{12}$
Orange	622 – 597	$(482 - 503) \times 10^{12}$
Yellow	597 – 577	$(503 - 520) \times 10^{12}$
Green	577 – 492	$(520 - 610) \times 10^{12}$
Blue	492 – 455	$(610 - 659) \times 10^{12}$
Violet	455 – 390	$(659 - 769) \times 10^{12}$

When visible light falls on an object, that object absorbs and reflects wavelengths in the visible spectrum. A white object, such as the white page forming the background to this writing, reflects all wavelengths, and our eyes see white. A black object, such as the letters printed on this page, absorbs all wavelengths, and no light is reflected.

Colours are seen because of special absorption and reflection of light by an object. The green grass and leaves are perceived as green because the chlorophyll in green plant matter strongly absorbs the wavelengths surrounding green and reflects wavelengths in the green region of the visible spectrum. A red traffic light is red when light passes through it because it passes wavelengths in the red part of the visible light spectrum and absorbs light in the remaining parts of the visible spectrum. If our eyes intercept predominantly red wavelengths coming from an object, the eye perceives that object as having a red colour. Colour is the result of a filtered white light spectrum, in cases where the illumination light is white.

1.6.5 Ultraviolet Light

Ultraviolet radiation is the portion of the electromagnetic spectrum that lies between x-rays and visible light. Ultraviolet radiation was discovered by Johann Wilhelm Ritter (1776–1810). It is not seen by the human eye, but some regions of UV act in much the same way as visible light. There is no sharp dividing line between UV and x-rays, or between UV and visible light. The approximate wavelength range of UV is between 4 and 400 nm. It is not often realized that UV radiation covers such a wide range of wavelengths; the ratio of the longest-to-shortest of these wavelengths is nearly 100 to 1. By comparison, the entire range of visible light seen by humans extends from approximately 380 to 760 nm, which is only a 1 to 2 ratio in wavelength. The portions of the UV spectra are commonly referred to as near UV (from approximately 300–400 nm) and far UV (from approximately 185–300 nm). Medical literature divides regions into UVA long wave (315–380 nm), UVB midrange (280–315 nm), and UVC shortwave (<280 nm).

Ultraviolet radiation is produced by special lamps and very hot bodies. The sun emits large amounts of UV radiation. Fortunately, most of the UV radiation received by the Earth is absorbed in the ozone (O_3) layer in the upper atmosphere at an altitude of about 40 to 50 km. Sunglasses are now labeled to indicate the UV protection standards they meet in shielding the eyes from this potentially harmful radiation.

Artificial sources of UV radiation include incandescent and arc lamps. Lamps used for the generation of UV light are usually enclosed arcs containing mercury. Other sources include argon, krypton, and xenon gases, as well as zinc and cadmium vapour.

UV may be produced directly by these arcs or indirectly by fluorescence of phosphors deposited in thin layers on the lamp envelope. These phosphors absorb UV and re-emit this energy at longer UV wavelengths or as visible light as follows. High frequency UV photons collide with atoms and part of the photon's energy is transferred to the atoms by boosting electrons to higher energy states. Upon de-excitation, as electrons fall back to lower energy states, energy is released as photons of light. Since only a portion of the incoming photon's energy was transferred to an electron, these emitted photons have less energy than the incoming UV photons, so their wavelengths are longer than the excitation photons.

In some phosphoric materials, the fluorescence lingers and disappears, slowing after the UV source is removed. Here, the electron returns slowly to its original state, and this delayed fluorescence is called phosphorescence.

1.6.6 X-Rays

X-rays were discovered in 1895 by the German physicist Wilhelm C. Röntgen (1845–1923) when he noted the glow of a piece of fluorescent paper caused by some mysterious radiation coming from a cathode ray tube. Because of the apparent mystery involved, these were named x-radiation or x-rays. Extending in frequency from roughly 2.4×10^{16} to 5×10^{19} Hz, they have extremely short wavelengths; most are smaller than an atom. The photon energies of x-rays (100 eV to 0.2 MeV) are large enough so that x-ray quanta can interact with particles of matter one at a time in a clearly granular fashion, almost like bullets of energy. One of the most practical mechanisms for producing x-rays is the rapid deceleration of high-speed charged particles. The resulting broad frequency braking radiation arises when a beam of energetic electrons is fired at a material target, such as a copper plate. Collisions with the Cu nuclei produce deflections of the beam electrons, which in turn radiate x-ray photons.

Typically, the x-rays used in medicine are generated by the rapid deceleration of high-speed electrons projected against a metal target. These energetic rays, which are only weakly absorbed by the skin and soft tissues, pass through our bodies rather freely, except when they encounter bones, teeth, or other relatively dense material. However, at low intensities, x-rays can be used with relative safety to view the internal structure of the human body and other opaque objects. X-rays can pass through materials that are opaque to other types of radiation. The denser the material is, the greater its absorption of x-rays, and the less intense the transmitted radiation will be. For example, as x-rays pass through the human body, many more x-rays are absorbed by bone than by tissue, as shown in Figure 1.7. If the transmitted radiation is directed onto a photographic plate or film, the exposed areas show variations in intensity and thus form a picture of internal structures. This property makes x-rays most valuable for medical diagnosis, research, and treatment. Still, x-rays can cause damage to human tissue, and it is desirable to reduce unnecessary exposure to these rays as much as possible.

X-ray processes take place in colour television picture tubes, which use high voltages and electron beams. When the high-speed electrons are stopped as they hit the screen, they can emit x-rays into the environment. Fortunately, all modern televisions have the shielding necessary to protect viewers from exposure to this radiation. Examples of optical instruments using x-rays include x-ray microscopes, picosecond x-ray streak cameras, x-ray diffraction grating, interferometers, and x-ray holography.

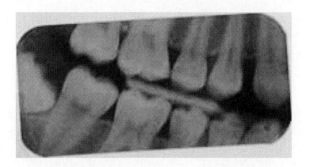

FIGURE 1.7 X-ray photograph of teeth.

1.6.7 GAMMA RAYS

Electromagnetic waves with frequencies above 10^{20} Hz are often referred to as gamma (γ) rays. They are the highest energy (10^4 eV to about 10^{19} eV), shortest wavelength electromagnetic waves. This high-frequency radiation is produced in nuclear reactions, in particle accelerators, and also in certain types of nuclear radioactivity. Gamma rays are emitted by particles undergoing transitions within the atomic nucleus. A single gamma ray carries so much energy that it can be detected with little difficulty. At the same time, its wavelength is so small that it is difficult to observe any wavelike properties. Gamma rays are also highly penetrating and destructive to living cells. It is for this reason that they are used to treat cancer cells and more recently to kill microorganisms in food processing.

1.7 PROPERTIES OF LIGHT

Light travels in a straight line through a uniform optical medium. The eye depends on back-projecting rays entering the eye that direct to the origin of the light rays. Eyes can locate objects, as long as the light from them has travelled to the eye in a straight line. However, the path or nature of light can be altered by optical materials, which can absorb, reflect, or transmit light, or do a combination of all three. Details of light properties are presented throughout this book.

1.7.1 ABSORPTION

Light falling on an optical material may be absorbed, transmitted, or reflected. Light that is neither reflected nor transmitted by an optical surface is absorbed and usually converted into heat energy.

The absorbed light is responsible for the heating effects that occur when light strikes a surface. The more light that is reflected or transmitted, the smaller the absorption and the less heating that takes place. This is most evident in the heating of common objects by sunlight. This depends on the colour of the optical material; a red material reflects red light and absorbs much of the rest of the other colours. The colour of a material is that colour which is reflected rather than absorbed. Black or dark-coloured painted objects or dyed fabrics have extremely low reflectance and transmit little or no light. Consequently, they absorb most of the radiant energy that falls on them. This energy is converted to heat, which explains why the dark materials tend to heat up much more rapidly than light-coloured ones when in the presence of bright sunlight. On the other hand, white and other colours that have high reflectance over large areas of the visible spectrum tend to reflect most of the incident radiation. Because little energy is absorbed by such colours, only a small amount of energy is converted into heat.

Greenhouses also utilize the same effect to capitalize on the energy from the sun. In cool periods of the year, visible light enters through the glass and is absorbed by the dark soils and vegetation.

1.7.2 TRANSMISSION

In addition to the colour that depends on the detailed nature of the source, light may take on a particular colour as it passes through a material that selectively absorbs some wavelengths. For example, a piece of red glass is a filter that selectively passes red light but absorbs shorter wavelengths. Similarly, a blue filter transmits blue but not green, yellow, or red. Thus, colour may be produced by selective absorption.

1.7.3 REFLECTION

Reflection is the most familiar property of light. Reflection enables us to see objects around us. Light from a source, such as the sun or a lamp, travels in a straight line until it strikes a material, at which point it may be either absorbed, transmitted, or reflected, or a combination of two or three processes.

Reflection occurs when waves encounter a boundary that does not absorb the light's waves and bounces the waves off the surface. The incoming light wave is referred to as an incident wave and the wave that is bounced from the material's surface is called the reflected wave. Surfaces that reflect the most light appear white, silver, or mirror-finished. A highly polished, smooth, flat silver surface acts as a mirror, reflecting a perfect image of the objects around it.

Colour is also produced by reflection as well as transmission. The perceived colour of most objects is due to the selective reflection of light. Thus, a red object reflects red light but absorbs green and blue. A red apple appears a vivid red when illuminated with red light, but it looks dark under blue light because it does not reflect blue light well.

The absorption, transmission, and reflection properties of light depend on the following major factors, which are discussed in detail throughout the book:

1. optical characteristics of the optical materials,
2. intensity of a light source,
3. wavelength of a light source,
4. angle of incidence of a light beam on an optical component surface,
5. mechanical characteristics of an optical component surface,
6. environmental and operating temperatures of optical components,
7. colours of the optical components,
8. thickness and density of the optical components, and
9. dopant and codopant percentage in the optical components.

1.7.4 REFRACTION

Light that is transmitted through an optical medium will usually deviate somewhat from the straight path it was previously following. This phenomenon is familiar with transparent optical materials, such as glass, plastic, water, and lenses. Objects seen through them appear larger, smaller, or distorted. Place half of a stick's length into water and the stick appears to be bent at the surface. Refraction is an important characteristic of lenses, and allows them to focus a beam of light onto a single point. Refraction occurs as light passes from one optical material to another when there is a difference in the index of refraction between the two optical materials.

1.7.5 INTERFERENCE

Interference is the net effect of the combination of two or more wave types moving on intersecting or coincident paths. The effect is the addition of the amplitudes of the individual waves at each point affected by more than one wave. There are two types of interference. Assume two waves have the same frequency and amplitude. If two of the waves are in phase with each other, the two waves are reinforced, producing constructive interference. If the two waves are out of phase, the result is destructive interference, producing complete cancellation. One of the best examples of interference is demonstrated by the light reflected from a film of oil floating on water or a soap bubble, or a computer compact disk (CD). The disk reflects a variety of rainbow colours when illuminated by natural or artificial white light sources.

1.7.6 DIFFRACTION

Diffraction occurs when light waves pass by a corner or through an opening or slit. The opening size must be smaller than the light's wavelength. Diffraction is a specialized case of light scattering, in which an object with regularly repeating features (such as a diffraction grating) produces an orderly diffraction of light, seen as a diffraction pattern. In the real world, most objects are very complex in shape and thus are considered to be composed of many individual diffraction features that can collectively produce a random scattering of light.

1.7.7 POLARIZATION

Natural sunlight and most forms of artificial illumination transmit light waves whose electric fields vibrate in all perpendicular planes with respect to the direction of propagation; this is called un-polarized light. The light has two components: the electric field vibrating in the vertical plane and the magnetic field vibrating in the horizontal plane. The two fields combine to form the shape of the electromagnetic wave. When the electric fields of light waves are restricted to a single plane by filtration, the light is polarized with respect to the direction of propagation so that all waves vibrate in the same plane.

FURTHER READING

Beiser, Arthur, *Physics*, 5th ed., Addison-Wesley Publishing Company, Reading, MA, 1991.

Born, M. and Wolf, E., *Principles of Optics: Electromagnetic Theory of Propagation, Interference, and Diffraction of Light*, 7th ed., Cambridge University Press, Cambridge, England, 1999.

Bouwkamp, C. J., Diffraction theory, *Rep. Prog. Phys.*, 17, 35–100, 1949.

Bromwich, T. J. I' A., Diffraction of waves by a wedge Proc. London Math. Soc. 14, 450–468, 1916.

Camperi-Ginestet, C., Young W, *Kim*, Micro-opto-mechanical devices and systems using epitaxial lift off. JPL, Proceedings of the Workshop on Microtechnologies and Applications to Space Systems., 305–316, 1993.

Cox, Arthur, *Photographic Optics*, 15th ed., Focal Press, London, New York, 1974.

Cutnell, John D. and Johnson, Kenneth W., *Physics*, 5th ed., Wiley, New York, 2001.

Cutnell, John D. and Johnson, Kenneth W., *Student Study Guide: Physics*, 5th ed., Wiley, New York, 2001.

Douglas, C., *Giancoli: Physics*, 5th ed., Prentice Hall, 1998.

Evans, R. D., *The Atomic Nucleus*, McGraw-Hill, New York, 1955.

Ewen, Dale, *Physics for Career Education*, 4th ed., Prentice Hall, Englewood Cliffs, NJ, 1996.

Ghatak, Ajoy K., *An Introduction to Modern Optics*, McGraw-Hill Book Company, New York, 1972.

Griot, Melles, *The Practical Application of Light*, Melles Griot Catalog, San Jose, CA, 2001.

Halliday, Resnick, and Walker, *Fundamental of Physics*, 6th ed., Wiley, New York, 1997.

Heath, Robert W., *Fundamentals of Physics*, D.C. Heath Canada Ltd, Toronto, 1979.

Hewitt, Paul G., *Conceptual Physics*, 8th ed., Addison-Wesley, Inc., Reading, MA, 1998.

Jones, Edwin and Richard, Childers, *Contemporary College Physics*, McGraw-Hill Higher Education, New York, NY, 2001.

Lehrman, Robert L., *Physics-The Easy Way*, 3rd ed., Barron's Educational Series, Inc., New York, 1998.

McDermott, Lillian C., *Introduction to Physics*. Preliminary Edition, Prentice Hall, Inc., Englewood Cliffs, NJ, 1988.

McDermott, Lillian C., *Tutorials in Introductory Physics*. Preliminary Edition, Prentice Hall, Inc., Englewood Cliffs, NJ, 1998.

Nichols, Daniel H., *Physics for Technology with Applications in Industrial Control Electronics*, Prentice Hall, Englewood Cliffs, NJ, 2002.

Nolan, Peter J., *Fundamentals of College Physics*, Wm. C. Brown Publishers, Boston, 1993.

Robinson, Paul, *Laboratory Manual to Accompany Conceptual Physics*, 8th ed., Addison-Wesley, Inc., Reading, MA, 1998.

Romine, Gregory S., *Applied Physics Concepts into Practice*, Prentice Hall, Inc., Englewood Cliffs, NJ, 2001.

Salah, B. E. A. and Teich, M. C., *Fundamentals of Photonics*, John Wiley and Sons, New York, 1991.

Sears, Francis W., *University Physics - Part II*, 6th ed., Addison-Wesley Publishing Company, Reading, MA, 1998.

Tippens, Paul E., *Physics*, 6th ed., Glencoe McGraw-Hill, Westerville, OH, U.S.A., 2001.

Urone, Paul Peter, *College Physics*, Brooks/Cole publishing Company, Pacific Grove, CA, 1998.

Walker, James S., *Physics*, Prentice Hall, Englewood Cliffs, NJ, 2002.

Warren, Mashuri L., *Introduction to Physics*, W.H. Freeman and Company, San Francisco, CA, 1979.

White, Harvey E., *Modern College Physics*, 6th ed., Van Nostrand Reinhold Company, New York, 1972.

Williams, John E., *Teacher's Edition Modern Physics*, Holt, Rinehart and Winston, Inc., New York, 1968.

Wilson, Jerry D., *Physics: A Practical and Conceptual Approach Saunders Golden Sunburst Series*, Saunders College Publishing, London, 1989.

Wilson, Jerry D. and Buffa, Anthony J., *College Physics*, 5th ed., Prentice Hall, Inc., Englewood Cliffs, NJ, 2000.

Yeh, C., *Applied Photonics*, Academic Press, New York, 1994.

2 Light and Shadows

2.1 INTRODUCTION

To cast a shadow, we need light from a concentrated source, such as the Sun during the day or the Moon at night. A shadow is also cast by light that comes from a point source, which gives parallel light rays. Consider a flat white screen located a short distance away from the light source and an object placed between the screen and the light source. Part of the light rays are blocked by the object making a dark area, and part of the rays that reach the screen make that part of the screen bright. The dark area is called shadow. The shadow's location can be figured out by drawing straight lines from the point source to the edge of the object and continuing them to the screen. The resulting shadow resembles the object and we can easily recognize the shape of the object in two dimensions.

Perhaps the most spectacular astronomical events that one can observe without a telescope are solar and lunar eclipses, which were considered omens of great fortune or complete disaster in ancient times. We now know that the occurrence of an eclipse is a consequence of the shadows cast by the orbits of the Earth and Moon with respect to the Sun.

X-ray pictures are simply shadows in x-ray light made visible by a fluorescent screen or photographic film.

This chapter presents the principles, types, and applications of shadows. Solar and lunar eclipses are also presented in detail. Two experimental studies of light shadow formation using different optical objects are included in this chapter.

2.2 SHADOWS

The electromagnetic wave of light emitted in all directions from point source S of light is shown in Figure 2.1. Light propagation can be represented by a series of spherical wavefronts moving away from the point light source at the speed of light. The point source of light can be defined as a source whose dimensions are small in comparison with the distances that light travels. Notice that the spherical wavefronts become essentially plane wavefronts in all directions at a large distance from

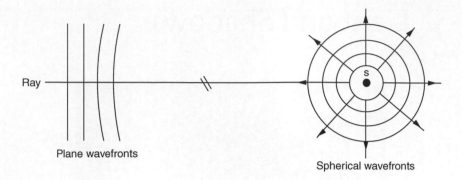

FIGURE 2.1 Point source of light emits rays in all directions.

the light source. More details on this subject will be presented later in this book. An imaginary straight line drawn perpendicular to the wavefront in the direction of the moving wavefronts is called a ray, as shown in Figure 2.1. There are an infinite number of rays starting from the point light source propagating in all directions.

Any dark object absorbs some wavelengths of light, as discussed in the properties of light in Chapter 1. A black object absorbs nearly all of the visible light wavelengths when they are incident on it. Light that is not absorbed when incident upon an object is either transmitted or reflected. If all light incident upon an object is transmitted, the object is called transmissive. If all light incident upon an object is reflected or absorbed, the object is called opaque. Since light cannot pass through an opaque object, a shadow will be produced in the space behind the object. The shadow formed by a point source S of light is shown in Figure 2.2. Since light is propagated in straight lines, rays drawn from the point source past the edges of the opaque object form a sharp shadow. The shape of the shadow is proportional to the shape of the object. The shape of the shadow depends on the opaque object location relative to the point source. The region in which no light has entered is called the umbra.

When the source of light is an extended type, the shadow will consist of two portions, as shown in Figure 2.3. The inner portion receives no light from the extended source; this portion is the umbra, as explained above. The outer portion is called the penumbra. An observer within the penumbra would see a portion of the source but not all of the source. An observer located outside both regions would see the source completely. Shadows are produced when objects are exposed to a light source or natural light. A light shadow occurs in the universe when solar and lunar eclipses occur by a similar formation of shadows, as explained above. One consequence of the Moon's orbit around the Earth is that the Moon can shadow the Sun's light as viewed from the

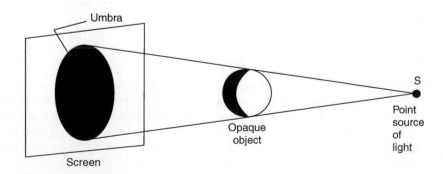

FIGURE 2.2 Shadow formed by an opaque object placed between a screen and a point light source.

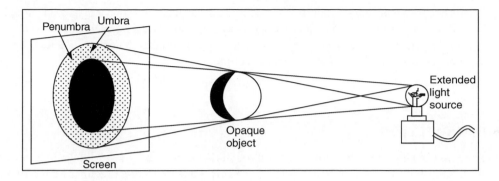

FIGURE 2.3 Shadow formed by an extended source of light.

Earth, or when the Moon passes through the shadow cast by the Earth. The former is called a solar eclipse, and the later is called a lunar eclipse.

2.3 SOLAR ECLIPSE

A solar eclipse occurs when the Moon passes between the Sun and the Earth, blocking the light from the Sun. This causes a shadow to be cast on a small area of the Earth. Due to the relative sizes of the Moon and the Sun and their distances from Earth, at times they appear to be the same size in the sky. When the Moon exactly covers the Sun, a total solar eclipse occurs. Three types of solar eclipses (as observed from any particular point on the Earth) can be defined as:

1. *Total Solar Eclipses* occur when the umbra of the Moon's shadow touches a region on the surface of the Earth. This occurs when the Moon exactly covers the Sun. The alignment of the Earth, Moon, and Sun have to be exact for a total eclipse to occur.
2. *Partial Solar Eclipses* occur when the penumbra of the Moon's shadow passes over a region of the Earth's surface. If the alignment is not exact, a partial eclipse may occur. This is when the Moon only partially overlaps the Sun and blocks only part of the Sun from view.
3. *Annular Solar Eclipses* occur when a region on the Earth's surface is in line with the umbra, but the distance is such that the tip of the umbra does not reach the Earth's surface, as shown in Figure 2.4. An annular eclipse occurs when the apparent size of the Moon is smaller than that of the Sun, and the Moon does not fully block the Sun from view.

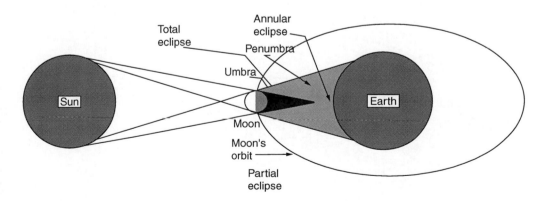

FIGURE 2.4 Geometry of annular solar eclipse.

This occurs because the Moon's orbit is elliptical, causing the Moon's distance from the Earth to vary. When the Moon is farther away from the Earth, it appears to be smaller. Therefore, there are times when the Moon appears to be smaller than the Sun. If an eclipse occurs at this time, it is called an annular eclipse.

2.4 LUNAR ECLIPSE

A lunar eclipse occurs when the Moon passes through the Earth's shadow, as shown in Figure 2.5. Because the Earth is much larger than the Moon, usually the entire Moon is eclipsed. Since the full phase can be seen from anywhere on the night side of the Earth, a lunar eclipse can be seen by more people than a solar eclipse. Similar to solar eclipses, lunar eclipses can be partial or total, depending on whether the light of the Sun is partially or completely blocked from reaching the Moon. Since the Moon is moving through the Earth's shadow, and the size of the Earth is much greater than the size of the Moon, a lunar eclipse lasts for about 3.5 h (as opposed to a solar eclipse, which lasts for about 7.5 min).

Although the Moon is always in a new phase during a solar eclipse, a solar eclipse does not occur every time the Moon is in the new phase. This is because the orbit of the Moon is tilted relative to the Earth's orbit around the Sun. Though the Moon tilt is only 5°, it is enough to allow the alignment of the Earth, Moon, and Sun to occur only about once every six months. This holds true for lunar eclipses as well. In fact, lunar and solar eclipses generally occur together; that is, if the alignment is correct for a lunar eclipse during the full phase of the Moon, it will also be correct for a solar eclipse during the next new phase of the Moon. There are three basic types of lunar eclipses that occur in the universe, as shown in Figure 2.6.

1. *Penumbral Lunar Eclipses* occur when the Moon passes through the Earth's penumbral shadow. These events are only of academic interest, since they are subtle and quite difficult to observe.
2. *Partial Lunar Eclipses* occur when a portion of the Moon passes through the Earth's umbral shadow.
3. *Total Lunar Eclipses* occur when the entire Moon passes through the Earth's umbral shadow. These events are quite striking because of the vibrant range of colors the Moon can display during the total phase.

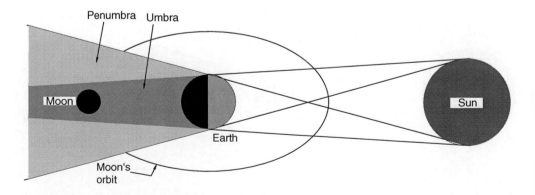

FIGURE 2.5 Geometry of a lunar eclipse—full moon.

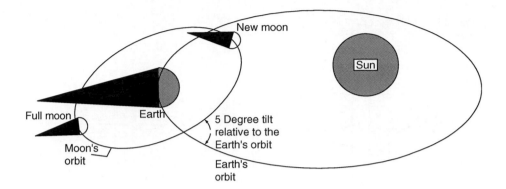

FIGURE 2.6 Types of lunar eclipse.

2.5 APPLICATIONS OF SHADOWS

The following are some of the applications of shadows:

1. The x-ray images of teeth and bones are shadows, made by a kind of light that goes through most things, but is scattered more by dense materials, such as metals.
2. A sundial is a way to measure the position of the Sun (and thus determine the time) by observing its shadow.
3. The first measurement of the size of the Earth was based on the length of shadows exactly at noon in different places. For the lengths of the shadows to be very different, you have to go north a significant fraction of the radius of the Earth. Observing the shadow lengths allows the size of the Earth to be calculated.
4. During an eclipse of the Sun, we are briefly in the shadow of the Moon; during an eclipse of the Moon, it is passing through the Earth's shadow. The duration of the eclipses was one of the first hints to early astronomers that the Moon is smaller than the Earth (but not much smaller), and that the Moon's distance from Earth is about 100 Earth-diameters, very much less than the distance to the Sun.
5. There is another kind of astronomical shadow that played an important role in the universe. About every 100 years, the planet Venus passes directly in front of the Sun, so that it appears as a small dark region on the Sun's disk. From different places on the surface of the Earth, different views are seen, and the path of Venus across the Sun's disk is slightly different. This provided the first means by which astronomers could measure the distance from the Earth to the Sun. The reason that the British Explorer Captain James Cook (1728–1780) sailed to the South Pacific (accidentally discovering New Zealand on the way) was to make the transit of Venus that occurred in 1769.
6. The results of Thomas Young's (1773–1829) experiment predicted the location of bright and dark fringes generated by the diffraction of light when light passes through an opening. The diffraction of light waves is of extreme importance in optical instruments because it sets the ultimate limit on the possible magnification. The interference of light waves is the principle of the American physicist A. Michelson (1852–1931), who invented the interferometer, which is used to make accurate length measurements.

2.6 EXPERIMENTAL WORK

This experiment studies the formation of a light shadow using an opaque object. This experimental work can be divided into two cases:

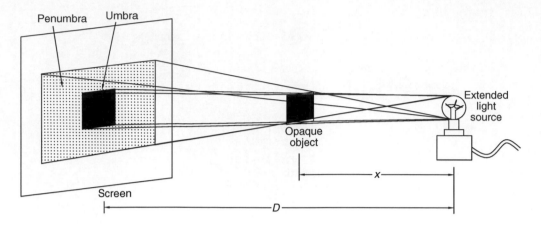

FIGURE 2.7 Shadow formed by an object located between a light source and a screen.

a. The student will practise forming shadows from a light source using an opaque object. The student will measure the sizes of the umbra and penumbra areas that form for at least five different distances from the light source. Then the size of the umbra and penumbra will be drawn as a function of a distance. A comparison among light shadows will be carried out. This case is illustrated in Figure 2.7.

b. The student will practise forming shadows from a light source using three different optical objects. The objects are an opaque object, which blocks light; a translucent object, which allows some light to pass through, but the light is scattered; and a transparent object, which allows light to pass through (relatively) undisturbed. The student will measure the size of the umbra and penumbra that are formed for at least five different distances from the light source, for each object. The student will also measure the light intensity at the umbra and penumbra areas for each object. Then the size of the umbra and penumbra and the light intensity will be drawn as a function of the distance for each object. A comparison among light shadows and light intensity will be carried out. This case is illustrated in Figure 2.8.

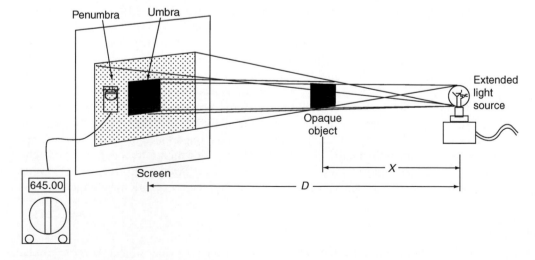

FIGURE 2.8 Shadow formation and light intensity measurements.

2.6.1 Technique and Apparatus

Appendix A presents the details of the devices, components, tools, and parts.

1. Opaque object and object holder, as shown in Figure 2.9
2. Unfrosted bulb in a light source, as shown in Figure 2.9
3. Opaque, translucent, and transparent objects and object holder
4. Hardware assembly (clamps, posts, positioners, etc.)
5. Screen
6. Light meter
7. Ruler

2.6.2 Procedure

Follow the laboratory procedures and instructions given by the professor and/or instructor.

2.6.3 Safety Procedures

Follow all safety procedures and regulations regarding the use of optical instruments and measurement devices, and light source devices.

2.6.4 Apparatus Set-Up

Case (a): Formation of Shadows

1. Figure 2.9 shows the experimental apparatus set-up.
2. Prepare an opaque optical object on the table.
3. Mount a screen on the table or wall.
4. Mount a light source at a distance D from the screen.
5. Mount an opaque object at a distance x_1 from the light source facing the screen.
6. Connect the light source to the power outlet.
7. Align the light source and the opaque object to face each other.
8. Measure the distance D, the size of the object, and the distance x_1. Fill out Table 2.1.
9. Switch on the power to the light source.
10. Turn off the lights in the lab before taking measurements.

FIGURE 2.9 Formation of shadows apparatus set-up.

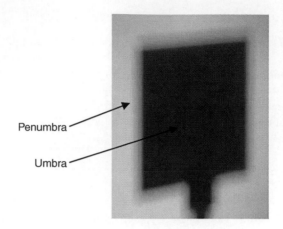

FIGURE 2.10 Umbra and penumbra of a shadow.

11. Measure the size of the umbra and penumbra, as shown in Figure 2.10. Fill out Table 2.1.
12. Switch off the power to the light source.
13. Turn on the lights in the lab.
14. Repeat steps 5 to 13 for four different distances between the object and light source.
15. Illustrate the location of the light source, object, and screen in a diagram.

Case (b): Measuring Light Intensity of Shadows Formed by Optical Materials

Repeat the procedure of Case (a) using three optical objects: opaque, translucent, and transparent. Figure 2.9 also shows the experimental apparatus set-up for this case. Measure the light intensity on the umbra and penumbra areas for each optical object. Fill out Table 2.2 for each optical object. Illustrate the location of the light source, object, and screen in a diagram.

2.6.5 DATA COLLECTION

Case (a): Formation of Shadows

Measure the distance D, size of the object, distance x, and the size of the umbra and penumbra areas when the object is located in front of the light source at five distances. Fill out Table 2.1.

TABLE 2.1
Shadows Formed by an Object

Distance D from Light Source to Screen (unit)	Size of Object (unit)	Distance x from Light Source to Object (unit)	Size of Umbra (unit)	Size of Penumbra (unit)
		$x_1 =$ ( )	( )	( )
		$x_2 =$ ( )	( )	( )
		$x_3 =$ ( )	( )	( )
		$x_4 =$ ( )	( )	( )
( )	( )	$x_5 =$ ( )	( )	( )

Case (b): Measuring Light Intensity of Shadows Formed by Optical Materials

Measure the distance D, the size of the object, the distance x, and the size of the umbra and penumbra areas when the three optical objects are located in front of the light source at five distances. Measure the light intensity at the umbra and penumbra areas for each optical object. Read the principles and procedure of light intensity, which are presented in this book. Fill out and use Table 2.2 for each optical object.

TABLE 2.2
Light Intensity Measurements

Distance D from Light Source to Screen (unit)	Size of Object (unit)	Distance x from Light Source to Object (unit)	Size of Umbra (unit)	Light Intesity on Umbra Area (unit)	Size of Penumbra (unit)	Light Intesity on Penumbra Area (unit)
		$x_1 = ($ $)$	$($ $)$	$($ $)$	$($ $)$	$($ $)$
		$x_2 = ($ $)$	$($ $)$	$($ $)$	$($ $)$	$($ $)$
		$x_3 = ($ $)$	$($ $)$	$($ $)$	$($ $)$	$($ $)$
		$x_4 = ($ $)$	$($ $)$	$($ $)$	$($ $)$	$($ $)$
$($ $)$	$($ $)$	$x_5 = ($ $)$	$($ $)$	$($ $)$	$($ $)$	$($ $)$

2.6.6 CALCULATIONS AND ANALYSIS

Calculate the areas of the umbra and penumbra for Cases (a) and (b).

2.6.7 RESULTS AND DISCUSSIONS

Case (a): Formation of Shadows

1. Illustrate the location of the light source, object, and screen in a diagram.
2. Present the measurements of the umbra and penumbra areas at five different distances.
3. Create a relation of the umbra and penumbra sizes as a function of the distances.
4. Discuss your results.

Case (b): Measuring Light Intensity of Shadows Formed by Optical Materials

1. Illustrate the location of the light source, object, and screen in a diagram.
2. Present the measurements of the umbra and penumbra areas at five different distances.
3. Present the measurements of light intensity on the umbra and penumbra areas at different distances from the light source.
4. Create a relation of the umbra and penumbra sizes as a function of the distances.
5. Create a relation of light intensity on the umbra and penumbra areas as a function of the distances.
6. Compare among light shadows and light intensities.
7. Discuss your results.

2.6.8 Conclusion

Summarize the important observations and findings obtained in this lab experiment.

2.6.9 Suggestions for Future Lab Work

List any suggestions for improvements using different experimental equipment, procedures, and techniques for any future lab work. These suggestions should be theoretically justified and technically feasible.

2.7 LIST OF REFERENCES

List any references that were used in the report. Use one format in writing the references. Never mix reference formats in a report.

2.8 APPENDICES

List all of the materials and information that are too detailed to be included in the body of the report.

FURTHER READING

Beiser, A., *Physics*, 5th ed., Addison-Wesley Publishing Company, Reading, MA, 1991.
Cutnell, J. D. and Johnson, K. W., *Physics*, 5th ed., Wiley, New York, 2001.
Ewen, D., *Physics for Career Education*, 4th ed., Prentice Hall, Englewood Cliffs, NJ, 1996.
Giancoli, D. C., *Physics*, 5th ed., Prentice Hall, Englewood Cliffs, NJ, 1998.
Halliday, R. W., *Fundamental of Physics*, 6th ed., Wiley, New York, 1997.
Jones, E. and Childers, R., *Contemporary College Physics*, McGraw-Hill Higher Education, New York, 2001.
McDermott, L. C., *Introduction to Physics*, Preliminary Edition, Prentice Hall, Inc., Englewood Cliffs, NJ, 1988.
Nolan, P. J., *Fundamentals of College Physics*, Wm. C. Brown Publishers, Dubuque, IA, 1993.
Robinson, P., *Laboratory Manual to Accompany Conceptual Physics*, 8th ed., Addison-Wesley, Inc., Reading, MA, 1998.
Salah, B. E. A. and Teich, M. C., *Fundamentals of Photonics*, Wiley, New York, 1991.
Sears, F. W., *University Physics—Part II*, 6th ed., Addison-Wesley Publishing Company, Reading, MA, 1998.
Tippens, P. E., *Physics*, 6th ed., Glencoe McGraw-Hill, Westerville, OH, U.S.A., 2001.
Urone, P. P., *College Physics*, Brooks/Cole Publishing Company, Pacific Grove, CA, 1998.
Walker, J. S., *Physics*, Prentice Hall, Englewood Cliffs, NJ, 2002.
White, H. E., *Modern College Physics*, 6th ed., Van Nostrand Reinhold Company, New York, 1972.
Wilson, J. D., *Physics: A Practical and Conceptual Approach*, Saunders College Publishing, Philadelphia, 1989.
Wilson, J. D. and Buffa, A. J., *College Physics*, 5th ed., Prentice Hall, Inc., Englewood Cliffs, NJ, 2000.
Young, H. D. and Freedman, R. A., *University Physics*, 9th ed., Addison-Wesley Publishing Company, Inc., Reading, MA, 1996.

3 Thermal Radiation

3.1 INTRODUCTION

In the absence of heat, life cannot be sustained in the universe. Thermal radiation serves as an essential ingredient for the formation (and existence) of life in the cosmos Thermal radiation represents a finite region of the electromagnetic spectrum.

Thermal imaging cameras permit the ability to observe the flow of thermal energy between two bodies. Thermal images from space have illustrated the flow of energy within the universe.

Thermal radiation is the transfer of energy by electromagnetic waves, which have a range of wavelengths. One example of thermal radiation heat transfer is that of energy transfer between the Sun and the Earth. The emitted radiant heat is the net flow of heat from a higher temperature material to a lower temperature material. The high temperature material emits energy and the low temperature material absorbs this energy. Many developing countries have exploited this principle as a means of renewable solar energy. This transfer (or flow) is dependent on a material's ability to transmit, reflect or absorb thermal energy. In this chapter, the nature of thermal radiation, its characteristics, and properties are presented.

The experiment included in this chapter requires an understanding of the concepts of solar radiation intensity levels and measurements when using a solar radiation meter. The solar radiation variation during the day will be measured and studied. Also the sun positions during the day hours will be studied.

3.2 THERMAL RADIATION

Thermal radiation is the electromagnetic radiation emitted by a body as a result of its temperature. For a net thermal radiation interchange of energy between two bodies to occur, there must be a temperature difference. The rate at which the thermal radiation interchange occurs is proportional to the difference of the fourth powers of their absolute temperatures. For a better understanding of thermal radiation, it will be necessary to study several different disciplines:

1. Electromagnetic Theory: Thermal radiation can be treated mathematically as if it was a traveling wave that contains energy and exerts a pressure. This wave-like nature is described by using electromagnetic wave theory.
2. Statistical Mechanics: Thermal radiation energy is transported over a range of wavelengths. Statistical mechanics help describe how this energy is distributed over the spectrum of wavelengths. For example, most energy coming from the Sun is radiated over many wavelengths.
3. Quantum Mechanics: All substances radiate and receive thermal radiation energy continuously. Associated with these processes are atomic and molecular activities that are modeled with the principles of quantum mechanics. Quantum mechanics gives a description of molecular processes associated with radiation.
4. Thermodynamics: Thermodynamics gives a description of how bulk properties of a material behave in the presence of thermal radiation.

There is only one type of thermal radiation; it propagates at the speed of light $c = 3.00 \times 10^8$ m/s. This speed is equal to the product of the frequency f and wavelength λ of the radiation as given in Equation (1.1). A portion of the electromagnetic spectrum is shown in Figure 3.1. This figure shows that the thermal radiation lies in the range from about 100 to 2000 nm and the very narrow portion of the spectrum is the visible light. Visible light lies in the range from about 450 to 780 nm. The propagation of thermal radiation takes place in the form of discrete quanta; each quantum having the energy of a photon can be defined by Equation (1.8).

By considering the thermal radiation of a gas, the principles of quantum-statistical thermodynamics can be applied to derive an expression for the energy density of radiation per unit volume and per unit wavelength as given in Equation (3.1):

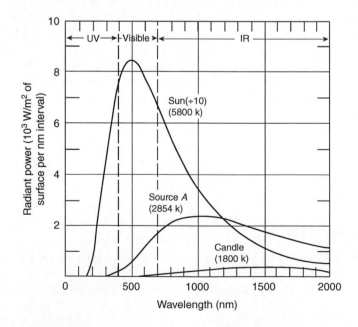

FIGURE 3.1 Blackbody radiation as a function of wavelength and temperature.

$$u_\lambda = \frac{8\pi hc\lambda^{-5}}{e^{hc/\lambda kT} - 1} \tag{3.1}$$

where

h is Planck's constant $= 6.626 \times 10^{-34}$ (J s),
k is Boltzmann's constant $= 1.38066 \times 10^{-23}$ J/molecule K, and
T is the temperature in degrees Kelvin.

When the energy density is integrated over all wavelengths, the total energy emitted is proportional to the absolute temperature to the fourth power, which is known as the Stefan–Boltzmann law as given in Equation (3.2):

$$E_b = \sigma T^4 \tag{3.2}$$

where

E_b is the energy radiated in W/m^2,
T is the temperature in degree K, and
σ is the Stefan–Boltzmann constant, $\sigma = 5.669 \times 10^{-8}$ W/m^2 K^4.

The total energy radiated by the surface increases rapidly with temperature. Figure 3.1 compares the spectrum of blackbody radiation at different temperatures for several light sources. Almost all of the energy at the temperatures shown lies in the infrared regions of the spectrum, where wavelengths are longer than visible light. The wavelength of peak (greatest) emission λ_{peak} (nm) decreases as temperature (T) increases, and is inversely proportional to the temperature in Kelvin. This is known as Wien's displacement law as given in Equation (3.3):

$$\lambda_{peak}T = 2.898 \times 10^{-3} \text{ m K} \tag{3.3}$$

If an object is heated to 2000 K, part of the blackbody radiation spectrum lies in the visible region and is seen as a dull glow, although the wavelength of maximum intensity remains in the infrared. If the temperature increases to 4000 K the glow is a bright orange. At 6000 K (comparable to the temperature of the Sun's surface atmosphere) the maximum peak is in the visible spectrum and the object glows white hot.

In thermodynamics and heat transfer analysis, the energy density is defined as the energy radiated from a surface per unit time and per unit area. Thus the heated interior surface of an enclosure produces a certain energy density of thermal radiation in the enclosure. The thermal radiation energy exchange from a surface is related to its temperature. The subscript b in Equation (3.2) denotes that this is the radiation from a blackbody. This is called blackbody radiation because materials, which obey this law, appear black to the eye. They appear black because they do not reflect any radiation. Thus a blackbody also is considered as a body that absorbs all radiation incident upon it. E_b is called the emissive power of a blackbody.

When radiant energy is incident on a material's surface, part of the radiation is reflected, part is absorbed, and part is transmitted, as shown in Figure 3.2. One defines the reflectivity ρ as the fraction reflected, the absorptivity α as the fraction absorbed, and the transmissivity τ as the fraction transmitted. The total radiant energy distribution can be written in the form:

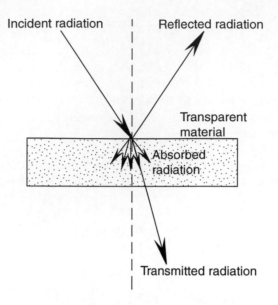

FIGURE 3.2 Incident thermal radiation on a transparent material.

$$\rho + \alpha + \tau = 1 \qquad (3.4)$$

Most solid bodies do not transmit thermal radiation; therefore, the transmissivity may be taken as zero. In this case, Equation (3.4) becomes:

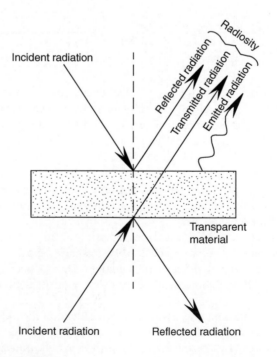

FIGURE 3.3 The definition of radiosity.

$$\rho + \alpha = 1 \qquad (3.5)$$

Two types of reflection phenomena may be observed when radiation strikes a surface. If the angle of incidence is equal to the angle of reflection, the reflection is referred to as specular. On the other hand, when the reflected radiation is distributed uniformly in all directions, the reflection is referred to as diffuse. The reflection phenomena are presented in details later in this book.

Figure 3.3 shows the definition of another radiation, referred to as the radiosity. Consider a transparent material receiving radiation on both sides. All the radiation leaving one of its surfaces, which includes emitted, reflected, and transmitted components, is defined as radiosity. Although Figure 3.3 shows the components of radiosity as single rays, it should be remembered that radiosity leaves the surface in all directions and over all wavelengths. Again, if the material is opaque, then the transmitted component is zero and Equation (3.5) applies.

3.3 LIGHT AND ENERGY

The idea that the energy in a light beam consists of numerous amounts of photons simply highlights the fact that light is a form of energy. Energy has the ability to cause transformations from one state to other. Thus ice can be changed to water by heat or a car can be set in motion by the power of fuel energy. Light is just one form of energy. There are many other kinds of energy as well, such as thermal, chemical, electrical, nuclear, and mechanical energy. Whenever a transformation takes place in nature, energy is being changed from one form into another. One particularly important type of transformation involves the conversion of light energy into chemical energy on the retina of the eye or in the film of a camera. In both cases, the energy contained in individual photons is used to produce chemical reactions that are necessary for visual or photographic processes. Both the eye and the camera use visible light photons to initiate the appropriate chemical reactions. Photons from infrared, microwave, and radio wave parts of the spectrum do not contain sufficient energy to be effective. On the other hand, ultraviolet, x-ray, and γ-ray photons carry too much energy, and would be damaging and hazardous, especially to the retina of the eye. The higher energy corresponds to the shorter wavelength of light.

3.4 SOLAR RADIATION ENERGY

The Sun supplies an inexhaustible amount of solar thermal energy to the universe. Solar radiation energy is one of the renewable engines available for many applications. Solar radiation energy reaches the Earth's surface in two forms: (i) direct radiation (the solar parallel rays in a clear sky); and (ii) diffuse radiation (the non-parallel rays of sky radiation, scattered in the atmosphere by a cloudy sky, gases, and atmospheric dust). The average intensity of solar radiation normal to the Sun's rays at the outer edge of the Earth's atmosphere is 1353 W/m^2. This energy is received at the surface of the Earth from the Sun and is subject to variations due to:

1. Variations in tilt of the Earth relative to the Sun. The Earth rotates around its own axis, one complete rotation in 24 h. The axis of this rotation is tilted at an angle of 23.5° to the plane of the Earth's orbit and the direction of this axis is constant, as shown in Figure 3.4. Maximum intensity of solar radiation is received on a plane normal to the direction of radiation. The equatorial regions of the Earth closest to the direction of solar radiation would always receive maximum radiation, if the axis of the Earth were perpendicular to the plane of the orbit. However, due to the tilt of the Earth's axis, the area receiving the maximum solar radiation moves north and south, between the Tropic of Cancer

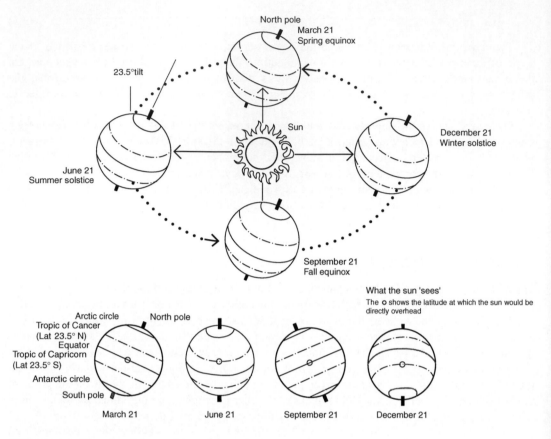

FIGURE 3.4 Diagram of the earth's path.

(latitude 23.5°N) and the Tropic of Capricorn (latitude 23.5°S). This is the primary cause of seasonal change.

2. Variations in atmospheric scattering by air molecules of water vapor, gases, and dust particles.

3. Variations in atmospheric absorption by O_2, O_3, H_2O, and CO_2. Solar radiation incident normal to the Earth's atmosphere has a special distribution. The x-ray and other very short-wave radiation of the solar spectrum are absorbed high in the ionosphere by nitrogen, oxygen, and other atmospheric components; most of the ultra-violet radiation is absorbed by ozone.

At wavelengths longer than 2.5 μm, a combination of low extra-terrestrial radiation and strong absorption by CO_2 and H_2O means that very little energy reaches the ground. For the application of solar energy, only radiation of wavelengths between 0.29 and 2.5 μm need to be considered, as shown in Figure 3.5. This figure shows the intensity of the light from the Sun as a function of wavelength.

Solar radiation is transmitted through the atmosphere, undergoing variations due to scattering and absorption. The other components of the atmosphere scatter a portion of the solar radiation reaching the ground. Thus there is always some diffuse radiation, even in periods of very clear skies. Particles of water and solids in clouds scatter radiation, and in periods of heavy clouds, all of the radiation reaching the ground will be diffused. According to the Rayleigh

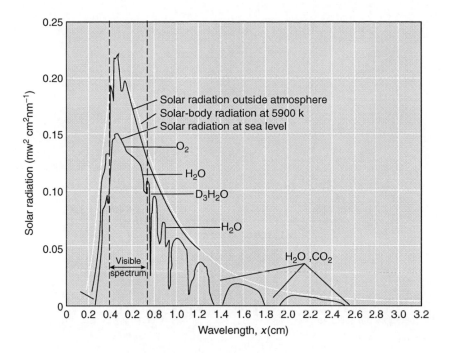

FIGURE 3.5 Spectral radiation distribution curves as a function of wavelength.

theory of scattering, the shorter wavelengths are scattered most and hence, diffuse radiation will tend to be at shorter wavelengths.

Solar radiation measurements often include total (beam and diffuse) radiation, in energy per unit time per unit area, on a horizontal surface. The measurements are made by a pyrheliometer using a collimated detector for measuring solar radiation for a small portion of the sky including the Sun (i.e., beam radiation) at normal incidence. Also the radiation measurements made by a pyranometer measure the total hemispherical solar (beam + diffuse) radiation, usually on a horizontal surface, as shown in Figure 3.6(a). If shaded from the beam radiation by a shade ring, it measures only the diffuse radiation, as shown in Figure 3.6(b). The primary meteorological measurements, done in virtually every weather network stations, are those of global solar radiation and of sunshine

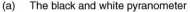

(a)   The black and white pyranometer   (b)   A shadow band stand to shade the
                                                              pyranometer

FIGURE 3.6 Solar radiation measurements. (a) The black and white pyranometer. (b) A shadow band stand to shade the pyranometer.

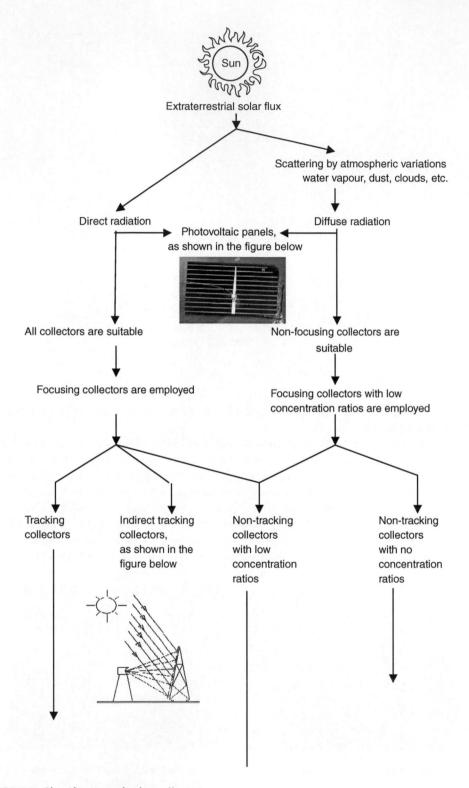

FIGURE 3.7. Classification of solar collectors.

Fresnel lens concentrators
STRA collector
Paraboloidal concentrators
Fresnel reflectors
Heliostat collectors
Parabolic cylindrical collectors,
as shown in the figure below

Honeycomb collectors
Evacuated collectors
Fined flat-plate collectors
Flat-plate collectors,
as shown in the figure
below

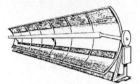
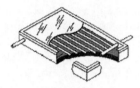

Trapezoidal concentrating collector
Spiral or sea shell collectors
Tabor circular cylinder collector
Axicon (conical) mirror concentrator
AKS composite axicon-lens collector
Trombe-meinel collector
Compound parabolic concentrator,
as shown in the figure below

FIGURE 3.7   (*continued*)

duration, using pyranometers. Pyranometers are radiometers designed for measuring the irradiance on a plane surface, normally from solar radiation and lamps, for measuring photosynthetically active radiation (PAR) in natural daylight or for measurement of illuminance, for the measurement of sunshine duration. There are many types of pyrgeometers designed to measure meteorological radiation parameters. Pyrgeometers are designed for IR (infrared) radiation measurement. Sunshine duration is defined by the World Metrological Organization (WMO) as when the solar radiation exceeds the level of 120 W/m^2.

3.5 CLASSIFICATION OF SOLAR COLLECTORS

Solar collectors may convert solar radiation directly into a useable energy form. A solar collector differs in several respects from more conventional heat exchangers. In solar collectors, energy transfer is from a distant source of radiant energy into thermal energy in a fluid or into electrical energy by photovoltaic panels. Thus, the analysis of solar collectors presents unique problems of low and variable energy fluxes. The type of device used to collect solar radiation energy depends primarily on the application. There are different types of collectors suitable for hot water, hot air, refrigeration, air-conditioning, and electricity generation.

Solar energy collectors are divided into two classes:

1. Liquid-heating solar collectors
2. Air-heating solar collectors

Collectors of solar energy are classified into two main categories. The non-concentrating solar collector category, such as flat plate collectors, is shown in Figure 3.7. In this category the received flux density is low, which makes this type of collector suitable for low temperature applications. On the other hand, in the concentrating solar collectors, such as cylindrical collectors shown in Figure 3.7, the incident flux on a large area is concentrated into a narrow zone with a high intensity of solar flux which generates a high temperature. This correspondingly reduces thermal losses (radiation and convection losses). Thus the concentrating collectors are suitable for high temperature applications. Some applications may require a higher temperature than obtainable by a flat plate collector. However, still a very high temperature is achievable by using solar concentrators with very high concentration ratios, such as cylindrical and heliostat solar concentrators.

Solar electrical collectors are called photovoltaic panels. They generate direct electrical power to customers, as shown in Figure 3.7. All types of solar collectors are a costly purchase, but the cost can be recovered in the long run. Solar collectors also can be used to concentrate solar radiation onto fibre optic cables, which are located in the collector's focusing line. In this arrangement, the fibre optic cables carry solar light into the interior spaces of a building for specular lighting via fibre cables. Fibre optic lighting technology is discussed in detail in this book. Figure 3.7 shows the types of solar collectors, which have been recently built.

3.6 FLAT-PLATE COLLECTORS

The flat-plate solar collector is the basic device used in solar heating and cooling systems. The operation of a flat-plate solar collector is simple. Most of the solar energy incident on the collector is absorbed by the surface, which is black to solar radiation. The essential parts of the collector are:

1. the absorber plate, generally made of metal with a non-reflective black finish to maximize the absorption of solar radiation,

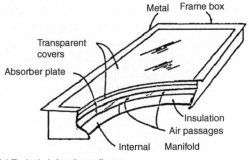

(a) Typical air-heating collector

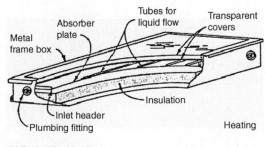

(b) Typical liquid-collector

FIGURE 3.8 Flat-plate collectors. (a) Typical air-heating collector. (b) Typical liquid-heating collector.

2. pipes or ducts to circulate either liquid or air in thermal contact with the absorber plate; thermal insulation for the back side and edges of the plate, and
3. one or more air spaces separated by transparent covers to provide insulation for the top of the plate.

The important parts of a typical flat-plate solar collector are shown in Figure 3.8. The transparent covers usually are made of glass. Glass has excellent weather durability and good mechanical properties. Plastics can be used; however, they generally do not have as high a resistance to weathering as glass; the plastic surface can be scratched and many plastics degrade, yellow with age, and reduce transitivity. The advantage of glass over plastics is that glass absorbs or reflects the entire long-wave radiation incident on it from the solar heated absorber plate. This reduces radiation losses from the absorber plate more effectively than plastics, since plastic transmits part of the long-wave radiation.

Flat-plate collectors absorb both beam and diffuse radiation. Flat-plate collectors can be designed for applications requiring energy delivery at moderate temperatures for such purposes as air heating, water heating for domestic uses, refrigerant systems and heating swimming pools.

Flat-plate collectors are usually mounted in a stationary position with an orientation dependent upon the location and the time of year in which the solar energy system is intended to operate. The cost of energy delivered from a solar collector will depend on the system's thermal performance, installation costs, and maintenance costs.

3.7 SOLAR HEATING SYSTEMS

A solar heating and/or cooling system can be defined as any system, which utilizes solar energy to heat and/or cool a building, although a distinction is often made between active and passive systems. A passive system can be defined as having no moving parts, but it may involve natural circulation of fluid to the heated space. A south-facing window or a skylight, which transmits sunlight can be considered a passive system if it admits more energy than it loses as heat. Another type of passive system may involve movable insulation material, which reduces heat loss from solar absorbing surfaces when there is no sunshine. In contrast to a passive system, an active system involves hardware that collects solar energy, stores heat, and distributes the heat to the rooms in the building. A solar heating system can be classified into a space heating system or a water heating system depending on the fluid employed. In the latter, a liquid is heated in the collector and transferred to a storage unit or directly to the heating loads. The two most common heat transfer working fluids are air and water.

3.7.1 SOLAR AIR HEATING SYSTEMS

An example of a basic solar air heating system is shown in Figure 3.9. This system is called an active solar air heating system. The components of this typical system include:

1. a solar air heating collector, as shown in Figure 3.8(a),
2. a pebble-bed heat storage unit to and from which heat is transferred by circulating air through the bed,
3. a control unit, which includes the sensors and control logic necessary to automatically maintain comfort conditions at all, times,
4. an air heating module comprising of automatic dampers, filters and blowers, and
5. an auxiliary heating unit (usually a warm-air furnace) to provide 100% back-up space heating when storage temperatures are insufficient to meet demands or when the solar system is not operating.

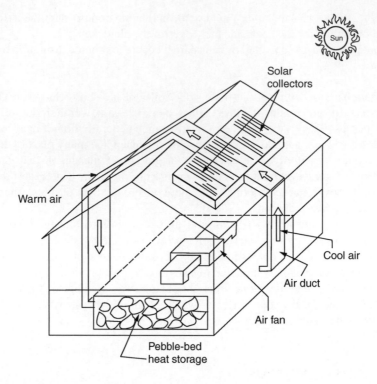

FIGURE 3.9 Basic active air-heating system.

Many solar air-heating designs (e.g., the Denver Solar House, MIT house IV), have been experimentally and commercially employed for space heating of residential buildings in different regions of the world.

There are many designs of passive air solar heating systems used in buildings, some of which are used to create natural ventilation to heat/cool buildings. Figure 3.10 shows two types of the passive air solar systems; direct solar energy collection through the glass roof of a room is one

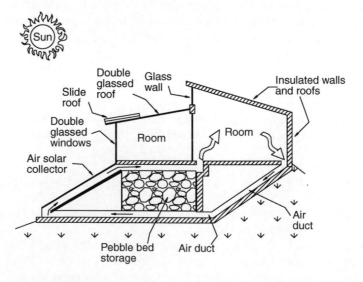

FIGURE 3.10 A passive air heating solar system.

system. The other system is called passive solar air heating through an air solar collector, which is connected to an air heating system.

3.7.2 Solar Water Heating Systems

Most of the solar water systems that have been experimentally and commercially used for domestic hot water can be placed in two main groups:

1. Circulating types involving the supply of solar heat to a fluid circulating through a collector, as shown in Figure 3.8(b), and storage of hot water in a separate tank.
2. Non-circulating types involving the use of water containers that serve both as solar collector and storage.

The circulating group may be further divided into the following types and sub-types:

1. Direct heating, single-fluid types in which the water is heated directly in the collector, by:
   (a) thermosiphon circulation between collector and storage and
   (b) pumped circulation between collector, load or storage.
2. Indirect heating, dual-fluid types in which a non-freezing medium is circulated through the collector for subsequent heat exchange with water when:
   (a) heat transfer medium is a non-freezing liquid and
   (b) heat transfer medium is air.

The most common type of direct heating, thermosiphon circulating solar water systems shown in Figure 3.11 utilizes the natural upward movement of heated fluids to circulate water from the collectors to the storage tank without the use of pumps and other components of "active" systems. Location of the storage tank higher than the top of the collector permits circulation of water from the bottom of the storage tank through the collector and back to the top of the storage tank.

Figure 3.12 shows the direct heating, pump circulation type of solar water heater. This arrangement is usually more practical than the thermosiphon type, since the collector would often be located on the roof with a storage tank in the basement. When solar energy is available, a temperature sensor activates a pump, which circulates water through the collector-storage loop. If the solar water heat is used in a cold climate, it may be protected from freeze damage by draining the

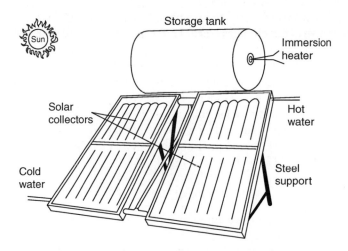

FIGURE 3.11 Passive circulation type of a solar water heater direct heating by thermal themosiphon.

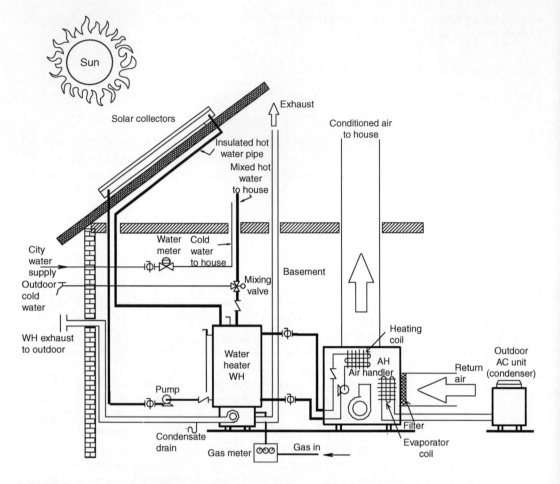

FIGURE 3.12 Active circulation type of a solar water heater direct heating system.

collector when sub-freezing temperatures are encountered. Drainage of the collector can be accomplished by automatic valves, which provide water outflow to a drain and the inflow of air to the collector.

Figure 3.13 illustrates a method for solar water heating using a liquid heat transfer medium in the solar collector used in freezing climates, such as in Canada. The most commonly used liquid is a solution of antifreeze (ethylene glycol) in water. A pump circulates this un-pressurized solution, as in the direct water heating system, and delivers the liquid to and through a liquid-to-liquid heat exchanger. Simultaneously, another pump circulates domestic water from the storage tank through the exchanger, then back to storage. The control system is essentially the same as any design employing water in a direct solar collector system. In a manner similar to that described above, solar energy can be employed in an air heating collector with subsequent transfer to domestic water in an air-to-water heat exchanger, as shown in Figure 3.14.

3.8 HOT WATER AND STEAM GENERATION SYSTEMS

Alternative versions of solar heating systems can be utilized for hot water and steam generation for industrial processes, as shown in Figure 3.15. Water is heated to 60°C in flat-plate solar collectors and then to 88°C in concentrating solar collectors. The hot water flows from the solar collector field into insulated water storage tanks to feed the system by a pump.

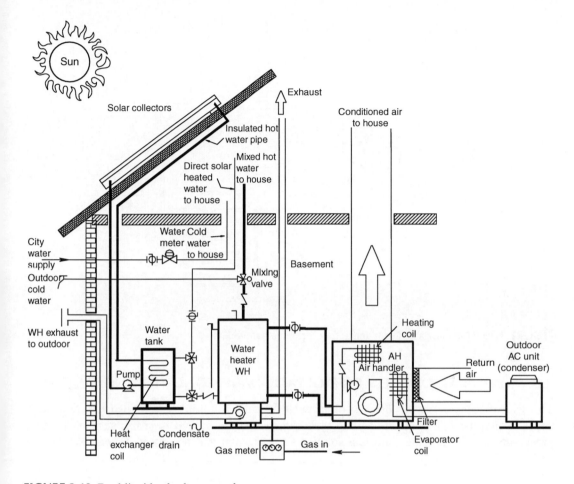

FIGURE 3.13 Dual liquid solar hot water heater.

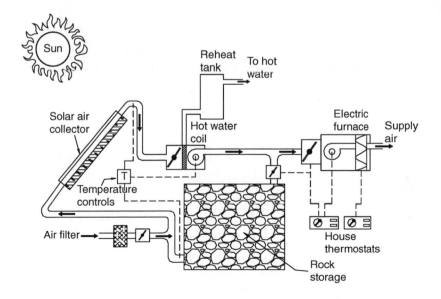

FIGURE 3.14 Hot water heater with solar air collector.

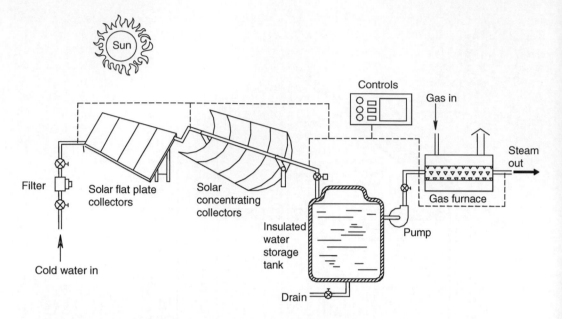

FIGURE 3.15 Hot water and steam generation for industrial processes.

3.9 VAPOUR ABSORPTION REFRIGERATION/AIR CONDITIONING SYSTEMS

The use of solar energy to drive cooling cycles has been considered for two different major applications; the first is in the cooling of buildings and the second is in refrigeration for food preservation. The working fluid is a solution of refrigerant and absorbent. When solar heating

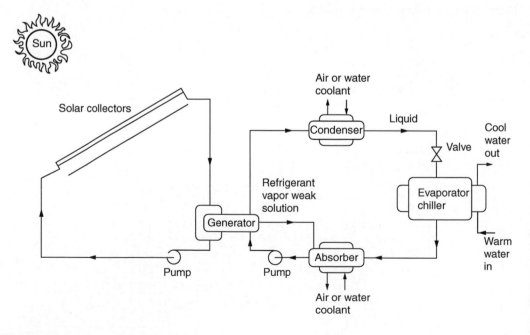

FIGURE 3.16 A solar air conditioner.

is supplied to the generator, some refrigerant is vaporized, and a weak mixture is left behind. The vapor then is condensed, and the liquid is expanded through a throttle valve to the lower pressure evaporator. The spent refrigerant leaving the evaporator next is returned to be absorbed by the weak solution returned from the generator. The resulting strong solution is transferred to the generator. A schematic diagram of one arrangement of such a system is shown in Figure 3.16. The most common refrigerant mixtures used in solar energy absorption refrigeration systems are: ammonia-water and lithiumbromide-water, the former for high temperatures required in the generator while the latter can be operated within the temperature range attainable with simple flat-plate collectors.

3.10 PHOTOVOLTAIC SYSTEMS

The word *photovoltaic* (PV) is a combination of the Greek word for light and the name of the physicist Allesandro Volta. Photovoltaic devices sometimes are called solar cells. They directly convert the incident solar radiation into electrical power. The conversion process is based on the photoelectric effect discovered by Alexander E. Becquerel in 1839. The photoelectric effect describes the release of positive and negative charge carriers in a solid state when light strikes its surface. The solar energy knocks electrons loose from their atoms, allowing the electrons to flow through the semiconductors to produce electricity. This process of converting light (photons) to electricity (voltage) is called the photovoltaic (PV) effect. Solar cells often are used to power calculators, watches, outdoor lighting, remote telecommunication systems, etc. Solar cells are combined typically into modules that hold about 40 cells; about 10 of these modules are mounted in solar cell arrays that can measure up to several metres per side. These flat-plate solar cell arrays can be mounted at a fixed angle facing south, or they can be mounted on a tracking device that follows the sun allowing them to capture the most sunlight over the

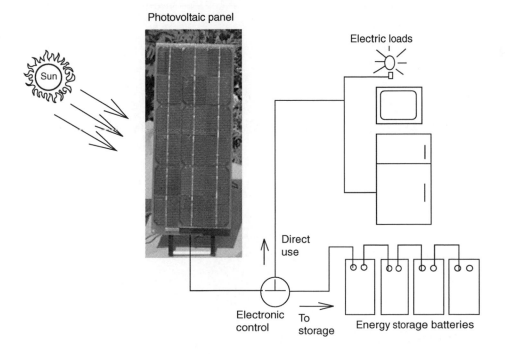

FIGURE 3.17 Major photovoltaic system components.

course of a day. The solar cell arrays are tilted at an angle equal to the geographical location of the site. About 10–20 solar cell arrays can provide enough power for a household. For large electric utility or industrial applications, hundreds of arrays can be interconnected to form a single, large solar cell system. A simplified schematic diagram of a typical photovoltaic system producing electricity is shown in Figure 3.17. More detailed materials and experiments are presented in the semiconductor chapter.

3.11 EXPERIMENTAL WORK

3.11.1 SOLAR RADIATION MEASUREMENTS

The purpose of this experiment is to measure the solar radiation intensity using a solar radiation meter (black and white pyranometer) as explained above. Students will measure daily solar radiation using a pyranometer connected to a data recorder and a monitor, which displays the data. The students will study the Sun positions during the day hours. Figure 3.18 illustrates outdoor solar radiation measurements arrangement.

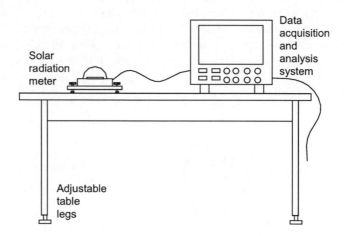

FIGURE 3.18 Solar radiation measurements arrangement.

3.11.2 TECHNIQUE AND APPARATUS

Appendix A presents the details of the devices, components, tools, and parts.

1. Solar radiation meter (black and white pyranometer), as shown in Figure 3.6(a).
2. Data acquisition and computer system with display.
3. Clock.
4. Table.

3.11.3 PROCEDURE

Follow the laboratory procedures and instructions given by the professor and/or instructor.

3.11.4 SAFETY PROCEDURE

Follow all safety procedures and regulations regarding the use of pyranometer and working under the sun.

3.11.5 APPARATUS SET-UP

Solar Radiation Measurements

1. Figure 3.19 shows the experimental apparatus set-up.
2. Place the table in a clear area free from shadows generated by trees and buildings during the day.
3. Adjust the legs of the table to level the top surface of the table.
4. Mount a solar radiation meter (black and white pyranometer) on the table. Normally, the solar radiation meter is positioned on a horizontal surface.
5. Place a data acquisition and computer system on the table.
6. Connect the solar radiation meter to the data acquisition system.
7. Measure the solar radiation intensity, time duration, and date using the data acquisition system. Fill out Table 3.1.
8. Keep track of the periods of sunshine during the day of the experiment. Fill out Table 3.2.
9. Repeat the steps 7 and 8 for a few days.

FIGURE 3.19 Solar radiation measurements set-up.

3.11.6 DATA COLLECTION

1. Measure solar radiation intensity, time duration, and date using the data acquisition system. Fill out Table 3.1.
2. Keep track of the periods of sunshine during the day of the experiment. Fill out Table 3.2.

TABLE 3.1
Solar Radiation Measurements

Day:	Month:	Year:
Hour	Solar radiation intensity (unit)	
8:00		
9:00		
10:00		
11:00		
12:00		
13:00		
14:00		
15:00		
16:00		
17:00		

TABLE 3.2
Sunshine Duration

Day:	Month:	Year:
	Sunshine duration	
Start time	End time	Duration

3.11.7 CALCULATIONS AND ANALYSIS

No calculations and analysis are required for this case.

3.11.8 RESULTS AND DISCUSSIONS

1. Present solar radiation intensity for each day in a graph.
2. Present sunshine duration for each day in a graph.
3. Discuss solar radiation data collection related to the hours during the day and weather conditions.

3.11.9 CONCLUSION

Summarize the important observations and findings obtained in this lab experiment.

3.11.10 SUGGESTIONS FOR FUTURE LAB WORK

List any suggestions for improvements using different experimental equipment, procedures, and techniques for any future lab work. These suggestions should be theoretically justified and technically feasible.

3.12 LIST OF REFERENCES

List any references that were used in the report. Use one format in writing the references. Never mix reference formats in a report.

3.13 APPENDICES

List all of the materials and information that are too detailed to be included in the body of the report.

3.14 WEATHER STATION

This is a proposal for experimental work using a weather station to measure and study some weather elements, such as solar radiation (direct and diffuse components), temperatures, precipitation, UV index, atmospheric pressure, and wind speed and direction.

FURTHER READING

Adams, R. A., Courage, M. L., and Mercer, M. E., Systematic measurements of human neonatal color vision, *Vision Research*, 34, 1691–1701, 1994.

Air-Conditioning and Refrigeration Institute, *Refrigeration and Air-Conditioning*, Prentice-Hall, Inc., Englewood Cliffs, NJ, 1979.

Al-Azzawi, A. R., The use of solar energy to provide air conditioning and hot water in buildings. M.Sc., Thesis, University of Strathclyde, Glasgow, Scotland, U.K., 1980.

Duffie, J. A. and Beckman, W. A., *Solar Energy Thermal Processes*, Wiley-Interscience Publication, New York, 1974.

EPLAB, *EPLAB Catalogue 2002*, Eppley Laboratory, Inc., Newport, RI, 2002.

Halliday, D., Resnick, R., and Walker, J., *Fundamentals of Physics*, 6th ed., Wiley, New York, 1997.

Holman, J. P., *Thermodynamics*, 4th ed., McGraw Hill, New York, 1988.

Holman, J. P., *Heat Transfer*, 9th ed., McGraw Hill Higher Education, New York, 2002.

Janna, W. S., *Engineering Heat Transfer*, PWS Publishers, Boston, MA, 1986.

Kreith, F. and Kreider, J. F., *Principles of Solar Energy*, McGraw-Hill Book Co., New York, 1980.

Lerner, R. G. and Trigg, L. G., *Encyclopedia of Physics*, 2nd ed., VCH Publishers, Inc., New York, 1991.

McDaniels, D. K., *The Sun: Our Future Energy Source*, 2nd ed., Wiley, New York, 1984.

McDermott, L. C. and Shaffer, P. S., *Tutorials in Introductory Physics, Preliminary Edition*, Prentice Hall Series in Educational Innovation, Upper Saddle River, New Jersey, U.S.A., 1988.

Melles Griot. The practical application of light. *Melles Griot Catalog, Rochester*, NY, U.S.A., 2001.

National Material Advisory Board, National Research Council, *Harnessing Light Optical Science and Engineering for the 21st Century*, National Academy Press, Washington, DC, 1998.

National Resources Canada NRC, *Advanced Houses Testing New Ideas for Energy-Efficient Environmentally Responsible Homes*, CANMET's Building Group, NRC, Ottawa, On, Canada, 1993.

National Resources Canada NRC, Canada advances on the home front. Canada's advance houses program: Leading the way in energy efficiency, indoor air quality, and environmental responsibility. *Home Energy Magazine, NRC, Ottawa, On, Canada*, 1996.

Ontario Ministry of Environment and Energy, *Home Heating and Cooling: A Consumer's Guide, Featuring the Energy Calculator*, Ontario Ministry of Environment and Energy, Toronto, On., 1994.

SEPA, *Solar Power*, Solar Electric Power Association, Washington, DC, 2002.

Smith, W. J., *Modern Optical Engineering*, McGraw-Hill Book Co., New York, 1966.

Tao, W. K. Y. and Janis, R. R., *Mechanical and Electrical Systems In Buildings*, Prentice Hall, Englewood Cliffs, NJ, 2001.

Wilson, J. D., *Physics-A Practical and Conceptual Approach*, Saunders College Publishing, London, 1989.

Young, H. D. and Freedman, R. A., *University Physics*, 9th ed., Addison-Wesley Publishing Co., Inc., Reading, MA, 1996.

4 Light Production

4.1 INTRODUCTION

For many thousands of years, humans relied on the sun as their only source of light. The ability to create fire, and its by-product, light, led to a profound change in the way humans managed their time. Today, there are many options for creating light.

Light can be produced through the rapid change of state of an electron from a state of relatively high energy to a ground state of lower energy. The energy of the electron has to leave the atom somehow, and it is often emitted in the form of a photon of light. This light emission takes place in some optical materials. A series of controlled rapid transitions will produce a stream of photons sufficient to provide illumination. This principle is utilized in the design of flashlights, light emitting diodes, and lasers.

Production of artificial light can be achieved by two types of emissions: either by spontaneous emission or stimulated emission. The spontaneous emission of light serves as the basis for most lighting systems. By passing an electric current through a metal wire (filament), the filament will begin to glow. The chemical composition and temperature of the wire will determine the wavelength of the light being generated. As such, light that is generated by this method is composed of many different wavelengths. The stimulated emission of light occurs in special materials when the energy transfer is stimulated by monochromatic wave light. As a result, the monochromatic light is amplified to produce the lasing light. Stimulated emission can be achieved using either light, electricity, or both as the catalyst. Lasers are the most common example of stimulated light emission. Gas and semiconductor lasers offer the coherent and monochromatic light required to re-shape a cornea, remove a pigment from skin, or transmit signals efficiently through optical fibre. Light production is explained in detail in this chapter.

This chapter also presents the types of energy sources used as a catalyst to produce light. The energy transfers involved vary in conversion efficiency. Experiments at the end of this chapter involve testing different light sources and calculating the conversion efficiency of the light sources.

4.2 SPONTANEOUS LIGHT EMISSION

Spontaneous light emission is the most common process of light production. When an electron is elevated to a high-energy state. This state is usually unstable. The electron will return spontaneously to the ground state, which is a more stable state, immediately emitting a photon. When light is emitted spontaneously, the light propagates in all directions. Light emission from an incandescent light bulb is an example of spontaneous emission. The glow of the thin wire filament of a light bulb is caused by the electrical current passing through it. The electrical current is transferred to thermal energy by the collisions of the excited electrons in the atoms of the wire. This causes the wire's temperature to elevate to a high level so that it emits light. There are many different types of materials with several energy states that can be used to produce spontaneous light at various wavelengths.

4.3 STIMULATED LIGHT EMISSION

Stimulated light emission is another process of light production. This process happens in laser light production. Once the lasing medium is pumped to a higher energy level by an external power source, it contains atoms with many electrons sitting in excited energy levels. The excited electrons occupy higher energy levels than the more relaxed electrons. The higher energy levels are meta-stable states. Just as the electron absorbed some amount of energy to reach this excited level, it can also release this energy. The electron can simply relax, emitting some energy in the form of a photon. This photon has a very specific wavelength that depends on the change of the electron's energy when the photon is released. While the excited electron is in the metastable state, an incoming photon can stimulate it to emit an identical photon with the same wavelength and phase. Thus, laser light is produced. This process is called stimulated light emission.

4.4 LIGHT PRODUCTION BY DIFFERENT ENERGY SOURCES

As explained in light emission processes, energy from an external source must be supplied to a material to boost the electron from its low energy state to a higher energy state before emitting a photon. The following are the types of energy sources used to produce light.

4.4.1 HEAT ENERGY

The most common way of providing an external energy to boost an electron into a higher energy state is to apply heat energy. The electron cannot stay in this excited state for very long. It immediately returns to a lower energy state and gives off its energy in the form of a photon. One example is light emission from a hot metal.

4.4.2 ELECTRICAL DISCHARGE

When an electrical current passes through a gas, such as neon, energy from the current ionizes the gas particles. This ionizing process injects energy into the electrons in the gas. When this ejected electron is reclaimed into a molecule, energy is given off in the form of a photon. One example is the emission of light from an ordinary fluorescent tube.

4.4.3 ELECTRICAL CURRENT

Another way to produce light is by applying electrical power to semiconductors, such as lasers and light emitting diodes. An electrical current is applied to a semiconductor's p–n junction to produce pairs of electrons and holes, as shown in Figure 4.1. When the positive pole is connected to the p-type material and the negative pole connected to the n-type; then the junction conducts. On the n-type side, free electrons are repelled from the electrical contact and pushed towards the junction.

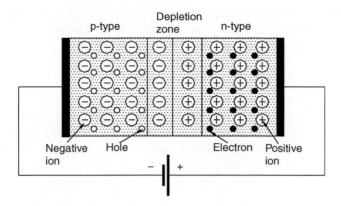

FIGURE 4.1 A P–N junction in a semiconductor.

FIGURE 4.2 Light emitting and laser diodes.

On the p-type side, holes are repelled from the positively charged contact towards the junction. At the junction, electrons will cross from the n-type side to the p-type side, and holes will cross from the p-type side to the n-type side. As soon as they cross the junction, most holes and electrons will recombine and eliminate each other. During this recombination process, each free electron loses a discrete packet of energy called a quantum of energy. This quantum of energy is radiated in the form of photons as electromagnetic waves.

   Light emitting diodes and laser diodes are used to produce light in various wavelengths. Light emitting diodes have a similar but simpler structure than laser diodes but they have a lot in common. Almost all light sources used in communication systems are made from semiconductors. Figure 4.2 shows different types of light emitting diodes, and a pigtail fibre connected to a side emitting laser diode commonly used in telecommunication systems.

4.4.4 ABSORPTION OF LIGHT

Some materials, such as phosphorus materials absorb light at a particular wavelength. During light absorption, the electrons move to an excited energy state. They then give off light at a different wavelength. Phosphorus material is used to coat the inside surface of the fluorescent tube. The phosphor absorbs the UV light, which is produced by electrical discharge within the tube. The UV light is then re-emitted as visible light.

4.4.5 Chemical Reaction

There are many chemical materials that emit light when they are mixed together during chemical reactions. This process occurs without adding heat energy; it can even occur at room temperature. During a chemical reaction, atoms and molecules are restructured. Often as a result of the restructuring process, electrons are left in high-energy unstable states. The electrons will emit some of the energy while returning to the lower energy state. This excess energy is either emitted in the form of light or heat. Figure 4.3 shows an example of a chemical reaction taking place between two components in plastic tubes.

4.4.6 Biological Reactions

Light production by biological reactions is simply chemical reactions taking place within living organisms. Many biological organisms use bioluminescence for attracting prey and mates or to scare predators. Within these organisms, light is generated by initiating chemical reactions that leave electrons in a high energy state. These excited electrons subsequently decay to a more stable energy state by emitting their energy as light. This occurs in luminous insects, such as the glowworm and the North American firefly. Many fish and sea animals are also capable of this light emission.

4.4.7 Nuclear Radiation

Light production by nuclear radiation can result when there is a nuclear reaction. A nuclear reaction occurs within the unstable nucleus of an atom of a nuclear-reactive (radioactive) material. Similar to chemical reactions, nuclear reactions are accompanied by a release of energy that was stored in the nucleus. This released energy is partially emitted as stream of photons of light in the form of electromagnetic radiation. The nuclear reaction produces a major fragmentation of the particles in the nucleus and releases high-energy nuclear radiations called alpha, beta, and gamma radiations.

FIGURE 4.3 Light production by chemical reaction.

An alpha (α) radiation particle (the nucleus of the He atom) is heavy and moves at very high speeds (high energy) after a nuclear reaction. A beta (β) radiation particle is simply an electron moving at high speeds after a nuclear reaction. Gamma (γ) radiation rays are electromagnetic radiation of very short wavelength (very high frequency) light; therefore, they have very high energy.

Light emission can be a by-product of a chain reaction of nuclear radiation. The main nuclear decay creates nuclear radiation moving at excessive speed. Nuclear radiation will eject electrons from their orbit in the atoms in almost any material it encounters. This process is called ionization. The encounter transfers energy to the ejected electron, exciting the electron to a higher energy state. When the electron in the ionized atom then relaxes to the lower energy state, it gives off the excess energy, which can be emitted as a photon of light. Gamma rays are capable of ionizing more atoms in their path than alpha or beta particles.

Depending on the material involved in the nuclear reaction, the emitted light can be visible. For example, cobalt (a radioactive material) placed in a storage pool of heavy water emits intense cobalt blue light.

4.4.8 Electrical Current

Electrical current can be applied through metal or gas to produce light. The following describes the methods of light production using different lamps. These are reasonably efficient light sources

4.4.8.1 Incandescent Light Lamps

When a solid material is heated, it emits radiant energy in the form of light. The light power at each wavelength depends on the temperature of the material. Figure 4.4 shows the various continuous spectra produced by a black body radiator at increasing temperatures. The amount of energy that falls within the visible portion of the spectrum is a function of temperature. The wavelengths of the visible light range from approximately 400–700 nm. The ideal light source would need to operate around 6000 K.

An incandescent light bulb has a filament of tungsten (a high melting point metal) that is heated to a high temperature by the passage of electric current and emits visible light of all wavelengths.

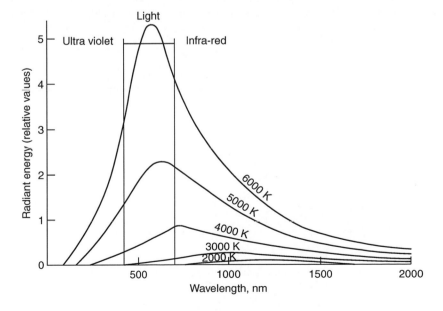

FIGURE 4.4 The radiant energy from a black body at various temperatures.

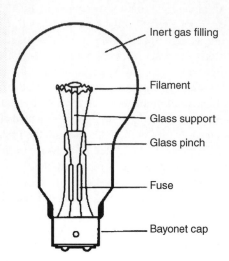

FIGURE 4.5 Incandescent lighting bulbs.

Figure 4.5 shows incandescent bulbs for general lighting service. The filaments are constructed of drawn tungsten wire, which is then coiled. The glass envelopes are spherical, elliptical or mushroom in shape and can be clear, pearl, or white. Clear and pearl bulbs have the same efficiency. The glass envelopes are filled with gases that are chemically inactive with hot tungsten filaments. At the bottom of the glass envelope, a lamp cap holds the electric terminal contacts.

Most of the power supplied to an incandescent light bulb is converted to light and heat. This bulb increases in efficiency with input electrical power, since the higher the filament temperature, the greater the proportion of visible light in the total radiation. The light emitted by an incandescent bulb is called white light, which is a mixture of all visible light wavelengths. Light with a single wavelength is called monochromatic or one colour. Figure 4.6 shows the radiant energy spectrum for an incandescent lamp. Incandescent bulbs are very inefficient light sources. Appendix C.1. presents shape codes and lamp designations for most common incandescent bulbs that are used in different lighting systems.

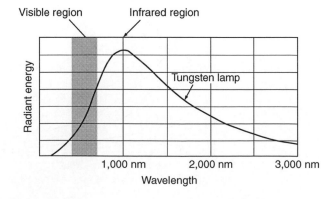

FIGURE 4.6 The radiant energy spectrum for an incandescent lamp.

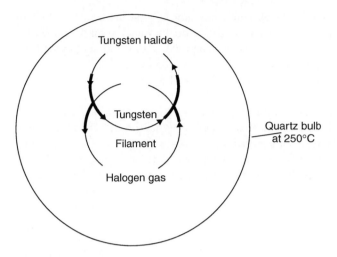

FIGURE 4.7 The halogen regenerative cycle.

4.4.8.2 Tungsten Halogen Lamps

The temperature of a filament is limited by the rate of gas evaporation. The introduction of a halogen (iodine, chlorine, bromine, or fluorine) gas sets up a regenerative cycle as illustrated in Figure 4.7. The evaporated tungsten associates with the halogen at the bulb wall temperature so that the tungsten does not get deposited on the bulb wall blackening it. Tungsten halide vapour forms and when it diffuses back to the hot filament, the vapour dissociates, and the tungsten is redeposited on the filament. This cycle enables the filament to run at temperatures above 3000 K. Appendix C.2. presents shape codes for most common tungsten lamps that are used in different lighting systems.

4.4.8.3 Fluorescent Light Lamps

The history of the electric gas-discharge lamp dates back to before the electric filament lamp, but it was only in 1939 that the fluorescent tube became generally available as an alternative to the filament

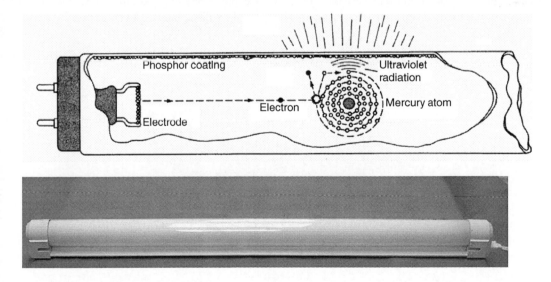

FIGURE 4.8 The tubular fluorescent lamp.

lamp. The fluorescent tube development has been so rapid since then that if a fitting installed in 1940 was relamped today, it could give three times the illumination for the same filament lamp wattage. In the same period, there has been very little equivalent development in the filament lamp output.

A fluorescent light lamp consists of a glass tube filled with a mixture of mercury vapour and an inert gas, such as argon, as shown in Figure 4.8. When a current passes through the mixture, ultraviolet radiation is produced. The inside of the tube is coated with a phosphorus material. This material absorbs the ultraviolet radiation and emits visible light.

Figure 4.9 shows that most of the emitted light is in the visible region of the light spectrum. Fluorescent light lamps produce little heat, so they are very efficient, more so than incandescent lamps. Appendix C.3. presents shape codes and lamp designations for most common fluorescent light lamps that are used in different lighting systems.

4.4.8.4 Black Lights

The fluorescent excitation by ultraviolet (UV) radiation has another common use. Black light used at nightclubs emits radiation in the violet and near-ultraviolet regions. Ultraviolet black lights cause fluorescent paints and dyes on signs, posters, and performers' clothes to fluoresce with brilliant colours. Many products, such as laundry detergents, are packaged in boxes that have bright colours containing phosphors in the ink. These fluoresce somewhat under the store's fluorescent lamps and appear brighter to get your attention. In the dark, when illuminated with black light, the boxes glow brilliantly. Some natural minerals are also fluorescent. Illumination with black lights is a method of identifying these materials.

4.4.8.5 Phosphorescent Materials

Phosphorescent materials, such as zinc sulfide products have numerous applications in safety signs, luminous watch dials, and military and novelty applications. They can be produced as paints, tapes, ropes, vinyl, plastisol, inks, pigments, and varnishes to fit many applications.

Phosphors absorb UV light and re-emit this energy at longer UV wavelengths or as visible light. When an atom is raised to an excited state, it drops back down to a relaxed state within a fraction of a second. However, in phosphorescent substances, atoms excited by photon absorption to high-energy states, called metastable states, can remain there for a few seconds. In a collection of such atoms, many of the atoms will descend to the lower energy state fairly soon, however many will remain in the excited state for over an hour. Hence, light will be emitted even after long periods of time.

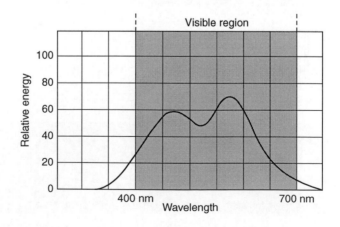

FIGURE 4.9 The radiant energy spectrum for a tubular fluorescent lamp.

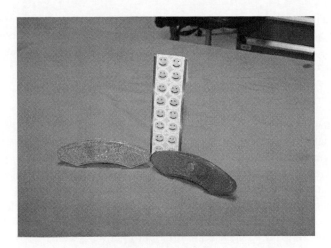

FIGURE 4.10 Phosphorescent materials emitting light.

Figure 4.10 shows phosphorescent materials emitting light. There are also super phosphorescent products based on strontium, an earth alkaline. This self-emitting light technology allows after glow time to exceed 12 h. Safety and novelty applications use this luminous pigment.

4.4.8.6 High-Pressure Mercury Discharge Lamps

A high-pressure discharge lamp contains a gas or vapour that causes light to radiate due to the passage of an electric current. The colour of the light depends on the nature of the gas or vapour. In a high-pressure mercury discharge lamp, the mercury gas discharge is increased, the resonant light radiation at 253.7 nm is absorbed and other radiations occur, many within the visible part of the spectrum. The colour of light from a high-pressure lamp is much whiter than that from a low-pressure one and the ultraviolet content is reduced. Figure 4.11 shows a high-pressure mercury

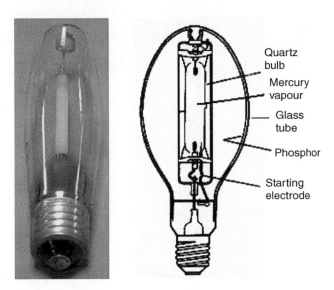

FIGURE 4.11 A high-pressure mercury discharge lamp.

discharge lamp. The lamp operates on a choke circuit and incorporates its own starting electrode; no circuit transformer or switch is necessary. High-pressure lamps are more efficient than incandescent lamps. Appendix C.4. presents shape codes for most common high-pressure mercury lamps that are used in different lighting systems.

4.4.8.7 Metal Halide Discharge Lamps

The high-pressure mercury discharge lamp emits a rather cold white light, as the colour distribution is strong in blues and greens but weak in reds. Better colour distribution can be achieved by introducing other metals in their halide form into the discharge tube. This produces additional radiation. The wavelength and colour depend on the metal used. Two examples of these metals are the iodides of thallium and sodium. Metal halide lamps require a higher starting voltage than the high-pressure mercury discharge lamps, but operate at higher efficiencies, and their colour appearance is suitable for industrial and commercial uses. Appendix C.5. presents shape code for most common metal halide lamps that are used in different lighting systems.

4.4.8.8 Sodium Lamps

Sodium lamps are a very efficient light source. However, the radiation is concentrated into two wavelengths around 589 nm resulting in a strong yellow light, monochromatic in nature. This is inappropriate where any reasonable colour is required. The lamp produces light by discharging an electrical current through sodium vapour. A control circuit limits the current and provides a high starting voltage. There are a number of different types of low-pressure sodium lamps which are illustrated in Figure 4.12. High-pressure sodium lamps are used in lighting systems. Appendix C.6. presents shape codes for most common high-pressure sodium lamps that are used in different lighting systems.

4.4.8.9 Energy Efficient Light Bulbs

An energy-efficient lighting system uses less natural resources and reduces the impact on the environment. Energy efficiency in lighting systems requires careful attention to the type of lighting and fixture purchased and to their location in the building. Energy efficient bulbs will provide energy savings and improve the quality of lighting.

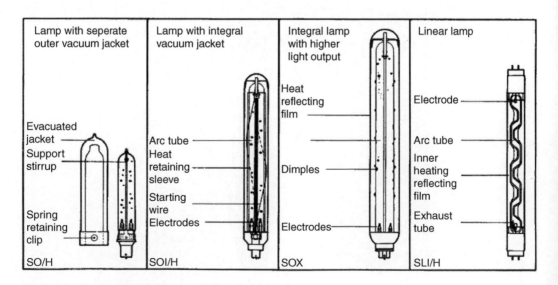

FIGURE 4.12 Low-pressure sodium lamps.

A bewildering array of light fixtures and lamps are available many of which have changed dramatically in recent years. The choices made when choosing new lighting products can lower the energy spending. Every lighting product can be thought of as having two costs: the initial cost is the purchase price, and the secondary cost is the ongoing cost of the energy used by the light bulb. If an energy efficient light bulb is chosen, the initial cost will be high, but the ongoing cost will be far less than any ordinary light bulb with the same power.

The first consideration is the amount of artificial light that is required. There is a natural tendency to provide too much light. The lighting selection also depends on the application. General lighting provides moderate light throughout a room. Several light sources can create uniform lighting and minimize glare and contrasts. Task lighting facilitates activities, such as reading and cooking; it requires more focused light in addition to general lighting. Protective and safety lighting helps prevent accidents, especially in stairwells, and discourages prowlers. Decorative lighting highlights room features, such as drapes, a fireplace, or a piece of art.

The second consideration is the type of fixture to be used. The three main types of household lighting available are standard incandescent, fluorescent, and tungsten-halogen fixtures. With incandescent bulbs, a very small percentage of the electricity used actually becomes light. Fluorescent lights and tungsten-halogen bulbs, on the other hand, are much more energy-efficient.

Whatever type of lighting is selected, the choice of fixtures and the range of efficiencies that they offer is wide. To select the most efficient fixture for the application, the following should be considered:

- Fixtures are specifically designed for the purpose of the application. For task lighting, such as for reading, the lamp should provide highly directional lighting. For general room lighting, choose a fixture that provides light over a broad area.
- Some fixtures have features that limit light output. Heavy shades and bowls, for example, can reduce light levels significantly. Features that will enhance useful light output, such as reflectors, will direct the light outwards.
- A fixture with a single bulb gives more useful light than one with several bulbs having the same total wattage. For example, four 25 W bulbs give little more than half the light of one 100 W bulb.

Energy Efficient Incandescent Bulbs

Energy efficient incandescent bulbs are regular incandescent light bulbs that have been improved to use less energy, but with slightly less light output. They do not offer energy savings as large as the compact fluorescents; however they are compatible with dimmers, work well outdoors, fit any light fixtures that take a regular incandescent bulb, and cost less. Long-life or extended-life incandescent bulbs last longer than regular bulbs, but output 30 per cent less light while using the same amount of energy. Bulbs with a higher than normal voltage rating are also available (typically, 130 V instead of the standard 120 V). These are intended for use where the voltage supply fluctuates (as in some rural areas). These bulbs are less efficient than standard incandescent bulbs.

Energy Efficient Fluorescent Tubes

Traditionally, fluorescent lighting has had limited use in most homes; the older fixtures were big, and the light quality was poor for most home needs. Today, however, new types of fluorescent tubes and bulbs produce light comparable to incandescent lighting, and a new generation of compact fluorescent tubes and fixtures are available. Fluorescent lighting now can be used in almost any area of a building and is best suited for utility areas. Though the initial cost may be higher, the greatly reduced energy costs and the very long tube life can make fluorescent fixtures an economical choice. Fluorescent tubes use 60–80% less energy and last 10–20 times longer than incandescent bulbs. They also work with conventional switches.

Energy Efficient Compact Fluorescent Tubes

Fluorescent lighting now can fit conveniently into most standard light fixtures by using compact fluorescent tubes. Compact fluorescents use about 25% of the energy that incandescent bulbs use and last up to 10 times longer. This makes them ideal for lights that are hard to reach or difficult to change, such as stairwells and recessed luminaires. To maximize the efficiencies of compact fluorescent tubes, they are best suited for lights that are used for long periods of time (three or more hours per day).

Compact fluorescents work most efficiently when the lamp is oriented downwards, with the base up. This is because the efficiency of the tube depends on the temperature of the coldest part of the lamp, which is the end most remote from the ballast. As heat rises, a base-up lamp will be coolest at the bottom, and therefore produce the greatest amount of light. Compact fluorescents emit the same kind of warm, natural light as regular incandescent bulbs. Two types can be installed together in the same room to produce a balanced, even illumination. Most fluorescent tubes are not compatible with dimmer switches and are not recommended for cold outdoor applications.

Energy Efficient Tungsten-Halogen Bulbs

Tungsten-halogen parabolic aluminum reflectors (PAR) type bulbs are low wattage floodlights (halogen-type incandescent bulbs). A standard 150 W incandescent spotlight can be replaced by a 90 W PAR bulb; this will cut down on electricity consumption by up to 40%. Tungsten-halogen bulbs can be used outdoors as well as indoors and are suitable for small parking lots, gardens, and marking pathways. No warm-up or waiting is required after a momentary power interruption.

Energy Efficient High Intensity Discharge (HID) Bulbs

Efficient HID light sources include high-pressure sodium (HPS) and metal halide types. HPS bulbs are an efficient source for exterior lighting. They provide bright light, which is ideal to ensure safety around buildings. HPS bulbs use 70% less energy than standard floodlights and last up to eight

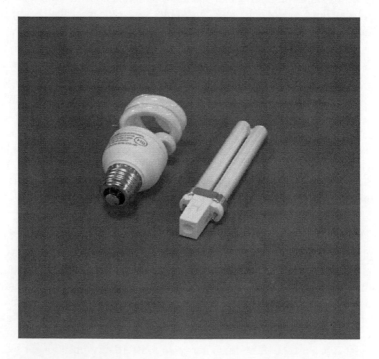

FIGURE 4.13 Energy efficient light lamps.

times longer. This type of fixture is estimated to have a lifetime of approximately 10 years. Metal halides provide a blue–white light source and can be used to highlight plants in gardens.

Figure 4.13 shows light output that is measured in lumens. A standard 110 W bulb produces about 1,680 lumens; whereas a 26 W compact fluorescent tube produces about 1,800 lumens. This figure can be used when replacing regular bulbs with energy efficient compact fluorescents.

Solar Energy Powered Lighting

Solar energy is an alternative energy that is useful for lighting to reduce the energy cost. Solar lighting systems are suitable for small garden and sign displays. Figure 4.14 shows a solar powered system. It's fixture has a re-chargeable battery to store energy generated by solar photovoltaic cells. The solar photovoltaic cells generate energy from solar radiation during day hours for use at night. The fixture uses a fluorescent lamp powered by the stored energy.

4.4.8.10 Lasers

One of the most fascinating inventions in the second half of the twentieth century is the laser. Laser is an acronym for Light Amplification by Stimulated Emission of Radiation. Laser devices produce a very narrow intense beam of monochromatic and coherent light. Laser beams are used in many applications such as for aligning subway tunnels, precisely adjusting integrated circuit chips, boring holes in steel, sending signals into fibre optic cables in communication systems, performing surgery on the retina of the eye, activating remote control devices, playing CDs, printing materials in laser printers, reading bar codes. Lasers send thousands of signals through telephone lines and many television signals over fibre optic cables. There are other applications listed in details in the laser chapter.

The beam emitted from the laser is coherent and directional, while an ordinary light source emits incoherent light in all directions. The action of laser production is based on electrons existing at specific energy levels or states characteristic of that particular atom. Electrons can be pumped up to higher energy levels by the external energy. These electrons may jump spontaneously to the lower state, resulting in the emission of energy in the form of a photon. However, if a photon with

FIGURE 4.14 Solar powered outdoor lighting.

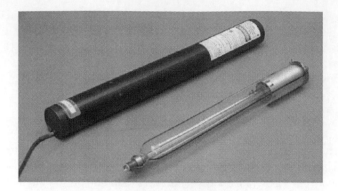

FIGURE 4.15 Helium–Neon (He–Ne) laser device.

this same energy strikes the excited atom, it can stimulate the atom to make the transition to the lower state. In this phenomenon, called stimulated emission, the new photon has the same frequency as the original one. Also, these two photons are in phase and moving in the same direction. This is how coherent light is produced in a laser. Depending on the particular lasing material being used, specific wavelengths of light are emitted.

There are many types of lasers, such as the gas laser, Ruby laser, Helium–Neon (He–Ne) laser, Fabry-Perot laser, CO_2 gas laser, and laser diode. Figure 4.15 shows the Helium-Neon (He–Ne) laser device with internal glass tube that has two parallel mirrors, one at each end. More details on laser principles, operations, and devices are presented laser chapter.

4.5 EXPERIMENTAL WORK

This experiment studies light emission from different types of light sources. The student will study power output from different light sources, such as a fluorescent tube, incandescent bulb, halogen lamp, mercury lamp, He–Ne laser, LED, and laser diode. The student will measure the power

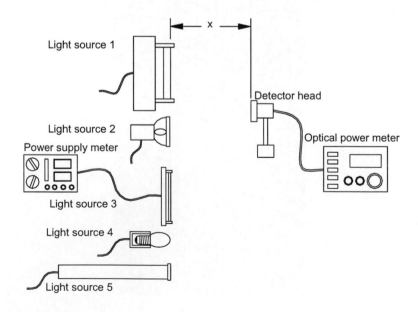

FIGURE 4.16 Light emission from different light sources.

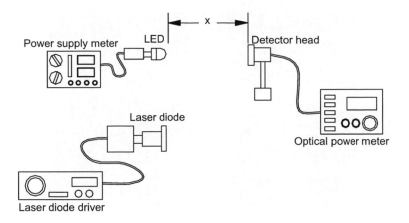

FIGURE 4.17 Light emission from laser diode and light emitting diode.

consumption and the energy output from these light sources. Then the efficiency of a light source will be calculated. An efficiency comparison among each light source will be carried out.

This experiment is arranged into two cases:

a. Light emission from five different light sources shown in Figure 4.16.
b. Light emission from an LED and laser diode shown in Figure 4.17.

4.5.1 TECHNIQUE AND APPARATUS

Appendix A presents the details of the devices, components, tools, and parts.

1. He–Ne laser source, as shown in Figure 4.16 and Figure 4.18. Always follow laser operation procedure and safety instructions when using a laser light source.
2. Laser power supply.
3. Fluorescent tube, incandescent lamp, halogen lamp, mercury lamp, as shown in Figure 4.18.
4. LED and laser diode/laser diode drive, as shown in Figure 4.2.
5. Hardware assembly (clamps, posts, screw kits, screwdriver kits, sundry positioners, etc.).

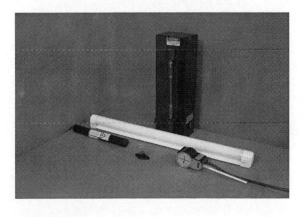

FIGURE 4.18 Five different light sources.

FIGURE 4.19 Light emission from five different light sources.

  6. Electrical power supply meter, as shown in Figure 4.19.
  7. Optical detector head and optical power meter, as shown in Figure 4.19.
  8. Ruler.

4.5.2 PROCEDURE

Follow the laboratory procedures and instructions given by the professor and/or instructor.

4.5.3 SAFETY PROCEDURE

Follow all safety procedures and regulations regarding the use of light source devices, measurement instruments, and power supplies.

4.5.4 APPARATUS SET-UP

4.5.4.1 Light Emission from Five Different Light Sources

  1. Figure 4.19 shows the experimental apparatus set-up.
  2. Prepare five types of light sources on the table.
  3. Connect light source 1 to the power supply meter.
  4. Mount an optical detector head connected to an optical power meter at a distance of x cm facing the light source.
  5. Align the optical detector and light source 1 to face each other.
  6. Measure the distance x between light source 1 and the optical detector head. Fill out Table 4.1.
  7. Turn on the power supply to provide electrical power to light source 1. Be sure not to exceed the maximum voltage and current for each light source.
  8. Turn off the lights of the lab.
  9. Measure the voltage (V, in Volts) and the current (I, in Amperes) to calculate the power (W, in Watts $= V \times I$) consumed by light source 1. Fill out Table 4.1.
  10. Measure light power emitted by light source 1 using the optical power meter. Fill out Table 4.1.
  11. Repeat steps 3–10 for the four remaining light sources.
  12. Illustrate the location of the light sources and the optical detector head in a diagram.

FIGURE 4.20 Light emission from a light emitting diode and laser diode.

4.5.4.2 Light Emission from an LED and Laser Diode

1. Figure 4.20 shows the experimental apparatus set-up.
2. Repeat the procedure explained in Case (a).
3. Use an LED and laser diode/laser diode drive as light sources. Be sure not to exceed the maximum voltage and current for the LED and laser.
4. Measure the voltage V and the current I for the LED and laser diode, and the distance x. Fill out Table 4.2.
5. Measure the light power emitted by the LED and laser diode. Fill out Table 4.2.
6. Illustrate the location of the LED and laser diode and the optical detector head in a diagram.

4.5.5 DATA COLLECTION

4.5.5.1 Light Emission from Five Different Light Sources

1. Measure the voltage (V), the current (I), and the distance (x). Fill out Table 4.1.
2. Measure the light power emitted by the light sources. Fill out Table 4.1.

TABLE 4.1
Light Emission from Five Different Light Sources

Light	Light Source	Volts	Amperes	Input Power = ($V \times I$)	Optical Power (Output Power)	Distance	Efficiency = (Output Power/Input Power)×100
Source	Type	(V)	(I)	(W)	(W)	(x)	(%)
No. 1							
No. 2							
No. 3							
No. 4							
No. 5							

4.5.5.2 Light Emission from an LED and Laser Diode

1. Measure the voltage (V), the current (I), and the distance (x). Fill out Table 4.2.
2. Measure the light power emitted by the LED and laser diode. Fill out Table 4.2.

TABLE 4.2
Light Emission from a Laser Diode and LED

Light Source	Light Source Type	Volts (V)	Amperes (I)	Input Power = (V × I) (W)	Optical Power (Output Power) (W)	Distance (x)	Efficiency = (Output Power/Input Power)×100 (%)
No. 1							
No. 2							

4.5.6 Calculations and Analysis

4.5.6.1 Light Emission from Five Different Light Sources

1. Calculate the power (W, in Watts $= V{\times}I$) consumed by the light sources. Fill out Table 4.1.
2. Calculate the efficiency of each light source, by dividing power output by power input. Fill out Table 4.1.
3. Illustrate the location of the light sources and the optical detector head in a diagram.

4.5.6.2 Light Emission from an LED and Laser Diode

1. Calculate the power (W, in Watts $= V{\times}I$) consumed by the LED and laser diode. Fill out Table 4.2.
2. Calculate the efficiency of the LED and laser diode, by dividing the power output by power input. Fill out Table 4.2.
3. Illustrate the location of the LED and laser diode and the optical detector head in a diagram.

4.5.7 Results and Discussions

4.5.7.1 Light Emission from Five Different Light Sources

1. Discuss the measurements of the voltage and current, the calculation of the power consumed by the light sources, and the optical power measurements.
2. Compare the efficiency of the light sources.
3. Verify the calculated efficiency with the efficiency provided by the manufacturer.

4.5.7.2 Light Emission from an LED and Laser Diode

1. Discuss the measurements of the voltage and current, the calculation of the power consumed by the LED and laser diode, and the optical power measurements.
2. Compare the efficiency of the LED and laser diode.
3. Verify the calculated efficiency with the efficiency provided by the manufacturer.

4.5.8 Conclusion

Summarize the important observations and findings obtained in this lab experiment.

4.5.9 SUGGESTIONS FOR FUTURE LAB WORK

List any suggestions for improvements using different experimental equipment, procedures, and techniques for any future lab work. These suggestions should be theoretically justified and technically feasible.

4.6 LIST OF REFERENCES

List any references that were used in the report. Use one format in writing the references. Never mix reference formats in a report.

4.7 APPENDICES

List all of the materials and information that are too detailed to be included in the body of the report.

FURTHER READING

Cornsweet, T. N., *Visual Perception*, Academic Press, New York, 1970.

Deveau, R. L., *Fiber Optic Lighting: A Guide for Specifiers*, 2nd ed., Prentice Hall PTR, Upper Saddle River, NJ, 2000.

Falk, D. et al., *Seeing the Light Optics in Nature, Photography, Color, Vision, and Holography*, Wiley, New York, 1986.

Grove, A. S., *Physics and Technology of Semiconductor Devices*, Wiley, New York, 1967.

Hood, D. C. and Finkelstein, M. A., Sensitivity to light handbook of perception and human performance, In *Sensory Processes and Perception*, Boff, K. R., Kaufman, L., and Thomas, J. P., Eds., Vol. 1, John Wiley & Sons, Toronto, 1986.

Illuminating Engineering Society of North America, *IESNA lighting education—Intermediate level*, IESNA ED-150.5A, 1993.

Kuhn, K., *Laser Engineering*, Prentice Hall, Inc, Upper Saddle River, NJ, 1998.

Lengyel, B., *Lasers*, Wiley, New York, 1971.

Lerner, R. G. and Trigg, G. L., *Encyclopedia of Physics*, 2nd ed., VCH Publishers, Inc, New York, 1991.

McComb, G., *The laser cookbook—88 practical projects Tab Book*, Division of McGraw-Hill, Inc, Blue Ridge Summit, PA, 1988.

Melles Griot, The practical application of light, *Melles Griot Catalog*, Irvine, California, U.S.A., 2001.

Overheim, R. D. and Wagner, D. L., *Light and Color*, John Wiley and Sons, Inc, New York, 1982.

Pritchard, D. C., *Environmental Physics: Lighting*, Longmans, Green & Co, London, 1969.

Product Knowledge, Lighting reference guide, *Product Development*. 2nd ed., Ontario Hydro, Toronto, ON, Canada, 1988.

Salah, B. E. A. and Teich, M. C., *Fundamentals of Photonics*, Wiley, New York, 1991.

Schildegen, T. E., *Pocket Guide to Color With Digital Applications*, Delmar Publishers, Albany, New York, U.S.A., 1998.

SCIENCETECH, *Designers and Manufacturers of Scientific Instruments Catalog*, SCIENCETECH, London, Ontario, Canada, 2003.

Serway, R. A., *Physics for Scientists and Engineers*, 3rd ed., Saunders Golden Sunburst Series, Philadelphia, 1990.

Weisskopf, V. F., How light interacts with matter, *Scientific American*, 219, (3), pp. 60–71, U.S.A., September, 1968.

Williamson, S. J. and Cummins, H. Z., *Light and Color in Nature and Art*, Wiley, New York, 1983.

Yariv, A., *Optical Electronics*, Wiley, New York, 1997.

Yeh, C., *Applied Photonics*, Academic Press, San Diego, CA, 1994.

5 Light Intensity

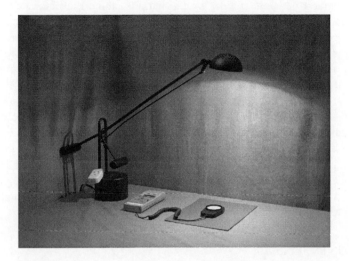

5.1 INTRODUCTION

The light from a fire or flame will propagate spherically outwards from the source. In general, the apparent intensity of the light will vary inversely with the distance from the observer to the source. The earliest unit for measuring light, the foot-candle, was based on the amount of light being emitted from a single candle at a distance of one foot. The more candles lit, the brighter the radiant light.

Since light waves propagate spherically from a candle, the number of photons per unit area will vary based on the distance to the candle. The measurement of light intensity is called luminous flux.

Experimental work presented in this chapter involves measuring light intensity of five different light sources at five distances. It also involves measuring the distribution of light intensity on a floor, to see if the light is evenly distributed throughout the floor and if it complies with the recommended indoor lighting levels.

5.2 LIGHT INTENSITY

Electromagnetic wave motion involves the propagation of energy from a source to a detector. The rate of energy transfer is expressed as intensity. The intensity is defined as the energy transported per time per unit area as given in the following equation:

$$\text{Intensity} = \frac{\frac{\text{Energy}}{\text{Time}}}{\text{Area}} = \frac{\text{Power}}{\text{Area}} \tag{5.1}$$

The standard units of intensity are watts per square metre (W/m^2).

5.3 LUMINOUS FLUX

Light sources (except coherent and directional light sources such as lasers) emit electromagnetic waves in all directions. A lamp emits radiation when electric energy is supplied to the lamp. The radiation energy emitted per unit of time by the lamp is called the radiant power or sometimes referred to as the radiant flux. Radiant energy covers a wide range of the light spectrum. The visible light is a small range of the light spectrum. The range of the visible light is approximately between wavelengths of 400 and 700 nm. The visual sensing of the eye perceives the visible light and luminous energy radiation. Therefore, the luminous flux is defined as that part of the total radiant power emitted from a light source that is capable of affecting the sense of sight.

The human eye is not equally sensitive to all colours. In other words, equal radiant power of different wavelengths does not produce equal perceived brightness. Each colour has perceived brightness different from other colours. For example, a 40-W green light bulb appears brighter than a 40-W blue light bulb. A relative sensitivity graph of the visible light spectrum is shown in Figure 5.1. The sensitivity curve is a bell-shape with the centre of the visible spectrum at wavelength of 555 nm. Under normal conditions, the human eye is most sensitive to yellow–green light of wavelength 555 nm. This curve also shows the sensitivity falling off rapidly for longer and shorter wavelengths.

5.4 LUMINOUS INTENSITY

The brightness of a light source is referred to as luminous intensity I, for which the candela (cd) is the SI unit (metric system). The candela is defined in terms of the light emitted by a small pool of platinum at its melting point. A candle has a luminous intensity of about 1 cd, and in fact the former standard of this quantity was an actual candle of specified composition and dimensions.

Consider the definition of a solid angle θ in terms of the radius and the circular arc length in two-dimensional analysis. Figure 5.2 shows the angle θ, measured in radians, is defined as:

$$\theta = \frac{S}{R} \tag{5.2}$$

where S is the arc length and R is the radius of the circle.

Similarly, the solid angle Ω in space analysis is explained in Figure 5.3. Point C represents the centre of the sphere. A steradian sr is defined as conical in shape. The solid angle (conical) Ω, representing one steradian, such that the area A of the subtended portion of the sphere is equal to R^2,

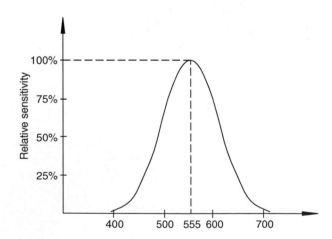

FIGURE 5.1 A relative sensitivity curve for visible light perceived by the eye.

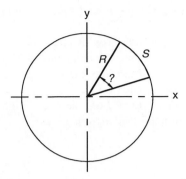

FIGURE 5.2 An angle θ expressed in radians in a plane.

where R is the radius of the sphere. The total solid angle of a sphere is 4π steradians. The number of steradians in a given solid angle can be determined by dividing the area on the surface of a sphere (lying within the intersection of that solid angle with the surface of the sphere) by the square of the radius R of the sphere.

Therefore, the solid angle in steradians sr is defined as:

$$\Omega = \frac{A}{R^2} \, (\text{sr}) \tag{5.3}$$

Notice that, the steradian is a unitless quantity in Equation (5.3). The solid angle is independent of the radius R of a sphere. There are 4π sr in a sphere, regardless of the length of its radius as given in the following equation:

$$\Omega = \frac{A}{R^2} = \frac{4\pi R^2}{R^2} = 4\pi \, (\text{sr}) \tag{5.4}$$

The luminous flux F emitted by a light source describes the total amount of visible light that it emits. A small bright source may give off less light than a large diameter source, just as a small hot object may give off less heat than a large warm one. The unit of luminous flux is the lumen lm. Consider a 1-candela light source at the centre of the sphere, as shown in Figure 5.3. Such a light source is referred to as isotropic because it radiates equally in all directions. One lm equals the luminous flux falling on each square metre of a 1 m radius sphere. A lumen of flux is equivalent to about 0.0015 W of yellow–green light of wavelength 555 nm. Since the area of a sphere of radius R

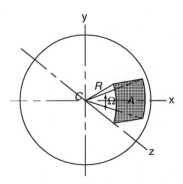

FIGURE 5.3 A solid angle Ω expressed in steradians in space.

is $4\pi R^2$, the above sphere has a total surface area of 4π m^2, and the total luminous flux radiated by a 1-cd source is thus 4π lm.

Light travels radially outward in straight lines from a point source that is small compared with its surroundings. For a point light source, the luminous flux included in a solid angle Ω remains the same in all directions from the source. Therefore, the luminous flux per unit solid angle is simply expressed as the total flux. The physical quantity that expresses this relationship is called the luminous intensity I. The luminous intensity I of a source of light is the luminous flux F emitted per unit solid angle Ω, as given in the following equations:

$$I = \frac{F}{\Omega}.$$ (5.5)

or,

$$F = I\Omega =$$ (5.6)

The unit for luminous intensity I is the lumen per steradian (lm/sr), which is called a candela (cd). The candela, or candle as it was sometimes called, originated when the international standard was defined in terms of the quantity of light emitted by the flame of a certain make of candle. This standard was found unsatisfactory and eventually was replaced by the platinum standard. The total solid angle Ω for an isotropic source is 4π sr. Thus, the luminous flux is given by:

$$F = 4\pi I$$ (5.7)

5.5 ILLUMINATION

If the intensity of a source is increased, the luminous flux F transmitted to each unit of surface area in the vicinity of the source is also increased. The surface appears brighter. The illumination E of a surface A is defined as the luminous flux F per unit area A, as given in the following equation:

$$E = \frac{F}{A}$$ (5.8)

When the flux F is measured in lumens and the area A in square metres, the illumination E has units of lumens per square metre or lux (lx). When the area A is expressed in square feet, the E is expressed in lumens per square foot. The lumen per square foot is sometimes referred to as the footcandle, where 1 footcandle $= 0.093$ lux.

Figure 5.3 shows a point light source emitting the light energy in all directions. To see the relations between intensity and illumination, consider a surface A at a distance R from a point source at C of intensity I, as shown in Figure 5.3. The solid angle subtended by the surface at the source is given by Equation (5.3) as where the area A is perpendicular to the emitted light. If the luminous flux makes an angle θ with the normal to the surface, then the projected area, $A \cos \theta$, is the perpendicular area to the normal. Therefore, the solid angle in Equation (5.3) can be written in general form as follows:

$$\Omega = \frac{A \cos \theta}{R^2}$$ (5.9)

Substitute Equation (5.9) into Equation (5.6) to obtain:

$$F = I\Omega = \frac{IA \cos \theta}{R^2}$$ (5.10)

In this equation, θ is the angle between the direction of the light and a normal to the surface where the light strikes. When $\theta = 0°$, then $\cos \theta = 1$. To express the illumination as a function of

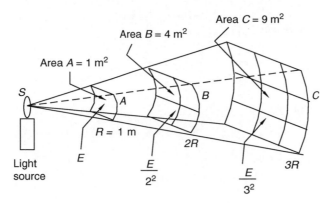

FIGURE 5.4 Inverse-square law of light.

intensity, substitute Equation (5.10) into Equation (5.8) to give the following equation:

$$E = \frac{F}{A} = \frac{IA \cos \theta}{AR^2} = \frac{I \cos \theta}{R^2} \tag{5.11}$$

For the special case in which the surface is normal to the flux, $\theta = 0°$, Equation (5.11) is simplified to:

$$E = \frac{I}{R^2} \tag{5.12}$$

Equation (5.12) involving illumination and luminous intensity is a mathematical formulation of the inverse-square law of light waves. This law can be stated as: The illumination of a surface is perpendicular to the luminous intensity of a point light source and is inversely proportional to the square of the distance. This describes the fundamental law of light intensity variation with distance from a point light source.

Figure 5.4 illustrates this inverse-square law of light. The figure shows light emitted from a point source S becomes distributed over 1 m^2 at the first distance R. If the distance is doubled to $2R$, the light is distributed over 4 m^2, and the illumination per square metre is one-quarter of what it was at the distance R. If the distance is tripled to $3R$, the illumination per square metre will be one-ninth of what is was at distance R. Therefore, the intensity of light emitted from the point light source varies inversely as the square of the distance from the light source. This is called the inverse square law for electromagnetic light waves.

5.6 EXPERIMENTAL WORK

This experiment studies the light intensity from different types of light sources. These different light sources could be a fluorescent tube, incandescent bulb, halogen lamp, mercury lamp, LED, and laser diode. The student will measure the light intensity for at least five different distances from these light sources. Then intensity of a light source will be drawn as a function of distance. Light intensity comparison among each light source will be carried out. The student will also measure light intensity on a flat area and compare it with the recommended lighting levels related to the area under consideration according to indoor lighting system codes.

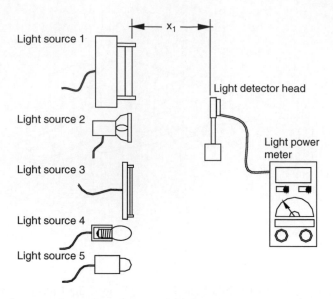

FIGURE 5.5 Light intensity measurements from different light sources.

This experiment covers two cases:

a. Light intensity measurements of five different light sources are shown in Figure 5.5. Each light source is placed in front of a light meter at five distances between them. Light intensity is measured using a light meter.

b. Distribution of light intensity on a flat surface is shown in Figure 5.6. A light meter is used to measure the distribution of the intensity of the light from the ceiling lighting system projected onto the floor of a room. The measurements will show if the light is evenly distributed throughout the floor and also if the measurements comply with the recommended indoor lighting levels.

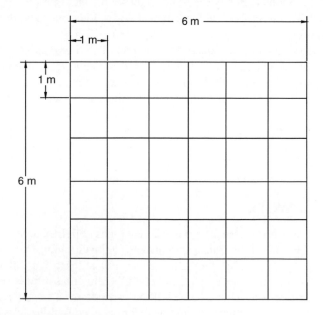

FIGURE 5.6 Distribution of light intensity on a flat surface.

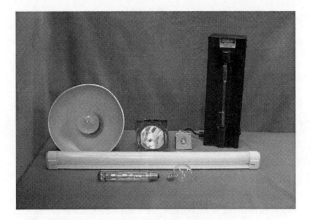

FIGURE 5.7 Five different light sources.

5.6.1 TECHNIQUE AND APPARATUS

Appendix A presents the details of the devices, components, tools, and parts.

1. Fluorescent tube, high power incandescent lamp, low power incandescent lamp, halogen lamp, and mercury lamp, as shown in Figure 5.7.
2. Hardware assembly (clamps, posts, positioners, etc.).
3. Light meter, as shown in Figure 5.6.2.
4. Ruler.

FIGURE 5.8 Light intensity from five different light sources set-up.

5.6.2 PROCEDURE

Follow the laboratory procedures and instructions given by the professor and/or instructor.

5.6.3 Safety Procedure

Follow all safety procedures and regulations regarding the use of optical instruments and measurements, and light source devices.

5.6.4 Apparatus Set-up

5.6.4.1 Light Intensity From Five Different Light Sources

1. Figure 5.8 shows the experimental apparatus set-up.
2. Prepare five types of light sources on the table.
3. Connect light source 1 to the power outlet.
4. Mount a light meter at a distance x_1 from light source 1.
5. Align the light meter and light source 1 to face each other.
6. Measure the distance x_1 between light source 1 and the light meter. Fill out Table 5.1.
7. Switch on the power supply of light source 1.
8. Turn off the lights of the lab.
9. Measure the intensity of light source 1 using the light meter. Fill out Table 5.1.
10. Switch off the power supply of light source 1.
11. Repeat steps 3 to 10 for four different light sources.
12. Turn on the lights of the lab.
13. Illustrate the location of the light sources and the light meter in a diagram.

5.6.4.2 Distribution of Light Intensity on a Flat Surface

1. Figure 5.9 shows the experimental set-up.
2. Find an unoccupied flat area of a floor below the ceiling lighting system.
3. Create a grid pattern on the flat area in 1 square metre sections. Use masking tape to map out the flat floor.
4. Draw the flat area layout on a paper using a suitable scale.
5. Measure the light intensity at the centre of each square in the grid. Be careful as not to shadow the light meter with your body during the measurements. Fill out Table 5.2.
6. Illustrate the distribution of the light intensity on the flat floor in a diagram.

FIGURE 5.9 Distribution of light intensity on a flat floor.

5.6.5 DATA COLLECTION

5.6.5.1 Light Intensity from Five Different Light Sources

Measure the intensity of each of the five light sources, at five distances. Fill out Table 5.1.

TABLE 5.1
Light Intensity from Five Different Light Sources

Distance between Light Source and Light Meter (unit:)	Intensity of the (unit:)				
	Light Source Type 1	Light Source Type 2	Light Source Type 3	Light Source Type 4	Light Source Type 5
$x_1 =$					
$x_2 =$					
$x_3 =$					
$x_4 =$					
$x_5 =$					

5.6.5.2 Distribution of Light Intensity on a Flat Surface

Measure the intensity of light at the centre of each grid on the floor. Fill out Table 5.2.

TABLE 5.2
Distribution of Light Intensity on a Flat Surface

	1	2	3	4	5	6 (meter)
1	× (unit:)	×	×	×	×	×
2	×	×	×	×	×	×
3	×	×	×	×	×	×
4	×	×	×	×	×	×
5	×	×	×	×	×	×
6 (meter)	×	×	×	×	×	×

5.6.6 CALCULATIONS AND ANALYSIS

5.6.6.1 Light Intensity from Five Different Light Sources

No calculations are required for this case.

5.6.6.2 Distribution of Light Intensity on a Flat Surface

No calculations are required for this case.

5.6.7 RESULTS AND DISCUSSIONS

5.6.7.1 Light Intensity from Five Different Light Sources

1. Discuss the measurements of the intensity of light sources at different distances.
2. Compare the intensity of the light sources.
3. See if the inverse square law of light is applicable with your measurements.
4. Verify the intensity of the light sources with the intensity that is provided by the manufacturer.

5.6.7.2 Distribution of Light Intensity on a Flat Surface

1. Discuss the measurements of the intensity of light at the centre of each grid of the flat floor under consideration.
2. Present the light intensity distribution of the flat floor under consideration.
3. Compare your measurements of the intensity of light with indoor lighting system codes.

5.6.8 CONCLUSION

Summarize the important observations and findings obtained in this lab experiment.

5.6.9 SUGGESTIONS FOR FUTURE LAB WORK

List any suggestions for improvements using different experimental equipment, procedures, and techniques for any future lab work. These suggestions should be theoretically justified and technically feasible.

5.7 LIST OF REFERENCES

List any references that were used in the report. Use one format in writing the references. Never mix reference formats in a report.

5.8 APPENDICES

List all of the materials and information that are too detailed to be included in the body of the report.

FURTHER READING

Adams, R. A., et al., Systematic measurements of human neonatal color vision, *Vision Research*, 34, 1691–1701, 1994.

Alda, J., Laser and Gaussian beam propagation and transformation, in *Encyclopedia of Optical Engineering*, Barry Johnson, R., Ed., Marcel Dekker, Inc., New York, 2002.

Beiser, A., *Physics*, 5th ed., Addison-Wesley Publishing Company, Reading, MA, 1991.

Chen, K. P., In-Fiber light powers active fiber optical components, *Phontonics Spectra*, 78–90, 2005.

Cutnell, J. D. and Johnson, K. W., *Physics*, 5th ed., Wiley, New York, 2001.

Digonnet, M. J. F., *Rare-Earth Fiber Lasers and Amplifiers*, Marcel Dekker, NY, U.S.A., 2001.

Falk, D., et al., *Seeing the Light Optics in Nature, Photography, Color, Vision, and Holography*, Wiley, New York, 1986.

Hood, D. C. and Finkelstein, M. A., Sensitivity to light handbook of perception and human performance, in *Sensory Processes and Perception*, Boff, K. R., Kaufman, L., and Thomas, J. P., Eds., Vol. 1, Wiley, Toronto, 1986.

Jameson, D. and Hurvich, L. M., Theory of brightness and color contrast in human vision, *Vision Research*, 4, 135–154, 1964.

Jenkins, F. W. and White, H. E., *Fundamentals of Optics*, McGraw Hill, New York, 1957.

Jones, E. R. and Childers, R., *Contemporary college physics*, McGraw-Hill Higher Education, Maidenhead, U.K., 2001.

Kuhn, K., *Laser engineering Englewood, Chffs*, Prentice Hall, Inc, NJ, 1998.

Lerner, R. G. and Trigg, G. L., *Encyclopedia of Physics*, 2nd ed., VCH Publishers, Inc, New York, 1991.

Melles Griot, The practical application of light, *Melles Griot Catalog*, Irvine, California, U.S.A., 2001.

Pritchard, D. C., *Environmental Physics: Lighting*, Longmans, Green & Co, London, 1969.

Product Knowledge, Lighting reference guide, *Product development*, 2nd ed., Feb. 1988.

Robinson, P., *Laboratory Manual to Accompany Conceptual Physics*, 8th ed., Addison-Wesley, Inc., Reading, MA, 1998.

Romine, G. S., *Applied physics concepts into practice Englewood, Chffs*, Prentice Hall, Inc, NJ, 2001.

Shen, L. P., Huang, W. P., Chen, G. X., et al., Design and optimization of photonic crystal fibers for broad-band dispersion compensation, *IEEE Photonics Technology Letter*, 15, 540–542, 2003.

Urone, P. P., *College Physics Belmont*, Brooks/Cole Publishing Company, CA, U.S.A., 1998.

Warren, M. L., *Introduction to Physics*, W.H. Freeman and Company, San Francisco, CA, U.S.A., 2001.

Weisskopf, V. F., How light interacts with matter, *Scientific American*, 60–71, 219(3), U.S.A., Sept. 1968.

6 Light and Colour

6.1 INTRODUCTION

Visible light radiating from the sun consists of a broad spectrum of wavelengths ranging from violet to red. When light strikes a material, some of the wavelengths are reflected and others are absorbed. The characteristics of a material and the incident light striking them determine the apparent colour of an object. Light can be used to measure an object's temperature; for example, blue stars are hotter than red stars.

Display devices mix red, green, and blue (RGB) light to form different coloured pixels in an image. Likewise, digital cameras separate and measure the RGB light from a scene to form digital images.

Experimental work presented in this chapter will demonstrate mixing coloured light to produce different colours. Newton's colour wheel and black and white colour wheels are used in the experiment.

6.2 COLOURS

Colour is a phenomenon of light, and a visual perception using the physical sense of human sight. As with most human senses, vision is susceptible to sensory adaptation. It has been known at least since the early days of the ancient Egyptians that fragments of clear, colourless glass and precious stones emit the colours of the rainbow when placed in the path of a beam of white light. It was not until 1666, however, that this phenomenon called desperation was systematically investigated by scientists. The refracting telescope was invented by a Dutch eyeglass maker named Hans Lippershey (1570–1619). Later, the English physicist and mathematician, Isaac Newton (1642–1727), was starting to search for a technique for removing colouration from the images seen through reflecting telescopes. In 1672, Newton described his experiments to the Royal Society in London. His theory, that white light was made up of many colours, was revolutionary and it was greeted with skepticism. Indeed Newton and another English physicist, Robert Hooke (1635–1703), became involved in a bitter debate, and Newton refused to publish his conclusions until after Hooke's death, 32 years later.

As with all human sense organs, the eye is sensitive to physical energy. The human eye is sensitive to a very small portion of the electromagnetic light spectrum. This portion of the spectrum is referred to as light energy covering the visible range. The human eye only sees colours starting from blue light waves that measure about 400 nm and ending with the longer red light waves of about 700 nm, as shown in Figure 6.1.

Light is one of the four essential elements in the visual experience of colour. For the human sensation of colour vision, the following four elements are required:

1. A coloured object
2. A light source to illuminate the coloured object
3. The human eye as a receptor of the light energy reflected from the object
4. The human brain to interpret the electrochemical neural impulses sent from the eye

There are many different pigments or colourants, and these can be organic or inorganic in nature. The coloured object either absorbs or reflects the incident light that illuminates it. Some objects, such as coloured glass, can allow light to be transmitted through them, filtering all colours other than their own.

The eye is the physical receptor of light energy. The major parts of the human eye are shown in Figure 6.2. As the eye focuses on a coloured object, light passes through the cornea and is regulated by the pupil opening. The lens of the eye focuses the light on the retina in the back of the eye. The retina is the light-sensitive surface located around the back of the eye. The retina is made up of rods and cones, which are the photosensitive cells that convert the light energy into different nerve impulses. The retina also has ganglion cells and bipolar cells. Vision is a function of light energy reaching the rods and cones of the eye. The bulbous cones can see chromatic vision or colour, such as red, green, and blue. The cones can also see achromatic colour, such as white, grey, and black. There are more than six million bulbous cones distributed in the retina of the eye. The cones need higher levels of illumination to produce colour vision. There are around 100 million rods in the retina, and they function in dim light conditions and produce monochromatic vision. The rods produce vision in times of low illumination. The rods see black and white or grey tonal values. If you look at a landscape during the daylight, the grass and shrubs are green and the sky is blue in

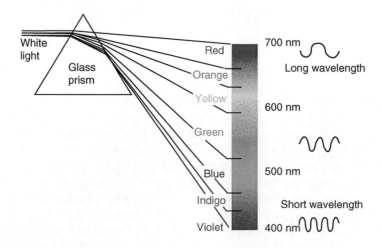

FIGURE 6.1 (See colour insert following page 512.) Dispersion of white light into the visible spectrum by a prism.

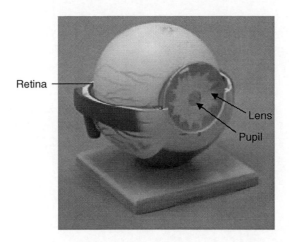

Retina

Lens

Pupil

FIGURE 6.2 (**See colour insert following page 512.**) The major parts of the human eye.

colour as a result of the different wavelengths of energy that stimulate the bulbous cones. Looking at the landscape at dusk requires the use of the rods in the eye and the green grass or blue sky appears to be a dark grey due to limited illumination. The fovea is the most sensitive part of the retina in daylight conditions. The fovea is made up of only bulbous cones and is centrally located in the retina. The light reflected from a coloured object is seen best when it strikes the fovea, because these cones are connected directly to the optic nerve which transmits the data to the brain. Other cones and rods in the retina require a relay of nerves to the brain. There are several theories regarding colour vision. The most common theory regarding colour vision was presented by the Young-Helmholtz theory, composed by Thomas Young (1773–1829), a British physician and Egyptologist and Hermann Helmholtz (1821–1894), a German physicist and physiologist. Their theory of colour vision proposed that the eye has three different colour receptors, each sensitive to different wavelengths of the spectrum. Because the eye receives light energy it was proposed that the bulbous cones are either sensitive to red, green, or blue light energy.

Another more modern colour theory includes elements from previous models of colour vision. In 1955, this theory was called either the zone or the opponent-process, and was proposed by Leo Hurvich and Dorothea Jameson. In the opponent-process theory, the eye has three colour receptors (cones with broad bands of sensitivity) that peak at three different portions of the visible spectrum. These different wavelength receptors in the eye are connected to the ganglion cells, some of which are always active regardless of light energy. The ganglion cells that receive light energy pulsate faster, and if the cell is inhibited in some way the pulsation diminishes. In this theory the ganglion cells can provide two opposing signals to the brain, which better explains differences in colour vision and negative afterimages that we experience.

Normal colour vision is often called trichromat, which refers to a person having the ability to sense or discriminate light from dark, red from green, and yellow from blue. Colour blindness is when an individual is limited in his/her ability to discriminate colour, such as red from green and yellow from blue. If a person has the ability to see contrast from light to dark and cannot discriminate one colour pattern from another, that individual has dichromat vision. If the individual has only the ability to see contrast of light to dark, the person has the vision of a monochromat.

Each colour of light covers a specific range of wavelengths. The spectral power distribution of the colour is measured by a spectroscope, which is an instrument that works on the principle of white light dispersion in a prism to indicate the power of the light's components at various wavelengths. Light entering the spectroscope is dispersed through a prism, such as in Newton's

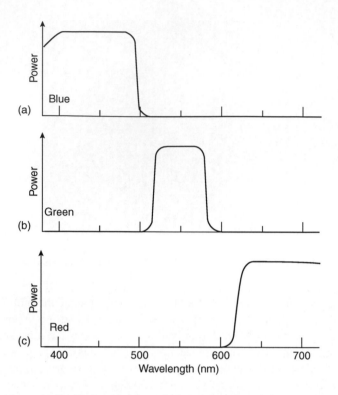

FIGURE 6.3 Spectral power distributions for the (a) blue, (b) green, and (c) red colours of the white light dispersion in a prism.

experiment. As a detector moves across the screen, for each wavelength, it measures the power of light received from the light source. The resulting curve giving the power to each wavelength is called the spectral power distribution. Figure 6.3 shows the spectral power distributions (SPD) for the blue, green, and red portions of the visible light spectrum. For blue, the power is found concentrated at short wavelengths; for green, at intermediate wavelengths; and for red, at long wavelengths.

6.3 MIXING LIGHT COLOURS

All colour reproduction is accomplished through either additive or subtractive colour theory. Additive theory is synonymous with light energy and the direct experience of light. Subtractive theory describes the experience of light reflected from a coloured object, such as paper coated with printing inks.

6.3.1 ADDITIVE METHOD OF COLOUR MIXING

Newton's edition of *Optics* in 1704 described a number of experiments in which he arranged mirrors to direct two different monochromatic components of the spectrum to the same location on a white screen. When he observed the colour of the mixed beams, Newton noticed that he could not always produce white light by mixing two monochromatic colours together. His systematic studies revealed the general rules of colour addition. As described above, each colour has a spectral power distribution. The spectral power distribution of one colour is added to the spectral power distribution of the other colour to produce the net spectral power distribution of the new colour. The rules of additive colour mixing are valid whether or not the light that produces a given colour is

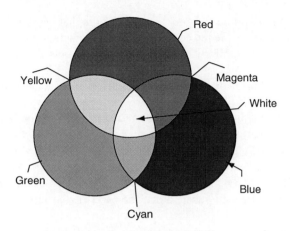

FIGURE 6.4 (**See colour insert following page 512**.) Additive colour mixing.

monochromatic. For simplicity, it is assumed that the brightness of each original is the same. Mixing green light with red light produces a yellow light. A second example of addition is when blue light and red light are mixed to produce magenta light (a purple). For a third example, blue light and green light are added together to produce cyan light (a turquoise). When the three colours blue, green, and red are mixed additively in various proportions a wider range of colours can be produced. If these three colours have the same proportion and brightness when added, they produce white colour. Therefore, the three colours the red, green, and blue (R, G, and B, respectively) are called primary colours for additive colour mixing. Adding the three primary colours of red, green, and blue lights will create white light, as shown in Figure 6.4.

Any two colours that produce white colour when added together are called complementary. The complement of a primary colour is called a secondary colour. Therefore, cyan is a secondary colour and is the complement of red colour. The complement of green is blue plus red colours, or magenta, so magenta is also a secondary colour. Also, the complementary of blue is green plus red colours, or yellow. Thus, yellow is a secondary colour and is the complementary of blue. As shown in Figure 6.4, the outside colours are the primary colours and the interior colours are the secondary colours.

6.3.2 SUBTRACTIVE METHOD OF COLOUR MIXING

The idea of primary colours can be applied to a different set of situations where the spectral power distribution is altered as light passes through a medium, which absorbs differing amounts of power across the spectrum. This occurs when a beam of light passes through one or more coloured filters (such as a piece of coloured glass or a plastic containing a dye). The more coloured filters we insert into the path of the beam, the more light is absorbed and the dimmer is the emerging light. A similar effect is produced when paints of various colours are mixed together. These are examples of subtractive colour mixing. The rules of subtractive mixing are more complicated than those of additive mixing. If several filters are placed in a beam of white light, only the first filter receives all of the white light. The others receive variously coloured light depending on the filters that precede them.

An understanding of the theory of filters is a good introduction to subtractive mixing of colours. By convention, a filter is named by the colour of the light that emerges from it when it is placed in a white beam. A red filter allows long wavelengths to pass and absorbs the remaining wavelengths. A common yellow filter allows medium through long wavelengths to pass and absorbs short wavelengths. The process where a material absorbs light in only certain regions of the spectrum is called selective absorption. It is the primary cause of most colours that we see. In terms of our additive primaries, a red filter allows only the red component of the incident light to pass. A common yellow filter transmits green and red. Similarly, a magenta filter transmits blue and red, and a cyan filter transmits blue and green.

Consider the combination of two filters, such as those of magenta and yellow. The magenta filter allows blue and red to pass. The yellow filter would transmit any incident green and red, but since green is absorbed by the magenta filter, only red emerges. The combination of the two filters therefore has the same effect as a red filter. It makes no difference which of the two filters is inserted into the beam first. To figure out what happens we must follow the path of light filtering, as shown in Figure 6.5. This figure shows the effect of (a) Magenta, (b) Yellow, (c) Cyan filters on white light. Also, (d–f) illustrates how the three additive primary colours can be produced from white light by using two filters in succession. The SPD of the emerging light are shown for each case.

The three subtractive primary colours are yellow, magenta, and cyan, as shown in Figure 6.6. The primary subtractive colours receive their name from the fact that each colour absorbs one-third of the white light spectrum. These subtractive primaries will make suitable ink colours

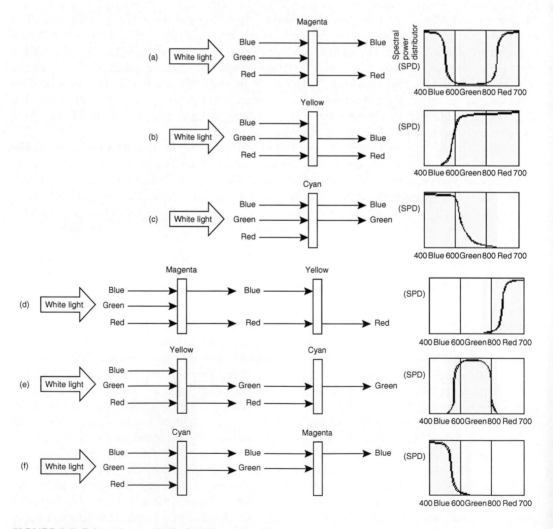

FIGURE 6.5 Subtractive method of mixing using filters.

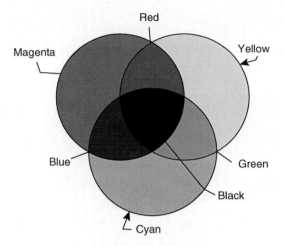

FIGURE 6.6 (**See colour insert following page 512**.) Subtractive colour mixing.

because when all are mixed they create black, or the total absence of colour. Printers subtract colour from the white sheet of paper. Basically, all the colours we see in a printed separation are contained in the red, green, and blue light energy reflected from a white sheet of paper. To produce the image, different degrees of red, green, and blue light are subtracted from the white light spectrum or paper.

The relationship between the additive primary colours and the subtractive primaries is considered complementary, as shown in Figure 6.7. The term complementary comes from the paired relationship where each subtractive colour absorbs its complementary one-third of the light spectrum. Yellow ink on white paper absorbs its complementary colour of blue light. Magenta ink on white paper will absorb one-third of the light spectrum, or its complementary colour of green. Finally, the cyan ink will absorb its complement of red light.

The colour wheel is used as a graphic aid in understanding basic colour theory. Though simple in design, the colour wheel represents a significant amount of information. The printing of colour reproductions involves creating colour separations of the primary colours contained in the original image, and reproducing them. When you mixed red and green tempera paints they created brown paint. The colour wheel used in graphic arts is built on the physical sciences; when red and green light are mixed we see the yellow. Using the colour wheel shown in Figure 6.8, note that the three additive primary colours are separated by the subtractive primary colours. Also, the

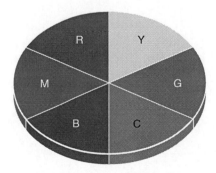

FIGURE 6.7 (**See colour insert following page 512**.) The Relationship between the additive and subtractive primary's complement.

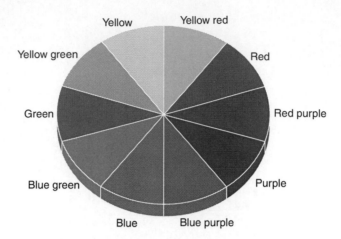

FIGURE 6.8 (See colour insert following page 512.) The colour wheel (the three primary additive colours sandwiched between variations of coloured light).

complementary colour pairs are opposite each other in the colour wheel. Finally, note that the two additive colours of light that make up a particular subtractive primary are located on each side of that ink colour.

6.4 THE COLOUR TRIANGLE

The colour triangle, as shown in Figure 6.9, is a triangular arrangement of the additive and subtractive primaries with white at the centre. Red, green, and blue are located at the corners, while magenta, yellow, and cyan are located at the sides. The order of the colours is such that the sum of any two additive primaries at the corners gives the subtractive primary between them on the sides, and the sum of all three primaries gives white at the centre.

Colours opposite each other on the colour triangle are complementary. Two colours are said to be complementary if, when added together, they produce white. Magenta and green are complementary, for when added together, they contain all of the spectrum colours of white light. Similarly red and cyan, as well as yellow and blue, are complementary.

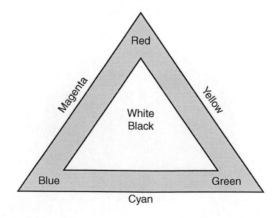

FIGURE 6.9 Diagram of the colour triangle, with the additive primaries at the corners and the subtractive primaries at the sides.

6.5 THE C.I.E. CHROMATICITY DIAGRAM

Well-planned steps toward a quantitative measurement of colour were taken by the Commission Internationale de I' Eclairage C.I.E. (International Commission on Illumination I.C.I.) in 1931. The three primaries—red, green, and blue—were adopted, in which the visible spectrum was divided into three overlapping spectral curves. It should be mentioned here that any given colour sample can be measured with a spectroscope in terms of the three adopted primaries, and the results of the measurements can be expressed by two numbers. These two numbers can then be plotted on a graph. When the pure spectrum colours (red R, orange O, yellow Y, green G, blue B, and violet V) are matched against a mixture of the standard primaries, a smooth curve is obtained, as shown in Figure 6.10. The figure shows white colour at the centre, the complete array of all possible colour mixtures lies within the enclosed area RGBVW, with the purples P and magentas M confined to the region RWV between the two ends of the spectrum.

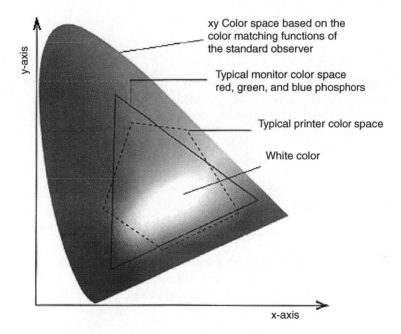

FIGURE 6.10 (See colour insert following page 512.) The C. I. E. Chromaticity diagram.

6.6 COLOUR TELEVISIONS

The use of artificial light sources to reproduce colour, such as a colour television, must also use the blue, green, and red portions of the spectrum to produce a white image. The television set has three cathode ray guns that project different degrees of red, green, and blue onto the phosphor coating on the inside of the picture tube to create different colours in the spectrum. The mixing of light energy to create a colour image is additive colour theory. In a dark room, the overlapping of red, green, and blue light beams will create additional colours, such as magenta, cyan, and yellow. Additive colour can also be seen in a television studio where coloured lights are used to create different effects on a studio backdrop. The various light adjustments on a control panel allow for different amounts of red, green, and blue filtered lights to create any colour desired, as shown in Figure 6.11.

The colour monitor on a desktop computer or colour workstation also uses red, green, and blue dots of phosphor on a black background to produce images. When electronically activated, the phosphor dots emit light energy. Unfortunately, the range of light energy emitted from various

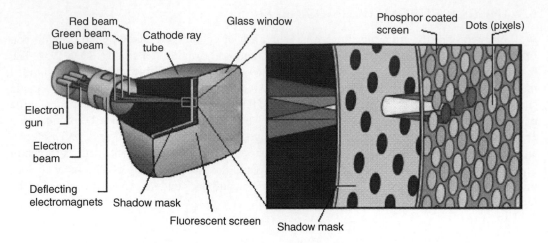

FIGURE 6.11 (**See colour insert following page 512.**) The additive theory is used in colour television.

monitors differ just like in various colour televisions for the home market. With this in mind, it becomes critical to understand that one colour monitor on a desktop workstation will vary from another. Furthermore, as a colour monitor and phosphor coating age the colour output will shift. There are some more expensive colour monitors that have been colour-calibrated and can be adjusted to a production standard every day. The need to calibrate colour devices is becoming more critical as the hardware and software become more sophisticated.

6.7 SPECTRAL TRANSMITTANCE CURVES

Figure 6.12 shows the spectral transmitted curve for different colours. The curve specifies the percentage of incident light that is transmitted for each wavelength across the spectrum. This percentage at a given wavelength is called the spectral transmittance for that wavelength. The spectral transmittance can range from 0 to 1 or, when expressed as a percentage, from 0 to 100%. The rest of the colour is absorbed by the filter or material, or is reflected back from one or both surfaces. The value of the spectral transmittance at a given wavelength is defined in Equation (6.1):

$$\text{Spectral Transmittance} = \frac{\text{Transmitted Power}}{\text{Incident Power}} \tag{6.1}$$

6.8 COLOUR TEMPERATURE

Colour temperature is a method of specifying the colour of a light source, but it should only be used when the source emits a continuous spectrum if it is to indicate the colour-rendering properties of the source. The colour temperature is measured in absolute temperature in Kelvin (K). Figure 6.13 shows some examples of colour temperature.

Although white light contains all the colours of the spectrum, they need not be present with the same intensity. For example, sunlight, often used as a reference standard for white light, does not have the same intensity at all wavelengths. The intensity of sunlight reaching the earth's surface peaks near a wavelength of 474 nm and falls off sharply in the ultraviolet. A comparison of the solar data with the Planck theory of blackbody radiation can be used to determine the sun's temperature, giving a result of about 6000 K. Incandescent lamps are good approximations of blackbody

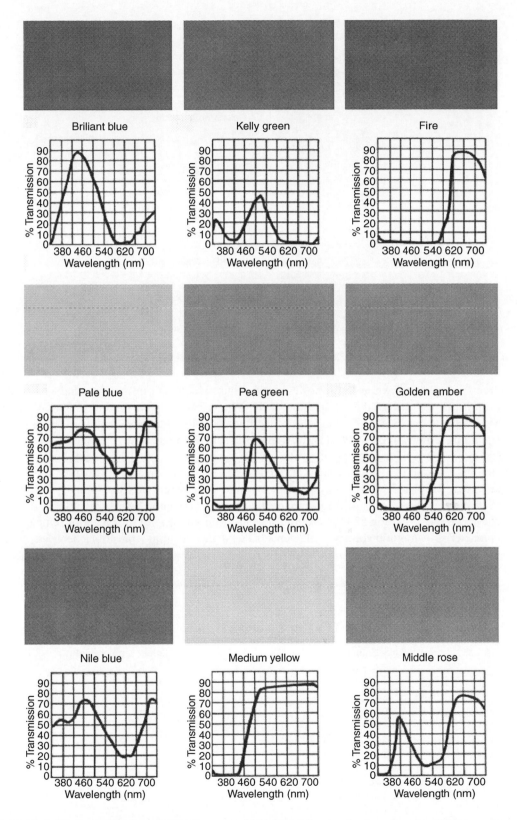

FIGURE 6.12 (See colour insert following page 512.) Spectral transmittance curve for different colours.

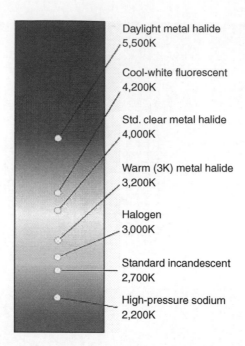

Daylight metal halide
5,500K

Cool-white fluorescent
4,200K

Std. clear metal halide
4,000K

Warm (3K) metal halide
3,200K

Halogen
3,000K

Standard incandescent
2,700K

High-pressure sodium
2,200K

FIGURE 6.13 (See colour insert following page 512.) Colour temperature chart.

radiations and give off white light; however, this light is deficient in the blue and violet. The spectrum is skewed toward the red because of the relatively low operating temperature of incandescent lamps, about 2900 K. The spectrum distribution of the light is described in terms of the corresponding temperature of a blackbody, termed the colour temperature. At a colour temperature of 2900 K, only 3% of the energy dissipated in the lamp emerges as visible light. Special high-temperature lamps designed for photography and television usually operate at one of two reference temperatures, 3000 or 3400 K. They produce considerably more blue than ordinary household lamps. Table 6.1 shows some of the characteristics of a few common light sources.

The spectral distribution of a light source depends on more than just the operating temperature. The characteristic emission spectra of elements is due to the electronic transitions between atomic energy levels. Thus, the light from each element has its own characteristic set of emission

TABLE 6.1
Colour Temperature of a Few Common Light Sources

Light Source	Colour Temperature (K)
Mercury arc	6000
Daylight	5500
High-intensity carbon arc	5500
Cool white fluorescent	4200
Incandescent tungsten (Photoflood 3400)	3400
Incandescent tungsten (Photoflood 3200)	3200
Warm white fluorescent	3000
Incandescent tungsten (100 W)	2900
Incandescent tungsten (40 W)	2650
High pressure sodium arc	2200

wavelengths and intensities, which give it a characteristic colour. An electric discharge in hydrogen produces light of the Balmer series. We perceive this mixture of wavelengths as pink. Sodium vapour lamps produce a characteristic yellow light. Similarly, neon lamps produce a red–orange light. The fluorescent lamp is a gas-discharge tube containing mercury vapour and a coating of fluorescent powder on the inside of the tube. Fluorescent materials absorb light at some wavelengths and then radiate light at longer wavelengths. Generally, both the absorption and emission states are broad bands rather than sharply-defined energy levels of the type observed in free atoms. When the mercury vapour in a fluorescent lamp is excited by an electric discharge, it emits its characteristic spectral radiation. Part of this radiation extends beyond the visible range into the ultraviolet. The fluorescent material absorbs the ultraviolet radiation and then radiates light in the visible spectrum. Fluorescent lamps of this type are considerably more efficient than incandescent lamps, converting about 20% of the electrical energy into visible light.

6.9 NEWTON'S COLOUR WHEEL

In 1672, Newton described his experiment to the Royal Society in London. His theory that white light was made up of many colours was revolutionary, and it was greeted with skepticism. The demonstration of recomposition with Newton's colour wheel is only possible because of the persistence of vision. The image of a colour produced on the retina of the eye is retained for a fraction of a second. If the wheel is rotated fast enough, the image of one colour is still present on the retina when the image of the next colour is formed. The brain sums up and blends together the rapidly changed coloured images on the retina, producing the effect of a white image.

The seven colours in Newton's optical spectrum (red, orange, yellow, green, blue, indigo, and violet) may be recombined, by rotating Newton's colour wheel, as shown Figure 6.14. When the wheel is rotating with enough speed, the colours blur together; the eye, unable to respond rapidly enough, sees the colours mixed together to form white colour.

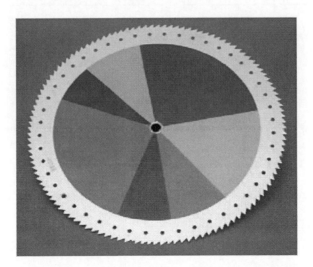

FIGURE 6.14 (See colour insert following page 512.) Newton's colour wheel.

6.10 BLACK AND WHITE COLOUR STRIP INTERSECTION WHEEL

The black and white colour strip intersection wheel works on the same principle as Newton's colour wheel. Figure 6.15 shows the picture of this wheel. If the wheel is rotated fast enough, the image produces

FIGURE 6.15 Black and white colour strip intersection wheel.

a combination of white and black bands of grey colours. The brain sums up and blends together the rapidly changing coloured images on the retina, producing the effect of a white, black, or grey image.

6.11 BLACK AND WHITE COLOUR STRIP WHEEL

The black and white colour strip wheel works as explained in Case (d). Figure 6.16 shows the picture of this wheel. If the wheel is rotated fast enough, the image produces a combination of white and black bands of grey colours.

FIGURE 6.16 Black and white colour strip wheel.

6.12 EXPERIMENTAL WORK

This experiment is a study of the colour mixing principles. The students will experience colour mixing for the following cases:

    a. Additive method of colour mixing
    b. Subtractive method of colour mixing
    c. Newton's colour wheel
    d. Black and white colour strip intersection wheel
    e. Black and white colour strip wheel

6.12.1 TECHNIQUE AND APPARATUS

Appendix A presents the details of the devices, components, tools, and parts comprised in the experiment:

1. Newton's colour wheel, as shown in Figure 6.14
2. Black and white colour strip intersection wheel, as shown in Figure 6.15
3. Black and white colour strip wheel, as shown in Figure 6.16
4. Blue, green, and red slides, as shown in Figure 6.17
5. Three slide projectors, as shown in Figure 6.18
6. White screen and stand
7. Subtractive colour filter set, as shown in Figure 6.19
8. Motor-driven rotator, as shown in Figure 6.20
9. Stroboscope with a filter in 35 mm slide format window, and power supply, as shown in Figure 6.21
10. White cardboard

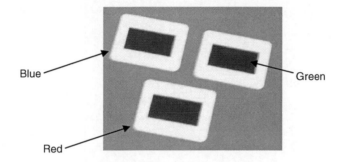

FIGURE 6.17 Blue, green, and red slides.

FIGURE 6.18 Slide projector.

FIGURE 6.19 (**See colour insert following page 512.**) Subtractive colour filter set.

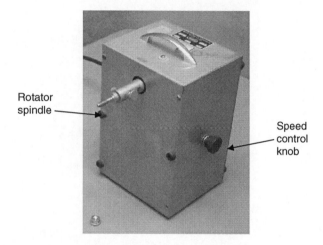

FIGURE 6.20 Motor-driven rotator.

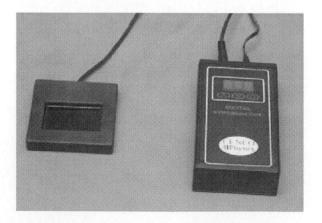

FIGURE 6.21 Stroboscope with a filter in 35 mm slide format window.

6.12.2 PROCEDURE

Follow the laboratory procedures and instructions given by the professor and/or instructor.

6.12.3 SAFETY PROCEDURE

Follow all safety procedures and regulations regarding the use of electric and optical devices and test measurement instruments.

6.12.4 APPARATUS SET-UP

6.12.4.1 Additive Method of Colour Mixing

1. Figure 6.22 shows the experimental apparatus set-up.
2. Place the white cardboard against the wall.
3. Mount three slide projectors on a working bench. Locate the three slide projectors facing the white cardboard.
4. Turn on the slide projectors individually and focus the projected images on the white cardboard.

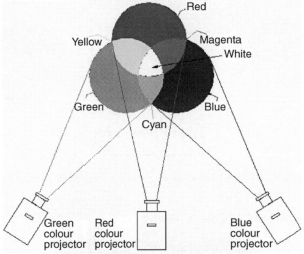

(a) Three slide projectors project lights onto the white cardboard.

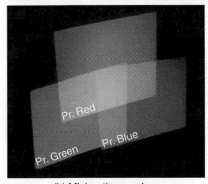

(b) Mixing three colors

FIGURE 6.22 (See colour insert following page 512.) Additive colour mixing. (a) Three slide projectors project lights onto the white cardboard. (b) Mixing three colours.

5. Turn off the laboratory light to be able to see the colours.
6. Insert the green, red, and blue slide colours into the slide projectors, with one-slide colour in each slide projector.
7. Align the images on the white cardboard. Try to mix three colours at a time, as shown in Figure 6.22.
8. Report colour observations from additive method of colour mixing when two or three colours are mixed at a time.
9. For more experience in mixing colours, try to insert two slide colours in one projector at the same time. Try to achieve the colour mixing shown in Figure 6.23.
10. Turn on the laboratory light.

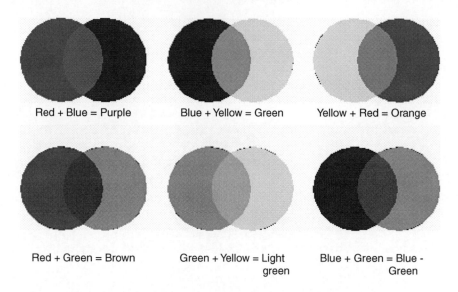

Red + Blue = Purple   Blue + Yellow = Green   Yellow + Red = Orange

Red + Green = Brown   Green + Yellow = Light green   Blue + Green = Blue - Green

FIGURE 6.23 (**See colour insert following page 512.**) Two colour additive mixing.

6.12.4.2 Subtractive Method of Colour Mixing

1. Hold the subtractive colour filter set between your hands, as shown in Figure 6.24.
2. Try to mix two or three colours at a time.
3. Report colour observations from subtractive method of colour mixing when two or three colours are mixed at a time.

6.12.4.3 Newton's Colour Wheel

1. Figure 6.25 shows the experimental apparatus set-up.
2. Take the colour wheel and mount it on the motor-driven rotator and plug it into a power source.
3. Turn on the colour wheel using the speed control knob and set it to its lowest rotational speed: "Speed 1."
4. Measure the rotational speed of the colour wheel (Speed 1) using the stroboscope with a filter in 35 mm slide format window, as shown in Figure 6.21.
5. Record what colour is observed on the wheel in Table 6.2.
6. Increase the rotational speed to setting "Speed 2" using speed control knob.
7. Repeat steps 4 and 5 for four different speeds and fill out Table 6.2.

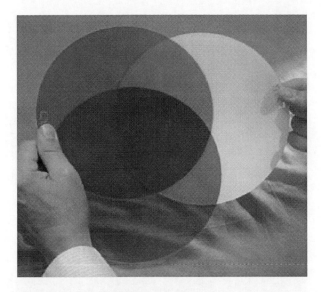

FIGURE 6.24 (See colour insert following page 512.) Subtractive colour mixing.

6.12.4.4 Black and White Colour Strip Intersection Wheel

1. Figure 6.26 shows the experimental apparatus set-up.
2. Take the black and white colour strip intersection wheel and mount it on the motor-driven rotator and plug it into a power source.

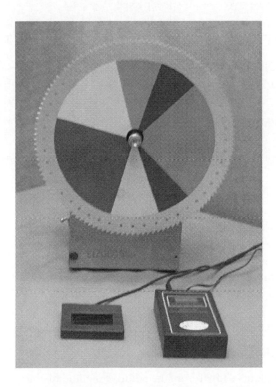

FIGURE 6.25 (See colour insert following page 512.) Newton's colour wheel apparatus set-up.

FIGURE 6.26 Black and white colour strip intersection wheel apparatus set-up.

3. Turn on the black and white colour strip intersection wheel using the speed control knob and set it to its lowest rotational speed: "Speed 1."
4. Measure the rotational speed of the black and white colour strip intersection wheel (Speed 1) using the stroboscope with a filter in 35 mm slide format window, as shown in Figure 6.21.
5. Record what colour is observed on the wheel in Table 6.3.
6. Increase the rotational speed to setting "Speed 2" using the speed control knob.
7. Repeat steps 4 and 5 for four different speeds and fill out Table 6.3.

6.12.4.5 Black and White Colour Strip Wheel

1. Figure 6.27 shows the experimental apparatus set-up.
2. Take the black and white colour strip wheel and mount it on the motor-driven rotator and plug it into a power source.
3. Turn on the black and white colour strip wheel using the speed control knob and set it to its lowest rotational speed: "Speed 1."
4. Measure the rotational speed of the black and white colour strip wheel (Speed 1) using the stroboscope with a filter in 35 mm slide format window, as shown in Figure 6.21.
5. Record what colour is observed on the wheel in Table 6.4.
6. Increase the rotational speed to setting "Speed 2" using the speed control knob.
7. Repeat steps 4 and 5 for four different speeds and fill out Table 6.4.

6.12.5 Data Collection

6.12.5.1 Additive Method of Colour Mixing

Observe and report the colours shown from the additive method of colour mixing when two or three colours are mixed at a time.

FIGURE 6.27 Black and white colour strip wheel apparatus set-up.

6.12.5.2 Subtractive Method of Colour Mixing

Observe and report the colours shown from the subtractive method of colour mixing when two or three colours are mixed at a time.

6.12.5.3 Newton's Colour Wheel

Fill out Table 6.2 with the observed colour and wheel rotational speed.

TABLE 6.2
Newton's Colour Wheel

Wheel Speed	Measured Wheel Speed (RPM)	Observed Colour
Speed 1		
Speed 2		
Speed 3		
Speed 4		
Speed 5		

6.12.5.4 Black and White Colour Strip Intersection Wheel

Fill out Table 6.3 with the observed colour and wheel rotational speed.

TABLE 6.3
Black and White Colour Strip Intersection Wheel

Wheel Speed	Measured Wheel Speed (RPM)	Observed Colour
Speed 1		
Speed 2		
Speed 3		
Speed 4		
Speed 5		

6.12.5.5 Black and White Colour Strip Wheel

Fill out Table 6.4 with the observed colour and the wheel rotational speed.

TABLE 6.4
Black and White Colour Strip Wheel

Wheel Speed	Measured Wheel Speed (RPM)	Observed Colour
Speed 1		
Speed 2		
Speed 3		
Speed 4		
Speed 5		

6.12.6 CALCULATIONS AND ANALYSIS

Not applicable in this lab.

6.12.7 RESULTS AND DISCUSSIONS

6.12.7.1 Additive Method of Colour Mixing

1. Report and discuss the additive method of colour mixing, when two or three colours are mixed at a time.

6.12.7.2 Subtractive Method of Colour Mixing

1. Report and discuss the subtractive method of colour mixing, when two or three colours are mixed at a time.

6.12.7.3 Newton's Colour Wheel

1. Report and discuss the observed colour when the Newton colour wheel rotates at different rotational speeds.

6.12.7.4 Black and White Colour Strip Intersection Wheel

1. Report and discuss the observed colour when the black and white colour strip intersection wheel rotates at different rotational speeds.

6.12.7.5 Black and White Colour Strip Wheel

1. Report and discuss the observed colour when the black and white colour strip wheel rotates at different rotational speeds.

6.12.8 CONCLUSION

Summarize the important observations and findings obtained in this lab experiment.

6.12.9 SUGGESTIONS FOR FUTURE LAB WORK

List any suggestions for improvements using different experimental equipment, procedures, and techniques for any future lab work. These suggestions should be theoretically justified and technically feasible.

6.13 LIST OF REFERENCES

List any references that were used in the report. Use one format in writing the references. Never mix reference formats in a report.

6.14 APPENDICES

List all of the materials and information that are too detailed to be included in the body of the report.

FURTHER READING

Adams, R. A., Courage, M. L., and Mercer, M. E., Systematic measurements of human neonatal color vision, *Vision Res.*, 34, 1691–1701, 1994.
Beiser, A., *Physics*, 5th ed., Addison-Wesley, New York, 1991.

Bise, R. T., Windeler, R. S., Kranz, K. S., Kerbage, C., Eggleton, B. J., and Trevor, D. J., Tunable Photonic Bandgap Fiber, Paper presented at Proc. Optical Fiber Communication Conference and Exhibit, Murray Hill, NJ, 466–468, March 17, 2002.

Bowmaker, J. K. and Dartnall, H. J. A., Visual pigments of rods and cones in human retina, *J. Physiol.*, 298, 501–511, 1980.

Cornsweet, T. N., *Visual Perception*, Academic Press, New York, 1970.

Cox, A., *Photographic Optics*, 15th ed., Focal Press, London, 1974.

Ewen, D., Nelson, R. J., and Schurter, D., *Physics for Career Education*, 4th ed., Prentice Hall, Upper Saddle River, NJ, 1996.

Falk, D., Brill, D., and Stork, D., *Seeing the Light Optics in Nature, Photography, Color, Vision, and Holography*, Wiley, New York, 1986.

Fehrman, K. R. and Fehrman, C., *Color the Secret Influence*, Prentice Hall, Inc., Upper Saddle River, 2000.

Ghatak, A. K., *An Introduction to Modern Optics*, McGraw-Hill, New York, 1972.

Giancoli, D. C., *Physics*, 5th ed., Prentice Hall, Upper Saddle River, NJ, 1998.

Griot, M., The practical application of light, *Melles Griot Catalog*, Wiley, New York, 2001.

Halliday, D., Resnick, R., and Walker, J., *Fundamental of Physics*, 6th ed., Wiley, New York, 1997.

Hood, D. C. and Finkelstein, M. A., Sensitivity to light handbook of perception and human performance, In *Sensory Processes and Perception*, Boff, K. R., Kaufman, L., and Thomas, J. P., Eds., Vol. 1, Wiley, Toronto, 1986.

Hurvich, L. M. and Jameson, D., A psychophysical study of white I natural adoption, *J. Opt. Soc. Am.*, 41, 521–527, 1951a.

Hurvich, L. M. and Jameson, D., A psychophysical study of white III adoption as variant, *J. Opt. Soc. Am.*, 41, 701–709, 1951b.

Hurvich, L. M. and Jameson, D., Some quantitative aspects of an opponent colors theory, II. Brightness, saturation, and hue in normal and dichromatic vision, *J. Opt. Soc. Am.*, 45, 602–616, 1955.

Jameson, D. and Hurvich, L. M., Theory of brightness and color contrast in human vision, *Vision Res.*, 4, 135–154, 1964.

Jenkins, F. W. and White, H. E., *Fundamentals of Optics*, McGraw Hill, New York, 1957.

Jones, E. and Childers, R., *Contemporary College Physics*, McGraw-Hill Higher Education, New York, 2001.

Kaiser, P. K. and Boynton, R. M., *Human Color Vision*, 2nd ed., Optical Society of America, Washington, DC, 1996.

Keuffel & Esser Co., *Physics*, Keuffel & Esser Audiovisual Educator—Approved Diazo Transparency Masters, Audiovisual Division, Keuffel & Esser Co., U.S.A., 1989.

Lerner, R. G. and Trigg, G. L., *Encyclopedia of Physics*, 2nd ed., VCH Publishers, Inc., New York, 1991.

Manngeim, L. A., *Photography Theory and Practice*, Focal Press Ltd., Boston, 1970.

Nassau, K., *Experimenting with Color*, Franklin Watts A Division of Grolier Publishing, New York, 1997.

Newport Corporation, *Photonics Section, Newport Resource 2004 Catalog*, Newport Corporation, Irvine, CA, 2004.

Nolan, P. J., *Fundamentals of College Physics*, Wm. C. Brown Publishers, Dubuque, Iowa, 1993.

Overheim, R. D. and Wagner, D. L., *Light and Color*, Wiley, New York, 1982.

Pritchard, D. C., *Environmental Physics: Lighting*, Longmans, Green & Co. Ltd., London and Harlow, Great Britain, 1969.

Product Knowledge, Lighting reference guide, *Product Development*, 2nd ed., Ontario Hydro, Toronto, ON, Canada, 1988.

Schildegen, T. E., *Pocket Guide to Color with Digital Applications*, Delmar Publishers, Albany, NY, 1998.

Stroebel, L., Compton, J., Current, I., and Zakia, R., *Photographic Materials and Processes*, Focal Press Ltd., Boston, 1986.

Tippens, P. E., *Physics*, 6th ed., Glencoe McGraw-Hill, Westerville, OH, 2001.

Williamson, S. J. and Cummins, H. Z., *Light and Color in Nature and Art*, Wiley, New York, 1983.

Wilson, J. D., *Physics—A Practical and Conceptual Approach Saunders Golden Sunburst Series*, Saunders College Publishing, Philadelphia, 1989.

Wilson, J. D. and Buffa, A. J., *College Physics*, 5th ed., Prentice Hall, Inc., Upper Saddle River, 2000.

7 Laws of Light

7.1 INTRODUCTION

Basic understanding of the reflection and refraction properties of light paved the way for many of the optical measuring devices in use today. When light strikes a surface, a portion of the incident ray is reflected from the surface. Depending on the surface characteristics, either specular reflection or diffuse reflection is observed. When light passes through an optically dense media, such as glass or water, the incident light ray bends as it enters the media. The characteristics of the material (air, glass, water, etc.) and the orientation of the ray determine the path of a reflected or refracted ray.

The laws of reflection and refraction govern the behaviour of light incident on a flat surface, separating two optical media. Light propagation through optical components is explained by these laws, as well as by critical and total internal reflection.

Four experimental lab cases presented in this chapter cover light passing through a water layer as an optical component. The experimental cases include observing laser light passing through a water layer, measuring the angle of incidence and angle of refraction, measuring and calculating the critical angle at the water–air interface, and measuring the total internal reflection occurring at the water–air interface.

7.2 LAW OF REFLECTION

Light travels in a straight line at a constant speed in a uniform optical medium. When a light ray traveling in a medium meets a boundary leading into a second medium, part of the incident ray reflects back to the first medium. Consider a light ray traveling in air incident at an angle (θ_i) on a flat mirror surface, as shown in Figure 7.1. A light ray incident on a mirror surface dividing two uniform media is described by an angle of incidence (θ_i). This angle is measured relative to a line perpendicular to the reflecting surface, as shown in Figure 7.1. Similarly, the reflected ray is

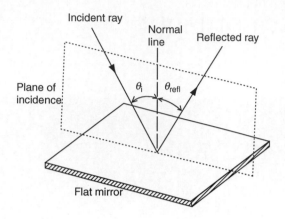

FIGURE 7.1 Light ray reflection.

described by an angle of reflection (θ_{refl}). The incident ray, the normal line, and the reflected ray lie in the plane of incidence.

Figure 7.2 shows three parallel light rays incident on and reflected off a reflecting surface. This figure also illustrates how to derive the law of reflection. The line AB lies along an ingoing wavefront, while CD lies on an outgoing wavefront. In effect, AB transforms upon reflection into CD. The wavelet emitted from A will arrive at C in-phase with the wavelet just being emitted from D, as long as the distances AC and BD are equal. In other words, if all the wavelets emitted from the surface overlap in-phase and form a single reflected plane wave, it must be that $AC = BD$. The two right triangles ABD and ACD are congruent, and they have the same hypotenuse AD. From Figure 7.2, the hypotenuse AD can be obtained:

$$\frac{\sin \theta_i}{BD} = \frac{\sin \theta_{refl}}{AC} \qquad (7.1)$$

All the waves travel in the incident medium with the same speed (v_i). In the same time (Δt) that it takes the wavelet from point B on the wavefront to reach point D on the surface, the wavelet emitted from A reaches point C. Therefore, $BD = v_i \Delta t = AC$, and Equation (7.1) can be

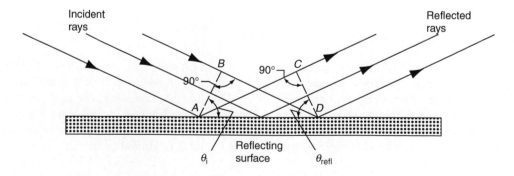

FIGURE 7.2 Light rays incident on and reflected off a reflecting surface.

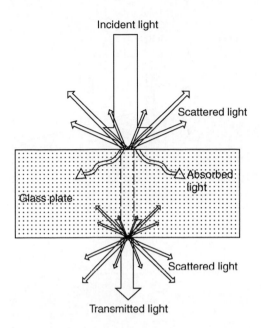

FIGURE 7.3 Transmission of light through a glass plate.

re-written as:

$$\sin \theta_i = \sin \theta_{refl} \qquad (7.2)$$

which means that

$$\theta_i = \theta_{refl} \qquad (7.3)$$

Therefore, the angle of incidence equals the angle of reflection. The relationship between these angles is given by the law of reflection.

Figure 7.3 shows light passing through an optical medium, for example, a glass plate. When light rays are incident perpendicular to an optical medium surface, part of the light is scattered at the surface of the medium. One part of the light rays is absorbed by the optical medium, whereas the other part is transmitted through the optical medium. Light reflection, scattering, and absorption are important factors in the calculation of light loss in an optical medium.

7.2.1 Fresnel Reflection

There are two common types of reflection that occur in optical components. Fresnel reflection occurs when a light ray passes through materials, which have different refractive indices, as shown in Figure 7.4. A portion of the incident light ray is subjected to multiple reflections between two parallel optical surfaces. The multiple reflections are called Fresnel reflection, named after the French physicist, Augustin Fresnel (1788–1827). The portion of the incident light ray reflected is dependent on the angle of incidence (measured from the normal line) and the polarization of the

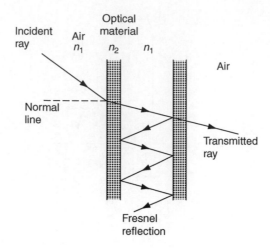

FIGURE 7.4 Fresnel reflection between parallel optical surface.

electric field relative to the plane of incidence. This kind of reflection normally takes place at the ends of mated fibre optic connectors.

7.2.2 BACK REFLECTION

Back reflection, which occurs at the end of the fibre optic cables, is shown in Figure 7.5. Additional losses can occur if the fibre optic cable ends are cut and polished at an angle that does not match the connector angle. Strong back reflection losses can provide feedback, which, in some cases, introduces a spurious signal and increased noise levels.

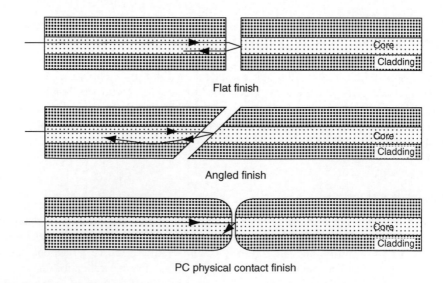

FIGURE 7.5 Back reflection in fibre optic cable ends.

7.3 LAW OF REFRACTION

Light travels in a straight line at a constant speed in a uniform optical medium. If the properties of the optical medium change, then the speed of light within the medium will also change. The velocity of light in an optical medium is less than the velocity of light in a vacuum of 3×10^8 m/s. The ratio of the velocity of light in a vacuum (c) to the velocity of light in a particular medium (v) is called the index of refraction (n) for that medium. The index of refraction of a medium is defined by Equation (7.4).

$$n = \frac{c}{v} \tag{7.4}$$

The index of refraction is a unitless quantity that is always greater than unity. Table 7.1 lists the index of refraction for several common substances.

Consider a ray of light traveling from one medium to a second medium, as shown in Figure 7.6. The ray that enters the second medium is bent at the interface between the two mediums; this phenomenon is called refraction of light. The incident ray, the normal line, and the refracted ray lie on the plane of incidence. The sine of the angle of refraction (θ_{refr}) is directly proportional to the sine of the angle of incidence (θ_i), as shown in Figure 7.6; this relation is called Snell's law of refraction.

Figure 7.7 shows how to derive Snell's law of refraction. Two parallel light rays in a medium with index of refraction n_1 meet at the interface of a medium with index of refraction n_2. It is assumed that the second medium has a greater index of refraction than the first ($n_2 > n_1$). Both rays bend toward the normal as they pass into the denser medium.

In the time interval Δt, ray light ray 1 propagates from A to B and light ray 2 propagates from C to D. Since the velocity v_2 in the second medium is less than the velocity v_1 in the first medium, the distance AB will be shorter than the distance CD. These lengths are given by:

$$AB = v_2 \Delta t \tag{7.5}$$
$$CD = v_1 \Delta t \tag{7.6}$$

TABLE 7.1
Index of Refraction for Several Common Substances

Substance	Index of Refraction (n)
Air	1.00029
Diamond	2.24
Ethyl alcohol	1.36
Fluorite	1.43
Fused quartz	1.46
Glass (by type)	
Crown	1.52
Flint	1.66
Glycerin	1.47
Ice	1.31
Polystyrene	1.49
Rock salt	1.54
Water	1.33

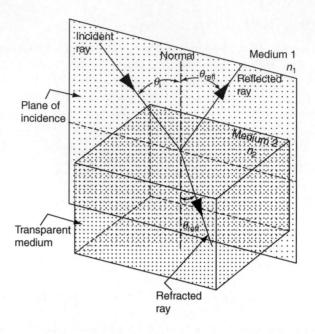

FIGURE 7.6 Refraction of a light ray.

The geometry in Figure 7.7 shows that the angle CAD is equal to θ_i, and the angle ADB is equal to θ_{refr}. The line AD forms a hypotenuse that is common to the two triangles, ACD and ABD. From these triangles, we find that

$$\sin \theta_i = \frac{v_1 \Delta t}{AD} \tag{7.7}$$

$$\sin \theta_{refr} = \frac{v_2 \Delta t}{AD} \tag{7.8}$$

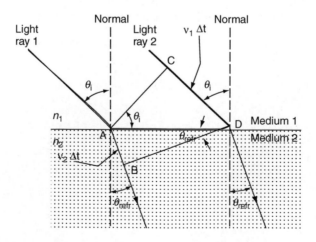

FIGURE 7.7 Deriving Snell's law.

Dividing Equation (7.7) by Equation (7.8) gives:

$$\frac{\sin \theta_i}{\sin \theta_{refr}} = \frac{v_1}{v_2} \tag{7.9}$$

Equation (7.9) shows that the ratio of the sine of the angle of incidence to the sine of the angle of refraction is equal to the ratio of the velocity of light in the incident medium to the velocity of light in the refracted medium.

Using Equation (7.4) in conjunction with Equation (7.9), gives the following equation:

$$n_1 \sin \theta_i = n_2 \sin \theta_{refr} \tag{7.10}$$

where n_1 and n_2 are the index of refraction for medium 1 and 2, respectively. The relationship between these angles is known as the law of refraction, or Snell's law.

More generally, the following relationships for light refraction in an optical medium can easily be determined from Snell's law:

1. When a light ray enters a denser medium, its speed decreases, and the refracted ray is bent towards the normal. For example, this occurs when the light ray moves from air into glass, as shown in Figure 7.8.
2. When a light ray enters a less dense medium, its speed increases, and the refracted ray is bent away from the normal. For example, this occurs when the light ray moves from glass into air, as shown in Figure 7.9.
3. There is no change in the direction of propagation, if there is no change in index of refraction. The greater the change is in the index of refraction the greater the change in the propagation direction will be.
4. If a light ray goes from one medium to another along the normal, it is undeflected, regardless of the change in the index of refraction, as shown in Figure 7.10. Although the angle does not change, the speed of light does change as the light travels from one medium to another.

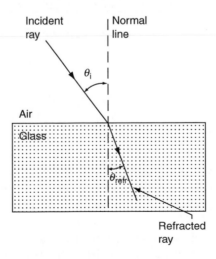

FIGURE 7.8 Light ray propagates from air into glass.

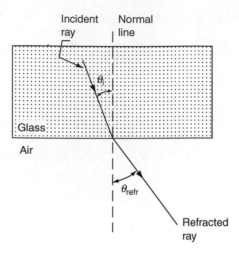

FIGURE 7.9 Light ray propagates from glass in to air.

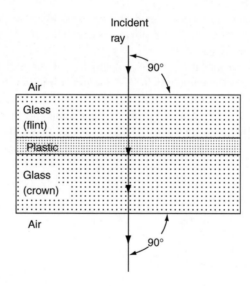

FIGURE 7.10 Normal incident light ray passes through differenct optical materials.

The principles of reflection and refraction are illustrated in Figure 7.6. An incident light ray is partially reflected and partially refracted, and then transmitted at the interface plane, separating two transparent media.

Figure 7.11 shows a light ray passing through two media (air and glass). The change in index of refraction at each interface causes the incident ray to bend twice. This figure also shows that the angle θ_1 equals the angle θ_3, and the ray exiting the glass plate is parallel to the incident ray.

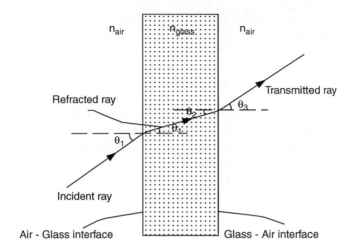

FIGURE 7.11 Light ray refraction through a glass plate.

7.3.1 CRITICAL ANGLE AND TOTAL INTERNAL REFLECTION

Total internal reflection can occur when light, traveling in a more optically dense medium, is incident on the boundary with a less optically dense medium, as shown in Figure 7.12. The index of refraction (n_1) of medium 1 is greater than index of refraction (n_2) of medium 2. A good example is when light goes from water into air. Consider a light beam traveling in medium 1 and incident on the boundary interface between medium 1 and medium 2. Various possible directions of the light ray are indicated by rays 1 through 5, as illustrated in

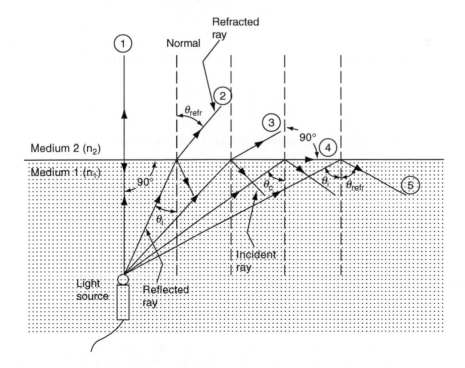

FIGURE 7.12 Critical angle and total internal reflection.

Figure 7.12. The refracted rays are bent away from the normal, according to Snell's law because n_1 is greater than n_2. Furthermore, when the light ray refracts at the interface between the two media, it partially reflects back into medium 1. Figure 7.12 shows that rays 1 through 4 are partially reflected back into medium 1, and ray 5 is totally reflected back into medium 1.

Consider individual light rays in Figure 7.12:

1. Ray 1 is incident perpendicular to the interface between the two media, and passes straight through from medium 1 to medium 2. Part of ray 1 reflects back in the same direction into medium 1; this type of reflection is called a back reflection. Back reflection is one type of loss in optical devices.
2. Ray 2 is an incident ray in medium 1 to medium 2, where $n_1 > n_2$. A portion of ray 2 refracts to medium 2, as given by Snell's law, and a portion of ray 2 reflects back into medium 1 (since the angle of incidence equals the angle of reflection).
3. Similarly, for ray 3, as the angle of incidence increases, the angle of refraction also increases.
4. For ray 4, as the angle of incidence increases, the angle of refraction increases until the angle of refraction is 90°. The angle of incidence is called the critical angle (θ_c), when the refracted angle is 90°. Ray 4 refracts and propagates parallel to the interface between the two media.
5. For angles of incidence greater than the critical angle (θ_c), ray 5 is entirely reflected at the interface. All the intensity of incident ray 5 goes into the reflected ray without any loss; this is called total internal reflection of light.

The critical angle for total internal reflection is found by setting $\theta_{\text{refr}} = 90°$ and applying Snell's law, Equation (7.10):

$$n_1 \sin \theta_c = n_2 \sin 90 = n_2 \times 1$$
$$\because \sin 90 = 1 \tag{7.11}$$

Therefore, the critical angle (θ_c) is given by the following equations:

$$\sin \theta_c = \frac{n_2}{n_1} \tag{7.12}$$

$$\theta_c = \sin^{-1}\left(\frac{n_2}{n_1}\right) \tag{7.13}$$

Since $\sin \theta_c$ is always less than or equal to 1, the index of refraction (n_1) of medium 1 must be larger than the index of refraction (n_2) of medium 2. Thus, total internal reflection can occur only when light in one medium encounters an interface with another medium that has a lower index of refraction (i.e., speed of light is greater in the second medium).

7.4 EXPERIMENTAL WORK

The cases in this experiment use the theoretical principles of light reflection, refraction, and total internal reflection for light passing through optical components. The following cases examine the behaviour of light passing through a water layer in a water tank; these principles are explained by the laws of the reflection and refraction of light. Students will observe the effects of a light ray as it propagates through a tank filled with water for the following cases:

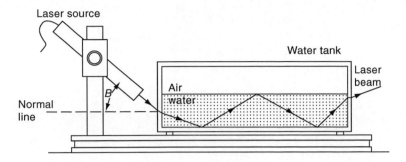

FIGURE 7.13 Incident laser beam of an angle θ into water layer.

7.4.1 LASER LIGHT PASSES THROUGH A WATER LAYER

The students will observe a laser beam bouncing through the water layer in a water tank. The students will observe the reflection of a laser beam incident at different angles on the side of the water tank, as illustrated in Figure 7.13.

7.4.1.1 Law of Refraction at Air–Water Interface

The students will calculate the angles of incidence and refraction of a laser beam incident on the top water surface of a water tank. Refraction occurs when light passes from air to water. Air has a lower index of refraction than water. Therefore, the refracted ray bends toward the normal as the ray passes through water. The students will calculate the angle of incidence (θ_i) and the angle of refraction (θ_{refr}) at the air–water interface using the measurement of horizontal and vertical distances, as illustrated in Figure 7.14.

Using the geometrical relations in this figure, the following equations can be written to calculate the angles of incidence (θ_i) and refraction (θ_{refr}).

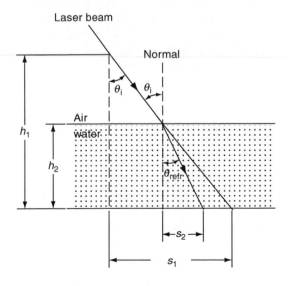

FIGURE 7.14 Refraction angle of light at air–water interface.

$$\tan \theta_i = \frac{s_1}{h_1} \qquad (7.14)$$

$$\theta_i = \tan^{-1} \frac{s_1}{h_1} \qquad (7.15)$$

$$\tan\theta_{\text{refr}} = \frac{s_2}{h_2} \qquad (7.16)$$

$$\theta_{\text{refr}} = \tan^{-1} \frac{s_2}{h_2} \qquad (7.17)$$

7.4.1.2 Critical Angle at Water–Air Interface

The students will perform an experiment to calculate the critical incident angle (θ_c) when the refracted angle is 90° with the normal, as illustrated in Figure 7.15. Total internal reflection can occur when light propagates from water to air and the incident angle is greater than the critical angle. In this case, the refracted ray reflects inside the water, and the law of reflection is applicable. When the critical angle occurs, light incident on the underside of a water–air interface produces a laser beam that lies on the water–air interface and shows a bright spot on the target card. The students will find the experimental and theoretical values of the critical angle at the water–air interface.

Using the geometrical relations in this figure, the following equations can be written to calculate the critical angles (θ_c):

$$\tan \theta_c = \frac{X}{Y} \qquad (7.18)$$

$$\theta_c = \tan^{-1} \frac{X}{Y} \qquad (7.19)$$

7.4.1.3 Total Internal Reflection at Water–Air Interface

The students will observe the phenomena of the total internal reflection of light rays when the incidence angle is greater than the critical angle. The students will observe light rays shining from the bottom of a water tank filled with water, as illustrated in Figure 7.16. Specifically, they will measure the distance from the light source to the bottom of the water tank in relation to the diameter of the light spots on the top and bottom surfaces of the water tank. The students will determine the critical angle experimentally and theoretically at the water–air interface.

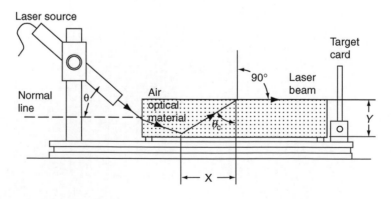

FIGURE 7.15 Critical angle of light at water–air interface.

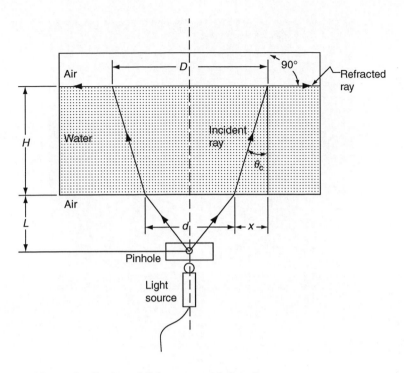

FIGURE 7.16 Total internal reflection of light at water-air interface.

Using the geometrical relations in Figure 7.16, the following equations can be written to calculate the critical angle (θ_c) and x:

$$x = \frac{(D-d)}{2} \tag{7.20}$$

$$\tan \theta_c = \frac{x}{H} \tag{7.21}$$

$$\theta_c = \tan^{-1}\left(\frac{x}{H}\right) \tag{7.22}$$

7.4.2 TECHNIQUE AND APPARATUS

Appendix A presents the details of the devices, components, tools, and parts.

1. 2×2 ft. optical breadboard
2. HeNe laser light source and power supply
3. Laser mount assembly
4. Incandescent or halogen light source
5. Light source positioners
6. A water tank, as shown in Figure 7.17
7. Hardware assembly (clamps, posts, screw kits, screwdriver kits, sundry positioners, etc.)
8. Adjustable pinhole, as shown in Figure 7.18
9. Rotation Stage
10. Black/white card and cardholder
11. Target card

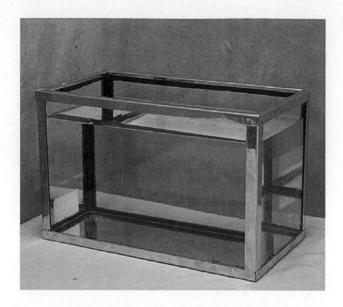

FIGURE 7.17 Water tank.

12. Vernier
13. Ruler

7.4.3 PROCEDURE

Follow the laboratory procedures and instructions given by the professor and/or instructor.

7.4.4 SAFETY PROCEDURE

Follow all safety procedures and regulations regarding the use of laser light and light source devices.

FIGURE 7.18 Adjustable pinhole.

7.4.5 APPARATUS SET-UP

7.4.5.1 Laser Light Passes through a Water Layer

1. Figure 7.19 shows the apparatus set-up.
2. Bolt the laser short rod to the breadboard.
3. Bolt the laser mount to the clamp using bolts from the screw kit.
4. Put the clamp on the short rod.
5. Place the HeNe laser into the laser mount and tighten the screw. Turn on the laser device. Follow the operation and safety procedures of the laser device in use.
6. Mount the water tank and fill it with water to a depth of about 10 cm.
7. Place the water tank near the laser source.
8. Turn off the laboratory light to be able to see the laser beam path.
9. Direct the laser beam in the perpendicular position to the left side of the water tank (set the first angle equal to zero degrees with the horizontal level).
10. Observe the laser beam path. Report your observation in Table 7.2.
11. Align the laser source at an angle so the laser beam enters the left side of the water tank. Observe the laser beam path when the laser beam bounces between the bottom of the water tank and water surface. Report your observation in Table 7.2.
12. Repeat step 11 for another two different angles of the laser beam entering the left side of the water tank. Report your observations in Table 7.2.
13. Turn on the laboratory light.

7.4.5.2 Law of Refraction at Air–Water Interface

1. Figure 7.20 shows the apparatus set-up.
2. Bolt the laser short rod to the breadboard.
3. Bolt the laser mount to the clamp using bolts from the screw kit.
4. Put the clamp on the short rod.
5. Place the HeNe laser into the laser mount and tighten the screw. Turn on the laser device. Follow the operation and safety procedures of the laser in use.

FIGURE 7.19 Laser light passes through a water layer.

6. Place the water tank on the breadboard near the laser source.
7. Place a lined paper under the water tank, for ease of marking the laser beam spots.
8. Turn off the laboratory light to be able to see the laser beam path.
9. Direct the laser beam to be perpendicular to the bottom of the water tank.

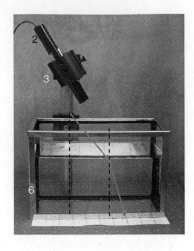

FIGURE 7.20 Refraction angle of light at air–water interface.

10. Mark the laser beam spot on the lined paper, when the laser beam is positioned vertically to the bottom of the water tank.
11. Rotate the laser source to direct the laser beam by an angle less the 90° with the normal. For example, 30°. Refer to Figure 7.14 for distances to be measured.
12. Measure the vertical distance (h_1) from the laser source to the spot where the laser beam is incident on the lined paper. This is the reference spot. Fill out Table 7.3.
13. Measure the distance (s_1) from the reference spot to where the laser beam is incident on the lined paper. Fill out Table 7.3.
14. Fill the water tank with water to about 5 cm below the top. (Note: It is preferable to add sugar or salt to the water to enhance the viewing of the light shining through the water tank. In this case, you will need to find the index of refraction (n) of the solution).
15. Mark the laser beam spot on the lined paper when the laser beam is incident on it.
16. Make a vertical line from where the laser beam intersects the water surface to the lined paper. Mark this spot on the lined paper.
17. Measure the vertical distance (h_2) and the horizontal distance (s_2), as shown in Default 7.14 and 7.20. Fill out Table 7.3.
18. Turn on the laboratory light.

7.4.5.3 Critical Angle at Water–Air Interface

1. Figure 7.21 shows the apparatus set-up.
2. Bolt the laser short rod to the breadboard.
3. Bolt the laser mount to the clamp using bolts from the screw kit.
4. Put the clamp on the short rod.
5. Place the HeNe laser into the laser mount and tighten the screw. Turn on the laser device. Follow the operation and safety procedures of the laser in use.
6. Mount the water tank on the breadboard and fill it with water to a depth of about eight to ten centimetres. (Note: It is preferable to add sugar or salt to the water to enhance the viewing of the light shining through the water tank. In this case, you will need to find the index of refraction (n) of the solution).
7. Place the water tank near the laser source.
8. Place a target card on the side of the water tank.
9. Turn off the laboratory light to be able to see the laser beam path.

FIGURE 7.21 Critical angle at water–air interface.

10. Align the laser source at an angle of incident on the side of the water tank so the laser beam enters the left side of the water tank.
11. Observe the laser beam path when it bounces between the bottom of the water tank and the water surface. Keep aligning the laser beam angle until you see the refracted laser beam lies on the water surface and produces a red spot on the target card (i.e. the aligned laser beam is at the critical angle and the refracted angle is 90°).
12. Measure the vertical and the horizontal distances (Y) and (X), as shown in Figure 7.15. Fill out Table 7.4.
13. Turn on the laboratory light.

7.4.5.4 Total Internal Reflection at Water–Air Interface

1. Figure 7.22 shows the apparatus set-up.
2. Mount the water tank on the stand. Fill it with water to within a few centimetres below the top of the water tank. (Note: It is preferable to add sugar or salt to the water to

FIGURE 7.22 Total internal reflection at water–air interface.

enhance the viewing of the light shining through the water tank. In this case, you will need to find the index of refraction (n) of the solution).
3. Measure water depth (H) in the water tank.
4. Place the light source underneath the water tank so the light shines up towards the bottom of the water tank. Place the light source at a distance (L_1) from the bottom of the water tank.
5. Place the pinhole on the top of the light source.
6. Connect the light source to the power supply.
7. Turn off the laboratory light to be able to see the light shining through the water.
8. Adjust the distance (L_1) until you see a bright circular pattern on the top water surface in the water tank.
9. Measure the distance (L_1), the diameter of the bright circle (D_1) on the top water surface, and the diameter (d_1) of the bright circle on the bottom water surface, as shown in Figure 7.16. Fill out Table 7.5.
10. Repeat steps 8 and 9 for another five distances. Fill out Table 7.5.
11. Turn on the laboratory light.

7.4.6 DATA COLLECTION

7.4.6.1 Laser Light Passes through a Water Layer

Observe the laser beam path for four angles of incidence (θ) on the left side of the water tank. Draw the laser beam path and report your observation in Table 7.2.

TABLE 7.2
Observation of the Laser Beam Path

Laser Beam Angle (Degrees)	Observation of the Laser Beam Path
$\theta_1 = 0^\circ$	
$\theta_2 =$	
$\theta_3 =$	
$\theta_4 =$	

7.4.6.2 Law of Refraction at Air–Water Interface

1. Measure the vertical distances (h_1 and h_2) and horizontal distances (s_1 and s_2).
2. Report your measurements in Table 7.3.

TABLE 7.3
Law of Refraction at Water–Air Interface

Vertical Distances (cm)		Horizontal Distances (cm)		Calculated Incident Angle (θ_i) (degrees)	Calculated Refraction Angle (θ_{refr}) (degrees)
h_1	h_2	s_1	s_2		

7.4.6.3 Critical Angle at Water–Air Interface

1. Measure the vertical and horizontal distances (Y and X).
2. Report your measurements in Table 7.4.

TABLE 7.4
Critical Angle at Water–Air Interface

Vertical Distance Y	Horizontal Distance X	Calculated Critical Angle (θ_c) Using Equation (32.18) (degrees)	Refraction Angle (θ_{refr}) (degrees)	Calculated Critical Angle (θ_c) Using Equation (32.12) (degrees)
			90°	

7.4.6.4 Total Internal Reflection at Water–Air Interface

1. Measure the distances (L) between the pinhole and the tank bottom, the light spot diameters (D) at the top water surface, and the light spot diameters (d) at the bottom water surface. Report your measurements in Table 7.5.
2. Measure water depth (H) in the water tank.

TABLE 7.5
Total Internal Reflection at Water–Air Interface

Measured Distance (L) (cm)	Measured Top Diameter (D) (cm)	Measured Bottom Diameter (d) (cm)	x (cm)	Determined Critical Angle (θ_c) (degree)	Calculated Critical Angle by Snell's Law (θ_c) (degree)
$L_1 =$	$D_1 =$	$d_1 =$	$x_1 =$	$\theta_{c\,1} =$	
$L_2 =$	$D_2 =$	$d_2 =$	$x_2 =$	$\theta_{c\,2} =$	
$L_3 =$	$D_3 =$	$d_3 =$	$x_3 =$	$\theta_{c\,3} =$	$\theta_c =$
$L_4 =$	$D_4 =$	$d_4 =$	$x_4 =$	$\theta_{c\,4} =$	
$L_5 =$	$D_5 =$	$d_5 =$	$x_5 =$	$\theta_{c\,5} =$	
$L_6 =$	$D_6 =$	$d_6 =$	$x_6 =$	$\theta_{c\,6} =$	

7.4.7 CALCULATIONS AND ANALYSIS

7.4.7.1 Laser Light Passes through a Water Layer

No calculations and analysis are required in this case.

7.4.7.2 Law of Refraction at Air–Water Interface

1. Calculate the angles of incidence and refraction, using Equation (7.15) and Equation (7.17). Report your results in Table 7.3.
2. Verify Snell's law, given by Equation (7.10), using your calculated angles and the index of refraction for air and water as 1.0003 and 1.33, respectively.

7.4.7.3 Critical Angle at Water–Air Interface

1. Calculate the experimental critical angle using Equation (7.19). Report your results in Table 7.4.
2. Calculate the theoretical critical angle using Equation (7.13) and the index of refraction for air and water as 1.0003 and 1.33, respectively. Report your results in Table 7.4.
3. Verify the experimental and theoretical critical angles.

7.4.7.4 Total Internal Reflection at Water–Air Interface

1. Draw the geometry of the light propagation from the light source through the water for each distance (L). Determine the angles of incidence (θ_{in}) of the light beam path through the water, from the geometry of the light propagation. Report your results in Table 7.5.
2. Plot a graph of distance (L) verses diameters (D) and (d).
3. Calculate the critical angle (θ_c) using Snell's Law in Equation (7.10).
4. Calculate the critical angles (θ_c) obtained from the geometry of the light propagation in the water using Equation (7.20) and Equation (7.22).
5. Compare the critical angle obtained from Snell's law to the critical angles obtained from the geometry of the light propagation in the water.

7.4.8 RESULTS AND DISCUSSIONS

7.4.8.1 Laser Light Passes through a Water Layer

Discuss your observations of the laser beam path at four angles of incidence on the left side of the water tank.

7.4.8.2 Law of Refraction at Air–Water Interface

Discuss Snell's law verification. Try to explain any discrepancies between your measurements and Snell's law.

7.4.8.3 Critical Angle at Water–Air Interface

1. Present a geometric drawing of the laser beam propagation through the water.
2. Discuss the calculated and measured critical angles.
3. Explain the variation between the critical angles obtained by Snell's law and the critical angle obtained from the geometry of the light beam passing through the water layer.

7.4.8.4 Total Internal Reflection at Water–Air Interface

1. Present a ray drawing of the light propagation from the light source through the water for each distance (L).

2. Find a relation between the distance (L) and diameters (D) and (d).
3. Discuss the critical angles for each distance (L).
4. Explain the variation between the critical angle obtained by Snell's law and the critical angles obtained from the geometry of the light propagation in the water.

7.4.9 Conclusion

Summarize the important observations and findings obtained in this lab experiment.

7.4.10 Suggestions for Future Lab Work

List any suggestions for improvements using different experimental equipment, procedures, and techniques for any future lab work. These suggestions should be theoretically justified and technically feasible.

7.4.11 List of References

List any references that were used in the report. Use one format in writing the references. Never mix reference formats in a report.

7.4.12 Appendices

List all of the materials and information that are too detailed to be included in the body of the report.

FURTHER READING

Beiser, Arthur, *Physics*, 5th ed., Addison-Wesley, Reading, MA, 1991.

Born, M. and, Wolf, E., Elements of the theory of diffraction and rigorous diffraction theory, *Principles of Optics: Electromagnetic Theory of Propagation, Interference, and Diffraction of Light*, Cambridge University Press, Cambridge, England, pp. 370–458 and 556–592, 1999.

Bouwkamp, C. J., Diffraction theory, *Rep. Prog. Phys.*, 17 35, 100, 1949.

Bromwich, T. J. I'A, Diffraction of waves by a wedge, *Proc. London Math. Soc.*, 14, 450–468, 1916.

Cutnell, J. D. and Johnson, K. W., *Physics*, 5th ed., John Wiley & Sons, Inc., New York, New York, U.S.A., 2001a.

Cutnell, J. D. and Johnson, K. W., *Student Study Guide—Physics*, 5th ed., John Wiley & Sons, Inc., New York, U.S.A., 2001b.

Francon, M., *Optical Interferometry*, Academic Press, New York, pp. 97–99, 1966.

Ghatak, K., *An Introduction to Modern Optics*, McGraw-Hill, New York, 1972.

Griot, Melles, *The Practical Application of Light*, Melles Griot Catalog, Rochester, NY, U.S.A., 2001.

Halliday, D., Resnick, R., and Walker, J., *Fundamental of Physics*, 6th ed., John Wiley & Sons, New York, U.S.A., 1997.

Heath, R. W., Macnaughton, R. R., and Martindale, D. G., *Fundamentals of Physics*, D.C. Heath Canada, Ltd., Canada, 1979.

Hecht, Eugene, *Optics*, 4th ed., Addison-Wesley Longman, Inc., Reading, MA, 2002.

Hewitt, P. G., *Conceptual Physics*, 8th ed., Addison-Wesley, Inc., Reading, MA, U.S.A., 1998.

Hoss, R. J., *Fiber Optic Communications—Design Handbook*, Prentice Hall Pub. Co., Englewood, Chffs. NJ, 1990.

Jenkins, F. W. and White, H. E., *Fundamentals of Optics*, McGraw Hill, New York, 1957.

Lambda, Lambda Catalog, Lambda, Research Optics, Inc., Costa Mesa, California, U.S.A., 2004.

Lehrman, R. L., *Physics-The Easy Way*, 3rd ed., Barron's Educational Series, Inc., U.S.A., 1998.

Lerner, R. G. and Trigg, G. L., *Encyclopedia of Physics*, 2nd ed., VCH Publishers, Inc., New York, 1991.

Levine, H. and Schwinger, J., On the theory of diffraction by an aperture in an infinite plane screen II, *Phys. Rev.*, 75, 1423–432, 1949

McDermott, L. C., *Introduction to Physics*, Prentice Hall, Inc., Englewood, Chffs. NJ, U.S.A., 1988.

Naess, R. O., *Optics for Technology Students*, Prentice Hall, Englewood, Chffs. NJ, 2001.

Newport Corporation, *Optics and Mechanics Section*, the Newport Resources 1999/2000 Catalog, Newport Corporation, Irvine, CA, U.S.A., 1999/2000.

Newport Corporation, *Photonics Section*, the Newport Resources 2004 Catalog, Newport Corporation, Irvine CA, U.S.A., 2004.

Nolan, P. J., *Fundamentals of College Physics*, Wm. C. Brown Publishers, Dubuque, IA, U.S.A., 1993.

Otto, F. B. and Wilson, J. D., *Physics—a Practical and Conceptual Approach*, 3rd ed., Saunders Goldern Sunburst Series, Saunders College Publishing, Harcourt Brace College Publishers, Orlando, Florida, U.S.A., 1993.

Pedrotti, F. L. and Pedrotti, L. S., *Introduction to Optics*, 2th ed., Prentice Hall, Inc., Englewood, Chffs. NJ, 1993.

Plamer, Christopher, *Diffraction Grating Handbook*, 5th ed., Thermo RGL, Richardson Grating Laboratory, New York, 2002.

Robinson, P., *Laboratory Manual to Accompany Conceptual Physics*, 8th ed., Addison-Wesley, Inc., Reading, MA, U.S.A., 1998.

Serway, R. A. and Jewett, J. W., *Physics for Scientists and Engineers with Modern Physics*, 6th ed., Vol. 2, Thomson Brooks/Cole, U.S.A., 2004.

Smith, W. J., *Modern Optical Engineering*, McGraw-Hill Book Co., New York City, NY, 1966.

Tippens, P. E., *Physics*, 6th ed., Glencoe McGraw-Hill, Westerville, OH, U.S.A., 2001.

Urone, P. P., *College Physics*, Thomson Brooks/Cole Publishing, U.S.A., 1993.

Walker, J. S., *Physics*, Prentice Hall, Upper Saddle River, New Jersey, U.S.A., 2002.

Warren, M. L., *Introduction to Physics*, W.H. Freeman and Company, San Francisco, CA, U.S.A., 1979.

White, H. E., *Modern College Physics*, 6th ed., Van Nostrand Reinhold Company, New York, 1972.

Williams, J. E., *Teacher's Edition Modern Physics*, Holt, Rinehart and Winston, Inc., New York, U.S.A., 1968.

Woods, N., *Instruction's Manual to Beiser Physics*, 5th ed., Addison-Wesley Publishing, Reading, MA, 1991.

Yeh, Chai, *Handbook of Fiber Optics: Theory and Applications*, Academic Press, San Diego, 1990.

Section II

Optics

8 Plane Mirrors

8.1 INTRODUCTION

Plane mirrors are made from a piece of glass polished on one side. If one stands before a plane mirror, one sees an upright image of oneself as far "behind" the mirror as one's position in front of it. All plane mirrors form an upright image equal to the height of the object placed in front them. If one raises one's right hand, the mirror image raises its left hand. Lettering read in a plane mirror is reversed right to left as well. This is the reason why writing on emergency vehicles, such as ambulances and police cars, is reversed right to left. Thus, when a driver looks into the mirror of his car, the writing can easily be read, and he can give room to emergency vehicles passing through.

Image formation by plane mirrors follows the law of reflection, as explained in the chapter on the laws of light. The ray-tracing method is used to show the image formation by a plane mirror. The reflection of light from a smooth shiny surface is more clear and focused than the reflection of light from a rough surface.

Plane mirrors are used in simple reflection applications, such as building optic and optical fibre devices.

Seven experimental lab cases presented in this chapter study the formation of images when an object or a laser beam source is placed in front of a plane mirror. The cases are: a candle placed in front of a fixed plane mirror and between two plane mirrors at right angles, tracing a laser beam passing between two plane mirrors at different angles, and tracing a laser beam incident on a rotating plane mirror.

8.2 THE REFLECTION OF LIGHT

When a light ray traveling in a medium comes across a boundary leading into a second medium, such as that between air and glass, part of the incident light ray is reflected back into the first

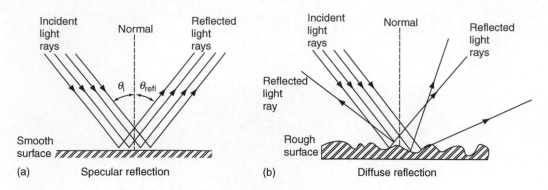

FIGURE 8.1 Light reflection from smooth and rough surfaces.

medium. The light entering the glass is partially absorbed and partially transmitted. The transmitted light is refracted. These phenomena are called the laws of reflection and refraction, and have been presented in detail in the laws of light chapter.

Figure 8.1(a) shows two types of reflection: specular and diffuse reflection. Light reflection from a smooth surface is called regular or specular reflection. The figure also shows that all the reflected light moves in parallel. Reflection of light from mirrors and smooth surfaces is an example of specular reflection. In contrast, if a surface is rough, as shown in Figure 8.1(b), the reflected light is spread out and scattered in all directions. Light reflection from a rough surface is called diffuse reflection. Reflection of light from brick, concrete, and other rough surfaces is an example of diffuse reflection.

Consider a point source of light placed a distance S in front of a plane mirror at a location O, as shown in Figure 8.2. The distance S is called the object distance. Light rays leaving the light source are incident on the mirror and reflect from the mirror. After reflection, the rays diverge, but they appear to the viewer to come from a point I located "behind" the mirror at distance S'. Point I is called the image of the object at O. The distance S' is called the image distance. Images are formed in the same way for all optical components, including mirrors and lenses. Images are formed at the point where rays of light actually intersect at the point from which they appear to originate. Solid lines represent the light coming from an object or a real image. Dashed lines are drawn to represent the light coming from an imaginary image. Figure 8.2 shows that the rays appear to originate at I, which is behind the mirror. This is the location of the image. Images are classified as real or virtual. A real image is one in which light reflects from a mirror or passes through a lens and the image can be captured on a screen. A virtual image is one

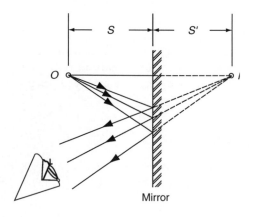

FIGURE 8.2 Image formation by a plane mirror.

in which the light does not really pass through a mirror or lens but appears either behind the mirror or in front of the lens. The virtual image cannot be displayed on a screen.

8.2.1 An Object Placed in Front of a Plane Mirror

Figure 8.3 shows a virtual image formed by a plane mirror. The images seen in plane mirrors are always virtual for real objects. Real images can usually be displayed on a screen, but virtual images cannot. Some of the properties of the images formed by plane mirrors can be examined using simple geometric techniques shown in Figure 8.3. In order to find out where an image is formed, it is always necessary to follow at least two rays of light as they reflect from the mirror. One of those rays starts at O, follows a horizontal path to the mirror, and reflects back on itself, OA. The second ray follows the oblique path OB and reflects, as shown in Figure 8.3. An observer to the left of the mirror, as shown in this figure, would trace the two reflected rays back to the point from which they appear to originate, point I. A continuation of this process for points on the object other than I would result in a virtual image to the right of the mirror, as shown in this figure. Since triangles OAB and IAB are congruent, $OA = IA$. Therefore, the image formed by an object placed in front of a plane mirror is the same distance behind the mirror as the object is in front of the mirror.

The geometry of Figure 8.3 also shows that the object height, h, equals the image height, h'. Let lateral magnification, M, of a plane mirror be defined as given in Equation (8.1):

$$M = \frac{h'}{h} \tag{8.1}$$

Equation (8.1) gives a general definition of the lateral magnification of any type of plane mirror. $M = 1$ for a plane mirror because $h' = h$, as shown in the geometry of Figure 8.3.

The image formed by a plane mirror has one more important property: that of right-left reversal between image and object. This reversal can be seen by standing in front of a mirror and raising one's right hand. The image seen raises its left hand.

In summary, the image formed by a plane mirror has the following properties:

1. The image is as far "behind" the plane mirror as the object location is in front of the mirror.
2. The image is unmagnified, virtual, and erect, as shown in Figure 8.3.
3. The image has right–left reversal.

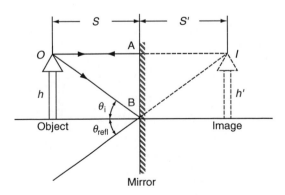

FIGURE 8.3 Geometric construction used to locate the image of an object placed in front of a plane mirror.

8.2.2 Multiple Images Formed by an Object Placed between Two Plane Mirrors at Right Angles

Figure 8.4 shows an object at point O placed between two plane mirrors at right angles to each other. In this situation, multiple images are formed. The positions of the multiple images that formed by the two mirrors can be located using the simple geometric techniques shown in Figure 8.4. The image of the object is at $image_1$, behind mirror 1, and at $image_2$, behind mirror 2. In addition, a third image is formed at $image_3$, which will be the image of $image_1$ in mirror 2 or, equivalently, the image of $image_2$ in mirror 1. That is, the image at $image_1$ (or $image_2$) serves as the object for $image_3$. When viewing $image_3$, note that the rays reflect twice after leaving the object point at O.

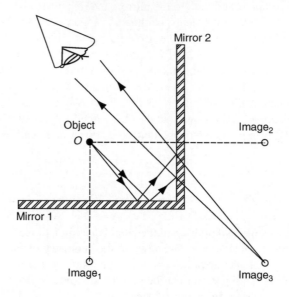

FIGURE 8.4 An object placed between two plane mirrors at right angles.

8.2.3 Tracing a Laser beam Passing between Two Plane Mirrors at an Acute Angle

This is sometimes called double reflected light ray between two mirrors. Two mirrors make an acute angle, β, with each other, as shown in Figure 8.5. A light ray is incident on mirror 1 at an angle θ_1 to the normal. From the law of reflection, the reflected ray also makes an angle θ_1 with the normal.

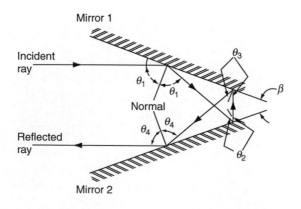

FIGURE 8.5 Laser beam path passes between two plane mirrors at an acute angle.

The reflected ray from mirror 1 is then incident on mirror 2, making an angle θ_2 with the normal. The reflected ray from mirror 2 also makes an angle θ_2 with the normal. Similarly, the reflected ray from mirror 2 is incident on mirror 1 with an angle θ_3 with the normal. The final reflected ray from mirror 1 onto mirror 2 makes angles of incidence and reflection θ_4 with the normal.

From the triangle made by the first reflected ray and the two mirrors, one can see the first reflected ray from mirror 1 and mirror 2 in relation to the acute angle between the mirrors. Figure 8.5 also shows that the first incident ray on mirror 1 is parallel to the last reflected ray from mirror 2. The first incident and last reflected rays are in opposite directions.

8.2.4 TRACING A LASER BEAM PASSING BETWEEN TWO PLANE MIRRORS AT RIGHT ANGLES

Two mirrors make a right angle with each other, as shown in Figure 8.6. A light ray is incident on mirror 1 at an angle 45° to the normal. From the law of reflection, the reflected ray also makes an angle 45° with the normal. The reflected ray from mirror 1 is then incident on mirror 2. The incident ray makes an angle of 45° with the normal, and the reflected ray from mirror 2 also makes an angle of 45° with the normal. From the triangle made by the first reflected ray and the two mirrors, it is possible to see the first reflected ray from mirror 1 and mirror 2 in relation to the right angle between the mirrors. Figure 8.6 also shows that the incident ray on mirror 1 is parallel to the reflected ray from mirror 2. The first incident and last reflected rays are in opposite directions.

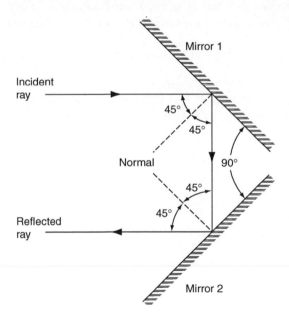

FIGURE 8.6 Laser beam path passes between two plane mirrors at right angles.

8.2.5 TRACING A LASER BEAM PASSING BETWEEN TWO PLANE MIRRORS AT AN OBTUSE ANGLE

Two mirrors make an obtuse angle β with each other, as shown in Figure 8.7. A light ray is incident on mirror 1 at an angle θ_1 to the normal. From the law of reflection, the reflected ray also makes an angle θ_1 with the normal. Thus, the reflected ray from mirror 1 is incident on mirror 2. The incident ray makes an angle θ_2 with the normal and the reflected ray from mirror 2 also makes an angle θ_2 with the normal. From the triangle made by the first reflected ray and the two mirrors, you can see the first reflected ray from mirror 1 and mirror 2 in relation to the obtuse angle between the mirrors. Figure 8.7 also shows that the incident ray on mirror 1 is not parallel to the reflected ray from mirror 2. The first incident and last

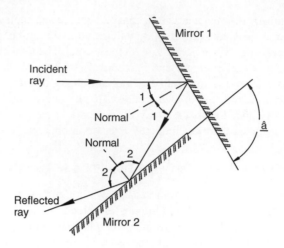

FIGURE 8.7 Laser beam path passes between two plane mirrors at an obtuse angle.

reflected rays are in opposite direction at an angle with each other. It is possible to direct the last reflected ray to any angle by rotating mirror 1 and/or mirror 2, i.e., changing the angle β.

8.2.6 Tracing a Laser Beam Passing between Three Plane Mirrors at Different Angles

Three mirrors make different angles with each other, as shown in Figure 8.8. A light ray is incident on mirror 1 at an angle θ_1 to the normal. From the law of reflection, the reflected ray also makes an angle θ_1 with the normal. The reflected ray from mirror 1 is then incident on mirror 2, making an

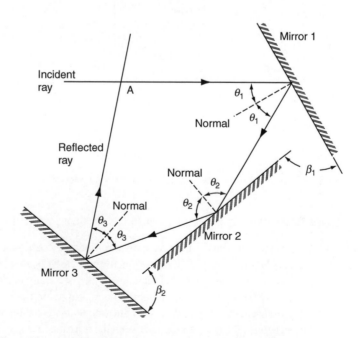

FIGURE 8.8 Laser beam path passes between three plane mirrors at different angles.

angle θ_2 with the normal. The reflected ray from mirror 2 also makes an angle θ_2 with the normal. Similarly, the reflected ray from mirror 2 is incident on mirror 3, with an angle θ_3 with the normal. The reflected ray from mirror 3 makes an angle θ_3 with the normal.

From the triangle made by the first reflected ray and the three mirrors, one can see the first reflected ray from mirror 1, mirror 2, and mirror 3 in relation to the angles between the mirrors. Figure 8.8 also shows that the incident ray on mirror 1 is not parallel to the reflected ray from mirror 3. Depending on the angles between the mirrors, the last reflected ray could be in any direction at an angle with the first incident ray. The last reflected and first incident rays can intercept each other at a point A. It is possible to direct the last reflected ray to any angle by rotating mirror 1, mirror 2, or mirror 3.

8.2.7 TRACING A LASER BEAM INCIDENT ON A ROTATING MIRROR

Figure 8.9(a) shows a single ray of light incident normally upon a plane mirror. The angle of incidence is zero to the mirror; thus, the angle of reflection is also zero, according to the law of reflection (the angle of incidence is equal to the angle of reflection). If the mirror is now rotated through an angle, α, without changing the direction of the incident light ray, the angle of incidence also becomes equal to α. Therefore, the angle of reflection NBC becomes α as well, as shown in Figure 8.9(b). Thus, if the mirror rotates through an angle α, the separation of the incident and reflected rays changes by 2α. Since the incident ray remains fixed throughout, it is equally true that the rotation of the mirror through an angle α causes a movement of the reflected ray through an angle of 2α.

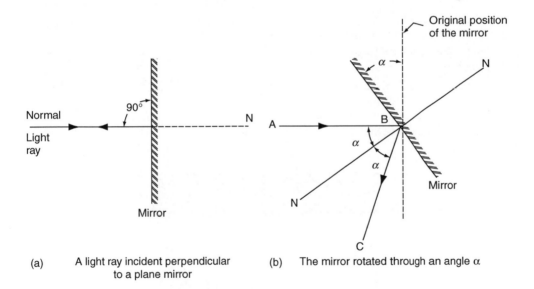

(a)   A light ray incident perpendicular   (b)   The mirror rotated through an angle α
       to a plane mirror

FIGURE 8.9 The rotating mirror.

8.3 EXPERIMENTAL WORK

The theory of this experiment is based on the first law of light, which is called the law of reflection. A light ray incident on a reflecting surface is described by an angle of incidence (θ_i). This angle is measured relative to a line normal (perpendicular) to the reflecting surface, as shown in Figure 8.1. Similarly, the reflected ray is described by an angle of reflection (θ_{refl}). The incident ray, the normal line, and the reflected ray lie in the plane of incidence. The relationship between these angles is given by the law of reflection: the angle of incidence is equal to the angle of reflection.

This experiment is a study of the formation of images when an object is placed in front of a plane mirror. The student will study the image formation by reflection. Mirrors are optical components that work on the basis of image formation by reflection. Such optical components are commonly used in manufacturing optical devices, instruments, and systems. In the theory of this experiment, the ray approximation assumes that light travels in straight lines. This corresponds to the field of geometric optical components. This experiment discusses the manner in which optical components, such as mirrors, form images. The plane mirrors are the simplest optical components to be considered in detail.

The purpose of this lab is to observe the image(s) and laser beam path when a candle or a laser beam source is placed in front of plane mirror arrangements for the following cases:

a. A candle placed in front of a fixed plane mirror.
b. A candle placed between two plane mirrors at right angles.
c. A laser beam passing between two plane mirrors at an acute angle.
d. A laser beam passing between two plane mirrors at right angles.
e. A laser beam passing between two plane mirrors at an obtuse angle.
f. A laser beam passing between three plane mirrors at different angles.
g. A laser beam incident on a rotating plane mirror.

8.3.1 Technique and Apparatus

Appendix A presents the details of the devices, components, tools, and parts.

1. 2×2 ft. optical breadboard.
2. HeNe laser light source and power supply.
3. Laser mount assembly.
4. Candles.
5. Hardware assembly (clamps, posts, screw kits, screwdriver kits, sundry positioners, etc.).
6. Mirrors and mirror holders, as shown in Figure 8.10.
7. Rotation stages, as shown in Figure 8.10.
8. Target card and cardholder, as shown in Figure 8.11.

FIGURE 8.10 A plane mirror mounted on a rotation stage.

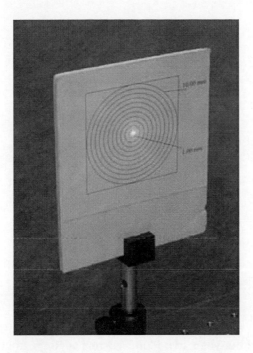

FIGURE 8.11 A target card and cardholder.

9. Protractor.
10. Ruler.

8.3.2 Procedure

Follow the laboratory procedures and instructions given by the instructor.

8.3.3 Safety Procedure

Follow all safety procedures and regulations regarding the use of optical components and instruments, light source devices, and optical cleaning chemicals.

8.3.4 Apparatus Setup

8.3.4.1 A Candle Placed in Front of a Fixed Plane Mirror

1. Figure 8.12 shows the experimental apparatus setup.
2. Mount a plane mirror on the breadboard so that the mirror lines up with the holes.
3. Place the candle in front of the plane mirror at a reasonable distance.
4. Measure the distance between the candle and the mirror by counting the holes on the breadboard. Fill out Table 8.1.
5. Measure the distance between the mirror and the image by counting the holes on the breadboard. Fill out Table 8.1.
6. Illustrate the locations of the candle, mirror, and image in a diagram.

FIGURE 8.12 A candle placed in front of a plane mirror.

8.3.4.2 A Candle Placed between Two Plane Mirrors at Right Angles

1. Figure 8.13 shows the experimental apparatus setup.
2. Mount two plane mirrors on the breadboard. Position the two mirrors to make a right angle with each other.
3. Place the candle between the mirrors.
4. Move the candle between the two mirrors and observe the formation of images.
5. Measure the distance between the candle and the mirrors by counting the holes on the breadboard. Fill out Table 8.2.
6. Measure the distance between the mirrors and the images by counting the holes on the breadboard. Fill out Table 8.2.
7. Count the number of images that are produced between the mirrors.
8. Illustrate the locations of the candle, mirrors, and images in a diagram.

FIGURE 8.13 A candle placed between two plane mirrors at right angles.

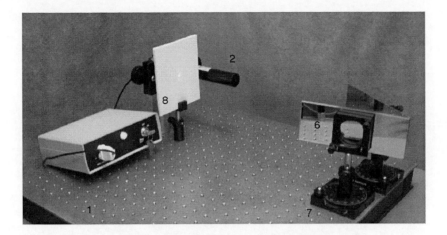

FIGURE 8.14 A laser beam passing between two plane mirrors at an acute angle.

8.3.4.3 A Laser Beam Passing between Two Plane Mirrors at an Acute Angle

1. Figure 8.14 shows the experimental apparatus setup.
2. Bolt the laser short rod to the breadboard.
3. Bolt the laser mount to the clamp using bolts from the screw kit.
4. Put the clamp on the short rod.
5. Place the HeNe laser into the laser mount and tighten the screw. Turn on the laser device. Follow the operation and safety procedures of the laser device in use.
6. Check the laser alignment with the line of bolt holes on the breadboard and adjust when necessary.

FIGURE 8.15 The path of the laser beam.

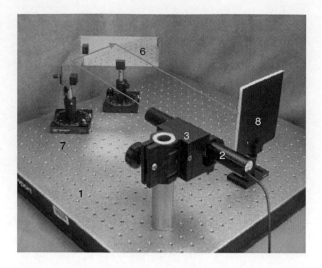

FIGURE 8.16 A laser beam passing between two plane mirrors at right angles.

7. Mount two plane mirrors on the breadboard.
8. Align the laser beam to be incident on the first mirror. Align the second mirror to make an acute angle with the first, as shown in Figure 8.15.
9. Observe the laser beam path between the two mirrors and intercept the laser beam with the target card, as shown in Default 8.14 and 8.15.
10. Measure the angles of the laser beam path relative to the mirrors. Fill out Table 8.3.
11. Illustrate the locations of the mirrors and the laser beam path in a diagram.

8.3.4.4 A Laser Beam Passing between Two Plane Mirrors at Right Angles

Repeat the procedure explained in Case (c) of this experiment. However, in Case (d), make a right angle between the two plane mirrors. Figure 8.16 shows the experimental apparatus setup and the laser beam path. Fill out Table 8.4.

8.3.4.5 A Laser Beam Passing between Two Plane Mirrors at an Obtuse Angle

Repeat the procedure explained in Case (c) of this experiment. However, in Case (e), make an obtuse angle between the two plane mirrors. Figure 8.17 shows the experimental apparatus setup and the laser beam path. Fill out Table 8.5.

8.3.4.6 A Beam Source Passing between Three Plane Mirrors at Different Angles

Repeat the procedure explained in Case (c) of this experiment. However, in Case (f), choose different angles between the three plane mirrors. Figure 8.18 shows the experimental apparatus setup and laser beam path. Fill out Table 8.6.

8.3.4.7 A Laser Beam Incident on a Rotating Mirror

1. Figure 8.19 shows the experimental apparatus setup.
2. Repeat steps 2 to 8 from Case (c).
3. Mount a plane mirror on a rotation stage and place it on the breadboard.
4. Set the rotation stage to zero degree or align the mirror to give you a back reflection into the laser source, as shown in Figure 8.19(a).

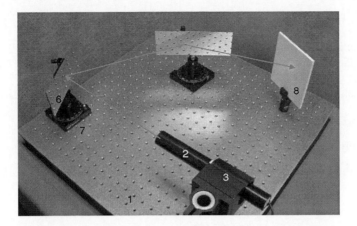

FIGURE 8.17 A laser beam passing between two plane mirrors at an obtuse angle.

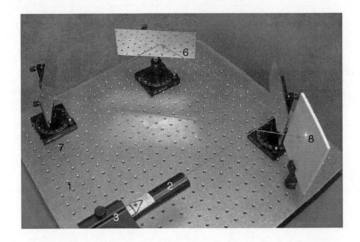

FIGURE 8.18 A laser beam passing between three plane mirrors at different angles.

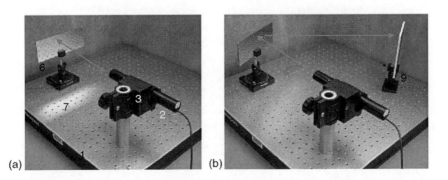

FIGURE 8.19 A laser beam incident on a rotating mirror.

5. Turn the rotation stage by an angle α (several degrees, say 20°), as shown in Figure 8.19(b).
6. Try to capture the reflected laser beam on the target card.
7. Measure the angle between the incident beam and the reflected beam; this is 2α. Fill out Table 8.7.
8. Illustrate the locations of the laser source, mirror, and the target card in a diagram.

8.3.5 DATA COLLECTION

8.3.5.1 A Candle Placed in Front of a Fixed Plane Mirror

Measure the locations of the candle, mirror, and image, and fill out Table 8.1.

TABLE 8.1
A Candle Placed in Front of a Plane Mirror

Location of		
Candle	Mirror	Image

8.3.5.2 A Candle Placed between Two Plane Mirrors at Right Angles

Measure the locations of the candle, mirrors, and images, and fill out Table 8.2.

TABLE 8.2
A Candle Placed between Two Plane Mirrors at Right Angles

Location of					
Candle	Mirror 1	Mirror 2	Image 1	Image 2	Image 3

8.3.5.3 A Laser Beam Passing between Two Plane Mirrors at an Acute Angle

Measure the locations of the laser source, mirrors, and target card, and fill out Table 8.3. A reference line can be used for the measurements of locations and angles.

TABLE 8.3
Laser Beam Passing between Two Plane Mirrors at an Acute Angle

Location of				Angle of	
Laser Source	Mirror 1	Mirror 2	Target Card	Mirror 1	Mirror 2

8.3.5.4 A Laser Beam Passing between Two Plane Mirrors at Right Angles

Measure the locations of the laser source, mirrors, and target card, and fill out Table 8.4. A reference line can be used for the measurements of locations and angles.

TABLE 8.4
Laser Beam Passing between Two Plane Mirrors at Right Angles

Location of				Angle of	
Laser Source	Mirror 1	Mirror 2	Target Card	Mirror 1	Mirror 2
				45°	45°

8.3.5.5 A Laser Beam Passing between Two Plane Mirrors at an Obtuse Angle

Measure the locations of the laser source, mirrors, and target card, and fill out Table 8.5. A reference line can be used for the measurements of locations and angles.

TABLE 8.5
Laser Beam Passing between Two Plane Mirrors at an Obtuse Angle

Location of				Angle of	
Laser Source	Mirror 1	Mirror 2	Target Card	Mirror 1	Mirror 2

8.3.5.6 A Laser Beam Passing between Three Plane Mirrors at Different Angles

Measure the locations of the laser source, mirrors, and target card, and fill out Table 8.6. A reference line can be used for the measurements of locations and angles.

TABLE 8.6
Laser Beam Passing between Three Plane Mirrors at Different Angles

Location of					Angle of		
Laser Source	Mirror 1	Mirror 2	Mirror 3	Target Card	Mirror 1	Mirror 2	Mirror 3

8.3.5.7 A Laser Beam Incident on a Rotating Mirror

Measure the locations of the laser source, mirror, and target card, and fill out Table 8.3.6. A reference line can be used for the measurements of locations and angles.

TABLE 8.7
A Laser Beam Incident on a Rotating Mirror

Location of			Angle of		
Laser Source	Mirror	Target Card	Mirror	Incident Ray	reflected Ray

8.3.6 CALCULATIONS AND ANALYSIS

8.3.6.1 A Candle Placed in Front of a Fixed Plane Mirror

Illustrate the locations of the candle, mirror, and image in a diagram; include your measurements.

8.3.6.2 A Candle Placed between Two Plane Mirrors at Right Angles

Illustrate the locations of the candle, mirror, and image in a diagram; include your measurements.

8.3.6.3 A Laser Beam Passing between Two Plane Mirrors at an Acute Angle

Illustrate the locations of the laser source, mirrors, target card, and laser beam path in a diagram; include your measurements.

8.3.6.4 A Laser Beam Passing between Two Plane Mirrors at Right Angles

Illustrate the locations of the laser source, mirror, target card, and laser beam path in a diagram; include your measurements.

8.3.6.5 A Laser Beam Passing between Two Plane Mirrors at an Obtuse Angle

Illustrate the locations of the laser source, mirrors, target card, and laser beam path in a diagram; include your measurements.

8.3.6.6 A Laser Beam Passing between Three Plane Mirrors at Different Angles

Illustrate the locations of the laser source, mirrors, target card, and laser beam path in a diagram; include your measurements.

8.3.6.7 A Laser Beam Incident on a Rotating Mirror

1. Illustrate the locations of the laser beam, mirror, target card, and laser beam path in a diagram; include your measurements.
2. Show that the reflected ray rotates 2α when the mirror rotates an angle α.

8.3.7 RESULTS AND DISCUSSIONS

8.3.7.1 A Candle Placed in Front of a Fixed Plane Mirror

1. Discuss your observation of the formation of the image in a diagram.
2. Confirm that the distance of the object from the mirror is equal to the distance of the image formed behind the mirror.

8.3.7.2 A Candle Placed between Two Plane Mirrors at Right Angles

1. Discuss your observation of the formation of the images.
2. Confirm that the distance of the object from the mirror is equal to the distance of the image formed behind the mirrors.
3. Confirm the number of images formed by the two plane mirrors at a right angle with each other.

8.3.7.3 A Laser Beam Passing between Two Plane Mirrors at an Acute Angle

1. Discuss your observation of the laser beam path.
2. Confirm the law of reflection of the laser beam when incident on a mirror.

For Cases (d) through (g), repeat the steps as explained in Case (c).

8.3.8 CONCLUSION

Summarize the important observations and findings obtained in this lab experiment.

8.3.9 SUGGESTIONS FOR FUTURE LAB WORK

List any suggestions for improvements using different experimental equipment, procedures, and techniques for any future lab work. These suggestions should be theoretically justified and technically feasible.

8.4 LIST OF REFERENCES

List any references that were used in the report. Use one format in writing the references. Never mix reference formats in a report.

8.5 APPENDICES

List all of the materials and information that are too detailed to be included in the body of the report.

FURTHER READING

Beiser, A., *Physics,* 5th ed., Addison-Wesley Publishing Company, Reading, MA, 1991.
Cutnell, J. D. and Johnson, K. W., *Physics* 5th ed., John Wiley and Sons, Inc., New York, 2001a.
Cutnell, J. D. and Johnson, K. W., *Student Study Guide—Physics*, 5th ed., John Wiley and Sons, Inc., New York, 2001b.

Davis, C. C., *Lasers and Electro-Optics: Fundamentals and Engineering*, Cambridge University Press, Cambridge, 1996.

Edmund Industrial Optics, *Optics and Optical Instruments Catalog*, Edmund Industrial Optics, Barrington, NJ, 2004.

Ghatak, A. K., *An Introduction to Modern Optics*, McGraw-Hill Book Company, New York, 1972.

Giancoli, D. C., *Physics,* 5th ed., Prentice Hall, Engle Wood, Cliffs, NJ, 1998.

Halliday, D., Resnick, R., and Walker, J., *Fundamentals of Physics*, 6th ed., John Wiley & Sons, Inc, USA, 1997.

Heath, R. W., Macaughton, R. R., and Martindale, D. G., *Fundamentals of Physics*, D.C. Heath Canada, Ltd., Canada, 1979.

Hecht, E., *Optics,* 4th ed., Addison-Wesley Longman, Inc, USA, 2002.

Jenkins, F. W. and White, H. E., *Fundamentals of Optics*, McGraw Hill, New York, 1957.

Keuffel & Esser Co., *Physics*, Keuffel & Esser Audiovisual Educator—Approved Diazo Transparency Masters, Audiovisual Division, Keuffel & Esser Co., U.S.A., 1989.

Lambda Research Optics, Inc., *Catalog 2004*, Lambda Research Optics, Costa Mesa, CA, 2004.

Lerner, R. G. and Trigg, G. L., *Encyclopedia of Physics*, 2nd ed., VCH Publishers, Inc, New York, NY, 1991.

McDermott, L. C., *Introduction to Physics*, Preliminary Edition, Prentice Hall, Inc., Engle Wood, Cliffs, NJ, 1988.

Melles, G., *The Practical Application of Light*, Melles Griot Catalog, Irvine, CA, 2001.

Naess, R. O., *Optics for Technology Students*, Prentice Hall, Englewood, Cliffs, NJ, 2001.

Nichols, D. H., *Physics for Technology with Applications in Industrial Control Electronics*, Prentice Hall, Englewood, Cliffs, NJ, 2002.

Nolan, P. J., *Fundamentals of College Physics*, Wm. C. Brown Publishers, Dubuque, Iowa, 1993.

Pedrotti, F. L. and Pedrotti, L. S., *Introduction to Optics*, 2nd ed., Prentice Hall, Inc, Engle Wood, Cliffs, NJ, 1993.

Serway, R. A. and Jewett, J. W., *Physics for Scientists and Engineers* with Modern Physics, 6th ed., volume 2, Thomson Books/Cole, U.S.A., 2004.

Smith, W. J., *Modern Optical Engineering*, McGraw-Hill Book Co., New York, 1966.

Warren, M. L., *Introduction to Physics*, W.H. Freeman and Company, San Francisco, 1979.

White, H. E., *Modern College Physics*, 6th ed., Van Nostrand Reinhold Company, New York, 1972.

Williams, J. E. et al., *Teacher's Edition—Modern Physics*, Holt, Rinehart and Winston, Inc, New York, 1968.

Woods, N., *Instruction's Manual to Beiser Physics*, 5th ed., Addison-Wesley Publishing Company, Reading, MA, 1991.

9 Spherical Mirrors

9.1 INTRODUCTION

Like plane mirrors, spherical mirrors are a common type of mirror used in many applications. Spherical mirrors are similar to plane mirrors in image formation, but instead of being made from a flat piece of glass, spherical mirrors have the shape of a section from the surface of a hollow sphere. If the inside surface of the mirror is reflective, it is called a concave mirror. If the outside surface of the mirror is reflective, it is called a convex mirror. The concave and convex mirrors are sometimes also called converging and diverging mirrors, respectively. Spherical mirrors are used in many applications, such as simple reflections, collimating, light convergence and divergence, and reflecting telescopes.

There are also curved mirrors with parabolic or elliptical shapes. These mirrors are used in many applications, including light reflectors in automobile headlights and solar ray concentrators. Spherical and curved mirrors are used mostly for concentrating or diverging light.

Three experimental lab cases presented in this chapter cover the formation of images of an object placed in front of a spherical mirror. The student will study the image formation using the ray-tracing method with experimental measurements and calculations. The cases are: image formation by concave mirrors; image formation by convex mirrors; and image formation by two spherical concave mirrors.

9.2 IMAGES FORMED BY SPHERICAL MIRRORS

The theory of this chapter is based on the first law of light, which is called the law of reflection. A light ray incident on a reflecting surface is described by an angle of incidence (θ_i) and an angle of reflection (θ_{refl}). The law of reflection has been presented in detail among the Laws of Light in Chapter 7.

Most curved mirrors used in practical optical applications are spherical. A spherical mirror is a portion of a reflecting sphere, which has a radius of curvature R and centre of curvature located at

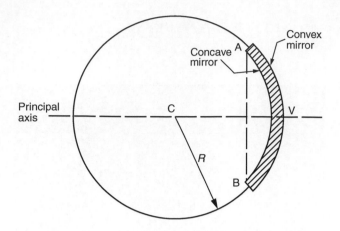

FIGURE 9.1 Definition of terms for spherical mirrors.

point C, as illustrated in Figure 9.1. The principal axis is an imaginary line that runs through C and the vertex V of the mirror. The opening AB, often used in solving optical problems, is called the linear aperture of the mirror.

There are two types of spherical mirrors: with the reflecting surface applied on the inner or on the outer surfaces. The reflecting surface can be created by polishing or reflective coating process. When the inner surface of the sphere is the reflecting surface, the mirror is called a concave mirror, as illustrated in Figure 9.2. When the outer surface of the sphere is the reflecting surface, the mirror is called a convex mirror, as illustrated in Figure 9.3.

Consider the reflection of parallel light rays from a spherical mirror. As illustrated in Figure 9.4, a light beam of parallel rays is incident on a concave mirror. Since the mirror is perpendicular to the principal axis of the mirror at the vertex V, the rays are reflected back and pass through a point F. This point is called the focal point of the mirror. The geometry of the reflection shows that the point F is located on the principal axis halfway between the centre of curvature C and the vertex V of the mirror. The distance from F to V is called the focal length f of the mirror. The focal length is defined

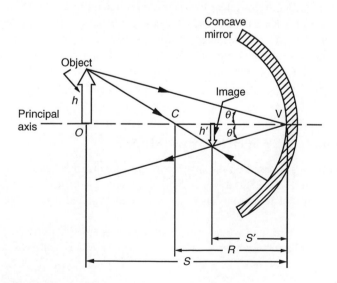

FIGURE 9.2 Formation of an image by a spherical concave mirror.

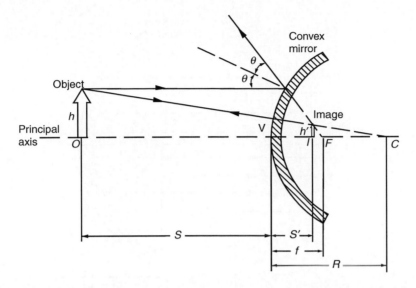

FIGURE 9.3 Formation of an image by a spherical convex mirror.

by Equation (9.1):

$$f = \frac{R}{2} \tag{9.1}$$

All light rays from a distant object, such as the sun, are parallel rays and will converge from the mirror at the focal point. For this reason, concave mirrors are frequently called converging mirrors. The focal point of a spherical mirror can be found experimentally by converging parallel light rays

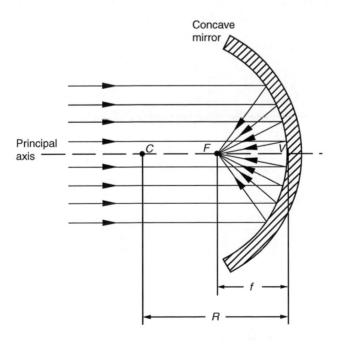

FIGURE 9.4 Light rays from a distant object reflect from a concave mirror through the focal point F.

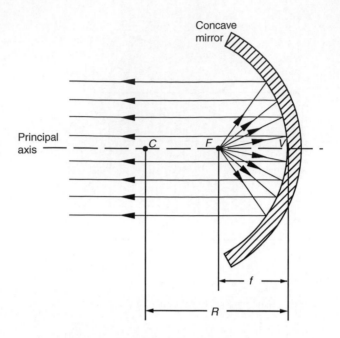

FIGURE 9.5 A light source at the focal point *F* of a concave mirror.

onto a target card. When sunlight is converged on a piece of paper at the focal point of the mirror, the paper can be burned by the intensity of the concentrated sunlight.

Since the light rays are reversible in direction, a source of light placed at the focal point of a concave mirror will form its image at infinity. In other words, the emerging light rays will be parallel to the principal axis of the mirror, as shown in Figure 9.5.

The same general principles can be applied for concave and convex mirrors. Light rays parallel to the principal axis, incident on the convex mirror surface, will diverge after reflecting off the mirror surface. The diverging reflected light rays appear to come from the focal point *F* located

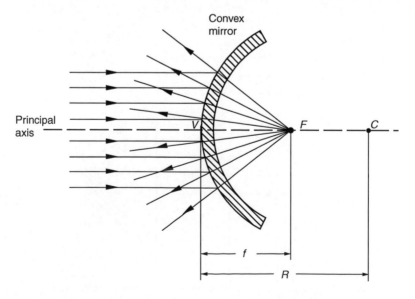

FIGURE 9.6 Light rays from a distant object reflect from a convex mirror and form the focal point *F* behind the mirror.

behind the mirror, but no light rays actually pass through it. Even though the focal point is virtual, the distance VF is called the focal length f, as shown in Figure 9.6. Since the actual light rays diverge when striking a convex mirror, convex mirrors are frequently called diverging mirrors. Equation (9.1) also applies in defining the focal length for a convex mirror.

9.2.1 CONCAVE MIRRORS

The geometrical optics method, sometimes called the ray-tracing method, is used for locating the images formed by spherical mirrors. This method traces the reflection of a few rays diverging from a point on an object O; this point must not be on the principal axis of the mirror. The point at which the reflected rays intersect will locate the image. Consider an object located on the principal axis away from the centre of curvature C of the concave mirror. Three rays originate from the point O on the top of the object. Figure 9.7 illustrates each of the three rays converging from the concave mirror.

Ray 1 is drawn from the top of the object from point O, parallel to the principal axis of the mirror, and then reflected back through the focal point F of a concave mirror.

Ray 2 is drawn from the top of the object from point O, through the focal point F of a concave mirror to the vertex V of the mirror, and then is reflected with the angle of incidence equal to the angle of reflection.

Ray 3 is drawn from the top of the object from point O, through the centre of curvature C, and then is reflected back along its original path.

The intersection of any two of these rays is enough to locate the image of an object. The third ray serves as a check on the image formation by the ray-tracing method. The image location obtained in the ray tracing method must always agree with the image location S' calculated from the spherical mirror equation. In Figure 9.7, the solid lines are used to identify real rays and real images.

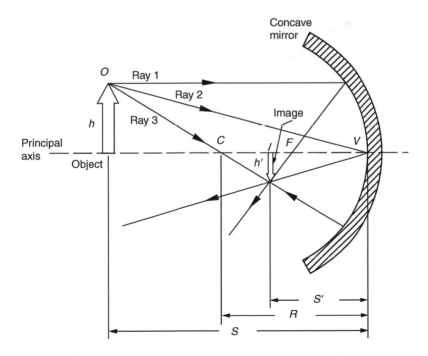

FIGURE 9.7 Formation of an image by a concave mirror using the ray tracing method.

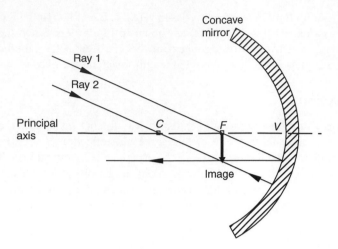

FIGURE 9.8 Image formed by a concave mirror for a distant object.

Consider the image formation by the ray-tracing method for an object which is placed in different positions in front of a concave mirror using two rays. In Figure 9.8, the parallel rays come from an object located at a distance far away from the centre of the concave mirror. The mirror forms an image at the focal point. The image is real, inverted, and smaller than the object.

In Figure 9.9, the object is located at the centre of curvature, C, of the concave mirror. The mirror forms an image at the centre of curvature. The image is real, inverted, and the same size as the object.

In Figure 9.10, the object is located at the focal point of F of the concave mirror. All reflected rays are parallel and will never intersect at any distance from the mirror. Therefore, no image will be formed in this situation.

In Figure 9.11, the object is located between the focal point, F, and the vertex, V, of the concave mirror. The rays form an image behind the mirror. The image can be located by extending the reflected rays to a point where they intersect behind the mirror. The image is virtual, erect, larger

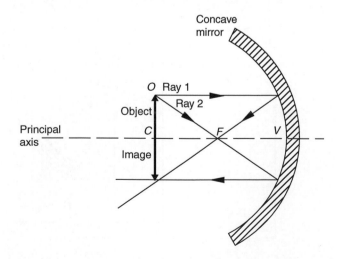

FIGURE 9.9 Image formed by a concave mirror for an object located at the centre of the curvature.

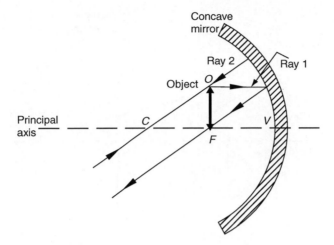

FIGURE 9.10 No image formed by a concave mirror for an object located at the focal point.

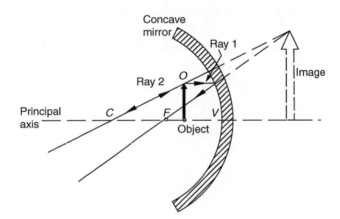

FIGURE 9.11 Image formed by a concave mirror for an object located inside the focal length.

than the object, and located behind the mirror. The magnification in this case follows the principle of spherical mirrors forming enlarged virtual images.

In summary, the characteristics of the image formed by a concave mirror depend on the location of the object. The images can be real or virtual, inverted or erect, and enlarged or reduced.

9.2.2 THE MIRROR EQUATION

The geometry shown in Figure 9.12 is used to calculate the image distance, S', knowing the object distance, S, and the radius of curvature, R, of a concave mirror. By convention, these distances are measured from the vertex point V of the concave mirror. Figure 9.12 shows two rays of light leaving from the point O on the top of the object. Ray 1 strikes the mirror at the vertex point V and reflects, as shown in the figure, obeying the law of reflection. Ray 1 is incident at an angle θ and reflected at an equal angle. Ray 2 passes through the centre of curvature C of the mirror, hits the mirror perpendicular to the surface, and reflects back on itself. The image is located where these two

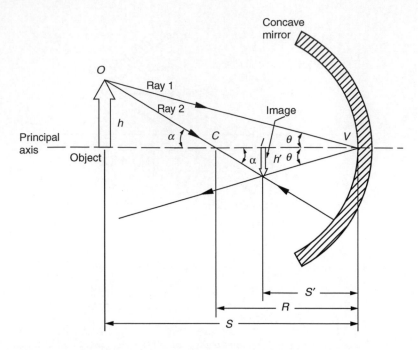

FIGURE 9.12 Deriving the mirror equation for an image formed by a concave mirror.

rays intersect. The image located at I is a real image, since actual light rays form it. The image is smaller than the object and is inverted.

From the largest triangle in Figure 9.12, one can write that $\tan \theta = h/s$, while the smallest triangle gives $\tan \theta = -h'/s'$. The negative sign signifies that the image is inverted. Thus, the magnification of the concave mirror can be written as:

$$M = \frac{h'}{h} = -\frac{S'}{S} \tag{9.2}$$

From the two triangles in Figure 9.12, one can also write the following relations:

$$\tan \alpha = \frac{h}{S-R} \quad \text{and} \quad \tan \alpha = -\frac{h'}{R-S'}$$

from which it is found that

$$\frac{h'}{h} = -\frac{R-S'}{S-R} \tag{9.3}$$

Combining Equation (9.2) and Equation (9.3), one has

$$\frac{S'}{S} = \frac{R-S'}{S-R} \tag{9.4}$$

Rearranging the terms using simple algebra, the following equation is obtained:

$$\frac{1}{S} + \frac{1}{S'} = \frac{2}{R} \tag{9.5}$$

Equation (9.5) is often written in terms of the focal length, f, of the mirror, instead of the radius of curvature. Again, the focal length of a mirror is given by Equation (9.1).

The mirror equation can be rewritten in terms of the focal length, by combining Equation (9.1) and Equation (9.5), to obtain:

$$\frac{1}{S} + \frac{1}{S'} = \frac{1}{f} \tag{9.6}$$

Equation (9.5) and Equation (9.6) are the two forms of the equation known as the spherical mirror equation. This equation is applicable to any spherical mirror, concave or convex.

9.2.3 CONVEX MIRRORS

Light is reflected from the silver coating of the outer convex surface of a spherical mirror. This type of mirror is sometimes called a diverging mirror. The rays from a point on a real object diverge after reflection as though they were coming from some point behind the convex mirror. Figure 9.13 illustrates an image formation by a convex mirror. In this figure, the image is virtual, since it lies behind the mirror at the location where the reflected rays appear to originate.

The ray-tracing method for the formation of images by spherical convex mirrors is applied and illustrated in Figure 9.13. The point at which the reflected rays intersect will locate the image. Consider an object located on the principal axis away from the centre of curvature C of the convex mirror. Three rays originate from the point O on the top of the object.

Ray 1 is drawn from the top of the object from point O, parallel to the principal axis of the mirror, and then reflected back to the same side. If the ray is extended, it seems to come from the focal point F of the convex mirror.

Ray 2 is drawn from the top of the object from point O, proceeds toward the focal point F of the convex mirror, and then is reflected parallel to the principal axis.

Ray 3 is drawn from the top of the object from point O, proceeds toward the centre of curvature C, and then is reflected back along its original path.

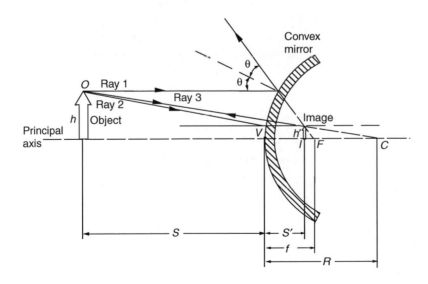

FIGURE 9.13 Ray tracing for a convex mirror.

The intersection of any two of these rays is enough to locate the image of an object. The third ray serves as a check on the image formation by the ray-tracing method.

The image location obtained in the ray-tracing method must always agree with the image location S' calculated from the spherical mirror equation. In Figure 9.13, the dashed lines are used to identify virtual rays and virtual images. Equation (9.6), the mirror equation, can be used to calculate the image location.

9.2.4 Sign Convention for Spherical Mirrors

S is *positive* if the object is in front of the mirror (real object).
S is *negative* if the object is in back of the mirror (virtual object).
S' is *positive* if the image is in front of the mirror (real image).
S' is *negative* if the image is in back of the mirror (virtual image).
Both f and R are positive if the centre of curvature is in front of the mirror (concave mirror).
Both f and R are negative if the centre of curvature is in back of the mirror (convex mirror).
If M is positive, the image is erect.
If M is negative, the image is inverted.

This convention applies only to the numerical values substituted into the mirror equations, Equation (9.5) and Equation (9.6). The quantities S, S', and f should maintain their signs unchanged until the substitution is made.

9.3 SPHERICAL ABERRATION

Spherical mirrors form sharp images as long as their apertures are small compared to their focal lengths. When large spherical mirrors are used, some of the rays from objects strike near the outer edges (away from the principal axis) and are focused to different points on the principal axis, producing a blurred image. This focusing defect is known as spherical aberration. The spherical aberration is illustrated in Figure 9.14. This defect can be corrected by using a parabolic mirror. Theoretically, parallel light rays incident on a parabolic mirror section will focus the rays at a single

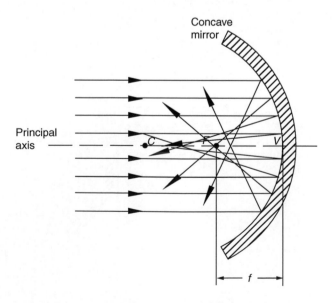

FIGURE 9.14 Spherical aberration occurring in a spherical mirror.

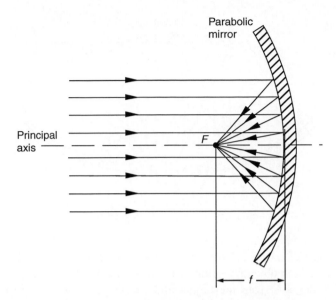

FIGURE 9.15 Parabolic mirror focuses all parallel light rays to the focal point.

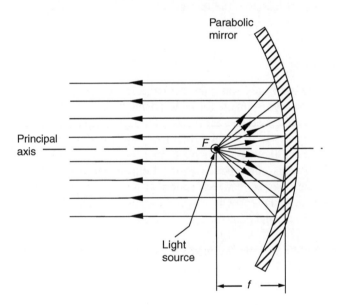

FIGURE 9.16 A light source at the focal point of a parabolic mirror.

point on the principal axis, as illustrated in Figure 9.15. Similarly, a small light source located at the focal point of a parabolic mirror is used in many optical devices, such as spotlights and searchlights. The light rays emitted from a light source are parallel to the principal axis of a parabolic mirror, as illustrated in Figure 9.16.

9.3.1 FORMATION OF IMAGE BY TWO SPHERICAL CONCAVE MIRRORS

Consider the image formation between two matching spherical concave mirrors, as illustrated in Figure 9.17. The two mirrors have back-silvered spherical glass surfaces, and each mirror has a hole

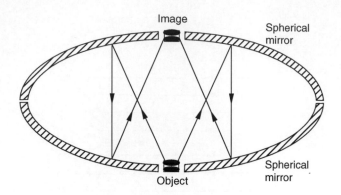

FIGURE 9.17 Forming a real three-dimensional image in the mirage.

at the centre of its spherical surface. The two mirrors are placed one on top of the other, with the silvered surfaces facing each other, and framed in a wooden box. An object is located at the hole of the lower mirror. Light rays from the object are first incident on the upper mirror, and then reflected onto the lower mirror. Next, the rays are all reflected from the lower mirror and intersect with each other to form the image in the hole of the upper mirror. The image appears as a real three-dimensional image in the mirage.

9.4 EXPERIMENTAL WORK

This experiment is a study of the formation of images of an object placed in front of a spherical mirror. The student will study the image formation using the ray-tracing method. Spherical mirrors are optical components that work on the basis of image formation by reflection. Such optical components are commonly used in manufacturing optical devices and instruments, such as reflecting telescopes. In the theory of this experiment, the ray approximation assumes that light travels in straight lines; this corresponds to the field of geometric optics. The theory of this experiment discusses the manner in which spherical mirrors form images. Spherical mirrors are similar to plane mirrors in image formation. The following cases deal with the image formation for different spherical mirror arrangements.

  a. Formation of images by concave mirrors.
  b. Formation of images by convex mirrors.
  c. Formation of image by two spherical concave mirrors.

9.4.1 TECHNIQUE AND APPARATUS

Appendix A presents the details of the devices, components, tools, and parts.

  1. Hardware assembly (clamps, posts, positioners, etc.).
  2. Spherical concave and convex mirrors, as shown in Figure 9.18.
  3. Spherical mirror holders.
  4. Two spherical concave mirrors, as shown in Figure 9.19.
  5. Optical rail.
  6. Candle.
  7. Black/white card and cardholder.
  8. Ruler.

FIGURE 9.18 Spherical concave and convex mirrors.

FIGURE 9.19 Spherical concave mirrors.

9.4.2 PROCEDURE

Follow the laboratory procedures and instructions given by the instructor.

9.4.3 SAFETY PROCEDURE

Follow all safety procedures and regulations regarding the use of light source, optical components, and optical cleaning chemicals.

9.4.4 APPARATUS SETUP

9.4.4.1 Formation of Images by Concave Mirrors

1. Figure 9.20 shows the experimental apparatus setup.
2. Place an optical rail on the table.

FIGURE 9.20 Formation of images by a concave mirror.

3. Mount a concave mirror on the table so that the centre of the mirror lines up perpendicular to the optical rail.
4. Place the candle on the optical rail in front of the concave mirror at a reasonable distance outside the centre of curvature of the concave mirror.
5. Measure the distance between the candle and the vertex of the mirror by counting the divisions on the optical rail. Fill out Table 9.1.
6. Measure the height of the candle and the height of the image. Fill out Table 9.1.
7. Choose three other positions for the candle in front of the concave mirror for this case. Repeat steps 5 and 6. Fill out Table 9.1.
8. Illustrate the locations of the candle, mirror, and image in diagrams using the ray-tracing method.

9.4.4.2 Formation of Images by Convex Mirrors

1. Figure 9.21 shows the experimental apparatus setup.
2. Place an optical rail on the table.
3. Mount a convex mirror on the table so that the centre of the mirror lines up perpendicular to the optical rail.
4. Place the candle on the optical rail in front of the convex mirror at a reasonable distance.
5. Measure the distance between the candle and the vertex of the mirror by counting the divisions on the optical rail. Fill out Table 9.2.
6. Measure the height of the candle and the height of the image. Fill out Table 9.2.
7. Choose three other positions for the candle in front of the convex mirror for this case. Repeat steps 5 and 6. Fill out Table 9.2.
8. Illustrate the locations of the candle, mirror, and image in diagrams using the ray-tracing method.

FIGURE 9.21 Formation of images by a convex mirror.

9.4.4.3 Formation of Image by Two Spherical Concave Mirrors

1. Mount two spherical mirrors, one on top of the other, on the table, as shown in Figure 9.22.
2. Place a paper clip in the centre hole of the lower spherical mirror.
3. Observe the image formation of the paper clip in the centre hole of the upper spherical mirror.
4. Illustrate the locations of the paper clip, mirrors, and image in a diagram using the ray-tracing method.

FIGURE 9.22 Formation of image by two spherical concave mirrors.

9.4.5 Data Collection

9.4.5.1 Formation of Images by Concave Mirrors

1. Measure the distance of the candle from the mirror and the heights of the candle and the image. Fill out Table 9.1.
2. Repeat step 1 for three additional candle positions. Fill out Table 9.1.

TABLE 9.1
A Candle Placed in Front of a Concave Mirror

Candle Position	Measured			Calculated		
	Candle Distance S (unit)	Candle Height h (unit)	Image Height h' (unit)	Magnification M	Image Distance S' (unit)	Mirror Focal Length f (unit)
1						
2						
3						
4						

9.4.5.2 Formation of Images by Convex Mirrors

1. Measure the distance of the candle from the mirror and the heights of the candle and the image. Fill out Table 9.2.
2. Repeat step 1 for three additional candle positions. Fill out Table 9.2.

TABLE 9.2
A Candle Placed in Front of a Convex Mirror

Candle Position	Measured			Calculated		
	Candle Distance S (unit)	Candle Height h (unit)	Image Height h' (unit)	Magnification M	Image Distance S' (unit)	Mirror Focal Length f (unit)
1						
2						
3						
4						

9.4.5.3 Formation of Image by Two Spherical Concave Mirrors

No data collection is required in this section.

9.4.6 CALCULATIONS AND ANALYSIS

9.4.6.1 Formation of Images by Concave Mirrors

1. For four candle positions, calculate the magnification of the mirror and the image distance using Equation (9.2), and the focal length of the mirror using Equation (9.6). Fill out Table 9.1.
2. For four candle positions, illustrate the locations of the candle, mirror, and image in diagrams, being sure to include dimensions.
3. Find the published value of the focal length of the mirror as provided by the manufacturer/supplier.

9.4.6.2 Formation of Images by Convex Mirrors

1. For four candle positions, calculate the magnification of the mirror and the image distance using Equation (9.2) and the focal length of the mirror using Equation (9.6). Fill out Table 9.2.
2. For four candle positions, illustrate the locations of the candle, mirror, and image in diagrams, being sure to include dimensions.
3. Find the published value of the focal length of the mirror as provided by the manufacturer/supplier.

9.4.6.3 Formation of Image by Two Spherical Concave Mirrors

Illustrate the locations of the paper clip, mirrors, and image in a diagram.

9.4.7 RESULTS AND DISCUSSIONS

9.4.7.1 Formation of Images by Concave Mirrors

1. Discuss your measurements and calculations of the formation of the images in diagrams.
2. Verify the calculated focal length of the mirror against the value published by the manufacturer/supplier.

9.4.7.2 Formation of Images by Convex Mirrors

1. Discuss your measurements and calculations of the formation of the images in diagrams.
2. Verify the calculated focal length of the mirror against the value published by the manufacturer/supplier.

9.4.7.3 Formation of Image by Two Spherical Concave Mirrors

Illustrate the locations of the paper clip, mirrors, and image in a diagram.

9.4.8 CONCLUSION

Summarize the important observations and findings obtained in this lab experiment.

9.4.9 SUGGESTIONS FOR FUTURE LAB WORK

List any suggestions for improvements using different experimental equipment, procedures, and techniques for any future lab work. These suggestions should be theoretically justified and technically feasible.

9.5 LIST OF REFERENCES

List any references that were used in the report. Use one format in writing the references. Never mix reference formats in a report.

9.6 APPENDIX

List all of the materials and information that are too detailed to be included in the body of the report.

FURTHER READING

Beiser, A., *Physics*, 5th ed., Addison-Wesley Publishing Company, Reading, MA, 1991.

Cutnell, J. D. and Johnson, K. W., *Physics*, 5th ed., Wiley, New York, 2001.

Cutnell, J. D. and Johnson, K. W., *Student Study Guide—Physics*, 5th ed., Wiley, New York, 2001.

Edmund Industrial Optics, *Optics and Optical Instruments Catalog*, Edmund Industrial Optics, Barrington, NJ, 2004.

Ewald, W. P., Young, W. A., and Roberts, R. H., *Practical Optics*, Makers of Pittsford, Pittsford, Rochester, New York, 1982.

Ewen, D. et al., *Physics for Career Education*, 4th ed., Prentice Hall, Englewood Cliffs, NJ, 1996.

Ghatak, A. K., *An Introduction to Modern Optics*, McGraw-Hill Book Company, New York, 1972.

Giancoli, D. C., *Physics*, 5th ed., Prentice Hall, Englewood Cliffs, NJ, 1998.

Halliday, D., Resnick, R., and Walker, J., *Fundamentals of Physics*, 6th ed., Wiley, New York, 1997.

Heath, R. W., Macnaughton, R. R., and Martindale, D. G., *Fundamentals of Physics*, D.C. Heath Canada Ltd., Canada, 1979.

Hecht, E., *Optics*, 4th ed., Addison-Wesley Longman, Inc, Reading, MA, 2002.

Jenkins, F. W. and White, E. H., *Fundamentals of Optics*, McGraw-Hill, New York, 1957.

Keuffel & Esser Co., Physics, Keuffel & Esser Audiovisual Educator-Approved Diazo Transparency Masters, Audiovisual Division, Keuffel & Esser Co., U.S.A., 1989.

Lambda Research Optics, Inc., *Catalog 2004*, Lambda Research Optics, Costa Mesa, CA, 2004.

Lehrman, R. L., *Physics—The Easy Way*, 3rd ed., Barron's Educational Series, Inc., Hauppauge, NY, 1998.

Lerner, R. G. and Trigg, G. L., *Encyclopedia of Physics*, 2nd ed., VCH Publishers, Inc., New York, 1991.

Malacara, D., *Geometrical and Instrumental Optics*, Academic Press, Boston, MA, 1988.

McDermott, L. C. et al., *Introduction to Physics*, Preliminary Edition, Prentice Hall, Inc., Englewood Cliffs, NJ, 1988.

Naess, R. O., *Optics for Technology Students*, Prentice Hall, Englewood Cliffs, NJ, 2001.

Newport Corporation, *Optics and Mechanics Section*, Newport Resources 1999/2000 Catalog, Newport Corporation, Irvine, CA, 1999.

Nolan, P. J., *Fundamentals of College Physics*, Wm. C. Brown Publishers, Dubuque, IA, 1993.

Pedrotti, F. L. and Pedrotti, L. S., *Introduction to Optics*, 2nd ed., Prentice Hall, Inc., Englewood Cliffs, NJ, 1993.

Sears, F. W., Zemansky, M. W., and Young, H. D., *University Physics—Part II*, 6th ed., Addison-Wesley Publishing Company, Reading, MA, 1998.

Serway, R. A., *Physics for Scientists and Engineers*, 3rd ed., Saunders Golden Sunburst Series, Philadelphia, PA, 1990.

Smith, W. J., *Modern Optical Engineering*, McGraw-Hill Book Co., New York, 1966.

Urone, P. P., *College Physics*, Brooks/Cole Publishing Company, Pacific Grove, CA, 1998.

Walker, J. S., *Physics*, Prentice Hall, Englewood Cliffs, NJ, 2002.

Warren, M. L., *Introduction to Physics*, W.H. Freeman and Company, San Francisco, CA, 1979.

White, H. E., *Modern College Physics*, 6th ed., Van Nostrand Reinhold Company, New York, 1972.

Williams, J., Metcalfe, H., Trinklein, F., and Lefler, R., *Teacher's Edition—Modern Physics*, Holt, Rinehart and Winston, Inc., New York, 1968.

Wilson, J. D., *Physics—A Practical and Conceptual Approach*, Saunders College Publishing, Philadelphia, PA, 1989.

Wilson, J. D. and Buffa, A. J., *College Physics*, 5th ed., Prentice Hall, Englewood Cliffs, NJ, 2000.

10 Lenses

10.1 INTRODUCTION

A typical lens is made of glass or plastic. Lenses are classified in two categories: thin lenses and thick lenses. Each category has its applications and equations for calculations. Lenses have two refraction surfaces; each surface is a segment of either a sphere or a plane. Lenses are commonly used to form images by the refraction of light. Light refraction in lenses is one application of the theory of the second law of light: the law of refraction. Lenses are used in building optic/optical fibre devices and instruments, such as cameras, microscopes, slide projectors, and fibre optical switches. The graphical method for locating images formed by mirrors will be used in this chapter as well. The experimental work in this chapter will enable students to practise tracing image formation by a lens or lens combination.

10.2 TYPES OF LENSES

The following sections introduce the types of lenses which are used in building optic and optical fibre devices and instruments.

10.2.1 CONVERGING AND DIVERGING LENSES

Figure 10.1 and Figure 10.2 show some common shapes of lenses. Typical spherical lenses have two surfaces defined by two spheres. The surfaces can be convex, concave, or planar. Lenses are divided into two types: (a) converging and (b) diverging. Converging lenses have positive focal lengths and are thickest at the middle. Common shapes of converging lenses are (1) biconvex, (2) convex–concave, and (3) plano-convex. Diverging lenses have negative focal lengths and are thickest at the edges. Common shapes of diverging lenses are (1) biconcave, (2) convex–concave, and (3) plano-concave.

FIGURE 10.1 Converging and diverging lenses.

Figure 10.3 illustrates the geometries of two common types of lenses: (a) a biconvex lens and (b) a biconcave lens. The type and thickness of the lens depends on the radius of curvatures R_1 and R_2, and the distance between the centres of curvature.

The principle of operation for a lens forming an image is explained by the second law of light, the law of refraction. When light rays pass through a lens, they are bent or deviated from their

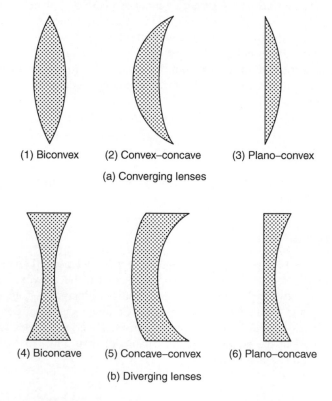

FIGURE 10.2 Various shapes of lenses.

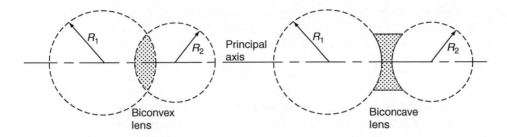

FIGURE 10.3 Two common types of lenses.

original paths, according to the law of refraction. The theory of light refraction through an optical medium is presented in Chapter 7, The Laws of Light. Refraction by a prism is addressed in Chapter 11, Prisms.

To study light refraction in a biconvex lens, two prisms can be placed base to base to approximate the convex lens operation, as shown in Figure 10.4(a). Parallel light rays that pass through the prisms are deviated so that the various rays intersect. They do not intersect or focus at a single point. However, if the surfaces of the prisms are curved rather than flat, then it becomes a converging lens,

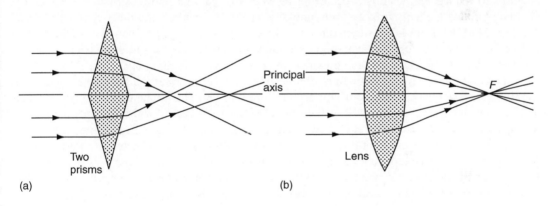

FIGURE 10.4 The principle of the converging lens.

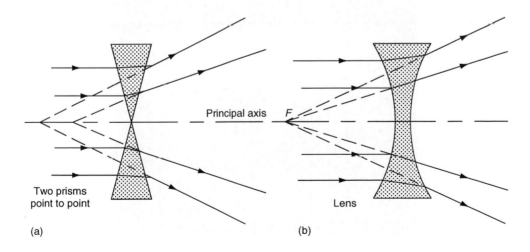

FIGURE 10.5 The principle of the diverging lens.

as shown in Figure 10.4(b). The converging lens brings incoming parallel rays of light to a single point *F*, called the focal point, at the principal axis. Because the refracted light rays pass through *F*, the focal point is real. This type of lens is often called a converging or convex lens.

Similarly, a biconcave lens can be approximated by two prisms with their apexes together, as shown in Figure 10.5(a). Parallel light rays that pass through the prisms are spread outward, but these diverging rays cannot be projected back to a single point. However, if the surfaces of the prisms are curved rather than flat, then it becomes a diverging lens, as shown in Figure 10.5(b). The diverging rays appear to originate from a single point on the incident side of the lens. The focal point *F* is not real; it is virtual because the rays do not actually pass through *F*. This type of lens is often called a diverging, or concave lens.

10.2.2 GRIN LENSES

GRIN is the acronym of the GRadient INdex lens. GRIN lenses have a cylindrical shape, as shown in Figure 10.6. One end is polished at an angle of 2, 6, 8, or 12 degrees, while the other end is polished at an angle of 2 or 90 degrees.

If the lens has one end polished, it is called a single angle lens; likewise, if both ends are polished, the lens is called a double angle lens. The choice of the angle for the polished face of the GRIN lens is dependent on the application. GRIN lenses are coated with anti-reflection (AR) coatings to reduce Fresnel losses occurring between two parallel optical surfaces. AR coatings also help protect the lens surfaces from humidity, chemical reaction, and other types of physical damage. The GRIN lens is constructed from multiple concentric layers, somewhat like the annular rings of a tree. Each concentric layer of glass, arranged radially from the central axis of the lens, has a lower index of refraction than the previous layer, as shown in Figure 10.7.

Figure 10.7 also shows the structure of a GRIN lens, and illustrates the paths of the light rays. Light travels faster in materials with a lower index of refraction. The further the light ray is from the lens centre, the greater its speed. The light traveling near the centre of the lens has the slowest average velocity. As a result, all rays tend to reach the end of the lens at the same time. Each layer within the lens refracts the light differently. Instead of being sharply reflected as in a step-index fibre, the light is now bent or continually refracted in a sinusoidal pattern. Those rays that follow the longest path by traveling near the outside of the lens have a faster average velocity. Collimated light entering from one end is focused on the focal point of the lens on the other end.

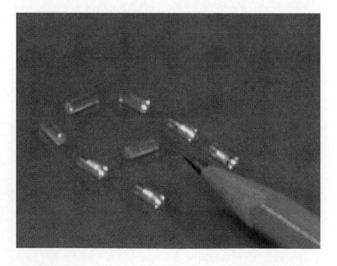

FIGURE 10.6 Actual GRIN lenses.

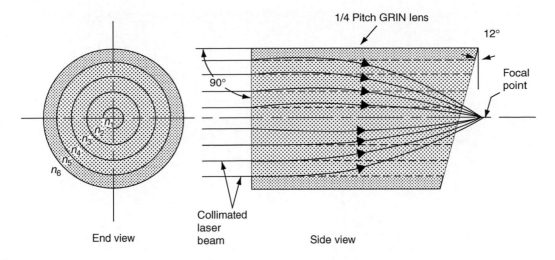

FIGURE 10.7 GRIN len's structure showing light rays.

Figure 10.8 illustrates a comparison between a GRIN lens and a convex lens. The GRIN lens equation is defined as:

$$P = \frac{Z\sqrt{A}}{2\pi}.$$

(10.1)

where

P the pitch
Z the lens length (mm); and
$\sqrt{A}$ the index gradient constant.

GRIN lenses are often specified in terms of their pitch lengths and operating wavelengths. Figure 10.9 illustrates GRIN lens of different pitch lengths. Figure 10.9(a) shows a GRIN lens with a length of more than one pitch. Where the length of a GRIN lens is very long and covered by cladding and jacket, a GRIN fibre optic cable is created. The principle of GRIN fibre optic cable will be discussed in detail later in this book. Figure 10.9(b) shows a one pitch GRIN lens. The image is erect and the same size as the object. In (c), which demonstrates a GRIN lens with a length of 1/2 pitch, the image is inverted and the same size as the object. And in (d), showing a GRIN lens

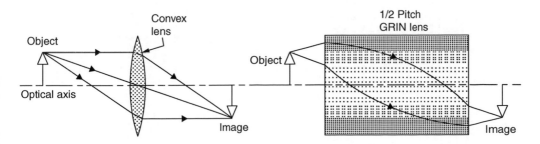

FIGURE 10.8 Comparison between a convex lens and a GRIN lens.

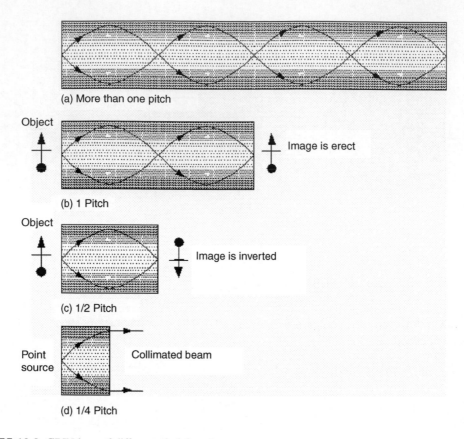

(a) More than one pitch

(b) 1 Pitch

(c) 1/2 Pitch

(d) 1/4 Pitch

FIGURE 10.9 GRIN lens of different pitch lengths.

with a length of 1/4 pitch, a point light source from one end is converted into a collimated beam exiting from the other end.

Coupling light into a fibre cable can be accomplished by using a quarter-pitch GRIN lens, which focuses the collimated beam from the laser source onto a small spot on the core of the fibre cable. Figure 10.10 is a schematic diagram showing the GRIN lens between the laser source and the fibre cable. The input power increases by inserting a GRIN lens. When the GRIN lens focuses the

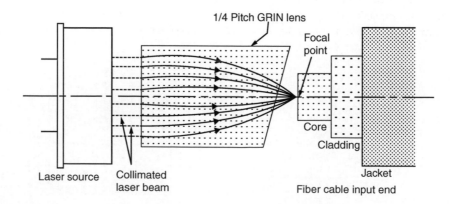

FIGURE 10.10 GRIN lens between a laser source and fibre cable input end.

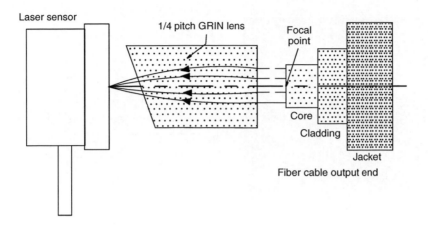

FIGURE 10.11 GRIN lens between a laser sensor and fibre cable output end.

power of the collimated laser beam onto the core of the fibre cable, the fibre input power is optimized.

Similarly, inserting a GRIN lens at the output of the fibre cable increases the output power, as shown in Figure 10.11.

GRIN lenses are used to focus and collimate light within a variety of fibre optical components. Passive component manufacturers use GRIN lenses in WDMs, optical switches, and attenuators. Active component manufacturers use GRIN lenses for fibre-to-detector and laser-to-fibre coupling.

GRIN lenses are useful for imaging and optical fibre communications systems because:

1. they are shaped for specific optical applications;
2. a completely flat piece of radial-GRIN material can act as a lens;
3. they can be used to image an object at one end and a detector at the other;
4. they are easy to use in both the design and alignment of optical fibre devices;
5. they can be used to produce micro-optical fibre devices; and
6. they are smaller than conventional lenses.

Radial graded-index (GRIN) lenses are useful in building fibre optical devices that are used in imaging and telecommunications systems. Made using a thermal diffusion technique, the new GRIN lenses are likely to be cheaper and of higher optical quality than existing GRIN lenses. This fabrication method offers flexibility in size and refractive index variation that is difficult to achieve with other methods.

10.2.3 BALL LENSES

Ball lenses, as shown in Figure 10.12(a), are used in manufacturing fibre optic devices for modern communication systems. Ball lenses couple fibres from micro-optic components to light sources, from laser diodes to fibre cables, and from fibre cables to detectors. They also collimate light from fibre cables in thin-film filters for dense wavelength division multiplexing. In a ball lens, trans-mitted light changes direction at the curved boundaries, traversing the interior in straight lines, as shown in Figure 10.12(b). An incoming parallel beam converges at the focal point on the other side of the ball lens. The effective focal length (EFL) depends upon the refractive index of the glass and the lens diameter. Placing a point light source at the focal point, F, of the lens results in collimated light.

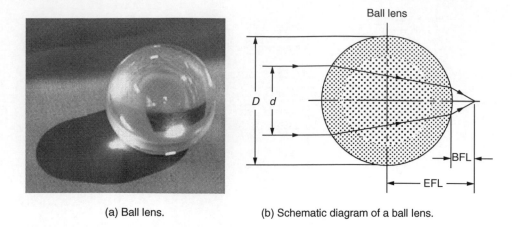

(a) Ball lens.   (b) Schematic diagram of a ball lens.

FIGURE 10.12 Ball lens (a) ball lens (b) schematic diagram of a ball lens.

Ball lenses benefit from cost and size advantages over gradient index, convex, and concave lenses; however, misperceptions concerning the relative ease of use of GRIN lenses have often led communications users to choose these over ball lenses. Due to recent improvements in ball lens antireflective coating technology, this preference for GRIN lenses has diminished. In some cases, ball lenses may even yield higher coupling efficiencies in smaller packages.

As shown in Figure 10.12, the EFL of a ball lens depends on two variables: the ball lens diameter (D) and the material index of refraction of the ball (n). The EFL is measured from the centre of the lens. The back focal length (BFL) is calculated as:

$$BFL = EFL - D/2 \tag{10.2}$$

$$EFL = nD/4(n-1) \tag{10.3}$$

The numerical aperture (NA) of a ball lens is dependent on the focal length of the ball and on the input beam diameter (d). The NA is defined in the following equation:

$$NA = 2d(n-1)/nD \tag{10.4}$$

When coupling light efficiently from a laser into a fibre optic cable, the choice of the ball is dependent on the NA of the fibre cable and the diameter of the laser beam. The diameter of the laser beam is used to determine the NA of the ball lens, as given by Equation 10.4. The NA of the

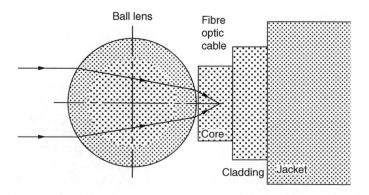

FIGURE 10.13 Ball lens coupling into a fibre optic cable.

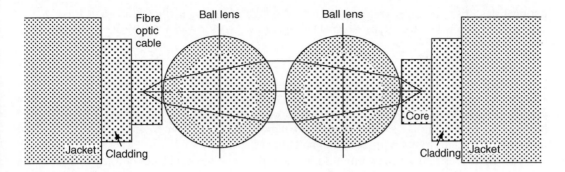

FIGURE 10.14 Coupling two fibre optic cables.

ball lens must be less than or equal to the NA of the fibre cable in order to couple all of the light into the fibre cable. This concept will be discussed in detail in the fibre optic cable chapter. The ball lens is placed directly onto the fibre cable for optimum coupling, as shown in Figure 10.13.

To couple light from one fibre cable to another fibre cable of similar NA, two identical ball lenses are used. The two lenses are placed in contact with the fibre cables, as shown in Figure 10.14. This figure also shows that light goes from one fibre cable to another in both directions.

Half-ball (hemispherical) lenses are available in the market. They are ideal for applications such as fibre communication, endoscopy, microscopy, optical pick-up devices, and laser interferometry.

10.2.4 FRESNEL LENSES

A Fresnel lens replaces the curved surface of a conventional lens with a series of concentric grooves molded into the surface of a thin, lightweight glass or plastic sheet, as shown in Figure 10.15. The grooves act as individual refracting surfaces, like tiny prisms, when viewed in cross section, bending parallel rays in a very close approximation to a common focal length. Because the lens is thin, very little light is lost by absorption compared to conventional lenses. Fresnel lenses are a compromise between efficiency and image quality. High groove density allows higher quality images (as needed in projection), while low groove density yields better efficiency (as needed in light gathering

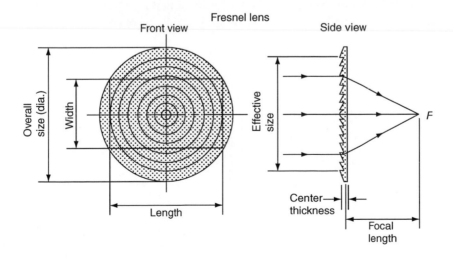

FIGURE 10.15 Fresnel lens.

applications). Fresnel lenses are most often used in light gathering applications, such as condenser systems or emitter/detector setups. Fresnel lenses can also be used as magnifiers or projection lenses.

10.2.5 LIQUID LENSES

Researchers have developed a unique lens system with a variable focus, which is able to function without mechanical moving parts. This development can be utilized in a wide range of optical applications, such as digital cameras, endoscopes, and security systems. The system simulates the human eye, which adjusts its focal length by changing its shape. The new lens overcomes the fixed focus disadvantages of many more economical image systems.

The liquid lens consists of two immiscible liquids with different indexes of refraction. Between these liquid layers are both an electrically conductive aqueous solution and an electrically nonconductive oil. This liquid lens is contained in a short tube with transparent end caps. The internal surface of the tube wall and one of the end caps are covered with a hydrophobic layer which forces the aqueous solution to take on a hemispherical shape at the opposite end of the tube.

The shape of the lens is controlled by applying an electric field across the hydrophobic coating. This causes the coating to become less hydrophobic as the surface tension changes. As a result, the aqueous solution starts to wet the side walls of the tube, and so changes the meniscus between the two liquids as well as the curve radius and focal length of the lens. In this way, the surface of the original convex lens can be made completely planar or concave by applying an electric field.

The liquid lens measures a few millimetres in diameter and length. The focusing range is from a few centimetres to infinite. The focal point switches quickly over the complete range. The lens is reliable, has very low loss, is shock-resistant, and functions over a wide temperature range.

10.3 GRAPHICAL METHOD TO LOCATE AN IMAGE FORMED BY CONVERGING AND DIVERGING LENSES

Graphical methods using ray diagrams are convenient for determining the location of an image formed by a lens. Figure 10.16 illustrates the method of locating the image of an object placed in front of a convex (converging) lens. Consider an object placed at location O and at distance S from the centre of the converging lens of focal length f. For a converging lens, the focal length is positive and is located on the right side of the lens (the side opposite to the object). To locate the image, the three rays are drawn from the top of the object as follow:

1. *Ray 1* is drawn from the top of the object parallel to the principal axis. After the ray is refracted through the lens, this ray passes through the focal point of the lens.

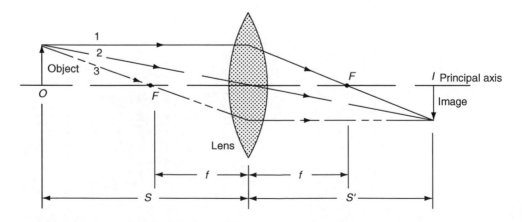

FIGURE 10.16 Graphical diagram to locate an image for a convex lens.

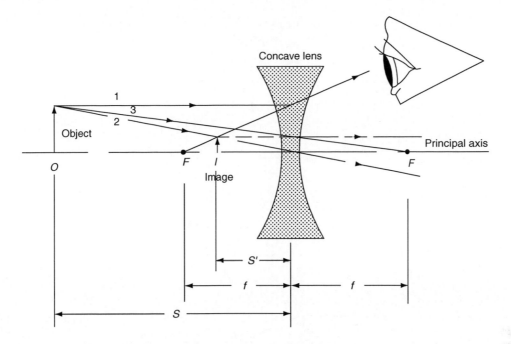

FIGURE 10.17 Graphical diagram to locate an image for a concave lens.

2. *Ray 2* is drawn from the top of the object passing through the centre of the lens. This ray continues in a straight line and intersects with ray 1 at a point at the top of the image. This intersection point of rays 1 and 2 shows the location of the image on the principal axis at point *I* at a distance S' away from the lens centre. Though the two rays can locate the image, it is better to confirm the image location using the third ray.
3. *Ray 3* is drawn from the top of the object passing through the focal point of the lens located on the left side (object side) of the lens. Ray 3 refracts from the lens parallel to the principal axis and intersects with rays 1 and 2 at the same point.

Figure 10.17 illustrates the method of locating the image of an object placed in front of a concave (diverging) lens. Consider an object placed at location O and at distance S from the centre of the diverging lens of focal length f. For a diverging lens, the focal length is negative and thus the focal point is on the left side of the lens (the object side).

A similar construction is used to locate the image from a diverging lens. In contrast, the deviated ray 1 appears to originate from the focal point and ray 3 heads towards the other focal point. The point of intersection of any two rays in the graphical diagram can be used to locate the image. Ray 3 serves as a check on the construction procedure of the image location. For diverging lenses, the rays appear to originate from a virtual image, as shown in dashed lines. This image is imaginary, located on the same side as the object, and is reduced in size.

10.4 IMAGE FORMATION BY CONVERGING LENSES

Figure 10.18 shows the cases of image formation by an object placed at different locations in front of a converging lens.

1. *Case (a)* occurs when the object is placed between the focal point F and the centre of the lens. The image is erect, enlarged, and virtual. The image is located on the object side of

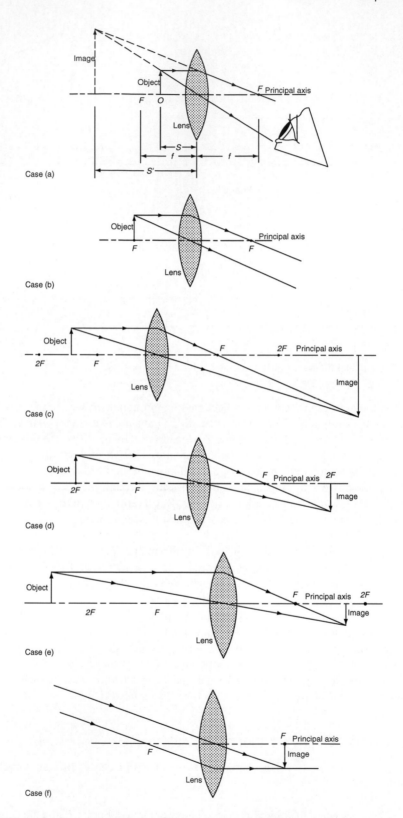

FIGURE 10.18 Images formation by a converging lens. Cases (a)–(f).

the lens because the refracted rays diverge as though coming from a point behind the lens. A magnifying glass is a simple example of this case.

2. *Case (b)* occurs when the object is located at the focal point *F* of the lens. All refracted rays are parallel and will never intersect at any distance from the lens. Therefore, no image will be formed; it seems the image is at an infinite distance from the lens. Light coming from a lighthouse is a common example.

3. *Case (c)* occurs when the object is placed beyond the focal point *F* of the lens. The image is inverted, real, and larger than the object. The image is located at a distance beyond 2*f*. This case illustrates the principle of image formation in an optical projector.

4. *Case (d)* occurs when the object is placed at a distance of 2*f*. The image is formed on the other side of the lens at location 2*f*. The image is real, inverted, and the same size as the object. This case is present in the operation of an office photocopier.

5. *Case (e)* occurs when the object is placed beyond 2*f*. The image is formed at a location between *f* and 2*f*. The image is real, inverted, and smaller than the object. This case shows the principle of the camera.

6. *Case (f)* occurs when the object is placed at infinity. The image is formed at the focal point of the lens. The image is real, inverted, and smaller than the object. This case presents the principle of the camera viewing a distant object.

In summary, a variety of image sizes, orientation, and locations are produced by a converging lens. In contrast, the image of a real object formed by a diverging lens is always virtual, erect, and smaller than the object, as shown in Figure 10.17. The further the object is located from the lens, the smaller the image.

10.5 THE LENS EQUATION

Consider defining the focal length *f* of a lens as a function of the object distance *S* and image distance *S'* from the lens centre, as shown in Figure 10.19. The lens equation can be derived geometrically as illustrated in Figure 10.19. The two triangles *ABO* and *CDO* are similar, which means that the corresponding sides of those triangles are proportional. Thus:

$$\frac{CD}{AB} = \frac{CO}{AO} = \frac{S'}{S}. \qquad (10.5)$$

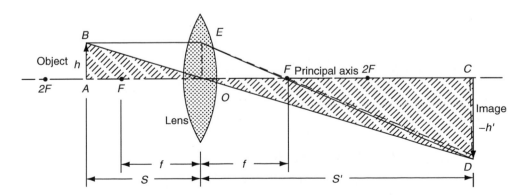

FIGURE 10.19 Ray diagram for deriving the lens equation.

The triangles *OEF* and *CDF* are also similar, so:

$$\frac{CD}{OE} = \frac{CF}{OF} = \frac{S'-f}{f} = \frac{S'}{f} - 1 \tag{10.6}$$

Since the light ray *BE* is parallel to the principal axis,

$$OE = AB \tag{10.7}$$

Therefore:

$$\frac{CD}{OE} = \frac{CD}{AB} \tag{10.8}$$

Which means the right side of Equation (10.5) and Equation (10.6) are equal, as shown in Equation 10.9:

$$\frac{S'}{f} - 1 = \frac{S'}{S} \tag{10.9}$$

Dividing each term in Equation (10.9) by S' gives:

$$\frac{1}{f} - \frac{1}{S'} = \frac{1}{S} \tag{10.10}$$

Rearranging terms gives the lens equation:

$$\frac{1}{S} + \frac{1}{S'} = \frac{1}{f} \tag{10.11}$$

Equation (10.11) is the basic equation for thin lenses. Thin lenses are defined as lenses whose thickness can be ignored. In the following sections, thin lenses are explained in detail.

The sign convention for thin lenses is given below:

S is *positive* if the object is in front of the lens (real object).
S is *negative* if the object is behind the lens (virtual object).
S' is *positive* if the image is behind the lens (real image).
S' is *negative* if the image is in front of the lens (virtual image).
f is *positive* if the focal point of the incident surface is in front of the lens (convex lens).
f is *negative* if the focal point of the incident surface is in back of the lens (concave lens).

This convention applies only to the numerical values substituted into the thin lens equation. The quantities S, S', and f should maintain their signs unchanged until the substitution is made into the lens equation.

10.6 MAGNIFICATION OF A THIN LENS

Figure 10.19 shows a converging lens that forms the image *CD* of the object *AB*. The object height is $AB = h$ and the image height $CD = -h'$ is negative because the image is inverted. The triangles *ABO* and *CDO* are similar, and their corresponding sides are proportional. Hence:

$$\frac{CD}{AB} = \frac{CO}{AO} \tag{10.12}$$

Since $CD = -h'$, $AB = h$, $CO = S'$, and $AO = S$, then Equation (10.12) becomes the lateral magnification M of the lens:

$$M = \frac{h'}{h} = -\frac{S'}{S} \tag{10.13}$$

In other words, the lateral magnification M of the thin lens is defined as the ratio of the image height h' to the object height h or also defined as the image distance S' to the object distance S, as given in Equation 10.14:

$$M = \frac{\text{Image height}}{\text{Object height}} = -\frac{\text{Image distance}}{\text{Object distance}} \tag{10.14}$$

When M is positive, the image is erect and on the same side of the lens as the object. When M is negative, the image is inverted and on the side of the lens opposite the object.

10.7 THE LENSMAKER'S EQUATION

Figure 10.20 shows a converging lens that has a focal length f, index of refraction n, and two radii with surface curvatures R_1 and R_2. The thin lens is a lens whose thickness is small compared with R_1 and R_2. The Lensmaker's equation, which is the relation between the focal length, the index of refraction, and the surface curvatures, is given by:

$$\frac{1}{f} = (n-1)\left(\frac{1}{R_1} - \frac{1}{R_2}\right) \tag{10.15}$$

Both f and R are *positive* if the centre of curvature of the incident surface is in front of the lens (convex lens).

Both f and R are *negative* if the centre of curvature of the incident surface is in back of the lens (concave lens).

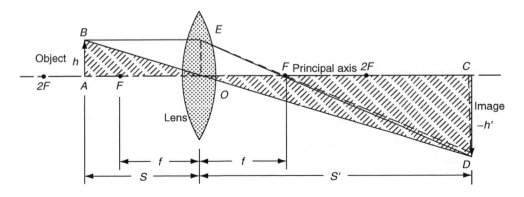

FIGURE 10.20 Biconvex lens.

10.8 COMBINATION OF THIN LENSES

Two lenses can be combined to form an image in optical devices and instruments, such as microscopes and telescopes. The graphical method can be used to locate the image, as explained previously. The first image of the first lens can be located by the graphical method. It can be calculated applying Equation 10.11 as if the second lens was not present. The light then approaches the second lens as if it had originated from the image formed by the first lens. Therefore, the image of the first lens is treated as the object to the second lens. The image of the second lens is the final image of the combination of two lenses. Draw and apply Equation 10.11 to the second lens as if the first lens were not present. The same procedure can be applied to any optical device or instrument with three or more lenses. The total magnification of the combination of lenses equals the product of the magnification of the separate lenses. Applications of lens combinations are presented in Chapter 14, covering optical instruments for viewing applications.

Suppose two thin lenses of focal lengths f_1 and f_2 are placed at a distance L between each other, as shown in Figure 10.21. If S_1 is the distance of the object from lens 1, then apply Equation 10.16 to find the location of the image S_1' formed by the first lens. The first image formed by the first lens becomes the object for the second lens. Apply the thin lens Equation 10.17 to find the location of the second image S_2' formed by the second lens. Now the distance between the two lenses can be calculated. Remember to consider the types of lens combinations when locating the images. The lenses involved in these combinations could be convex–convex, convex–concave, or any other combination. Lens combinations can form real or virtual images.

$$\frac{1}{S_1} + \frac{1}{S'_1} = \frac{1}{f_1} \qquad (10.16)$$

$$\frac{1}{S_2} + \frac{1}{S'_2} = \frac{1}{f_2} \qquad (10.17)$$

The distance L between the lenses can be calculated by Equation 10.17:

$$L = S'_1 + S_2 \qquad (10.18)$$

The distances S'_1 and S_2 could be real or virtual, i.e., positive and negative. The total magnification for the combination of the two lenses can be found by multiplying the magnification of first and second lenses, which are calculated using Equation 10.14. The total magnification M_{total} for the

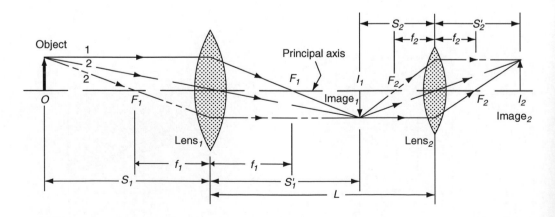

FIGURE 10.21 Combination of two types of lenses.

combinations of lenses is given by:

$$M_{\text{total}} = M_1 M_2. \tag{10.19}$$

The conventional signs for M_1 and M_2 carry through the product to indicate, from the sign of M_{total}, whether the final image is erect or inverted.

10.9 LENS ABERRATIONS

Aberration is a deviation from the ideal focusing of light by a lens. As a result, the image is not sharply focused. Some of the problems may lie with the geometry of the lens in use. Other aberrations may occur because different wavelengths of light are not refracted equally at the same time. The latter problem is called dispersion of light. This is clearly visible when white light passes through a prism. White light dispersion in a prism is discussed in detail in Chapter 11, Prisms. Lenses, like spherical mirrors, also have aberrations. Here are some common types of aberrations.

10.9.1 SPHERICAL ABERRATION

As explained in lenses chapter, the spherical mirrors, aberration occurring in spherical mirrors is the same phenomena that occurs in lenses. Spherical aberration results because the light rays which are far from the principal axis focus at different focal points than the rays which are passing near the axis. Figure 10.22 illustrates spherical aberration for parallel rays passing through a converging lens. Rays near the middle of the lens are imaged farther from the lens than rays near the edges. Therefore, there is no single focal point for a spherical lens.

Spherical aberration can be minimized by using an aperture to reduce the effective area of the lens, so that only light rays near the axis are transmitted. This happens in most cameras equipped with an adjustable aperture to control the light intensity, and when possible, reduce spherical aberration. To compensate for the light loss of a smaller aperture, a longer exposure time is used. Another way to minimize the effect of aberration is by combinations of converging and diverging lenses. The aberration of one lens can be compensated by the optical properties of another lens.

10.9.2 CHROMATIC ABERRATION

As described in the previous chapters, the index of refraction of a material varies slightly with different wavelengths; therefore, different wavelengths of light refracted by a lens focus at different points, giving rise to chromatic aberration. When white light passes through a lens, the transmitted rays of different wavelengths (colours) do not refract at the same angle. As a result, there is no

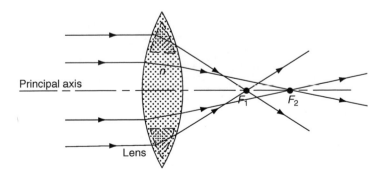

FIGURE 10.22 Spherical lens aberration.

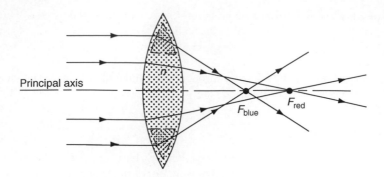

FIGURE 10.23 Chromatic lens aberration.

common focal point for all the wavelengths. Figure 10.23 shows that the images of different colours are produced at different locations on the principal axis. The chromatic aberration for a diverging lens is the opposite of that for a converging lens. The red colour focuses before the blue colour. Chromatic aberration can be greatly reduced by the use of a combination of converging and diverging lenses made from two different types of glass with differing indexes of refraction. The lenses are chosen so that the dispersion produced by one lens is compensated for by opposite dispersion produced by the other lens.

10.9.3 ASTIGMATISM

Astigmatism is the imaging of a point off-set from the axis as two perpendicular lines in different planes. This effect is shown in Figure 10.24. In astigmatism, the horizontal rays from a point object converge at a certain distance from the lens to a line called the primary image. The primary image is perpendicular to the plane defined by the optic axis and the object point. At a somewhat different distance from the lens, the vertical rays converge to a second line, called the secondary image. The secondary image is parallel to the vertical plane. The circle of least confusion (greatest convergence) appears between these two positions.

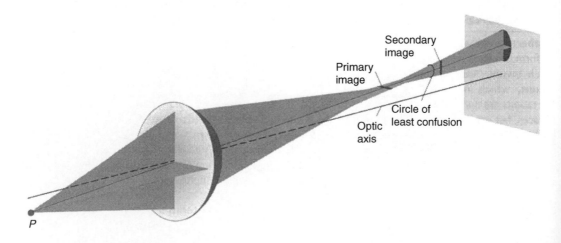

FIGURE 10.24 Astigmatism of a lens.

The location of the circle of least confusion depends on both the object point's transverse distance from the axis, as well as its longitudinal distance from the lens. As a result, object points lying in a plane are generally not imaged in a plane, but are imaged in some curved surface. This effect is called curvature of field.

Finally, the image of a straight line that does not pass through the optic axis may be curved. As a result, the image of a square with the axis through its centre may resemble a barrel (sides bent outward) or a pincushion (sides bent inward). This effect, called distortion, is not related to the lack of sharpness of the image, but results from a change in lateral magnification with distance from the axis. In high-quality optical instruments, astigmatism can be minimized by using properly designed, nonspherical surfaces or specific lens combinations.

10.10 LENS POLISHING TECHNOLOGY

Many lens polishing techniques are available in the industry. The technique chosen depends on the lens performance requirements and the cost. A new fabrication technology called magnetorheological finishing has been developed, which provides some advantages over traditional techniques. The main advantage of this technology is that it offers a way to produce high precision asphere lenses in moderate volume.

The spherical shape of any lens surface causes focus error, which is called a spherical aberration, as explained above. The manufacturers of the lenses use multiple surfaces to overcome spherical aberration. The problem lies with the traditional techniques used for optics fabrication shown in Figure 10.25. To produce spherical surfaces, the lens blanks are polished between convex and concave surfaces of the intended radius of curvature, using an abrasive slurry.

The semi-random movement of a convex spherical surface against a concave spherical surface tends to produce a spherical wear pattern. Originally, the lens blanks are not a perfect spherical shape. But over time, the wear from the polishing process causes them to take on a spherical shape. With process control, this method can yield an extremely high degree of precision. For example, lenses with $\lambda/4$ or better surface accuracy at 632.8 nm (HeNe laser) have a surface that is spherical to within a fraction of micrometre.

To produce aspheric surfaces, the movement of the polishing tool must be constrained. For lenses in which the aspheric profile departs from a sphere by only a small amount, typical lens

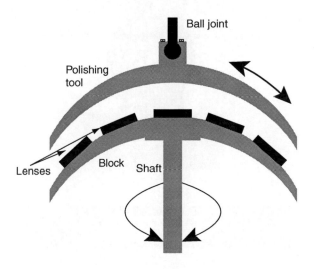

FIGURE 10.25 Traditional polishing technique.

polishing starts with the closest-fit sphere using standard techniques. The final aspheric shape is then achieved with a polishing tool that is much smaller than the optic (a sub-aperture tool) on another polishing machine. Controlling the amount of time that the tool spends working a particular zone of the lens surface can produce a specific aspheric profile.

If the aspheric shape departs significantly from spherical, a computer numerical control machine would typically be used to generate the aspheric shape directly into the lens blank. Alternatively, a special grinding tool can be used to transform the closest-fit sphere into a close approximation of the final desired lens shape. Once the blank closely resembles the desired asphere, the goal is to polish the blank without removing a significant amount of material. This is accomplished using a modified form of a traditional polishing apparatus along with sub-aperture polishing.

Unfortunately, because there are many environmental and operator dependent variables in this type of polishing, successful operation of this process requires a high degree of operator skill. Given the mechanical limitations of the machinery commonly used, it is hard to achieve surface precision near $\lambda/20$. Process times for most common lens polishing methods are measured in hours, and even in days or weeks in some cases. Thus, high-quality aspheres produced in this way can be prohibitively expensive compared with spherical lenses.

Magnetorheological finishing technology was developed to overcome the limitations of traditional lens polishing techniques. It is a controlled way of producing high precision aspheres that can be used reliably and in high-volume production. It also potentially requires less operator know-how than traditional approaches, reduces process times significantly, and can yield surface precision to $\lambda/40$.

Figure 10.26 shows the basic elements of the magnetorheological system including a pump that continuously supplies a small amount of magnetorheological fluid from a reservoir. This fluid is sprayed onto a rotating wheel that draws out the fluid into a thin ribbon. The work piece (lens) to be

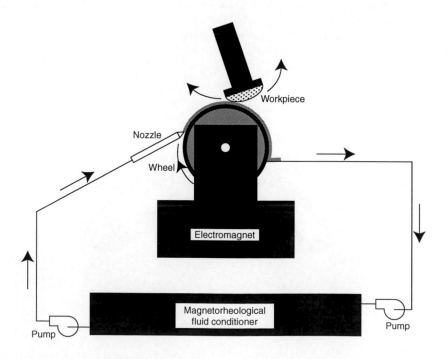

FIGURE 10.26 Magnetorheological system.

polished is partially immersed in this moving ribbon of fluid. A specially designed electromagnet beneath the polishing wheel produces a strong local magnetic field gradient. When the moving fluid enters this magnetic field, it stiffens substantially in a matter of milliseconds; it then returns to its fluid state as it leaves the field. This zone of stiffened fluid is the polishing tool. After the fluid passes the lens, it is removed by suction and pumped back into the reservoir.

A polishing session begins by running a test routine that characterizes the exact polishing shape and properties of the polishing fluid. To polish a given lens, the initial surface error (the difference between the present and desired shape) of the lens is supplied to the system software. The software generates a set of instructions for the computer numerical control machine, which then starts polishing the lens. During the polishing process, the machine rotates the lens on the spindle, as shown in Figure 10.27. At the same time, the machine slowly tilts the lens so that different zones in the lens are exposed to the fluid. During this process, the machine maintains a constant gap and surface normal for the lens in the fluid. The computer program controls the speeds of these movements to remove only the necessary amount of material. In other words, the system will spend more time polishing a zone with a high spot than it will spend polishing a low spot. The machine continuously monitors and controls the properties of the polishing fluid, such as temperature, viscosity, and evaporation rate, to maintain process invariance.

After the lens is completely polished, it has the correct specifications and is entirely transparent. The lens material is very sensitive and can scratch easily. Therefore, a harder coating must be applied to the surfaces of the lens. There are many types of coating materials. One coating is an acrylic monomer solution that is spin-coated onto the lens. The coating is then cured under a UV lamp for a specific time. This coating helps protect the lens from scratches and wear.

Most of the lenses and other optical components are coated with at least one antireflective (AR) layer. AR coating is used to reduce back reflection and to filter undesirable light waves. An antireflection coating material is a composite thin film. There are between four to six layers that have several purposes. Some layers help bond the other layers to the lens while other layers provide the antireflection properties. Each layer of the composite coating depends on the properties of the surrounding layers in order to achieve the desired specifications. AR coatings made of multi-layer

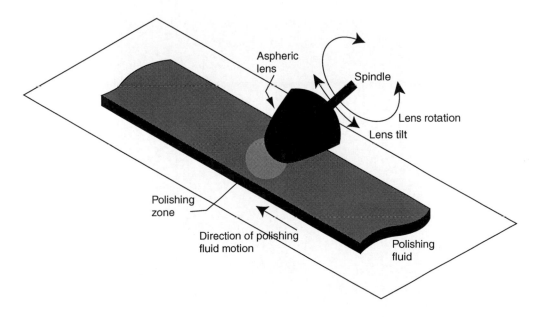

FIGURE 10.27 Polishing process.

thin film metallic oxide can increase the light transmittance to nearly 100%. Typically, titania and zircona are used because they have a high index of refraction. Silica is usually used with material that has a low index of refraction. The thickness of the layers determines the optical performance of the coating, thus the specifications need to be precisely calculated and carefully controlled.

10.11 EXPERIMENTAL WORK

The theory of this experiment is based on the second law of light, which is called the law of refraction. Students will practise tracing image formation by lenses. Lenses are optical components that work on the basis of image formation by refraction. Such optical components are commonly used in manufacturing optical devices, instruments, and systems. The purpose of this experiment is to observe the image formed when an object is placed in front of a lens or combination of two lenses. The lens applications are shown in the following cases:

a. Image formed by a lens.
b. Image formed by a combination of two lenses.

10.11.1 TECHNIQUE AND APPARATUS

Appendix A presents the details of the devices, components, tools, and parts.

1. Two thin convex lenses, as shown in Figure 10.1, Figure 10.29, and Figure 10.31.
2. Lens holders and positioners, as shown in Figure 10.29 and Figure 10.31.
3. Hardware assembly (clamps, posts, screw kits, screwdriver kits, sundry positioners, etc.).
4. Optical rail.
5. Light source/object and positioner, as shown in Figure 10.28.
6. Black/white card and cardholder.
7. Ruler.

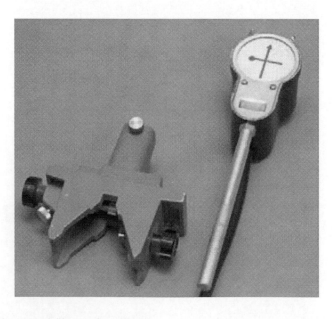

FIGURE 10.28 Light source/object and positioner.

10.11.2 PROCEDURE

Follow the laboratory procedures and instructions given by the professor and/or instructor.

10.11.3 SAFETY PROCEDURE

Follow all safety procedures and regulations regarding the use of optical components, light source devices, and optical cleaning chemicals.

10.11.4 APPARATUS SET-UP

10.11.4.1 Image Formed by a Lens

1. Figure 10.29 shows the experimental apparatus set-up.
2. Place an optical rail on the table.
3. Mount a light source on one end of the optical rail.
4. Mount a convex lens and lens holder on the optical rail.
5. Mount the black/white card on the optical rail.
6. Connect the light source to the power supply.
7. Turn off the lights of the lab.
8. Move the lens back and forth until you capture a focused image on the card.
9. Make a fine adjustment to the lens position to get a clear and focused image, as shown in Figure 10.30. Describe the characteristics of the image.
10. Measure the object and image distances. Record in Table 10.1.
11. Measure the height of the object and the height of the image. Record in Table 10.1.
12. Turn on the lights of the lab.
13. Illustrate the locations of the light source/object, lens, and image, in a diagram using the ray tracing method.

FIGURE 10.29 Image formation by a lens.

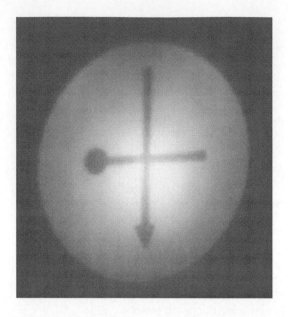

FIGURE 10.30 Image formation by a convex lens.

10.11.4.2 Image Formed by a Combination of Two Lenses

1. Figure 10.31 shows the experimental set-up.
2. Repeat the steps in Case (a) to find the image formed by the lens.
3. Mount another lens between the first lens and the card.
4. Find the image formed by lens 2 as explained in Case (a).
5. Make a fine adjustment to the position of the lenses to get a clear and focused image, as shown in Figure 10.32. Describe the characteristics of the image.

FIGURE 10.31 Image formation by a combination of two lenses set-up.

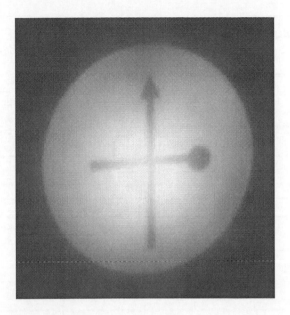

FIGURE 10.32 Image formation by a combination of two lenses.

6. Measure the distances between the light source, lens 1, lens 2, and the card, to determine the object and image distances. Record in Table 10.2.
7. Turn on the lights of the lab.
8. Illustrate the locations of the light source/object, lens 1, lens 2, and image, in a diagram using the ray tracing method.

10.11.5 DATA COLLECTION

10.11.5.1 Image Formed by a Lens

Measure the object and imaged distances. Record in Table 10.1.

TABLE 1
10.1 Image Formed by a Lens

Distance of (unit)		Lens Focal Length f (unit)		Magnification
Object S	Image S'	Given	Calculated	M

10.11.5.2 Image Formed by a Combination of Two Lenses

Measure the distances between the light source, lens 1, lens 2, and the card, to determine the object and image distances. Record in Table 10.2.

TABLE 2
10.2 Image Formed by a Combination of Two Lenses

Lens 1				
Distance of (unit)		Lens Focal Length f (unit)		Magnification
Object S	Image S'	Given	Calculated	M

Lens 2				
Distance of (unit)		Lens Focal Length f (unit)		Magnification
Object S	Image S'	Given	Calculated	M
Distance between the lenses L				
Total magnification M_{total}				

10.11.6 CALCULATIONS AND ANALYSIS

10.11.6.1 Image Formed by a Lens

1. Calculate the focal length of the lens, using Equation 10.11. Calculate the lateral magnification M of the lens, using Equation 10.13. Record in Table 10.1.
2. Find the published value of the focal length of the lens, as provided by the manufacturer/supplier. Record in Table 10.1.
3. Illustrate the locations of the light source/object, lens, and image in a diagram; include the dimensions.

10.11.6.2 Image Formed by a Combination of Two Lenses

1. Calculate the focal length of lenses 1 and 2, using Equation 10.16 and Equation 10.17, respectively. Record in Table 10.2.
2. Calculate the distance L between the lenses, using Equation 10.18. Record in Table 10.2.
3. Calculate the lateral magnification M of lens 1 and lens 2, using Equation 10.14. Record in Table 10.2.
4. Calculate the total magnification M_{total}, using Equation 10.19. Record in Table 10.2.
5. Find the published value of the focal length for each of the lenses, as provided by the manufacturer/supplier. Record in Table 10.2.
6. Illustrate the locations of the light source/object, lens 1, lens 2, and image, in a diagram; include the dimensions.

10.11.7 RESULTS AND DISCUSSIONS

10.11.7.1 Image Formed by a Lens

1. Discuss your measurements and calculations of the formation of the image by a thin lens.
2. Discuss the characteristics of the image.

3. Verify the calculated focal length of the lens against the value published by the manufacturer/supplier.

10.11.7.2 Image Formed by a Combination of Two Lenses

1. Discuss the measurements and calculations of the formation of the image by combinations of two thin lenses.
2. Discuss the characteristics of the image
3. Verify the calculated focal length of the lenses against the values published by the manufacturer/supplier.

10.11.8 CONCLUSION

Summarize the important observations and findings obtained in this lab experiment.

10.11.9 SUGGESTIONS FOR FUTURE LAB WORK

List any suggestions for improvements using different experimental equipment, procedures, and techniques for any future lab work. These suggestions should be theoretically justified and technically feasible.

10.12 LIST OF REFERENCES

List any references that were used in the report. Use one format in writing the references. Never mix reference formats in a report.

10.13 APPENDICES

List all of the materials and information that are too detailed to be included in the body of the report.

FURTHER READING

Beiser, Arthur, *Physics*, 5th ed., Addison-Wesley, Reading, MA, 1991.

Blaker, J. Warren and Peter, Schaeffer, *Optics an Introduction for Technicians and Technologists*, Prentice Hall, Inc., Englewood Cliffs, NJ, 2000.

Born, M. and Wolf, E., *Elements of the Theory of Diffraction Principles of Optics: Electromagnetic Theory of Propagation, Interference, and Diffraction of Light*, 7th Ed., Cambridge University Press, Cambridge, pp. 370–458, 1999.

Born, M. and Wolf, E., *Rigorous Diffraction Theory Principles of Optics: Electromagnetic Theory of Propagation, Interference, and Diffraction of Light*, 7th Ed., Cambridge University Press, Cambridge, pp. 556–592, 1999.

Clark, Timothy and Wanser, Keith, Ball vs gradient index lenses, *Photonics Spectra*, February, 94–96, 2001.

Cox, Arthur, *Photographic Optics*, 15th ed., Focal Press, London, New York, 1974.

Cutnell, John D. and Johnson, Kenneth W., *Physics*, 5th Ed., Wiley, New York, 2001.

Cutnell, John D. and Johnson, Kenneth W., *Student Study Guide-Physics*, 5th Ed., Wiley, New York, 2001.

DiCon Fibre Optics, Inc. *GRIN Lenses Catalog*, Rev. D. Dicon Fibre Optics, Inc., Richmond, CA, U.S.A., 2002.

Drisoll, Walter G. and Vaughan, William, *Handbook of Optics*, McGraw Hill Book Company, New York, 1978.

Drollette, Dan, Liquid lenses make better beams,*Photonics Spectra*, pp. 25–26, August, U.S.A., 2001.

Edmund Industrial Optics, *Optics and Optical instruments Catalog*, Edmund Industrial Optics, New Jersey, 2004.

EPLAB, *EPLAB Catalogue 2002*. The Eppley Laboratory, Inc., 12 Sheffield Avenue, P.O. Box 419, Newport, Rhode Island 02840, U.S.A.

Ewald, Warren J. et al., *Optics: The Matrix Theory*, Marcel Dekker, New York, 1972.

Ewald, William P. et al., *Practical Optics*, 1982.

Francon, M., *Optical Interferometry*, Academic Press, New York, 1966.

Ghatak, Ajoy K., *An Introduction to Modern Optics*, McGraw-Hill, New York, 1972.

Giancoli, Douglas C., *Physics*, 5th Ed., Prentice Hall, Englewood Cliffs, NJ, 1998.

Halliday, R., Resnick, D., and Walker, J., *Fundamental of Physics*, 6th Ed., Wiley, New York, 1997.

Heath, R. W., Macnaughton, R. R., and Martindale, D. G., *Fundamentals of Physics*, D.C. Heath Canada, Ltd, Canada, 1979.

Hecht, Eugene, *Optics*, 4th Ed., Addison-Wesley Longman, Inc., New York, 2002.

Hewitt, Paul G., *Conceptual Physics*, 8th Ed., Addison-Wesley, Inc., Reading, MA, 1998.

Jenkins, F. W. and White, H. E., *Fundamentals of Optics*, McGraw Hill, New York, 1957.

Jones, Edwin and Richard, Childers, *Contemporary College Physics*, McGraw-Hill Higher Education, New York, 2001.

Kennedy, T. P., Understanding Ball Lenses. *Optics and Optical Instruments Catalog*, Edmund Industrial Optics Co., U.S.A., 2004.

Keuffel & Esser Co., *Physics*, Keuffel & Esser Audiovisual Educator—Approved Diazo Transparency Masters, Audiovisual Division, Keuffel & Esser Co., U.S.A., 1989.

Lambda Research Optics, Inc., *Catalog 2004*, Lambda Research Optics, Inc., California, 2004.

Lehrman, Robert L., *Physics-The Easy Way*, 3rd Ed., Barron's Educational Series, Inc., Hauppage, NY, 1998.

Lerner, Rita G. and Trigg, George L., *Encyclopedia of Physics*, 2nd Ed., VCH Publishers, Inc., New York, 1991.

McDermott, L. C. and Shaffer, P. S., *Tutorials in Introductory Physics*, Preliminary Edition, Prentice Hall Series in Educational Innovation, Upper Saddle River, New Jersey, U.S.A., 1988.

Melles Griot, *The Practical Application of Light Melles Griot Catalog, 2001*, Melles Griot, Rochester, NY, 2001.

Newport Corporation. Optics and mechanics. *Newport 1999/2000 catalog*, Newport Corporation.

Nichols, Daniel H, *Physics for Technology with Applications in Industrial Control Electronics*, Prentice Hall, Englewood Cliffs, NJ, 2002.

Nolan, P. J., *Fundamentals of College Physics*, Wm. C. Brown Communications, Inc., Dubuque, Iowa, U.S.A., 1993.

Ocean Optics, Inc. *Product Catalog, 2003*, Florida.

Pedrotti, Frank L. and Pedrotti, Leno S., *Introduction to Optics*, 2nd Ed., Prentice Hall, Inc., Englewood Cliffs, NJ, 1993.

Robinson, Paul, *Laboratory Manual to Accompany Conceptual Physics*, 8th Ed., Addison-Wesley, Inc., Reading, MA, 1998.

Romine, Gregory S., *Applied Physics Concepts into Practice*, Prentice Hall, Inc., Englewood Cliffs, NJ, 2001.

Sears, F. W., Zemansky, M. W., and Young, H. D., *University Physics- Part II*, 6th Ed., Addison-Wesley Publishing Company, Reading, MA, U.S.A., 1998.

Serway, R. A., and Jewett, J. W., *Physics for Scientists and Engineers*, with Modern Physics, 6th Ed., Volume 2, Thomson Books/Cole, U.S.A., 2004.

Sciencetech, *Designers and Manufacturers of Scientific Instruments Catalog*, Sciencetech, London, Ont., Canada, 2003.

Smith, W. J., *Modern Optical Engineering*, McGraw-Hill Book Co., New York, 1966.

Sterling, Donald J. Jr. *Technician's Guide to Fibre Optics*, 2nd Ed., Delmar Publishers Inc., New York, 1993.

Tippens, P. E., *Physics*, 6th Ed., Glencoe McGraw-Hill, Westerville, OH, U.S.A., 2001.

Tom, Miller, Aspherics come of age, *Photonics Spectra*, February, 76–81, 2004.

Urone, Paul Peter, *College Physics*, Brooks/Cole Publishing Company, Belmont, CA, 1998.

Walker, James S., *Physics*, Prentice Hall, Englewood Cliffs, NJ, 2002.

White, Harvey E., *Modern College Physics*, 6th Ed., Van Nostrand Reinhold Company, New York, 1972.

Wilson, Jerry D., *Physics—A Practical and Conceptual Approach Saunders Golden Sunburst Series*, Saunders College Publishing, London, 1989.

Wilson, Jerry D. and Buffa, Anthony J., *College Physics*, 5th Ed., Prentice Hall, Inc., Englewood Cliffs, NJ, 2000.

11 Prisms

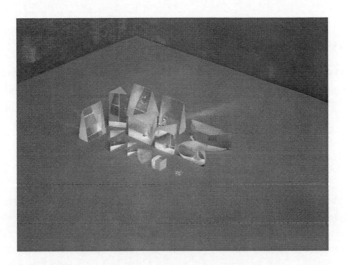

11.1 INTRODUCTION

The purpose of this chapter is to explain the basic principles that govern light passing through a prism or a combination of prisms. Types of prisms and image formation are also presented. The concept of light passing through prisms is very important to photonics and has significant use in image formation and building optical devices. Particular emphasis will be given to calculating the index of refraction of a prism, producing a rainbow of colours, and mixing a rainbow of colours using a glass rod or tube. Also in this chapter, along with the theoretical presentation, five experimental cases demonstrate the principles of light passing through prisms.

11.2 PRISMS

Prisms are blocks of optical material with flat polished faces arranged at precisely controlled angles, as shown in the figure above. In many situations it is necessary to direct a beam of light entering from one side and exiting from the other side of the prism. Light passing through prisms is governed by the laws of light.

Prisms are widely used in building optical devices, such as a prism spectrometer, which is commonly used to study the wavelengths emitted by a light source. Prisms are also used in building optical fibre devices, such as an opt-mechanical switch, which deflects or deviates an optical signal through a telecommunication system. Prisms can invert or rotate an image, deviate a light beam, disperse light into its component wavelengths, and separate states of polarization. The orientation of a prism (with respect to the incident light beam) determines its effect on the beam. Prisms can be designed for specific applications. The most popular prisms are right angle prisms, Brewster's angle dispersing prisms, Penta prisms, solid glass retro-reflectors, equilateral dispersing prisms, littrow dispersion prisms, wedge prisms, roof prisms, and Dove prisms. Some of these commonly used prisms are discussed in detail in this chapter.

The light beam can be arranged to exit the prism either parallel or horizontal to the input light and could also exit on the same or the opposite side of the input light. Depending on the desired characteristics of the image, the distance between the input and output beams can be determined using a combination of two prisms. This is usually done with different types of prisms.

11.3 PRISM TYPES

There are many types of prisms that are used in different applications. They are commonly used in building optic/optical fibre devices. In the following sections, the types of prisms and applications are presented in detail.

11.3.1 RIGHT ANGLE PRISMS

Right angle prisms are generally used to achieve a 90° bend in a light path and to change the orientation of an image. Depending on the prism orientation, images viewed through the prism will be inverted while maintaining correct left-to-right orientation, as shown in Figure 11.1(a). If the prism is rotated by 90°, images viewed through the prism will be erect, but they will be reversed left-to-right. Prisms can also be used in combination for image/beam displacement. A right angle prism may be used as a front surface mirror. Right angle prisms are also used to reverse the light beam, as shown in Figure 11.1(b). The input and output light beams are on the same side of the prism. It is possible to achieve a fixed distance between the input and the output in a right angle prism.

Figure 11.2 illustrates the arrangement of a combination of two right angle prisms. The light beam is incident on the first prism and exits from the second prism. The light beam exiting from the second prism is parallel to the incident light beam at the first prism. In this arrangement, the

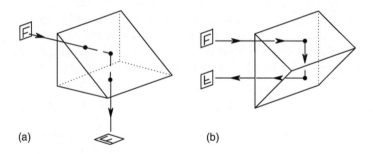

(a)   (b)

FIGURE 11.1 (a) Right angle prism as a mirror. (b) The direction of the light beam is reversed.

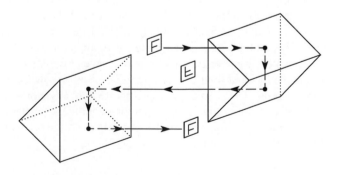

FIGURE 11.2 Light beam passing through two prisms.

distance between the incident and the exit beams can be varied. The common application of a two-prism combination is use in a submarine periscope.

11.3.2 Equilateral Prisms

Equilateral prisms, also called dispersing prisms, have three equal 60° angles, as shown in Figure 11.3. They are used for wavelength separation applications. A light beam is twice refracted while passing through the prism with a deviation angle (δ). Deviation is a function of the index of refraction and the wavelength of the light beam. These prisms are used within a laser cavity to compensate for dispersion of fixed laser cavity optics, such as a Ti:Sapphire crystal, or external to the cavity to manipulate the pulse characteristics.

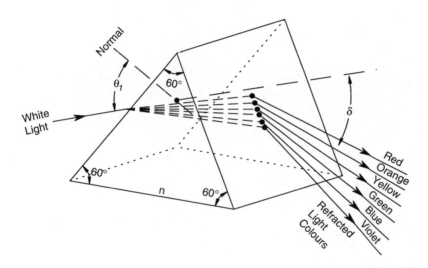

FIGURE 11.3 White light dispersing through an equilateral prism.

11.3.3 Dove Prisms

Dove prisms are commonly used as image rotators, as shown in Figure 11.4. As the prism is rotated, the image will rotate at twice the angular rate of the prism. Dove prisms can also be used with parallel or collimated light.

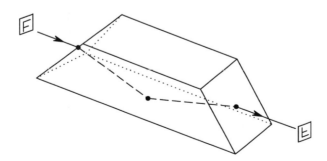

FIGURE 11.4 The Dove prism.

11.3.4 Roof Prisms

The roof or Amici prism is essentially a truncated right-angle prism whose hypotenuse has been replaced by a 90° total internal reflection roof, as shown in Figure 11.5. The prism deviates, or deflects, the image through an angle of 90°. Roof prisms are commonly used to split the image down the middle of the prism and interchange the right and left portions. They are often used in building simple telescope systems to correct for the image reversal introduced by the telescope lenses.

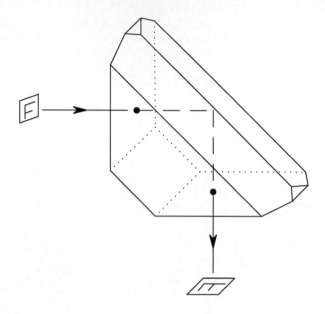

FIGURE 11.5 The Roof prism.

11.3.5 Penta Prisms

The Penta prism deviates the beam by precisely 90° without affecting the orientation of the image (neither inverting nor reversing the image), as shown in Figure 11.6. Penta prisms will

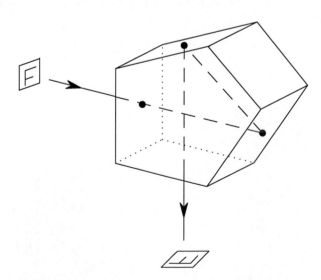

FIGURE 11.6 The Penta prism.

direct a beam through the same angle regardless of the prism's orientation to the beam. They are used in applications requiring an exact 90° deviation without having to orient the prism precisely. They are also often used as end reflectors in small range finders. The design of a Penta prism also makes it inherently more stable than a system consisting of two mirrors.

11.3.6 Double Porro Prisms

The double Porro prism, as shown in Figure 11.7, consists of two right-angle prisms. The Porro prism is made with rounded corners to reduce weight and size. They are relatively easy to manufacture. A small slot is often cut in the hypotenuse face of the prism to obstruct rays that are internally reflected at glancing angles. Since there are four reflections between the two Porro prisms, the exiting image will be right-handed. For example, two matching Porro prisms are used in binoculars to produce erect final images. At the same time, they permit the distance between the object-viewing lenses to be greater than the normal eye-to-eye distance, thereby enhancing the stereoscopic effect produced by ordinary binocular vision.

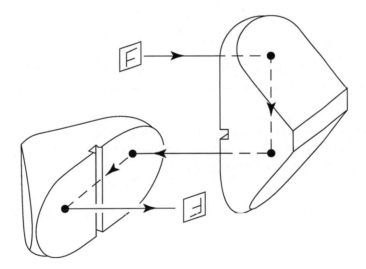

FIGURE 11.7 The double Porro prism.

11.3.7 Leman–Springer Prisms

Figure 11.8 shows that the Leman–Springer prism has a 90° roof like the roof prism. The input beam is displaced without being deviated, but the emerging image is right-handed and rotated through 180°. This prism can be used to correct image orientation in telescope systems.

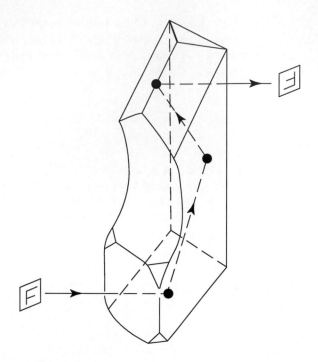

FIGURE 11.8 The Leman–Springer prism.

11.3.8 WEDGE PRISMS

Wedge prisms can be used individually to deviate a laser beam by a set angle (θ_d), as shown in Figure 11.9. Also, two wedge prisms can be combined together as an anamorphic pair. A single prism's ability to deviate the angle of an incident beam is measured in dioptres. One dioptre is defined as the angular deviation of the beam of 1 cm at a distance of one metre from the wedge prism. Wedge prisms are used as beamsteering elements in optical systems.

The apex angle of a wedge prism necessary to produce a given minimum deviation angle (θ_d) or deflection is determined by the wedge angle (θ_w):

$$\theta_w = \arctan\left[\frac{\sin \theta_d}{(n - \cos \theta_d)}\right] \tag{11.1}$$

where n is the index of refraction of a wedge prism.

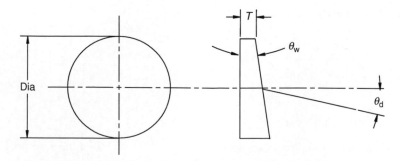

FIGURE 11.9 A Wedge prism.

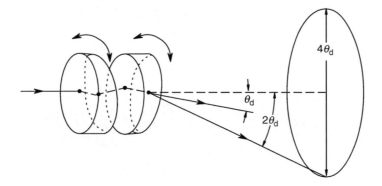

FIGURE 11.10 Wedge prisms used in beamsteering applications.

Figure 11.10 shows a pair of matching wedge prisms, which can steer a beam anywhere within a circle described by the full angle ($4\theta_d$), where θ_d is the deviation angle from a single prism as defined in Equation 11.2. The equation to determine the deviation angle (θ_d) for the same input direction but other wavelengths is:

$$\theta_d = \text{arc} \sin(n \sin \theta_w) - \theta_w \tag{11.2}$$

This beam steering is accomplished by rotating the two wedge prisms independently of each other. Beam steering is typically used to scan a beam to different locations in imaging applications. By combining two wedge prisms of equal power in near contact, and by independently rotating them about an axis parallel to the normals of their adjacent faces, a ray can be steered in any direction within the narrow cone. The deviation will change with the input angle.

11.3.9 PRISMS WITH SPECIAL APPLICATIONS

Prisms can be combined to produce achromatic overall behaviour. The achromatic prism has a net dispersion of zero for two given wavelengths, even though the deviation is not zero. On the other hand, the direct vision prism accomplishes zero deviation for a particular wavelength while providing chromatic dispersion. Figure 11.11 illustrates special combinations that form these two prism types. The arrangement of prisms, as illustrated in Figure 11.11(a), is combined so that one prism cancels the dispersion of the other. These can also be reversed so that the dispersion is additive, providing double dispersion. Figure 11.11(b) shows a direct vision prism when light with a wavelength of λ passes through. These prisms can be used in laser beam alignment.

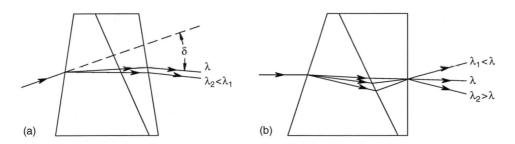

FIGURE 11.11 Nondispersive and nondeviating prisms. (a) Achromatic prism. (b) Direct vision prism for wavelength λ.

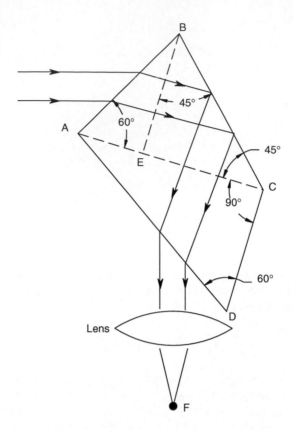

FIGURE 11.12 Pellin–Broca prism of constant deviation.

A prism design useful in spectrometers is one that produces a constant deviation for all wavelengths as they are observed or detected. One example is the Pellin–Broca prism illustrated in Figure 11.12. A collimated beam of light enters the prism at face AB and departs at face AD, making an angle of 90° with the incident direction. The dashed lines are merely added to assist in analyzing the operation of the prism. Of the incident wavelengths, only one will refract at the precise angle of deviation, as shown, with the light rays parallel to the prism base AE. At face BC total internal reflection occurs to direct the light beam into the prism section ACD. Since the prism section BEC acts only as a mirror, the beam passes through sections ABE and ACD, which together constitute a prism of 60° apex angle. The beams are focused at F, the focal point of lens.

An observing telescope may be rigidly mounted, as the prism is rotated on its prism table (or about an axis normal to the page). As the prism rotates, various wavelengths in the incident beam deviate and follow the path indicated and focus at F.

11.3.10 OTHER TYPES OF PRISMS

There are many other types of prisms that are used in building various optical devices and systems. Many reflecting prisms perform specific functions. Non-standard prism designs can be made according to customer specifications. Prisms made from birefringent crystals are useful for producing a highly polarized light wave or polarization splitting of light, such as the Wollaston prism. This prism will be addressed in detail in Chapter 15.

11.4 PRISMS IN DIFFERENT COMBINATIONS

Prism combinations are commonly used in optical instruments for purposes such as inversion of an image in prism binoculars. Laser light can be passed through two or more prisms aligned one beside the other, as shown in Figure 11.13. Some prism combinations are described above in the prism types. The laser light can be directed to a point by moving one prism in the prism combination. Consider the orientation of the final image when designing any prism combination.

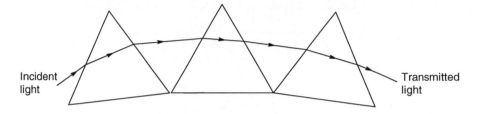

FIGURE 11.13 Laser beam passing through prism combinations.

11.5 LIGHT BEAM PASSING THROUGH A PRISM

Refraction of light passing through a prism, as shown in Figure 11.14, is understood using light rays and Snell's law. This figure shows how a single light ray that is incident on the prism from the left emerges bent away from its original direction of travel. When the ray exits the prism, the emerged ray is bent by an angle of δ, called the angle of deviation (δ). The angle of deviation will change as the angle (θ_1) of the incident ray changes. The magnitude of the deviation angle also depends on the apex angle (A) of the prism, the index of refraction of the prism material, and the wavelength of the incident light. The angle of deviation reaches its minimum value when the light passing through the prism is symmetrical; i.e., when $\theta_1 = \theta_2$, as shown in Figure 11.14. Then, the angle is called the minimum deviation angle (δ_m).

The minimum angle of deviation (δ_m) is related to the apex angle of the prism (A) and the index of refraction (n) of the prism. The index of refraction of a prism in Equation 11.3 is calculated by applying Snell's law and using simple geometry.

$$n = \frac{\sin\left(\frac{A+\delta_m}{2}\right)}{\sin\left(\frac{A}{2}\right)} \tag{11.3}$$

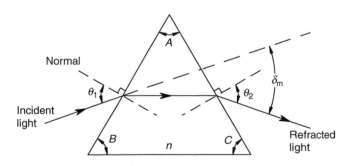

FIGURE 11.14 Light ray incident on the prism surface.

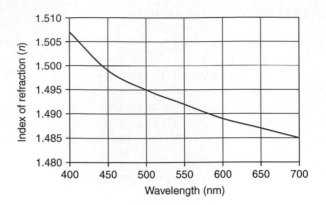

FIGURE 11.15 Index of refraction as a function of wavelength for a transparent plastic material.

White light is a uniform mixture of all visible light. Natural light, like sunlight, is considered to be white light and contains the entire visible spectrum.

Chromatic dispersion is the spreading of white light into its spectrum of wavelengths. This type of dispersion occurs when the refraction process deviates the light. The angle of deviation depends on the wavelength of the light, since the index of refraction of any optical material is a function of wavelength.

The light of shorter wavelengths travels with slightly smaller wave velocities than that of longer wavelengths. This means that the prism's index of refraction is not constant across the visible spectrum, but decreases continuously as the wavelengths increase from violet to red light. Figure 11.15 shows the relation between the light wavelength and the index of refraction of a transparent plastic material, which has the typical dispersive behaviour of decreasing the index of refraction in conjunction with increasing the wavelength.

Water, glass, transparent plastics, and quartz are all optically dispersive materials. Table 11.1 lists the indices of refraction for such materials. Again, these indices are dependent on the wavelength of light.

The index of refraction is greater for violet light than for red light, thus violet light deviates through a greater angle than does red light, as shown in Figure 11.16.

TABLE 11.1
Index of Refraction for Different Materials at Various Wavelengths

Material	Red $\lambda=660$	Orange $\lambda=610$	Yellow $\lambda=580$	Green $\lambda=550$	Blue $\lambda=470$	Violet $\lambda=410$
Water	1.331	1.332	1.333	1.335	1.338	1.342
Diamond	2.410	2.415	2.417	2.426	2.444	2.458
Glass (crown)	1.512	1.514	1.518	1.519	1.524	1.530
Glass (flint)	1.662	1.665	1.667	1.674	1.684	1.698
Polystyrene	1.488	1.490	1.492	1.493	1.499	1.506
Quartz (fused)	1.455	1.456	1.458	1.459	1.462	1.468

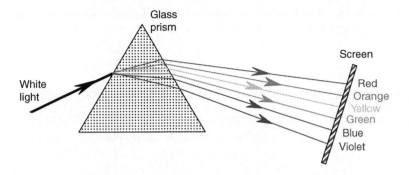

FIGURE 11.16 White light is dispersed by a prism.

11.6 FACTORS GOVERNING DISPERSION OF LIGHT BY A PRISM

Dispersion of white light by a prism depends on the following factors:

1. The optical properties of the prism (index of refraction n)
2. The angle of incidence (θ) of the light ray at the prism surface
3. The apex angle (A) of the prism
4. The wavelength of light (λ)

11.7 DISPERSION OF WHITE LIGHT BY A PRISM

Isaac Newton's first scientific paper, published in 1672, described his experiments with light and colour. Newton passed a beam of sunlight through a prism, spreading the light into a spectrum of colours. A second prism turned the opposite way re-combined the spectrum of colours back into a narrow beam of white light. Newton found that when white light is passed through a prism, a spectrum of colours emerges as shown in Figure 11.16. This separation of white light into its constituent spectral wavelengths is called chromatic dispersion. In real life, dispersion can be observed in such examples as rainbows, the surface of CD's, and an oil layer on water.

700 nm	The visible light spectrum	400 nm
Red		Violet
long λ		short λ

FIGURE 11.17 The visible spectrum of white light.

TABLE 11.2
Wavelength of the Colours in the Visible Spectrum

Colour	Wavelength Range λ (nm)
Red	630–700
Orange	590–630
Yellow	570–590
Green	500–570
Blue	450–500
Violet	400–450

Colours are associated with different wavelengths of light, as shown in Figure 11.17. It is possible to assign an approximate wavelength range to each colour of the spectrum, as given in Table 11.2.

11.8 MIXING SPECTRUM COLOURS USING A GLASS ROD AND TUBE

Colour mixing theory has been presented earlier in this book. A rainbow of colours can be mixed using a glass rod or tube, as shown in Figure 11.18. A narrow beam of white light from a light source passes through a lens and enters a glass prism where it is dispersed into the complete visible spectrum. If a mirror is placed after the prism, all colours (after reflection) can be directed onto a translucent glass rod or tube. A card held in front of the mirror can control the colours that are permitted to mix by the rod or tube. For example, if the violet, blue, and green colours are blocked, then the remaining colours of red, orange, and yellow combine, and the rod appears orange.

The visible spectrum can be divided into three equal parts, as shown in Figure 11.19. These colours (red, green, and blue) are the additive primaries and appear as three large circular areas in Figure 11.19.

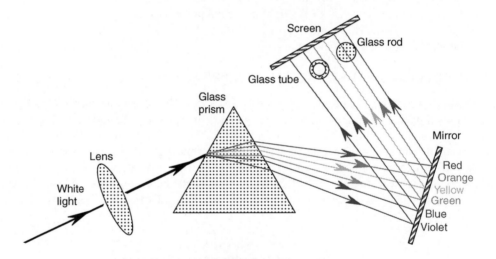

FIGURE 11.18 Experimental arrangement for mixing spectrum colours using a translucent rod.

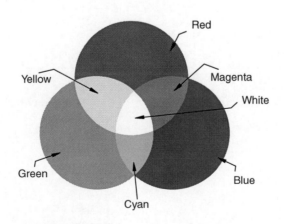

FIGURE 11.19 Additive colour mixing.

Consider mixing the primary colours of light (two at a time) and observing their resultant colour mixture, which is called a secondary colour. When primary red and primary green mix, they produce yellow; red and blue produce magenta; and green and blue produce cyan, which is a light blue–green. When red light and orange light mix, the rod appears as bright red; when yellow and green mix, the rod appears bright green; and when blue and violet mix, the rod appears blue–violet.

11.9 EXPERIMENTAL WORK

This experiment is designed to demonstrate the theory of light by passing light through both a single prism and through prism combinations. It also demonstrates light dispersion into a spectrum by a prism. Since the index of refraction varies with the wavelength, the angles of refraction are wavelength dependent. When white light enters a prism, the colour rainbow (a sequence of red, orange, yellow, green, blue, and violet colours) exits the prism, because the refractive deviation increases steadily with decreasing wavelength. In this experiment the student will perform the following cases:

a. Observe a laser beam passing through a right angle prism and draw the laser beam path.
b. Observe a laser beam passing through a Dove prism and draw the laser beam path.
c. Observe a laser beam passing through a Porro prism and draw the laser beam path.
d. Observe a laser beam passing through a prism and draw the laser beam path.
e. Observe a laser beam passing through prisms in different arrangements and draw the laser beam path through the prism combinations.
f. Use a laser beam to calculate the index of refraction of a prism.
g. Use white light passing through a prism to observe a rainbow of colours. Measure the refractive deviation angle of the rainbow colours and calculate the index of refraction of the prism.
h. Mix the spectrum colours using a glass rod and tube.

11.9.1 TECHNIQUE AND APPARATUS

Appendix A presents the details of the devices, components, tools, and parts.

1. 2×2 ft. optical breadboard
2. HeNe laser source and power supply

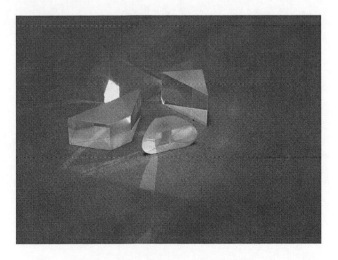

FIGURE 11.20 Prisms.

3. Laser mount assembly
4. Light source (white light)
5. Lab jack
6. Multi-axis translation stage
7. 360° rotational stage
8. Hardware assembly (clamps, posts, screw kits, screwdriver kit, positioners, post holder, laser holder/clamp, etc.)
9. Different types of prisms, as shown in Figure 11.20
10. Glass rod and tube, as shown in Figure 11.21
11. Lens and lens holder/positioner, as shown in Figure 11.22
12. Mirror and mirror holder, as shown in Figure 11.23
13. Black/white card and cardholder (large and small sizes)
14. Protractor
15. Ruler

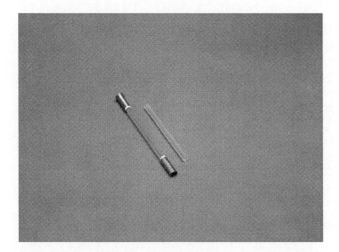

FIGURE 11.21 A glass rod and tube.

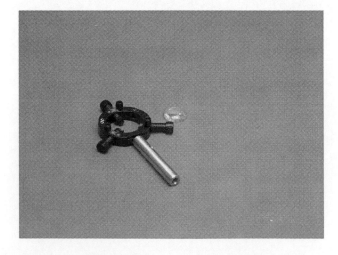

FIGURE 11.22 A lens and lens holder/positioner.

FIGURE 11.23 A mirror and mirror holder.

11.9.2 PROCEDURE

Follow the laboratory procedures and instructions given by the professor and/or instructor.

11.9.3 SAFETY PROCEDURE

Follow all safety procedures and regulations regarding the use of optical components, electrical and optical devices, and optical cleaning chemicals.

11.9.4 APPARATUS SET-UP

Case (a): Laser Beam Passing through a Right Angle Prism

1. Figure 11.24 shows the apparatus set-up.
2. Bolt the laser short rod to the breadboard.

FIGURE 11.24 Laser beam passing through a right angle prism.

3. Bolt the laser mount to the clamp using bolts from the screw kit.
4. Put the clamp on the short rod.
5. Place the HeNe laser into the laser mount and tighten the screw. Turn on the laser device. Follow the operation and safety procedures of the laser device in use.
6. Mount a 360° rotational stage to the breadboard so that the stage is approximately 10 cm in front of the laser source. Secure the stage to the breadboard.
7. Mount a right angle prism on top of the rotational stage platform and ensure that one side of the prism is facing the laser beam. The right angle prism should be orthogonal to the laser beam and parallel to the optical breadboard.
8. Arrange the location of the black/white card to capture the laser beam from the right angle prism.
9. Make the necessary adjustments so that the laser beam passes through the right angle prism, exits from the other side of the prism, and projects onto the white/black card.
10. Finely align the laser beam, using the rotational stage under the right angle prism, to direct the laser beam passing through the prism.
11. Observe and draw the laser beam path in a diagram.

Case (b): Laser Beam Passing through a Dove Prism

Figure 11.25 shows the apparatus set-up. Using a Dove prism, repeat the procedure explained in Case (a).

FIGURE 11.25 Laser beam passing through a Dove prism.

Case (c): Laser Beam Passing through a Porro Prism

Figure 11.26 shows the apparatus set-up. Using a Porro prism, repeat the procedure explained in Case (a).

FIGURE 11.26 Laser beam passing through a Porro prism.

Case (d): Laser Beam Passing through a Prism

Figure 11.27 shows the apparatus set-up. Using a prism repeat the procedure explained in Case (a).

FIGURE 11.27 A laser beam passing through a prism.

Case (e): Laser Beam Passing through Prism Combination

Figure 11.28 shows the apparatus set-up. Repeat the procedure as explained in Case (a) from step 2 through step 5. Add the following steps to complete the lab set-up for a laser beam passing through a combination of three prisms:

1. Mount three 360° rotational stages to the breadboard so that the first stage is approximately 10 cm in front of the light source. Mount the second and the third rotation stages offset as show in the Figure 11.28. Secure the stages to the breadboard.
2. Mount a prism on top of each rotational stage platform and ensure that one side of the prism is facing the laser beam. The prism should be orthogonal to the laser beam and parallel to the optical breadboard.

FIGURE 11.28 A laser beam passing through prism combination.

3. Arrange the location of the black/white card to capture the laser beam from the third prism.
4. Make the necessary adjustments so that the laser beam passes through the first prism, exits from the other side of the first prism, is incident on the second prism, and the transmitted light is incident on the third prism. Try to project the transmitted laser beam from the third prism onto the white/black card.
5. Finely align the laser beam, using rotational stages under the prisms, to direct the laser beam passing through the prisms.
6. Observe and draw the laser beam path in a diagram.

Case (f): Laser Beam Passing through a Prism to Calculate the Index of Refraction

Figure 11.27 also shows the apparatus set-up. Repeat the procedure explained in Case (a) from step 2 through step 11. Add the following steps to complete the lab set-up for a laser beam passing through a prism:

1. Use a protractor to measure the prism angles, the angle of incidence of the laser beam to the prism, the angle of deviation (δ), and the angle of refraction of the laser beam exiting the prism.
2. Record the measured angles in Table 11.3.

Case (g): Dispersion of White Light by a Prism

1. Figure 11.29 shows the apparatus set-up.
2. Mount the white light source on the lab jack to the optical breadboard.
3. Mount the lens and lens holder in front the white light beam to focus the light.
4. Use the lab jack to adjust the height of the light beam to the centre of the lens.
5. Mount a multi-axis translation stage on the optical breadboard.
6. Mount a prism on top of the multi-axis translation stage platform and ensure that one side of the prism is facing the light source. The prism should be orthogonal to the light source and parallel to the optical breadboard.
7. Turn off the laboratory light to be able to see the light exiting from the prism.

FIGURE 11.29 Apparatus set-up for light dispersion by a prism.

8. Make the necessary adjustments so that the light passes through the lens, prism, exits from the other side of the prism, and projects onto the white/black card.
9. Arrange the location of the black/white card to capture the rainbow colours exiting from the prism.
10. Finely align the light, using multi-axis translation stages under the prism, to maximize the amount of light passing through the prism. This will brighten the spectrum on the black/white card. Make a very fine alignment to the stage to ensure that the light exiting the prism appears as a clear rainbow on the black/white card, as shown in Figure 11.30.
11. Note that the rainbow of colours will be different from one prism to other. This depends on the prisms optical properties (index of refraction), apex angles, and wavelength of the light source.
12. Use a protractor to measure the prism angles, the angle of incidence of the light source to the prism, the angle of deviation (δ), and the angle of refraction of each of the six colours exiting the prism.

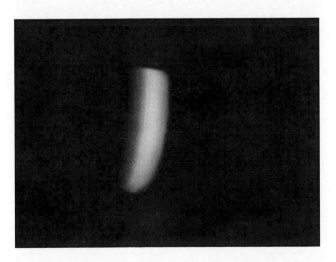

FIGURE 11.30 Rainbow colour exiting from the prism captured on the card.

13. Turn on the laboratory light.
14. Record the measured angles in Table 11.4.

Case (h): Mixing the Spectrum Colours Using a Glass Rod and Tube

Figure 11.31 shows the apparatus set-up. Repeat the procedure explained in Case (g) from step 1 through step 10. Add the following steps to mix the colours by using a glass rod and tube:

1. Mount a mirror in place of the black/white card.
2. Mount the black/white card to the optical breadboard to capture the rainbow colours reflected from the mirror, as shown in Figure 11.31.
3. Place a glass rod in front of the black/white card, as shown in Figure 11.32.
4. Place the glass rod between two adjacent colours of the rainbow starting from red towards violet.

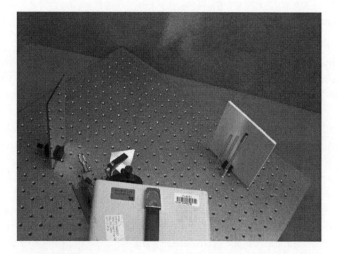

FIGURE 11.31 Apparatus set-up for light dispersion by a prism and mixing spectrum colours.

FIGURE 11.32 A glass rod and tube intercepting the rainbow colours.

5. Observe the mixing of spectrum colours exiting from the glass rod. When primary red and primary green mix at the glass rod, they produce yellow; red and blue produce magenta; and green and blue produce cyan.
6. Repeat step 3 through step 5 using a glass tube in front of the black/white card, as shown in Figure 11.32.
7. Report your observations when mixing spectrum colours.

11.9.5 DATA COLLECTION

11.9.5.1 Laser Beam Passing through a Right Angle Prism

No data collection is required for this case.

11.9.5.2 Laser Beam Passing through a Dove Prism

No data collection is required for this case.

11.9.5.3 Laser Beam Passing through a Porro Prism

No data collection is required for this case.

11.9.5.4 Laser Beam Passing through a Prism

No data collection is required for this case.

11.9.5.5 Laser Beam Passing through Prism Combination

No data collection is required for this case.

11.9.5.6 Laser Beam Passing through a Prism to Calculate the Index of Refraction

Measure the angles of the prism: the apex angle (A) and the base angles (B) and (C). Measure the angle of incidence of the laser beam incident on the prism, the angle of deviation (δ), and the refracted angle of the laser beam that leaves the prism. Record the measured angles in Table 11.3.

TABLE 11.3
Prism Angles, and Incidence and Refracted Angles of the
Laser Beam

Prism Angles	Incidence Angle of Light	Angle of Deviation	Calculated Refractive Index
$A = ($ $^{0})$			
$B = ($ $^{0})$	$\theta_1 = ($ $^{0})$	$\delta = ($ $^{0})$	$n =$
$C = ($ $^{0})$			

11.9.5.7 Dispersion of White Light by a Prism

1. Measure the angles of the prism: the apex angle (A) and the base angles (B) and (C). Measure the angle of incidence of light incident on the prism, the angle of deviation (δ), and the refracted angles of the rainbow colours that leave the prism.
2. Record the measured angles in Table 11.4.

TABLE 11.4
Prism Angles, and Incidence and Refracted Angles of the Light

Prism Angles

$A = (\quad ^{o})$

$B = (\quad ^{o})$

$C = (\quad ^{o})$

Incidence Angle of Light

$\theta_1 = (\quad ^{o})$

Angle of Deviation

$\delta = (\quad ^{o})$

Rainbow Colours	Rainbow Colours Refracted Angle from Prism (^{o})
Red	
Orange	
Yellow	
Green	
Blue	
Violet	

11.9.5.8 Mixing the Spectrum Colours Using a Glass Rod and Tube

No data collection is required for this case.

11.9.6 Calculations and Analysis

11.9.6.1 Laser Beam Passing through a Right Angle Prism

No calculations and analysis are required for this case.

11.9.6.2 Laser Beam Passing through a Dove Prism

No calculations and analysis are required for this case.

11.9.6.3 Laser Beam Passing through a Porro Prism

No calculations and analysis are required for this case.

11.9.6.4 Laser Beam Passing through a Prism

No calculations and analysis are required for this case.

11.9.6.5 Laser Beam Passing through Prism Combination

No calculations and analysis are required for this case.

11.9.6.6 Laser Beam Passing through a Prism to Calculate the Index of Refraction

1. Calculate the index of refraction (n) of the prism, using Equation 11.3.
2. Record the calculated value for the index of refraction of the prism in Table 11.3.
3. Find the published value of the index of refraction of the prism, as provided by the manufacturer.

11.9.6.7 Dispersion of White Light by a Prism

1. Calculate the index of refraction (n) of the prism for one colour, using Equation 11.3.
2. Record the calculated value for the index of refraction of the prism in Table 11.4.
3. Find the published value of the index of refraction of the prism, as provided by the manufacturer.

11.9.6.8 Mixing the Spectrum Colours Using a Glass Rod and Tube

No calculations and analysis are required for this case.

11.9.7 Results and Discussions

11.9.7.1 Laser Beam Passing through a Right Angle Prism

Report your observations when a laser beam passes through a right angle prism. Compare the laser beam diagram with the theory.

11.9.7.2 Laser Beam Passing through a Dove Prism

Report your observations when a laser beam passes through a Dove prism. Compare the laser beam diagram with the theory.

11.9.7.3 Laser Beam Passing through a Porro Prism

Report your observations when a laser beam passes through a Porro prism. Compare the laser beam diagram with the theory.

11.9.7.4 Laser Beam Passing through a Prism

Report your observations when a laser beam passes through a prism. Compare the laser beam diagram with the theory.

11.9.7.5 Laser Beam Passing through Prism Combination

Report your observations when a laser beam passes through the combination of prisms. Compare the laser beam diagram with the theory.

11.9.7.6 Laser Beam Passing through a Prism to Calculate the Index of Refraction

1. Report the angle measurements for the prism.
2. Explain how a prism works and provide examples of optical devices that use prisms combined with other optical components.
3. Compare the calculated index of refraction with the published value of the prism as provided by the manufacturer. Explain the possible reasons for any differences.
4. List the most important factors to be considered when choosing the specifications of a prism.

11.9.7.7 Dispersion of White Light by a Prism

1. Report the colours of the rainbow.
2. Report the angle measurements for the prism.
3. Explain how a prism works and provide examples of optical devices that use prisms with other optical components.
4. Compare the calculated index of refraction with the published value of the prism as provided by the manufacturer. Explain the possible reasons for any differences.
5. List the most important factors to be considered when choosing the specifications of a prism.

11.9.7.8 Mixing the Spectrum Colours Using a Glass Rod and Tube

Report your observations when mixing spectrum colours using a glass rod and tube.

11.9.8 Conclusion

Summarize the important observations and findings obtained in this lab experiment.

11.9.9 Suggestions for Future Lab Work

List any suggestions for improvements using different experimental equipment, procedures, and techniques for any future lab work. These suggestions should be theoretically justified and technically feasible.

11.10 LIST OF REFERENCES

List any references that were used in the report. Use one format in writing the references. Never mix reference formats in a report.

11.11 APPENDICES

List all of the materials and information that are too detailed to be included in the body of the report.

FURTHER READING

Beiser, A., *Physics*, 5th ed., Addison-Wesley Publishing Company, Inc., USA, 1991.

Blaker, J. W. and Schaeffer, P., *Optics: An Introduction for Technicians and Technologists*, Prentice Hall, Inc., Upper Saddle River, NJ, 2000.

Born, M. and Wolf, E., Elements of the Theory of Diffraction, In *Principles of Optics: Electromagnetic Theory of Propagation, Interference, and Diffraction of Light*, 7th ed., Cambridge University Press, Cambridge, pp. 370–458, 1999.

Born, M. and Wolf, E., *Rigorous Diffraction Theory Principles of Optics: Electromagnetic Theory of Propagation, Interference, and Diffraction of Light*, 7th ed., Cambridge University Press, Cambridge pp. 556–592, 1999.

Cox, A., *Photographic Optics*, 15th ed., Focal Press, London, 1974.

Cutnell, J. D. and Johnson, K. W., *Physics*, 5th ed., Wiley, New York, 2001.

Cutnell, J. D. and Johnson, K. W., *Student Study Guide—Physics*, 5th ed., Wiley, New York, 2001.

Drisoll, W. G. and Vaughan, W., *Handbook of Optics*, McGraw Hill Book Company, New York, 1978.

Edmund Industrial Optics, *Optics and Optical Instruments Catalog*, Edmund Optics, Inc., New Jersey, 2004.

EPLAB, *EPLAB Catalogue 2002*, The Eppley Laboratory, Inc., 12 Sheffield Avenue, P.O. Box 419, Newport, Rhode Island 02840, USA, 2002.

Ewald, W. P., Young, W. A., and Roberts, R. H., *Practical Optics*, Makers of Pittsford, Rochester, New York, 1982.

Francon, M., *Optical Interferometry*, Academic Press, New York pp. 97–99, 1966.

Giancoli, D. C., *Physics*, 5th ed., Prentice Hall, Upper Saddle River, NJ, 1998.

Halliday, D., Resnick, R., and Walker, J., *Fundamental of Physics*, 6th ed., Wiley, New York, 1997.

Heath, R. W. et al., *Fundamentals of Physics*, Heath Canada Ltd., Toronto, DC, 1979.

Hecht, E., *Optics*, 4th ed., Addison-Wesley Longman, Inc., New York, 2002.

Jenkins, F. W. and White, H. E., *Fundamentals of Optics*, McGraw Hill, New York, 1957.

Jones, E. and Childers, R., *Contemporary College Physics*, McGraw-Hill Higher Education, New York, 2001.

Keuffel & Esser Co., *Physics Approved Diazo Transparency Masters*, Keuffel & Esser Audiovisual Educator, New York, 1989.

Lambda, *Catalog 2004*, Research Optics, Inc., California, 2004.

Lehrman, R. L., *Physics—The Easy Way*, 3rd ed., Barron's Educational Series, Inc., Hauppauge, NY, 1998.

Lerner, R. G. and George, L. T., *Encyclopedia of Physics*, 2nd ed., VCH Publishers, Inc., New York, NY, 1991.

McDermott, L. C. et al., *Introduction to Physics*, Preliminary Edition, Prentice Hall, Inc., Upper Saddle River, NJ, 1988.

Melles, G., *The Practical Application of Light Melles Griot Catalog*, Melles Griot, Rochester, NY, 2001.

Newport Corporation, Optics and Mechanics Section, the *Newport Resources 1999/2000 Catalog*, Newport Corporation, Irvine, CA, USA, 1999/2000.

Nolan, P. J., *Fundamentals of College Physics*, Wm. C. Brown Publishers, Dubuque, Iowa, 1993.

Ocean Optics, *Product Catalog*, Ocean Optics, Inc., Florida, 2003.

Pedrotti, F. L. and Pedrotti, L. S., *Introduction to Optics*, 2nd ed., Prentice Hall, Inc., Upper Saddle River, NJ, 1993.

Romine, G. S., *Applied Physics Concepts into Practice*, Prentice Hall, Inc., Upper Saddle River, NJ, 2001.

Sears, F. W., Zemansky, M. W., and Young, H. D., *University Physics—Part II*, 6th ed., Addison-Wesley Publishing Company, Massachusetts, 1998.

Serway, R. A. and Jewett, J. W., *Physics for Scientists and Engineers*, with Modern Physics, 6th ed., Vol. 2, Thomson Books/Cole, USA, 2004.

Sterling, D. J. Jr., *Technician's Guide to Fiber Optics*, 2nd ed., Delmar Publishers Inc., Albany, NY, 1993.

Tippens, P. E., *Physics*, 6th ed., Glencoe McGraw-Hill, Westerville, OH, U.S.A., 2001.

Smith, W. J., *Modern Optical Engineering*, McGraw-Hill Book Co., New York, 1966.

Urone, P. P., *College Physics*, Brooks/Cole Publishing Company, Florence, KY, 1998.

Walker, J. S., *Physics*, Prentice Hall, Englewood Cliffs, NJ, 2002.

White, H. E., *Modern College Physics*, 6th ed., Van Nostrand Reinhold Company, New York, 1972.

Wilson, J. D., *Physics—A Practical And Conceptual Approach Saunders Golden Sunburst Series*, Saunders College Publishing, Philadelphia, PA, 1989.

Wilson, J. D. and Buffa, A. J., *College Physics*, 5th ed., Prentice Hall, Inc., Upper Saddle River, NJ, 2000.

12 Beamsplitters

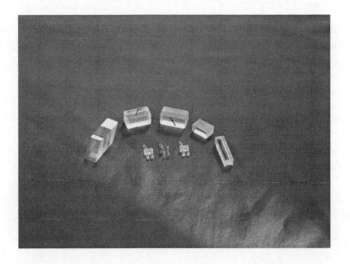

12.1 INTRODUCTION

A beamsplitter is a common optical component that partially transmits and partially reflects an incident light beam; this splitting usually occurs in unequal proportions. This chapter explains the basic principles that govern light passing through a beamsplitter. Various types of beamsplitters and their applications are presented. In addition to the function of dividing light, beamsplitters can be employed to recombine two separate light beams or images into a single path. The concept of light passing through beamsplitters is applicable to image formation and building optic/optical devices. Three experimental cases presented in this chapter demonstrate the principles of the light passing through beamsplitters in optical fibre devices.

12.2 BEAMSPLITTERS

As explained in the prisms chapter, the beamsplitters are also common optical components. Beamsplitters are blocks of optical material with flat polished faces. Beamsplitters partially transmit and partially reflect an incident light beam, usually in unequal ratios. They can also be used to recombine two separate light beams into a single light beam. Beamsplitters are used where it is necessary to direct a beam of light entering from one side and exiting from the other sides of the beamsplitter. Light passing through beamsplitters is governed by the laws of light. Many types of beamsplitters are used to build optical devices. The most common types of beamsplitters split the incoming light beam into two components of a 50/50 beamsplitting ratio. Rectangular beamsplitters are used to manufacture polarization beamsplitter devices with operating wavelengths of 1480 and 1550 nm, which are then used in communications systems. Some beamsplitters are shown in the figure above.

The light beam output components can be arranged exiting the beamsplitters either parallel or perpendicular to the input light. Each output light component exits from one side of the beamsplitter. Depending on the characteristics of the input light and the desired output light components, a specific beamsplitter can be chosen to accomplish the design requirements.

12.3 BEAMSPLITTER TYPES

There are many types of beamsplitters that are used in different applications. Some are used with polarizing and non-polarizing light, or with visible and near infrared waves. In the following sections, some beamsplitters and their applications are presented in detail.

12.3.1 STANDARD CUBE BEAMSPLITTERS

Standard cube beamsplitters consist of matched pairs of right angle prisms cemented together along their hypotenuses, as shown in Figure 12.1. The hypotenuse of one prism has a partial reflective coating. A black dot on the bottom side of the prism is used to identify which prism has the partial reflectore. The incident beam must enter the prism containing the partial reflector first. This prism splits the ray incident at the normal into two orthogonal components on the surface of the prism hypotenuse. It splits the incident beam into a 50/50 beamsplitting ratio. One component is called transmission, while the other is called reflection. Transmission and reflection approach 50%, though the output is partially polarized.

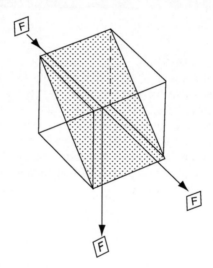

FIGURE 12.1 Standard cube beamsplitter.

The non-polarizing cube beamsplitters are constructed from a pair of precision, high-tolerance right angle prisms cemented together with a metallic-dielectric coating on the hypotenuse of one of the prisms. A broadband antireflective coating is applied to each face of the beamsplitter in order to produce maximum transmission efficiency.

Cube beamsplittes are available in two categories: polarizing and non-polarizing. If polarization sensitivity is critical in an application, then using the polarizing cube beamsplitters is recommended. Cube beamsplitters are available in three types: broadband hybrid, broadband dielectric, and laser-line non-polarizing.

12.3.2 POLARIZING CUBE BEAMSPLITTERS

Polarizing cube beamsplitters split randomly-polarized light into two orthogonal, linearly polarized components; S-polarized light is reflected at a 90° angle (perpendicular), while P-polarized light is transmitted (horizontal).

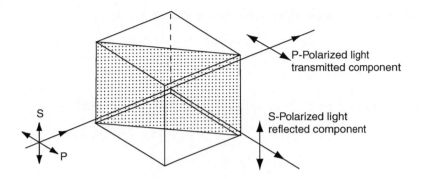

FIGURE 12.2 Polarizing cube beamsplitter.

Polarizing cube beamsplitters are constructed similarly to the standard cube beamsplitters, as shown in Figure 12.2. The beamsplitter consists of a pair of precision, high-tolerance right angle prisms cemented together with a dielectric coating on the hypotenuse of one of the prisms. A multi-layer antireflective coating is applied to each face of the beamsplitter in order to produce maximum transmission efficiency.

12.3.3 Rectangular Polarizing Beamsplitters

Rectangular polarizing beamsplitters perform the same function as cube beamsplitters. They consist of three prisms carefully cemented together along their hypotenuses, as shown in Figure 12.3. Rectangular polarization beamsplitters are used in optical devices where both of the output components are required to exit the side opposite from the input signal. This produces a lateral displacement between the two output components. Figure 12.3 shows a beamsplitter splitting the input signal into its two orthogonal 50/50 components. Rectangular beamsplitters are used in building polarizing beamsplitter devices for 1480 and 1550 nm wavelengths that are employed in communication systems.

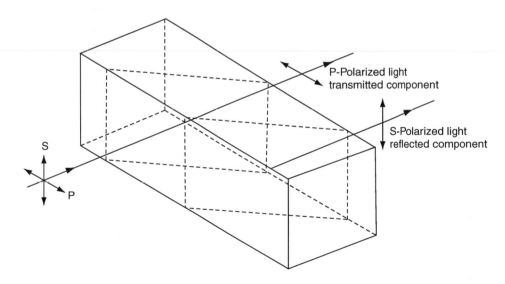

FIGURE 12.3 Rectangular polarizing beamsplitter.

12.3.4 Lateral Displacement Polarizing Beamsplitters

As explained in rectangular beamsplitters, lateral displacement beamsplitters output two parallel beams separated by a distance that depends on the size chosen, as shown in Figure 12.4. These beamsplitters consist of a precision rhomboid prism cemented to a 1/8-wavelength right angle prism. The entrance and exit faces have an antireflective coating layer to increase efficiency. Polarizing and non-polarizing lateral displacement beamsplitters are available.

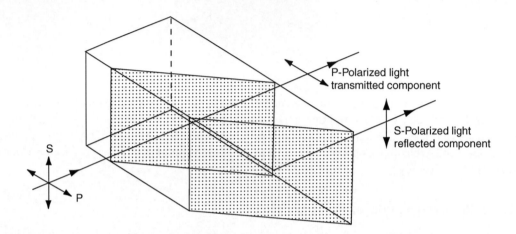

FIGURE 12.4 Lateral displacement polarizing beamsplitter.

12.3.5 Glan Thompson Polarizing Beamsplitters

The Glan Thompson polarizing beamsplitter is designed to permit the output of the S-polarized beam at 45° from the straight-through P-polarized beam, as shown in Figure 12.5. These are useful for utilizing both linear polarization states. They are commonly sold mounted in a rectangular metal cell.

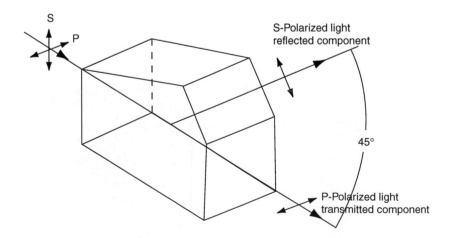

FIGURE 12.5 Glan Thompson polarizing beamsplitter.

12.3.6 POLKA-DOT BEAMSPLITTERS

Polka-dot beamsplitters offer a distinct advantage over standard dielectric beamsplitters; they have a constant 50/50 reflection-to-transmission ratio over a large spectral range. They are ideal for use with broadband, extended sources, such as tungsten, halogen, deuterium, and xenon lamps, and in monochromators, spectrophotometers, and other optical systems. The beamsplitter has an aluminum coating cell. The coated-to-uncoated surface areas resemble polka-dots. Input beams are split evenly: 50% of the incident light is reflected by the coating and 50% is transmitted through the clear glass, as shown in Figure 12.6. Since the polka-dot beamsplitters are not sensitive to the incident angle, wavelength, or polarization, they are ideal for splitting energy from a radiant light source.

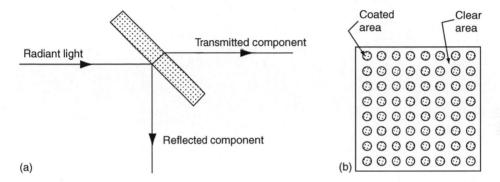

FIGURE 12.6 Polka-dot beamsplitter.

12.3.7 ELLIPTICAL PLATE BEAMSPLITTERS

The elliptical plate beamsplitter creates a circular aperture equal to the diameter of the minor axis, when oriented at 45°, as shown in Figure 12.7. They maximize beamsplitting efficiency while minimizing required mounting space. The elliptical plate beamsplitters are ideal for diffuse axial and in-line illumination. They operate in either visible light or near infrared light regions. They are a 50/50 beamsplitter coated with a high-efficiency multi-layer antireflection coating to reduce back reflections.

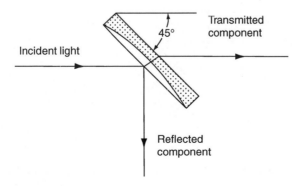

FIGURE 12.7 Elliptical plate beamsplitter.

12.3.8 Mirror-Type Beamsplitters

A mirror-type (plate) beamsplitter is an optical window with a semi-transparent mirrored coating that breaks a light beam into two beams, as shown in Figure 12.8. This beamsplitter will reflect a portion of the incident light, absorb a relatively small portion, and transmit the remaining light.

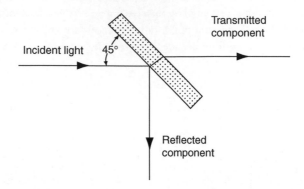

FIGURE 12.8 Mirror-type beamsplitter.

12.3.9 Pellicle Beamsplitters

Pellicle beamsplitters are very thin, nitrocellulose membranes (pellicle) bonded to lapped aluminum frames. Figure 12.9 shows a comparison between a pellicle beamsplitter and a glass plate beamsplitter. Ghost images occur in the glass beamsplitter shown in Figure 12.9(b). In a pellicle beamsplitter, ghost images are eliminated by the thinness of the membrane, since the second surface's reflection superimposes on the first surface reflection. The uncoated pellicle reflects 8% and transmits 92% of light in the visible and near infrared regions. Pellicle beamsplitters can be coated with various thin films to change the reflectivity and transitivity, and thus the splitting ratio. Pellicles are usually mounted on a precision-lapped, hard-black, anodized aluminum alloy frame. Pellicle beamsplitters are not affected by mechanical shock or variations in temperature and humidity. Pellicle membranes are extremely delicate and can be easily punctured.

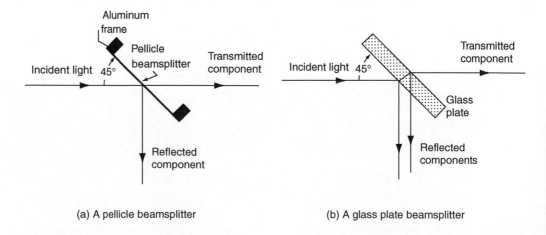

FIGURE 12.9 Comparison between a pellicle beamsplitter and a glass plate beamsplitter.

12.3.10 Visible and Near Infrared Region Plate Beamsplitters

Plate beamsplitters are designed for a 45° angle of incidence and random polarization, as shown in Figure 12.10. Plate beamsplitters are available in 30% reflection/70% transmission, as well as the standard 50% reflection/50% transmission ratios, for both the visible (400–700 nm) and near infrared (700–1100 nm) wavelengths. These plate beamsplitters have the optimum combination of polished, high quality, optical-grade glass with minimum substrate thickness. The broadband dielectric coating can be chosen to cover spectral ranges to meet the application needs. Plate beamsplitters with low-absorption coatings allow maximum throughput to minimize ghost images and have much less light loss than metallic coatings. The back surfaces of these plate beamsplitters are multi-layer antireflection coated, which reduces back reflections to less than about 1% for each wavelength range. Typical applications include use in dual magnification imaging systems and in combining low-power laser beams.

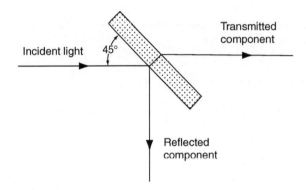

FIGURE 12.10 Plate beamsplitter.

12.3.11 Quartz Beamsplitters

Quartz beamsplitters are used to attenuate laser light. They attach directly to a power-meter probe and expose the probe to less than 10% of the laser light. The laser attenuators are ideal when pulsed laser light density levels approach damage limits of the power-meter probe coatings.

12.3.12 Dichroic Plate Beamsplliters

The dichroic manufacturing process gives the dichroic plate beamsplitters the steepest edges, and the flattest and the highest reflection and transmission bands. More complete transmission and reflection means that less stray light gets through the imaging system, yielding higher-contrast images and a better signal-to-noise ratio. Dichroic beamsplitters are critical components in fluorescence microscopy, genomics, proteomics, gel plate, and microplate readers, as well as in original equipment manufacturer instrumentation requiring beam separation, beam combination, or multispectral detection. These beamsplitters can also be used as highly efficient bandpass filters that cover several colours.

12.3.13 Other Types of Beamsplitters

There are many other types of beamsplitters used in building optical devices and systems, such as low- and high-laser line polarizing cube beamsplitters, medium and high extinction broadband polarizing cube beamsplitters, circular variable metallic beamsplitters, variable beamsplitter/attenuators, and electronic variable beamsplitters. These beamsplitters are useful for producing polarized and non-polarized light. Other polarizing optical components, such as polarizing cubes and polarizing plate beamsplitters, and polarizing rectangular beamsplitters, will be explained in polarization of light in Chapter 15. Non-standard beamsplitter design can be made according to the customers' required specifications.

12.4 EXPERIMENTAL WORK

This experiment is designed to demonstrate the theory of light passing through beamsplitters. It also demonstrates light splitting into two components by a cube beamsplitter and a rectangular beamsplitter. In this experiment the student will perform the following cases:

a. Observe a laser beam passing through a cube beamsplitter and draw the laser beam components.
b. Observe a laser beam passing through a rectangular beamsplitter and draw the laser beam components.
c. Observe a laser beam passing through the Glan Thompson polarizing beamsplitter and draw the laser beam components.

12.4.1 TECHNIQUE AND APPARATUS

Appendix A presents the details of the devices, components, tools, and parts.

1. 2×2 ft. optical breadboard
2. HeNe laser source and power supply
3. Laser light sensor
4. Laser light power meter
5. Laser mount assembly
6. Hardware assembly (clamps, posts, screw kits, screwdriver kits, sundry positioners, etc.)
7. Beamsplitter holder/positioner assembly
8. Cube beamsplitter, as shown in Figure 12.11
9. Rectangular beamsplitter, as shown in Figure 12.12
10. Glan Thompson polarizing beamsplitter, as shown in Figure 12.13

FIGURE 12.11 Cube beamsplitter.

FIGURE 12.12 Rectangular beamsplitter.

FIGURE 12.13 Glan Thompson polarizing beamsplitter.

11. Black/white card and cardholder
12. Protractor
13. Ruler

12.4.2 PROCEDURE

Follow the laboratory procedures and instructions given by the professor and/or instructor.

12.4.3 SAFETY PROCEDURE

Follow all safety procedures and regulations regarding the use of optical components and instruments, light source devices, and optical cleaning chemicals.

12.4.4 Apparatus Set-Up

12.4.4.1 Cube Beamsplitter

1. Figure 12.14 shows the apparatus set-up.
2. Bolt the laser short rod to the breadboard.
3. Bolt the laser mount to the clamp using bolts from the screw kit.
4. Put the clamp on the short rod.
5. Place the HeNe laser into the laser mount and tighten the screw. Turn on the laser device. Follow the operation and safety procedures of the laser device in use.
6. Check the laser alignment with the line of bolt holes and adjust when necessary.
7. Mount a beamsplitter holder/positioner to the breadboard.
8. Mount the cube beamsplitter on the beamsplitter holder/positioner, as shown in Figure 12.14.
9. Carefully align the laser beam so that the laser beam follows the direction of the normal to the centre of cube beamsplitter face.
10. Try to capture the laser beam exiting the cube beamsplitter on the black/white card, as shown in Figure 12.14.
11. Turn off the lights of the lab before taking measurements.
12. Measure the power of the laser beam at the input face and output face of the cube beamsplitter. Fill out Table 12.1 for Case (a).
13. Turn on the lights of the laboratory.

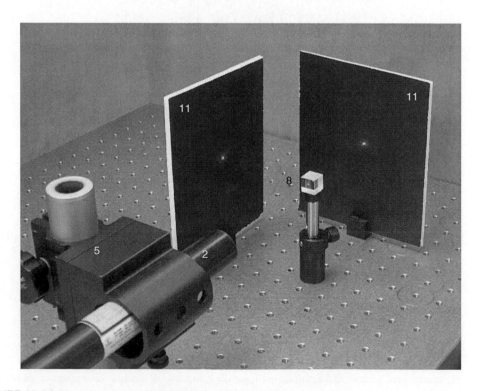

FIGURE 12.14 Laser beam passing through a cube beamsplitter.

12.4.4.2 Rectangular Beamsplitter

Figure 12.15 shows the apparatus set-up. Repeat the procedure from steps 1–7 in Case (a) using a rectangular beamsplitter; then add the following steps:

1. Mount a rectangular beamsplitter on the beamsplitter holder/positioner, as shown in Figure 12.15.
2. Carefully align the laser beam so that the laser beam impacts normally to the centre of the wide face of the rectangular beamsplitter.
3. Try to capture the laser beam exiting the rectangular beamsplitter on the black/white card, as shown in Figure 12.15. Two laser spots can be caught on the black/white card of the laser beam components exiting from the output face of the rectangular beamsplitter, as shown in Figure 12.16.
4. Turn off the lights of the laboratory before taking measurements.
5. Measure the laser beam power at the input face and output face of the rectangular beamsplitter. Fill out Table 12.1 for Case (b).
6. Turn on the lights of the laboratory.

FIGURE 12.15 Laser beam passing through a rectangular beamsplitter.

12.4.4.3 Glan Thompson Polarizing Beamsplitter

Figure 12.17 shows the apparatus set-up. Repeat the procedure from step 1–7 in Case (a) using a Glen Thompson polarizing beamsplitter; then add the following steps:

1. Mount a Glan Thompson polarizing beamsplitter on the beamsplitter holder/positioner, as shown in Figure 12.17.
2. Carefully align the laser beam so that the laser beam impacts normally to the centre of the face Glan Thompson polarizing beamsplitter.

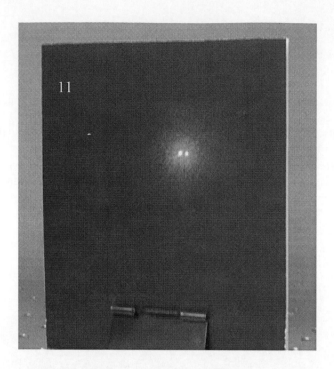

FIGURE 12.16 Laser beam outputs exiting the rectangular beamsplitter.

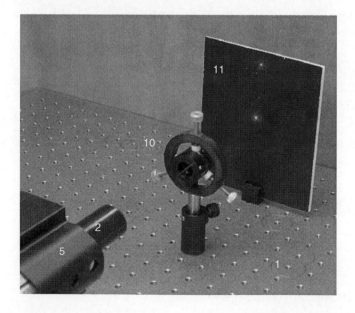

FIGURE 12.17 Laser beam passing through the Glan Thompson polarizing beamsplitter.

3. Try to capture the laser beam exiting from the face of the Glan Thompson polarizing beamsplitter on the black/white card, as shown in Figure 12.17. Two laser spots can be caught on the black/white card of the laser beam components exiting from the output faces of the Glan Thompson polarizing beamsplitter.

4. Turn off the lights of the laboratory before taking measurements.
5. Measure the laser power at the input face and output faces of the Glan Thompson polarizing beamsplitter. Fill out Table 12.1 for Case (c).
6. Turn on the lights of the laboratory.

12.4.5 DATA COLLECTION

12.4.5.1 Cube Beamsplitter

1. Measure the laser beam power at the input face and output faces of the cube beamsplitter.
2. Fill out Table 12.1 with the collected data for Case (a).

12.4.5.2 Rectangular Beamsplitter

1. Measure the laser beam power at the input face and output face of the rectangular beamsplitter.
2. Fill out Table 12.1 with the collected data for Case (b).

12.4.5.3 Glan Thompson Polarizing Beamsplitter

1. Measure the laser beam component power at the input face and output faces of the Glan Thompson polarizing beamsplitter.
2. Fill out Table 12.1 with the collected data for Case (c).

TABLE 12.1
Beamsplitters Data Collection

Case	Beamsplitter	Laser Power Input (unit)	P Component Power (unit)	S Component Power (unit)	Splitting Ratio
(a)	Polarizing cube beamsplitter				
(b)	Rectangular polarizing beamsplitter				
(c)	Glan Thompson Polarizing beamsplitter				

12.4.6 CALCULATIONS AND ANALYSIS

12.4.6.1 Cube Beamsplitter

1. Compare the laser beam power at the input face and output faces of the cube beamsplitter.
2. Calculate the splitting ratio of the cube beamsplitter. Fill out Table 12.1 with the collected data for Case (a).

12.4.6.2 Rectangular Beamsplitter

1. Compare the laser beam power at the input face and output face of the rectangular beamsplitter.
2. Calculate the laser beam splitting ratio of the rectangular beamsplitter. Fill out Table 12.1 with the collected data for Case (b).

12.4.6.3 Glan Thompson Polarizing Beamsplitter

1. Compare the laser beam power at the input face and output faces of the Glan Thompson polarizing beamsplitter.
2. Calculate the laser beam splitting ratio of the Glan Thompson polarizing beamsplitter. Fill out Table 12.1 with the collected data for Case (c).

12.4.7 RESULTS AND DISCUSSIONS

12.4.7.1 Cube Beamsplitter

1. Report the laser beam power measurements at the input face and output faces of the cube beamsplitter.
2. Verify the results with the technical specifications provided by the manufacturer.

12.4.7.2 Rectangular Beamsplitter

1. Report the laser beam power measurements at the input face and output face of the rectangular beamsplitter.
2. Verify the results with the technical specifications provided by the manufacturer.

12.4.7.3 Glan Thompson Polarizing Beamsplitter

1. Report the laser beam power measurements at the input face and output faces of the Glan Thompson polarizing beamsplitter.
2. Verify the results with the technical specifications provided by the manufacturer.

12.4.8 CONCLUSION

Summarize the important observations and findings obtained in this lab experiment.

12.4.9 SUGGESTIONS FOR FUTURE LAB WORK

List any suggestions for improvements using different experimental equipment, procedures, and techniques for any future lab work. These suggestions should be theoretically justified and technically feasible.

12.5 LIST OF REFERENCES

List any references that were used in the report. Use one format in writing the references. Never mix reference formats in a report.

12.6 APPENDICES

List all of the materials and information that are too detailed to be included in the body of the report.

FURTHER READING

Al-Azzawi, A. and Casey, R. P., *Fiber Optics Principles and Practices*, Algonquin Publishing Centre, Ontario, 2002.

Beiser, A., *Physics*, 5th ed., Addison-Wesley, Publishing Company, Inc., USA, 1991.

Blaker, J. W. and Schaeffer, P., *Optics: An Introduction for Technicians and Technologists*, Prentice Hall, Englewood, Cliffs, NJ, 2002.

Cox, A., *Photographic Optics*, 15th ed., Focal Press, Doncaster, UK, 1974.

Cutnell, J. D. and Johnson, K. W., *Physics*, 5th ed., John Wiley & Sons, New York, USA, 2001.

Cutnell, J. D. and Johnson, K. W., *Student Study Guide: Physics*, 5th ed., Wiley, New York, 2001.

Drisoll, W. G. and William, V., *Handbook of Optics*, McGraw Hill, New York, 1978.

Edmund Industrial Optics, *Optics and Optical Instruments Catalog 2004, Edmund Industrial*, New Jersey, USA, 2004.

Eppley Laboratory, Inc., *EPLAB Catalogue 2002*, Epplay Laboratory, Inc., Newport, Rhode Island, USA, 2002.

Ewald, W. P., Young, W. A., and Roberts, R. H., *Practical Optics*, Image Makers of Pittsford, USA, 1982.

Francon, M., Optical Interferometry,, Academic Press, New York, pp. 97–99, 1966.

Ghatak, A. K., *An Introduction to Modern Optics*, McGraw-Hill, New York, 1972.

Giancoli, D. C., *Physics*, 5th ed., Prentice Hall, Englewood, Cliffs, NJ, 1998.

Halliday, D., Resnick, R., and Walker, J., *Fundamentals of Physics*, 6th ed., Wiley, New York, 1997.

Heath, R. W., Macnaughton, R. R., and Martindale, D. G., *Fundamentals of Physics*, D.C. Heath Canada Ltd., Canada, 1979.

Hecht, E., *Optics*, 4th ed., Addison-Wesley Longman, Boston, MA, 2002.

Jenkins, F. W. and White, H. E., *Fundamentals of Optics*, McGraw Hill, New York, 1957.

Keuffel & Esser Co., *Physics*. Keuffel & Esser Audiovisual Educator—Approved Diazo Transparency Masters, Audiovisual Division, Keuffel & Esser Co., USA, 1989.

Lambda, Research Optics, Inc., *Catalog 2004*. California, USA, 2004.

Lehrman, R. L., *Physics-The Easy Way*, 3rd ed., Barron's Educational Series, Inc., USA, 1998.

Lerner, R. G. and Trigg, G. L., *Encyclopedia of Physics*, 2nd ed., VCH Publishers, Inc., New York, USA, 1991.

McDermott, L. C. et al., *Introduction to Physics*, Preliminary Edition, Prentice Hall, Englewood, Cliffs, NJ, 1988.

Melles Griot. The Practical application of light, *Melles Griot catalog Melles Griot, Inc., Rochester, NY*, 2001.

Newport Corporation, Optics and Mechanics. Section, the Newport Resources 1999/2000 Catalog, Newport Corporation, Irvine, CA, USA, 1999/2000.

Nolan, P. J., *Fundamentals of College Physics*, Wm. C. Brown Publishers, Inc., Dubugue IA, USA, 1993.

Ocean Optics, Inc., *Product Catalog 2003*. Ocean Optics, Inc., Florida, USA, 2003.

Pedrotti, F. L. and Pedrotti, L. S., *Introduction to Optics*, 2nd ed., Prentice Hall, Englewood, Cliffs, NJ, 1993.

Romine, G. S., *Applied Physics Concepts Into Practice*, Prentice Hall, Englewood, Cliffs, NJ, 2001.

Sears, F. W. et al., *University Physics—Part II*, 6th ed., Addison-Wesley, Wokingham, UK, 1998.

Smith, W. J., *Modern Optical Engineering*, McGraw-Hill, New York, 1966.

Serway, R. A., *Physics for Scientists and Engineers*, 3rd ed., Saunders College Publishing, London, 1990.

Sterling, D. J. Jr., *Technician's Guide to Fiber Optics*, 2nd ed., Delmar, Publishing Inc., Albany, New York, USA, 1993.

Tippens, P. E., *Physics*, 6th ed., Glencoe McGraw-Hill, Westerville, OH, USA, 2001, 1999.

Urone, P. P., *College Physics*, Brooks/Cole publishing Company, New York, 1998.

Walker, J. S., *Physics*, Prentice Hall, Upper Saddle River New Jersey, USA, 2002.

White, H. E., *Modern College Physics*, 6th ed., Van Nostrand Reinhold Company, New York, 1972.

Wilson, J. D., *Physics—A Practical and Conceptual Approach*, Saunders College Publishing, London, 1989.

Wilson, J. D. and Buffa, A. J., *College Physics*, 5th ed., Prentice Hall, Englewood, Cliffs, NJ, 2000.

13 Light Passing through Optical Components

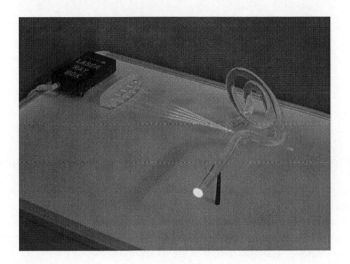

13.1 INTRODUCTION

This chapter demonstrates the behaviour of light passing through various optical components. Optical components can be lenses, mirrors, prisms, beamsplitters, glass rods and tubes, filters, polarizers, gratings, etc. These components are used in the manufacture of optical devices and systems for many applications, such as telecommunications, imaging, scanning, and microscopy. Theoretical principles of light propagation through optical components are explained in the previous chapters. This chapter presents a few cases involving different light sources passing through various optical components.

13.2 EXPERIMENTAL WORK

Students will practise aligning and observing the effects of several sources of light propagating through optical components in the following cases:

a. Laser, incandescent, or halogen light passing through various optical components from the laser optics kit
b. Laser light passing through various optical components from the ray optics laser set
c. Laser, incandescent, or halogen light passing through a glass rod and tube
d. Laser, incandescent, or halogen light passing through a spiral bar
e. Laser, incandescent, or halogen light passing through a fibre-optic cable bundle

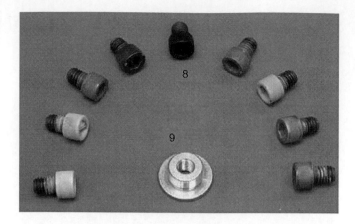

FIGURE 13.1 Laser optics component kit.

13.2.1 LIGHT PASSING THROUGH OPTICAL COMPONENTS FROM THE LASER OPTICS KIT

This case shows that the light exiting each different optical component produces a different image. The image produced is also dependent on the light source. Three types of light sources can be used in this experiment: laser, incandescent light, and halogen light. A comparison can be made between the light sources and the light images produced from the optical components. Figure 13.1 shows the laser optics components from the laser optics kit as follows:

1. Dark-blue-coloured optical component creates a spider web effect.
2. Green-coloured optical component creates a star pattern.
3. Light-blue-coloured optical component produces bright and dark area of constructive and destructive interference of the light waves.
4. Orange-coloured optical component produces an offset Fresnel pattern.
5. Red-coloured optical component creates a diverging and dispersing laser beam pattern.
6. Violet-coloured optical component creates a laser scanner effect.
7. Yellow-coloured optical component projects a curved slice of light.
8. Gold-coloured optical component displays a swirling light show when rotated.
9. Black-coloured optical component polarizes the laser light beam.

13.2.2 LASER LIGHT PASSING THROUGH OPTICAL COMPONENTS FROM THE RAY OPTICS LASER SET

The ray optics laser set contains a laser ray box, a variety of optical components, and templates to simulate actual optical devices. The optical components are lenses, reflecting surfaces, rods, and prisms. The ray optics laser set uses a laser ray box that projects five sharp parallel laser beams of 635 nm monochromatic light at 2 cm apart. The laser beams are easily visible in a normally lit room. In a moderately darkened room, it is easy to see secondary phenomena, such as front-surface and total internal reflections. The bottom surface of the laser ray box has a magnetized coating, so it can be mounted on the magnetic whiteboard when upright, as shown in Figure 13.7. The ray optics set includes thirteen large acrylic optical components with a magnetized coating on their back surfaces. The laser beams can pass through any optical components and make many combinations with two or more optical components. Additionally, the set includes six full-size template sheets, each 30×40 cm, printed on flexible white plastic with magnetized strips. These templates can be used to simulate a camera, the Galilean telescope, the Kepler telescope, the human eye and eyeglasses, spherical aberration of lenses and its correction, an angle scale for reflection and refraction, and any combination of optical components.

13.2.3 Light Passing through a Glass Rod and Tube

Imagine a glass rod that is simply a core with a large diameter, as shown in Figure 13.3. Light enters the glass rod at an angle, as shown in Figure 13.8, that is perpendicular to the entrance face. This angle can be increased from perpendicular to its maximum value that is equal to the acceptance angle. Light will propagate in the glass rod by the principles of total internal reflection when the incidence angle of the light inside the rod is less than or equal to the critical angle. When the incident angle at the glass-rod face becomes greater than the acceptance angle, most of the light will reflect away. In this case, the light cannot enter the glass rod. Therefore, no light will propagate through the glass rod. Some optical materials absorb most of the light when a small amount of light enters. As a result, the optical material glows. This principle is applicable to the glass tube and it is even more evident when the glass rod and tube are joined together. The length of the optical material has a significant effect on the amount of light absorption. Similarly, when light propagating in a fibre-optic cable is subjected to many types of light losses. Equation 13.1 is the general loss equation that can be used to calculate light loss (in dB) in optical components. The common types of light losses in optical components, devices, and systems are represented as light attenuation in optical components and fibre-optic cables theory.

$$dB = -10 \log_{10} \frac{P_{\text{out}}}{P_{\text{in}}}, \tag{13.1}$$

where P_{out} and P_{in} are the power of light at the output and input, respectively.

13.2.4 Light Passing through a Spiral Bar

Similarly, as explained in Case (a), imagine a spiral bar that is simply a core with a large diameter, as shown in Figure 13.4. Light propagates from the entrance to the exit of the spiral bar by the principle of total internal reflection, as shown in Figure 13.9. When light passes through the spiral section of the bar, some of the light escapes to the outside when the incident angle is greater than the critical angle. Therefore, part of the light is lost to the outside of the bar. Light propagation in a spiral bar is similar to that in a fibre-optic cable with a large bend.

13.2.5 Light Passing through a Fibre-Optic Cable Bundle

The principles of this case are similar to Cases (c) and (d). Figure 13.10 shows the experimental set-up for this case.

13.2.6 Technique and Apparatus

Appendix A presents the details of the devices, components, tools, and parts.

1. 2×2 ft. optical breadboard
2. HeNe laser source and power supply
3. Laser mount assembly
4. Laser light detector
5. Incandescent light source
6. Halogen light source
7. Light-source positioners

8. Laser optics component kit, as shown in Figure 13.1
9. Optical component mounting adapter, as shown in Figure 13.1
10. Optical component mount (holder/positioner assembly), as shown in Figure 13.2
11. Types of glass rods and tubes, as shown in Figure 13.3
12. Spiral bar, as shown in Figure 13.4
13. Ray optics laser set, as shown in Figure 13.7
14. Fibre-optic cable bundle, as shown in Figure 13.10
15. Hardware assembly (clamps, posts, screw kits, screwdriver kits, sundry positioners, etc.)
16. Black/white card and cardholder
17. Ruler

FIGURE 13.2 Optical component mount.

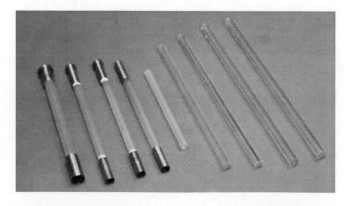

FIGURE 13.3 Types of glass rods and tubes.

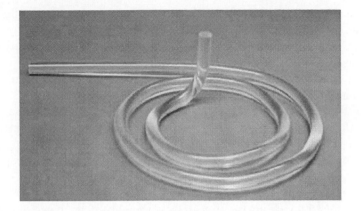

FIGURE 13.4 A spiral bar.

13.2.7 PROCEDURE

Follow the laboratory procedures and instructions given by the professor and/or instructor.

13.2.8 SAFETY PROCEDURE

Follow all safety procedures and regulations regarding the use of laser light, light source devices, optical components, and optical cleaning chemicals.

13.2.9 APPARATUS SET-UP

13.2.9.1 Light Passing through Optical Components from the Laser Optics Kit

1. Figure 13.5 shows the apparatus set-up.
2. Bolt the laser short rod to the breadboard.
3. Bolt the laser mount to the clamp using bolts from the screw kit.
4. Put the clamp on the short rod.

FIGURE 13.5 Laser light and coloured optical components apparatus set-up.

5. Place the HeNe laser into the laser mount and tighten the screw. Turn on the laser device. Follow the operation and safety procedures of the laser device in use.
6. Check the laser alignment with the line of bolt holes on the breadboard and adjust if necessary.
7. Place the optical component mount in front the laser source beam at a reasonable distance.
8. Align the laser source and the optical component mount so that the laser beam passes through the centre hole. Place the optical component adaptor into the optical component mount.
9. Inspect the dark-blue-coloured optical component under the microscope. Screw the component into the optical component mounting adapter.
10. Place the black/white card and cardholder at the output of the dark-blue-coloured component to display the image on the black/white board, as shown in Figure 13.6.
11. Turn off the laboratory light to be able to observe clear images.
12. Observe the output of the dark-blue-coloured component and report your observation in Table 13.1.
13. Unscrew the dark-blue-coloured optical component and replace it with the next coloured optical component. Report your observation in Table 13.1 for all colours.
14. Repeat steps 9–13 using incandescent light and halogen light sources with all coloured optical components. Report your observations in Table 13.1.
15. Turn on the lights of the laboratory.

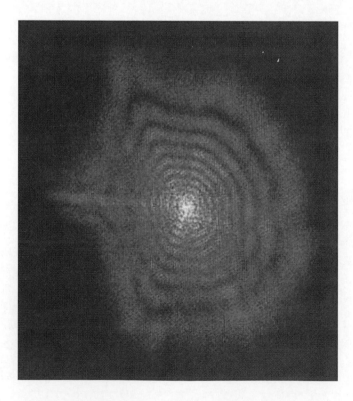

FIGURE 13.6 Blue-coloured component creates a spider-web effect.

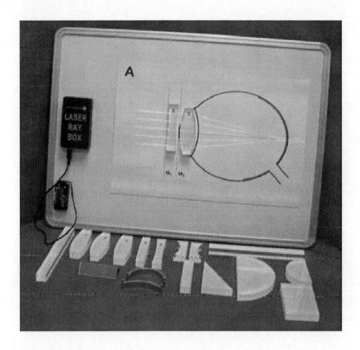

FIGURE 13.7 Ray optics laser set.

13.2.9.2 Laser Light Passing through Optical Components from the Ray Optics Laser Set

1. Figure 13.7 shows the apparatus set-up.
2. Place the whiteboard on a table using the built-in swiveling foot, which allows the whiteboard to stand vertically on the table.
3. Attach a paper to the whiteboard to trace the rays' paths.
4. Mount the laser ray box on the whiteboard. Connect the power cord of the laser ray box to the power supply.
5. Turn off the laboratory light to be able to observe clear laser rays reflecting, refracting, bending, and bouncing back and forth between the optical components.
6. Use the optical components and templates to simulate real optical devices.
7. Build three different optical devices, such as a camera, a Galilean telescope, a Kepler telescope, the human eye and eyeglass, and spherical aberration of lenses and its correction, using the setup template sheets.
8. Try to conduct three combinations between different optical components set-ups using two or more optical components.
9. Document your work in a laboratory report.

13.2.9.3 Light Passing through a Glass Rod and Tube

This experiment has three parts: measuring the light-beam output power after passing through a glass rod alone, a glass tube alone, and a glass rod and tube together. The output power of the light beam is measured when the light beam enters the glass rod and tube individually or together at different angles. Light-beam output power is measured by a light sensor that is located at a fixed distance from the end of the glass rod and tube. The output power of the light beam in these cases depends on the angle at which the glass rod, the glass tube, or the glass rod and tube are arranged. Most likely, there is an angle at which the output power is the greatest; at all other angles the output power will be less. If there is an angle at which the light beam is not passing through the glass rod and tube, the light beam is reflected back to the incoming side of the light beam. At a certain angle, the light beam is completely absorbed by the glass rod and tube. As the glass rod and tube are combined, the output power will decrease because the light is being absorbed more as the length of the glass rod and tube increases.

The following steps illustrate the experimental lab setup:

1. Figure 13.8 shows the apparatus setup using a laser light source. (Note: the glass rod and tube are placed together in this figure).
2. Bolt the laser short rod to the breadboard.
3. Bolt the laser mount to the clamp using bolts from the screw kit.
4. Put the clamp on the short rod.
5. Place the HeNe laser into the laser mount and tighten the screw. Turn on the laser device. Follow the operation and safety procedures of the laser device.
6. Check the laser alignment with the line of bolt holes and adjust if necessary.
7. Place the laser sensor in front of the laser source at a fixed distance for all cases.
8. Turn off the lights of the laboratory.
9. Measure the laser input power (P_{in}). Fill out Table 13.2.
10. Prepare a glass rod by cleaning it using a swab dampened with denatured ethanol. Follow the optical components' cleaning procedure.
11. Mount the glass rod into the slide holder/positioner.
12. Place the glass rod and glass-rod holder/positioner in front of the laser source at a fixed distance between the laser source and the laser sensor, as shown in Figure 13.8. Make certain that the glass rod is perpendicular to the laser beam. Keep the same distance for all cases.
13. Measure the laser output power (P_{out}) from the glass rod. Fill out Table 13.2.
14. Rotate the laser source 15 degrees from the normal. Rearrange the laser beam positioner to direct the laser beam at the centre of the glass rod. Measure the laser output power from the glass rod. Fill out Table 13.2.

FIGURE 13.8 Laser light loss in a glass rod and/or glass tube.

15. Repeat step 14 and rotate the laser source 15 degrees each time. Measure the laser output power. Fill out Table 13.2.
16. Repeat steps 10–15 to prepare a glass tube. Measure the laser output power at each angle. Fill out Table 13.3.
17. Repeat steps 10–15 to prepare a glass rod and glass tube together. Measure the laser output power at each angle. Fill out Table 13.4.
18. Repeat the above steps for the incandescent and halogen light sources. Table 13.2, Table 13.3, and Table 13.4.
19. Turn on the lights of the laboratory.

13.2.9.4 Light Passing through a Spiral Bar

The following steps illustrate the experimental lab setup:

1. Figure 13.9 shows the apparatus setup using an incandescent light source.
2. Get three types of light sources ready on the table.
3. Turn off the lights of the laboratory.
4. Turn on the incandescent source power. Measure the incandescent input power (P_{in}). Fill out Table 13.5.
5. Prepare a spiral bar by cleaning it using a swab dampened with denatured ethanol. Follow the optical components' cleaning procedure.
6. Mount the spiral bar into the bar holder/positioner.
7. Place the spiral bar and spiral bar holder/positioner in front of the incandescent light source at as close a distance as possible.
8. Place the light sensor at the output end of the spiral bar. Keep the distance between the end of the spiral bar and the light sensor the same for other light sources.
9. Measure the incandescent light output power (P_{out}) from the spiral bar. Fill out Table 13.5.
10. Repeat steps 4–9 using two types of light sources. Measure the light input and output powers. Fill out Table 13.5.
11. Turn on the lights of the laboratory.

FIGURE 13.9 Visible light passing through a spiral bar.

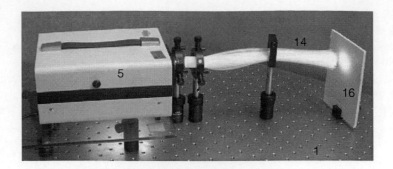

FIGURE 13.10 Visible light passing through a fibre-optic cable bundle.

13.2.9.5 Light Passing through a Fibre-Optic Cable Bundle

The following steps illustrate the experimental lab setup:

1. Figure 13.10 shows the apparatus setup using an incandescent light source.
2. Get three types of light sources ready on the table.
3. Turn off the lights of the laboratory.
4. Turn on the incandescent source power. Measure the incandescent input power (P_{in}). Fill out Table 13.6.
5. Prepare a fibre-optic cable bundle by cleaning its input and output ends using a swab dampened with denaturated ethanol. Follow the optical components' cleaning procedure.
6. Mount the fibre-optic cable bundle into the fibre-optic cable bundle holder/positioner.
7. Place the fibre-optic cable bundle and fibre-optic cable bundle holder/positioner in front of the incandescent light source at a close distance as possible.
8. Place the light sensor at the output end of the fibre-optic cable bundle. Keep the distance between the output end of the fibre-optic cable bundle and the light sensor the same for other light sources.
9. Measure the incandescent light output power (P_{out}) from the fibre-optic cable bundle. Fill out Table 13.6.
10. Repeat steps 4–9 using two types of light sources. Measure the light input and output powers. Fill out Table 13.6.
11. Turn on the lights of the laboratory.

13.2.10 DATA COLLECTION

13.2.10.1 Light Passing through Optical Components from the Laser Optics Kit

1. Observe image output of all coloured optical components using laser, incandescent, and halogen light sources. Report your observations in Table 13.1.

13.2.10.2 Laser Light Passing through Optical Components from the Ray Optics Laser Set

1. Draw the rays' paths on a paper for each case of the three combinations using two or more optical components.

TABLE 13.1
Image Comparison

	Colored Optical Components								
Light Source	**Dark Blue**	**Green**	**Light Blue**	**Orange**	**Red**	**Violet**	**Yellow**	**Gold**	**Black**
Laser Light									
Incandescent Light									
Halogen Light									

13.2.10.3 Light Passing through a Glass Rod and Tube

1. Record the input power (P_{in}) without a glass rod or tube in front of the laser light.
2. Record the output power (P_{out}) of the laser light exiting from the glass rod for every 15 degrees.
3. Repeat steps 1–2 for each of the incandescent and halogen light sources.
4. Fill out Table 13.2 for part one of Case (c).
5. Repeat the steps 1–5 for a glass rod. Fill out Table 13.3 for part two of Case (c).
6. Repeat the steps 1–5 for a glass rod and tube together. Fill out Table 13.4 for part three of Case (c).

TABLE 13.2
Light Passing through a Glass Rod and Tube

Laser Light Power Input P_{in} (unit)	
Incandescent Light Power Input P_{in} (unit)	
Halogen Light Power Input P_{in} (unit)	

	Laser Light		Incandescent Light		Halogen Light	
Angle of Glass Microscope Slide with the Normal (degrees)	**Power Output** P_{out} **(unit)**	**Loss (dB)**	**Power Output** P_{out} **(unit)**	**Loss (dB)**	**Power Output** P_{out} **(unit)**	**Loss (dB)**
0						
15						
30						
45						
60						
75						
90						

TABLE 13.3
Light Passing through a Glass Tube

Laser Light Power Input P_{in} (unit)						
Incandescent Light Power Input P_{in} **(unit)**						
Halogen Light Power Input P_{in} **(unit)**						
	Laser Light		**Incandescent Light**		**Halogen Light**	
Angle of Glass Microscope Slide with the Normal (degrees)	**Power Output P_{out} (unit)**	**Loss (dB)**	**Power Output P_{out} (unit)**	**Loss (dB)**	**Power Output P_{out} (unit)**	**Loss (dB)**
0						
15						
30						
45						
60						
75						
90						

TABLE 13.4
Light Passing through the Glass Rod and Tube Together

Laser Light Power Input P_{in} **(unit)**						
Incandescent Light Power Input P_{in} **(unit)**						
Halogen Light Power Input P_{in} **(unit)**						
	Laser Light		**Incandescent Light**		**Halogen Light**	
Angle of Glass Microscope Slide with the Normal (degrees)	**Power Output P_{out} (unit)**	**Loss (dB)**	**Power Output P_{out} (unit)**	**Loss (dB)**	**Power Output P_{out} (unit)**	**Loss (dB)**
0						
15						
30						
45						
60						
75						
90						

13.2.10.4 Light Passing through a Spiral Bar

1. Record the input power (P_{in}) without a spiral bar in front of the incandescent light source.
2. Record the output power (P_{out}) of the incandescent light exiting from the spiral bar.
3. Repeat steps 1–2 for each of the laser and halogen light sources.
4. Fill out Table 13.5.

TABLE 13.5
Light Passing through a Spiral Bar

Laser Light Power Input (unit)	P_{in}	
Incadescent Light Power Input (unit)	P_{in}	
Halogen Light Power Input (unit)	P_{in}	

Laser Light		Incandescent Light		Halogen Light	
Power Output P_{out} (unit)	Loss (dB)	Power Output P_{out} (unit)	Loss (dB)	Power Output P_{out} (unit)	Loss (dB)

13.2.10.5 Light Passing through a Fibre-Optic Cable Bundle

1. Record the input power (P_{in}) without a fibre-optic cable bundle in front of the incandescent light source.

TABLE 13.6
Light Passing through a Fibre-Optic Cable Bundle

Laser Light Power Input (unit)	P_{in}	
Incandescent Light Power Input (unit)	P_{in}	
Halogen Light Power Input (unit)	P_{in}	

Laser Light		Incandescent Light		Halogen Light	
Power Output P_{out} (unit)	Loss (dB)	Power Output P_{out} (unit)	Loss (dB)	Power Output P_{out} (unit)	Loss (dB)

2. Record the output power (P_{out}) of the incandescent light exiting from the fibre-optic cable bundle.
3. Repeat steps 1–2 for each of the laser and halogen light sources.
4. Fill out Table 13.6.

13.2.11 Calculations and Analysis

13.2.11.1 Light Passing through Optical Components from the Laser Optics Kit

No calculations and analysis are required in this case.

13.2.11.2 Laser Light Passing through Optical Components from the Ray Optics Laser Set

No calculations and analysis are required in this case.

13.2.11.3 Light Passing through a Glass Rod and Tube

1. Calculate the power loss (dB) of the laser, incandescent, and halogen light sources when they are passing through a glass rod, a tube, or both, using Equation 13.1. Fill out Table 13.2, Table 13.3, and Table 13.4.
2. Compare power loss (dB) of the three types of light sources among the three parts involved in this case.

13.2.11.4 Light Passing through a Spiral Bar

1. Calculate the power loss (dB) of the laser, incandescent, and halogen light sources when they are passing through a spiral bar using Equation 13.1. Fill out Table 13.5.
2. Compare power loss (dB) of the three types of light sources involved in this case.

13.2.11.5 Light Passing through a Fibre-Optic Cable Bundle

1. Calculate the power loss (dB) of the laser, incandescent, and halogen light sources when they are passing through a fibre-optic cable bundle using Equation 13.1. Fill out Table 13.6.
2. Compare power loss (dB) of the three types of light sources involved in this case.

13.2.12 Results and Discussions

13.2.12.1 Light Passing through Optical Components from the Laser Optics Kit

1. Discuss your observations when using the three types of light sources with all the coloured optical components.
2. Discuss the outputs of each coloured optical component when using different light sources.

13.2.12.2 Laser Light Passing through Optical Components from the Ray Optics Laser Set

1. Discuss your observations for each of the three combinations using two or more optical components.
2. Describe the reflections, refractions, and intersections of the rays that occurred through the optical components.

3. Discuss the combinations of the optical components that were used in the laboratory work.

13.2.12.3 Light Passing through a Glass Rod and Tube

1. Report the power loss (dB) of the light sources.
2. Report the incident angle of the light where no power output is recorded.
3. Compare and discuss the power loss in each component.

13.2.12.4 Light Passing through a Spiral Bar

1. Report the power loss (dB) of the light sources.
2. Compare and discuss the power loss for each light source.

13.2.12.5 Light Passing through a Fibre-Optic Cable Bundle

1. Report the power loss (dB) of the light sources.
2. Compare and discuss the power loss for each light source.

13.2.13 CONCLUSION

Summarize the important observations and findings obtained in this laboratory experiment.

13.2.14 SUGGESTIONS FOR FUTURE LAB WORK

List any suggestions for improvements using different experimental equipment, procedures, and techniques for any future laboratory work. These suggestions should be theoretically justified and technically feasible.

13.3 LIST OF REFERENCES

List any references that were used in the report. Use one format in writing the references. Never mix reference formats in a report.

13.4 APPENDIX

List all of the materials and information that are too detailed to be included in the body of the report.

FURTHER READING

Beiser, A., *Physics*, 5th ed., Addison-Wesley, Reading, MA, 1991.

Blaker, W. J., *Optics: The Matrix Theory*, Marcel Dekker, New York, 1972.

Blaker, J. W. and Schaeffer, P., *Optics an Introduction for Technicians and Technologists*, Prentice-Hall, Upper Saddle River, NJ, 2000.

Chen, K. P., In-fiber light powers active fiber optical components, *Photonics Spectra*, April, 78–90, 2005.

Cutnell, J. D. and Johnson, K. W., *Physics*, 5th ed., Wiley, New York, 2001.

Dutton, H. J. R., *Understanding Optical Communications*, IBM/Prentice-Hall, Englewood Cliffs, NJ/Research Triangle Park, NC, 1998.

Edmund Industrial Optics, *Optics and Optical Instruments Catalog*, 2004.

EPLAB, *EPLAB Catalogue 2002*, The Eppley Laboratory, Inc., 12 Sheffield Avenue, P.O. Box 419, Newport, RI 02840, USA, 2002.

Ghatak, A. K., *An Introduction to Modern Optics*, McGraw-Hill, New York, 1972.

Halliday, D., Resnick, R., and Walker, J., *Fundamental of Physics*, 6th ed., Wiley, New York, 1997.

Hecht, J., *Understanding Fiber Optics*, 3rd ed., Prentice-Hall, Upper Saddle River, NJ, 1999.

11. Hecht, E., *Optics*, 4th ed., Addison-Wesley/Longman, Reading, MA, 2002.

Hewitt, P. G., *Conceptual Physics*, 8th ed., Addison-Wesley, Reading, MA, 1998.

Jenkins, F. W. and White, H. E., *Fundamentals of Optics*, McGraw-Hill, New York, 1957.

Jones, E. and Childers, R., *Contemporary College Physics*, McGraw-Hill Higher Education, Boston, MA, 2001.

Kao, C. K., *Optical Fiber Systems: Technology, Design and Applications*, McGraw-Hill, New York, 1982.

Kolimbiris, H., *Fiber Optics Communications*, Prentice-Hall, Upper Saddle River, NJ, 2004.

Naess, R. O., *Optics for Technology Students*, Prentice-Hall, Upper Saddle River, NJ, 2001.

Newport Corporation, *Projects in Fiber Optics Applications Handbook*, Newport Corporation, 1986.

Newport Corporation, *Optics and Mechanics Section, the* Newport Resources 1999/2000 Catalog, Newport Corporation Irvine, CA, 1999/2000.

Nolan, P. J., *Fundamentals of College Physics*, Wm. C. Brown Publishers, Inc., Dubuque, IA, 1993.

Okamoto, K., *Fundamentals of Optical Waveguides*, Academic Press, San Diego, 2000.

Pedrotti, F. L. and Pedrotti, L. S., *Introduction to Optics*, 2nd ed., Prentice-Hall, Englewood Cliffs, NJ, 1993.

Pritchard, D. C., *Environmental Physics: Lighting*, Longmans, Green, London, 1969.

Salah, B. E. A. and Teich, M. C., *Fundamentals of Photonics*, Wiley, New York, 1991.

Serway, R. A. and Jewett, J. W., *Physics for Scientists and Engineers with Modern Physics*, 6th ed., volume 2, Thomson Books/Cole, USA, 2004.

Shamir, J., *Optical Systems and Processes*, SPIE Optical Engineering Press, Bellingham, WA, 1999.

Sterling, D. J. Jr., *Technician's Guide to Fiber Optics*, 2nd ed., Delmar Publishers, Albany, NY, 1993.

Tippens, P. E., *Physics*, 6th ed., Glencoe/McGraw-Hill, Westerville, OH, 2001.

White, H. E., *Modern College Physics*, 6th ed., Van Nostrand Reinhold Company, New York, 1972.

Woods, N., *Instruction's Manual to Beiser Physics*, 5th ed., Addison-Wesley, Reading, MA, 1991.

Yeh, C., *Handbook of Fiber Optics: Theory and Applications*, Academic Press, San Diego, 1990.

14 Optical Instruments for Viewing Applications

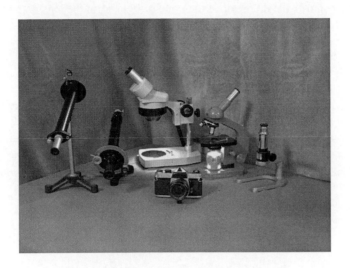

14.1 INTRODUCTION

The previous chapters presented the principles of image formation by optical components, such as mirrors and lenses. This chapter presents the use of these optical components in building common optical instruments. Such optical instruments include the eye, camera, projector, microscope, telescope, and binoculars, which are used for different vision enhancement applications. These optical instruments help people to perform ordinary and sophisticated tasks associated with vision, magnification, and image formation. The ray diagram method is used in image formation in these devices. Image formation by the mirrors, lenses, lenses combinations, and prisms is discussed in detail in the previous chapters. There are many optical devices and instruments using basic optical components that are covered in other chapters.

This chapter includes an experiment involving the operation, image formation, functionality, disassembly, and reassembly of the optical components of an optical viewing instrument.

14.2 OPTICAL INSTRUMENTS

There is a wide variety of optical devices, instruments, and systems that use mirrors and lenses, for example: cameras, microscopes, projectors, and telescopes. There are also optical instruments that use prisms, gratings, and fibre optics, along with one type of lens or mirror. These instruments will be described later in the optics and optical fibre sections of this book. Such optical components are commonly used in manufacturing optical fibre devices, which are used in building communication systems, medical instruments, scanning and imaging processors, optical spectrum analysers, etc.

14.3 THE CAMERA

Cameras are among the most common of optical instruments. The camera shown in Figure 14.1 is a simple optical instrument. Figure 14.2 illustrates a cross-sectional view of a simple camera. The basic elements of a camera are: a light-tight box, a converging lens, an aperture, a shutter, and a light sensitive film. The converging lens produces a real and an inverted image of the object being photographed. The shutter allows the light from the lens to strike the film for a prescribed length of time. The aperture controls the diameter of the cone of light from the lens onto the film. The light-sensitive film records the image, which is formed by the lens.

Focusing of the image is accomplished by varying the distance between lens and film using a mechanical arrangement in most cameras. When the camera is in proper focus, the position of the film coincides with the position of the real image formed by the lens, as shown in Figure 14.2. The resulting photograph then will be as sharp as possible. When using a converging lens, the image distance increases as the object distance decreases. Hence, to focus the camera, the lens is moved closer to the film for a distant object and farther from the film for a nearby object. This is achieved by rotating the threaded ring, which holds the lens. This rotation controls the linear movement of the lens. The shutter, located behind the lens, is a mechanical device that is opened for selected time intervals. With this arrangement, one can photograph moving objects by using short exposure times, or dark scenes by using long exposure times.

FIGURE 14.1 Camera.

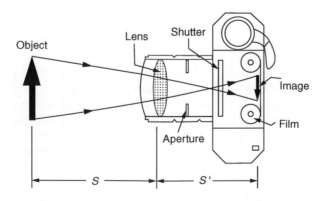

FIGURE 14.2 Cross-sectional view of a simple camera.

The major cause of blurred images is the movement of the camera or the object while the shutter is open. To avoid blurred images, a short time exposure and/or a tripod can be used, even for stationary objects. Most expensive cameras have compound lenses to correct the aberration that normally occurs in lenses. Most cameras have an aperture of adjustable diameter (either behind or in between the lenses) to provide further control of the intensity of the light reaching the film. When an aperture of small diameter is used, only light from the central portion of the lens reaches the film and so the aberration is reduced to a minimum.

The choice of the focal length f for a camera lens depends on the film size and desired angle of view. A lens of long focal length captures a small angle of view and produces a large image of a distant object; such a lens is called a telephoto lens. A lens of short focal length captures a small image and produces a wide angle of view; such a lens is called a wide-angle lens. Again, the focal length of a lens is defined as the distance from the lens to the image, when the object is infinitely far way from the lens. In general, for any distant object, using a lens of longer focal length produces a small image distance. This also increases the height of the image. In other words, the lens produces a magnified image of the object. As was discussed in the chapters on spherical mirrors and lenses the lateral magnification is defined as the ratio of the image height h' to the object height h. The lateral magnification is also equal to the ratio of the image distance S to the object distance S', as given in the following equation:

$$M = \frac{h'}{h} = -\frac{S'}{S} \tag{14.1}$$

When using a lens of short focal length, the lateral magnification is small, and a distant object produces only a small image. When a lens with a long focal length is used, the image of this same object may entirely cover the area of the film.

The intensity of light reaching the film is proportional to the area viewed by the camera lens and the effective area of the lens. The light intensity I will be proportional to the area of the lens. The area of the lens is proportional to the square of the angle of view of the lens so it also is nearly proportional to $1/f^2$. The effective area of the lens is controlled by means of an adjustable lens aperture, or diaphragm, a nearly circular hole with variable diameter D. The effective area is proportional to D^2. The intensity of light reaching the film is proportional to f^2/D^2. Thus, it follows that the light intensity is also proportional to $1/f^2$. The ratio f/D is defined as the f-number of the lens and given by the following equation:

$$f\text{-number} = \frac{\text{Focal Length}}{\text{Aperture Diameter}} = \frac{f}{D} \tag{14.2}$$

For a lens with a variable-diameter aperture, increasing the diameter by a factor of $\sqrt{2}$ changes the f-number by $1/\sqrt{2}$ and increases the intensity at the film by a factor of 2. Adjustable apertures usually have scales labeled with successive numbers related by factors of $\sqrt{2}$, such as: $f/2, f/2.8, f/4, f/5.6, f/8, f/11,$ and $f/16$.

A useful consequence of the f-number method of classifying lens openings is that for the same shutter speed, lenses of different focal lengths give proper exposure at the same f-numbers. This applies independent of the focal length for the lens in place. This result is due to two compensating factors. First, the amount of the light that passes through the aperture is proportional to its area and thus to the square of its diameter, D^2. Second, the light per unit area that reaches the film depends inversely on the area of the image.

For the usual situation in which the object distance is large compared with the focal length, the magnification is proportional to the focal length f of the lens, so the area of the image is proportional to f^2. The rate at which a photographic image is formed, or the speed of the lens, is then given by the following equation:

$$\text{Speed of Lens} \propto \frac{1}{(f/D)^2} \propto \frac{1}{(f\text{-number})^2} \tag{14.3}$$

The f-number is a measure of the light intensity and the speed of the lens. A fast lens has a small f-number. Fast lenses, with an f-number as low as about 1.4, are more expensive because it is more difficult to keep aberrations acceptably small. Cameras lenses often are marked with various f-numbers, such as: $f/2$, $f/2.8$, $f/4$, $f/5.6$, $f/8$, $f/11$, and $f/16$. The various f-numbers are obtained by adjusting the aperture, which effectively changes D. The smallest f-number corresponds to the case where the aperture is wide open and the full lens area is in use.

14.4 THE EYE

Vision is perhaps the most important sense of living beings. The eye of living creatures is one of the most familiar optical instruments. It is also one of the most complex parts of a living creature. The optical behaviour of the eye is similar to the optical principle of a simple camera. The essential external and internal parts of the human eye are shown in Figure 14.3 and Figure 14.4, respectively. The eye is nearly spherical in shape, and its diameter is typically less than 3 cm. There are a number of similarities between a human eyeball and a simple camera. The eye is basically a light-tight box, whose is outer walls are formed by the hard white sclera. There is a two-element lens system, consisting of the outer cornea and the inner crystalline lens or eyelens. The eyelens is acting as a converging lens. The cornea is a curved transparent tissue. The retina is a light sensitive lining, which corresponds to the film in the camera. The lens system forms an inverted image on the retina at the back of the eyeball, as shown in Figure 14.4. The coloured iris corresponds to the diaphragm in a camera. Its pupil, the dark and circular hole through which the light enters, corresponds to the camera aperture. The iris can change the pupil diameter, which controls the amount of light entering the eye. The eyelid is like a lens cover, protecting the cornea from dirt and objects. The eye or tear fluid cleans and moistens the cornea and can be compared to a lens cloth or brush. The volume of the eye is not empty, like the eye of the camera, but instead is filled with two transparent jellylike liquids, the vitreous humour and aqueous humour, which provide nourishment to the eyelens and

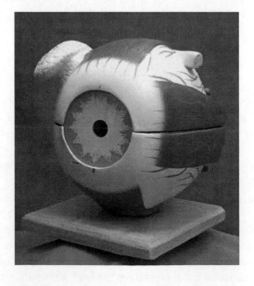

FIGURE 14.3 External parts of the human eye.

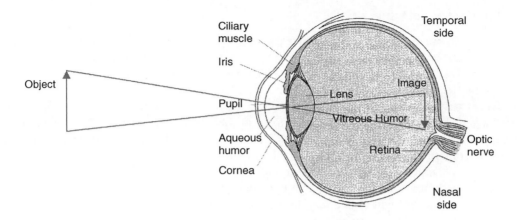

FIGURE 14.4 Cross section of the human eye.

cornea. Additionally, their internal pressure helps to hold the shape of the eyeball, which is necessary for stable vision. The lack of blood vessels in the cornea and eyelens means that, when surgically transplanted, these elements are less likely to be rejected by the immune system, which resides in the blood. Nowadays, corneal transplants are performed routinely.

The eye focuses on an object by varying the shape of the flexible crystalline lens through an amazing process called accommodation. An important component in accommodation is the ciliary muscle, which is attached to the lens. When the eye is focused on distant objects, the ciliary muscle is relaxed. For an object distance of infinity, the focal length of the eye (the distance between the lens and the retina) is about 1.7 cm. The eye focuses on nearby objects by tensing the ciliary muscle. This action effectively decreases the focal length of the eyelens by slightly decreasing the radius of the curvature of the lens, which allows the image to be focused on the retina. This lens adjustment takes place inside the eye automatically.

The retina of the eye is composed of two types of photosensitive cells, called rods and cones numbering about a hundred million. These cells are packed together like the chambers of a honeycomb. Near the centre of the retina, in a small region called the fovea, there are a great number of cones but no rods. This region is responsible for most precise vision and also for colour vision. Away from the fovea, the rods predominate. At the extremes of the retina, corresponding to peripheral vision, there are only a few cones. The more numerous rods have a greater sensitivity to light and can distinguish between low light intensities for twilight (black and white) vision. The cones respond selectively to certain colours of light, some to one colour and others to other colours. Cones are considerably less sensitive to light than are rods. This is why humans cannot see colour in very dim light. The rods and cones of the retina are connected to optic nerve fibres, from which the light stimulates signals to the brain.

In the region where the optic nerve enters the eyeball, there are no rods and cones. As a result, the human eye has a blind spot for which there is no optical response. Binocular (two-eyed) vision compensates for the blind spots. Binocular vision also accounts for some depth perception, which depends on slight differences between the shapes and positions of the images on the retina of the two eyes. This occurs because the eyes are set several centimetres apart and get slightly different views of objects.

The extremes of the range over which eye vision is possible are known as the far point and near point. The far point is the farthest distance that the normal eye can see clearly. The far point is assumed to be infinity. The near point is the position closest to the eye at which objects can be seen clearly. The near point depends on the extent that the eyelens can be deformed by accommodation. The range of accommodation decreases with age as the eyelens loses its elasticity.

14.4.1 Defects of Vision

Figure 14.5(a) shows the image formation in a normal eye. As mentioned above, the focusing of the image is controlled by the ciliary muscle, which is attached to the lens. When the eye is focused on distant objects, the ciliary muscle is relaxed. The eye focuses on nearby objects by tensing the ciliary muscle.

Three common defects of vision are myopia (nearsightedness), hyperopia (farsightedness), and astigmatism. In myopia, the eyeball is too long, or the cornea is too curved. Light from a very distant object comes to a focus in front of the retina, as shown in Figure 14.5(b). This occurs even though the ciliary muscle is completely tensed. Accommodation permits nearby objects to be seen clearly but not more distant ones. A diverging lens of the proper focal length can correct this condition, as shown in Figure 14.5(b). In hyperopia, the eyeball is too short or the cornea has insufficient curvature. Light from a very distant object does not come to a focus within the eyeball, even when the ciliary muscle is relaxed. Its power of accommodation permits a hyperopic eye to focus on distance objects, but the range of accommodation is not enough for nearby objects to be seen clearly. In this condition, light from a distant object is focused behind the retina. The correction for hyperopia is a converging lens, as shown in Figure 14.5(c).

Another common eye defect, astigmatism, occurs when the cornea is not spherical but is more curved in one plane then in another. As a result, the focal length of the astigmatism eye is different in one plane than in the perpendicular plane. When light rays that lie in one plane are in focus on the

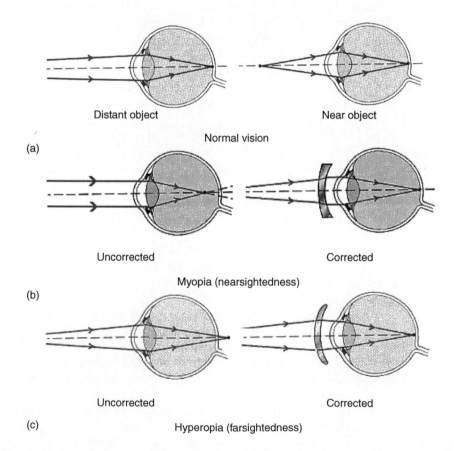

FIGURE 14.5 Common defects of vision. (a) Normal vision, (b) Myopia (nearsightedness), (c) Hyperopia (farsightedness).

retina of an astigmatism eye, those in the perpendicular plane will be in focus either in front or in back of the retina. Astigmatism is corrected by a lens that converges (or diverges) rays in one plane, while not effecting rays in the perpendicular plane. Such a lens is a cylindrical lens, which is curved in one direction but not in the perpendicular direction, as if cut out of a cylinder of glass. More modern techniques, such as laser surgery, are being used to correct common vision defects.

The eye is also subject to several diseases. One disease, which usually occurs in old age, is the formation of cataracts occurring when the lens becomes partially or totally opaque. The common remedy for cataracts is surgery performed on the lens. Another disease, called glaucoma, arises from an abnormal increase in the fluid pressure inside the eyeball. The pressure increase can lead to a swelling of the lens and also to strong myopia. A chronic form of glaucoma can lead to blindness. If the disease is discovered at an early stage, it can be treated with medicine or surgery.

14.4.2 COLOUR VISION

Colour vision is actually a physiological sensation of the brain in response to the light excitation of the cones reception in the retina. The cones in human eye are sensitive to light with wavelengths between about 400–700 nm (7.5×10^{14}–4.3×10^{14} Hz frequencies). Different wavelengths of light are perceived by the brain as having different colours. The association of colours with particular light wavelengths is subjective. Monochromatic light has a particular wavelength, but perception of its colour may vary from one person to another.

14.5 THE MAGNIFYING GLASS

Many optical devices are used to produce magnified images of objects. The simplest optical magnifying device is the magnifying lens, or simple microscope with an objective lens, as shown in Figure 14.6. The magnifying glass is a single converging lens. The lens held near the eye produces an image whose projected size on the retina is larger than the real size of the object (when compared to unaided eye). Maximum magnification of the object can be found by adjusting the distance between the lens and object. Varying this distance, one can find a clear image. The magnifying glass is considered a passive optical device. The magnifying glass is not only used to enlarge a fine object; it is also used in lens compensation, which is used in a compound microscope. In this arrangement, the magnifying glass is called the eyepiece, and the other lens near to an object is called the objective lens. The compound microscope instrument will be explained in detail in the next section. The primary function of magnifying glasses is to increase the angular size of the image while viewing with a relaxed eye.

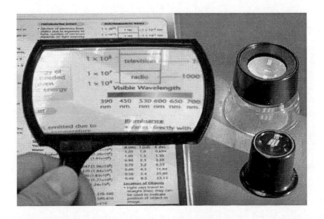

FIGURE 14.6 Simple magnifying lenses.

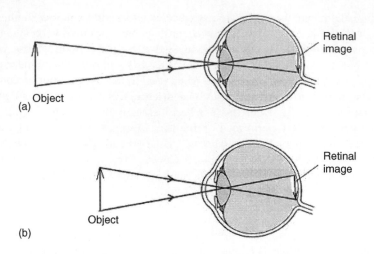

FIGURE 14.7 (a) The object placed a distance from the eye. (b) The object placed closer to the eye.

The apparent size of an object is determined by the size of its image on the retina. To an unaided eye, an object placed a distance from the eye creates an image covering part of the retina, as shown in Figure 14.7(a). If the object comes closer to the eye, its image covers a larger part of the retina, as shown in Figure 14.7(b). The greater the angle viewed, the larger the image appears.

In Figure 14.8(a) the object is located at the near point, where it subtends an angle θ at the eye. In Figure 14.8(b), a converging lens is used to form a virtual image that is larger and farther from

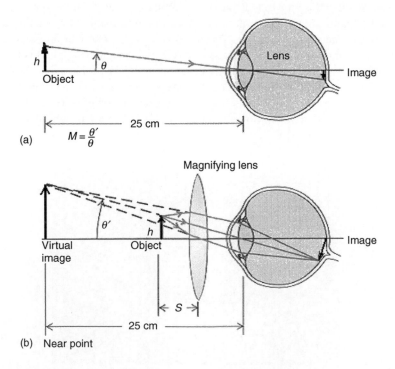

FIGURE 14.8 (a) An object placed at the near point of an unaided eye. (b) A magnifying glass placed close to the eye.

the eye than the object. The image formation by converging lenses is widely explained in the lenses chapter. A lens used to enlarge an object is called a simple magnifier, or sometimes is called a magnifying glass. In Figure 14.8(b) a magnifier in front of the eye forms an image at the near point, with the angle θ' subtended at the magnifier. The magnification power of the magnifier is defined by the ratio of the angle θ' (viewed with the magnifier) to the angle θ (viewed without the magnifier). This ratio is called the angular magnification M_a, as defined by:

$$M_a = \frac{\theta'}{\theta} \qquad (14.4)$$

Notice here the difference between the definition of angular magnification M_a and the lateral magnification M. The angular magnification M_a is defined in Equation (14.4). The lateral magnification M is defined as the ratio of the height of an image to the height of the corresponding object. Lenses chapter presents a detailed definition of lateral magnification.

To find the value of the angular magnification, consider angles small enough that each angle measured in radians is equal to its sine and its tangent values. Figure 14.8(a) shows an object of height h placed at the standard near the point of 25 cm from the eye, where it subtends at an angle of θ. In Figure 14.8(b) the object is moved close to the eye.

If the angle θ is small, it can be defined as:

$$\tan \theta \cong \theta = \frac{h}{25} \qquad (14.5)$$

where both the near point and the object height h are expressed in centimetres.

Similarly, Figure 14.8(b) shows the angle θ' subtended by the virtual image is approximately (ignoring the small distance between the magnifier lens and the eye) as:

$$\tan \theta' \cong \theta' = \frac{h}{S} \qquad (14.6)$$

To calculate the object distance S, using the thin lens equation with focal length f and virtual image at a distance of -25 cm, obtains S by:

$$\frac{1}{S} + \frac{1}{-25} = \frac{1}{f} \qquad (14.7)$$

$$S = \frac{25f}{(25 + f)} \qquad (14.8)$$

Substituting Equation (14.8) for S in Equation (14.6), from Equation (14.4) gives:

$$M_a = \frac{\theta'}{\theta} = \frac{25}{25f/(25 + f)} \qquad (14.9)$$

which simplifies for the case of an image at the near point to:

$$M_a = 1 + \frac{25}{f} \qquad (14.10)$$

If the object is held at or just inside the focal point of the lens, the image forms very far away, essentially at infinity, rather than at the near point. This far point corresponds to the most comfortable viewing distance because the eye is relaxed. In this case, $\theta' = h/f$, and Equation (14.6) can be rewritten so the angular magnification M_a for image at infinity is given by:

$$M_a = \frac{25}{f} \tag{14.11}$$

From Equation (14.11), a maximum angular magnification power can be obtained by using lenses with short focal lengths. In fact, the lens aberrations limit the practical range of a single magnifying glass to about 3X and 4X, or a sharp image magnification of three or four times the size of the object when used normally. The manufacturers of the magnifying glasses commonly use Equation (14.11) to specify the magnification power of a magnifying glass.

14.6 THE COMPOUND MICROSCOPE

In many magnification applications, a system using two or more lenses is better than a single lens. For example, an ordinary magnifying glass cannot produce sharp images enlarged more than 3X and 4X. However, two converging lenses combined together as a microscope can produce sharp images magnified hundreds of times. Such an optical instrument, called a compound microscope, is shown in Figure 14.9.

Figure 14.10 shows image formation in a compound microscope explained by the ray method. The objective lens is a converging lens of short focal length that forms a real and enlarged image of the object. An object to be viewed is placed at O just beyond the focal length of the objective lens. The objective lens forms a real and enlarged image at I_1. This image is enlarged further by the eyepiece lens, which acts as a simple magnifier to form an enlarged virtual image at I_2. In this lens combination, the first image at I_1 produced by the objective acts as an object of the eyepiece, and the second image at I_2 has been magnified twice. As shown in Figure 14.10, in a properly designed instrument the first image at I_1 lies just inside the focal length of the eyepiece. The position of the second image at I_2 may be anywhere between the near and far points of the eye.

The total angular magnification M_{total} of a lens combination in a microscope is the product of magnifications produced by two lenses. The first magnification factor is the lateral magnification M_1 of the objective; M_1 determines the linear size of the real image I_1. The magnification factor M_2 is the angular magnification of the eyepiece. M_2 relates the angular size of the virtual image I_2 (seen through the eyepiece) to the angular size that the real image I_1 would have (if viewed without the eyepiece). The lateral magnification factor of the objective is given by:

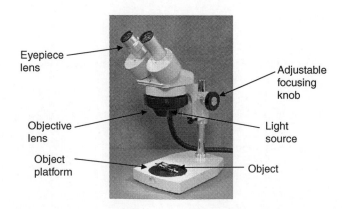

FIGURE 14.9 Compound microscope.

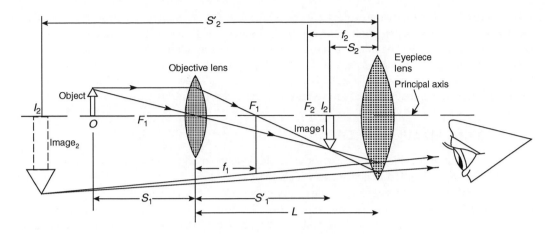

FIGURE 14.10 Image formation by a compound microscope.

$$M_1 = -\frac{S_1'}{S_1} \tag{14.12}$$

where S_1' and S_1 are the object and image distances, respectively, for the objective lens.

Under normal operation, the object location is adjusted to be very close to the focal point f_1 of the objective lens. In this case, the image distance S_1' is very great in comparison to the focal length f_1 of the objective lens. Thus S_1 is approximately equal to f_1, and the lateral magnification M_1 of the objective lens can be written as:

$$M_1 = -\frac{S_1'}{f_1} \tag{14.13}$$

Under normal operation, the eyepiece is adjusted so that the real image is just inside the focal point F_2 of the eyepiece lens. The eyepiece angular magnification M_2, can be written as:

$$M_2 = \frac{25}{f_2} \tag{14.14}$$

where f_2 is the focal length of the eyepiece.

Again, the total angular magnification M_{total} of a lens combination in a microscope is the product of magnifications produced by two lenses.

$$M_{\text{total}} = M_1 \times M_2 = -\frac{25S_1'}{f_1 f_2} \tag{14.15}$$

where S_1', f_1, and f_2 are measured in centimetres. The second image is inverted with respect to the object.

Microscope manufacturers usually specify the values of the magnification powers M_1 and M_2 rather than the focal lengths of the objective and eyepiece lenses, respectively. These magnification values assume that the object and first image I_1 is near to the focal point of the objective and eyepiece, respectively. This is the normal operation of the microscope. Interchangeable eyepieces with different magnification power from about 5X to over 100X are available. Most microscopes often are equipped with rotating turrets, which usually contain three or more objectives for different focal lengths, so that the same object can be viewed at different magnification powers. These objectives and eyepieces can be used in various combinations to provide common magnifications from 50X to about 900X. The maximum magnification power available for compound microscopes is about 2000X. All microscopes are equipped with a light source to illuminate the object with the required light intensity to view focused images. There are advanced compound microscopes that produce higher magnifications that are required by researchers and investigators.

14.7 ADVANCED MICROSCOPES

Many types of advanced microscopes are now available in the market. Most of them are operated by a laser source and controlled by software, which analyses and displays the data captured by a set of optical components. These microscopes are very expensive and sophisticated but also easy to use. They are used for viewing microscopic materials and particles, such as human cells, DNA, micro-organisms, and biological and medical test samples. Some of the highly sophisticated advanced microscopes are used in research and development activities.

Figure 14.11 shows the main parts and attachments of a confocal scanning laser microscope (CSLM). Figure 14.12 illustrates the basic components and operation of the CSLM.

The beam of approximately 1 mm in diameter emerges from the HeNe laser source mounted under the optical plate and then is deflected by a fixed mirror upwards towards a mirror, mounted on a kinematic mount. The beam is deflected horizontally from left-to-right towards a 15X beam expander. A beam 5 mm in diameter emerges at the output of the beam expander. The 15X beam expander consists of two positive lenses of 10 mm and 150 mm focal lengths, respectively. The beam then is defected by a mirror on a kinematic mount towards the front of the instrument. A broadband (range 400–700 nm) beamsplitter (50% reflection and 50% transmission) deflects the beam to the left towards an XY set of galvanometric mirrors. The beam is deflected by the galvo scanners to the front towards a unitary (1X) beam expander that consists of two identical 100 mm focal length positive lenses. The unitary telescope is designed to include a mirror at 45 degrees on a fixed mount between the two lenses. The beam passes through the first lens and is deflected by the mirror vertically downwards through the second lens of the unitary beam expander into the objective lens. The unitary telescope's function is to transfer a stationary beam at the galvos, as it is

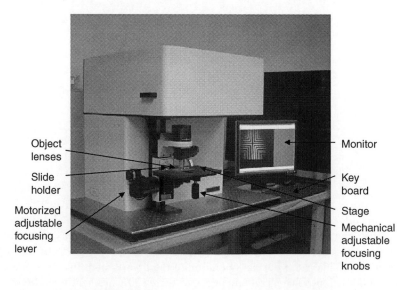

FIGURE 14.11 Confocal microscope.

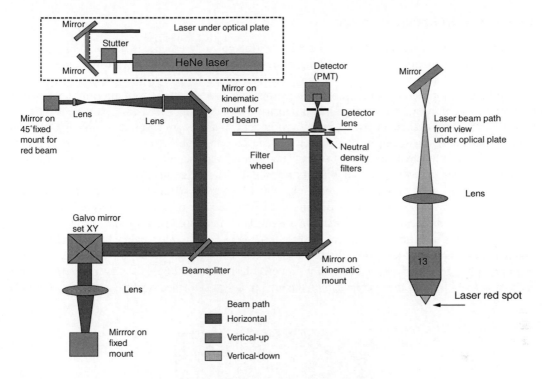

FIGURE 14.12 Basic components and operation of a confocal microscope.

scanned along the *x* and *y* directions, into a stationary beam at the entrance pupil of the objective lens. The objective lens focuses the laser beam onto a sample mounted on a motorized stage.

Light from the sample is reflected back, collected by the objective lens, passes through the unitary beam expander, and de-scanned by the XY galvo mirror set towards the beam splitter. The reflected light beam that passes through the beamsplitter is deflected by a mirror on a kinematic mount towards a neutral density (ND) filter mounted on a motorized filter wheel. The reflected light that passes through the ND filter is focused by the detector lens (100 mm focal length) onto a pinhole. Light that passes through the pinhole is incident on a detector type called a photomultiplier tube (PMT) and is measured. The PMT is a current generating device.

A current-to-voltage preamplifier converts the current to voltage and amplifies it. The voltage is transferred from the microscope to the computer by a coaxial cable where it is converted into a 16-bit digital number, by an analogue-to-Digital converter board that resides at one of the PCI slots of the host computer, and is displayed on the screen.

14.8 THE TELESCOPE

The optical instrument known as the telescope applies the principles of lenses and mirrors. The telescope is designed to aid in viewing distant objects, such as the planets and stars in the solar system. Basically, there are two types of telescopes: the refracting telescope, which uses a combination of lenses to form an image, and the reflecting telescope, which uses a curved mirror and a lens to form an image. The refracting telescope depends on the converging and reflecting of light by lens combinations. The reflecting telescope depends on the converging and reflecting of light, by lens and mirror combinations.

14.8.1 The Refracting Telescope

A simple refracting telescope, sometimes called astronomical telescope, is shown in Figure 14.13. The refracting telescope consists of a long tube with an objective lens toward the object and an adjustable eyepiece lens toward the viewer. The final image formed by this telescope is inverted, but the inverted image poses no problem in astronomical work. Image formation in the telescope is similar to a compound microscope. In both instruments, the image formed by an objective lens is viewed through an eyepiece lens. As in a microscope, two lenses are involved, with an eyepiece to enlarge the image produced by the objective. The telescope objective has a long focal length, whereas that of a microscope is very short. Therefore, the telescope is used to view large objects at long distances in outer space, such as stars and galaxies in the universe.

Figure 14.14 shows image formation, explained by the ray method, in a refracting telescope. The objective is a converging lens of long focal length. The object being viewed is far away compared with the focal length of the converging lens. Incoming rays from a distant object are nearly parallel. These rays form a real, inverted image at I_1, which is smaller than the object. The image appears near the focal point of the objective. The eyepiece lens enlarges this image further to form an enlarged virtual image at I_2. In a properly designed telescope, the first image I_1 must be at the first focal point of the eyepiece, then the final image at I_2 formed by the eyepiece is at infinity.

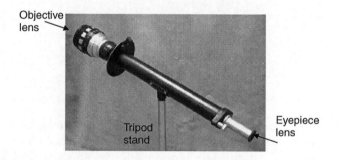

FIGURE 14.13 Simple refracting telescope.

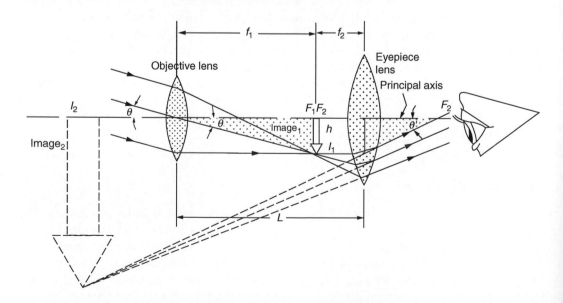

FIGURE 14.14 Image formation by a refracting telescope.

The eyepiece can be adjusted to move the image closer to the viewer's eyes. The distance between the objective lens and eyepiece lens, which is the length of the telescope L, is the sum of the focal length of objective and eyepiece lenses, f_1 and f_2.

The angular magnification M of a telescope is defined as the ratio of the angle θ' subtended by the object (when viewed through the telescope) to the angle θ subtended (when the object is viewed with the unaided eye). From Figure 14.14, the tangent of the angles θ and θ' can be written as:

$$\tan \theta \cong \theta = \frac{-h}{f_1} \tag{14.16}$$

$$\tan \theta' \cong \theta' = \frac{h}{f_2} \tag{14.17}$$

and the angular magnification M_a of a telescope is defined as the ratio of the focal length of the objective lens to the focal length of the eyepiece lens. The negative sign shows that the final image is inverted.

$$M_a = \frac{\theta'}{\theta} \cong -\frac{f_1}{f_2} \tag{14.18}$$

14.8.2 Terrestrial Telescopes

A telescope in which the final image is formed is erect is called a terrestrial telescope. Terrestrial telescopes are more convenient than astronomical telescopes, for viewing an upright image. This is useful for viewing objects on Earth. There are two types of terrestrial telescopes. Figure 14.15 shows one type of terrestrial telescope, which is called a Galilean telescope after the Italian astronomer and physicist Galileo Galilei (1564–1642), who built one in 1609.

In the Galilean telescope, the converging eyepiece is replaced by a diverging lens, so that an erect image is formed, as shown in Figure 14.16. Incoming parallel rays from a distant object will form a real inverted image from the objective lens. A diverging eyepiece lens is adjusted to be just within the focal length of the objective lens, so that the rays striking the eyepiece emerge parallel to each other. In this case an erect, virtual image is formed. The angular magnification M_a is again defined by Equation (14.18) as the ratio of the focal lengths. Note in this case, the focal length of the diverging lens is negative ($-f_2$). Therefore, the angular magnification of a Galilean telescope is positive, which indicates the final image is erected and enlarged.

Galilean telescopes have disadvantages, such as very narrow fields of view and limited magnification, compared to other telescope types. A better type of terrestrial telescope is shown in Figure 14.17. In this type, a third lens, called the intermediate erecting lens, is used between the objective and the eyepiece to produce an erect final image, as illustrated in Figure 14.17. If the first image formed by the objective at a distance is twice the focal length of the intermediate erecting lens (at $2f_e$), then the erecting lens inverts the image without magnification. The angular magnification M_a of this telescope type is still defined by Equation (14.18) as the ratio of the angles.

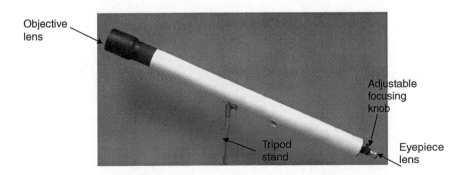

FIGURE 14.15 Galilean telescope.

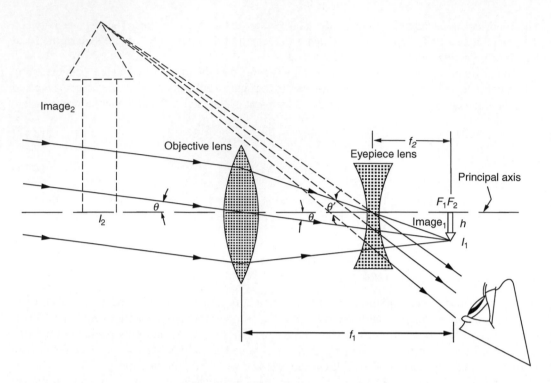

FIGURE 14.16 Image formation by a Galilean telescope.

The disadvantage of this type of telescopes that the telescope tube is longer than a telescope with similar magnification. A better design shortens the instrument length by using a pair of prisms that both invert the image so that the image is erect. This shorter arrangement is used in the design of binoculars. The principles of binoculars will be explained in a later section.

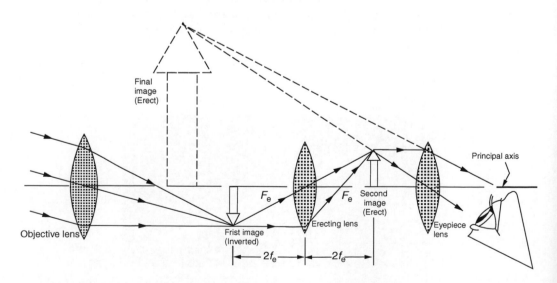

FIGURE 14.17 In a terrestrial telescope an intermediate lens is used to form an erect image.

14.8.3 THE REFLECTING TELESCOPE

In astronomical applications, such as observing nearby astronomical objects (such as the sun, moon, and planets), the magnification achieved by most refracting telescopes is sufficient. However, stars are so far away that they always appear as small points of light. Large research telescopes with very high magnification are used to study very distant objects. The higher the magnification of a telescope, the greater in diameter the objective lens must be, in order to gather in enough light for the image to be visible. Large lenses for refraction telescopes are difficult and expensive to manufacture. Another difficulty with large lenses is that their large weight requires a big telescope structure. This causes an additional source of lens aberration, dispersion, and misalignment. These problems can be solved by replacing the objective lenses with reflecting, concave parabolic mirrors. Such a mirror produces a real image of a distant object and is subject to less aberration.

Figure 14.18 illustrates a design for a typical reflecting astronomical telescope. This type of telescope is called Newtonian focus, because it was Newton who developed and constructed it in 1670. Incoming light rays pass down the barrel of the telescope and are reflected by a concave parabolic mirror at the base. These rays converge toward a small secondary flat mirror that reflects the light toward an opening in the side of the tube. Then, the rays pass into an eyepiece for viewing or more often for photographing.

Figure 14.19 shows another arrangement, which is called a Cassegrain focus reflecting telescope. This telescope uses a small mirror to form the image below the main parabolic mirror at the base. The mirror reflects the light toward an opening in the base. Then, the rays pass into an eyepiece for viewing or more often for photographing.

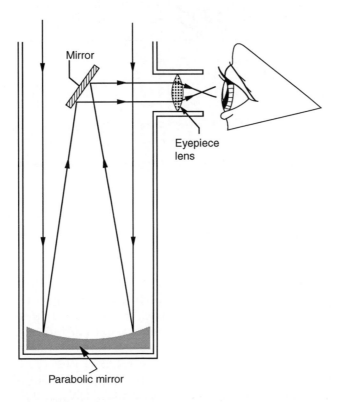

FIGURE 14.18 Newtonian focus reflecting telescope.

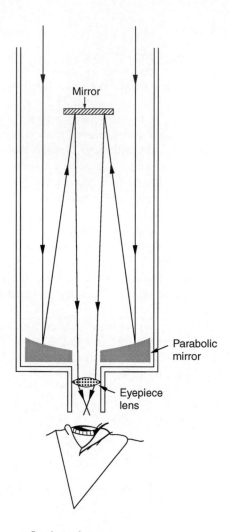

FIGURE 14.19 Cassegrain focus reflecting telescope.

14.8.4 FAMOUS TELESCOPES

The Hubble Space Telescope is in orbit outside of the Earth's atmosphere to enable observations without the distortion caused by refraction. In this way the telescope can be diffraction limited, and used for coverage in the ultraviolet (UV) and infrared ranges.

The Very Large Telescope (VLT) is currently the record holder in size; it has four telescopes, each 8 m in diameter. The four telescopes, belonging to the European Southern Observatory (ESO) and located in the Atacama desert in Chile, can operate independently or together.

The Largest refracting telescope in the world, which is located at the Yerkes Observatory in Williams Bay, Wisconsin in the United State of America, has a of 1.02-m diameter.

In contrast, the largest single-mirror optical telescope in the world is the 6-m diameter reflecting telescope on Mount Pastukhov in the Caucasus Mountains of Russia. At the same site is the RATAN 600 radio telescope, consisting of radio reflectors in a circle of 600-m diameter. Both instruments are operated by the Special Astrophysical Observatory of the Russian Academy of Sciences in St Petersburg.

Next in size is a 5-m reflector at Mount Palomar in California that enlarged the horizons of astronomy since 1948. The 5-m Hale telescope at Mt. Palomar is a conventional research telescope that was the largest for many years. It has a single borosilicate mirror that was famously difficult to construct. The equatorial mounting is also unique, permitting the telescope to image near the north celestial pole.

The 2.5-m Mt. Wilson telescope was used by Edwin Hubble to discover galaxies, and the redshift. It is now part of a synthetic aperture array with several other Mt. Wilson telescopes, and is still useful for advanced research.

The 60-cm telescope currently utilized by the Telescope In Education (TIE) programme has a distinguished history. Designed and built by Caltech in the early 1960s. This telescope was used to investigate the moon and prove that the lunar surface was solid rather than covered by a thick layer of dust. A thick layer of fine dust would have been a hazard to astronauts landing on the moon during the Apollo missions. This telescope was designed as a 60-cm (f/16) Cassegrain telescope. There were two secondary mirrors made: One was larger and used for visual and photographic observations. The second mirror was smaller because it was housed in a special fixture that could move the mirror for accurate infrared observations. The first secondary mirror was destroyed in an accident during an observing run. The second mirror, though slightly undersized, continues to function well on the instrument. This 60-cm telescope has been thoroughly refurbished and reconfigured to f/3.5 with a Newtonian focus, in order to operate with digital cameras.

The latest generation of astronomical telescopes do not rely on single large mirrors; practical difficulties limit the mirror diameter to at most 8 m. Instead, a number of individual mirrors are linked to produce a single image. In one approach the Keck telescope in Hawaii in the United State of America uses hexagonal segments to give a collecting surface 10 m across. In another approach separate circular mirrors are used. A proposed new American telescope will have four 7.5-m mirrors mounted in one structure for the equivalent of a 15-m mirror. A third scheme, which is being developed in Europe, uses four individual 8-m telescopes whose images are added together electronically.

14.8.5 Research Telescopes

In recent years, some technologies have been adapted to overcome the negative effect of atmosphere on ground-based telescopes. These technologies have achieved successful results. Some research telescopes have several instruments to choose from for different applications, such as imagers of different spectral responses, spectrographs useful in different regions of the spectrum, and polarimeters that detect light polarization.

Most large research telescopes can operate as either a Cassegrainian (longer focal length, and a narrower field with higher magnification) or Newtonian (brighter field). In a new era of telescope making, a synthetic aperture composed of six segments synthesizes a mirror of 4.5 m in diameter. Another company has developed a telescope with a synthetic-aperture 10 m in diameter.

The current generation of ground-based telescopes being constructed has a primary mirror of between 6 and 8 m in diameter. In this generation of telescopes, the mirror is usually very thin and is kept in an optimal shape by an array of actuators. This technology has driven new designs for future telescopes with diameters of 30, 50 and even 100 m.

Initially the detector used in telescopes was the human eye. Later, the sensitized photographic plate took its place. Thus, the spectrograph was introduced, allowing the gathering of spectral information. After the photographic plate, successive generations of electronic detectors, such as CCDs, have been perfected, each with more sensitivity and resolution.

The phenomenon of optical diffraction sets a limit to the resolution and image quality that a telescope can achieve, which is the effective area of the Airy disc, which limits how close to place two such discs. This absolute limit is called Sparrow's resolution limit. This limit depends on the wavelength of the studied light (so that the limit for red light comes much earlier than the limit for blue light) and on the diameter of the telescope mirror. This means that a telescope with a certain mirror diameter can resolve up to a certain limit at a certain wavelength, so if more resolution is needed at that very wavelength, a wider mirror must be built.

There are many plans for even larger telescopes; one of them is the Overwhelmingly Large Telescope (OWL), which is intended to have a single aperture of 100 m in diameter.

14.9 THE BINOCULARS

A pair of common binoculars shown in Figure 14.20 is shorter in length than any type of telescope. To shorten the optical instrument length, the common binoculars design uses a pair of prisms that both invert the image so that the image is erect to the viewer.

FIGURE 14.20 Pair of binoculars.

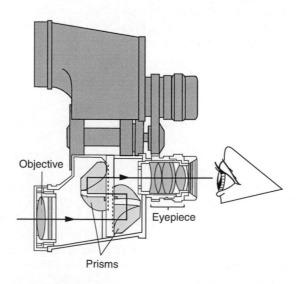

FIGURE 14.21 Optical components of binoculars.

The arrangement of the prisms and lenses in binoculars is shown in Figure 14.21. Common binoculars have three major parts. An objective lens produces an inverted image and focuses the image onto a prism. A set of prisms turns the image upright, and an eyepiece magnifies the image.

There are two types of prisms that are used in making binoculars: Porro prisms and roof prisms. There are advantages to each prism type. The Porro prism design is simpler and more light efficient, and its images show better contrast. Roof prism binoculars appear simpler than Porro prism binoculars. But inside, they have a more complex light path and require much greater optical precision in manufacturing. As a result, roof prism binoculars cost more to make. Though modern eyepieces and objective lenses are each comprised of multiple elements, their basic functions remain unchanged.

14.10 THE SLIDE PROJECTOR

A slide projector is an opto-mechanical device, as shown in Figure 14.22, used to view photographic slides, as shown in Figure 14.22. It has four main elements: a fan-cooled electric light bulb or other light source, a focusing lens to direct the light to cover the area of the slide, a holder for the slide, and magnifying lenses, as shown in Figure 14.23. Light passes through the transparent slide and magnifying lenses; the resulting image is enlarged and projected onto a perpendicular flat screen so the audience can view its reflection. Alternatively, the image may be projected onto a translucent rear projection screen using continuous automatic display for close viewing.

Common in the 1950s and 1960s households as an alternate to television or movie entertainment, family members and friends would gather, darken the living room and show slides of recent holidays or vacations. In-home photographic slides and slide projectors have largely been replaced by low cost paper prints, digital cameras, DVD media, video display monitors, and digital projectors.

FIGURE 14.22 Slide projector.

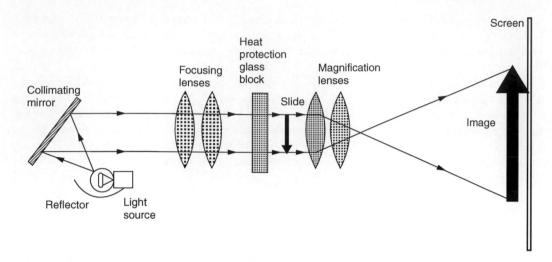

FIGURE 14.23 Image formation by a slide projector.

14.11 THE OVERHEAD PROJECTOR

The overhead projector is another common type of viewing device, as shown in Figure 14.24. The overhead projector is a display system used to display images to an audience. It typically consists of a large box containing a very bright lamp and a fan to cool it, on top of which is a large Fresnel lens that collimates the light. Above the box, typically on a long arm, is a magnifyer convex lens and a

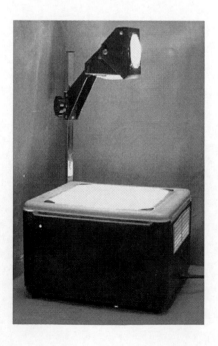

FIGURE 14.24 Overhead projector.

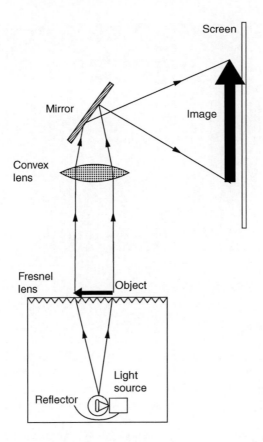

FIGURE 14.25 Image formation by an overhead projector.

mirror that redirects the light forward onto the screen. Figure 14.25 shows an image formation by an overhead projector.

Overhead projectors were a common fixture in most classrooms and business conference rooms, but today are being replaced by larger computer monitors and dedicated computer projection systems.

14.12 THE LIQUID CRYSTAL DISPLAY (LCD) PROJECTOR

A liquid crystal display (LCD) projector is a device for giving presentations generated on a computer. They are the modern equivalent to the slide projector and overhead projector. LCD projectors place a small LCD panel, almost always colour, in front of the bright lamp used in all overhead systems. The imagery on the LCD is provided by an attached computer. Currently, LCD projectors tend to be smaller and much more portable than older systems.

Figure 14.26 shows a computer display crystal screen. The LCD units can be used as a display panel in cellular phones, calculators, pocket television screens, computer screens, and diagnostic and measuring devices. LCDs give excellent readability from any angular position to within a few degrees of the horizontal plane, even in the most intense, direct sunlight.

FIGURE 14.26 Liquid crystal display.

14.13 THE LIGHT BOX

A light box is made of hardwood with a translucent acrylic diffuser panel and double-thick glass for durability, as shown in Figure 14.27. A special light diffuser illuminates the working surface with uniform colour-shifted light for bright, sharp images. The light box features a metal side rail for a T-square. Ventilation channels on the side keep the unit cool. This box is used for making engineering drawings.

Another design for a light box is used for slide viewing, as shown in Figure 14.28. Fluorescent illumination is used for tracing, opaquing, stripping, transparency and slide viewing, and slide sorting. The surface viewing area stays cool. The light box features durable, lightweight construction, and a frosted acrylic top.

A standard x-ray illuminator is another type of light box. They are used for x-ray viewing by medical staff in clinics and hospitals, as shown in Figure 14.29. All standard x-ray illuminators are constructed from cold rolled steel and finished inside and out with a non-yellowing baked white enamel finish. The x-ray illuminators provide the best viewing conditions for diagnostic purposes. The viewing area of the light box is one-piece, white, translucent, shatterproof plastic.

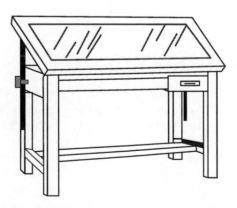

FIGURE 14.27 Light box with a support frame.

FIGURE 14.28 Light box for slide viewing.

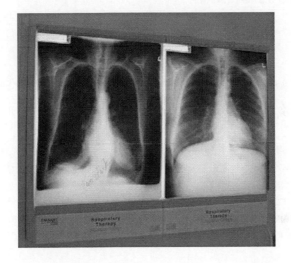

FIGURE 14.29 Standard x-ray illuminator.

14.14 EXPERIMENTAL WORK

The theory behind this experiment is based on the laws of light. Optical components, such as mirrors, lenses, and prisms, work on the basis of image formation by reflection and refraction. These optical components are used in building different optical viewing instruments.

In this experiment, the student will examine, operate, dismantle, and reassemble a variety of optical viewing instruments. These instruments are composed of different types of optical components that are used for image formation. First, the student will operate and observe image formation by optical instruments. Second, the student will examine the internal components and determine the principle operation of the instrument.

The student will operate, observe image formation, disassemble, examine the functionality of the optical components, and reassemble an optical viewing instrument. The following cases of optical viewing instruments are considered for this experiment:

a. a slide projector;
b. a film projector.

   c. an overhead projector.

   d. a camera.

   e. a compound microscope.

   f. a telescope.

14.14.1 Technique and Apparatus

Appendix A presents the details of the devices, components, tools, and parts.

1. Slide projector, film projector, overhead projector, camera, compound microscope, and telescope.
2. Slides, transparent film, overhead transparency, and a small object.
3. Screen.
4. Toolbox (screwdriver kits, etc.).
5. Black/white card and cardholder.
6. Ruler.

14.14.2 Procedure

Follow the laboratory procedures and instructions given by the professor and/or instructor.

14.14.3 Safety Procedure

Follow all safety procedures and regulations regarding the use of optical viewing instrument.

14.14.4 Apparatus Set-Up

14.14.4.1 A Slide Projector: Operate, Observe Image Formation, Disassemble to Examine the Functionality of the Optical Components, and Reassemble

1. Place a slide projector on the table.
2. Have the toolbox ready for use.
3. Inspect the slide projector condition and optical parts.
4. Load the slide tray with a few slides.
5. Connect the slide projector to the power supply.
6. Turn on the slide projector.
7. Turn off the lights of the lab.
8. Try to project and focus the image of the slide on the screen.
9. Turn on the lights of the lab.
10. Turn off the slide projector and disconnect the power cable from the wall outlet.
11. Take a screwdriver and unscrew the screws on the side/back cover of the slide projector, as shown in Figure 14.30.
12. Examine the optical components that are used in building the slide projector.
13. Illustrate the locations of the light source, mirror, lenses, slide, and image in a diagram using the ray tracing method.
14. Put back the side/back cover and screw the screws back in place.
15. Test the slide projector again.
16. Put the slide projector and the parts back into the slide projector case.

For Cases (b), (c), (d), (e), and (f), repeat the procedure described in Case (a), using the appropriate optical viewing instrument.

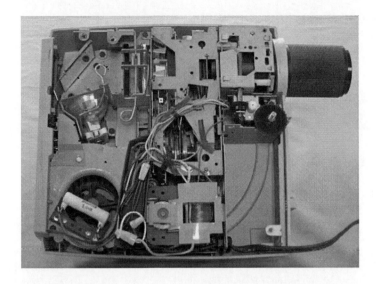

FIGURE 14.30 Disassembled slide projector.

14.14.5 DATA COLLECTION

No data collection is required.

14.14.6 CALCULATIONS AND ANALYSIS

No calculations and analysis are required for any case.

14.14.7 RESULTS AND DISCUSSIONS

14.14.7.1 A Slide Projector: Operate, Observe Image Formation, Disassemble to Examine the Functionality of the Optical Components, and Reassemble

1. Illustrate the locations of the light source, lens, object, and image in a diagram using the ray tracing method.
2. Discuss the characteristics of the image.

For Cases (b), (c), (d), (e), and (f), repeat the steps for the results and discussions in Case (a).

14.14.8 CONCLUSION

Summarize the important observations and findings obtained in this lab experiment.

14.14.9 SUGGESTIONS FOR FUTURE LAB WORK

List any suggestions for improvements using different experimental equipment, procedures, and techniques for any future lab work. These suggestions should be theoretically justified and technically feasible.

14.15 LIST OF REFERENCES

List any references that were used in the report. Use one format in writing the references. Never mix reference formats in a report.

14.16 APPENDICES

List all of the materials and information that are too detailed to be included in the body of the report.

FURTHER READING

Beiser, A., *Physics*, 5th ed., Addison-Wesley Publishing Company, Inc., USA, 1991.

Blaker, W. J., *Optics: The Matrix Theory*, Marcel Dekker, New York, 1972.

Blaker, J. W. and Schaeffer, P., *Optics: An Introduction for Technicians and Technologists*, Prentice Hall, Englewood Cliffs, NJ, 2000.

Cutnell, J. D. and K, W., *Physics Physics*, 5th ed., John Wiley and Sons, Inc., New york, 2001.

Damaskinos, S., *Confocal Scanning Laser Microscope*, MACROscope®—the widefield confocal™, 550 Parkside Dr., Unit A12, Waterloo, Ontario, Canada, N2L 5V4, 2005.

Edmund Industrial Optics, *Optics and Optical Instruments Catalog, 2004*, Edmund Industrial Optics, Barrington, NJ, 2004.

Ghatak, A. K., *An Introduction to Modern Optics*, McGraw-Hill Book Company, New York, 1972.

Halliday, D., Resnick, R., and Walker, J., *Fundamental of Physics*, 6th ed., John Wiley and Sons, Inc., New York, 1997.

Hecht, E., *Optics*, 4th ed., Addison-Wesley Longman, Inc., Reading, MA, 2002.

Hewitt, P. G., *Conceptual Physics*, 8th ed., Addison-Wesley, Inc., Reading, MA, 1998.

Jenkins, F. W., H., E., and White, *Fundamentals of Optics*, McGraw Hill, New York, 1957.

Jones, E. and Childers, R., *Contemporary College Physics*, 3rd ed., McGraw-Hill Higher Education, USA, 2001.

Naess, R. O., *Optics for Technology Students*, Prentice Hall, Englewood Cliffs, NJ, 2001.

Newport Corporation, Optics and Mechanics Section, *the Newport Resources 1999/2000 Catalog*, Newport Corporation, Irvine, CA, USA, 1999/2000.

Nolan, P. J., *Fundamentals of College Physics*, Wm. C. Brown Publishers, Inc., Dubuque, IA, USA, 1993.

Pedrotti, F. L., L., S., and Pedrotti, *Introduction to Optics*, 2nd ed., Prentice Hall, Englewood Cliffs, NJ, 1993.

Pritchard, D. C., *Environmental Physics: Lighting*, Longmans, Green and Co. Ltd., London, 1969.

Serway, R. A. and Jewett, J. W., *Physics for Scientists and Engineers with Modern Physics*, 6th ed., volume 2, Thomson Books/Cole, USA, 2004.

Shamir, J., *Optical Systems and Processes*, SPIE-The International Society for Optical Engineering, SPIE Press, USA, 1999.

Sterling, D. J. Jr., *Technician's Guide to Fiber Optics*, 2nd ed., Delmar Publishers Inc., Albany, New York, 1993.

Tippens, P. E., *Physics*, 6th ed., Glencoe McGraw-Hill, Westerville, OH, USA, 2001.

White, H. E., *Modern College Physics*, 6th ed., Van Nostrand Reinhold Company, New York, USA, 1972.

Woods, N., *Instruction's Manual to Beiser Physics*, 5th ed., Addison-Wesley Publishing Company, Reading, MA, 1991.

15 Polarization of Light

15.1 INTRODUCTION

Polarization is another interesting wave property of transverse light waves. Polarization is the principle applied to polarized sunglasses used in protecting the eyes from the sun's rays under a clear sky. Light waves are propagated in an electromagnetic wave as vibrating electric and magnetic fields, which are perpendicular to the direction of the wave propagation. The human eye cannot distinguish between polarized and unpolarized light. Therefore, an analyser is needed to detect polarized light.

Polarization states are linear, circular, or elliptical according to the paths traced by electric field vectors in a propagating wave. Unpolarized light, such as light from an incandescent fluorescent tube, is a combination of all polarization states. Randomly polarized light, in reference to laser output, is composed of two orthogonally linearly polarized beams.

The polarization phenomenon is used in many optical applications, such as testing plastic or glass under mechanical stress. Polarizer materials, such as beamsplitters and prisms, are used as components in polarization beamsplitter devices that are used in fibre communication systems. In this chapter, along with the theoretical presentation, four experimental cases demonstrate the principles of the polarization of light.

15.2 POLARIZATION OF LIGHT

Light can be considered to have a dual nature. In some cases light acts like a wave, and in others light acts like a particle. Classical electromagnetic wave theory provides an adequate explanation of light propagation and of the effects of interference. On the other hand, the photoelectric effect, involving the interaction of light with matter, is best explained by considering light as a particle.

An electromagnetic wave consists of an electric field and magnetic field oriented perpendicular to one another. Both fields have a direction and strength (or amplitude). The propagation axis of

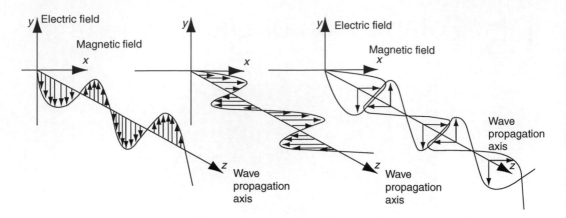

FIGURE 15.1 Electric and magnetic fields in a light wave.

the wave is perpendicular to the plane containing the electric and magnetic fields. As shown in Figure 15.1, if the electric field is oriented along the y-axis, and the magnetic field along the x-axis, then the wave propagation is along the z-axis. This means that the vibrations in the electric field are parallel to one another at all points in the wave, so that the electric field forms a plane called the plane of vibration. Similarly, all points in the magnetic field component of the wave lie in a plane that is perpendicular to the electric field plane.

The electric and magnetic fields vary sinusoidally with time. The curved line represents field strength or amplitude. A field starting at a maximum in one direction decays to zero and then builds up in the other direction until it reaches a maximum again. The electric and magnetic fields oscillate in phase; this means that the electric and magnetic fields reach their peaks (and their troughs) at exactly the same time. The rate of oscillation represents the frequency of the wave. The distance traveled during one period of oscillation represents the wavelength.

15.3 FORMS OF POLARIZATION OF LIGHT

Light from an ordinary light source consists of many waves emitted by the atoms or molecules of the light source. Each atom produces a wave with its own orientation of the electric and magnetic vibration; the orientation corresponds to the direction of the atomic vibration. Figure 15.2 shows a light source with electric and magnetic field vectors. The direction of propagation is coming out of the page in this figure. When the field vectors of a light source are randomly oriented, the light is called unpolarized, as shown in Figure 15.2(a). If there is some partial preferential orientation of the field vectors, the light is called partially polarized, as shown in Figure 15.2(b). If the field vectors are laid on the plane of polarization, the light is called linearly polarized, plane polarized, or sometimes called simply polarized. The plane of the linearly polarized light could be vertical or horizontal, as shown in Figure 15.2(c).

There are three forms of polarization in light: linear polarization, elliptical polarization, and circular polarization, as shown in Figure 15.3. In linearly polarized light, the field vectors vibrate in one axis. In the case of elliptical polarization, both electric and magnetic field vectors do not have the same amplitudes; while with circular polarization, the two fields do have equal amplitudes.

The analyser absorbs or transmits the polarized light, depending on the orientation of the analyser. When the analyser has the same orientation as the polarizer, light is transmitted. When the polarizer is rotated 90°, the analyser absorbs the polarized light, and then no light is transmitted.

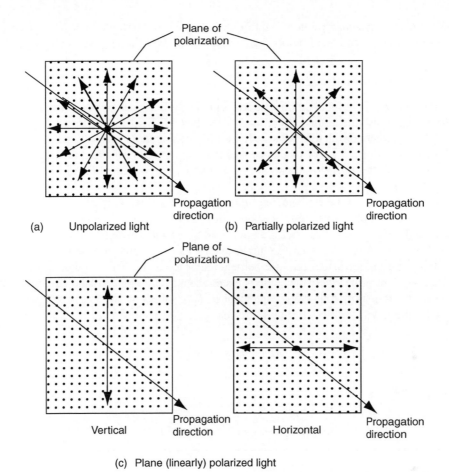

(a)   Unpolarized light   (b) Partially polarized light

(c) Plane (linearly) polarized light

FIGURE 15.2 An unpolarized and a polarized light of transverse waves.

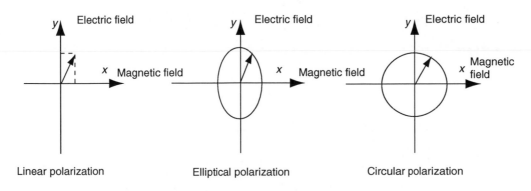

Linear polarization   Elliptical polarization   Circular polarization

FIGURE 15.3 Forms of polarization of light.

15.4 OCCURRENCE OF POLARIZATION

Ordinary light (unpolarized) can become polarized in several ways. Polarization may result from the reflection of light from various surfaces, from refraction and double refraction of light through a crystal, from selective absorption of light in a crystal, and scattering by particles of the medium. Polarization by these means will be discussed in detail in the following subsections.

15.4.1 POLARIZATION BY REFLECTION

Consider unpolarized light incident on a smooth optical surface, such as glass. The light is partially reflected and partially transmitted according to the laws of light, as shown in Figure 15.4. The reflected light is either completely polarized, partially polarized, or unpolarized, depending on the angle of incidence. If the angle of incidence is either 0 or 90° to the surface, then the reflected light is unpolarized. If the angle of incidence is between 0 and 90°, then the reflected light is partially polarized.

For one particular angle of incidence, the reflected light is completely polarized while the refracted light is partially polarized, as shown in Figure 15.5. This angle of incidence is called polarizing angle (θ_p) or the Brewster's angle after Sir David Brewster (1781–1868), a Scottish physicist. He discovered that when the angle of incidence is equal to this polarizing angle, the reflected light and the refracted light are perpendicular to each other. In other words, Brewster's angle is that incident angle where the angle between the reflected light and the refracted light is 90°. The angle of refraction (θ_{refr}) is the complement of (θ_p), so $\theta_{refr} = 90° - \theta_p$.

Applying the law of refraction (Snell's Law):

$$n_1 \sin \theta_p = n_2 \sin \theta_{refr} \tag{15.1}$$

Substituting for θ_{refr} gives:

$$n_1 \sin \theta_p = n_2 \sin(90° - \theta_p) = n_2 \cos \theta_p \tag{15.2}$$

Dividing the Equation 15.2 by $\cos \theta$ gives:

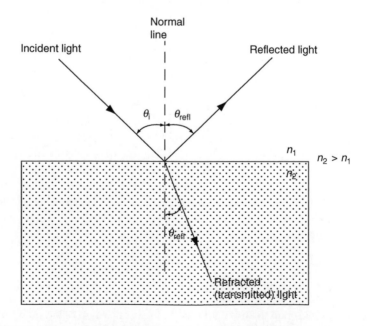

FIGURE 15.4 Light incident, reflected, and refracted by an optical material.

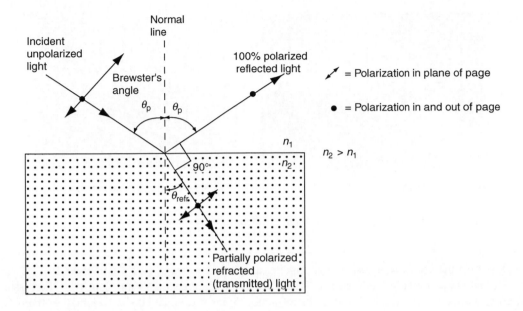

FIGURE 15.5 Polarization at Brewster's angle.

$$\frac{\sin \theta_p}{\cos \theta_p} = \frac{n_2}{n_1} = \tan \theta_p \tag{15.3}$$

For light incidence in air $(n_1) \approx 1.00$, then:

$$\tan \theta_p = n \tag{15.4}$$

Then, Brewster's angle is:

$$\theta_p = \tan^{-1} n \tag{15.5}$$

This relation is known as Brewster's law. For example, Brewster's angle for crown glass $(n = 1.52)$ has the value $\theta_p = \tan^{-1}(1.52) = 56.7°$. Brewster's angle is a function of the index of refraction (n) and varies to some degree with the wavelength of the incident light. This is because (n) varies with wavelength for a given optical material, as explained in the laws of light chapter.

Polarization by reflection is a common phenomenon. Sunlight that is reflected from water, glass, smooth asphalt, or snow surfaces is partially polarized. When the sunlight is reflected from a horizontal surface, the electric field vector of the reflected light has a strong horizontal component. Sunglass manufacturers make the polarizing axis of the lens material vertical, to absorb the strong horizontal component of the reflected light. The sunglasses can reduce the overall intensity of the light transmitted through the lenses to a certain value depending on the quality of the lens material. Other types of polarizing filters are widely used in different industrial applications.

15.4.2 POLARIZATION BY DOUBLE REFRACTION

An Iceland spar (calcite) crystal placed on a page shows two refracted images of the print, as shown in Figure 15.6. This double refraction of light by calcite was discovered by Erasmus Bartholinus, a Scottish Physician, in 1669, and later studied by Huygens and Newton.

(a) Calcite crystal   (b) Double refraction phenomenon

FIGURE 15.6 Calcite crystal. (a) Calcite crystal; (b) double refraction phenomenon.

Most crystalline substances, such as calcite, quartz, mica, sugar, topaz, aragonite, and ice, are now known to exhibit double refraction. The double refraction depends on the purity of the crystal. In other words, it depends on the index of refraction of the crystal and the wavelength of the light. Calcite (Calcium Carbonate $CaCo_3$) and quartz (Silicon Dioxide SiO_2) are the most important crystals used in manufacturing optical devices and instruments. Calcite is a natural, birefringent material and always has the shape shown in Figure 15.6. The calcite must be cut, ground, and polished to exact angles with respect to its optical axis. Each face of the calcite crystal is a parallelogram whose angles are 78 and 102°. The two opposite surfaces of the calcite crystal are parallel to each other.

When a beam of unpolarized light passes at an angle to the calcite crystal's optical axis, the beam is doubly refracted, dividing into two light beams as it enters the surface of the crystal. Both refracted light beams are found to be plane polarized and their planes of polarization perpendicular to each other, as shown in Figure 15.7. One light beam, called the ordinary light beam, is linearly polarized with its vibrations in one plane. The ordinary beam passes straight through without deviation. This beam is the horizontal component of the light. The second light beam, called the extraordinary light beam, is linearly polarized with its vibrations in a plane

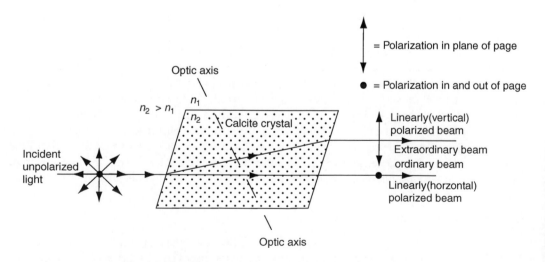

FIGURE 15.7 Polarization by double refraction in a calcite crystal.

perpendicular to the first plane. This beam is the vertical component of the light. The ordinary beam follows the law of refraction, Snell's law, and the extraordinary beam does not. Because the two opposite surfaces of the calcite crystal are always parallel to each other, the two refracted beams always emerge parallel to the incident light beam and are parallel to each other. One important property of the calcite crystal is that there is only one direction through the crystal in which there is no double refraction. This particular direction, called the optic axis, is indicted by dashed lines shown in Figure 15.7.

15.4.3 POLARIZATION BY SCATTERING

When light is incident on atmospheric particles, such as the molecules of air (nitrogen and oxygen) and dust, some of the light can be absorbed and reradiated. This process, called scattering of light is shown in Figure 15.8. Atmospheric scattering causes the sky's light to be polarized. In order for the scattering to occur, the sizes of the air molecules present in the atmosphere must be smaller than the wavelengths of visible light. The molecules scatter light inversely proportional to the fourth power of the wavelength ($1/\lambda^4$). This wavelength relation discovered by Lord Rayleigh (1842–1919), a British physicist, is due to Rayleigh scattering. Blue light has a shorter wavelength than red light. Therefore, blue light is scattered toward the ground more than red light. This is the reason why the sky appears blue. At sunset, the sun's rays pass through a maximum length of atmosphere. Much of the blue has been removed by scattering. The light that reaches the surface of the Earth is thus lacking of blue, which is why sunsets appear reddish.

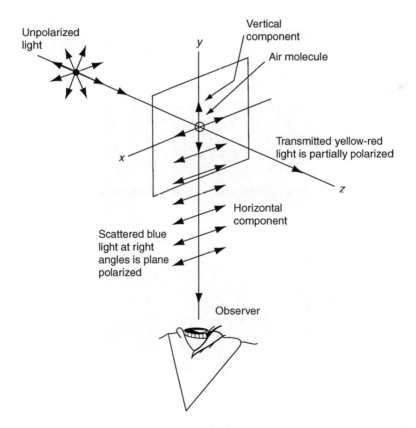

FIGURE 15.8 Polarization by scattering.

It is possible to produce polarized light by filtering ordinary light through a polarizer, which is a material that transmits only those waves that vibrate in a single plane. Polarizers will be explained in the next sections. Polarizers can be used as analysers or filters. When an ordinary beam of light strikes a polarizer, part of the light is absorbed and part is transmitted, as shown in Figure 15.9(a). The light emerging from this polarizer is completely polarized along the direction of propagation. If a second polarizer acting as analyser is placed behind the first and aligned in the same orientation, light is still transmitted. However, there is less transmission because of the light absorbed in the two polarizers. If the second polarizer is rotated by 90° about the axis of the light beam, then no light passes through, as shown in Figure 15.9(b). When the emerging polarized light from the first polarizer falls on the second polarizer, all the light is absorbed. This is because the second polarizer is turned so that its polarization axis is 90° away from that the first polarizer.

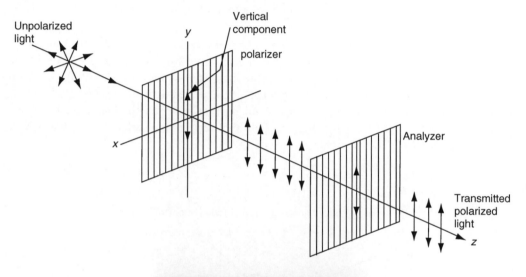

(a)

FIGURE 15.9. Polarization of light by two polarizers.

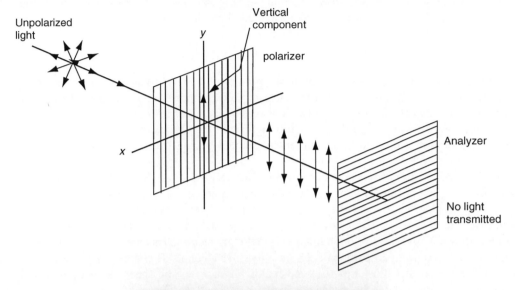

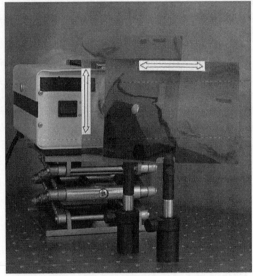

(b)

FIGURE 15.9 (*continued*)

15.4.4 POLARIZATION BY SELECTIVE ABSORPTION

When ordinary light enters a crystal of tourmaline, double refraction takes place in much the same way as with a calcite crystal, as explained above, but with one difference. The tourmaline crystal entirely absorbs the ordinary beam while the extraordinary beam passes through. This phenomenon is called selective absorption because the crystal absorbs light waves vibrating in one plane and absorbs light waves vibrating in the other plane. The common type of tourmaline crystal is the Nicol prism. When two such crystals are lined up parallel, with one behind the other, the plane polarized light from the first crystal passes through the second with a very low light loss. If either crystal is rotated to be perpendicular to the other, then the light is completely absorbed and none of the light passes through. The result is the same as explained above in the polarizer case.

15.5 POLARIZING MATERIALS

The most common technique for obtaining polarized light is to use a polarizing material. These materials transmit waves when the electric field components vibrate in a plane parallel to a specific direction. The material absorbs the waves whose electric field components vibrate in the other direction. Any substance that transmits light with the electric field component vibrating in only one direction and absorbs the other component during transmission is referred to as a dichroic substance.

Another dichroic crystal is quinine sulfide periodide, also called herapathite, after W. Herapath, an English physician who discovered its polarizing properties in 1852. In 1928, Edwin Land, an American scientist, discovered such a dichroic material, which is now called Polaroid. In 1938, Land invented H-sheet, which is now the most widely used linear polarizer. The H-sheet polarizes light through selective absorption in linearly oriented molecules. This material is fabricated in thin sheets of long-chain hydrocarbons such as polyvinyl alcohol. The sheets are stretched during manufacturing, so that the long-chain molecules align. After the sheet has been dipped into a solution containing iodine, the molecules become conductive. The conduction takes place along the hydrocarbon chains because the valence electrons of the molecules can move easily only along the chains (recall that valence electrons are free electrons that can readily move through the conductor). As a result, the molecules readily absorb light in which the electric field component is parallel to the chain. Likewise, the polarizing effect also causes the molecules to transmit light when the electric field component is perpendicular to the chain. It is common to refer to the direction perpendicular to the molecular chains, as the transmission axis. In an ideal polarizer, all light with the electrical field parallel to the transmission axis is transmitted, and all light with the electrical field perpendicular to the transmission axis is absorbed. Figure 15.10 shows a polarizing sheet, which is used in experiments to identify the polarization axis of a light beam.

FIGURE 15.10 Polarizing sheet and sheet holder.

15.6 POLARIZING OPTICAL COMPONENTS

There are many types of polarizing optical components that are used in splitting light into two beams. These optical components are prisms, rectangular beamsplitters, polarizers, etc. They work in different wavelength ranges. These components are explained in detail in prisms and beamsplitters in chapters. Figure 15.11 shows polarized light passing through a polarizing cube beamsplitter.

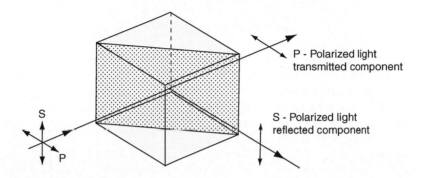

FIGURE 15.11 Polarizing cube beamsplitter.

15.7 THE LAW OF MALUS

Figure 15.12 shows a polarized light beam of intensity I_o passing through a polarizer whose axis is at an angle θ relative to the polarized light axis. The transmitted light intensity varies as the square of the transmitted light amplitude. The intensity of the transmitted polarized light varies as:

$$I = I_o \cos^2\theta \qquad (15.6)$$

where

I_o is the intensity of the polarized light incident on the polarizer in (W/m^2) and
θ is the angle between two transmission axes of the polarized light and the polarizer in (degrees).

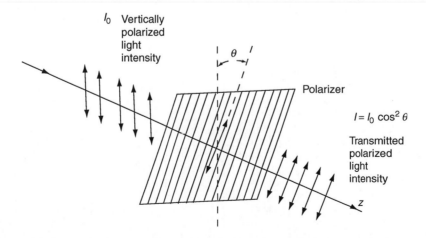

FIGURE 15.12 Transmission of polarized light through a polarizer—Malus's law.

This formula is known as the law of Malus, after the French Engineer Etienne-Louis Malus (1775–1812). Applying this formula, the transmitted intensity is unchanged if $\theta = 0$, and is zero if $\theta = 90°$. The orientation of the polarizer can be adjusted to give a beam of transmitted light of variable intensity and polarization.

Consider vertically polarized light with an intensity (I_o) of 500 W/m^2 passing through a polarizer oriented at an angle of 45° to the vertical. The transmitted light intensity, (I), from the polarizer calculated by Malus's law, Equation (15.6), gives:

$$I = I_o \cos^2\theta = 500 \times \cos^2(45) = 250 \text{ W/m}^2$$

Figure 15.13 shows an unpolarized light beam with intensity, (I_o), incident on a polarizer. The light that passes through the polarizer is vertically polarized. In this case, there is no single angle θ; instead, to obtain the transmitted intensity, the $\cos^2\theta$ over all angles must be averaged. The average of $\cos^2\theta$ is one-half. Thus, the transmitted light intensity varies as the square of the transmitted light amplitude. The transmitted intensity of a polarized light is half of the intensity of an unpolarized light, as given:

$$I = \frac{I_o}{2} \tag{15.7}$$

Figure 15.14 shows an unpolarized light beam incident on the first polarizer, where the transmission axis is vertically indicated. The light that passes through the polarizer is vertically polarized. A second polarizer is lined up after the first polarizer, called the analyser. The analyser axis is at an angle θ to the polarizer axis. The intensity of the transmitted polarized light varies as Malus's law.

Consider unpolarized light with an intensity (I_o) of 600 W/m^2 passing through a polarizer and analyser. The axis of the polarizer is vertical. The analyser axis is at an angle 45° to the polarizer axis. The transmitted intensity (I_1) through the polarizer calculated using Equation (15.7) gives:

$$I_1 = \frac{I_o}{2} = \frac{600}{2} = 300 \text{ W/m}^2$$

The transmitted intensity through the analyser calculated by Malus's law, Equation (15.6), gives:

$$I_2 = I_1 \cos^2\theta = 300 \times \cos^2(45) = 150 \text{ W/m}^2$$

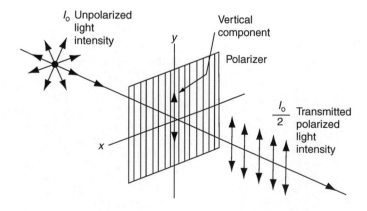

FIGURE 15.13 Transmission of unpolarized light through a polarizer.

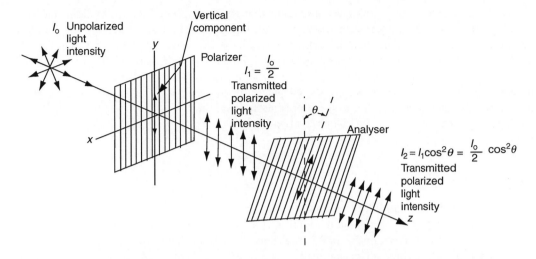

FIGURE 15.14 Transmission of unpolarized light through a polarizer and analyser.

15.8 OPTICAL ACTIVITY

Some optical materials, such as quartz, sugar, tartaric acid, and turpentine, are known as optically active because they rotate the plane of polarized light. The French physicist Dominique F.J. Arago first observed this phenomenon in 1811. The rotation of the polarized axis is due to the molecular structure of the optically active material. Figure 15.15 shows a linearly polarized light undergoing a continuous rotation as it propagates along the optic axis of a quartz material. The rotation may be clockwise or counterclockwise, depending on the molecular orientation. A water solution of cane sugar (sucrose) rotates the plane of polarization to clockwise. For a given path length through the solution, the angle of rotation is proportional to the concentration of the solution. Scientists have found numerous applications for the optically active materials in analytical procedures. For example, instruments for measuring the angle of rotation under standard conditions are known as polarimeters. These instruments are used specially for measuring the concentration of sugar solution and are known as saccharimeters. A standard method for determining the concentration of sugar solutions is to measure the rotation angle produced by a fixed length of the solution.

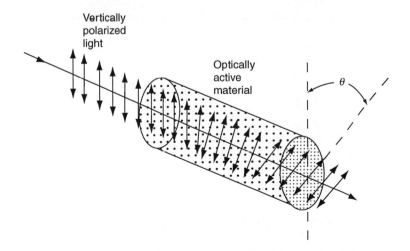

FIGURE 15.15 An optically active material rotates the direction of the polarization.

A Faraday rotator is an optical component that rotates the polarization of light due to the Faraday effect. The Faraday rotator works because one polarization of the input light is in ferromagnetic resonance with the material, which causes its phase velocity to be higher than the other. Specifically, given two rays of circularly polarized light, one with left-hand and the other with right-hand polarization, the phase velocity of the one with the polarization in the same sense as the magnetization is greater. In other words, the plane of linearly polarized light is rotated, when a magnetic field is applied parallel to the propagation direction.

The angle of rotation can be calculated by:

$$\theta = vBd \tag{15.8}$$

where

θ is the angle of rotation in (radians),
v is the Verdet constant for the material in (arc minutes cm^{-1} Gauss^{-1}),
B is the magnetic flux density in the direction of the propagation in (teslas), and
d is the length of the path in (metres).

This equation is proportionally constant (in radians per tesla per metre) varies with wavelength of the light and temperature. The angle θ also varies with various materials.

Faraday rotators (for fixed and tunable wavelenghts) are used in optical isolators to prevent undesired back reflection of light in reverse direction from disrupting or damaging an optical system.

15.9 PHOTOELASTICITY

Some optical materials become birefringent when they are subjected to mechanical stress. Materials, such as glass and plastic, become optically active when placed under any kind of

(a) Specimen under low stress.   (b) Specimen under high stress.

FIGURE 15.16 Photoelastic stress analysis for a photoelastic specimen. (a) Specimen under low stress. (b) Specimen under high stress.

stress. Photoelasticity is the basis of a technique for studying the stress analysis in both transparent and opaque materials, which was discovered by Sir David Brewster in 1816. If polarized light passes through an unstressed piece of plastic and then through an analyser with an axis perpendicular to that of the polarizer, none of the polarized light is transmitted. However, if the plastic is subjected to a stress, the regions of greatest stress produce the largest angles of rotation of polarized light. Therefore, a series of light and dark bands can be observed, as shown in Figure 15.16. Manufacturers of automobiles, bridges, machine parts, and building materials use models made of photoeleastic materials to study the effects of stresses on various structures, because the strains in the structures show clearly when analysed by polarized light. Very complicated stress distributions can be studied by these optical techniques.

15.10 LIQUID CRYSTAL DISPLAY

As briefly explained in the optical instruments for viewing chapter, the optical liquid crystal display (LCD) is used for display units. The LCD units can be used as a display panel in cellular phones, calculators, pocket television screens, computer screens, diagnostic and measuring devices, and so forth, as shown in Figure 15.17.

In 1888, the Austrian botanist named Friedrich Reinitzer (1857–1927) observed that cholesterol benzoate has a two-melting-point phonomena, one at which the crystal changed into a cloudy liquid and another where the crystal became transparent at high temperature. This crystal is know now as a liquid crystal (LC). Liquid crystals have long cigar-shaped molecules that can move about like ordinary liquids. There are three types of liquid crystal distinguished by the way in which its molecules align. The common liquid crystal is the nematic, where the molecules tend to be more or less parallel, even though their positions are fairly random.

To prepare a parallel nematic cell, one face of each to two pieces of flat glass is coated with a transparent electrically conducting metallic film. These two windows will also serve as the electrodes where a controlled voltage can be applied. These windows are etched carefully by indium tin oxide producing parallel microgrooves. This allows the LC molecules in contact with the windows to be oriented parallel to the glass and to each other. When a thin space about a few micrometers between two each prepared glass windows is filled with nematic LC, the molecules

(a) LCD for numberic display.

(b) LCD for colour display.

FIGURE 15.17 Liquid crystal display. (a) LCD for numberic display. (b) LCD for color display.

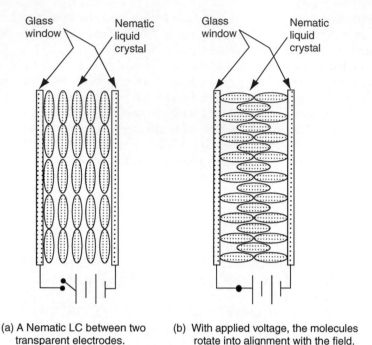

(a) A Nematic LC between two   (b) With applied voltage, the molecules
    transparent electrodes.   rotate into alignment with the field.

FIGURE 15.18 Liquid crystal operation. (a) A Nematic LC between two tranparent electrodes (b) With applied voltage, the molecules rotate into alignment with the field.

in contact with the mircogrooves anchor themselves parallel to the ridges. The LC molecules then essentially drag each other into alignment. Soon the entire liquid is similarly oriented, as shown in Figure 15.18(a). The direction in which the molecules of a liquid crystal are aligned is called the director.

When a voltage is applied across the LC, an electric field is created perpendicular to the glass windows. Electric diploes are either present or induced, and the LC molecules experience torques that cause them to rotate into alignment with the field. As the the voltage increases more and more molecules turn toward the direction of the field, as shown in Figure 15.18(b).

Figure 15.19 illustrates the operation of an LCD. The LCD consists of a liquid crystal sandwiched between crossed polarizers sheets (first vertical axis and second horizontal axis) and backed by a mirror. Figure 15.19(a) shows that unpolarized light enters the first polarizer (vertical axis), immediately linearly vertical polarized light exits, is rotated 90°, passes the second polarizer, is incident on the mirror and reflected, and again rotated 90° by liquid crystal, and leaves the LCD. After the return trip through the liquid crystal, the polarization direction of the light is the same of the first polarizer. Thus, the light is transmitted and leaves the diplay unit. Hence, the display appears lighted when illumimated.

When a voltage is applied to the liquid crystal (or heat is applied in the case of coloured LC), the liquid crystal reorients itself and loses its ability to rotate the plane of polarization, as shown in Figure 15.19(b). Vertical polarized light enters and leaves the liquid cell unchanged, only to be completely absorbed by the second polarizer. Thus, the entrance window is now black and no light emerges.

Since very little energy is required to give the voltage necessary to turn a liquid crystal cell on, a LCD is very energy efficient. In addition, the LCD uses light already present in the environment. A LCD does not need to produce its own light, as do similar displays.

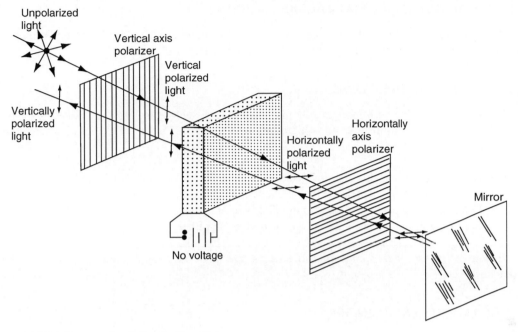

(a) No voltage applied to the liquid crystal

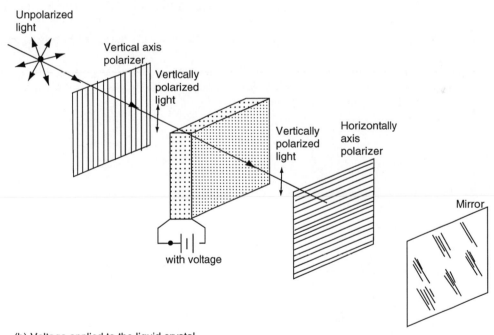

(b) Voltage applied to the liquid crystal

FIGURE 15.19 Liquid crystal display operation. (a) No voltage applied to the liquid crystal. (b) Voltage applied to the liquid crystal.

15.11 POLARIZATION MAINTAINING FIBRES

Design, construction, and applications of the polarization maintaining (PM) fibres are very import-ant in fibre optic cable technology and communication systems.

15.12 POLARIZATION LOSS

Polarization losses play an important role in many devices, such as isolators, modulators, polariz-ation beamsplitters, amplifiers, and polarization maintaining fibres. The polarization losses include:

- Polarization mode dispersion (PMD) in optical fibres,
- Polarization dependent loss (PDL) in passive optical components,
- Polarization dependent modulation (PDM) in electro-optic modulators, and
- Polarization dependent gain (PDG) in optical amplifiers.

Some of these losses and measurements are generally explained in detail in fibre optic testing.

15.13 EXPERIMENTAL WORK

In this experiment the student will perform the following cases:

- a. Light passing through a polarizing cube beamsplitter. Refer to the materials presented in prisms and beamsplitters chapters.
- b. Glan Thompson polarizing beamsplitter. Refer to the materials presented in prisms and beamsplitters chapters.
- c. Light passing through a calcite material. The students will test light passing through a birefringence of calcite material, as shown in Figure 15.20. They will also measure the power of the laser beams exiting from the calcite.
- d. The law of Malus.

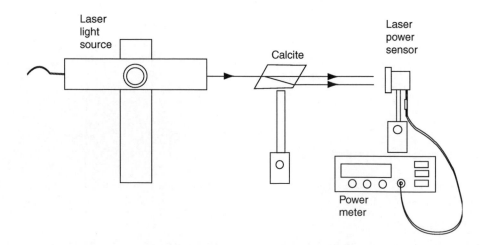

FIGURE 15.20 Light passes through a calcite.

In this case, the student will verify the law of Malus. This law states that when linearly polarized light goes through a polarizer, the transmitted light intensity is dependent on the axis angle of the polarizer with respect to the direction of the polarized light axis, as shown in Figure 15.12. This relation is given by Equation (15.6).

15.13.1 TECHNIQUE AND APPARATUS

Appendix A presents the details of the devices, components, tools, and parts.

1. 2×2 ft. optical breadboard.
2. HeNe laser light source and power supply.
3. Laser light sensor.
4. Laser light power meter.
5. Laser mount assembly.
6. Hardware assembly (clamps, posts, screw kits, screwdriver kits, sundry positioners, etc.).
7. Polarizing sheet and sheet holder, as shown in Figure 15.10.
8. Polarizing cube beamsplitter and holder/positioner assembly, as shown in Figures 15.21 and 15.24.
9. Glan Thompson polarizing beamsplitter and holder/positioner assembly, as shown in Figure 15.22.
10. Two disc polarizers and holder/positioner assemblies, as shown in Figure 15.23.
11. Calcite material and calcite holder/positioner assembly, as shown in Figure 15.26.
12. Black/white card and cardholder.
13. Protractor.
14. Ruler.

FIGURE 15.21 Polarizing cube beamsplitter.

FIGURE 15.22 Glan Thompson polarizing beamsplitter.

FIGURE 15.23 Disc polarizer and holder.

15.13.2 PROCEDURE

Follow the laboratory procedures and instructions given by the professor and/or instructor.

15.13.3 SAFETY PROCEDURE

Follow all safety procedures and regulations regarding the use of optical components and instruments, light source devices, and optical cleaning chemicals.

15.13.4 APPARATUS SET-UP

15.13.4.1 Light Passing through a Polarizing Cube Beamsplitter

1. Figure 15.24 shows the apparatus set-up.
2. Bolt the laser short rod to the breadboard.
3. Bolt the laser mount to the clamp using bolts from the screw kit.
4. Put the clamp on the short rod.
5. Place the HeNe laser into the laser mount and tighten the screw.
6. Turn on the laser device. Follow the operation and safety procedures of the laser device in use. Allow sufficient time to stabilize the laser beam polarization.
7. Check the laser alignment with the line of bolt holes and adjust when necessary.
8. Mount a cube beamsplitter holder/positioner to the breadboard.
9. Mount the polarizing cube beamsplitter on the beamsplitter holder/positioner, as shown in Figure 15.24.
10. Carefully align the laser beam so that the laser beam impacts normally to the centre of the centre of the cube beamsplitter face.
11. Try to capture the laser beam exiting the cube beamsplitter on the black/white card, as shown in Figure 15.24.
12. Turn off the lights of the lab before taking measurements.
13. Measure the power of the laser beam at the input face and output faces of the cube beamsplitter. Fill out Table 15.1 for Case (a).
14. Place the polarizing sheet to intersect the laser beam at the input face and output faces of the prism.
15. Observe the laser beam orientation using the polarizing sheet at the input face and output faces of the prism. Fill out Table 15.1 for Case (a).

FIGURE 15.24 Polarizing cube beamsplitter apparatus set-up.

16. Rotate the polarizing sheet 90° and observe the laser beam orientation exiting from the polarizing sheet at the input face and output faces of the prism.
17. Turn on the lights of the lab.

15.13.4.2 Glan Thompson Polarizing Beamsplitter

Figure 15.25 shows the apparatus set-up for Case (b). Repeat the procedure from step 1 to 8 in Case (a) using a Glan Thompson polarizing beamsplitter; add the following steps:

1. Mount a Glan Thompson polarizing beamsplitter holder/positioner to the breadboard.
2. Mount a Glan Thompson polarizing beamsplitter on the beamsplitter holder/positioner, as shown in Figure 15.25.
3. Carefully align the laser beam so that the laser beam impacts normally to the centre of the face of the Glan Thompson polarizing beamsplitter.
4. Try to capture the laser beam exiting from the face of the Glan Thompson polarizing beamsplitter on the black/white card. Two laser spots can be caught on the black/white card of the laser beam components exiting from the output faces of the Glan Thompson polarizing beamsplitter.
5. Turn off the lights of the lab before taking measurements.
6. Measure the laser power at the input face and output faces of the Glan Thompson polarizing beamsplitter. Fill out Table 15.1 for Case (b).
7. Place the polarizing sheet to intersect the laser beam at the input face and output faces of the prism and observe the laser beam on the card.
8. Observe the laser beam orientation using the polarizing sheet at the input face and output faces of the beamsplitter. Fill out Table 15.1 for Case (b).
9. Rotate the polarizing sheet 90° and observe the laser beams on the card.
10. Turn on the lights of the lab.

FIGURE 15.25 Glan Thompson polarizing beamsplitter apparatus set-up.

FIGURE 15.26 Calcite material apparatus set-up.

15.13.4.3 Light Passing through a Calcite Material

Figure 15.26 shows the apparatus set-up for Case (c). Repeat the procedure from step 1 to 8 in Case (a) using a calcite material; add the following steps:

1. Mount a calcite holder/positioner to the breadboard.
2. Mount the calcite on the calcite holder/positioner, as shown in Figure 15.26.
3. Carefully align the laser beam so that the laser beam impacts normally to the centre of the calcite face.
4. Try to capture the laser beams exiting the calcite on the black/white card, as shown in Figure 15.27.
5. Turn off the lights of the lab before taking measurements.

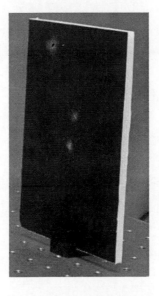

FIGURE 15.27 Laser beams exiting from the calcite material.

6. Measure the power of the laser beam at the input face and the two laser beams at the output face of the calcite.
7. Mount the polarizing sheet in a sheet holder.
8. Place the polarizing sheet to intersect the laser beam at the input face and output face of the calcite.
9. Observe the laser beam orientation using the polarizing sheet at the input face and output face of the calcite. Fill out Table 15.1 for Case (c).
10. Rotate the polarizing sheet 90° and observe the laser beam orientation exiting from the polarizing sheet at the input face and output face of the calcite.
11. Turn on the lights of the lab.

15.13.4.4 The Law of Malus

Figure 15.28 shows the apparatus set-up for Case (d). Repeat the procedure from step 1 to 8 in Case (a), using unpolarized laser light source or any unpolarized light source. Add the following steps:

1. Mount two polarizer holders/positioners to the breadboard at a distance of 15 cm apart. The two polarizers are lined up parallel to each other and perpendicular to the laser beam.
2. Mount one polarizer on each polarizer holder/positioner.
3. Mount unpolarized laser light source to the breadboard.
4. Carefully align the laser beam so that the laser beam impacts normally to the centre of the first and second polarizer face.
5. Turn off the lights of the lab before taking measurements.
6. Turn on the laser source power.
7. Measure the laser beam power from the laser source. Fill out Table 15.2.
8. Place the laser sensor after the first polarizer. Rotate the polarizer slowly and watching the laser power meter reading of the laser light.

FIGURE 15.28 The law of Malus apparatus set-up.

9. Rotate the polarizer until the power meter reading is at the maximum. Fix the position of the polarizer. Measure the power of the laser light beam. Fill out Table 15.2.
10. Measure the power of the laser beam exiting form the second polarizer. Fill out Table 15.2.
11. Rotate the second polarizer in $10°$ increments for a total of $180°$, using the rotating knob of the polarizer holder/positioner. Measure the power of the laser beam at each angle. Fill out Table 15.2.
12. Turn on the lights of the lab.

15.13.5 DATA COLLECTION

15.13.5.1 Light Passing through a Polarizing Cube Beamsplitter

1. Measure the laser beam power at the input face and output faces of the cube beamsplitter.
2. Observe the laser beam orientation using the polarizing sheet at the input face and output faces of the cube beamsplitter.
3. Fill out Table 15.1 with the collected data for Case (a).

TABLE 15.1
Polarizing Beamsplitters and Calcite

Case	Beamsplitter	Laser Power Input (unit)	P Component Power (unit)	P Component Orientation Axis	S Component Power (unit)	S Component Orientation Axis	Splitting Ratio
(a)	Polarizing Cube Beamsplitter						
(b)	Glan Thompson Polarizing						
(c)	Calcite Material						

15.13.5.2 Glan Thompson Polarizing Beamsplitter

1. Measure the laser beam component power at the input face and output faces of the Glan Thompson polarizing beamsplitter.
2. Observe the laser beam orientation, using the polarizing sheet at the input face and output faces of the Glan Thompson polarizing beamsplitter.
3. Fill out Table 15.1 with the collected data for Case (b).

15.13.5.3 Light Passing through a Calcite Material

1. Measure the laser beam component power at the input face and the two laser beams at the output face of the calcite.
2. Observe the laser beam orientation, using the polarizing sheet at the input face and output face of the calcite.
3. Fill out Table 15.1 with the collected data for Case (c).

TABLE 15.2
Application of the Law of Malus

Laser Light Beam Power I_o (unit)	Measured Laser Light Beam Power I_1 (unit)	Calculated Laser Light Beam Power I_1 (unit)	Rotation Angle θ (degrees)	Measured Laser Light Beam Power I_2 (unit)	$\cos^2\theta$	Calculated Laser Light Beam Power $I_2 = I_1 \cos^2\theta$ (unit)
			0			
			10			
			20			
			30			
			40			
			50			
			60			
			70			
			80			
			90			
			100			
			110			
			120			
			130			
			140			
			150			
			160			
			170			
			180			

15.13.5.4 The Law of Malus

1. Measure the power of the laser beam before and after exiting from the first and second polarizer.
2. Fill out Table 15.2 with the collected data for this case.

15.13.6 CALCULATIONS AND ANALYSIS

15.13.6.1 Light Passing through a Polarizing Cube Beamsplitter

1. Compare the laser beam power at the input face and output faces of the cube beamsplitter.
2. Calculate the splitting ratio of the cube beamsplitter. Fill out Table 15.1 with the collected data for Case (a).

15.13.6.2 Glan Thompson Polarizing Beamsplitter Cube

1. Compare the laser beam power at the input face and output faces of the Glan Thompson polarizing beamsplitter.
2. Calculate the splitting ratio of the Glan Thompson polarizing beamsplitter. Fill out Table 15.1 with the collected data for Case (b).

15.13.6.3 Light Passing through a Calcite Material

1. Compare the laser beam power at the input face and the two laser beams at the output face of the calcite.
2. Calculate the splitting ratio of the calcite material. Fill out Table 15.1 with the collected data for Case (c).

15.13.6.4 The Law of Malus

1. Calculate the power of the light beam exiting from the first polarizer. Fill out Table 15.2.
2. Calculate $\cos^2\theta$ and the transmitted power I_2 from the second polarizer using Malus's law, Equation (15.6). Fill out Table 15.2.
3. Plot I_2 the measured transmitted power of the laser beam against angle θ.
4. Plot I_2 the calculated transmitted power of the laser beam against angle θ.

15.13.7 RESULTS AND DISCUSSIONS

15.13.7.1 Light Passing through a Polarizing Cube Beamsplitter

1. Report the laser beam power measurements at the input face and output faces of the cube beamsplitter and the splitting ratio.
2. Verify the results with the technical specifications provided by the manufacturer.
3. Determine the polarization orientation of the laser beam at the input face and output faces of the cube beamsplitter.

15.13.7.2 Glan Thompson Polarizing Beamsplitter

1. Report the laser beam power measurements at the input face and output faces of the Glan Thompson polarizing beamsplitter and the splitting ratio.
2. Verify the results with the technical specifications provided by the manufacturer.
3. Determine the polarization orientation of the laser beam at the input face and output faces of the Glan Thompson polarizing beamsplitter.

15.13.7.3 Light Passing through a Calcite Material

Report the laser beam power measurements at the input face and output faces of the calcite material and the splitting ratio.

Verify the results with the technical specifications provided by the manufacturer.

Determine the polarization orientation of the laser beam at the input face and output faces of calcite material.

15.13.7.4 The Law of Malus

Report the power of the light beam exiting from the first polarizer using Equation (15.7) with the measured data.

Present the plotted curves with the measured and calculated data. Comment on the shape of the curves.

15.13.8 CONCLUSION

Summarize the important observations and findings obtained in this lab experiment.

15.13.9 SUGGESTIONS FOR FUTURE LAB WORK

List any suggestions for improvements using different experimental equipment, procedures, and techniques for any future lab work. These suggestions should be theoretically justified and technically feasible.

15.14 LIST OF REFERENCES

List any references that were used in the report. Use one format in writing the references. Never mix reference formats in a report.

15.15 APPENDICES

List all of the materials and information that are too detailed to be included in the body of the report.

FURTHER READING

Astle, T.B., Gilbert, A. R., Ahmed, A., and Fox, S., Optical components—The planar revolution? Merrill Lynch & Co., The Global Telecom Equipment—Wireline, In-depth Report, U.S.A., May 17, 2000.

Alkeskjold, T. T., Laegsggaard, J., Bjarklev, A., Hermann, D., Anawati, A., Broeng, J., Li, J., and Wu, S. T., All optical modulation in dye-doped nematic liquid crystal photonic bandgap fibers, *Optics Express*. vol. 12(24), pp. 5857–5871, November 20, 2004.

Blaker, W. J., *Optics: The Matrix Theory*, Marcel Dekker, New York, 1972.

Brower, D. L., Ding, W. X., and Deng, B. H., *Fizeau Interferometer for Measurement of Plasma Electron Current*, Electrical Engineering Department, University of California, Los Angeles, U.S.A., vol. 75(10), pp. 3399–3401, October 2004.

Callister, W. D. Jr., *Materials Science and Engineering—An Introduction*, 3rd ed., John Wiley & Sons, New York, U.S.A., 1994.

Chee, J. K. and Liu, J. M., Polarization-dependent parametric and raman processes in a birefringent optical fiber, *IEEE J. Quantum Electron.*, 26, 541–549, 1990.

Chen, J. H., Su, D. C., and Su, J. C., Holographic spatial walk-off polarizer and its application to a 4-port polarization independent optical circulator, *Opt. Express*, 11 (17), 2001–2006, 2003.

Chen, J. H., Su, D. C., and Su, J. C., Multi-port polarization-independent optical circulators by using a pair of holographic spatial- and polarization- modules, *Opt. Express*, Vol. 12 (4), 601–608, 2004.

Drisoll, W. G. and Vaughan, W., *Handbook of Optics*, McGraw Hill Book Company, New York, 1978.

EPLAB, *EPLAB Catalogue 2002*, The Eppley Laboratory, Newport, RI, 2002.

Fedder, G. K. and Howe, R. T., Multimode digital control of a suspended polysilicon microstructure, *IEEE J. MEMS*, 5 (4), 283–297, 1996.

Francon, M., *Optical Interferometry*, Academic Press, New York pp. 97–99, 1966.

Hariharan, P., Modified Mach-Zehnder interferometer, *Appl. Opt.*, 8 (9), 1925–1926, 1969.

Hecht, J., *Understanding Fiber Optics*, 3rd ed., Prentice Hall, Inc., Englewood Cliffs, NJ, U.S.A., 1999.

Hecht, E., *Optics*, 4th ed., Addison-Wesley Longman, Inc., San Diego, CA, U.S.A., 2002.

Hood, D. C. and Finkelstein, M. A., In *Sensitivity to Light Handbook of Perception and Human Performance Vol. 1 of Sensory Processes and Perception*, Boff, K. R., Kaufman, L., and Thomas, J. P., Eds., Wiley, Toronto, 1986.

Lambda Research Optics, Inc., *Lambda Catalog 2004*, Lambda Research Optics, Inc., Costa Mesa, California, 2004.

Lerner, R. G. and Trigg, G. L., *Encyclopedia of Physics*, 2nd ed., VCH Publishers, Inc., New York, 1991.

Melles Griot, The practical application of light, Melles Griot, Inc., *Melles Griot Catalog*, Rochester, NY, 2001.

Newport Corporation, *Projects in Fiber Optics Applications Handbook*, Newport Corporation, Newport, RI, U.S.A., 1986.

Newport Corporation. Optics and mechanics section, the Newport Resources 1999/2000 Catalog, Newport Corporation, Irvine, CA, 1999/2000.

Newport Corporation, Photonics Section, the Newport Resources 2004 Catalog, Newport Corporation, Irvine, CA, U.S.A., 2004.

Ocean Optics, Inc., *Product Catalog 2003*, Ocean Optics, Inc., Florida, U.S.A., 2003.

Palais, J. C., *Fiber Optic Communications*, 4th ed., Prentice Hall, Inc., Englewood Cliffs, NJ, 1998.

Salah, B. E. A. and Teich, M. C., *Fundamentals of Photonics*, Wiley, New York, 1991.

Shamir, J., *Optical Systems and Processes*, SPIE Optical Engineering Press, Washington, DC, 1999.

Shen, L. P., Huang, W. P., Chen, G. X. et al., Design and optimization of photonic crystal fibers for broad-band dispersion compensation, *IEEE Photon. Technol. Lett.*, 15, 540–542, 2003.

Smith, W. J., *Modern Optical Engineering*, McGraw-Hill Book Company, New York, 1966.

Von Weller, E. L., Photonics crystal fibers advances in fiber optics, *Appl. Opt.*, March 1, 2005.

Woods, N., *Instruction's Manual to Beiser Physics*, 5th ed., Addison-Wesley Publishing Company, San Diego, CA, 1991.

Yariv, A., *Optical Electronics*, Wiley, New York, 1997.

Zhang, L. and Yang, C., Polarization celective coupling in the holey fiber with Aasymmetric dual cores. LEOS2003, Tucson, U.S.A., paper ThP1, Oct 26–30, 2003.

Zhang, L. and Yang, C., Polarization splitter based on photonic crystal fibers, *Opt. Express*, 11 (9), 1015–1020, 2003.

Zhang, L. and Yang, C., Polarization splitting with dual-core holey fibers., CLEO/PACIFIC RIM 2003, Taipei Taiwan, paper W4C-(14)-3, 2003.

16 Optical Materials

16.1 INTRODUCTION

Optical materials are used in the construction of devices that guide light in some way. These materials refract light. This can be observed when light passes through a prism or a lens. The angle through which light is refracted is a function of the wavelength of the light, so that when more than one frequency is present the light can be dispersed into different components. Most of us have observed the rainbow produced when white light is passed through a prism.

Optical materials can also be used to attenuate, or decrease the intensity of light. Under the right conditions, some can also amplify, increasing the intensity of light. Optical materials are used to transmit light, not only visible, but also light in the infrared and ultraviolet ranges. Historically, glass was the first optical material used in the fabrication of imaging systems, such as telescopes and microscopes. As modern applications have evolved, other materials better suited to extended wavelength ranges and other design considerations have been developed.

16.2 CLASSES OF MATERIALS

The following sections introduce different classes of materials used in the fabrication of optical devices and instruments.

16.2.1 GLASS

Glass is characterized by an amorphous molecular structure, where the chemical bonds holding the material together do not have a regular lattice, such as is found in crystals. Most glass used in optical applications is based on silica (i.e., silicon dioxide) with other materials mixed in to control various optical properties. This creates a new glass type with different index of refraction and sometimes a different colour. Pure silica has a melting point closer to 2000°C. Often soda or potash is added to lower the melting point to about 1000°C. When soda is added to glass, the mixture becomes slightly soluble in water. Lime (calcium oxide) needs to be added to the mix to return the material to an insoluble form.

Most glass types will transmit light well through the visible range, into the near infrared. In order to build optical systems with good transmission in the ultraviolet, it is necessary to employ more exotic materials, such as quartz, fused silica, and sapphire, as shown in Figure 16.1. Some of

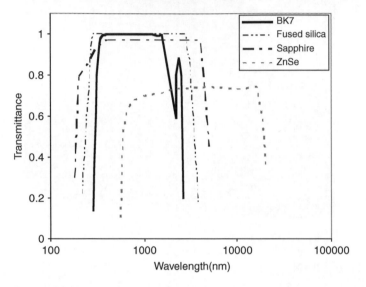

FIGURE 16.1 Transmission of various optical materials.

FIGURE 16.2 Glass lens formed by grinding and polishing with diamond tools.

the motivation for developing materials with improved ultraviolet transmission comes from the photolithography industry, which needs to work at shorter and shorter wavelengths.

At longer wavelengths, it is sometimes necessary to use other materials, which have transmission through to the far infrared. These materials include zinc selenide (ZnSe), zinc sulphide (ZnS), and germanium (Ge).

Usually, optical components are fabricated by grinding and polishing these optical materials. This process involves rotating the optic on a spindle and bringing a grinding or polishing tool in contact with it, as shown in Figure 16.2. The tool also rotates, and an abrasive is delivered in the form of a slurry. By reducing the grit size of the abrasive at various stages, the optical surface of the part can be made smoother.

16.2.2 DOPANTS

As discussed in the Section 16.2.1, dopants can be added to glass to change its optical, mechanical and thermal properties. Dopants can also be added to semiconductor materials to change the conductivity by selectively increasing the number of free electrons or holes in the material. For example, when silicon is doped with arsenic, there is an electron left over after the atom forms bonds with the silicon crystal. This electron can be easily promoted into the valence band by thermal energy, thus becoming a free carrier. In this case, the arsenic is a donor, and the material is called n-type.

If silicon is doped with boron, boron leaves one unpaired electron when it forms a bond in the silicon lattice. This unpaired electron can accept an electron from the valence band, which leaves behind a hole. A hole can move around in the lattice under the influence of an applied electric field. Here boron is an acceptor, and the material is termed p-type. Some dopants are shown in Figure 16.3.

FIGURE 16.3 Dopant materials.

A designer can manipulate the regions where different kinds of doping occurs, to design devices, such as light emitting diodes, diode lasers, as well as photo-detectors.

16.2.3 CO-DOPANTS

In some instances, both electrons donors and acceptors are doped into the optical material. Co-dopants can be added to glass to change its optical, mechanical and thermal properties.

Co-dopants can also be added to semiconductor materials to change the conductivity by selectively increasing the number of free electrons or holes in the material. Some of the free electrons will become trapped in the holes, and the free carriers will tend to cancel each other out. The net effect being that the dopant with the higher concentration will determine whether the material is n-type or p-type. In this case, the dopant with the lower concentration is called the co-dopant.

16.2.4 POLYMERS IN PHOTONICS

Optical polymers are transparent plastics, which can be used for lenses, optical storage media, such as compact discs, and for lightguides, metallized reflectors, optical fibres, and couplers. Materials commonly used include polycarbonates (PC), acrylics (polymethyl methacrylate (PMMA)), polystyrene (PS), allyl diglycol carbonate (ADC), and cyclic olefin copolymer (COC). These materials are lighter and cheaper to fabricate than glass, but have some additional challenges to overcome for demanding applications. A summary of some of the properties of these materials is given in Table 16.1.

TABLE 16.1
Properties of Some of the Optical Polymers

Property	PC	PMMA	PS	ADC	COC
Transmittance (%) in the 400–700 nm range of a 3 mm thick sample	85–91	92	87–90	89–91	92
Index of refraction	1.586	1.491	1.589	1.5	1.533
Abbe number	30.0–30.3	57.4	31	58	58
Density g/cm^3	1.2	1.18	1.06	1.31	1.01

These materials are usually injection-molded or cast, but can be diamond turned as well. Their operating temperature range is much smaller than that of glass, and extra care must be taken in applications where large optical power is transmitted.

16.3 APPLICATIONS

The following sections introduce the types of lenses used in building optic and optical fibre devices and instruments.

16.3.1 REFLECTORS

Reflections typically occur at the interface between two dissimilar materials, a common example is air and a coated or uncoated mirror. In the applications of reflectors, the bulk optical properties of the reflector material are unimportant. The properties considered will be the environmental properties of the material and its manufacturability.

Typically the surface accuracy of an optic is measured in terms of the wavelength of light. The HeNe laser's wavelength at 632 nm is most commonly used. In some applications, the optic is required to perform over a wide range of temperatures. In order to achieve the required performance of the optic, it is not only necessary to manufacture the optic within its required surface accuracy, but also to use a material that maintains that surface accuracy over the specified temperature range. Some materials undergo thermal expansion when heated, and in the case of reflectors, may undergo

FIGURE 16.4 Glass is often delivered as rough cut pieces or blanks, which must be ground or pressed to the approximate shape before polishing is started.

surface distortion, which may compromise the performance of the optic. Clearly, this is more of an issue for a reflecting mirror in a space telescope than for a flashlight reflector.

Figure 16.4 shows that glass is often delivered as roughly cut pieces or blanks, which must be ground or pressed to the approximate shape before polishing.

A variety of techniques can be used to fabricate a reflector. An inexpensive method is by molding. This process is suitable for plastics and glass. Glass may also be formed by pressing it in a metal mold at a temperature where it has softened.

Molding and pressing are not capable of producing high surface accuracy. Molded plastic optics will typically shrink by up to about 1%, after being released from the mold. Manufacturers can take this into account by using a mold with larger dimensions than required for the finished optic. Even so, surface irregularities of a few waves are typical. This means that the difference between the desired surface shape and the actual surface shape of the optic can be a few times the HeNe laser wavelength.

To fabricate optics with higher surface accuracy, other processes are used. Conventional grinding and polishing of glass will yield surfaces well below one wave in accuracy, and in some instances better than 1/50 wave.

Single point diamond turning is another process used to manufacture accurate surfaces. In this process, the optic substrate is rotated on a spindle, while a computer controlled diamond cutting tool is moved to cut the desired surface. Some metals, such as nickel, are suitable for diamond turning.

16.3.2 LENSES

The refractive indices of all materials change with wavelength. This can lead to optical distortions, such as chromatic aberration, occurring when the light source in the system is not monochromatic. An optical designer can reduce these distortions by using a combination of different glass types. The lens combination balances these refractive deviations in such a way that light of different

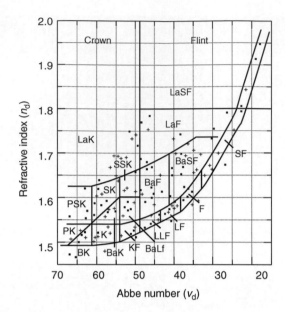

FIGURE 16.5 Index of refraction versus the Abbe number for different glasses.

wavelengths will focus at the correct position in the image plane. The glass map available to optical designers, shown in Figure 16.5, is a plot of the index of refraction versus the Abbe number for different glasses. The Abbe number is a measure of the chromatic dispersion of the glass: larger values correspond to lower dispersion.

16.3.3 FIBRE OPTICS

Fibre optics used in communication systems suffer from chromatic aberration effects. In an optical fibre transmitting digital information in the form of light pulses, chromatic aberration manifests itself as dispersion. It is impossible to form light pulses from purely monochromatic light. In fact, there is an inverse relationship between how short a pulse can be made, and how much bandwidth or spread in frequencies is required from light source. In order to increase the data rate travelling through a fibre optic system, it is desirable to operate with shorter pulses, which in turn require a larger range of frequencies in a pulse. As these pulses travel through the fibre, the longer wavelength components tend to travel slightly faster, since they encounter a slightly lower index of refraction. This causes the pulse to be stretched in time, and the amount it stretches increases with fibre length. At some point in a long fibre, the tail end of one pulse will overlap the leading edge of the next pulse, which limits the ability of the system to transmit information accurately.

In order to build systems that are able to transmit shorter pulses over long distances, fibre manufacturers attempt to use materials with low dispersion. In order to achieve low dispersion, dopants are added to the glass, to reduce dispersion in the wavelength range in which the fibre will be used.

Another important requirement for fibres used in long haul communication applications is low absorption. Most of the intrinsic absorption that occurs in fibres results from hydroxyl impurities in the glass core. Fibre manufacturers work hard to control their manufacturing processes, in order to minimize these impurities.

Light is guided in fibres by engineering an index of refraction difference between the core and the cladding of the fibre. In general, the core of the fibre needs to have a higher index of refraction in order to achieve this effect. Germanium, titanium or phosphorus can be added to glass, to increase the index of refraction. Boron or fluorine may be added to decrease the index of refraction.

Optical fibres can also be doped for other purposes. The fibre may be doped to provide a mechanism for the fibre to act as an optical amplifier. A fibre doped with erbium is used to act as an amplifier in the important communication wavelength range around 1550 nm.

16.3.4 MECHANICAL COMPONENTS

The materials used to make up the mechanical components of an optical system should also receive careful consideration. Strength and stiffness have ramifications in applications where the optical assembly may be subject to vibration and mechanical shock.

Some applications may have a weight limit; sometimes significant weight savings can be made by utilizing, for example, aluminum housings instead of steel. In order to reduce weight further, sometimes designers must use more exotic materials, such as titanium and carbon fibre composites.

Some systems must be designed to perform over large temperature ranges. This can be a particular challenge for infrared systems, as the index of refraction of many materials can be very temperature dependent. To compensate for the resulting change in focal lengths of components in these systems, one method uses materials with predictable thermal expansion properties. With careful mechanical design, it is possible to compensate for temperature dependent optical variations, by introducing temperature dependent mechanical changes. The optical and mechanical variables cancel each other out.

There are materials available to move optical components with the application of a bias voltage. One class of these materials is called piezoelectric. These materials are crystals, which acquire a charge when they are mechanically compressed. Conversely, a piezoelectric undergoes a dimensional change when a voltage is applied across it. Piezoelectric can be used to act as transducers for adjusting the position of optics. They have also been used to build telescopes with improved performance when viewing through distortions caused by atmospheric turbulence. Piezoelectric transducers (PZTs) are installed on the backs of mirrors used in these telescopes in such a way that by adjusting voltages the optical surface of the mirror can be distorted. This distortion can be used to compensate for distortion in the atmosphere, such that the two distortions cancel each other out, and a clearer view of celestial objects can be made. This field of adaptive optics has revolutionized the use of earth-based telescopes for astronomy.

16.4 EXPERIMENTAL WORK

In this experiment, the voltage coefficient of a piezoelectric transducer will be measured. Hysteresis will also be observed.

16.4.1 TECHNIQUE AND APPARATUS

Appendix A presents the details of the devices, components, tools, and parts.

1. 2×2 ft. optical breadboard.
2. HeNe laser source and power supply.
3. Laser mount assembly.
4. Hardware assembly (clamps, posts, screw kits, screwdriver kits, sundry positioners, etc.).
5. A Michelson interferometer will be required. The end mirror on one of the arms should be mounted on a spring-loaded single axis translation stage, as shown in Figure 16.6.
6. Beamsplitter and Beamsplitter holder/positioner assembly.
7. Diverging lens and lens holder/positioner assembly, as shown in Figure 16.6.
8. Piezoelectric stack and translation stage, as shown in Figure 16.7.

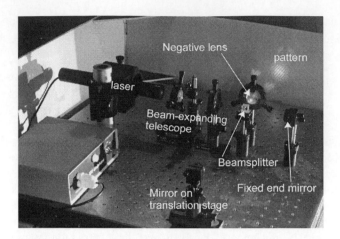

FIGURE 16.6 Interferometer set-up.

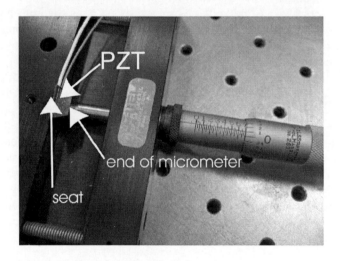

FIGURE 16.7 Installation of PZT into a translation stage.

9. DC voltage supply, with sufficient range to adjust the PZT stack through several microns of travel, as shown in Figure 16.7.
10. Voltmeter.
11. Black/white card and cardholder.
12. Protractor.
13. Viewing screen.
14. Ruler.

16.4.2 PROCEDURE

Follow the laboratory procedures and instructions given by the professor and/or instructor.

16.4.3 SAFETY PROCEDURE

Follow all safety procedures and regulations regarding the use of optical instruments and measurements, and light source devices.

16.4.4 APPARATUS SET-UP

16.4.4.1 Set Up of Interferometer

1. Figure 16.6 shows the experimental set-up for the interferometer.
2. Connect the PZT stack to the DC voltage supply.
3. Install the PZT stack in the spring-loaded single axis translation stage, as shown in Figure 16.7. The PZT should be placed in between the end of the spindle of the micrometre and the seat of the translation stage. Make sure that the wires do not interfere with the motion of the stage.
4. Connect the voltmeter to the DC voltage supply, to monitor the voltage applied to the PZT.
5. Install the single axis stage in the interferometer such that one of the arms could be lengthened by adjusting it.
6. Align the interferometer to the monochromatic source. Adjust the tilt of the end mirrors to get a few fringes at the output.
7. Install the diverging lens such that the output projected onto the viewing screen has a fringe spacing of a couple of centimetres.
8. Place the ruler perpendicular to the fringes on the screen.
9. Record the fringe spacing as measured on the viewing screen.
10. Setup a table for recording fringe displacement as a function of power supply voltage. Plan to take about 20 measurements through the voltage range of the DC voltage supply. Do not exceed the maximum voltage that the PZT can handle.
11. Mark the starting position of one of the fringes on the viewing screen. Turn on the power supply and ramp the voltage to the first point in the table. Do not reverse the direction of voltage adjustment. Record the distance the fringe has shifted from its original position and record the voltage.
12. Repeat for all of the voltages, up to the maximum range of the DC supply or the PZT, which ever is less.
13. Reverse the direction of voltage adjustment. Measure displacements at the same intervals, all the way down to zero voltage.
14. Reverse the positive and negative connections to the PZT at the power supply.
15. Repeat the measurements of fringe displacement through the range of the power supply in both the upward and downward direction. Remember that these voltages should be recorded with the opposite sign compared to the earlier measurements.

16.4.5 DATA COLLECTION

1. Record the distance the fringe has shifted from its original position and record the voltage. Repeat for all voltages.
2. Convert the displacement column from screen displacement to PZT displacement. Remember that the fringe spacing corresponds to one wavelength of extra travel in the arm of the interferometer (and that the light travels double pass through that arm).

16.4.6 CALCULATIONS AND ANALYSIS

1. Plot the data. Is there any observed hysteresis in the data?
2. Calculate the slope of the curve and determine the voltage coefficient.

16.4.7 RESULTS AND DISCUSSIONS

1. Present and discuss the plot of the data and the observation of a hysteresis in the data.
2. Present the slope of the curve and the voltage coefficient.

16.4.8 CONCLUSION

Summarize the important observations and findings obtained in this lab experiment.

16.4.9 SUGGESTIONS FOR FUTURE LAB WORK

List any suggestions for improvements using different experimental equipment, procedures, and techniques for any future lab work. These suggestions should be theoretically justified and technically feasible.

16.5 LIST OF REFERENCES

List any references that were used in the report. Use one format in writing the references. Never mix reference formats in a report.

16.6 APPENDIX

List all of the materials and information that are too detailed to be included in the body of the report.

FURTHER READING

Agrawal, G. P., *Nonlinear Fiber Optics Optics and Photonics*, 2nd ed., Academic Press, New York, 1995.

Becker, P. M., et al., *Erbium-Doped Fiber Amplifiers: Fundamentals and Technology*, Elsevier, Amsterdam, 1999.

Charschan, S., *Lasers in Industry*, Van Nostrand, New York, 1972.

Davis, C. C., *Lasers and Electro-Optics, Fundamental and Engineering*, Cambridge University Press, New York/Melbourne, 1996.

Digonnet, M. J. F., *Rare-Earth Fiber Lasers and Amplifiers*, Marcel Dekker, New York, 2001.

Edmund Industrial Optics, *Optics and Optical Instruments Catalog*, New Jersey, USA, 2004.

Eggleton, B. J., et al., Microstructured optical fiber devices, *Opt. Express*, December 17, 698–713, 2001.

Fedder, G. K., Iyer, S., and Mukherjee, T., Automated optimal synthesis of microresonators, In *Technical Digestion of the IEEE International Conference on Solid-State Sensors and Actuators (Transducers 97)*, Vol. 2, Chicago, IL, pp. 1109–1112, June 16–19, 1997.

Ghatak, A. K., *An Introduction to Modern Optics*, McGraw-Hill Book Company, New York, 1972.

Griffel, G., et al., Low-threshold InGaAsP ring lasers fabricated using bi-level dry etching, *IEEE Photon. Technol. Lett.*, 12, 146–148, 2000.

Hecht, E., *Optics*, 4th ed., Addison-Wesley Longman, Inc., Reading, MA, 2002.

Heidrich, H., et al., Passive mode converter with a periodically tilted InP/GaInAsP rib waveguide, *IEEE Photon. Technol. Lett.*, 4, 34–36, 1992.

Hurvich, L. M. and Jameson, D., A psychophysical study of white I adoption as variant, *J. Opt. Soc. Am.*, 41, 521–527, 1951.

Hurvich, L. M. and Jameson, D., A psychophysical study of white III adoption as variant, *J. Opt. Soc. Am.*, 41, 701–709, 1951.

Jeong, Y., et al., Ytterbium-doped double-clad large-core fibers for pulsed and CW lasers and amplifiers, *SPIE*, 140–150, 2004.

Kao, C. K., *Optical Fiber Systems: Technology, Design, and Applications*, McGraw-Hill, New York, 1982.

Koga, M., Compact quartzless optical quasi-circulator, *Electron. Lett.*, 30, 1438–1440, 1994.

Lerner, R. G. and Trigg, G. L., *Encyclopedia of Physics*, 2nd ed., VCH Publishers, Inc., New York, 1991.

Malacara, D., *Geometrical and Instrumental Optics*, Academic Press Co, Boston, MA, 1988.

McComb, G., *The Laser Cookbook- 88 Practical Projects*, Tab Book, Division of McGraw-Hill, Inc., New York, 1988.

Newport Corporation, *Optics and Mechanics*, Newport, 1999/2000 Catalog.

Salah, B. E. A. and Teich, M. C., *Fundamentals of Photonics*, Wiley, New York, 1991.

SCIENCETECH, *Designers and Manufacturers of Scientific Instruments Catalog*, London, Ontario, Canada, 2003.

Serway, R. S., *Serway: Physics for Scientists and Engineers*, 3rd ed., Saunders Golden Sunburst Series, London, 1990.

Shashidhar, N., *Lensing Technology*, Corning Incorporated, Fiber Product News, pp. 14–15, 2004.

Shen, L. P., Huang, W. P., Chen, G. X., and Jian, S. S., Design and optimization of photonic crystal fibers for broad-band dispersion compensation, *IEEE Photon. Technol. Lett.*, 15, 540–542, 2003.

Smith, W. J., *Modern Optical Engineering*, McGraw-Hill Book Co., New York, 1966.

Thompson, G. H. B., *Physics of Semiconductor Laser Device*, Wiley, Chichester, 1980.

Vail, E., et al., GaAs micromachined widely tunable Fabry–Perot filters, *Electron. Lett.*, 31 (3), 228–229, 1995.

Weisskopf, V. F., How Light Interacts with Matter, Scientific American, pp. 60–71, 1968.

Yariv, A., *Optical Electronics*, Wiley, New York, 1997.

Yariv, A., Universal relations for coupling of optical power between microresonators and dielectric waveguides, *Electron. Lett.*, 36, 321–322, 2000.

Yeh, C., *Handbook of Fiber Optics: Theory and Applications*, Academic Press, San Diego, CA, 1990.

Yeh, C., *Applied Photonics*, Academic Press, New York, 1994.

Zirngibl, M., Joyner, C.II., and Glance, B., Digitally tunable channel dropping filter/equalizer based on waveguide grating router and optical amplifier integration, *IEEE Photon. Technol. Lett.*, 6, 513–515, 1994.

Section III

Waves and Diffraction

17 Waves

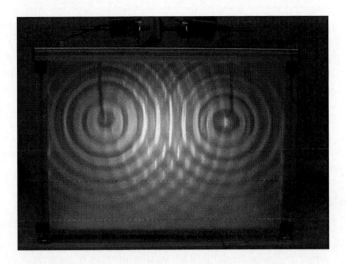

17.1 INTRODUCTION

Previous chapters explained that light acts as a stream of particles that allow the light to transfer from one point to another. The coming chapters will deal with light as a wave. The purpose of this chapter is to explain the basic principles of light when transmitted, reflected, or refracted through an optical material as a wave.

Energy can be transmitted from one place to another by vibrating objects, such as water waves that travel hundreds of kilometres over the ocean. The water particles move up and down as the wave passes. Similarly, when you shake a spiral spring, your energy is transferred from coil to coil down the spring. A wave is a transfer of energy in the form of vibrating particles in a medium. We live in a world surrounded by waves; some are visible and others are not. Water waves and the waves generated by a rope or a spring can be seen. Sound waves and radio waves cannot be seen. Waves also occur in light, sound, heat, microwaves, and in the ultra-microscopic world of atoms. Several types of waves and their applications will be presented in this chapter. Also in this chapter, along with the theoretical presentation, three experimental cases demonstrate the principles of Hook's law, wave generation, and the simple pendulum.

17.2 THE NATURE OF WAVES

17.2.1 ENERGY TRANSFER

There are various ways in which energy can be transferred from one place to another. The flow of heat through a metal from a region of high temperature to one of low temperature represents one method of transferring energy. The flow of electricity through a metal is somewhat analogous to heat flow. The conduction of heat energy and electric energy through metals depends upon the motion of particles that compose the metal.

The transfer of energy by the gross movement of materials or objects from one place to another is the basic principle of the nature of waves. Winds, tides, and projectiles in flight are examples for this type. Another method of energy transfer involves wave motion when a stone thrown into a quiet pond produces a familiar wave pattern on the surface of the water. A sound is heard because a wave disturbance travels from the source through the intervening atmosphere. The energy released by a great explosion can shatter windows far from its source. The shock wave of a supersonic airplane can have similar destructive effects.

A wave is a disturbance that moves through a medium. Previously, all chapters explained the properties of light by means of a particle theory. This chapter explains the properties of light by means of a wave theory. Physicists believe that light waves, radio waves, infrared and ultraviolet waves, x-rays, and gamma rays are fundamentally similar in their behaviour. These are all electromagnetic waves.

Light transmits through space as an electromagnetic wave. One of the most significant developments in physics during the last century has been the discovery that light has wave properties, such as wave interference.

17.2.2 Mechanical Waves

A mechanical wave is a disturbance that moves through a material. A source of energy is needed to produce mechanical waves. Energy produces the disturbance, and an elastic medium is needed to transmit the disturbance. An elastic medium behaves like an array of particles connected by springs, with each particle at equilibrium, as shown in Figure 17.1.

FIGURE 17.1 An elastic medium behaves like an array of particles connected by springs.

If particle 1 is displaced from its equilibrium position by being pulled away from particle 2, particle 1 is immediately subjected to a force (pulling right) from particle 2. This force attempts to restore particle 1 to its original position. At the same time, particle 1 exerts an equal but opposite force (pulling left) on particle 2, which attempts to displace particle 2 from its equilibrium position.

Similar events occur, but in opposite directions, if particle 1 is displaced from its equilibrium position by being pushed toward particle 2. If particle 1 is displaced permanently by an external source, particle 1 exerts a force, which displaces particle 2. Particle 2, in turn, being displaced from its equilibrium position, exerts a force on particle 3, which is in turn displaced. In this way, the displacement travels along from particle to particle.

Because the particles have inertia, the displacements do not all occur at the same time, but successively as the particles farther and farther from the source move. The kinetic energy imparted to the first particle by the source is transmitted from particle to particle in the medium. This mode of transmission of energy creates mechanical waves.

17.2.3 Elastic Potential Energy

When one pulls out a spring, the spring resists being stretched, and it returns to its original position if it is released. As explained above, this is known as elastic behaviour. If a spring is stretched, a restoring force comes to return the spring to its original length, as shown in Figure 17.2. The more the spring is stretched, the greater the restoring force that can be obtained. Similarly, when a spring is compressed,

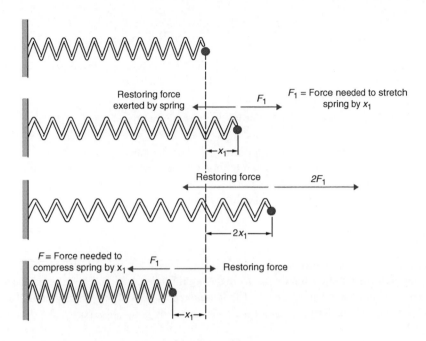

FIGURE 17.2 Hook's law.

a restoring force returns the spring to its original length, as shown in Figure 17.2. Again, a restoring force arises, and the more the compression, the greater the restoring force to be overcome.

The displacement, x, by which a spring is stretched or compressed by a force, is directly proportional to the magnitude of the force, F. The relation between the displacement and restoring force is called Hook's law, described by the following equation:

$$F = -kx \tag{17.1}$$

where k is the spring constant.

For many materials, k is constant if the displacement x is not too large. The spring constant is also known as the stiffness constant. The negative sign in Hook's law equation indicates that the restoring force due to the spring is in the opposite direction of the displacement.

The work needed to stretch or compress a spring that obeys Hook's law can be calculated. The work done by a force is the product of the magnitude of the force and the distance through which the force acts, as shown in Figure 17.3. Because the force F is proportional to the displacement x, the average force $\bar{F}$ applied while the spring is stretched from its normal length of $x = 0$ to its final length $x = x$. The average force is given by:

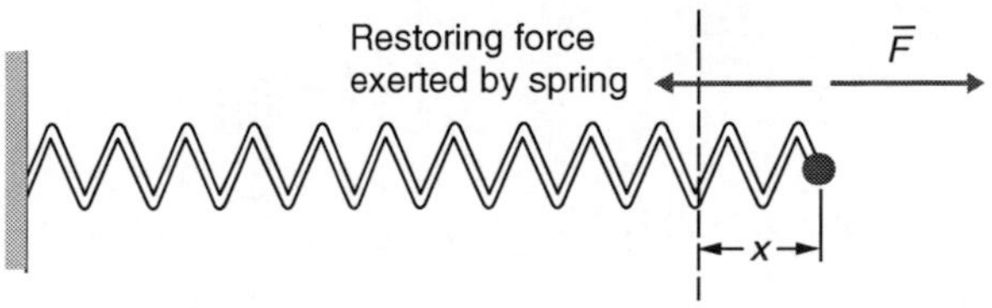

FIGURE 17.3 Elastic potential energy.

$$\bar{F} = \frac{F_{x=0} - F_{x=x}}{2} = \frac{0 + kx}{2} = \frac{1}{2}kx \qquad (17.2)$$

The work done in stretching the spring is the product of the average force and the total distance (displacement $= x$) in the direction of the force. Thus, the work done is given by:

$$W = \text{Potential Energy} = \frac{1}{2}kx^2 \qquad (17.3)$$

Equation 17.3 is most often used in calculating the work done by a spring when the spring is stretched or compressed. This work of stretching or compressing converts into elastic potential energy. When the spring is released, its potential energy is transferred into kinetic energy. This kinetic energy can then be applied to a mechanical system.

17.2.4 VIBRATING SPRING

Consider a block of mass m placed on a frictionless surface, as in Figure 17.4(a). The block is displaced, due to an external force F, by a distance of x from the equilibrium position. This external force keeps the block in the new position shown in Figure 17.4(b). If the force is released, the elastic potential energy in the spring accelerates the block. The acceleration a is described by Newton's second law of motion:

$$F = -kx = ma \qquad (17.4)$$

The acceleration a is found by combining Newton's second law of motion with Hook's law:

$$a = -\frac{kx}{m} = \frac{F}{m} \qquad (17.5)$$

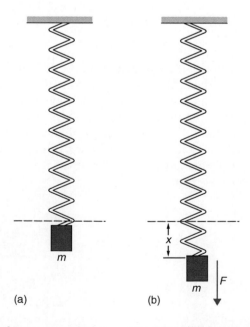

(a)   (b)

FIGURE 17.4 (a) A block of mass m attached to a spring at equilibrium position; (b) Block displaced by distance x due to force F.

The acceleration is proportional to the displacement and in the opposite direction. The proportionality between acceleration and displacement occurs in many types of vibrating objects, such as clock pendulums, children's swings, and rotating arms.

17.3 TYPES OF WAVES

There are two basic types of wave motion for mechanical waves: longitudinal and transverse. The following sections demonstrate these types of waves and illustrate the difference between the motion of the wave and the motion of the particles in the medium through which the wave is travelling.

17.3.1 TRANSVERSE WAVES

In a transverse wave, the particle displacement is perpendicular to the direction of wave propagation. Figure 17.5 shows a one-dimensional transverse plane wave propagating from left to right by a spring. The particles do not move along with the wave; they simply oscillate up and down about their individual equilibrium positions as the wave propagates.

As shown in this figure, the spring is attached to a wall. To generate a wave, pull on the free end with your hand, producing a tension in the spring, and then move your hand up and down. This action generates a wave pulse that will travel along the spring towards the wall. When the hand moves up and down with simple harmonic motion, the wave on the spring will have the shape of a sine or cosine wave. The motion of these waves is known as a simple harmonic motion. More details on the sine and cosine waves will be presented in the following sections.

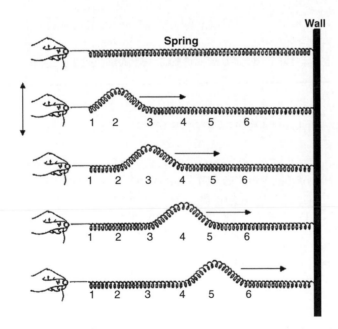

FIGURE 17.5 A transverse pulse travelling along the spring.

17.3.2 LONGITUDINAL WAVES

A longitudinal wave is easer to see when a spring has a large diameter. The spring is tied to a wall, as shown in Figure 17.6. Compress the spring several coils closer together at one end. Such a distortion is called compression. If these compressed coils are released, they attempt to spread out to their equilibrium positions, compressing the coils immediately to the right. In this way, the

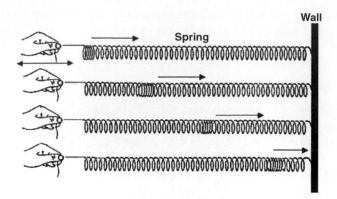

FIGURE 17.6 A compression pulse travelling along the spring.

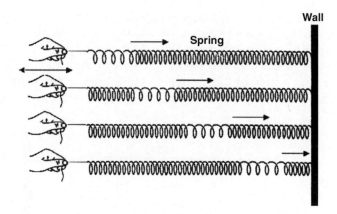

FIGURE 17.7 A rarefaction pulse travelling along the spring.

compression moves toward the wall, generating pulses. These pulses travel along the spring by displacing the particles of the spring parallel to the direction that the wave is travelling. Again, the wave transfers kinetic energy from particle to particle along the medium, without motion of the medium as a whole. This is called longitudinal wave motion.

If the coils at the left end of the spring are stretched apart, instead of compressed, a rarefaction is formed, as shown in Figure 17.7. Such a distortion is called a tension. When the coil compression is released, the rarefaction will travel along the spring.

17.3.3 WATER WAVES

Water waves are an example of waves that involve a combination of both longitudinal and transverse waves. Figure 17.8 shows the movement of the water as a wave carries a small piece of cork. While the water wave passes, the motion of the cork will trace the motion of the water surface. As a wave travels through the water, the cork travels in clockwise circles. The radius of these circles decreases as the depth of the wave increases. Notice that the cork moves in a roughly circular path, easily returning to approximately its starting point. Thus, each particle of the water moves both vertically and horizontally as the wave propagates in the horizontal direction.

From this demonstration, it can be seen that the water wave is a combination of both longitudinal and transverse waves. The water wave is difficult to analyse using a simple formula.

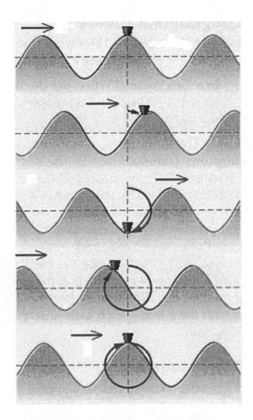

FIGURE 17.8 The movement of a water wave carrying a piece of cork.

Computer software is needed to solve the continuity, momentum, and energy equations for the variables of the water or other fluid in use.

17.3.4 RAYLEIGH SURFACE WAVES

Another example of waves having both longitudinal and transverse motion, one that may be found in solids, is known as Rayleigh surface waves. The particles in a solid through which a Rayleigh surface wave passes move in elliptical paths, with the major axis of the ellipse perpendicular to the surface of the solid. As the depth into the solid increases, the width of the elliptical path decreases. Rayleigh waves are different from water waves in one important way: In water waves, all particles travel in clockwise circles, as explained above. However, in a Rayleigh surface wave, particles at the surface trace out a counter-clockwise ellipses, while particles at a depth of more than 1/5th of a wavelength trace out clockwise ellipses. Rayleigh surface waves are more complicated to analyse than other wave types.

17.4 CHARACTERISTICS OF WAVES

All waves have several common characteristics. A wave has a finite speed, v, within a given transmitting medium. Wave speed may be quite slow, as with water waves; or it may be moderately fast, as with sound waves, which travel at speeds of 10^2–10^3 m/s. Waves may travel at the ultimate speed, that of light or radio waves, at 3×10^8 m/s.

The speed of a wave depends primarily on the nature of the wave disturbance and on the medium through which it passes. Under very close examination, the speed of wave propagation in certain media, called dispersive media, is also found to depend slightly on wave frequency. As a wave travels through a medium, the particles of the medium vibrate about their equilibrium positions in an identical fashion. However, the particles are in corresponding positions of their vibratory motion at different times. The position and motion of a particle indicate the phase of the wave. Particles that have the same displacement and are moving in the same direction are said to be in phase.

17.5 TRIGONOMETRIC NOTATION OF WAVES

An angle θ is generated by rotating a line about its fixed endpoint from an initial position to a terminal position. If the rotation of the line from the initial side is counterclockwise, the angle is defined as positive. The angles are measured in degrees (°) and radians (rad). The measurement unit of an angle can be changed from one measure to the other, by using the definition of π rad = 180°.

Figure 17.9 shows a particle P moving at constant speed in a circle of radius R having a projection, or shadow, that undergoes simple harmonic motion. The position of P is calculated by the velocity and acceleration as functions of time. The shadow of P on the x-axis oscillates back and forth between distances at $+x$ and $-x$. Because $\cos \theta$ is always between $+1$ and -1, the maximum displacement of the projection is equal to R. The position of the particle P at a specific time along the x-axis is given by:

$$x = R \cos \theta \qquad (17.6)$$

For uniform circular motion, the angle θ increases steadily with time t. Therefore, θ is proportional to t. The angular displacement of the particle at any time t is $\theta = \omega t$, and ω is the angular frequency. The particle linear speed in the circle is ωR, and centripetal acceleration is $\omega^2 R$. The time required to make one complete revolution ($\theta = 2\pi$) is the period T. Thus, the relationship between the angle in radians and the time t is given by:

$$\theta = 2\pi \frac{t}{T} \qquad (17.7)$$

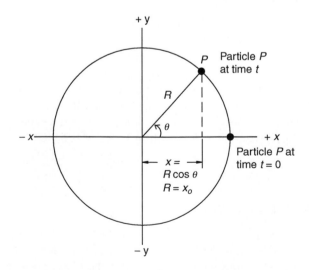

FIGURE 17.9 A particle P moving around a circular path.

Using the definition of frequency $f = 1/T$ and angular frequency $\omega = 2\pi f$ from the uniform circular motion definition, Equation 17.7 can be written as:

$$\theta = 2\pi ft = \omega t \tag{17.8}$$

Therefore, the projection distance along the x-axis of the circular motion given in Equation 17.6 can be written as:

$$x = R \cos 2\pi ft \tag{17.9}$$

The quantity x is the maximum displacement from the centre of the circle and is called the amplitude (A) of the displacement. When using $R = x_0$, Equation 17.9 can be rewritten as:

$$x = x_0 \cos 2\pi ft \tag{17.10}$$

The velocity of the particle P is also a function of time. Figure 17.10 shows the particle P with an amplitude value of velocity v_0. This figure also shows the velocity components in the x-axis and y-axis directions. The velocity in the x-axis is given as:

$$v_x = -v_0 \sin \theta = -v_0 \sin 2\pi ft \tag{17.11}$$

The velocity can be expressed in terms of ω as $v = -v_0 \sin \omega t$.

Figure 17.11 shows the particle P moving in a circle with constant speed and its acceleration radially directed toward the centre of rotation. The acceleration in the x-axis is given as:

$$a_x = -a_0 \cos \theta = -a_0 \cos 2\pi ft \tag{17.12}$$

where the acceleration a_0 is constant and positive and is the total acceleration of the particle P. The negative sign of the x and y components of the acceleration occur because the acceleration vectors are in the direction opposite to the position vector direction from the centre to the particle P.

Substituting Equation 17.9 into Equation 17.12 gives:

$$a_x = -a_0 \frac{x}{R} \tag{17.13}$$

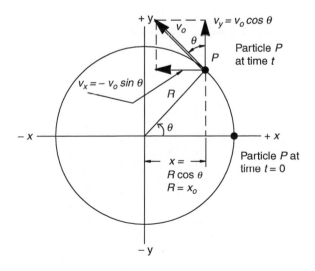

FIGURE 17.10 A particle P moving in a circular path with constant speed v_0.

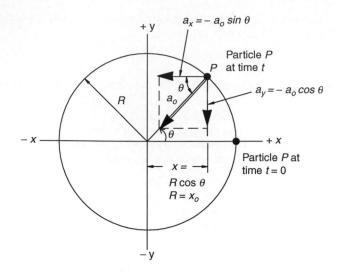

FIGURE 17.11 Particle P moving in a circular path with constant acceleration a_0.

This equation explains the oscillating projection of uniform circular motion on the x-axis. It also shows that the acceleration is proportional to the negative of the displacement.

17.6 SIMPLE HARMONIC MOTION

One of the simplest forms of periodic motion is the up-and-down motion of a mass m suspended from a spring. There are many oscillatory systems having motion similar to that of a spring. Any system in which acceleration is proportional to the negative of displacement undergoes simple harmonic motion.

Figure 17.12 shows a block of a mass m attached to a spring that obeys Hook's law is displaced by an initial amount x_0 from the equilibrium position and released from rest. The motion of the block is described by the curve in Figure 17.12. When the block is released at time $t = 0$ with initial

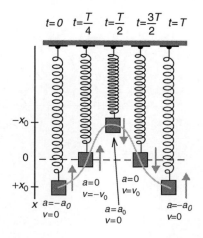

FIGURE 17.12 Simple harmonic motion.

velocity $v = 0$, but with the initial displacement and an initial acceleration opposite in direction to the displacement, the block moves toward the equilibrium position $x = 0$, gaining speed as it moves. As the block reaches the position of zero displacement, its momentum keeps it moving, even though the restoring force is zero at that point. The block is then displaced in the opposite direction from its initial displacement. A restoring force proportional to this new displacement gradually slows the block to a stop position and then accelerates it back toward the initial position. In this way, the block oscillates periodically with time.

If there were no frictional forces (internal and external) to slow the motion, the block would oscillate indefinitely. This oscillation of the block is called simple harmonic motion. The simple harmonic motion of the block generates a sinusoidal curve, as shown in Figure 17.12. Sinusoidal curves or waves can be represented by a trigonometric function. The most common types of sinusoidal curves are the sine and the cosine.

A sinusoidal wave can be produced in the lab using an experimental apparatus set-up shown in Figure 17.13. A pen attached to an oscillating block hung from a spring can be used to trace out a graph of the position of the block versus time on a sheet of paper pulled past the pen at a constant speed.

By applying the principle of energy conservation—kinetic energy equals potential energy—the relation between the maximum displacement x_0 and the maximum velocity v_0 is:

$$\text{Kinetic Energy} = \frac{1}{2}mv_0^2 = \text{Potential Energy} = \frac{1}{2}kx_0^2 \tag{17.14}$$

or

$$v_0 = x_0\sqrt{\frac{k}{m}} \tag{17.15}$$

The total energy, E, of the block motion is the sum of kinetic and potential energies at any time, and it is constant if the frictional forces in the spring are neglected. The total energy is written as:

$$E = \frac{1}{2}mv_x^2 + \frac{1}{2}kx^2 \tag{17.16}$$

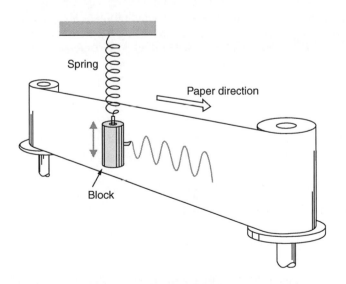

FIGURE 17.13 A set-up for demonstrating simple harmonic motion.

where the instantaneous values of displacement x and velocity v_x are given by Equation 17.10 and Equation 17.11. Submitting Equation 17.10 and Equation 17.11 into Equation 17.16 gives the instantaneous velocity v_x at any displacement x as:

$$v_x = \pm\sqrt{\frac{k}{m}(x_0^2 - x^2)}$$

(17.17)

17.7 PERIOD AND FREQUENCY OF SIMPLE HARMONIC MOTION

As explained in Section 17.6, the velocity of a simple harmonic motion depends on the spring constant and the mass of the block. Therefore, the period—the time required to complete one cycle of the motion—also depends on them. As shown in Figure 17.10, the particle moves with constant speed v_0. In one period, T, the block traverses a circular path of length of $2\pi x_0$. Thus:

$$v_0 T = 2\pi x_0$$

(17.18)

From this equation, the period is then:

$$T = \frac{2\pi x_0}{v_0}$$

(17.19)

Substituting Equation 17.19 into Equation 17.15 gives the period (in seconds) of all types of simple harmonic motion as:

$$T = 2\pi\sqrt{\frac{m}{k}}$$

(17.20)

Again, as given in Equation 17.8, a quantity often used in describing harmonic motion is frequency f (1/s). The frequency is the number of cycles that occur per unit of time. Hence, frequency is the reciprocal of period T.

$$f = \frac{1}{T}$$

(7.21)

The unit of frequency is the hertz, named after the German physicist Heinrich Hertz (1857–1894) in honour of his pioneering studies of radio waves. By definition, 1 Hz = 1 cycle/s.

The maximum displacement of a particle undergoing harmonic motion on either side of its equilibrium position is called the amplitude of the motion. In Figure 17.10, it can be seen that the displacement x is defined by Equation 17.6. Since the particle P moves with constant speed, the radius vector rotates with constant angular velocity ω, and the angle θ is changing at a constant rate, giving:

$$x = R\cos\omega t$$

(17.22)

As given in Equation 17.19, $R = x_0$, and the period is then:

$$T = \frac{2\pi R}{v_0}$$

(17.23)

In terms of the period of vibration T, the angular velocity ω of the particle is given by:

$$\omega = \frac{2\pi}{T}$$

(17.24)

The particle moving in a circle with uniform speed, shown in Figure 17.11, has a constant acceleration toward the centre given by:

$$a_0 = \frac{v_0^2}{R} \qquad (17.25)$$

As this acceleration (a_0) changes in direction, its component a_x along the x-axis changes in magnitude, as given by Equation 17.12. Then, Equation 17.25 can be written as:

$$a_x = \frac{v_0^2}{R} \cos \theta \qquad (17.26)$$

Since $\cos \theta = x/R$, substituting into Equation 17.26 gives:

$$a_x = \frac{v_0^2}{R} \times \frac{x}{R} = \frac{v_0^2}{R^2} x \qquad (17.27)$$

Multiplying both sides of Equation 17.27 by $R^2/a_x v_0^2$ and taking the square root gives:

$$\frac{R}{v_0} = \sqrt{\frac{x}{a_x}} \qquad (17.28)$$

Substituting Equation 17.28 into Equation 17.23 gives:

$$T = 2\pi \sqrt{\frac{x}{a_x}} \qquad (17.29)$$

When the displacement x is positive, the acceleration a_0 is toward the centre of rotation and therefore should be negative. Conversely, when x is negative, acceleration is positive. For this reason, the period should always written as:

$$T = 2\pi \sqrt{-\frac{x}{a_x}} = 2\pi \sqrt{-\frac{\text{displacement}}{\text{acceleration}}} \qquad (17.30)$$

17.8 THE SIMPLE PENDULUM

The pendulum was discovered by Galileo Galilei (1564–1642), an Italian scientist and philosopher. While he attended services in a cathedral in Pisa, Italy, he noticed that a chandelier hanging from the ceiling was swinging with a constant period, as timed by his pulse, regardless of its amplitude. He began to study the pendulum in 1581, as well as the motion of bodies. Continuing to use his pulse as a stopwatch, he observed that the period of a pendulum varies with its length, but is independent of the weight attached to the string. His experiments showed that the longer the pendulum, the longer the period of its swing. Later, he constructed the first crude pendulum clock and medical device, known as the pulsilogium, to measure a patient's pulse.

A simple pendulum consists of a mass m hanging at the end of a string of length L, with the other end of the string suspended from a fixed point, as shown in Figure 17.14. (Please see also http://library.thinkquest.org/16600/intermediate/pendulum.gif.) The motion of a pendulum can be considered as simple harmonic motion; the mass m moves in a curved path. The curve made by the mass is close enough to a straight line to make a small amplitude. The pendulum works almost like the spring. The force is always pointing opposite to the displacement. The mass is moving the fastest when it passes its lowest point at the equilibrium point, as shown in Figure 17.14. If there were no external frictional forces (air resistance) to slow the motion, the pendulum would continuously swing back and forth.

Consider the forces acting on the mass m in Figure 17.14 in terms of tangential and radial components. When the mass is displaced to the side by an angle θ from the vertical, the gravitational force mg has a component along the direction of the string and a component perpendicular to the string.

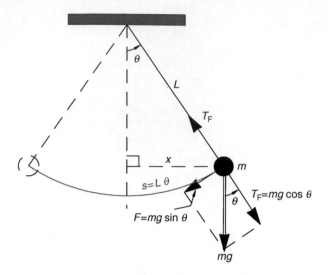

FIGURE 17.14 A simple pendulum consists of a mass hanging at the end of a string.

The force of gravity mg and the tension force ($T_F = mg \cos \theta$) in the string are acting on the mass. The tension force acts in the radial direction and supplies the force needed to keep the mass moving along its circular path.

The net tangential force F acting on m is simply the tangential component of its weight, as given by:

$$F = -mg \sin \theta \qquad (17.31)$$

where g is the gravity constant $= 9.80$ m/s^2.

The tangential force always points toward the equilibrium point. Thus, F is a restoring force to keep the mass in oscillating motion. When the angle θ is sufficiently small, $\sin \theta$ is approximately equal to the angle θ in radians. Then, Equation 17.31 becomes:

$$F = -mg\theta \qquad (17.32)$$

The angle θ is related to the displacement s along the arc through which the pendulum swings by $s = L\theta$, giving:

$$F = -\left(\frac{mg}{L}\right)s \qquad (17.33)$$

This equation has the same form as the spring Equation 17.1, except that here the spring constant k is replaced by mg/L. Thus, the simple pendulum is a harmonic oscillator. The period of the pendulum is given by Equation 17.20, with the substitution of mg/L for k. Thus, the period of a simple pendulum is:

$$T = 2\pi\sqrt{\frac{L}{g}} \qquad (17.34)$$

This equation shows that the period of the simple pendulum is independent of the mass m, as Galileo observed. The period is also independent of the amplitude of the motion for small amplitudes. However, the period of the pendulum does depend on its length L.

17.9 EXPERIMENTAL WORK

This experiment demonstrates the theory of waves. In this experiment the student will investigate the following cases:

17.9.1 Hook's Law for Springs

The student will measure the spring constant (k) for Hook's law. This is done by measuring the extension (x) of the spring, which is proportional to the mass $(F = mg$, force in Newtons) hung from it. A plot can be drawn for the extension of the spring versus applied force. The slope of the line is given by the following equation:

$$\text{Slope} = \frac{\Delta x}{\Delta F} = \frac{1}{k} \tag{17.35}$$

17.9.2 Generating Different Types of Waves

Students will generate and observe different types of waves generated from different wave demonstrators. The wave demonstrators used in this lab are: slinky spring, longitudinal wave model, transverse wave kit, ripple tank, etc.

17.9.3 Period of a Simple Pendulum

The student will measure the period of a simple pendulum for five different displacements and verify the period using Equation 17.34.

17.9.4 Technique and Apparatus

Appendix A presents the details of the devices, components, tools, and parts.
1. Spring kit and masses, as shown in Figure 17.15
2. Wave demonstrators, such as (a) longitudinal wave model, (b) transverse wave kit, (c) ripple tank, (d) tuning forks, etc., as shown in Figure 17.16
3. Pendulum apparatus, as shown in Figure 17.17
4. Stop watch

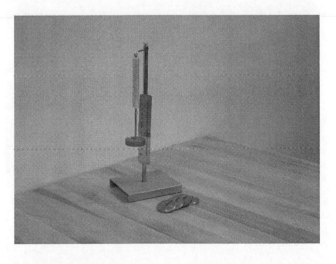

FIGURE 17.15 Hook's law apparatus set-up.

FIGURE 17.16. Different wave simulator apparatuses: (a) Longitudinal wave model, (b) Transverse wave kit, (c) Ripple tank, (d) Tuning forks.

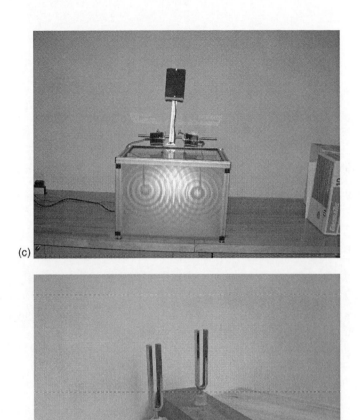

(c)

(d)

FIGURE 17.16 (*continued*)

FIGURE 17.17 Period of a simple pendulum.

5. Protractor
6. Ruler

17.9.5 PROCEDURE

Follow the laboratory procedures and instructions given by the instructor.

17.9.6 SAFETY PROCEDURE

Follow all safety procedures and regulations regarding the use of mechanical and electrical devices.

17.9.7 APPARATUS SET-UP

17.9.7.1 Hook's Law

1. Figure 17.15 shows the apparatus set-up.
2. Mount the spring kit on the table. Mark the initial position of the spring without any weight.
3. Add a small mass (m_1) to the weight hanger (Note: you have to choose the masses related to the spring stiffness). Measure the extension of the spring (x_1). Fill out Table 17.1.
4. Keep adding a mass and measure the extension of the spring. Repeat this step for seven masses. Fill out Table 17.1.

17.9.7.2 Generating Different Types of Waves

Figure 17.16 shows the apparatus set-up. Use only one of the wave demonstrators at a time. Read the operation manual of each wave demonstrator. The wave demonstrators shown in Figure 17.16 are: (a) longitudinal wave model, (b) transverse wave kit, (c) ripple tank, and (d) tuning forks.

17.9.7.3 Period of a Simple Pendulum

1. Figure 17.17 shows the apparatus set-up.
2. Mount the pendulum kit on the table.
3. Measure the length and mass of the pendulum. Fill out Table 17.2.
4. Draw the five angles with the normal (the stand) on a paper, as given in Table 17.2. Stick the paper on the wall behind the pendulum. These five angles represent five displacements of the pendulum.
5. Displace and release the mass for the first displacement and measure the time to complete full oscillation. Fill out Table 17.2.
6. Repeat step 4 for four displacements as marked on the paper.

17.9.8 DATA COLLECTION

17.9.8.1 Hook's Law

Measure the extension of the spring for each added mass. Fill out Table 17.1.

TABLE 17.1
Hook's Law

Mass Applied (kg)	Force F = mg (N)	Extension of Spring (m)
m_1		x_1
m_2		x_2
m_3		x_3
m_4		x_4
m_5		x_5
m_6		x_6
m_7		x_7
m_8		x_8

17.9.8.2 Generating Different Types of Waves

No data collection is required for this case.

17.9.8.3 Period of a Simple Pendulum

1. Measure the length of the pendulum.
2. Measure the mass of the pendulum.
3. Measure the time for one complete oscillation, for each displacement.
4. Fill out Table 17.2 for all the measured data.

TABLE 17.2
The Simple Pendulum Data

Length of Pendulum $L =$ (m)

Mass of Pendulum $m =$ (kg)

Calculated Period Using Equation (17.34) = (s)

Displacement (degree)	Measured Time for One Period (s)
$\theta_1 = 15°$	
$\theta_2 = 30°$	
$\theta_3 = 45°$	
$\theta_4 = 60°$	
$\theta_5 = 75°$	

17.9.9 CALCULATIONS AND ANALYSIS

17.9.9.1 Hook's Law

1. Calculate applied force $F = mg$ (N). Fill out Table 17.1.
2. Plot a graph of extension of the spring versus applied force.
3. Find spring constant k from the graph, using Equation 17.4.

17.9.9.2 Generating Different Types of Waves

No calculations or analysis are required for this case.

17.9.9.3 Period of a Simple Pendulum

1. Calculate the period using Equation 17.34. Fill out Table 17.1.
2. Verify the measured and calculated values of the period.

17.9.10 RESULTS AND DISCUSSIONS

17.9.10.1 Hook's Law

1. Present the measurements and calculations in a graph.
2. Present the spring constant k.
3. Compare the measured spring constant k with the actual manufacturer value.

17.9.10.2 Generating Different Types of Waves

Report your observations when generating different types of waves. If possible, take a picture for each wave demonstration and compare it with the theory.

17.9.10.3 Period of a Simple Pendulum

Present the measured and calculated values of the period.

17.9.11 CONCLUSION

Summarize the important observations and findings obtained in this lab experiment.

17.9.12 SUGGESTIONS FOR FUTURE LAB WORK

List any suggestions for improvements using different experimental equipment, procedures, and techniques for any future lab work. These suggestions should be theoretically justified and technically feasible.

17.10 LIST OF REFERENCES

List any references that were used in the report. Use one format in writing the references. Never mix reference formats in a report.

17.11 APPENDICES

List all of the materials and information that are too detailed to be included in the body of the report.

FURTHER READING

Beiser, A., *Physics*, 5th ed., Addison-Wesley Publishing Company, Reading, MA, 1991.
Blaker, J. W. and Schaeffer, P., *Optics: An Introduction for Technicians and Technologists*, Prentice Hall, Inc., Upper Saddle River, NJ, 2000.

Born, M., Wolf, E., Element of the theory of diffraction, In *Principles of Optics: Electromagnetic Theory of Propagation, Interference, and Diffraction of Light*, 7th ed., Cambridge University Press, Cambridge, UK, pp. 370–458, 1999a.

Born, M., Wolf, E., Rigorous diffraction theory, In *Principles of Optics: Electromagnetic Theory of Propagation, Interference, and Diffraction of Light*, 7th ed., Cambridge University Press, Cambridge, UK, pp. 556–592, 1999b.

Bromwich, T. J. IA., Diffraction of waves by a wedge, *Proc. London Math. Soc.*, 14, 450–468, 1916.

Cox, A., *Photographic Optics*, 5th ed., Focal Press, London, 1974.

Cutnell, J. D. and Johnson, K. W., *Physics*, 5th ed., Wiley, New York, 2001.

Ewen, D., Nelson, R., and Schurter, N., *Physics for Career Education*, 5th ed., Prentice Hall, Englewood Cliffs, NJ, U.S.A, 1996.

Ghatak, A. K., *An Introduction to Modern Optics*, McGraw-Hill Book Company, New York, 1972.

Giancoli, D. C., *Physics*, 5th ed., Prentice Hall, Upper Saddle River, NJ, 1998.

Halliday, D., Resnick, R., and Walker, J., *Fundamentals of Physics*, 6th ed., Wiley, New York, 1997.

Hecht, E., *Optics*, 4th ed., Addison-Wesley Longman, Inc., Reading, MA, 2002.

Jenkins, F. W. and White, H. E., *Fundamentals of Optics*, McGraw-Hill, New York, 1957.

Keuffel & Esser Co., *Physics*, Keuffel & Esser Audiovisual Educator-Approved Diazo Transparency Masters, Mid South, US, 1989.

Lehrman, R. L., *Physics—The Easy Way*, 3rd ed., Barron's Educational Series, Inc., Hauppauge, NY, 1998.

Lerner, R. G. and Trigg, G. L., *Encyclopedia of Physics*, 2nd ed., VCH Publishers, Inc., New York, 1991.

Levine, H. and Schwinger, J., On the theory of diffraction by an aperture in an infinite plane screen, part II, *Phys. Rev.*, 75, 1423–1432, 1949.

Loewen, E. G. and Popov, E., *Diffraction Gratings and Applications*, Marcel Dekker, New York, 1997.

McDermott, L. C. and Shaffer, P. S., *Introduction to Physics Preliminary Edition*, Prentice Hall, Inc., Upper Saddle River, NJ, 1988.

McDermott, L. C. and Shaffer, P. S., *Tutorials in introductory Physics Preliminary Edition*, Prentice Hall, Inc., Upper Saddle River, NJ, 1988.

Naess, R. O., *Optics for Technology Students*, Prentice Hall, Englewood Cliffs, NJ, 2001.

Nolan, P. J., *Fundamentals of College Physics*, Wm. C. Brown Publishers, Inc., Dubuque, IA, U.S.A., 1993.

Pedrotti, F. L. and Pedrotti, L. S., *Introduction to Optics*, 2nd ed., Prentice Hall, Inc., Upper Saddle River, NJ, 1993.

Plamer, C., *Diffraction Grating Handbook*, 5th ed., Thermo Richardson Grating Laboratory, New York, 2002.

Robinson, P., *Laboratory manual to accompany Conceptual Physics*, 8th ed., Addison-Wesley, Inc., Reading, MA, 1998.

Sears, F. W., Zemansky, M. W., and Young, H. D., *University Physics—Part II*, 6th ed., Addison-Wesley Publishing Company, Reading, MA, 1998.

Serway, R. A., *Physics for scientists and Engineers Sunders Golden Sunburst Series*, 3rd ed., Saunders College Publishing, Philadelphia, PA, 1990.

Shamir, J., *Optical Systems and Processes*, SPIE Optical Engineering Press, Bellingham, WA, 1999.

Silverman, M. P., *Waves and Grains—Reflection on Light and Learning*, Princeton University Press, Princeton, NJ, 1998.

Tippens, P. E., *Physics*, 6th ed., Glencoe/McGraw-Hill, Westerville, OH, U.S.A., 2001.

Urone, P. P., *College Physics*, Brooks/Cole Publishing Company, Florence, KY, 1998.

Walker, J. S., *Physics*, Prentice Hall, Englewood Cliffs, NJ, 2002.

Warren, M. L., *Introduction to Physics*, W. H. Freeman and Company, San Francisco, CA, 1979.

White, H. E., *Modern College Physics*, 6th ed., Van Nostrand Reinhold Company, New York, 1972.

Wilson, J. D., *Physics—A Practical and Conceptual Approach Saunders Golden Sunburst Series*, Saunders College Publishing, Philadelphia, PA, 1989.

Wilson, J. D. and Buffa, A. J., *College Physics*, 5th ed., Prentice Hall, Inc., Upper Saddle River, NJ, 2000.

18 Interference and Diffraction

18.1 INTRODUCTION

The particle nature of light is well known and has important applications. There are two accepted behaviours of light to explain the dual nature of light. In some cases, light acts like a particle, and in others, it acts like a wave. The proof of the wave nature of light came with the discovery of interference of light and diffraction. In this chapter, the wave nature of light will be studied with emphasis on two important wave phenomena, the diffraction and interference of light.

Also in this chapter are experiments designed to observe the diffraction patterns generated by objects, such as a blade, disk, washer, single-slit and double-slit holes, and grating, will be described. Students will practise light alignment techniques to generate diffraction patterns from the different geometrical objects.

18.2 INTERFERENCE OF LIGHT

When light waves from two light sources are mixed, the waves are said to interfere. This interference can be explained by the principle of superposition. When two or more waves of the same phase and direction go past a point at the same time, the instantaneous amplitude at that point is the sum of the instantaneous amplitudes of the two waves. If the waves are in phase, then they add together, resulting in a larger amplitude. This is referred to constructive interference, as shown in Figure 18.1(a). If the waves are out of phase with one another, then they cancel each other. This is referred to destructive interference, as shown in Figure 18.1(b). If the waves differ in amplitude and are out of phase with one another, then they add to give a partial cancellation or elimination. This is referred to as partial cancellation or elimination interference, as shown in Figure 18.1(c).

Interference occurs with monochromatic light. This light has a single colour and, hence, a single frequency. In addition, if two or more sources of light are to show interference, they must

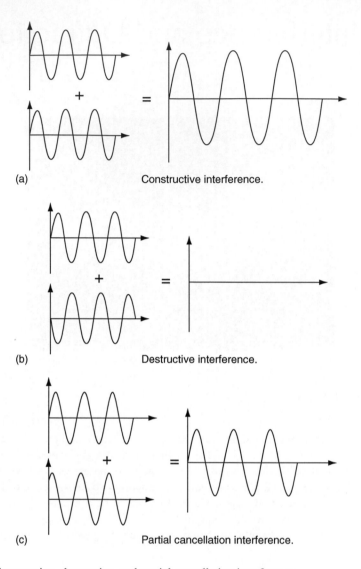

(a)   Constructive interference.

(b)   Destructive interference.

(c)   Partial cancellation interference.

FIGURE 18.1 Constructive, destructive, and partial cancellation interference.

maintain a constant phase relationship with one another. Light sources whose relative phases vary randomly with time show no discernible interference patterns and are called incoherent. Incoherent light sources include incandescent and fluorescent lights. In contrast, lasers emit light that is both monochromatic and coherent.

The following conditions must be met to sustain an interference pattern between two sources of light:

1. The light sources must be coherent. They must maintain a constant phase with respect to each other.
2. The light sources must have identical wavelengths.
3. The principle of superposition must apply.

Figure 18.2 shows interference patterns involving water waves produced by two vibrating sources in a ripple tank at the water's surface. If the two vibrating sources are made to vibrate with the same frequency, each of them causes circular waves to spread out from the point of contact. The waves from these two

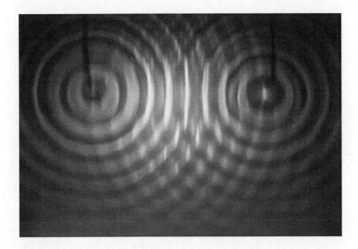

FIGURE 18.2 A wave interference pattern produced by two coherent sources in a ripple tank.

sources interfere with one another. At some points, the waves combine constructively, making waves of larger amplitude. At others, they combine destructively, so that there is little or no wave amplitude.

18.3 YOUNG'S DOUBLE-SLIT EXPERIMENT

The interference of light waves was first demonstrated in 1801 by the English physician and physicist Thomas Young (1773–1829). Young's double-slit experiment not only demonstrates the wave nature of light, but also allows measuring its wavelength. Figure 18.3(a) shows an arrangement of Young's double-slit experiment. A source of monochromatic light is placed behind a barrier with a single narrow slit. Another screen with two similar slits, S_1 and S_2, is placed on the other side. Light from the single-slit passes through both S_1 and S_2, and then to the viewing screen, where an interference pattern of bright and dark fringes is observed, as shown in Figure 18.3(b). This pattern of bright and dark fringes is called interference fringes. When the light from slits S_1 and S_2 arrives at a point on the screen so that constructive interference occurs at that location, a bright line appears. When the light from slits S_1 and S_2 combines destructively at any location on the screen, a dark line results.

If light has wave properties, two sources of light emitting light waves should produce a result similar to that just noted above for waves in a ripple tank, as shown in Figure 18.2.

Examples of constructive and destructive interference are shown in Figure 18.4. In Figure 18.4(a), the two waves, which leave the two slits S_1 and S_2 in phase, strike at the central point, O. Because these waves travel equal distances, they arrive in phase at O. As a result constructive interference occurs at this point and a bright fringe is observed. In Figure 18.4(b), the two light waves start in phase, but one wave has to travel an extra distance of one wavelength to reach point P on the screen. Because the lower wave falls behind the upper one by exactly one wavelength, the two waves still arrive in phase at P, and so a second bright fringe appears at P. In Figure 18.4(c), the point R is located midway between O and P. If one wave travels an extra distance of one-half a wavelength (or $3/2\lambda$, $5/2\lambda$, and so on), the two waves are exactly out of phase when they arrive at the screen. The crests of one wave arrive at the same time as the troughs of the other wave, and they combine to produce zero amplitude. This is destructive interference at R. As a consequence, a dark fringe is observed at R. Figure 18.4(d) shows the intensity distribution on the screen. This figure also shows that the central fringe is the most intense fringe, and

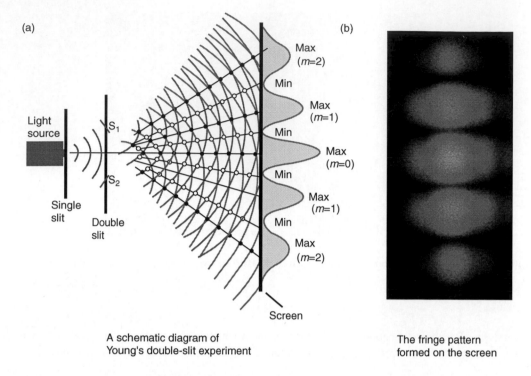

(a)

Light
source

Single
slit

Double
slit

S_1

S_2

Max
(m=2)

Min

Max
(m=1)

Min

Max
(m=0)

Min

Max
(m=1)

Min

Max
(m=2)

Screen

A schematic diagram of
Young's double-slit experiment

(b)

The fringe pattern
formed on the screen

FIGURE 18.3 (**See colour insert following page 512.**) Young's double-slit experiment.

that the intensity decreases for higher-order fringes. Constructive and destructive interferences occur in similar orders above and below the central line, as shown in Figure 18.3(b) and (c).

As explained above, Young's double-slit experiment allows measurement of the wavelength of light, as shown in Figure 18.5. Consider point O on the screen, which is positioned a perpendicular distance L from the screen. The barrier contains slits S_1 and S_2, which are separated by a distance d. The two waves from slits S_1 and S_2 travel distances r_1 and r_2, respectively, to reach the screen. Assume that the two waves have the same frequency, amplitude, and phase. The light intensity on the screen at P is the resultant of the constructive wave interference from both slits. The wave from slit S_1 travels a distance than the wave from slit S_2 by the amount of d sin θ. This distance is called the path difference, Δr, where:

$$\Delta r = r_2 - r_1 = d \sin \theta \qquad (18.1)$$

If the path difference is either zero or some integral multiple of the wavelength, the waves are in phase at P and constructive interference results. Therefore, the condition for bright fringes, or constructive interference, at P is given by:

$$\Delta r = d \sin \theta = m\lambda \quad m = 0, \pm 1, \pm 2, \pm 3, \ldots \qquad (18.2)$$

The number m is called the order number of the fringe. The zero-order fringe ($m = 0$) is the central bright fringe at $\theta = 0$, which corresponds to the central maximum, the first-order fringe ($m = 1$) is the first bright fringe on either side of the central maximum, and so on.

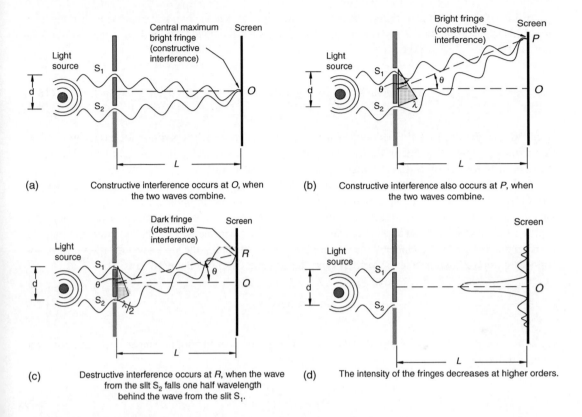

FIGURE 18.4 Light interference: (a) constructive interference occurs at O, when the two waves combine; (b) constructive interference also occurs at P, when the two waves combine; (c) destructive interference occurs at R, when the wave from the slit S_2 falls one half wavelength behind the wave from the slit S_1; (d) the intensity of the fringes decreases at higher orders.

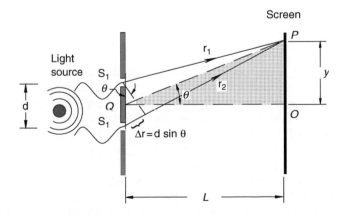

FIGURE 18.5 Geometry of Young's double-slit experiment.

Similarly, when the path difference is an odd multiple of $\lambda/2$, the two waves arriving at P are $180°$ out of phase, resulting in destructive interference. Therefore, the condition for dark fringes, or destructive interference, at P is given by:

$$\Delta r = d \sin \theta = \left(m + \frac{1}{2} \right) \lambda \quad m = 0, \pm 1, \pm 2, \pm 3, \ldots \tag{18.3}$$

If $m = 0$ in Equation 18.3, the path difference is $\Delta r = \lambda/2$, which is the condition for the location of the first dark fringe on either side of the central bright maximum. Similarly, for $m = 1$, $\Delta r = 3\lambda/2$, which is the condition for the second dark fringe on either side, and so on.

In this section, the locations of bright and dark fringes will be determined. Consider the geometry of Young's double-slit experiment shown in Figure 18.5. The first assumption is that the distance from the slits to the screen L is much greater than the distance between the two slits (i.e., $L \gg d$), as shown in Figure 18.5. The second assumption is that the distance between the two slits is much greater than the wavelength (i.e., $d \gg \lambda$). Under these assumptions, θ is small for the first several orders. Thus, the following approximations can be written from the triangle OPQ in Figure 18.5, as:

$$\sin \theta \approx \tan \theta = \frac{y}{L} \tag{18.4}$$

By substituting Equation 18.4 into Equation 18.2, the positions of the bright fringes, measured from the central point, O are:

$$y_{\text{bright}} = \frac{\lambda L}{d} m \tag{18.5}$$

Similarly, by substituting Equation 18.4 into Equation 18.3, the positions of the dark fringes, measured from the central point O, are:

$$y_{\text{dark}} = \frac{\lambda L}{d} \left(m + \frac{1}{2} \right) \tag{18.6}$$

By measuring the bright and dark fringes, the wavelength of the light can be determined. Young used this technique to make the first measurement of the wavelength of light.

18.4 WAVE PHASE CHANGES DUE TO REFLECTION

Changes in the phase of a wave due to reflection are used in many applications. The phase change due to reflection of light waves at a boundary depends on the optical densities, or the indices of refraction, of the two materials. There are two cases of wave phase change due to reflection. Figure 18.6(a) shows a light wave travelling in one medium, reflected from the boundary of the second medium, whose index of reflection is greater than that of the first medium ($n_2 > n_1$). The wave undergoes a $180°$ phase change. This case is similar to the reflected pulse on a string that undergoes a phase change of $180°$ when the pulse is reflected from a fixed end, as shown in Figure 18.6(b). Figure 18.6(c) shows a light wave travelling in one medium, reflected from the boundary of the second medium, whose index of reflection is lower than that of the first medium ($n_1 > n_2$). In this case, there is no phase change. This case is similar to the reflected pulse on a string that has a phase change shift of zero when the pulse is reflected from a free end, as shown in Figure 18.6(d).

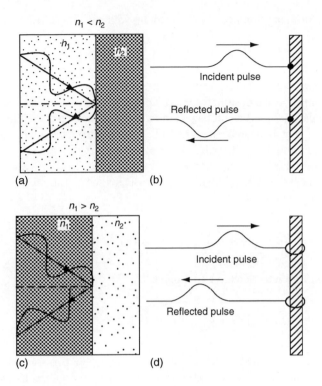

FIGURE 18.6 Reflection and phase changes.

18.5 INTERFERENCE IN THIN FILMS

Interference effects are commonly observed in thin films, such as soap bubbles and a thin layer of oil on the surface of water. The oil floats on the water's surface, because its density is less than that of the water. Consider a beam of monochromatic light that strikes a thin film of oil on water. The oil has a uniform thickness t and index of refraction n, as shown in Figure 18.7. As explained before, an electromagnetic wave of light travelling from a medium of index of refraction n_1 towards a medium of index of refraction n_2 undergoes a 180° phase change on reflection when $n_2 > n_1$. There is no phase change in the reflected

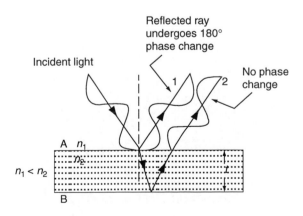

FIGURE 18.7 Reflection of light from a thin film.

wave if $n_1 > n_2$. The wavelength of light λ_n in a medium with index of refraction n is:

$$\lambda_n = \frac{\lambda}{n} \tag{18.7}$$

where λ is the wavelength of light in a vacuum.

The rules of constructive and destructive interference can be applied to the film in Figure 18.7. Light strikes the surface of the oil film. Ray 1 reflected from the upper surface of the film A undergoes a phase change of 180° with respect to the incident light. Ray 2 reflected from the lower surface B undergoes no phase change with respect to the incident light. Therefore, ray 1 is 180° out of phase with respect to ray 2, a situation that is equivalent to a path difference of $\lambda_n/2$. Thus, ray 2 travels an extra distance of $2t$ before the waves recombine. The required film thickness, t, for the constructive interference condition is given in the following equation:

$$2t = \left(m + \frac{1}{2} \right) \lambda_n \quad m = 0, 1, 2, 3, \ldots \quad \text{(maxima)} \tag{18.8}$$

Substituting Equation 18.7 into Equation 18.8 gives:

$$2nt = \left(m + \frac{1}{2} \right) \lambda \quad m = 0, 1, 2, 3\ldots \quad \text{(maxima)} \tag{18.9}$$

If the extra distance $2t$ traveled by ray 2 is a multiple of λ_n, the two waves combine out of phase, resulting in destructive interface. The general equation for destructive interference condition is given by:

$$2nt = m\lambda \quad m = 0, 1, 2, 3, \ldots \quad \text{(minima)} \tag{18.10}$$

An interesting case occurs when the thickness of the film changes along the length, giving rise to alternating regions of constructive and destructive interference.

Consider a thin film of index n surrounded by a medium of lower index of refraction on one side, and a medium of higher index on the other side, as shown in Figure 18.8. Interference of reflected light occurs in many thin film situations involving different indices of refraction. For example, a thin film of

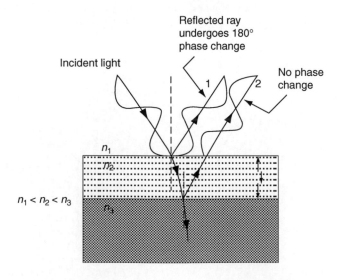

FIGURE 18.8 Reflection of light from a thin film on a material with a higher index of refraction.

oil ($n = 1.36$) on a glass plate ($n = 1.55$) involves two phase changes, one reflection from each interface. The conditions for constructive and destructive interference are the same as given in Equation 18.9
and Equation 18.10, except that the locations of maxima and minima are reversed. As explained, a
phase change occurs upon reflection at both surfaces, since both surfaces are low-index to high-index
boundaries. The condition for constructive interference in the reflected ray is given by:

$$2n_2t = m\lambda \quad m = 0, 1, 2, 3, \ldots \quad (\text{maxima}) \tag{18.11}$$

where n_2 is the index of refraction of the thin film in the middle.

When monochromatic light strikes the surface of a thin film, a phase change occurs between the two
reflected rays as a result of the difference in their paths. In this case, destructive interference occurs. The
condition for destructive interference in the reflected ray is given by:

$$2n_2t = \left(m + \frac{1}{2}\right)\lambda \quad m = 0, 1, 2, 3, \ldots \quad (\text{minima}) \tag{18.12}$$

18.6 NEWTON'S RINGS

Interference rings are observed when a curved lens is placed on a flat glass plate and illuminated with
white light, as shown in Figure 18.9(a). These rings are called Newton's rings, after Sir Isaac Newton
(1642–1727), who first described this interference. Again, as explained above, reflection takes place at
both the top and bottom surfaces of the air film between the lens and the glass plate (i.e., the lower
surface of the lens and upper surface of the glass plate), resulting in a series of constructive (bright) and
destructive (dark) interference bands. Because the thickness of the air film increases with distance from
the central point of contact, the pattern of the bright and dark bands consists of concentric circles with
different diameters, as shown in Figure 18.9(b). One of the important uses of Newton's rings is in the
testing of optical lenses. If there are irregularities in the lens surface, a distorted ring pattern is observed.

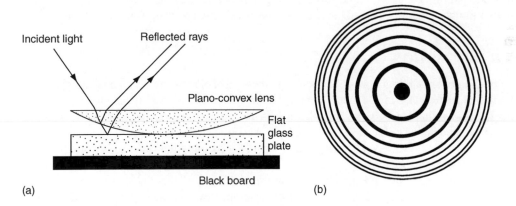

FIGURE 18.9 Newton's rings formed by the interference in the air gap between a lens and a flat glass.

18.7 THIN FILM APPLICATIONS

Thin films are used to make nonreflective glass lenses, such as those used on cameras and binoculars.
Interference properties allow for the reduction of reflected light from the lenses, or the greater transmission of light, which is needed for exposing photographic film or for binocular viewing.
A nonreflective lens has a thin film coating with a thickness of $\lambda/4$, corresponding to $m = 0$, since
there is no reflection at that wavelength. If a lens is coated with a thin film that has the exact thickness
and an index of refraction intermediate between those of air and the lens, reflection from the lens can be

FIGURE 18.10 Lens coated with an antireflection coating.

minimized. Such a coating is called an antireflection coating. Because reflections are reduced, more light is transmitted through lenses that have antireflection coatings. The coating is usually chosen in the yellow–green region of the visible light spectrum (about 550 nm), to which the human eye is most sensitive. Other wavelengths are still reflected and give the glass a somewhat bluish appearance. The bluish colour of the coated camera lens can be seen in Figure 18.10.

The film coating serves a double purpose: not only does it reduce reflection from the front of the lens, but is also cuts down on back reflection. Some of the light transmitted through the lens is reflected from the back surface. This could be reflected again from the front surface of an uncoated lens and produce a poor image, for example, on the photographic film in a camera. However, the reflections from the two film surfaces of a coated lens interfere destructively, and there is no reflection.

18.8 DIFFRACTION

The phenomenon of diffraction was explained in the time of Newton by assuming that light is composed of small particles. This idea—that a source of light is a source of high-speed particles radiated in all directions—was held by Newton and other scientists for many years. After many more years, this idea was abandoned in favor of a wave theory of light. According to the wave theory of light, a beam of light is made up of many waves and propagates in space. By adopting the wave theory of light, a complete understanding of the phenomena of reflection, refraction, diffraction, interference, and polarization was finally formulated on a mathematical basis at the beginning of the nineteenth century by Augustin Fresnel. The wave theory of light was further developed by the Dutch scientist and mathematician Christian Huygens (1629–1695).

An experimental demonstration of Huygens' principle is shown in Figure 18.11. Plane waves pass through a circular aperture, S, in a barrier. When light waves pass through such a small aperture and fall on a screen, an interference pattern of light and dark rings is observed on the screen. Examination of the rings sizes shows that light spreads in various directions beyond the aperture into regions where a shadow would be expected if light travelled in straight lines. This phenomenon is known as diffraction. It can be regarded as a consequence of interference from many coherent wave sources. At times, the phenomena of diffraction and interference are basically treated equally. Figure 18.11 illustrates a beam of light with a wavelength λ passing through an aperture of different diameters d. Different diffraction patterns are observed when light exits from the aperture. Figure 18.11(a) shows that when wavelength $\lambda \ll d$, no diffraction is observed, and the light rays remain collimated. Whereas in Figure 18.11(b),

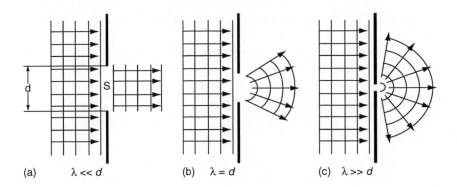

(a) $\lambda \ll d$ (b) $\lambda = d$ (c) $\lambda \gg d$

FIGURE 18.11 Diffraction patterns of light passes through different aperture diameters.

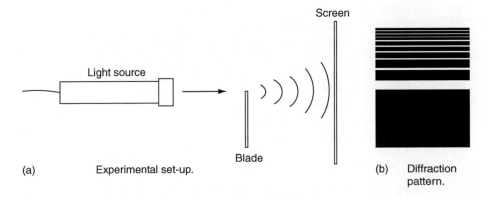

(a)   Experimental set-up.   (b)   Diffraction pattern.

FIGURE 18.12 Diffraction pattern of a blade.

when $\lambda = d$, diffraction becomes significant. In Figure 18.11(c), when $\lambda \gg d$, the diffraction pattern looks like it was made by a point source of light emitting spherical light waves. The diffraction patterns depend on the size of the aperture, the wavelength of the light source, and the location of the screen.

Diffraction occurs when a sharp blade is placed in front of a source of light, as shown in Figure 18.12(a). The diffraction pattern is shown on a screen, as seen in Figure 18.12(b). The boundary between the dark and illuminated regions on the screen is not sharp. The boundary shows that a small amount of light bends into the shadowed region. The region outside the shadow contains alternating

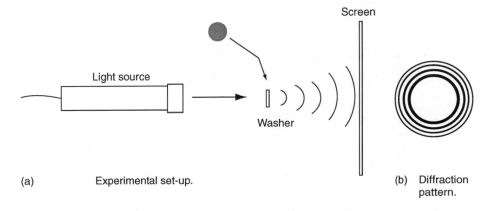

(a)   Experimental set-up.   (b)   Diffraction pattern.

FIGURE 18.13 Diffraction pattern of a small disk.

light and dark bands, as shown in Figure 18.12(b). This is certainly a most dramatic experimental proof of the wave nature of light.

The diffraction pattern of a small disk is shown in Figure 18.13(a). The diffraction pattern is shown on a screen, as shown in Figure 18.13(b). Circular fringes near the disk's circumference can be seen.

Similarly, diffraction can be created by an object of any shape when it is placed between a light source and a screen.

18.9 EXPERIMENTAL WORK

This experiment is designed to show the diffraction patterns generated by a blade, disk, washer, opening, slits, and gratings. Students will practise light alignment techniques for different geometrical shapes in the following experimental setups:

a. Diffraction pattern of a blade
b. Diffraction pattern of a small disk
c. Diffraction pattern of a small washer
d. Diffraction pattern of an arrow shape
e. Diffraction pattern of a grating
f. Diffraction pattern of Fresnel grating
g. Diffraction pattern of a single-slit
h. Diffraction pattern of a double-slit
i. Diffraction pattern of a circular aperture
j. Diffraction pattern of a two-light source through an aperture

18.9.1 TECHNIQUE AND APPARATUS

Appendix A presents the details of the devices, components, tools, and parts.

a. 2×2 ft. optical breadboard.
b. Laser source and power supply.
c. Laser clamps.
d. Hardware assembly (clamps, posts, screw kits, screwdriver kits, positioners, post holders, laser holder/clamp, etc.).
e. Single and double-slit, pinhole, Fresnel, grating, small disk, small washer, arrow shape, and blade targets, as shown in Figure 18.14.

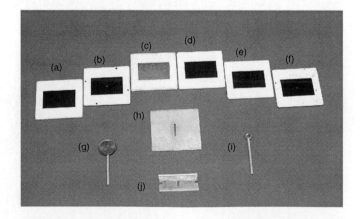

FIGURE 18.14 Targets: (a) single-slit; (b) double-slit; (c) vertical grating; (d) Fresnel grating; (e) pinhole; (f) linear diffraction grating; (g) small disk; (h) arrow shape; (i) small washer; and (j) blade.

FIGURE 18.15 Card with a hole and card holder/stage assembly.

f. Target holder/positioners.
g. Card with a hole and cardholder, as shown in Figure 18.15. Make the hole diameter slightly larger than the laser beam diameter (about 2 mm), so that the laser beam will pass through and back reflections from the mirrors can be easily seen.
h. Two lenses (-25.0 and 200 mm focal length) and lens holder/positioner assemblies, as shown in Figure 18.16.
i. Mirrors and mirror holder/assemblies, as shown in Figure 18.16.
j. Black/white card and cardholder.
k. Ruler.

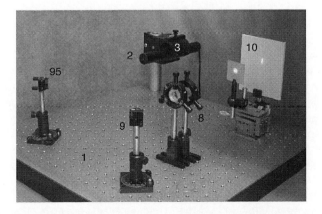

FIGURE 18.16 Diffraction experimental setup.

18.9.2 Procedure

Follow the laboratory procedures and instructions given by the instructor.

18.9.3 Safety Procedure

Follow all safety procedures and regulations regarding the use of laser sources, optical components and instruments, light source devices, and optical cleaning chemicals.

18.9.4 Apparatus Setup

18.9.4.1 Diffraction Pattern of a Blade

a. Figure 18.16 shows the experimental setup.
b. Bolt the laser short rod to the breadboard.
c. Bolt the laser mount to the clamp using bolts from the screw kit.
d. Put the clamp on the short rod.
e. Place the HeNe laser into the laser mount and tighten the screw. Turn on the laser device. Follow the operation and safety procedures of the laser device in use.
f. Align the laser beam so that it is parallel to the edge of the breadboard.
g. Mount a mirror and mirror holder assembly (M_1) to the breadboard at the corner facing the laser beam.
h. Mount a card with a hole-and-card holder assembly between the laser assembly and mirror M_1.
i. Adjust the position of the laser assembly such that the laser beam passes through the hole and is parallel to the edge of the breadboard.
j. Adjust the height of mirror M_1 until the laser beam intersects the centre of mirror M_1. Finely rotate the mirror M_1 post to make sure the laser beam is perpendicular to mirror M_1. The laser beam reflects should reflect back from the mirror through the hole to the laser source. Rotate mirror M_1 an angle of 45° away from the laser beam towards the right corner of the breadboard.
k. Place a second mirror and mirror holder assembly (M_2) to the breadboard at the corner that is facing mirror M_1.
l. Rotate mirror M_2 an angle of 45° away from mirror M_1 towards the right corner of the bread-board. After this step, alignment is achieved from the laser source to mirror M_1 and mirror M_2. The laser beam is at the same height, and is parallel to the breadboard, as shown in Figure 18.16.
m. Place the target card at the corner of the breadboard that is facing mirror M_2. Adjust the position of the target card so that the laser beam is incident on the centre of the target card, as shown in Figure 18.16.
n. The laser beam can be expanded by two ways: a Galilean telescope arrangement and by a Keplerian telescope arrangement.
o. To expand the laser beam by a Galilean telescope arrangement, insert a short-focal-length (-25.0 mm) negative lens (Lens 1) into a lens holder/positioner assembly and mount it 127 mm from the mirror M_2. Align the lens by raising or lowering the post in the lens holder and sliding the lens holder/positioner so that the diverging beam is centred on the black/white card.
p. Insert a longer-focal-length (200 mm) positive lens (Lens 2) into a holder/positioner. Place the lens about 175 mm (the sum of the focal lengths of the two lenses, remembering that the first lens is a negative lens) from the first lens in the diverging laser beam path.
q. Align the lens (Lens 2) by raising or lowering the post in the lens holder and sliding the lens holder/positioner so that the diverging beam is centred on the black/white card.
r. Carefully adjust the position of Lens 2 by moving it back and forth along the beam until the laser beam is expanded and incident on the black/white card.

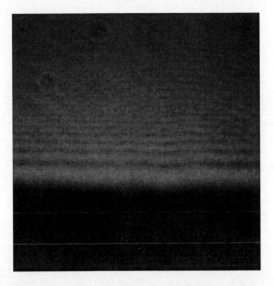

FIGURE 18.17 (See colour insert following page 512.) Diffraction patterns produced by a blade.

s. Mount a blade and blade holder assembly on the breadboard between Lens 2 and the black/-white card, as shown in Figure 18.16. (Note: in Figure 18.16, a slit with an arrow shape is used instead of a blade.)

t. Observe the diffraction pattern produced on the black/white card, as shown in Figure 18.17.

As you have experienced in Case (a), diffraction patterns can also be created for all the other cases: a small disk (2), a small washer (3), an arrow shape (4), a grating (5), a Fresnel grating (6), a single-slit (7), a double-slit (8), a circular aperture (9), and two light sources through an aperture (10), by placing one of the objects in the above list in place of the blade. The following figures show the diffraction patterns for some of the above cases (Figure 18.18 through Figure 18.24).

FIGURE 18.18 (See colour insert following page 512.) Diffraction patterns produced by a small disk.

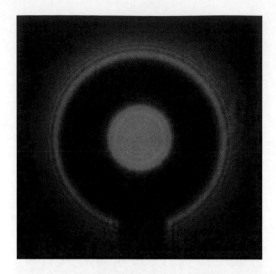

FIGURE 18.19 **(See colour insert following page 512.)** Diffraction patterns produced by a small washer.

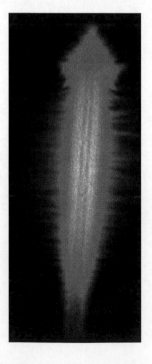

FIGURE 18.20 **(See colour insert following page 512.)** Diffraction patterns produced by an arrow shape.

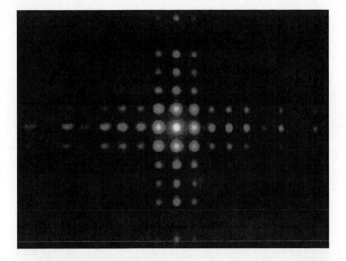

FIGURE 18.21 (**See colour insert following page 512.**) Diffraction patterns produced by a Fresnel grating.

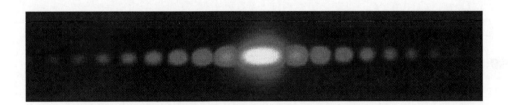

FIGURE 18.22 (**See colour insert following page 512.**) Diffraction patterns produced by a single-slit.

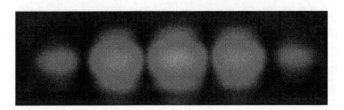

FIGURE 18.23 (**See colour insert following page 512.**) Diffraction patterns produced by a double-slit.

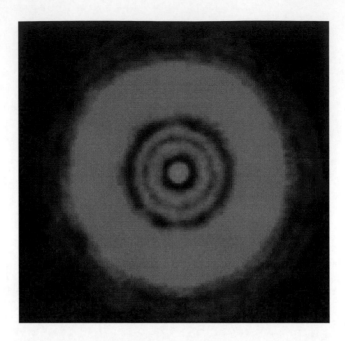

FIGURE 18.24 **(See colour insert following page 512.)** Diffraction patterns produced by a pinhole.

18.9.5 Data Collection

No data collection is required in these cases.

18.9.6 Calculations and Analysis

No calculations or analysis are required in these cases.

18.9.7 Results and Discussions

1. Discuss your observation for each case.
2. Compare the experimental interference patterns with the theoretical patterns for each case.

18.9.8 Conclusion

Summarize the important observations and findings obtained in this lab experiment.

18.9.9 Suggestions for Future Lab Work

List any suggestions for improvements using different experimental equipment, procedures, and techniques for any future lab work. These suggestions should be theoretically justified and technically feasible.

18.10 LIST OF REFERENCES

List any references that were used in the report. Use one format in writing the references. Never mix reference formats in a report.

18.11 APPENDIX

List all of the materials and information that are too detailed to be included in the body of the report.

FURTHER READING

Beiser, A., *Physics*, 5th ed., Addison-Wesley Publishing Company, Reading, MA, 1991.

Blaker, W. J., *Optics: The Matrix Theory*, Marcel Dekker, New York, 1972.

Born, M. and Wolf, E., Elements of the theory of diffraction, In *Principles of Optics: Electromagnetic Theory of Propagation, Interference, and Diffraction of Light*, 7th ed., Cambridge University Press, Cambridge, UK, pp. 370–458, 1999.

Born, M. and Wolf, E., Rigorous diffraction theory, In *Principles of Optics: Electromagnetic Theory of Propagation, Interference, and Diffraction of Light*, 7th ed., Cambridge University Press, Cambridge, UK, pp. 556–592, 1999.

Bouwkamp, C. J., Diffraction theory, *Rep. Prog. Phys.*, 17, 35–100, 1949.

Bromwich, T. J. I'A., Diffraction of waves by a wedge, *Proc. Lond. Math. Soc.*, 14, 450–468, 1916.

Cutnell, J. D. and Kenneth, W. J., *Physics*, 5th ed., Wiley, New York, 2001.

Cutnell, J. D. and Kenneth, W. J., *Student Study Guide—Physics*, 5th ed., Wiley, New York, 2001.

Drisoll, W. G. and William, V., *Handbook of Optics*, McGraw Hill Book Company, New York, 1978.

Ewald, W. P., Young, W. A., and Roberts, R. H., *Practical Optics*, Makers of Pittsford, Rochester, New York, 1982.

Ewen, D., Nelson, R., and Schurter, N., *Physics for Career Education*, 4th ed., Prentice Hall, Upper Saddle River, NJ, 1996.

Giancoli, D. C., *Physics*, 5th ed., Prentice Hall, Upper Saddle River, NJ, 1998.

Halliday, D., Resnick, A., and Walker, J., *Fundamentals of Physics*, 6th ed., Wiley, New York, 1997.

Hewitt, P. G., *Conceptual Physics*, 8th ed., Addison-Wesley, Inc., Reading, MA, 1998.

Jenkins, F. W. and White, H. E., *Fundamentals of Optics*, McGraw-Hill, New York, 1957.

Jones, E. and Richard, C., *Contemporary College Physics*, McGraw-Hill Higher Education, Australia, 2001.

Keuffel and Esser Co, *Physics*, Keuffel and Esser Audiovisual Educator-Approved Diazo Transparency Masters, Audiovisual Division, Keuffel & Esser Co., U.S.A., 1989.

Lehrman, R. L., *Physics—the Easy Way*, 3rd ed., Barron's Educational Series, Inc., Hauppauge, NY, 1998.

Lerner, R. G. and George, L. T., *Encyclopedia of Physics*, 2nd ed., VCH Publishers, Inc., New York, 1991.

Loewen, E. G. and Popov, E., *Diffraction Gratings and Applications*, Marcel Dekker, New York, 1997.

McDermott, L. C. and Shaffer, P. S., *Tutorials in Introductory Physics*, Preliminary Edition, Prentice Hall Series in Educational Innovation, Upper Saddle River, New Jersey, U.S.A., 1988.

Naess, R. O., *Optics for Technology Students*, Prentice Hall, Upper Saddle River, NJ, 2001.

Nolan, P. J., *Fundamentals of College Physics*, Wm.C. Brown Publishers, Dubuque, IA, 1993.

Okamoto, K., *Fundamentals of Optical Waveguides*, Academic Press, San Diego, CA, 2000.

Pedrotti, F. L. and Leno, S. P., *Introduction to Optics*, 2nd ed., Prentice Hall, Inc., Englewood Cliffs, NJ, 1993.

Plamer, C., *Diffraction Grating Handbook*, 5th ed., Thermo Richardson Grating Laboratory, New York, 2002.

Robinson, P., *Laboratory Manual to Accompany Conceptual Physics*, 8th ed., Addison-Wesley, Inc., Reading, MA, 1998.

Romine, G. S., *Applied Physics Concepts into Practice*, Prentice Hall, Inc., Upper Saddle River, NJ, 2001.

Salah, B. E. A. and Teich, M. C., *Fundamentals of Photonics*, Wiley, New York, 1991.

Sears, F. W., Zemansky, M. W., and Young, H. D., *University Physics—Part I*, 6th ed., Addison-Wesley Publishing Company, Reading, MA, 1998.

Serway, R. A., *Physics for Scientists and Engineers Saunders Golden Sunburst Series*, 3rd ed., Saunders College Publishing, Philadelphia, PA, 1990.

Shamir, J., *Optical Systems and Processes*, SPIE Optical Engineering Press, Bellingham, WA, 1999.

Silverman, M. P., *Waves and Grains—Reflections on Light and Learning*, Princeton University Press, Princeton, NJ, 1998.

Smith, W. J., *Modern Optical Engineering*, McGraw-Hill Book Co., New York, 1966.

Tippens, P. E., *Physics*, 6th ed., Glencoe/McGraw-Hill, Westerville, OH, U.S.A., 2001.

Tolansky, S., *An Introduction to Interferometry*, Longmans, Green and Co., London, 1955.

Urone, P. P., *College Physics*, Brooks/Cole Publishing Company, Florence, KY, 1998.

Walker, J. S., *Physics*, Prentice Hall, Upper Saddle River, NJ, 2002.

Warren, M. L., *Introduction to Physics*, W.H. Freeman and Company, New York, 1979.

White, H. E., *Modern College Physics*, 6th ed., Van Nostrand Reinhold Company, New York, 1972.

Wilson, J. D., *Physics—a Practical and Conceptual Approach*, Saunders Golden Sunburst Series, Saunders College Publishing, Philadelphia, PA, 1989.

Wilson, J. D. and Anthony, J. B., *College Physics*, 5th ed., Prentice Hall, Inc., Upper Saddle River, NJ, 2000.

Woods, N., *Instruction Manual to Beiser Physics*, 5th ed., Addison-Wesley Publishing Company, Reading, MA, 1991.

Young, H. D. and Roger, A. F., *University Physics*, 9th ed., Addison-Wesley Publishing Company, Inc., Reading, MA, 1996.

19 The Diffraction Grating

19.1 INTRODUCTION

A diffraction grating is typically made from a piece of glass or plastic upon which parallel grooves are ruled very close together, many thousands per centimetre. This device reflects or refracts light through an angle that depends on the wavelength. For example, if sunlight falls on a diffraction grating (at the correct angle), then the sunlight will be broken up into its component colours to form a rainbow. This phenomenon is similar to the dispersion of white light through a glass prism. Gratings come in many different types and are used as building blocks in optical fibre devices. Each type has its applications and advantages over the others. Instruments such as wave division multiplexers (WDMs), light spectrum analysers, fibre amplifiers, etc. use diffraction gratings as components.

This chapter will examine how gratings work based on the wave theory of light. The different types of gratings will be described and their characteristics explored. The various ways of mounting a grating in order to use it in many different types of applications will be discussed, along with their advantages and disadvantages. A laboratory experiment in which the student uses a standard diffraction grating to measure the wavelength of light using an articulating spectrometer then is detailed.

19.2 DIFFRACTION GRATINGS

A diffraction grating is an optical component that separates (diffracts) polychromatic (white) light into its component wavelengths. A grating is fabricated from an optical material that transmits or reflects certain wavelengths depending on whether it is a transmission or reflection grating. The grating has rows of fine lines or grooves etched on a surface of the optical material, as shown in Figure 19.1. The lines and grooves can be parallel or curved. The grooves have different shapes. Gratings vary in size, shape, groove spacing (groove density), groove angle, and index of refraction.

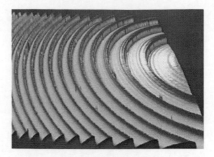

FIGURE 19.1 A diffraction grating.

The high-quality and high-precision process required for grating manufacturing also gives the flexibility to design various diffraction grating instruments.

There are two categories of gratings that describe the method of manufacturing gratings: ruled and holographic gratings. Diffraction ruled gratings are used in a variety of monochromators for research, teaching, and industry use, as shown in Figure 19.2. Almost all commercially available spectrophotometers (ultraviolet, visible, infrared, fluorescence, Raman, atomic absorption) utilize diffraction gratings to select specific wavelengths or scan over a wavelength range. In general, ruled gratings should be used when high volume of manufacturing is required. Holographic gratings are used when minimizing stray light is critical and high resolution is needed.

Holographic gratings are formed by the interference fringes of two laser beams when the standing wave pattern is exposed to a polished substrate coated with photo resist. Processing of the exposed medium results in a pattern of straight lines with a sinusoidal cross section, as shown in Figure 19.3. Holographic gratings produce less stray light than ruled gratings. They also can be produced with different numbers of grooves per millimetre for greater resolving power. Due to their sinusoidal cross section, holographic gratings cannot be easily blazed, and their efficiency is usually considerably less than a comparable ruled grating. There are, however, special exceptions, which should be noted. When the ratio of groove spacing to wavelength is near one, a holographic grating has virtually the same efficiency as the ruled version. A holographic grating with 1800 grooves per millimetre has the same efficiency at 0.500 μm as a blazed ruled grating.

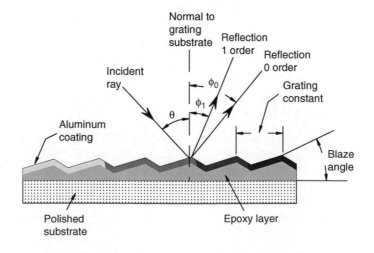

FIGURE 19.2 A ruled diffraction grating.

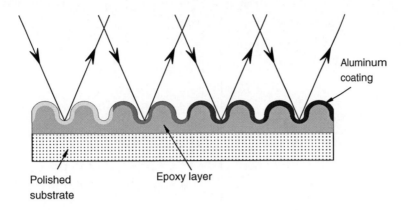

FIGURE 19.3 A holographic grating replica.

19.3 PROFILES OF GRATINGS

Figure 19.4 illustrates the cross-sections of basic grating profiles. There are many designs for the shapes of the grooves in a grating. The groove profile determines the relative strengths of diffracted orders produced. Types (a) and (b) in Figure 19.4 are called blazed grating (triangular and sawtooth). The wavelength dispersion of the blazed grating depends on the blaze angle of the profile. Grating grooves having two or more different blaze angles can be combined on a single diffraction grating, as shown in Figure 19.1. This structure allows for wavelength dispersion over a wider range. Type (c) in Figure 19.4 is called an unblazed profile. Type (d) in Figure 19.4 is called a rectangular profile. The blazed gratings are the most popular groove profiles because they allow a very high proportion of power to be transferred into the first order mode. However, a particular blazed grating will operate efficiently over only a very restricted range of wavelengths.

The common types of gratings are: planar gratings, imaging gratings, concave gratings, laminar gratings, laser gratings, and customized gratings.

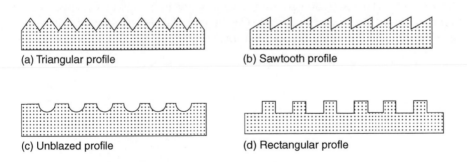

(a) Triangular profile

(b) Sawtooth profile

(c) Unblazed profile

(d) Rectangular profle

FIGURE 19.4 Different grating profiles.

19.4 PLANAR DIFFRACTION GRATINGS

A planar grating is one whose surface is flat. Gratings work in both transmission and reflection of light waves. Fraunhofer, a German physicist, first made transmission gratings in 1819. H.A. Rowland, an American physicist made the first reflection gratings in 1882. Although Rowland's first gratings were ruled on flat surfaces, his best ones were ruled upon the polished surfaces of concave mirrors.

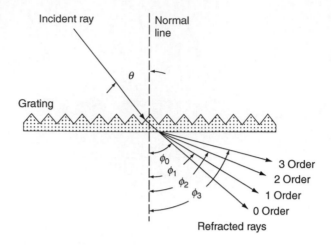

FIGURE 19.5 Transmission diffraction gratings.

In transmission gratings, the light passes through a material which has gratings written on its surface, as shown in Figure 19.5. The transmitted beam is diffracted into multiple orders. Each order has a different wavelength and intensity. Transmission gratings are commonly used for laser beam division and multiple laser line separation of the visible wavelengths. Several gratings are commercially available, offering different wavelength dispersion and power distributions.

A reflective diffraction grating consists of very closely spaced sets of parallel or circular lines or grooves made in the mirror surface of an optical material. A schematic diagram of the operation of a reflective grating is shown in Figure 19.6. A grating can be formed in almost any material where the optical properties (such as refractive index) can vary in a regular way with a period close to the wavelength. An incident ray (at an angle of θ to the normal) is projected onto the grating. The lines or grooves reflect light waves. A number of reflected rays are produced corresponding to different wavelengths. Interference amongst light waves with different wavelengths results in multiple orders of the reflected light.

Both reflection and transmission gratings separate white light into the spectrum of colours. This wavelength separation also occurs in prisms. The theoretical dispersion power of gratings is greater than that of prisms. This power is proportional to the total number of lines on the grating.

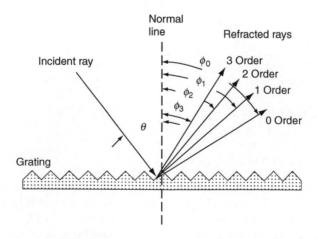

FIGURE 19.6 Reflective diffraction gratings.

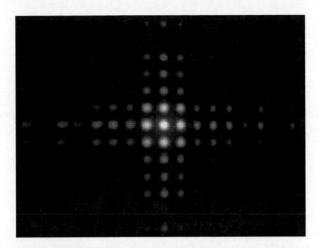

FIGURE 19.7 Laser beam passing through a grating.

Reflective gratings are called wavelength-selective filters. Reflective gratings are widely used in building optical fibre communication devices. In optical communications they are used for splitting and/or combining optical signals in WDM systems and as reflectors in external cavity of the Distributed Bragg Reflector (DBR) lasers. The transmission and reflection gratings are available in two shapes: rectangular and circular.

The number of orders of reflected rays produced depends on the relationship between the grating geometry (size, shape, and groove spacing), index of refraction of the grating, and the wavelength of the light that is incident on the grating. Figure 19.7 shows a pattern of the orders produced from an expanded laser beam passing through a grating and viewed on a black/white board. This figure is from one of the experiments in the Photonics Engineering laboratory.

19.5 CONCAVE GRATINGS

Other instruments do not use secondary focusing lenses or mirrors and rely instead on concave gratings both to focus and to disperse the light. The grooves ruled on a concave grating are equally spaced relative to a plane projection of the surface, not relative to the concave surface itself. In this way, spherical aberration is eliminated. There are many applications of the concave gratings in optical devices and instruments. The most common ones are described in a later section in this chapter.

19.6 CHARACTERISTICS OF GRATINGS

A wide selection of gratings allows the user flexibility in configuring the optical system. Customers can easily choose the groove density (which determines the resolution), starting wavelength (which, when combined with the groove density, determines the spectral range), and blaze wavelength (which determines the wavelength of highest reflection efficiency).

The following is a glossary of key terms used in the grating characteristics selection:

- Lines/mm
  Groove density (ruled or holographically etched) of the grating; the greater the groove density, the better the optical resolution that will result, but the shorter the spectral range.
- Spectral Range
  The dispersion of the grating across the linear array; also expressed as the size of the spectra on the array. When selecting gratings, a wavelength range with a width within the

spectral range must be chosen. The grating's highest efficiency is within the range listed in the best efficiency ($>30\%$) column.

- Blaze Wavelength

  The peak wavelength in the typical efficiency curve for a ruled grating. Also, for a holographic grating, the blaze wavelength is the most efficient wavelength region.
- Efficiency

  All ruled or holographically etched gratings optimize first-order spectra at certain wavelength regions; the best or most efficient region is the range where efficiency is $>30\%$.

19.7 EFFICIENCY OF DIFFRACTION GRATINGS

The blaze angle and groove spacing determines the specific wavelength and spectral region where the grating has its maximum efficiency. The various types of diffraction gratings have different efficiency versus wavelength characteristics. Classical ruled gratings usually peak with very high efficiency at a certain wavelength and become rapidly less efficient as one deviates from that wavelength. Blazed holographic gratings have similar properties. On the other hand, standard holographic gratings have very little variation in efficiency over the spectral range.

Classical ruled gratings (those ruled mechanically by cutting groove after groove into a substrate with a diamond tool) are characterized by a blaze wavelength λ_{Blaze} and a groove density. The useful range of a grating can be described by the 2/3–3/2 rule, which simply states that a grating lower limit equals 2/3 of λ_{Blaze} and upper limit equals 3/2 of λ_{Blaze}. Thus for a grating with a blaze of 0.4 µm, its useful range is 0.266–0.600 µm. It is not unusual to be able to operate the grating with reasonable efficiency above the magic 3/2 value. However, this is not suggested on the short wavelength side below the magic 2/3 value.

When considering efficiency, confusion often results in the definition of the terms absolute and relative efficiency. Absolute efficiency refers to the ratio of diffracted flux in a given order to that of the incident flux. Relative or groove efficiency is defined as the absolute efficiency divided by the reflectivity of the grating surface layer.

It often is required that the grating operates over a broad wavelength range without large fluctuations in grating efficiency or without having to exchange gratings. Such a uniform efficiency performance can be seen in efficiency curves for a 1200 gr/mm holographic grating designed for broadband performance from 0.2 to 1.0 µm. The manufacturers provide the efficiency curves.

When evaluating efficiency requirements there are several general rules concerning holographic gratings that can be presented:

1. In the region where the ratio of $\lambda/d < 0.4$ the diffraction behaviour of the grating acts according to pure geometrical optics (scalar wave theory). The maximum efficiency obtained in the first order is 34% for a given wavelength.
2. In the region approximately $0.4 < \lambda/d < 0.6$ grating anomalies can occur. This makes it very difficult to generalize on the efficiency behaviour of the grating over a range of wavelengths.
3. In the regions where the groove period ratio (λ/d) is between 0.6 and 1.8, sinusoidal groove profiles with deep modulation depth (>0.1 µm) can yield high efficiency (greater than 65% for unpolarized light and as high as 95% for preferred polarization.

19.8 MANUFACTURING OF DIFFRACTION GRATINGS

There are many techniques used to manufacture a grating. In one common method a diamond-tipped cutting tool is used to produce a grating by scratching parallel lines on a piece of transparent glass plate with a precision machining technique. The spaces between the scratches are transparent to the light and hence act as separate slits. Commercial gratings are made by depositing a thin film of aluminum on an optically flat surface and then removing some of the reflecting metal by cutting regularly spaced parallel lines. Precision diffraction gratings are made using two coherent laser beams, intersecting at an angle. The beams expose a layer of photosensitive material, which then is etched. The spacing of the grating lines is determined by the intersection angle of the beams. A typical plane grating consists of several thousand lines per centimetre, for example a precision grating may have 300,000 lines per centimetre or more.

Ruled diffraction gratings are produced by ruling a series of closely spaced, straight parallel grooves into an optically flat aluminum coated substrate known as the master grating. Precise, interferometrically-controlled ruling engines utilize a very fine diamond tool to form a sawtooth-shaped groove profile at the blaze angle on the surface of a prepared substrate. Each grading is replicated from the highly accurate master grating. The replication process begins with the surface contour of a ruled master grating being vacuum deposition-coated with an extremely thin separation layer. An aluminum coating then is deposited on top of this separation layer. Next, an epoxy-coated flat glass substrate is placed on top of the layer-covered master, duplicating the grooved surface. The combination is cured, and the process is finished when the replicated grating is separated from the master grating. The holographic master grating is replicated to become many gratings by a process identical to that used for ruled gratings.

Gratings with gold coating provide a 15–20% increase in efficiency in the near infrared (0.7–1.1 µm) wavelength range. This makes the gold-coated gratings an excellent choice for applications such as fibre optic pulse compression and spectroscopy setups using silicon detectors.

19.9 DIFFRACTION GRATING INSTRUMENTS

A diffraction grating instrument uses a grating as a special dispersing element. The instrument is designed around the type of grating for a particular application. There are many diffraction grating instruments capable of analyzing a variety of radiations ranging from soft x-Rays to far infrared. Diffraction grating instruments have been used in various optic applications and optical fibre applications. Some devices have now been combined with laser interferometric technology. There are a number of designs of diffraction grating instruments that are available for many applications. The next section presents a brief description of a few of the common ones.

The function of separating the spectrum into its spectral line components is performed by a monochromator. The name is from the Greek roots mono (single) and chroma (colour). A mono-chromator is an optical device that transmits a mechanically selectable narrow band of wavelengths of light chosen from a wider range of wavelengths available at the input. A monochromator is a spectrometer that images a single wavelength or wavelength band onto an exit slit; the spectrum is scanned by moving the entrance (and/or exit) slits with respect to the grating. A monochromator device that can produce monochromatic light has many uses in science and in optics, since many phenomena of interest vary with changes in colour.

A spectrograph is a spectrometer that images a range of wavelengths simultaneously, either onto photographic film or a series of detector elements, or through several exit slits (sometimes called a polychromator). The defining characteristic of a spectrograph is that an entire section of the spectrum is recorded at once.

A monochromator's adjustment range might cover the visible spectrum and/or part of the nearby UV and IR spectra, although monochromators are built for a great variety of optical ranges.

It is common for two monochromators to be connected in series, with their mechanical systems operating in tandem so that they both select the same colour. This arrangement is not intended to improve the narrowness of the spectrum but rather to lower the wavelength cutoff level. A double monochromator may have a cutoff about one millionth of the peak value, the product of the two cutoffs of the individual sections. The intensity of the light of other colours in the exit beam is referred to as the stray light level and is the most critical specification of a monochromator. Achieving a low stray light level is a large part of the art of making a practical monochromator.

The narrowness of the band of colours that a monochromator can generate is related to the focal length of the monochromator collimators. The longer focal length gives a narrower wavelength band. Using a longer focal length optical system also unfortunately decreases the amount of light that can be accepted from the source. Very high resolution monochromators might have a focal length of 2 metres. Building such monochromators requires exceptional attention to mechanical and thermal stability. For many applications a monochromator of about 0.4 metre focal length is considered to have excellent resolution. Many monochromators have a focal length less than 0.1 metre.

Monochromators often are calibrated in units of wavelength. Uniform rotation of a grating produces a linear change in wavelength, so such an instrument is easy to build. Many of the underlying physical phenomena being studied are linear in energy. Because wavelength and energy have a reciprocal relationship, spectral patterns that are simple and predictable when plotted as a function of energy are distorted when plotted as a function of wavelength. Some monochromators are calibrated in units of reciprocal centimetres (called wavenumbers) or some other energy units, but the scale may not be linear.

Monochromators are used in many optical measuring instruments and in other applications where tunable monochromatic light is wanted. Sometimes the monochromatic light is directed at a sample, and the reflected or transmitted light is measured. Also at times white light is directed at a sample, and then the monochromator is used to analyse the reflected or transmitted light.

Two monochromators are used in many fluorometers. One monochromator is used to select from the light source the wavelength needed to excite the florescent material. A second monochromator is used to analyse the emitted light. An automatic scanning spectrometer includes a mechanism to change the wavelength selected by the monochromator and to record the resulting changes in the measured quantity as a function of the wavelength.

A monochromator separates all the colours but only makes one colour available. If an imaging device replaces the exit slit, the result is the basic configuration of a spectrograph. This configuration allows the simultaneous analysis of the intensities of a wide band of colours. Photographic film or an array of photo detectors can be used, for instance to collect the light. Such an instrument can record a spectral function without mechanical scanning, although there may be tradeoffs in terms or resolution or sensitivity.

An absorption spectrophotometer measures the absorption of light through a sample as a function of wavelength. Sometimes the result is expressed as percent transmission, and sometimes it is expressed as the inverse logarithm of the transmission. The old nomenclature for absorption was optical density (OD); current nomenclature is absorbance units (AU). One AU is a tenfold reduction in light intensity; six AU is a million fold reduction. Absorption spectrophotometers often contain a monochromator to supply light to the sample. Some absorption spectrophotometers have automatic spectral analysis capabilities.

Absorption spectrophotometers have many everyday uses in chemistry, biochemistry, and biology. For example, they are used to measure the concentration or change in concentration of many substances that absorb light. Critical characteristics of many biological materials (enzymes for example) are measured by starting a chemical reaction that produces a colour change that depends on the presence or activity of the material being studied. Optical thermometers are created by calibrating the change in absorbance of a material against temperature.

Spectrophotometers are used to measure the specular reflectance of mirrors and the diffuse reflectance of coloured objects. They are used to characterise the performance of sunglasses, laser protective glasses, and other optical filters.

In the UV, visible and near IR regions, absorbance and reflectance spectrophotometers usually illuminate the sample with monochromatic light. In the corresponding IR instruments, the monochromator is used usually to analyse the light coming from the sample.

Monochromators also are used in optical instruments that measure other phenomena besides simple absorption or reflection, wherever the colour of the light is a significant variable. For example, circular dichroism spectrometers contain a monochromator. Lasers produce light, which is much more monochromatic than light from the optical monochromators discussed here, but most lasers are not easily tunable.

Atomic absorption spectrometers use hollow cathode lamps as a light source. These lamps emit light generated by ions of a specific element, such as iron or lead or calcium. The available colours are fixed but are very monochromatic and are excellent for measuring the concentration of specific chemical elements in a sample. Atomic absorption spectrometers behave as if they contained a very high quality monochromator, but their use is limited to analyzing only the chemical elements they are equipped for.

19.9.1 The Czerny–Turner Monochromator

Figure 19.8 shows the basic principles of a Czerny–Tuner system used as grating spectrometer. In the common Czerny–Turner design, a light source from an entrance slit is directed to the first concave collimating mirror. This mirror collimates the reflected light to be incident on the grating. The amount of light energy available for use depends on the intensity of the source in the space defined by the slit and the acceptance angle of the optical system. Each ray of the collimated light is diffracted from the grating into several wavelengths and then is incident on a second concave mirror. The second mirror focuses the spectrum at the exit slit. At the exit slit, the colours of the light are spread out. Because each colour arrives at a separate point in the exit slit plane, there are a series of images from the entrance slit focused on the plane. Due to the fact that the entrance slit is finite in width, parts of nearby images overlap. The light leaving the exit slit contains the entire image of the entrance slit of the selected colour plus parts of the entrance slit images of nearby colours. A rotation of the grating element causes the band of colours to move relative to the exit slit, so that the desired entrance slit image is centred on the exit slit. The range of colours leaving the exit slit is a function of the width of the slits. The entrance and exit slit widths are adjusted together.

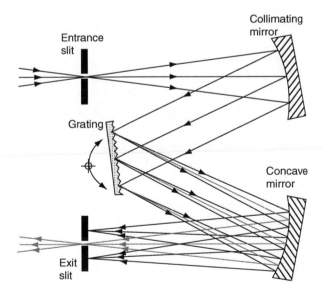

FIGURE 19.8 Basic principles of a Czerny–Turner monochromator.

19.9.2 The Ebert–Fastie Monochromator

Figure 19.9 shows the Ebert–Fastie mount. This design is a special case of a Czerny–Turner mount in which a single relatively large concave mirror serves to collimate and focus. Its use is limited because stray light and aberrations are difficult to control.

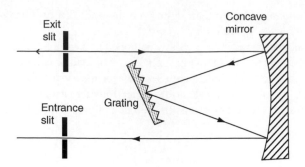

FIGURE 19.9 The Ebert–Fastie mount.

19.9.3 THE MONK–GILLIESON MONOCHROMATOR

Figure 19.10 shows the Monk–Gillieson mount. In this mount, a plane grating is illuminated by converging light. Usually light diverging from an entrance slit or a fibre connector is rendered converging by off-axis reflection from a concave mirror. The grating diffracts the light, which converges toward the exit slit. The spectrum is scanned by rotating the grating to bring different wavelengths into focus at or near the exit slit. If the light incident on the grating is not collimated, the grating introduces wavelength-dependent aberrations into the diffracted wavefronts. Consequently the spectrum cannot remain in focus at a fixed exit slit when the grating is rotated. For low-resolution applications, the Monk–Gillieson mount enjoys a certain amount of popularity, since it represents the simplest and least expensive spectrometric system imaginable.

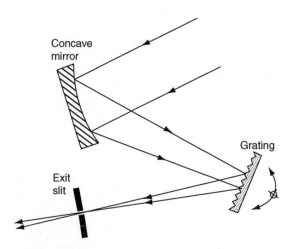

FIGURE 19.10 The Monk–Gillieson mounting.

19.9.4 THE LITTROW MOUNTING

Figure 19.11 shows the basic principles of the Littrow mounted plane grating. A grating used in the Littrow or autocollimating configuration diffracts light of various wavelengths back along the incident light direction. In a Littrow monochromator, the spectrum is scanned by rotating the grating. The same concave mirror can be used to collimate and focus because the diffracted rays retrace the incident rays. Usually the entrance slit and exit slit or photographic plate will be offset slightly along the direction

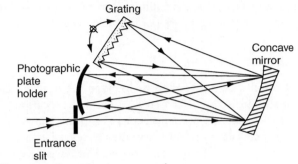

(a) Littrow mounting using a concave mirror.

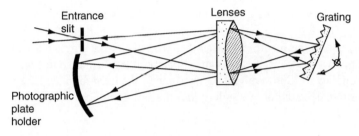

(b) Littrow mounting using lens combination.

FIGURE 19.11 Basic principles of the Littrow mounted plane grating. (a) Littrow mounting using concave mirror. (b) Littrow mounting using a lens combination.

parallel to the grooves so that they do not coincide; of course, this will generally introduce out-of-plane aberrations. True Littrow monochromators are quite popular in laser tuning applications.

19.9.5 THE ECHELLE GRATING

The Littrow condition also is used in the Echelle spectrograph, as shown in Figure 19.12. This design gives the high dispersion and resolution attainable with large angles of incidence on a blazed

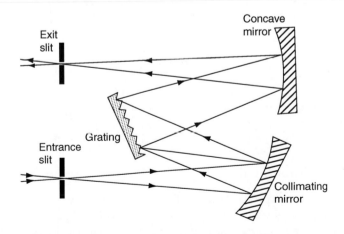

FIGURE 19.12 The Echelle grating.

plane grating. Light from the entrance slit is collimated by the mirror onto the grating. A concave mirror is used to direct the overlapping grating orders towards the exit slit.

19.9.6 The Paschen–Runge Mounting

One of the most useful of all spectrographs to be found in the research laboratory today is one whose design was originally devised by F. Paschen. A concave reflection grating with a radius of curvature (r) is mounted in one corner of an instrument box. The slit and light source are located in another corner with a long curved plateholder directly opposite to the grating, as shown in Figure 19.13. The slit, grating, and photographic plateholder are all located with high precision on the periphery of a circle whose diameter is equal to the radius of curvature of the grating. Such an arrangement brings the different wavelengths to focus all along the plateholder. Light from the source to be studied passes through the narrow slit and then falls on the grating to be diffracted. Note that the grating performs the double function of dispersing the light into a spectrum and of focusing it as well. The design makes use of the Rowland circle named after H.A. Rowland. When a photograph of any part of the spectrum is desired, a strip of plate or film is placed in the proper place in the plateholder and exposed to the spectrum. Spectral lines formed in this way may suffer rather severely from astigmatism. Concave-grating instruments are used for wavelengths in the soft x-ray (1–25 nm) and ultraviolet regions, extending into the visible region.

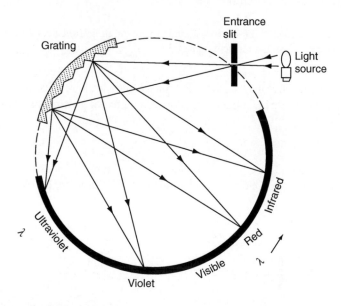

FIGURE 19.13 The Paschen–Runge mounting.

19.9.7 The Rowland Mounting

Another useful type of mounting for a concave reflection grating first was developed by Rowland and is commonly referred to as the Rowland mounting. As shown in Figure 19.14, the grating G and plateholder P are fixed to opposite ends of a rigid beam of length R. The two ends of this beam rest on swivel trucks, which are free to move along two tracks at right angles to each other. The slit S is

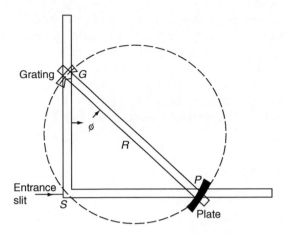

FIGURE 19.14 The Rowland mounting.

mounted just above the intersection of the two tracks. With this arrangement, the portion of the spectrum reaching the plate may be varied by sliding the beam one way or the other, thus varying the angle of incidence. It will be seen that this effectively moves S around the Rowland circle. For any setting the spectrum will be in focus at P, and it will be nearly a normal spectrum because the angle of diffraction is zero. The track SP is usually graduated in wavelengths since, as may be easily shown from the grating equation, the wavelength in a given order arriving at P is proportional to the distance SP.

19.9.8 THE VACUUM SPECTROGRAPH

Another useful type of mounting for a concave reflection grating, called the Eagle mounting, is shown in Figure 19.15. Different wavelength regions of the spectrum are brought to focus on the

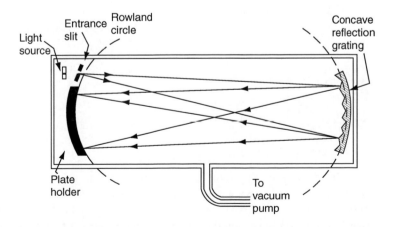

FIGURE 19.15 The vacuum spectrograph.

photographic plate by turning and moving the grating and plateholder by fine adjusting screws. Adjustments are made so that the slit, grating, and plateholder lie on the Rowland circle.

Since the light, from source to photographic plate, never traverses glass elements, this spectrograph can be mounted in a package and highly evacuated for the study of the ultraviolet and extreme ultraviolet spectrum of any light source. Oxygen and nitrogen gases absorb broad wavelength regions of the ultraviolet, from 190 nm to approximately 5 nm. Spectrum lines throughout this entire region are best photographed only by highly evacuated instruments of this kind.

19.9.9 THE WADSWORTH MOUNTING

Figure 19.16 shows the Wadsworth spectrograph configuration. This configuration eliminates astigmatism and spherical aberration by adding a primary mirror to collimate the light incident on the grating. Spectra are observed over a range, making small angles to the grating normal, perhaps 10° to either side. To record different regions of the spectrum, the grating can be rotated and higher grating orders can be used. This version of a grating spectrograph can be of more compact construction than the Paschen–Runge.

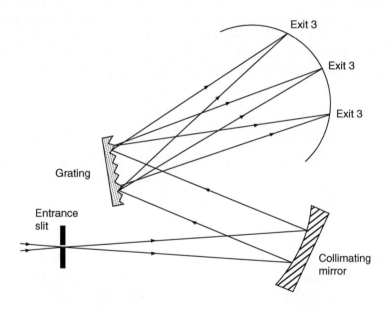

FIGURE 19.16 The Wadsworth mounting.

19.9.10 THE SEYA NAMIOKA MOUNTING

Figure 19.17 shows the Seya Namioka mounting design. In the Seya Namioka mounting the entrance and the exit collimators are fixed at an angle of about 70° 15′. A wavelength variation is produced by rotation of the grating. This type of grating is mainly used in the vacuum UV region. Applications include vacuum UV laser interaction experiments and general photochemical, photolysis, and harmonic generation research.

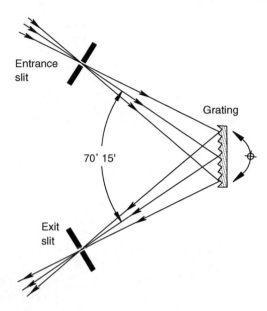

FIGURE 19.17 The Seya Namioka mounting.

19.9.11 THE ROBIN MOUNTING

A normal incidence mounting where δ for wavelength adjustment, the grating is rotated and transported along the bisector of the angle subtended by the entrance and exit axis, is called the Robin Mounting.

19.9.12 FLAT FIELD MOUNTING

A mounting of a specifically corrected interferometric grating or flat field grating, where for a considerable length of the spectrum a focal plane is obtained, is called a flat field mounting.

19.9.13 TRANSMISSION GRATING BEAMSPLITTERS

Transmission grating beamsplitters are commonly used for laser beam division and multiple laser line separation in visible wavelengths. The transmitted beam is diffracted into multiple orders. Transmission grating beamsplitters consist of an index matched epoxy replica on a polished glass substrate, for a high total efficiency. Several gratings are available, offering different dispersion and power distributions.

19.9.14 DOUBLE AND TRIPLE MONOCHROMATORS

Two monochromator mounts used in series form a double monochromator. The exit slit of the first monochromator usually serves as the entrance slit for the second monochromator, as shown in Figure 19.18. Stray light in a double monochromator is much lower than in a single monochromator. The stray light efficiency is the product of the ratios of stray light intensity to incident line intensity for each system. Also, the reciprocal linear dispersion of the entire system is the sum of the reciprocal linear dispersions of each monochromator. A triple monochromator mount consists of three monochromators in series. These mounts are used only when the demands to reduce stray light are extraordinarily severe such as in Raman spectroscopy.

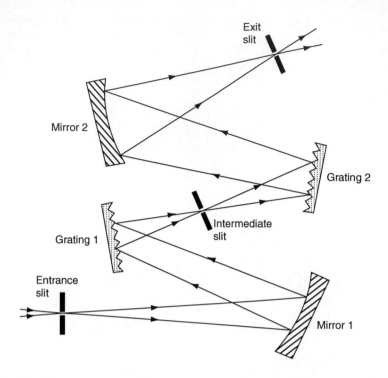

FIGURE 19.18 A double monochromator mounting.

19.9.15 THE GRATING SPECTROMETER

Figure 19.19 shows a grating spectrometer, which has no moving parts. Light from a light source enters the optical spectrometer through a fibre cable connector. The light passes through a filter installed in the connector holder assembly. The light passes through the installed slit, which acts as the entrance aperture. The light is incident on the collimating mirror. The light reflects from the collimating mirror as a collimated beam towards the diffraction grating. The light diffracts from the

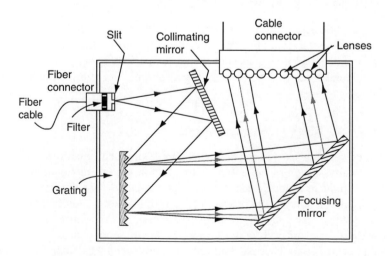

FIGURE 19.19 A grating spectrometer.

fixed grating and proceeds to the focusing mirror. When the diffracted light hits the focusing mirror, it reflects and focuses the light onto the detector collection lenses. The lenses focus the light onto the short detector elements to increase the light collection efficiency. The detector responds to each wavelength of the light that strikes the detector. An external electronic circuit transforms the complete spectrum for the software, which displays the spectrum on a screen.

19.10 EXPERIMENTAL WORK

The theory of this experiment is based on the behaviour of light transmission and reflection from gratings. Gratings are commonly used in manufacturing optical spectrometer devices. The basic arrangement of this experiment is a form of diffraction grating spectrometer. The experiment work will be conducted on the diffraction pattern of a light source by a plane diffraction grating. Figure 19.20 shows a plane diffraction grating of five slits placed perpendicular to the incident light wave from a light source. The light to be analysed passes through the grating. Diffraction patterns of bright and dark fringes occur when the light exits the diffraction grating. Exiting light travels to a distant viewing screen from each of five grating slits and forms the central bright fringe with the first-order bright fringes on either side. Higher-order bright fringes also are formed but are not shown in Figure 19.20. The first order bright fringe is located by an angle θ relative to the central fringe. These bright fringes sometimes are called the principal fringes or principal maxima because they are located where the light intensity is a maximum. The term principal distinguishes them from other dimmer fringes that are referred to as secondary fringes or secondary maxima.

Constructive interference creates the principal fringes. The light rays shown exiting the grating remain nearly parallel while the light travels toward the screen, as shown in Figure 19.21. In reaching the place on the screen where the first-order maximum is located, light from slit 2 travels a distance of one wavelength farther than light from slit 1. Similarly, light from slit 3 travels one wavelength farther than light from slit 2, and so on. For the first-order maximum, the enlarged view of one of these right triangles, in Figure 19.21, shows that constructive interference occurs when $\sin \theta = \lambda/d$, where d is the distance between the centres of the slits. The second-order maximum forms when the extra distance travelled by light from adjacent slits is two wavelengths, so that $\sin \theta = 2 \lambda/d$. The diffracted light that leaves the grating at angle θ satisfies Equation (19.1),

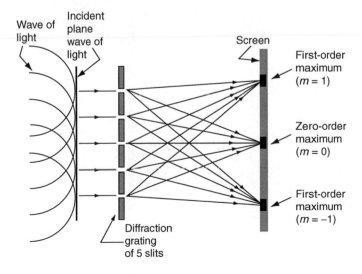

FIGURE 19.20 A side view of a diffraction grating.

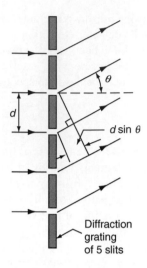

FIGURE 19.21 Path light difference in a diffraction grating.

called maxima of a diffraction grating.

$$d \sin\theta = m\lambda \quad m = 0, \pm 1, \pm 2,\ldots \tag{19.1}$$

The separation d between the slits can be calculated from N, the number of slits (lines) per centimetre of a grating. For example, a grating with 4000 slits per centimetre has a slit separation of $d = (1/4000)\text{cm} = 2.5 \times 10^{-4}$ cm. The number of lines per centimetre N is calculated by Equation (19.2).

$$N = \frac{1}{d} \tag{19.2}$$

Figure 19.22 shows the bright fringes that are produced by the diffraction grating of five silts. A diffraction grating produces bright fringes that are much narrower or sharper than those from a single

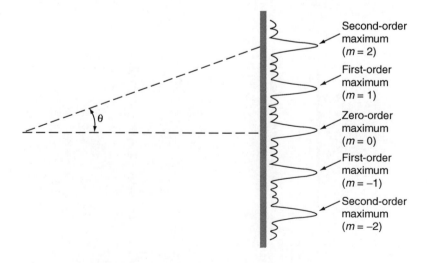

FIGURE 19.22 The bright fringes produced by a diffraction grating.

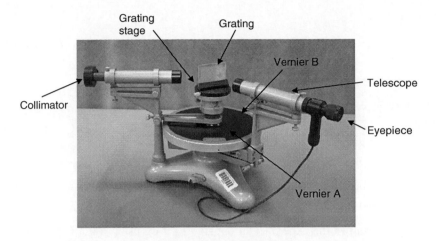

FIGURE 19.23 A grating and a conventional articulating spectrometer.

or double slit. This figure also shows that between the principal maxima of the diffraction grating there are secondary maxima fringes with much smaller intensities.

The number of orders m produced by a grating depends on the wavelength of the light and on the grating's distance d. From Equation (19.1), since $\sin\theta$ cannot exceed 1 ($\sin\theta \le 1$), the number of orders m is limited by the following equation:

$$m \le \frac{\lambda}{d} \qquad (19.3)$$

A simple arrangement that can be used to measure the angles at which the principal maxima of a grating occur, using a plane grating with a conventional articulating spectrometer, is shown in Figure 19.23. This arrangement can be called a grating spectroscope. From the measured values of the angle θ, calculations using Equation (19.1) can be turned around to provide the corresponding value of the wavelength.

Figure 19.23 shows a grating spectroscope that will be used in this experiment. The slit that admits light from the source (e.g., a hot gas tube) is located at the focal point of the collimating lens, so the light rays striking the grating are parallel. The telescope is used to view the bright fringes of the grating. The telescope sits on the attached articulating scale, which measures the angle θ. The wavelength can be determined by measuring the angle θ at which the bright fringes of the slits appear for the various orders.

The student will practise to construct a diffraction pattern and observe the light path passing through the grating. The student will identify line emission wavelengths from gas discharges and thus identify the gas in the discharge lamp. This is done using a grating with a known number of lines per centimetre and the articulating spectrometer. Since gas discharge emissions are at discrete wavelengths, this light passing through the grating creates coloured images of the illuminated slit at different angles with respect to the normal to the grating. By measuring this angle of deviation of the diffracted beam and knowing its order, the grating equation can be used to calculate the wavelength.

19.10.1 Technique and Apparatus

Appendix A presents the details of the devices, components, tools, and parts.

1. Grating and grating holder.
2. Conventional articulating spectrometer, as shown in Figure 19.23.
3. Light source (gas discharge lamp).

19.10.2 PROCEDURE

Follow the laboratory procedures and instructions given by the professor and/or instructor.

19.10.3 SAFETY PROCEDURE

Follow all safety procedures and regulations regarding the use of optical instruments and light source devices.

19.10.4 APPARATUS SET-UP

19.10.4.1 A Grating with a Conventional Articulating Spectrometer Experiment Set-Up

1. Figure 19.24 shows the experimental apparatus set-up using a grating, a conventional articulating spectrometer, and a light source.
2. First the spectrometer has to be set up and the grating mounted so that the plane of the grating is perpendicular to the collimator.
3. Adjust the spectrometer by following the instructions in the instruction manual. (Note: Appendix B walks you through the alignment of each part of the conventional articulating spectrometer that is shown in Figure 19.23).
4. Turn off the lights of the lab.
5. Turn on the light source (gas discharge lamp).
6. Illuminate the slit with the lamp.
7. Align the slit image with the telescope's vertical cross hair. Lock the rotary stage and the telescope arm. Fine adjust the telescope to exactly align the slit image with the vertical cross hair. Note the vernier readings. From these readings determine the telescope angle.
8. This is the reference angle (Zero).
9. Loosen the telescope locking screw and rotate the telescope through exactly 90 degrees

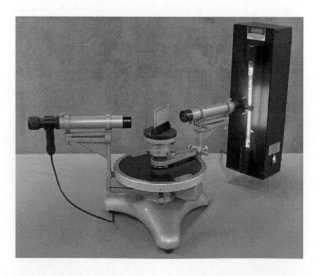

FIGURE 19.24 Spectrum test using a grating with a conventional articulating spectrometer.

so that the axis of the telescope is perpendicular to the axis of the collimator. Lock the telescope arm again.

10. Mount the grating in the holder onto the rotary stage; loosen the rotary stage lock so that the stage can be rotated by hand.
11. Rotate the stage until a reflected image of the slit off the surface of the grating can be seen in the telescope. Adjust the stage leveling screws to make this image vertical. Make sure the image coincides with the vertical cross hair in the telescope.
12. Rotate the stage by exactly 45 degrees so that the plane of the grating is perpendicular to the axis of the collimator. The set up is now ready for use.
13. In order to find out the wavelength of a line in the emission spectrum of the lamp provided, loosen the telescope arm and align the telescope to the reference angle. Observe the central spectral line that is visible when looking into the spectrometer.
14. Align the vertical cross hair with the central slit image (diffraction order $m = 0$).
15. On both sides of this order are several bright lines of different colours. Loosen the telescope arm and move the telescope to the next spectrum line. Observe the line spectra and record where the coloured lines are visible when looking into the spectrometer. Use the same technique as in step 7 for aligning the telescope cross hair with the line image. Take down readings of the verniers for two to four lines on both sides of the $m = 0$ line. From these readings determine the telescope angles.
16. Turn on the lights of the lab.

19.10.5 DATA COLLECTION

1. Record the number of lines per cm of the grating: $N =$ cm^{-1}.
2. Find $d = 1/N =$ cm.
3. Fill out Table 19.1.

TABLE 19.1
Spectrum Test Using Grating

Line	Left Reading Vernier A (LA)	Left Reading Vernier B (LB)	Right Reading Vernier A (RA)	Right Reading Vernier B (RB)	RA - LA $(2\theta_1)$	RB - LB $(2\theta_2)$	$\theta =$ $1/4\,(2\theta_1 + 2\theta_2)$	$\lambda = d\sin\theta$

19.10.6 CALCULATIONS AND ANALYSIS

1. Find θ and λ. Fill out Table 19.1.

19.10.7 RESULTS AND DISCUSSIONS

1. Compare the calculated values with the values for the various lamps already tabulated in the spectral chart.
2. From this comparison, figure out the type of the gas in the lamp.

19.10.8 CONCLUSION

Summarize the important observations and findings obtained in this lab experiment.

19.10.9 SUGGESTIONS FOR FUTURE LAB WORK

List any suggestions for improvements using different experimental equipment, procedures, and techniques for any future lab work. These suggestions should be theoretically justified and technically feasible.

19.11 LIST OF REFERENCES

List any references that were used in the report. Use one format in writing the references. Never mix reference formats in a report.

19.12 APPENDICES

List all of the materials and information that are too detailed to be included in the body of the report.

FURTHER READING

Beiser, A., *Physics*, 5th ed., Addison-Wesley Publishing Company, Reading, MA, 1991.
Blaker, J. W. and Schaeffer, P., *Optics: An Introduction for Technicians and Technologists*, Prentice Hall, Inc, NJ, U.S.A., 2000.
Born, M. and Wolf, E., Elements of the theory of diffraction and rigorous diffraction theory, *Principles of Optics: Electromagnetic Theory of Propagation, Interference, and Diffraction of Light*, 7th ed., Cambridge University Press, Cambridge, England, Born, 1949 pp. 370–458, 1999, See also pp. 556–59.
Bouwkamp, C. J., Diffraction theory, *Rep. Prog. Phys.*, 17, 35–100, 1949.
Bromwich, T. J. I'A. ., Diffraction of waves by a wedge, *Proc. London Math. Soc.*, 14, 450–468, 1916.
Christopher Plamer, *Diffraction Grating Handbook*, 5th ed., Thermo RGL, Richardson Grating Laboratory, New York, 2002.
Cutnell, J. D. and Johnson, K. W., *Physics*, 5th ed., Wiley, New York, 2001.
Cutnell, J. D. and Johnson, K. W., *Student Study Guide—Physics*, 5th ed., Wiley, New York, 2001.
Drisoll, W. G. and Vaughan, W., *Handbook of Optics*, McGraw Hill Book Company, New York, 1978.
Duarte, F. J. and Piper, J. A., Dispersion theory of multiple-prisms beam expanders for pulsed dye lasers, *Opt. Commun.*, 43, 303–307, 1982.
Duarte, F. J. and Piper, J. A., Multi-pass dispersion theory of prismatic pulsed dye lasers, *Opt. Acta*, 33, 331–335, 1984.
Duarte, F. J., Note on achromatic multi-prism beam expanders, *Opt. Commun.*, 53, 259–262, 1985.
Edmund Industrial Optics, *Optics and Optical Instruments Catalog*, Edmund Industrial Optics, New Jersey, 2004.
EPLAB, *EPLAB Catalogue 2002*, The Eppley Laboratory, Inc., New Port, RI, 2002.
Ewald, W. P. et al., *Practical Optics*, Image Makers of Pittsford, Rochester, 1982.
Francon, M., *Optical Interferometry*, Academic Press, New York, Francon, 2002 pp. 97–99, 966.
Giancoli, D. C., *Physics*, 5th ed., Prentice Hall, NJ, U.S.A., 1998.
Goralski, W. J., *SONET*, 2nd ed., McGraw-Hill, New York, 2000.
Guy, M. and Trepanier, F., Fiber bragg gratings: better manufacturing—better performance, *Photonics Spectra*, March, 106–110, 2002.
Guy, M. and Painchaud, Y., Fiber bragg gratings: a versatile approach to dispersion compensation, *Photonics Spectra*, August, 96–101, 2004.
Halliday, Resnick, and Walker, *Fundamental of Physics*, 6th cd., Wiley, New York, 1997.
Hariharan, P., Modified mach-zehnder interferometer, *Appl. Opt.*, 8 (9), 1925–1926, 1969.

Hecht, E., *Optics*, 4th ed., Addison-Wesley Longman, Reading, MA, 2002.

Hewitt, P. G., *Conceptual Physics*, 8th ed., Addison-Wesley, Inc., Reading, MA, 1998.

Jenkins, F. W. and White, H. E., *Fundamentals of Optics*, McGraw Hill, New York, 1957.

Jones, E. and Childers, R., *Contemporary College Physics*, McGraw-Hill Higher Education, New York, 2001.

Keuffel & Esser Co., *Physics, Keuffel & Esser Audiovisual Educator—Approved Diazo Transparency Masters*, Audiovisual Divison, Keuffel & Esser Co., U.S.A., 1989.

Lambda, Research Optics, Inc., *Catalog 2004*, Lambda, Reseach Optics, Inc., California, 2004.

Lehrman, R. L., *Physics-the Easy Way*, 3rd ed., Barron's Educational Series, Inc., New York, 1998.

Lerner, R. G. and Trigg, G. L., *Encyclopedia of Physics*, 2nd ed., VCH Publishers, Inc., New York, 1991.

Levine, H. and Schwinger, J., On the theory of diffraction by an aperture in an infinite plane screen I, *Phys. Rev.*, 74, 958–974, 1948.

Levine, H. and Schwinger, J., On the theory of diffraction by an aperture in an infinite plane screen II, *Phys. Rev.*, 75, 1423–1432, 1949.

Loewen, E. G. and Popov, E., *Diffraction Gratings and Applications*, Marcel Dekker, New York, 1997.

McDermott, L. C. et al., *Introduction to Physics*, Preliminary Edition, Prentice Hall, Inc., NJ, U.S.A., 1988.

McDermott, L. C. et al., *Tutorials in Introductory Physics*, Preliminary Edition, Prentice Hall, Inc., NJ, U.S.A., 1998.

Melles Griot, The practical application of light, Melles Griot, Inc., *Melles Griot catalog*, Rochester, NY, 2001.

Naess, R. O., *Optics for Technology Students*, Prentice Hall, NJ, U.S.A., 2001.

Newport Corporation, *Photonics Section, the Newport Resources 2004 Catalog*, Newport Corporation, Irvine, CA, U.S.A., 2004.

Newport Corporation, Optics and mechanics Section, the Newport Resources, *1999/2000 Catalog*, Newport Corporation, Irvine, CA, U.S.A., 1999/2000

Nolan, P. J., *Fundamentals of College Physics*, Wm. C. Brown Publishers, Inc., Dubuque, IA, U.S.A., 1993.

Ocean Optics, Inc., *Product Catalog 2003*, Ocean Optics, Inc., Florida, USA, 2003.

Okamoto, K., *Fundamentals of Optical Waveguides*, Academic Press, San Diego, 2000.

Page, D. and Routledge, I., Using interferometer of quality monitoring, *Photonics Spectra*, November, 147–153, 2001.

Pedrotti, F. L. and Pedrotti, L. S., *Introduction to Optics*, 2nd ed., Prentice Hall, Inc., NJ, U.S.A., 1993.

Romine, G. S., *Applied Physics Concepts into Practice*, Prentice Hall, Inc., NJ, U.S.A., 2001.

Sears, F. W. et al., *University Physics—Part II*, 6th ed., Addison-Wesley Publishing Company, Reading, MA, 1998.

Serway, R. A., *Physics for Scientists and Engineers*, 3rd ed., Saunders Golden Sunburst Series, London, 1990.

Shamir, J., *Optical Systems and Processes*, SPIE Optical Engineering Press, Washington, 1999.

Silverman, M. P., *Waves and Grains—Reflections on Light and Learning*, Princeton University Press, Princeton, 1998.

Smith, W. J., *Modern Optical Engineering*, McGraw-Hill Book Co, New York, 1966.

Tippens, P. E., *Physics*, 6th ed., Glencoe McGraw-Hill, Westerville, OH, U.S.A., 2001.

Urone, P. P., *College Physics*, Brooks/Cole Publishing Company, Belmont, 1998.

Walker, J. S., *Physics*, Prentice Hall, NJ, U.S.A., 2002.

Warren, M. L., *Introduction to Physics*, W.H. Freeman and Company, San Francisco, 1979.

Weisskopf, V. F., How light interacts with matter, *Sci. Am.*, September, 60–71, 1968.

White, H. E., *Modern College Physics*, 6th ed., Van Nostrand Reinhold Company, New York, 1972.

Williams, J. E. et al., *Teacher's Edition Modern Physics*, Holt, Rinehart and Winston, Inc., Austin, 1968.

Wilson, J. D. and Buffa, A. J., *College Physics*, 5th ed., Prentice Hall, Inc., NJ, U.S.A., 2000.

Woods, N., *Instruction's Manual to Beiser Physics*, 5th ed., Addison-Wesley Publishing Company, Massachusetts, 1991.

20 Interferometers

20.1 INTRODUCTION

The interferometer is designed to exploit, in any of a variety of ways, the interference of light and the fringe patterns that result from optical path differences. The interferometer is also called an optical interferometer. This general description of the interferometer reflects its wide variety of designs and uses. Applications of interferometers can also extend to acoustic and radio waves. This chapter explains the basic principles of the Michelson, Mach–Zehnder, and the Fabry–Pérot interferometers.

To achieve interference between two coherent beams of light, an interferometer divides an initial beam into two parts that travel diverse optical paths and then reunites the beams to produce an interference pattern. Wave-front division interferometers sample portions of the same wave-front of a coherent beam of light. In contrast, amplitude-division interferometers use one type of beam splitter that divides the initial beam into two parts. The Michelson interferometer is an amplitude-division interferometer. The beam splitting is usually controlled by a semireflecting metallic or dielectric film; it can also occur by frustrated total internal reflection at the interface of two prisms forming a cube or by means of double refraction or diffraction. Another means of classification distinguishing between those interferometers that make use of the interference of two beams, as in the case of the Michelson interferometer, and those that operate with multiple beams, as in the Fabry–Pérot interferometer.

Also in this chapter, along with the theoretical presentation, Michelson and Mach–Zehnder interferometers experimental setups will demonstrate the principles of light passing through beamsplitters (BSs) in optical fibre devices.

20.2 THE MICHELSON INTERFEROMETER

The interferometer was invented in 1881 by the American physicist Albert A. Michelson (1852–1931). The interferometer is now called the Michelson Interferometer, or the amplitude-splitting interferometer. The Michelson interferometer splits a beam of monochromatic light into two parts, so that one beam strikes a fixed mirror and the other a movable mirror. When the reflected beams are brought back together, an interference pattern results.

A schematic drawing of the Michelson interferometer is shown in Figure 20.1. Light beam 1 from a monochromatic source of light S is incident on a BS. Beam 1 is split into two parts by the BS by means of a thin, semitransparent-front-surface metallic or dielectric film deposited on glass. The interferometer is, therefore, of the amplitude-splitting type. The reflected part 2 and transmitted part 3, of roughly equal amplitudes, continue to the fully reflecting mirrors M_2 and (fixed) M_1, respectively, where their directions are reversed. M_2 and M_1 are movable and fixed mirrors, respectively. On returning to the BS, part 2 is now transmitted, and part 3 is reflected by the semitransparent film, so that the two parts come together and recombine by the BS as beam 4. This double-beam interferometer is such that all rays striking M_1 and M_2 will be exactly normal. The mirror M_2 is equipped with tilting adjustment screws that allow the surface of M_1 to be made exactly perpendicular to that of M_2. The mirror M_2 is also movable along the direction of the path by means of a very accurate track and micrometre screw. The compensator plate ensures that part 1 and part 2 pass through the same thickness of glass. This plate has the same index of refraction as the BS. Thus, beam 4 includes rays that have travelled different optical paths and will demonstrate interference. When monochromatic light is used as a source of light and the mirrors are in exact adjustment, circular interference fringes are observed, as shown in Figure 20.2. Thus, an observer who views the superposition of the beams sees constructive or destructive interference, depending only on the difference in path lengths d_1 and d_2 travelled by the two beams.

Figure 20.1 also shows that the mirrors are perpendicular to each other, the BS makes a 45° angle with each, and the distances d_1 and d_2 are equal. Beam parts 2 and 3 travel the same distance, and the field of view is uniformly bright due to constructive interference. However, if the adjustable mirror is

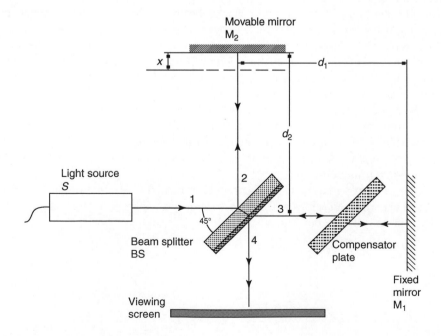

FIGURE 20.1 The Michelson interferometer.

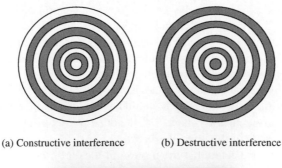

(a) Constructive interference   (b) Destructive interference

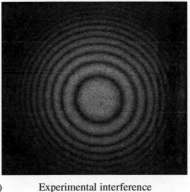

(c)   Experimental interference

FIGURE 20.2 Patterns of circular interference fringes: (a) Constructive interference, (b) Destructive interference, (c) Experimental interference.

moved away from the viewing screen by a distance of 1/4 λ, part 2 travels back and forth by an amount that is twice this value, leading to an extra distance of 1/2 λ. In this case, the parts are out of phase when they reach the viewing screen, destructive interference occurs, and the viewer sees a dark field. If the adjustable mirror is moved farther, full brightness returns as soon as the parts are in phase and interfere constructively. This in-phase condition occurs when part 2 travels a total extra distance of λ relative to part 3. Thus, as the mirror is continuously moved, the viewer sees the field of view change from bright to dark, then back to bright, and so on. The amount by which d_2 has been changed can be measured and related to the wavelength λ of the light, since a bright field changes into a dark field and back again to bright each time d_2 is changed by a half-wavelength (the back-and-forth change in distance is λ). If a sufficiently large number of wavelengths is counted in this manner, the Michelson interferometer can be used to obtain a very accurate value for the wavelength from the measured changes in d_2. Precise distance measurements (x) can be made with the Michelson interferometer by moving the mirror M_2 and counting the number of interference fringes (m) that appear, which move by a reference point. The distance x associated with m fringes is given by:

$$x = \frac{m\lambda}{2} \quad \text{or} \quad \lambda = \frac{2x}{m} \tag{20.1}$$

Alternatively, if the number of fringes m is found, and a distance x can be measured when the movable mirror M_2 is moved by this distance, an accurate value for the wavelength λ of the light source can be obtained using Equation 20.1.

The original application of the Michelson interferometer was in the historic Michelson–Morley experiment. Before the electromagnetic theory of light and Einstein's special theory of relativity became established, most physicists believed that the propagation of light waves occurred in a medium, called ether, which was believed to permeate all space. In 1887, the American scientists Albert Michelson and Edward Morley used the Michelson interferometer in an attempt to detect the motion of the earth through the ether. Suppose the interferometer in Figure 20.1 is moving from left to right relative to the ether. According to the ether theory, this would lead to changes in the speed of light in the portions of the path, shown as the distance d_1 in Figure 20.1. There would be fringe shifts, relative to the positions that the fringes would have if the instrument were at rest in the ether. Then, when the entire instrument was rotated 90°, the other portions of the paths would be similarly affected, giving a fringe shift in the opposite direction.

Michelson and Morley expected that the motion of the earth through the ether would cause a fringe shift of about four-tenths of a fringe when the instrument was rotated. The shift that was actually observed was less than a hundredth of a fringe and, within the limits of experimental uncertainty, appeared to be exactly zero. Despite its orbital motion around the sun, the earth appeared to be at rest relative to the ether. This negative result baffled physicists until Einstein developed the special theory of relativity in 1905. Einstein postulated that the speed of a light wave in a vacuum has the same magnitude c relative to all inertial reference frames, no matter what their velocity may be relative to each other. Therefore, the presumed ether plays no role, and the concept of ether was abandoned.

Besides the historical importance of the Michelson interferometer in the origin of the special theory of relativity, it is a very convenient tool. The Michelson interferometer features large free spaces in its two arms, between the BS and the mirrors. The Michelson interferometer is used to detect the changes in the temperature or composition of the atmosphere in one of the arms, the other one serving as a reference. A two-wave interferometer has many other possible uses. A very important one is as a Fourier-transform spectrometer that analyzes the spectrum of a light source or the absorption spectrum of a sample. It can also be used to measure the refractive index of air and investigate its dependence on pressure.

The interference fringe patterns that are generated by the interferometer can be used to analyze any vibration sources, such as vibration occurring on a table surface or any movement of the optical components. If the pattern moves rapidly and then settles down, the table is receiving ground vibrations. These types of vibrations may be caused by moving vehicles, people walking, elevators, dishwashers, etc.

Thermal expansion or contraction can be detected by the changes in the fringe pattern that move slowly in one direction. Thermal expansion or contraction occurs due to room temperature changes. The interference fringe patterns can also be used to detect any air movements in an enclosed space.

There are many ways in which a beam of light may be split into two parts and reunited after traversing diverse paths, as explained in the section regarding the Michelson interferometer. The operation of all other interferometers is based on the basic operation principle of the Michelson interferometer with simple or radical variations. The following sections present the principle of the designs and applications of some common interferometers.

20.3 THE MACH–ZEHNDER INTERFEROMETER

In 1891, another type of interferometer was introduced by Ernst Mach and Ludwig Zehnder. This interferometer, shown in Figure 20.3, is called the Mach–Zehnder interferometer in honour of their work. The incident beam of light is split into two orthogonal parts by beamsplitter BS_1. The two components are perpendicular to one other. Sometimes the BS is called a semitransparent mirror or one-way glass. Each part is totally reflected by mirrors M_1 and M_2, and the parts are made coincident again by the second beamsplitter BS_2. The path lengths of parts 1 and 2 around the rectangular system and through the BSs are identical. Thus, beam 4 includes rays that will

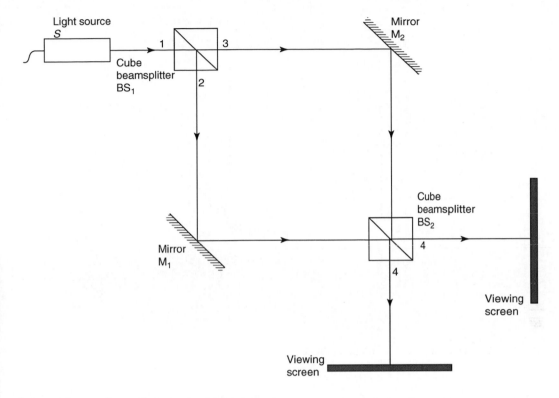

FIGURE 20.3 The Mach–Zehnder interferometer.

demonstrate interference on the viewing screen. When monochromatic light is used as a source of light and the mirrors are in exact adjustment, circular interference fringes will be observed.

The Mach–Zehnder interferometer is used in many applications; for example, in aerodynamic research, the geometry of airflow around an object in a wind tunnel is revealed through local variations of pressure and refractive index. A windowed test chamber, into which the model and a streamline flow of air are introduced, is placed in path 1. An identical chamber is placed in path 2 to maintain equality of the optical paths. The airflow pattern is revealed by the fringe pattern. For such applications, the interferometer must be constructed on a rather large scale. An advantage of the Mach–Zehnder over the Michelson interferometer is that, by appropriate small rotations of the mirrors, the fringes may be made to appear at the object being tested, so that both can be viewed or photographed together. In the Michelson interferometer, fringes appear localized on the mirror. Thus, they cannot be seen in sharp focus at the same time as a test object placed in one of its arms.

20.4 THE TWYMAN–GREEN INTERFEROMETER

A slight modification to the Michelson interferometer was made by Twyman and Green, as shown in Figure 20.4. Instead of using an extended source, this interferometer uses a point source together with a collimating lens L_1, so that all rays enter the BS parallel to the optical axis. The parallel rays emerging from the BS go through a second lens L_2 and brought to a focus at F, where the detector is placed. The circular fringes of equal inclination no longer appear; in their place are seen fringes of equal thickness. These fringes reveal imperfections in the optical system that cause variations in optical path length. When no distortions appear in the plane wave-fronts through the interferometer, uniform illumination is seen near F. If the interferometer components are of high quality, this

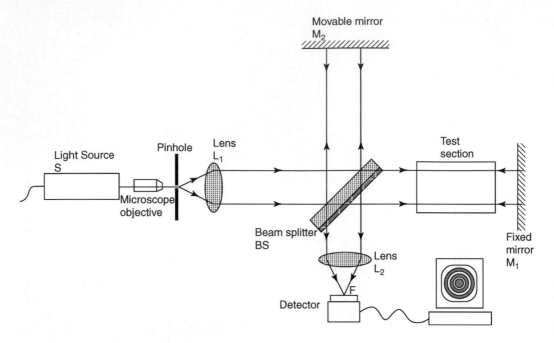

FIGURE 20.4 The Twyman–Green interferometer.

system can be used to test the optical quality of another optical component, such as a prism. Any surface imperfection or internal variation in refractive index shows up as a distortion of the fringe pattern. Lenses are tested for aberrations in the same way; one plane mirror, M_1, is replaced by a convex spherical surface that can reflect the refracted rays back along themselves.

The Twyman–Green interferometer is used to study transparent objects and is particularly useful in studying fluid flow in a wind tunnel. The Twyman–Green interferometer with distance measurement and image processing system is a very accurate method for measuring the topographic properties of ophthalmic surfaces. This method is also useful for studying the distribution of the tear layer on the contact lens or the corneal topography.

20.5 THE FIZEAU INTERFEROMETER

In 1859, French physicist Armand Fizeau (1819–1896) developed still another type of interferometer. In 1868, he suggested using an interferometric method to measure stellar diameters by placing a mask with two holes in front of a telescope aperture. He calculated that the interference fringes would vanish at a separation related to the size of the star. Fizeau's ideas were pursued unsuccessfully by Stephan and, using a different concept, successfully by Michelson.

The Fizeau interferometer is one of the simplest and most versatile interferometers; hence its popularity for measurement of both flat and spherical surfaces. It is used for measuring the surface height differences between two surfaces and to measure the radius of curvature of curved surfaces.

Figure 20.5 shows a schematic diagram of the Fizeau interferometer using a point light source. All light rays enter the BS and collimator. Light rays emerging from the collimator lens are parallel to the optical axis. Light rays are next incident on the divider plate (also called transmission plate), where amplitude division takes place. The bottom surface of the divider plate is the reference surface of a known flatness. This is the surface to which each test plate will be compared. Interference is created when the path of the incident light interacts with both reference surface

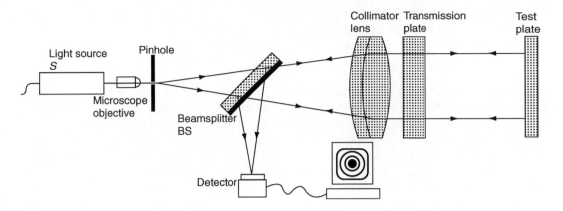

FIGURE 20.5 The Fizeau interferometer.

of the divider plate and test plate. Light rays reflect back through the collimator lens and reflect at a 90° angle into a detector connected to a computer for analysis.

The Fizeau interferometer is suitable for metrology application in fields, such as civil engineering, geology, biotechnology, life sciences, and vibration-insensitive field. The Fizeau interferometer features an instantaneous phase-shifting module. It is also used for testing infrared windows, focal systems, and image-forming optics to a high precision. The Fizeau interferometer is designed to develop and test technologies that will be needed for future interferometric spacecraft missions. It is also used to measure plasma electron density and toroidal current density through Faraday rotation. In phase-shifting Fizeau interferometers, nonlinear motion of the phase shifter and multiple-beam interference are the most common sources of systematic errors affecting high-precision phase measurement.

20.6 THE FABRY–PÉROT INTERFEROMETER

The Fabry–Pérot interferometer was designed in 1899 by Charles Fabry (1867–1945) and Alfred Pérot (1863–1925). It represents a significant improvement over the Michelson interferometer. A schematic diagram of the Fabry–Pérot interferometer is shown in Figure 20.6. The difference

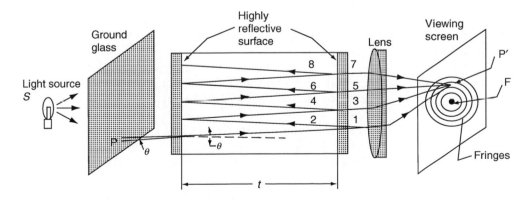

FIGURE 20.6 The Fabry–Pérot interferometer.

between the two is that the Fabry–Pérot design contains plane surfaces that all partially reflect light, so that multiple rays of light are responsible for creating interference patterns. The general theory behind interferometry still applies to the Fabry–Pérot model. However, these multiple reflections reinforce the areas where constructive and destructive effects occur, making the resulting interference fringes much more clearly defined.

Unlike the Michelson interferometer, which produces interference fringe patterns with two coherent beams of light, the Fabry–Pérot interferometer produces interference with a large number of coherent beams. Two optically-flat glass or quartz plates, each partially silvered on one face, are mounted in rigid frames. The fine screws enable the plates to be adjusted until their two silvered surfaces are precisely parallel. Light from an extended source S, passing through the interferometer, undergoes reflection back and forth, and the emerging parallel rays are brought together to interfere in the focal plane of a lens. The spacing, or thickness t, of the air layer between the two reflecting surfaces is an important performance parameter of the interferometer. When this spacing is fixed, the interferometer is often referred to as an etalon.

In Figure 20.6, a ray of light from the point P is shown incident on the first surface at an angle θ. Part of this light is reflected and part is transmitted. Part of the transmitted ray 1 is reflected at the second surface, and part is transmitted. Repeating this behaviour at each reflective surface, it can be seen that rays 1, 3, 5, 7, etc. emerge as parallel rays that have travelled successively greater distances. These numbers specify the number of times each ray traverses the gap of width t. The path difference between successive rays 1, 3, 5, 7, etc. is $2t \cos \theta$. If this distance is exactly equal to a whole number of wavelengths, the emergent rays, when brought together at the point P' on the viewing screen, will all be in phase and produce a bright spot. For such a bright spot, the distance t associated with m fringes can be written as:

$$t = \frac{m\lambda}{2 \cos \theta} \quad \text{or} \quad \lambda = \frac{2t \cos \theta}{m} \qquad (20.2)$$

where

m is a whole number and
λ is the wavelength.

Alternatively, if the number of fringes m is found and a thickness t can be measured, an accurate value for the wavelength λ of the light source can be obtained using Equation 20.2.

For all rays from all points of the source incident at the same angle θ, identical phase relations exist, and the lens will bring all sets of parallel rays to a focus on a circle in the focal plane of the lens. For those rays in which the angle θ is such that the path difference between successively reflected rays equals $(m+\frac{1}{2})\lambda$, the waves in alternate rays will destructively interfere to produce darkness. Hence, the existence of bright and dark concentric rings on the viewing screen indicates coherence as well as interference.

The interference ring patterns in Figure 20.7 illustrate the difference between the kinds of fringes observed (a) when two beams of light are brought together, as in the Michelson interferometer, and (b) when a large number of beams are brought together, as in the Fabry–Pérot interferometer.

Note that in Figure 20.6, when the rays entering the interferometer are parallel to the principal axis (when $\theta = 0$), the multiple reflected rays will be brought together at the focal point F, and if these waves arrive in phase, a bright spot will be produced. Equation 20.2 can then be written as:

$$t = \frac{m\lambda}{2} \quad \text{or} \quad \lambda = \frac{2t}{m} \qquad (20.3)$$

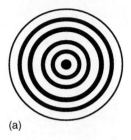
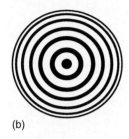

(a)   (b)

FIGURE 20.7 Comparison of the types of fringes produced with (a) the Michelson interferometer and (b) the Fabry–Pérot interferometer.

The number m is called the order of interference, or in microwave terminology, the principal oscillation mode of the cavity.

The Fabry–Pérot interferometer is probably the most adaptable of all interferometers; for example, in precision wavelength measurements, analysis of hyperfine spectral line structure, determination of refractive indices of gasses, and the calibration of the standard metre in terms of wavelengths. This interferometer is also used in high-resolution applications, such as atomic spectroscopy or measurement of narrow-band laser linewidths. The Fabry–Pérot interferometer is simple in structure, but it is a high-resolution instrument that has proved to be a powerful tool in a wide variety of applications.

20.7 OTHER TYPES OF INTERFEROMETERS

There are many other types of interferometers that are used in different optical applications. Scatterplate interferometers are used to test the curvature of mirrors, lenses, and prisms, and for creating microscopic images. A Smart Point Diffraction interferometer is used to study lateral shear interferometry, measure wavefront slope, and create interferograms.

20.8 EXPERIMENTAL WORK

This experiment is based on the Michelson interferometer principle. The student will practise creating interference fringes using prisms. Laser beams, prisms, and lenses are optical components that are used in fringe formation. Such optical components are commonly used in interferometry. The purpose of this experiment is to observe the fringe interference when a monochromatic light is passed through one or two prisms. Lenses are used to expand the laser beam for easy viewing of the interference fringes. The interferometer experimental set-ups are presented in the following cases:

20.8.1 THE MICHELSON INTERFEROMETER

As explained in the section on the Michelson interferometer, this part of the experiment will be conducted using a cube BS, as shown in Figure 20.8. The laser beam is reflected and directed by a mirror (M_1) toward and two lenses (L_1 and L_2). The two lenses expand the laser beam when they are acting as a Galilean telescope or Keplerian telescope arrangement. The cube BS splits the beam incident at the normal into two orthogonal components exiting from the BS into a 50/50 ratio. One component is called transmission, while the other is called reflection. Transmission and reflection

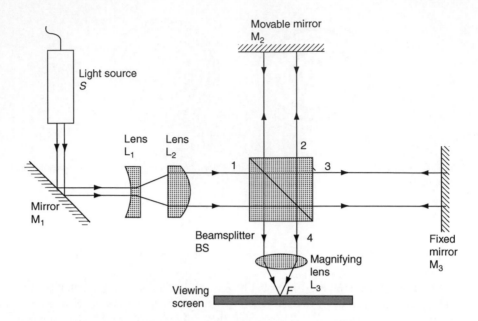

FIGURE 20.8 Diagram of the Michelson interferometer.

approach 50% through the output that is partially polarized. One component is incident on the movable mirror (M_2) and the second component is incident on the fixed mirror (M_3). The two mirrors are located at equal distances from the BS and orthogonal to each other. After reflecting the components from the mirrors (M_2 and M_3), the two parts are recombined by the BS to produce the light interference fringes. A magnifying lens (L_3) can be used to enlarge the interference fringes for easy viewing on the target card. The use of the magnifying lens is shown in two cases in this experiment.

20.8.2 THE MACH–ZEHNDER INTERFEROMETER

As explained with the Mach–Zehnder interferometer, this part of the experiment will be conducted using two cube BSs. A schematic diagram of the arrangement for the splitting and recombining a light beam using two cube BSs is shown in Figure 20.9.

The laser beam is split into its two orthogonal components (of equal strength) by a cube beamsplitter (BS_1). The two components are perpendicular to one other. One component is incident on the mirror (M_2) inclined at 45° relative to the incident beam component. The second component is incident on the mirror (M_3) inclined at 45° relative to the incident beam component. The two mirrors are located equal distances from the beamsplitters (BS_1) and (BS_2) and in a parallel position facing one other. After reflecting the components from the mirrors M_2 and M_3, the two components are eventually recombined by the beamsplitter (BS_2) to produce the laser interference fringes on the target card.

Two magnifying lenses (L_3 and L_4) can be used to enlarge the interference fringes for easy viewing on the target cards. The use of the magnifying lens is shown in two cases in this experiment.

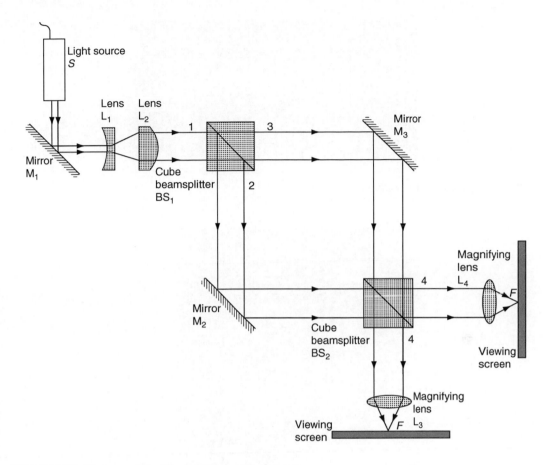

FIGURE 20.9 Diagram of the Mach–Zehnder interferometer.

20.8.3 Technique and Apparatus

Appendix A presents the details of the devices, components, tools, and parts.

1. 2×2 ft. optical breadboard.
2. HeNe laser source and power supply.
3. Laser light sensor.
4. Laser light power metre.
5. Laser mount assembly.
6. Hardware assembly (clamps, posts, screw kits, screwdriver kits, sundry positioners, etc.).
7. Two cube BSs, as shown in Figure 20.10.
8. Three Mirrors and mirror holders/positioners, as shown in Figure 20.11.
9. Lenses (200–25 mm, and 25.4-mm focal lens and magnifying lenses) and lens holders/positioners, as shown in Figure 20.12.
10. Beamsplitters holder/positioner assembly, as shown in Figure 20.13.
11. Target card and cardholder.
12. Rubber-tipped tweezers.
13. Ruler.

FIGURE 20.10 Two cube beamsplitters.

FIGURE 20.11 Mirror and mirror holder/positioner.

FIGURE 20.12 Two lenses and holders/positioners.

FIGURE 20.13 The Michelson interferometer setup.

20.8.4 PROCEDURE

Follow the laboratory procedures and instructions given by the professor and/or instructor.

20.8.5 SAFETY PROCEDURE

Follow all safety procedures and regulations regarding the use of optical components and instruments, light source devices, and optical cleaning chemicals.

20.8.6 APPARATUS SETUP

20.8.6.1 The Michelson Interferometer

1. Figure 20.13 shows the experimental apparatus setup.
2. Bolt the laser short rod to the breadboard.
3. Bolt the laser mount to the clamp using bolts from the screw kit.
4. Put the clamp on the short rod.
5. Place the HeNe laser into the laser mount and tighten the screw. Turn on the laser device. Follow the operation and safety procedures of the laser device in use.
6. Align the laser beam to be parallel to the edge of the breadboard.
7. Mount a mirror (M_1) and mirror holder assembly to the breadboard facing the laser beam at an angle of 45°.
8. Insert a short-focal-length (-25.0 mm) negative lens (L_1) into a lens holder/positioner assembly and mount it 127 mm from the mirror (M_1). Align the lens by raising or lowering the post in the lens holder and sliding the lens holder/positioner so that the diverging beam is centred on the mirror (M_2).
9. Mount a cube BS and BS holder assembly on the breadboard at a reasonable distance from the lens (L_1).
10. Insert a longer-focal-length (200 mm) positive lens (L_2) into a holder/positioner. Place it about 175 mm (the sum of the focal lengths of the two lenses, remembering that the first lens is a negative lens) from the first lens in the diverging laser beam path.

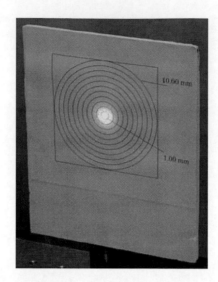

FIGURE 20.14 Expanded laser beam on the target card.

11. Align lens L_2 by raising or lowering the post in the lens holder and sliding the lens holder/positioner so that the diverging beam is centred on the cube BS.

12. Carefully adjust the position of the lens (L_2) by moving it back and forth along the beam until the laser beam is collimated and incident on the target card, as shown in Figure 20.14. The card should be placed between the lens (L_2) and the cube BS.

13. Mount two mirrors (M_2 and M_3) on two mirror holder/positioners (one each), as shown in Figure 20.13. The two mirrors must be located the same distance from the BS and perpendicular to each other. The centre of these two mirrors must be at the same height of the cube BS centre.

14. Place the target card on the breadboard facing the mirror (M_2). Adjust the position of the target card, so that the laser beam is incident on the centre of the target card, as shown in Figure 20.14.

15. Carefully recombine the laser beam components by finely adjusting the alignment of the cube BS and mirrors.

16. Try to capture the laser beam exiting the cube BS on the target card, as shown in Figure 20.15.

17. Move the mirror (M_2) a measured distance x. Find the wavelength λ of the laser beam from the manufacturer's specifications. Count the number of interference fringes (m). Fill out Table 20.1.

In order to see the interference fringes clearly, it is recommended to project the fringes on a wall at a distance far away from the BS. Due to the lack of distance in a lab, a magnifying lens can be used to enlarge the fringes. Figure 20.16 shows the apparatus setup for this case. Figure 20.17 shows the magnified interference fringes on the target card.

20.8.6.2 The Mach–Zehnder Interferometer

Continue the procedure explained in the Michelson interferometer using two cube BSs and two mirrors in the Mach–Zehnder interferometer arrangement. Figure 20.18 shows the apparatus setup for the Mach-Zehnder interferometer with magnifying lens. Add the following steps:

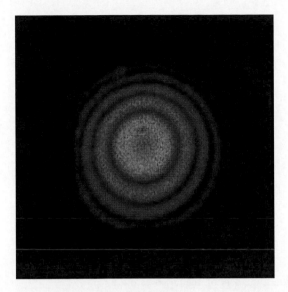

FIGURE 20.15 Laser pattern incident on the target card.

1. Mount the second cube BSs on a prism holder/positioners (one each), as shown in Figure 20.18. The two cubes should be lined up diagonally.
2. Carefully align the laser beam so that it is incident perpendicular at the centre of the face of the cube beamsplitter (BS_1).
3. Mount two mirrors on two mirror holder/positioners (one each), as shown in Figure 20.18. Rotate the two mirrors by 45°; the two mirrors should be lined up diagonally. The two mirrors must be located the same distances from the beamsplitters (BS_1) and (BS_2) and parallel, facing each other. The two mirrors centres must be at the same height of the cube BS centres.
4. Carefully recombine the laser beam components by finely adjusting the alignment of the cube BSs and mirrors.

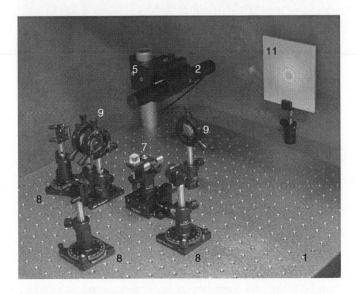

FIGURE 20.16 The Michelson interferometer with magnifying lens setup.

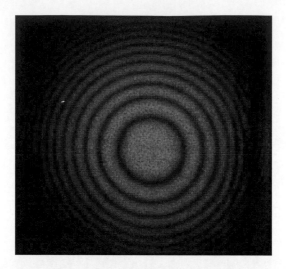

FIGURE 20.17 Laser pattern incident on the target card.

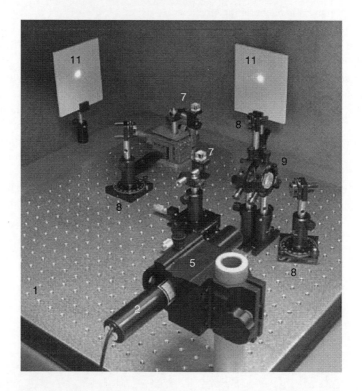

FIGURE 20.18 The Mach–Zehnder interferometer setup.

5. Try to capture the recombined laser beams exiting from the cube beamsplitter (BS$_2$) on the target cards, as shown in Figure 20.19.

In order to see the interference fringes clearly, it is recommended to project the fringes on a wall at a distance faraway from the BSs. Due to the lack of distance in a lab, magnifying lenses can be used to enlarge the fringes. Figure 20.20 shows the apparatus setup for this case. Figure 20.21 shows the magnified interference fringes on the target card.

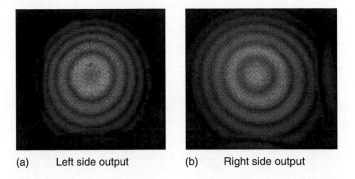

(a)   Left side output   (b)   Right side output

FIGURE 20.19 Laser pattern incident on the target card: (a) Left side output, (b) Right side output.

FIGURE 20.20 The Mach–Zehnder interferometer with magnifying lens setup.

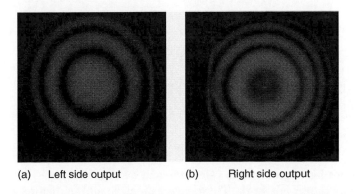

(a)   Left side output   (b)   Right side output

FIGURE 20.21 Laser pattern incident on the target card: (a) left side output, (b) right side output.

20.8.7 DATA COLLECTION

20.8.7.1 The Michelson Interferometer

Measure the distance (x), find the wavelength (λ) of the laser beam, and find the number of interference fringes (m). Fill out Table 20.1.

TABLE 20.1
The Michelson Interferometer Data Collection

Measured Distance x (unit)	Specified Light Wavelength λ (unit)	Experimental Number of Fringes m	Calculated Wavelength λ (unit)

20.8.7.2 The Mach–Zehnder Interferometer

No data collection is required for this case.

20.8.8 CALCULATIONS AND ANALYSIS

20.8.8.1 The Michelson Interferometer

Calculate the wavelength of the laser beam, using Equation 20.1. Fill out Table 20.1.

20.8.8.2 The Mach–Zehnder Interferometer

No calculations or analysis are required for this case.

20.8.9 RESULTS AND DISCUSSIONS

20.8.9.1 The Michelson Interferometer

1. Report the distance (x), the wavelength (λ) of the laser beam, and the number of the interference fringes (m).
2. Compare the calculated wavelength (λ) of the laser beam and the value that is provided by the manufacturer.

20.8.9.2 The Mach–Zehnder Interferometer

Report your observations and discoveries.

20.8.10 CONCLUSION

Summarize the important observations and findings obtained in this lab experiment.

20.8.11 SUGGESTIONS FOR FUTURE LAB WORK

List any suggestions for improvements using different experimental equipment, procedures, and techniques for any future lab work. These suggestions should be theoretically justified and technically feasible.

20.9 LIST OF REFERENCES

List any references that were used in the report. Use one format in writing the references. Never mix reference formats in a report.

20.10 APPENDIX

List all of the materials and information that are too detailed to be included in the body of the report.

FURTHER READING

Agrawal, Govind P., *Nonlinear Fiber Optics*, 2nd ed., Academic Press, Academic Press Optics and Photonics series, London, 1995.

Beiser, Arthur, *Physics*, 5th ed., Addison-Wesley Publishing Company, Reading, MA, 1991.

Blaker, Warren J., *Optics: The Matrix Theory*, Marcel Dekker, New York, 1972.

Born, M., Wolf, E., Elements of the theory of diffraction, *Principles of Optics: Electromagnetic Theory of Propagation, Interference, and Diffraction of Light*, 7th ed., pp. 370-458, Margaret Farley-Born and Emil Wolf, Cambridge University Press, Cambridge, UK, 1999.

Born, M., Wolf, E., Rigorous diffraction theory, *Principles of Optics: Electromagnetic Theory of Propagation, Interference, and Diffraction of Light*, 7th ed., pp. 556–592, Margaret Farley-Born and Emil Wolf Cambridge University Press, Cambridge, UK, 1999.

Bouwkamp, C. J., Diffraction theory, *Rep. Prog. Phys.*, 17, 35–100, 1949.

Bromwich, T.J.I'A., Diffraction of waves by a wedge, *Proc. London Math. Soc.*, 14, 450–468, 1916.

Chen, Y. K., et al., Low-crosstalk and compact optical add-drop multiplexer using a multiport circulator and fiber bragg gratings, *IEEE Photon. Technol. Lett.*, 12, 1394–1396, 2000.

Cutnell, John D. and Johnson, Kenneth W., *Physics*, 5th ed., John Wiley and Sons, Inc., New York, 2001.

Cutnell, John D. and Johnson, Kenneth W., *Student Study Guide—Physics*, 5th ed., John Wiley and Sons, Inc., New York, 2001.

Duarte, F. J. and Piper, J. A., Dispersion theory of multiple-prisms beam expanders for pulsed dye lasers, *Opt. Commun.*, 43, 303–307, 1982.

Edmund Industrial Optics, *Optics and Optical Instruments Catalog, 2004*, Edmund Industrial Optics, Barrington, NJ, 2004.

EPLAB, *EPLAB Catalogue 2002*, The Eppley Laboratory, Inc., Newport, RI, 2002.

Francon, M., *Optical Interferometry*, Academic Press, New York, 1966.

Ghatak, Ajoy K., *An Introduction to Modern Optics*, McGraw-Hill Book Company, New York, 1972.

Giancoli, Douglas C., *Physics*, 5th ed., Prentice Hall, New Jersey.

Halliday, D., Resnick, R., and Walker, J., *Fundamentals of Physics*, 6th ed., John Wiley & Sons, Inc., New York, 1997.

Hariharan, P., *1969. Modified Mach–Zehnder interferometer. Appl. Opt.*, 8(9), 1925–1926, 1969.

Hecht, Jeff, *Understanding Fiber Optics*, 3rd ed., Prentice Hall, Inc., New Jersey, 1999.

Hecht, Eugene, *Optics*, 4th ed., Addison-Wesley Longman, Inc., Reading, MA, 2002.

Hibino, Kenichi, Error-compensating phase measuring algorithms in a Fizeau interferometer, *Opt. Rev.*, 6(6), 529–538, 1999.

Jackson, R. A., The laser as a light source for the Mach–Zehnder interferometer, *J. Sci. Instrum.*, 42, 282–283, 1965.

Jenkins, F. W. and White, H. E., *Fundamentals of Optics*, McGraw-Hill, New York, 1957.

Johnstone, R. D. M. and Smith, W., A design for a 6 in., Mach–Zehnder interferometer, *J. Sci. Instrum.*, 42, 231–235, 1965.

Jones, Edwin and Childers, Richard, *Contemporary College Physics*, McGraw-Hill Higher Education, New York, 2001.

Kashyap, R., *Fiber Bragg Gratings*, Academic Press, New York, 1999.

Keuffel & Esser Co., *Physics*, Keuffel & Esser Audiovisual Educator—Approved Diazo Transparency Masters Avdiovisual Division, Keuffel & Esser Co., U.S.A., 1989.

Lambda Research Optics, Inc., *Catalog 2004*, Lambda Research Optics, San Jose, CA, 2004.

Lengyel, B., *Lasers*, John Wiley & Sons, New York, 1971.

Lerner, Rita G. and Trigg, George L., *Encyclopedia of Physics*, 2nd ed., VCH Publishers, Inc., New York, U.S.A., 1991.

Levine, H. and Schwinger, J., On the theory of diffraction by an aperture in an infinite plane screen, part I, *Phys. Rev.*, 74, 958–974, 1948.

Levine, H. and Schwinger, J., On the theory of diffraction by an aperture in an infinite plane screen, part II, *Phys. Rev.*, 75, 1423–1432, 1949.

Loewen, E. G. and Popov, E., *Diffraction Gratings and Applications*, Marcel Dekker, New York, 1997.

Malacara, Daniel, *Geometrical and Instrumental Optics*, Academic Press Co., Boston, 1988.

McDermott, Lillian C., et al., *Introduction to Physics*, Preliminary Edition, Prentice Hall, Inc., Upper Saddle River, NJ, 1988.

Melles Griot, *The Practical Application of Light*, Melles Griot Catalog, U.S.A., 2001.

Naess, Robert O., *Optics for Technology Students*, Prentice Hall, Upper Saddle River, NJ, 2001.

Newport Corporation, *Optics and Mechanics Section*, the Newport Resources, 1999/2000 Catalog, Newport, U.S.A., 1999/2000.

Newport Corporation, *Photonics Section, the Newport Resources 2004 Catalog*, Newport Corporation, U.S.A., 2004.

Nolan, Peter J., *Fundamentals of College Physics*, Wm. C. Brown Publishers, Inc., Dubuque, IA, U.S.A., 1993.

Ocean Optics, Inc., *Product Catalog 2003*, Ocean Optics, Inc., Florida, U.S.A., 2003.

Page, David and Routledge, Ian, Using interferometer of quality monitoring, *Photonics Spectra*, 147–153, November 2001.

Palais, Joseph C., *Fiber Optic Communications*, 4th ed., Prentice Hall, Inc., Upper Saddle River, NJ, 1998.

Pedrotti, Frank L. and Pedrotti, Leno S., *Introduction to Optics*, 2nd ed., Prentice Hall, Inc., New Jersey, 1993.

Plamer, Christopher, *Diffraction Grating Handbook*, 5th ed., Thermo Richardson Grating Laboratory, New York, 2002.

Salah, B. E. A. and Teich, M. C., *Fundamentals of Photonics*, John Wiley and Sons, New York, 1991.

Sciencetech, Modular optical spectroscopy, *Designers and Manufacturers of Scientific Instruments Catalog*, SCIENCETECH, London, Ontario, Canada, 2005.

Sears, Francis W., et al., *University Physics—Part II*, 6th ed., Addison-Wesley Publishing Company, Reading, MA, 1998.

Serway, Raymond A., *Physics for Scientists and Engineers*, 3rd ed., Saunders College Publishing, Saunders Golden Sunburst Series, London, 1990.

Shamir, Joseph, *Optical Systems and Processes*, SPIE Optical Engineering Press, Bellingham, 1999.

Silverman, Mark P., *Waves and Grains—Reflections on Light and Learning*, Princeton University Press, Princeton, NJ, 1998.

Smith, W. J., *Modern Optical Engineering*, McGraw-Hill Book Co., New York, 1966.

Tippens, Paul E., *Physics*, 6th ed., Glencoe McGraw-Hill, Westerville, OH, U.S.A., 2001.

Tolansky, S., *An Introduction to Interferometry*, Longmans, Green and Co., London, New York, Toronto, 1955.

Walker, James S., *Physics*, Prentice Hall, New Jersey, 2002.

Warren, Mashuri L., *Introduction to Physics*, W.H. Freeman and Company, San Francisco, CA, 1979.

Watanabe, T., et al., Polymeric arrayed-waveguide grating multiplexer with wide tuning range, *Electron. Lett.*, 33(18), 1547–1548, 1997.

White, Harvey E., *Modern College Physics*, 6th ed., Van Nostrand Reinhold Company, New York, 1972.

Wilson, Jerry D., *Physics—A Practical and Conceptual Approach*, Saunders College Publishing, Saunders Golden Sunburst Series. London, 1989.

Wilson, Jerry D. and Buffa, Anthony J., *College Physics*, 5th ed., Prentice Hall, Inc., New Jersey, 2000.

Woods, Nancy, *Instruction Manual to Beiser Physics*, 5th ed., Addison-Wesley Publishing Company, Reading, MA, 1991.

21 Spectrometers and Spectroscopes

21.1 INTRODUCTION

Spectrometers are optical devices, and often are called prism spectrometers because they often use a prism for light dispersion. As explained in the grating chapter, some devices use a diffraction grating in place of a prism because the dispersion from a grating can be much greater than that of a prism. Light dispersion occurs in a prism and in water particles in the atmosphere. Spectrometers are commonly used to analyse the wavelengths emitted by a light source. These devices are very useful to firefighters in identifying the fire coming from a burning object. Once they identify the type of the burning object, they easily can put out the fire with an appropriate fire extinguisher.

Common spectrum devices analyse light from a source. Light passes through a narrow slit and is focused into a parallel beam by a collimating lens; the beam then passes through a prism. Coloured beams emerge at a different angle according to the wavelength. A telescope focuses the colour beams and allows an observer to see an image of the spectrum. The spectrum can be matched with a chart of light source types to determine the type of the light source.

If the spectrum of a light source is recorded on a film or computer graph, rather than viewed by the eye, the device is called a spectrometer or spectrograph, as compared to a spectroscope, which is used for viewing only. Devices used to measure the intensity of light of a given wavelength are called spectrophotometers.

Spectrometers are widely used for determining the presence of certain types of molecules in laboratory specimens where chemical analysis would be difficult. Biological DNA identification is possible, since DNA and different types of protein absorb light in particular regions of the spectrum, such as in the UV region. Many applications, such as chemical analysis, metallurgy, biology, and medicine, use spectrometers.

This chapter presents five experimental cases that use different types of spectrometers and spectroscopes to identify a light source.

21.2 SPECTRAL ANALYSIS INSTRUMENTS

Although the individual design type may vary, all spectrometers perform the same function: they take an electromagnetic source and break it up into its constituent wavelengths. The following experiment performs this function for the visible light region of the electromagnetic spectrum. However, it is important to note that spectrometers also divide the source into the constituent IR and UV regions, which remain invisible to the human eye.

Spectrometers, spectroscopes, and spectrophotometers all perform the same function. The difference in suffix describes the method of detection: scope for observing with the eye and light meter for wavelength measurement. These instruments mentioned above may be split into two types, prism and grating. Spectrometers and spectroscopes use the principle of refraction and diffraction to disperse wavelengths.

21.2.1 SPECTRA

When a substance is heated, it will spontaneously emit its own particular type of light, depending on the atomic properties of the element, ion, or molecule of which it is made. This light emission results in a specific "fingerprint" of characteristic wavelengths for every material known, which is useful in discovering the chemical or elemental species, temperature, and other physical attributes of the substance.

Gasses have the unique property that, when excited, they produce discrete line spectra as opposed to the continuum that is produced from a solid or liquid. This is due to the fact that atoms in a gaseous state are loosely bound and, therefore, do not interact with the other surrounding atoms. This atomic freedom while in the gaseous state gives rise to the discrete energy levels that are responsible for the line spectra observed. Figure 21.1 shows a spectrum for a neon light source with lines in the visible region.

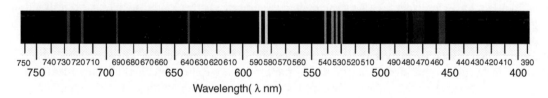

FIGURE 21.1 The neon spectrum.

The range of human vision covers wavelengths from roughly 400 to 700 nm (red to violet). According to the basic colour theory, it is understood that there are three additive primary colours and three subtractive primary colours. The three additive primaries are red, green, and blue, which represent the broad band of light energy, due to the fact that adding these three colours of light will create white light. The three subtractive primary colours are yellow, magenta, and cyan. The primary subtractive colours receive their name from the fact that each colour absorbs one-third of the white light spectrum.

21.2.2 THE PRISM

When white light passes through a prism, a spectrum of colours in the visible region is produced (red, orange, yellow, green, blue, and violet). Each wavelength has a different speed in an optical medium, such as glass or plastic. The prism-type spectrometer uses a prism to divide the light into its constituent wavelengths. Because each light wavelength has a different speed in the optical

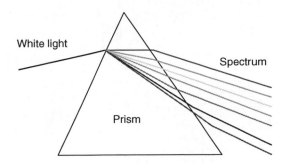

FIGURE 21.2 White light is dispersed into a spectrum by a prism.

medium, this causes the light beams to be refracted at a particular angle when they exit. The angle is often called the angle of deviation.

A prism produces a complete spectral range. However, because the angle of deviation increases steadily with decreasing wavelength, the red end is bent less than the blue, as shown in Figure 21.2.

21.2.3 THE DIFFRACTION GRATING

A diffraction grating is also used to break up light into its component wavelengths. The diffraction grating can be reflecting or transmitting. The transmitting grating uses many fine evenly-spaced lines or grooves to achieve the dispersion of light. Gratings are explained in detail in the diffraction gratings chapter.

21.2.4 NIGHT SPECTRA QUEST

This is the simplest and most inexpensive device that can be used by students to adequately identify common light sources. The diffraction grating in Night Spectra Quest consists of a clear plastic sheet on which thousands of fine parallel lines have been ruled. These lines are so fine that they are barely visible under a low-magnification microscope. Because they cannot be seen by the naked eye, they can only be oriented by the effect they produce. When a beam of white light passes through a diffraction grating, the different component colours interfere constructively in two directions, as shown in Figure 21.3.

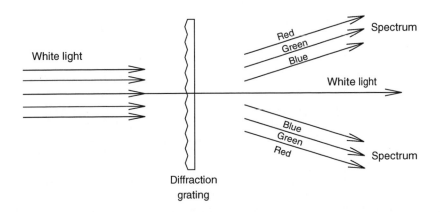

FIGURE 21.3 Night spectra quest operation.

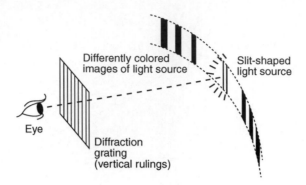

FIGURE 21.4 Looking through a diffraction grating.

When a distant light source is viewed through a diffraction grating, its spectrum appears as long and narrow coloured lines, if the grating lines are oriented parallel to the source length, as shown in Figure 21.4.

21.2.5 BLACK PLASTIC CASE SPECTROMETER

This is a simple spectroscope that can be used to analyse light sources in the visible range. It uses the principles of the diffraction grating, presented previously in the grating section. Figure 21.9 shows a black plastic case spectroscope.

21.2.6 PRISM SPECTROMETER

The prism spectrometer is another simple type of spectrometer, one that uses a prism to create the spectrum of the light source. As explained above in the prism section, when light enters a prism, a sequence of colours exits the prism. Figure 21.10 shows a prism spectroscope.

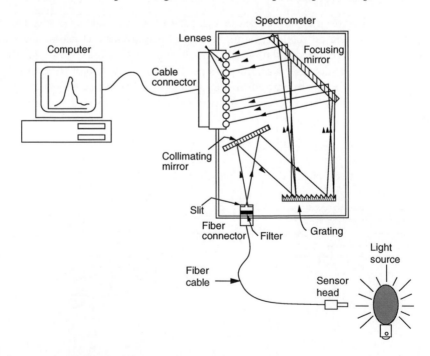

FIGURE 21.5 The instrumental aspects of a digital spectrometer.

21.2.7 DIGITAL SPECTROMETER

Figure 21.5 shows the optical components and the functionality of a typical digital spectrometer. The light source enters a fibre cable and is transmitted to the spectrometer. As the light exits the fibre cable, it is collimated by a mirror, and then is incident on a reflective diffraction grating. The light reflects from the grating spread out by colour. The colour beams then hit a focusing mirror, which projects them onto an array of lenses or Charge-coupled devices (CCDs) (e.g., most digital cameras use CCD to capture images. The CCD array consists of many photo-diodes that discharge a particular capacitor at a rate proportional to the photon flux incident on them. After a given sampling period, the capacitors release their charge, which is interpreted by a digital shift register. An analog-to-digital converter transfers the information to the computer, where it is output to the screen by the spectrometer software package.

21.2.8 CONVENTIONAL ARTICULATING SPECTROMETER

Figure 21.6 shows the components and operation of a conventional articulating spectrometer. This is a precision optical instrument that is often used to determine the composition and intensity of light. It can also be used to measure the refractive index of the prism or grating. The spectrometer has a heavy cast-iron base for stability and an adjustable prism table marked with grooved lines to assist the positioning of the prism. The turntable, mounted on heavy bearings, can be freely rotated about the vertical axis. The turntable has a circular scale graduated from 0° to 360° in one-half degree divisions and a vernier scale for beam angle measurement. The telescope is mounted on a movable pillar and allows fine adjustments along its axis. The instrument has an achromatic objective, eyepiece, rack and pinion focusing, and also includes prism and prism clamp, grating holder, and magnifier lens for reading verniers.

Light from the source is sent through a narrow, adjustable slit. The collimator is adjusted to produce a parallel and collimated beam. The light then passes through the prism and is dispersed into a spectrum. The refracted light is focused through the telescope objective and eyepiece lenses and observed by the eye. The angle, through which the telescope arm rotates to view a spectral line, relates to the wavelength of the spectral line. The colours of the light source spectrum are easily identified and diagnosed with the spectrum chart.

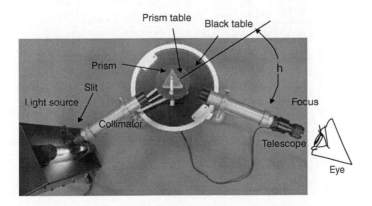

FIGURE 21.6 The components and operation of a conventional articulating spectrometer.

21.3 EXPERIMENTAL WORK

The student will gain theoretical and practical knowledge of the types of spectrometers and spectro-scopes that are used to identify light sources. The spectrometers to be used in this lab are:

a. A night spectra quest chart,
b. A black plastic case spectrometer,
c. A prism spectroscope,
d. A digital spectrometer, and
e. A conventional articulating spectrometer.

21.3.1 TECHNIQUE AND APPARATUS

Appendix A presents the details of the devices, components, tools, and parts.

1. Three types of light sources.
2. Spectrum chart for different light sources, as shown in Figure 21.7.
3. Night spectra quest chart, as shown in Figure 21.8.
4. Black plastic case spectrometer, as shown in Figure 21.9.

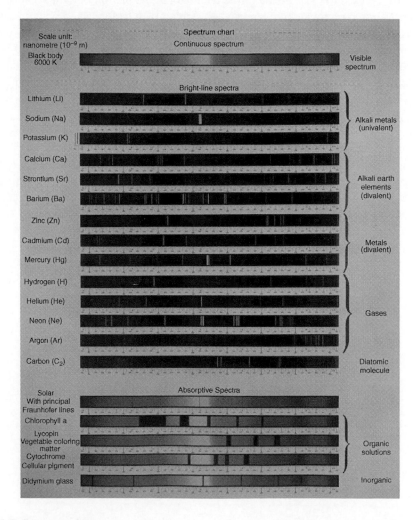

FIGURE 21.7 Spectrum chart.

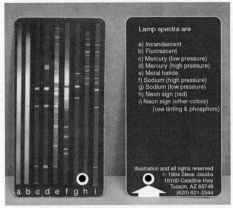

Front Side   Back Side

FIGURE 21.8 Night spectra quest spectral chart: Front side, back side.

FIGURE 21.9 Black plastic case spectrometer.

FIGURE 21.10 Prism spectroscope.

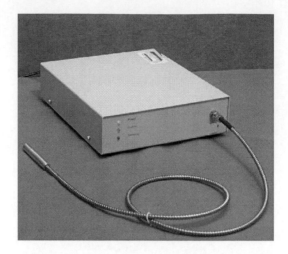

FIGURE 21.11 Digital spectrometer.

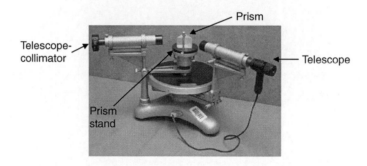

FIGURE 21.12 Conventional articulating spectrometer.

  5. Prism spectroscope, as shown in Figure 21.10.
  6. Digital spectrometer, as shown in Figure 21.11.
  7. Conventional articulating spectrometer, as shown in Figure 21.12.
  8. Black/white card and cardholder.
  9. Protractor.
  10. Ruler.

21.3.2 Procedure

Follow the laboratory procedures and instructions given by the instructor.

21.3.2.1 Safety Procedure

Follow all safety procedures and regulations regarding the use of optical components and devices, light source devices, electronic equipment, and optical cleaning chemicals.

21.3.2.2 Apparatus Setup

The following sections show the procedures that are used to identify light sources.

21.3.2.3 Night Spectra Quest Chart

1. Choose a light source and switch it on.
2. Choose a suitable distance from which to view the light source.
3. Hold a night spectra quest chart vertically parallel to the light source, as shown in Figure 21.13.
4. Observe the coloured spectrum and record what is seen.
5. Match the spectrum with the front side of the night spectra quest chart and determine the type of light source. The spectrum chart, as shown in Figure 21.7, can also be used to identify the light source.

21.3.2.4 Black Plastic Case Spectrometer

1. Use the same distance as in the preceding case.
2. Orient the spectrometer so that it is parallel with the table or other flat surface.
3. Align the slit as vertically to the source as possible, as shown in Figure 21.14.
4. Look through the diffraction grating at the narrow end of the spectrometer, pointing the slit towards the light source to be analysed.
5. The spectrum will appear on the right side of the slit, below the scale, where the wavelengths of the absorption or emission lines can be read.
6. Observe the line spectra and record where the lines fall on the ruled scale that is visible when looking into the device.

FIGURE 21.13 Light spectrum test using a night spectra quest chart.

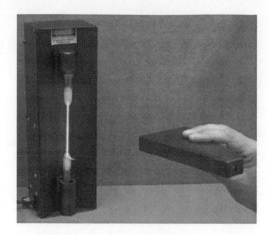

FIGURE 21.14 Light spectrum test using a black plastic case spectrometer.

21.3.2.5 Prism Spectroscope

1. Use the same distance as in the preceding case.
2. Orient and align the spectroscope so that it is facing the light source, as shown in Figure 21.15.
3. Observe the line spectra when looking into the spectroscope.

FIGURE 21.15 Light spectrum test using a prism spectroscope.

(a)

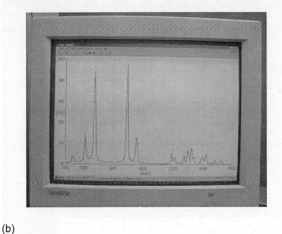

(b)

FIGURE 21.16 Light Spectrum test using a digital spectrometer: (a) Light spectrum test set-up, (b) Screenshot of the spectrum peaks.

21.3.2.6 Digital Spectrometer

1. Arrange the spectrometer sensor and setup, as shown in Figure 21.16(a).
2. Use the same distance as in the preceding case.
3. Observe the output on the screen noting where the peaks of the spectrum fall, as shown in Figure 21.16(b).
4. Capture the screen shot by hitting the Print Screen button.
5. Open up Microsoft Paint.
6. Press the paste key under the file tab at the top.
7. Save the image to disk or send as an email to yourself for the report.

21.3.2.7 Conventional Articulating Spectrometer

1. Figure 21.17 shows the experimental apparatus setup using a conventional articulating spectrometer and a light source.

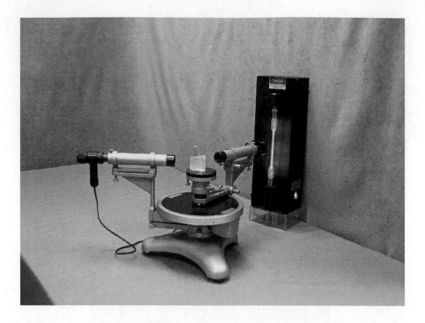

FIGURE 21.17 Spectrum test using a conventional articulating spectrometer.

2. The conventional articulating spectrometer uses a prism to find the spectrum of a light source. The alignment of the conventional articulating spectrometer and the prism is a very sensitive procedure. Appendix B explains the alignment of each part of the conventional articulating spectrometer.

21.3.3 Data Collection

21.3.3.1 Night Spectra Quest Chart

1. Record the colours observed while viewing the light source.

21.3.3.2 Black Plastic Case Spectrometer

1. Record the colours observed while viewing the light source.

21.3.3.3 Prism Spectroscope

1. Record the colours observed while viewing the light source.

21.3.3.4 Digital Spectrometer

1. Observe the output on the screen, noting where the peaks of the spectrum fall.
2. Capture the screen shot by hitting the Print Screen button.

21.3.3.5 Conventional Articulating Spectrometer

1. Find the apex angle (A) and incident angle.
2. Measure the angles of deviation (δ) for each spectral line.
3. Record the colours observed while viewing the light source.

21.3.4 CALCULATIONS AND ANALYSIS

21.3.4.1 Night Spectra Quest Chart

Match the spectrum of the light source with the spectrum chart and determine the type of light source.

21.3.4.2 Black Plastic Case Spectrometer

Match the spectrum of the light source with the spectrum chart and determine the type of light source.

21.3.4.3 Prism Spectroscope

Match the spectrum of the light source with the spectrum chart and determine the type of light source.

21.3.4.4 Digital Spectrometer

Determine the wavelengths of the light source and the type of light source.

21.3.4.5 Conventional Articulating Spectrometer

1. Report the apex angle, incident angle, and angles of deviation.
2. Determine the index of refraction for your prism, using Equation 21.1.
3. Match the spectrum of the light source with the spectrum chart and determine the type of light source.

$$n = \frac{\sin((A + \delta)/2)}{\sin(A/2)} \qquad (21.1)$$

where
A   is the apex angle of the prism, and
δ   is the angle of deviation.

21.3.5 RESULTS AND DISCUSSIONS

21.3.5.1 Night Spectra Quest Chart

Present the type of light source that was under observation.

21.3.5.2 Black Plastic Case Spectrometer

Present the type of light source that was under observation.

21.3.5.3 Prism Spectroscope

Present the type of light source that was under observation.

21.3.5.4 Digital Spectrometer

Present the type of light source that was under observation.

21.3.5.5 Conventional Articulating Spectrometer

1. Present the apex angle, incident angle, angle of deviation, and the index of refraction of the prism.
2. Present the type of light source that was under observation.

21.3.6 CONCLUSION

Summarize the important observations and findings obtained in this lab experiment.

21.3.7 SUGGESTIONS FOR FUTURE LAB WORK

List any suggestions for improvements using different experimental equipment, procedures, and techniques for any future lab work. These suggestions should be theoretically justified and technically feasible.

21.4 LIST OF REFERENCES

List any references that were used in the report. Use one format in writing the references. Never mix reference formats in a report.

21.5 APPENDICES

List all of the materials and information that are too detailed to be included in the body of the report.

FURTHER READING

Beiser, A., *Physics*, 5th ed., Addison-Wesley Publishing Company, Reading, MA, 1991.

Born, M. and Wolf, E., *Elements of the Theory of Diffraction Principles of Optics: Electromagnetic Theory of Propagation, Interference, and Diffraction of Light*, 7th ed., Cambridge University Press, Cambridge, UK, pp. 370–458, 1999.

Born, M. and Wolf, E., *Rigorous diffraction theory Principles of Optics: Electromagnetic Theory of Propagation, Interference, and Diffraction of Light*, 7th ed., Cambridge University Press, Cambridge, UK, pp. 556–592, 1999

Bouwkamp, C. J., Diffraction theory, *Rep. Prog. Phys.*, 17, 35–100, 1949.

Bromwich, T. J. I'A. ., Diffraction of waves by a wedge, *Proc. London Math. Soc.*, 14, 450–468, 1916.

Cutnell, John D. and Johnson, Kenneth W., *Physics*, 5th ed., Wiley, New York, 2001.

Desmarais, Louis, *Applied Electro Optics*, Prentice Hall, Inc, Upper Saddle River, NJ, 1998.

Drisoll, Walter G. and Vaughan, William, *Handbook of Optics*, McGraw-Hill Book Co, New York, 1978.

Edmund Industrial Optics, *Optics and Optical Instruments Catalog, 2004*, Edmund Industrial Optics, Barrington, NJ, 2004.

EPLAB, *EPLAB Catalogue 2002*, The Eppley Laboratory, Inc., Newport, RI, 2002.

Francon, M., *Optical Interferometry*, Academic Press, New York, 1966.

Giancoli, Douglas C., *Physics*, 5th ed., Prentice Hall, Upper Saddle River, NJ, 1998.

Griot, Melles, *The Practical Application of Light*, Melles Griot Catalog, Rochester, 2001.

Guy, Martin and Trepanier, Francois, Fiber Bragg gratings: Better manufacturing—better performance, *Photonics Spectra*, 106–110, 2002.

Guy, Martin and Painchaud, Yves, Fiber Bragg gratings: A versatile approach to dispersion compensation, *Photonics Spectra*, 96–101, 2004.

Halliday, D., Resnick, R., and Walker, J., *Fundamentals of Physics*, 6th ed., Wiley, New York, 1997.

Hariharan, P., Modified Mach–Zehnder interferometer, *Appl. Opt.*, 8 (9), 1925–1926, 1969.

Hecht, Jeff, *City of Light: The Story of Fiber Optics*, Oxford University Press, New York, 1999.

Hecht, Eugene, *Optics*, 4th ed., Addison-Wesley Longman, Inc., Reading, MA, 2002.

Jackson, R. A., The laser as a light source for the Mach–Zehnder interferometer, *J. Sci. Instrum.*, 42, 282–283, 1965.

Jacobs, Stephen F., *Night Spectra Quest*, University of Arizona Optical Sciences Center, Tucson, AZ, 2000.

Jenkins, F. W. and White, H. E., *Fundamentals of Optics*, McGraw Hill, New York, 1957.

Johnstone, R. D. M. and Smith, W., A design for a 6 in. Mach–Zehnder interferometer, *J. Sci. Instrum.*, 42, 231–235, 1965.

Jones, Edwin and Childers, Richard, *Contemporary College Physics*, McGraw-Hill Higher Education, New York, 2001.

Keuffel & Esser Co., *Physics*, Keuffel & Esser Audiovisual Educator-Approved Diazo Transparency Masters, Mid South, U.S., 1989.

Lambda Research Optics, Inc, *Catalog 2004*, Lambda Research Optics, Costa Mesa, CA, 2004.

Lerner, R. G. and Trigg, George L., *Encyclopedia of Physics*, 2nd ed., VCH Publishers, Inc., New York, 1991.

Levine, H. and Schwinger, J., On the theory of diffraction by an aperture in an infinite plane screen, part I, *Phys. Rev.*, 74, 958–974, 1948.

Levine, H. and Schwinger, J., On the theory of diffraction by an aperture in an infinite plane screen, part II, *Phys. Rev.*, 75, 1423–1432, 1949.

Loewen, E. G. and Popov, E., *Diffraction Gratings and Applications*, Marcel Dekker, New York, 1997.

Loreggia, David, Gardiol, D., Gai, M., Lattanzi, M. G., and Busonero, D., Fizeau interferometer for global astrometry in space, *Appl. Opt.*, 43 (4), 721–728, 2004.

Malacara, Daniel, *Geometrical and Instrumental Optics*, Academic Press Co., Boston, 1988.

McDermott, L. C. and Shaffer, P. S., *Introduction to Physics*, Preliminary edition, Prentice Hall, Inc., Upper Saddle River, NJ, 1988.

Naess, R. O., *Optics for Technology Students*, Prentice Hall, Upper Saddle River, NJ, 2001.

Newport Corporation, *Optics and Mechanics, 1999/2000 Catalog*, Newport Corporation, Irvine, CA, 1999.

Ocean Optics, Inc., *Product Catalog 2003*, Ocean Optics, Inc., Dunedin, FL, 2003.

Page, David and Routledge, Ian, Using interferometer for global astrometry in space, *Photonics Spectra*, November, 147–153, 2001.

Pedrotti, Frank L. and Pedrotti, Leno S., *Introduction to Optics*, 2nd ed., Prentice Hall, Inc, Upper Saddle River, NJ, 1993.

Plamer, Christopher, *Diffraction Grating Handbook*, 5th ed., Thermo Richardson Grating Laboratory, New York, 2002.

Pritchard, D. C., *Environmental Physics: Lighting*, Longmans, Green & Co., London, 1969.

Product Knowledge, *Lighting Reference Guide, Product Development,* 2nd ed., Ontario Hydro, Toronto, ON, Canada, 1988.

Romine, Gregory S., *Applied Physics Concepts into Practice*, Prentice Hall, Inc., Upper Saddle River, NJ, 2001.

Sciencetech, *Designers and Manufacturers of Scientific Instruments Catalog*, Sciencetech, London, 2003.

Sears, F. W., Zemansky, M. W., and Young, H. D., *University Physics—Part II*, 6th ed., Addison-Wesley Publishing Company, Reading, MA, 1998.

Serway, Raymond A., *Physics for Scientists and Engineers Saunders Golden Sunburst Series*, 3rd ed., Saunders College Publishing, Philadelphia, PA, 1990.

Smith, W. J., *Modern Optical Engineering*, McGraw-Hill Book Co, New York, 1966.

Silverman, M. P., *Waves and Grains—Reflections on Light and Learning*, Princeton University Press, Princeton, NJ, 1998.

Walker, James S., *Physics*, Prentice Hall, Upper Saddle River, NJ, 2002.

Warren, Mashuri L., *Introduction to Physics*, W.H. Freeman and Company, San Francisco, CA, 1979.

Weisskopf, V. F., How light interacts with matter, *Sci. Am.*, 219 (3), 60–71, 1968.

White, Harvey E., *Modern College Physics*, 6th ed., Van Nostrand Reinhold Company, New York, 1972.

Wilson, J. D., *Physics—a Practical and Conceptual Approach Saunders Golden Sunburst Series*, Saunders College Publishing, Philadelphia, PA, 1989.

Woods, Nancy, *Instruction manual to Beiser Physics*, 5th ed., Addison-Wesley Publishing Company, Reading, MA, 1991.

Young, Hugh D. and Freedman, Roger A., *University Physics*, 9th ed., Addison-Wesley Publishing Company, Inc., Reading, MA, 1996.

Section IV

Optical Fibres

22 Fibre Optic Cables

22.1 INTRODUCTION

Fibre optic cables transmit data through very small cores at the speed of light. Significantly different from copper cables, fibre optic cables offer high bandwidths and low losses, which allow high data-transmission rates over long distances. Light propagates throughout the fibre cables according to the principle of total internal reflection.

There are three common types of fibre optic cables: single-mode, multimode, and graded-index. Each has its advantages and disadvantages. There also are several different designs of fibre optic cables, each made for different applications. In addition, new fibre optic cables with different core and cladding designs have been emerging; these are faster and can carry more modes. While fibre optic cables are used mostly in communication systems, they also have established medical, military, scanning, imaging, and sensing applications. They are also used in optical fibre devices and fibre optic lighting.

This chapter will discuss the fabrication processes used in manufacturing fibre cables. The processes produce a thin flexible glass strand with a diameter even smaller than that of human hair. The chapter will also detail methods of coupling a light source with a fibre cable in the manufacturing of optical fibre devices. It will also compare fibre and copper cables and describe the applications of fibre optic cables in many fields and sectors of modern society. Finally, this chapter will present four experimental cases, including fibre cable inspection and handling, fibre cable end preparation, numerical aperture measurements and calculations, and fibre cable power output intensity measurements and calculations.

22.2 THE EVOLUTION OF FIBRE OPTIC CABLES

The evolution of optical communication systems dates back to the early 1790s, when the French engineer Claude Chappe (1763–1805) invented the optical telegraph. His system involved a series of semaphores mounted on towers, where human operators relayed messages from one tower to the next. This was certainly an improvement over hand-delivered messages. But by the mid-nineteenth century, the optical telegraph was replaced by the electric telegraph, leaving behind a legacy of Telegraph Hills.

In 1841, Swiss physicist Daniel Colladen (1802–1893) and French physicist Jacques Babinet (1794–1872) showed in their popular science lectures that light could be guided along jets of water for fountain displays. Then in 1870, Irish physicist John Tyndall (1820–1893) demonstrated the light-pipe phenomenon at the Royal Society in England. Tyndall directed a beam of sunlight into a container of water and opened the spout. Water flowed out in a jet, and the pull of gravity bent the water into a parabolic shape, shown in Figure 22.1. Light was trapped inside the water jet by the total internal reflection phenomena. The light beam bounced off the top surface, then off the lower surface of the jet, until turbulence occurred in the flowing water and broke up the beam. This experiment marked the first research into the guided transmission of light by an interface between two optical materials.

In 1880, Alexander Graham Bell patented an optical telephone system that he called the Photophone, but his earlier invention—the telephone—proved far more practical. While Bell dreamed of sending signals through the air, the atmosphere did not transmit light as reliably as wires carried electricity. For the next several decades, though light was used for a few special applications such as signaling between ships, inventions using optical communication gathered dust on the shelf. Bell donated his Photophone to the Smithsonian Institution.

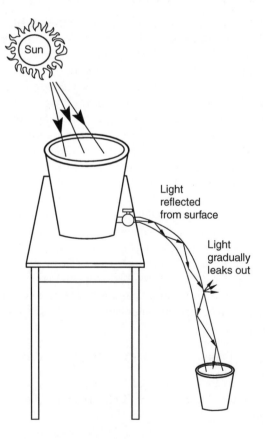

FIGURE 22.1 Total internal reflection of light in a water jet.

Ultimately, a new light-guiding technology that slowly took root solved the problem of optical transmission, The technology depended on the phenomenon of total internal reflection, which can confine light in an optical material that is surrounded by another optical material with a lower refractive index, such as glass in air. However, it was a long time before this method was adapted for communications.

Optical fibres went a step further. These are essentially transparent rods of glass or plastic stretched until they are long and flexible. During the 1920s, John Logie Baird in England and Clarence W. Hansell in the United States patented the idea of using arrays of hollow pipes or transparent rods to transmit images for television or facsimile systems. British Patent Spec 20,969/27 was registered to J. L. Baird, and US Patent 1,751,584 was granted to C. W. Hansell in 1930 for the scanning and transmission of a television image via fibres. Also in 1930, H. Lamm, in Germany, demonstrated light transmission through fibres. The next reported activity in this field took place in 1951, when A. C. S. van Heel in Holland and Harold H. Hopkins and Narinder S. Kapany of Imperial College in London investigated light transmission through bundles of fibres. While van Heel coated his fibres with plastic, Kapany explored fibre alignment, and as reported in his book *Fibre Optics*, produced the first undistorted image through an aligned bundle of uncoated glass fibres.

Neither van Heel nor Hopkins and Kapany made bundles that could carry light far, but their reports sparked the fibre optics revolution. The crucial innovation was made by van Heel, stimulated by a conversation with the American optical physicist Brian O'Brien. While all earlier fibres were bare, with total internal reflection at a glass–air interface, Heel covered a bare fibre of glass or plastic with a transparent cladding of lower refractive index. This protected the total reflection surface from losses and greatly reduced crosstalk between fibre cables. The next key step was the development of glass-clad fibres by Lawrence Curtiss at the University of Michigan. By 1960, glass-clad fibres had reached an attenuation of about one decibel per metre, which was fine for medical imaging though much too high for communications.

In 1958, higher optical frequencies seemed the logical next step to Alec Reeves, the forward-looking engineer at Britain's Standard Telecommunications Laboratories who invented digital pulse-code modulation before World War II. Other people climbed on the optical communications bandwagon when the laser was invented in 1960. Then the July 22, 1960, issue of *Electronics* magazine introduced its report on Theodore Maiman's demonstration of the first laser by saying, "Usable communications channels in the electromagnetic spectrum may be extended by development of an experimental optical frequency amplifier."

All was quiet in the new science until 1967, when Charles K. Kao and George A. Hockham of Britain's Standard Telecommunications Laboratories suggested the impending development of a new communications medium using cladded fibre, with wavelengths in the millimetre range. Optical fibres had been attracting attention because they were analogous in theory to the plastic dielectric waveguides used in certain microwave applications.

Since then, the technology has grown enormously—slowly at first, but almost exponentially during the 1970s.

The Corning Company's breakthrough in fibre optic cable manufacturing was among the most dramatic of the many developments that opened the door to fibre optic communications. In that same year, Bell Labs and a team at the Ioffe Physical Institute in St. Petersburg made the first semiconductor diode lasers that were able to emit continuous light waves at room temperature. Over the next several years, fibre losses dropped dramatically, aided both by improved fabrication methods and the shift to longer wavelengths at which fibres have inherently lower attenuation.

Early single-mode fibres had cores several micrometres in diameter. During the early 1970s, this bothered developers, who doubted that it would be possible to achieve the micrometre-scale tolerances needed to couple light efficiently into the tiny cores from light sources, or in splices or connectors. Not satisfied with the low bandwidth of step-index multimode fibre, they concentrated on multimode fibres with a refractive-index gradient between core and cladding, and core diameters of 50 or 62.5 µm. The first generation of telephone field trials in 1977 used these fibres to transmit light at 850 nm from gallium–aluminum–arsenide laser diodes.

Government and defence agencies, telephone companies, and a widening range of other private companies turned to fibre optics for telecommunications and other uses, many of which were protected as military or proprietary secrets. Each new user was attracted by the fact that optical transmission does not simply reduce but entirely avoids the risks of short circuits, wire tapping, and electromagnetic cross talk. Among the converts were General Telephone of Indiana, which had a 3-mile link between two switching centres in Fort Wayne, Indiana, carrying 5000 telephone circuits through only 14 fibres. A similar system was installed in San Angelo, Texas, before the end of 1981. Bell Canada also used fibre technology, as did Alberta Government Telephones of Canada. The list grew as companies recognized the superiority of fibre optic cable over the old copper wire technology.

Overseas, England completed nearly 450 km of optical fibre cable under water. Two grades of fibre cables were used, one for higher bit-rate systems of 140 Mbps, which used laser injection techniques, and another for systems of 8 to 34 Mbps. In addition, France, Germany, and other European countries were entering the field, primarily in telephone work, but also in other applications.

Those first-generation systems could transmit light over several kilometres but were limited by a loss of about 2 dB/km in the fibre cable. Thus, a second generation soon appeared that used new lasers that emitted at 1300 nm, with fibre attenuation as low as 0.5 dB/km. The development of hardware for the first transatlantic fibre cable showed that single-mode systems were feasible, so when deregulation opened the long-distance phone market in the early 1980s, carriers built national backbone systems of single-mode fibre with 1300 nm light sources. That technology spread into other telecommunication applications. Meanwhile, a newer generation of single-mode systems found applications in submarine cables and systems serving large numbers of subscribers; these operate at 1550 nm.

More important was the development of the erbium doped optical fibres, which serve as optical amplifiers. Submarine cables with these optical amplifiers operate at high speeds. Optical amplifiers are also attractive for fibre systems delivering the same signals to many terminals, because the fibre amplifiers can compensate for losses in dividing the signals among many terminals. The biggest challenge remaining for fibre optics now is one of economics.

By 1990, AT&T had laid a fibre cable that spanned the Atlantic Ocean. Cables between other countries were also completed, and electric utilities were installing fibre cable links for communicating with substations, replacing the venerable carrier telephone equipment that multiplexes via the power lines and through the air. Fibre cables have also been laid underground as security links.

Today, telephone and cable television companies can justify the cost of installing fibre links to remote sites that serve tens to hundreds of customers. However, terminal equipment remains too expensive to install the fibres all the way to customers's homes, at least for present services. Instead, cable and phone companies run twisted wire pairs or coaxial cables from optical network units to individual houses.

The wireless telecommunication market is growing quickly. Systems using wavelengths in the range 1625–675 nm are also available for very fast data transmission. Recent years have seen a drastic increase in the advanced types of optic cables available. These cables are very low loss and operate at very high speeds for a wide range of wavelengths. Time will tell how long fibre optic technology will last.

22.3 FIBRE OPTIC CABLES

Fibre optic cable is a filament of transparent material used to transmit light, as shown in Figure 22.2. Virtually all fibre optic cables share the same fundamental structure. The centre of the cable is referred to as the core. It has a higher refractive index than the cladding, which surrounds the core. The contact surface between the core and the cladding creates an interface surface that guides the light; the difference between the refractive index of the core and cladding is what causes the mirror-like interface surface, which guides light along the core. Light bounces through the core from one end to the other according to the principle of total internal reflection, as explained by the laws of light. The cladding is then covered with a protective plastic or PVC jacket. The diameters of the core, cladding, and jacket can vary widely; for example, a single fibre optic cable can have core, cladding, and jacket diameters of 9, 125, and 250 μm, respectively.

Figure 22.3 shows the structure of a typical fibre optic cable. The cores of most fibre optic cables are made from pure glass, while the claddings are made from less pure glass. Glass fibre optic cable has the lowest attenuation over long distances but comes at the highest cost. A pure glass

FIGURE 22.2 A fibre optic cable.

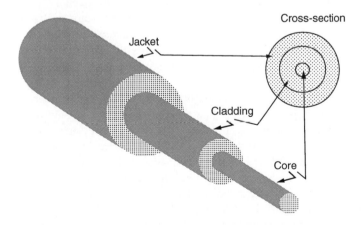

FIGURE 22.3 Schematic view and cross section of a fibre optic cable.

TABLE 22.1
Common Diameters of Fibre Optic Cables

Core (μm)	Standard Diameter of	
	Cladding (μm)	Jacket (μm)
8	125	250
50	125	250
62.5	125	250
100	140	250

fibre optic cable has a glass core and a glass cladding. Fibre optic cable cores and claddings may be made from plastic, which is not as clear as glass but is more flexible and easier to handle. Compared with other fibre cables, plastic fibre cables are limited in power loss and bandwidth. However, they are more affordable, easy to use, and attractive in applications where high bandwidth or low loss is not a concern. A few glass fibre cable cores are clad with plastic. Their performance, though not as good as all-glass fibre cables, is quite respectable. More details on plastic cables will be presented later in this chapter.

The jacket is made from polymer (PVC, plastic, etc.) to protect the core and the cladding from mechanical damage. The jacket has several major attributes, including bending ability, abrasion resistance, static fatigue protection, toughness, moisture resistance, and the ability to be stripped. Fibre optic cable jackets are made in different colours for colour-coding identification. Some optical fibres are coated with a copper-based alloy that allows operation at up to 700 and 500°C for short and long periods, respectively.

Table 22.1 shows the common diameters of the core, cladding, and jacket of four commonly used fibre optic cables in manufacturing optical fibre devices. Different types and cross sections are also available for building advanced fibre optic devices. These types will be presented in the coming sections of this chapter.

22.4 PLASTIC FIBRE CABLES

Plastic optical fibre (POF) has the highest attenuation over short distances, but it comes at the lowest cost. A plastic fibre optic cable has a plastic core and plastic cladding. It is also quite thick, with typical core/cladding diameters of 480/500, 735/750, and 980/1000 μ. The core generally consists of polymethylmethacrylate (PMMA) coated with a fluropolymer. Plastic fibre optic cables are used in small optical devices, lighting applications, automobiles, music systems, and other electronic systems. The cables are also used in communication systems where high bandwidth or low loss are not a concern. The increased interest in plastic optic fibre is due to two reasons: (1) the higher attenuation relative to glass, which may not be a serious obstacle with the short cable runs often required in premise networks; and (2) the cost advantage, which appeals to network architects faced with budget decisions. Plastic fibre optic cables do, however, have a problem with flammability. Because of this, they may not be appropriate for certain environments, and care must be given when they are run through a plenum. Otherwise, plastic fibre is considered extremely rugged, with a tight bend radius and the ability to withstand mechanical stress.

Plastic clad silica (PCS) fibre optic cable has an attenuation—and cost—that lie between those of glass and plastic. Plastic clad silica (PCS) fibre optic cable has a glass core that is often made of vitreous silica; the cladding is often plastic, usually a silicone elastomer with a lower refractive index. In 1984, the International Electrotechnical Commission (IEC) standardized PCS fibre optic cable to have the following dimensions: a core of 200 μ, a silicone elastomer cladding 380 of microns, and a jacket of 600 μ.

Plastic fibre cables are fabricated using the same principles as glass fibre cables. A core with a higher index of refraction is surrounded by a cladding with a lower index of refraction. The cladding is then coated with a coloured jacket for coding purposes; glass and plastic cables are similarly colour coded. POF cables are available in single- and multi-step index, as well as graded index.

Recent developments in the polymer industry have led to improvements in plastic fibre optic cables. Plastic fibre cables will eventually replace glass fibre cables because of their many advantages, including their ease in connection using epoxy as well as their lower price, durability, lower weight, and smaller bending radii.

22.5 LIGHT PROPAGATION IN FIBRE OPTIC CABLES

Figure 22.4 shows the principle of light propagation through a fibre optic cable. To understand the principle, consider when light is injected into an optical material that is surrounded by another optical material with a different index of refraction. The first and second optical materials will have a higher and lower index of refraction, respectively. Figure 22.4(a) shows light incident at an angle

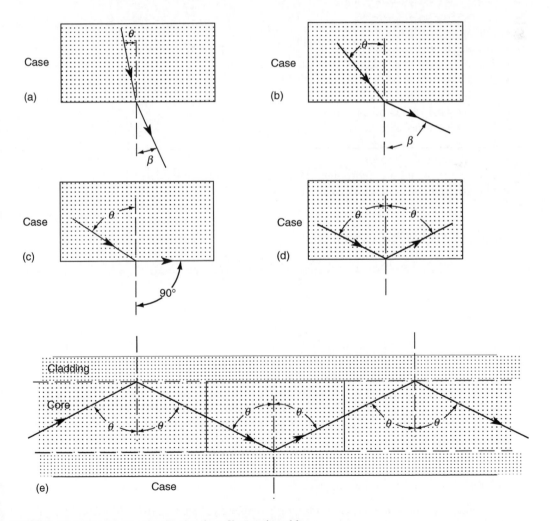

FIGURE 22.4 Total internal reflection in a fibre optic cable.

θ to the normal. The light refracts by an angle β from the normal. If the angle of incidence is increased, the angle of refraction will also increase, as shown in Figure 22.4(b). In Figure 22.4(c), light incident at an angle θ equal to the critical angle θ_c gives an angle of refraction β of 90 degrees. The refracted light lies on the interface of the first-second optical materials. Snell's Law is applicable here. In Figure 22.4(d), when light lands incident at an angle θ greater than the critical angle, the light will reflect by the total internal reflection phenomena. In this case, the angle of incidence equals the angle of reflection, which is defined by the first law of light.

Assume the first optical material is very long. It has a circular cross section with a very small diameter like the core, and at the same time is surrounded by another optical material resembling the cladding in a fibre optic cable, as shown in Figure 22.4(e). Now, light injected into a fibre optic cable and striking the core-to-cladding interface at a greater angle than the critical angle reflects back into the core. Since the angles of incidence and reflection are equal, the reflected light will again be reflected. The light will continue to bounce through the length of the fibre cable. Light in Figure 22.4 Cases (a), (b), and (c) show that the light passes into the cladding. The cladding is usually inefficient as a light carrier compared to the core of the fibre cable. Light in the cladding becomes part of the losses, some of which will be presented in coming sections of the chapter, that usually occur in any fibre optic cable. Therefore, light propagation in a fibre optic cable is governed by the following:

- The wavelength of light
- The angle of incidence of the light at the input of the fibre cable
- The indices of refraction of the core and cladding
- The composition of the core and cladding
- The length of the cable
- The bending radius of the cable
- The sizes of the core and cladding
- The design of the core and cladding.
- The transmission modes
- The temperature and environmental conditions of the fibre cable
- The strength and flexibility of the fibre cable

22.6 REFRACTIVE-INDEX PROFILE

As explained above, the fibre optic cable's core and cladding are made from different optical materials, each of which has a different index of refraction. The refractive-index profile describes the relationship between the indices of refraction of the core and cladding. When the core is made from uniform optical material, the refractive-index profile is flat. This is called a step-index profile, as shown in Figure 22.5(a) and (b). The step-index fibre cable has a core with a uniform index of refraction throughout the cable. The profile shows a sharp step at the junction of the core and cladding. When the core is made from multilayered optical materials and each layer has a differing index of refraction, the refractive-index profile has a curved shape. This type of profile is called a graded-index (GRIN), as shown in Figure 22.5(c). In contrast to the step-index profile, the graded-index profile has a non-uniform core. The index is highest at the centre and gradually decreases until it matches that of the cladding. There is no sharp step between the core and the cladding in graded-index fibre cables.

22.7 TYPES OF FIBRE OPTIC CABLES

There are, thus, three common types of fibre optic cables, as listed below. The suitability of each type for a particular application depends on the fibre optic cable's characteristics.

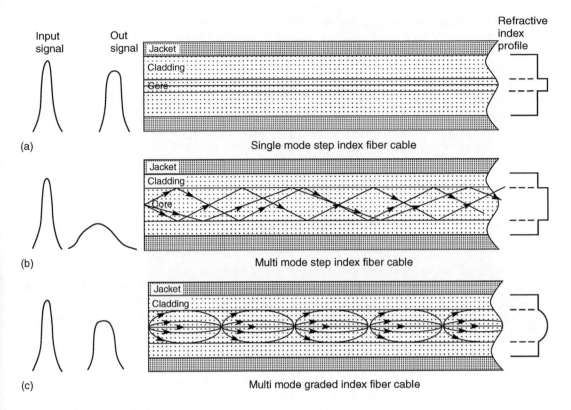

FIGURE 22.5 Types of fibre optic cables.

22.7.1 SINGLE-MODE STEP-INDEX FIBRE CABLE

The single-mode step-index fibre cable, sometimes called a single-mode fibre cable, is shown in Figure 22.5(a). The single and multimode step-index fibre cables are the simplest types. Single-mode fibre cables have extremely small core diameters, ranging from 5 to 9.5 μm. The core is surrounded by a standard cladding diameter of 125 μm. The jacket is applied on the cladding to provide mechanical protection, as shown in Figure 22.3. Jackets are made of one type of polymer in different colours for colour-coding purposes. Single-mode fibres have the potential to carry signals for long distances with low loss, and are mainly used in communication systems. The number of modes that propagate in a single-mode fibre depends on the wavelength of light carried. The number of modes will be given in Equation (22.9). A wavelength of 980 nm results in multimode operation. As the wavelength is increased, the fibre carries fewer and fewer modes until only one mode remains. Single-mode operation begins when the wavelength approaches the core diameter. At 1310 nm, for example, the fibre cable permits only one mode. It then operates as a single-mode fibre cable.

22.7.2 MULTIMODE STEP-INDEX FIBRE CABLE (MULTIMODE FIBRE CABLE)

The multimode step-index fibre cable, sometimes called a multimode fibre cable, is shown in Figure 22.5(b). Multimode fibre cables have bigger diameters than their single-mode counterparts, with core diameters ranging from 100 to 970 μm. They are available as glass fibres (a glass core and glass cladding), plastic-clad silica (a glass core and plastic cladding), and plastic fibres (a plastic core and cladding). They are also the widest ranging, although not the most efficient in long distances, and they experience higher losses than the single-mode fibre cables. Multimode fibre cables have the potential to carry signals for moderate and long distances with low loss (when optical amplifiers are used to boost the signals to the required power).

Since light rays bounded through a fibre cable reflect at different angles for different ray paths, the path lengths of different modes will also be different. Thus, different rays take a shorter or longer time to travel the length of the fibre cable. The ray that goes straight down the centre of the core without reflecting arrives at the other end faster. Other rays take slightly longer and thus arrive later. Accordingly, light rays entering a fibre at the same time will exit at the other end at different times. In time, the light will spread out because of the different modes. This is called modal dispersion. Dispersion describes the spreading of light rays by various mechanisms. Modal dispersion is that type of dispersion that results from the varying modal path lengths in the fibre cable.

22.7.3 MULTIMODE GRADED-INDEX FIBRE (GRADED-INDEX FIBRE CABLE)

Multimode graded-index fibres are sometimes called graded-index fibre cables (GRIN), as shown in Figure 22.5(c). Graded-index and multimode fibre cables have similar diameters. Common graded-index fibres have core diameters of 50, 62.5, or 85 μm, with a cladding diameter of 125 μm. The core consists of numerous concentric layers of glass, somewhat like the annular rings of a tree or a piece of onion. Each successive layer expanding outward from the central axis of the core until the inner diameter of the cladding has a lower index of refraction. Light travels faster in an optical material that has a lower index of refraction. Thus, the further the light is from the centre axis, the greater its speed. Each layer of the core refracts the light according to Snell's Law. Instead of being sharply reflected as it is in a step-index fibre, the light is now bent or continually refracted in an almost sinusoidal pattern. Those light rays that follow the longest path by traveling near the outside of the core have a faster average velocity. The light ray traveling near the centre of the core has the slowest average velocity. As a result, all rays tend to reach the end of the fibre at the same time. Thus, one way to reduce modal dispersion is to use graded-index fibres. This type of fibre optic cable is popular in applications that require a wide range of wavelengths, in particular telecommunication, scanning, imaging, and data processing systems.

Fibre cables are designed for a specific wavelength, called the cutoff wavelength, above which the fibre carries only one mode. A fibre designed for single-mode operation at 1310 nm has a cutoff wavelength of around 1200 nm. Although optical power is confined to the core in a multimode fibre, it is not so confined in a single-mode fibre. This diameter of optical power is called the "mode field diameter." It is usually more important to know the mode field diameter than the core diameter.

22.8 POLARIZATION MAINTAINING FIBRE CABLES

Polarization maintaining (PM) fibres are constructed by placing specially designed asymmetries into the core. PM fibres guide only one possible mode of propagation. They also maintain the electromagnetic field vector direction. This type of single-mode fibre is used in building optical fibre devices that work with polarized light, such as polarization beamsplitters, couplers, modulators, and interferometric sensors. PM fibres are desirable, since most lasers emit highly polarized light and the polarized properties are highly desirable in many measurement applications.

There are four common designs for PM fibres, as shown in Figure 22.6. Figure 22.6(a) shows the PANDA fibre. The PANDA fibre employs a stress technique that stresses the core of the fibre to create two propagation paths within the fibre core, with two stress rods placed within the cladding in the same plane on opposite sides of the core. Linearly polarized light aligned to either the slow or fast axis of the fibre will remain linearly polarized as it propagates through the fibre.

Figure 22.6(b) shows a cross section of the Bow-tie fibre. A pair of wedges on opposite sides of the core generates the optimum stress distribution within the fibre. This patented design offers the best in both in terms of performance and handling, with minimum stress breakout when cleaved, connectorized, or polished.

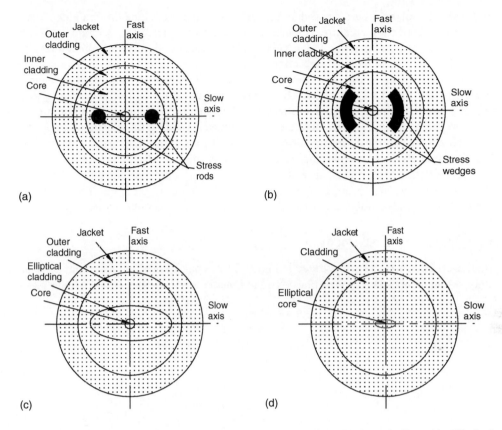

FIGURE 22.6 Polarization maintaining fibre cables. (a) PANDA fibre. (b) Bow-tie fibre. (c) Elliptical core fibre. (d) elliptical core.

Figure 22.6(c) shows a circular core surrounded by an elliptical boron-doped cladding. Figure 22.6(d) shows a PM fibre with a very high level of doping in an elliptical core, which causes waves polarized along the major and minor axes of the ellipse to have different effective indices of refraction. Thus, the fibres contain nonsymmetrical stress-production parts.

22.9 SPECIALTY FIBRE CABLES

Previous sections presented common fibre cables that are used to guide light over relatively long distances in communication systems, imaging, scanning, and medical applications. Fibre optic cable technology and its applications have experienced a diversity of technological advancements that make optical fibres able to fulfill every possible application. Other types of fibres are optimized for a variety of specific applications and for research and development. The next chapter will discuss the advanced fibre optic cables used in many applications requiring a low loss, low chromatic dispersion, and high transmission rate over long distances.

22.10 FIBRE CABLE FABRICATION TECHNIQUES

A variety of fabrication techniques are used to produce fibre optic cables. This section will discuss two common methods used to manufacture optical glass fibre cables. The first method involves directly drawing the fibre from two molten glass rods or "preforms," which are placed in two concentric crucibles; this is called the double crucible method. The second method involves forming a preform.

There are several processes for producing preforms to be used in the fibre-drawing method. The manufacture of fibre cable requires sophisticated and highly accurate techniques. The biggest challenge is the purification of the materials used in the construction of the core and cladding. In particular, the value of the index of refraction must be very precise, especially when manufacturing graded-index fibres. During the glass fibre optic cable fabrication process, impurities are intentionally added to the pure glass to obtain the desired indices of refraction needed to guide light. Germanium or phosphorous are added to increase the index of refraction, while boron or fluorine are added to decrease the index of refraction. Unfortunately, adding these residual impurities may increase the attenuation by either scattering or absorbing the light. The fabrication methods also need to be extremely precise regarding fibre dimensions and tolerances. The diameter of the core must be controlled to high precision, and the core must be located at the centre of the cladding. These same issues are also present when applying the cladding to the core. In the final step, the cladding is coated with a polymer jacket layer for mechanical and environmental protection.

22.10.1 Double Crucible Method

The double crucible method is illustrated in Figure 22.7. A pair of platinum crucibles sits one inside the other. Molten core glass is then placed in the inner crucible, and molten cladding glass is fed

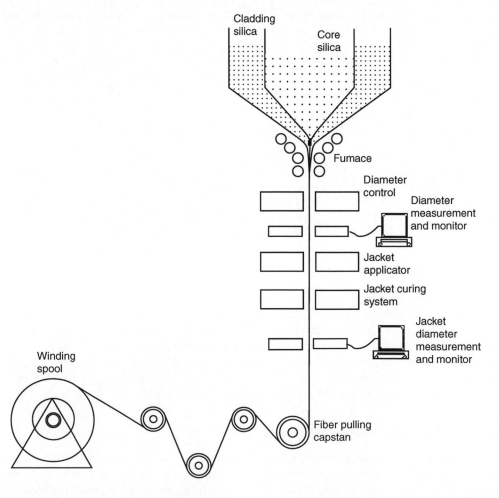

FIGURE 22.7 Double crucible fabrication method.

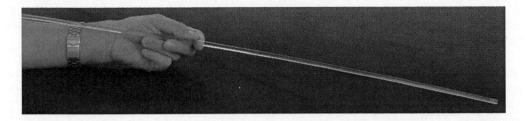

FIGURE 22.8 A preform.

into the outer crucible. The crucibles are kept at a high temperature, typically between 1850 and 2,000°C. Using a precision-feed mechanism, the two glasses come together at the base of the outer crucible, forming a core-cladding fibre. The fibre is drawn out from the crucibles. Then the fibre passes through high-precision diameter measurement and control equipment and is monitored by imaging or x-rays. The control equipment detects any non-homogeneity or bubbles in the drawn fibre. The fibre is then covered by a colored layer of jacket. Again, the jacket layer goes through diameter control and monitoring equipment. The end of the fibre cable is attached to a rotating spool, which turns steadily. The fibre is then tested for the attenuation (dB per kilometre), dispersion, and any other requirements specified by the customers or industry. Industry typically requires fibre cables to be a few hundred kilometres in length. Fibre cables are also available in different types and working wavelengths.

The rod-tube is one of the simplest methods of fibre fabrication. In the rod-tube procedure, a rod of core glass is placed inside a tube of cladding glass. This arrangement forms the preform that is required for the drawing process of the fibre. A preform is shown in Figure 22.8.

22.10.2 CHEMICAL VAPOUR DEPOSITION PROCESSES

The preforms used in the fibre drawing method are fabricated using chemical vapour deposition (CVD) processes. The CVD method is similar to the fabrication steps explained in the double crucible method. A preform is again needed for the fibre drawing process. The method for preparing the preform is called the rod-tube procedure. A rod of core glass is placed inside a tube of cladding glass.

All these processes are based on thermal chemical vapour reactions that form oxides. These oxides are deposited as layers of glass particles called soot, which is deposited on the outer rotating rod or inside glass tube to produce the preforms. Starting materials are solutions of O_2 mixed with $SiCl_4$, $GeCl_4$, $4POCl_3$, or gaseous BCl_3. These liquids are evaporated within an oxygen stream at a high temperature to produce silicon dioxide (SiO_2) and other oxides, known as dopants. Chemical reactions proceed as follows:

$$SiCl_4 + O_2 \rightarrow SiO_2 + 2Cl_2$$

$$GeCl_4 + O_2 \rightarrow GeO_2 + 2Cl_2$$

$$4POCl_3 + 3O_2 \rightarrow 2P_2O_5 + 6Cl_2$$

$$4BCl_3 + 3O_2 \rightarrow 2B_2O_3 + 6Cl_2$$

Silicon dioxide (SiO_2), or pure silica, is usually obtained in the form of small (submicron) particles called soot. This soot is deposited on the target rod or tube. The deposition of the silica soot, layer upon layer, forms a homogeneous transparent material. The manufacturer can control the exact amount of dopant added to each layer, thus controlling the refractive-index profile. For example, germanium dioxide (GeO_2) and phosphorus pentoxide (P_2O_5) increase the refractive index of glass, while boron oxide (B_2O_3) decreases it. Changing the composition of the mixture during the process influences the refractive-index profile of the preform. To change the value of a cladding's refractive index, some dopants are used. For example, fluorine (F) is used to decrease the cladding's refractive index in a depressed-cladding material.

The vapour process produces an extremely pure material whose characteristics are under the absolute control of the manufacturer. The preforms prepared in the vapour deposition processes will be explained next.

22.10.3 OUTSIDE VAPOUR DEPOSITION

This was the first successful mass-fabrication process that produced preforms used by the fibre drawing method. The outside vapour deposition (OVD) process, also called the soot process, was developed by Corning Company in 1972. This process consists of four phases: lay-down, consolidation, drawing, and measurement. During the lay-down phase, the materials that make up the core and cladding are vapour deposited around the rotating target rod. The result of this process is a soot preform. The refractive-index profile and fibre geometry are formed during this phase, as shown in Figure 22.9.

In the consolidation phase, the target rod is removed and the soot preform is placed inside a consolidation furnace. Here, the soot preform is consolidated into a solid, clear glass preform, and the centre hole is closed. During consolidation, a drying gas flows through the preform to remove residual moisture.

Then, in the drawing method, the preform is attached to a precision feed mechanism that feeds it into a furnace at a controlled speed, producing a fibre with the required diameter. Later, a colour jacket layer is applied. Diameter measurement and control equipment are also constantly checking the core, cladding, and jacket for diameter sizes and quality.

Finally, in the measurement phase, each fibre reel is tested for compliance with the fibre characteristics given in the data sheets. Specifically, the fibre is tested for strength, attenuation, and dimensional characteristics. Fibres are also tested for bandwidth, numerical aperture (NA), dispersion, and cutoff wavelength.

22.10.4 VAPOUR AXIAL DEPOSITION

The vapour axial deposition (VAD) is another form of outside deposition. This method was developed in 1977 by Japanese scientists. Figure 22.10 illustrates the vapour axial deposition

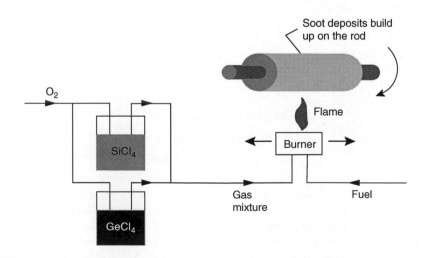

FIGURE 22.9 Outside vapour deposition.

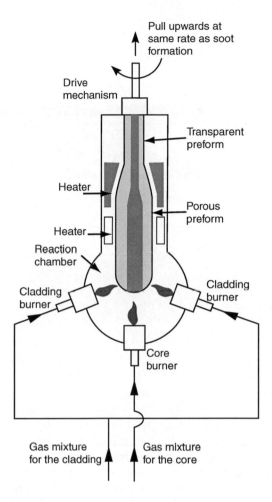

FIGURE 22.10 Vapour axial deposition.

(VAD) process. Silica particles obtained from a reaction among gases in a heated zone are deposited at the bottom end of a target, or seed rod, that rotates and moves upward. This deposition forms a porous preform, the upper end of which is heated in a ring furnace to produce a silica preform. The drawing and measurement steps are similar to those of the other deposition processes; however, the VAD process does not involve a central hole. The profile of the refractive index is formed by using many burners, a technique that allows the manufacturer to change the direction of the flow of a specific gas mixture. The VAD process produces both step-index and graded-index fibre profiles, as the deposited particle density varies due to temperature gradients produced in the plane perpendicular to the core axis.

22.10.5 MODIFIED CHEMICAL VAPOUR DEPOSITION

The modified chemical vapour deposition process was developed by Bell Laboratories in 1974 and has since been widely used in the production of graded-index fibre. Early in the development, the process was called Inside Vapour Deposition (IVD). Later, this was significantly improved and renamed as modified chemical vapour deposition (MCVD). This is the major process used in fibre production throughout the world. It is illustrated in Figure 22.11.

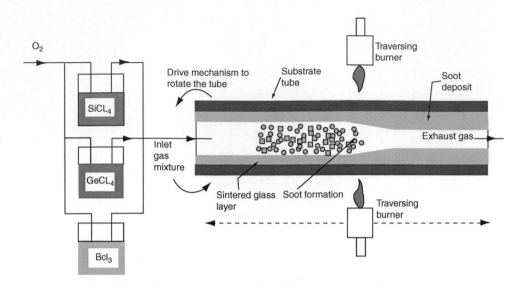

FIGURE 22.11 Modified chemical vapour deposition (MCVD).

In MCVD, a mixture of $SiCl_4$, $GeCl_4$, $POCl_3$, O_2, and H_2 gases flow though the inside of the tube, which is surrounded on the outside by a heat source. The heat source converts the gases into snow-like, high-surface-area silica soot inside the tube, as shown in Figure 22.11. The soot deposits on the tube downstream of the flame. The burner moves along the outside of the quartz tube, creating the fine soot particles and sintering the soot into a thin layer of doped glass on the inside. After the layer is deposited, the mixture of reactive gases is changed and the burner is brought back to the starting position. The above step is repeated and a subsequent layer is deposited. This process is continued, layer by layer, to construct the complex core structure of the optical fibre. Varying the concentration of dopants with each layer changes the refractive index, creating a graded-index profile. Once the glass is deposited, the tube is collapsed into a solid rod called a preform. The preform manufactured on the MCVD lathe is heated and drawn down to the standard diameter of 125 μ. Each preform generates many kilometres of fibre.

This operation is performed on a draw tower. The tower has a furnace at the top to melt the silica preform. Gauges are used to measure and control the diameter of the glass fibre to within submicrons as the fibre is pulled from the preform. Very fine control of the profile can be obtained using this technique. These MCVD layers are designed to be much thinner than the wavelength of light traveling down the fibre. An acrylate coating is applied during the draw process, which protects the pristine silica fibre from the environment.

After the fibre is manufactured, each spool of fibre is tested to ensure that it meets strict industry and internal specifications. These tests include the measurement of mechanical strength, geometric properties, and optical properties. One key test for laser-optimized multimode fibres is the high-resolution differential mode delay (HRDMD).

22.10.6 PLASMA CHEMICAL VAPOUR DEPOSITION

The plasma activated chemical deposition (PCVD) process was developed in 1975 by Phillips Research Laboratories, a Dutch consumer electronics and telecommunications company. The PCVD process is very similar to the MCVD process. However, instead of heating the outside of the silica tube, the energy source here is provided by a high-power microwave, resulting in ionized gas-plasma inside the tube. Figure 22.12 illustrates the PCVD process. Non-isothermal

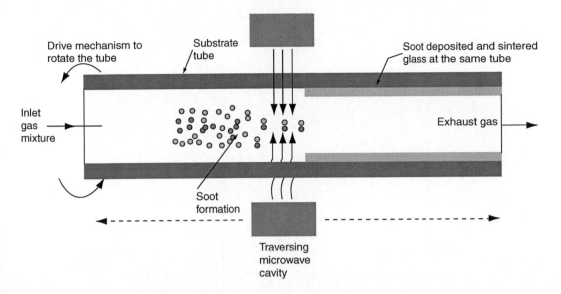

FIGURE 22.12 Plasma activated chemical vapour deposition (PCVD).

plasma in the microwave frequency range is used instead of a torch or flame. The plasma makes the reaction proceed at about 1,000 to 1,200°C. This results in very thin layers deposited inside the tube. Although this method allows layers to be grown at relatively low temperatures, the deposition rate is rather slow compared to other methods. However, this process can produce large preforms capable of producing a few hundred kilometres of fibre.

22.11 FIBRE DRAWING

Optical fibres are obtained by drawing from the preform at a high temperature. The drawing process must be integrated with the jacket-coating process to avoid contamination of the fibre surface. These two processes are shown in Figure 22.13. The preform is heated in a furnace to a molten state. Then it is attached to a precision feed mechanism that pushes the preform at the proper speed into the furnace at a high temperature. The drawing process is very precise and continuously controlled to check the diameter of the fibre; diameter drift cannot exceed 0.1%. The filament fibre then passes through a series of coating and jacket applicators and curing processes, depending on the customer's requirements. The outside diameter of the jacket is measured and monitored for defects. The fibre, in its final shape, is pulled down and wound around a winding drum or spool.

22.12 NUMERICAL APERTURE

Figure 22.14 illustrates a light ray incident on the fibre cable input passing through the fibre core. The cone is known as the acceptance cone for the fibre cable. Any ray outside this cone will not propagate through the fibre core, since the incident angle is too large. The angle (θ) is known as the half cone angle for this fibre cable. The full cone angle is also known as the acceptance angle (β) of the cable. Thus, the full-acceptance angle (β) can be calculated by:

$$\beta = 2\theta \qquad (22.1)$$

The numerical aperture (NA) is a measurement of the amount of light that can be collected by, or that emerges from, the core of a fibre optic cable. The NA is the product of the index of refraction

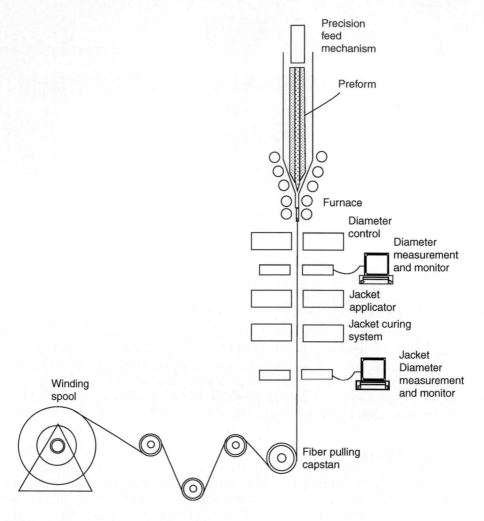

FIGURE 22.13 Fibre drawing and jacket coating.

of the incident medium (n_i) and the sine of the maximum ray angle (θ) of the light incident on the core, as shown in Figure 22.14.

$$\mathrm{NA} = n_i \sin(\theta) \tag{22.2}$$

In air, the sine of this half cone angle is the value of the numerical aperture (NA) of the cable.

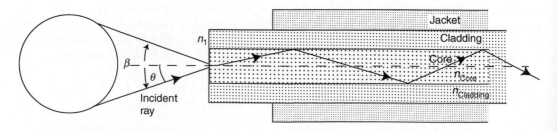

FIGURE 22.14 Light ray in a fibre optic cable.

Therefore, the half-acceptance angle (θ) is:

$$\theta = \sin^{-1}(NA) \tag{22.3}$$

Recall that the refractive index of a material (n) is defined as the ratio between the speed of light in a vacuum (c) and the speed of light in the medium (v):

$$n = \frac{c}{v} \tag{22.4}$$

For example, the refractive index for air, water, and glass crown are 1.0003, 1.33, and 1.52, respectively.

It is convenient to define the fractional refractive index change as:

$$\Delta = \frac{n_{core} - n_{cladding}}{n_{core}} \tag{22.5}$$

We can also calculate the NA from the following formula:

$$NA = n_{core}\sqrt{2\Delta} \tag{22.6}$$

To understand ray tracing in a fibre cable, recall the principles of Snell's Law, as explained by the law of refraction. The standard measure of acceptance angle is the numerical aperture (NA), which is the sine of the half-acceptance angle (θ). Therefore, the NA can be calculated by:

$$NA = \sqrt{n_{core}^2 - n_{cladding}^2} = \sin\theta \tag{22.7}$$

where n_{core} and $n_{cladding}$ are the refractive indices for the core and cladding, respectively.

22.13 MODES IN A FIBRE OPTIC CABLE

A mode is a mathematical and physical concept that describes the propagation of electromagnetic waves through an optical medium. In its mathematical form, mode theory derives from Maxwell's equations. James Maxwell, a Scottish physicist (1831–1879), first gave mathematical expression to the relationship between electric and magnetic energy, showing that they were both a single form of electromagnetic energy, not two different forms, as was then believed. His equations also showed that the propagation of this energy followed strict rules. Maxwell's equations form the basis of the electromagnetic theory of light.

A mode is an allowed solution to Maxwell's equations. A mode is simply a path that a light ray can follow when traveling through a fibre cable. Currently, the number of modes supported by a fibre cable ranges from one to hundreds of thousands. Thus, a fibre cable provides a path for one or thousands of light rays, depending on its size, design, and properties.

Each mode also carries a characteristic amount of optical power. Most fibre cables today support many modes. When light is first injected into the fibre cable, a mode may carry too much or too little power, depending on the injected power of light. Over the length of the fibre cable, power transfers between modes until all modes carry their characteristic power. When a fibre cable reaches this point, it is said to have reached the steady state, or equilibrium mode distribution (EMD). Achieving EMD in plastic fibres requires only a few hundred metres. For high-performance glass fibre cables, EMD often requires hundreds of kilometres.

The number of modes supported by a fibre cable can be calculated from the V number or normalized frequency. The V number is a fibre cable parameter that takes into account the core diameter (d), wavelength of the propagated light (λ), and numerical aperture (NA) of the fibre cable. The V number is calculated by

$$V = \frac{2\pi d}{\lambda} (NA) \qquad (22.8)$$

From the V number, the number of modes (N) in a fibre optic cable can be calculated for the following cases:

(a) For a single-mode step-index fibre cable, the number of modes (N) can be approximated by

$$N = \frac{V^2}{2} \qquad (22.9)$$

(b) For a multimode step-index and graded-index fibre cable, the number of modes (N) can be approximated by

$$N = \frac{V^2}{4} \qquad (22.10)$$

22.14 LIGHT SOURCE COUPLING TO A FIBRE CABLE

There are several techniques used for coupling a light source into a fibre optic cable. The coupling process can be done during the manufacture of a laser light source, which is connected to a fibre cable at one end and a pig tail connector at the other, as shown in Figure 22.15. Light-emitting diodes and laser diodes have edge or top-surface light emission. Therefore, coupling of a cable to the emitting side of a light source depends on many factors, such as the design and the packaging size of the device. There are also similar processes used in coupling a fibre cable to an optical component, such as a prism or a GRIN lens.

For an efficient coupling process, the emitting area of the light source should be equal to or slightly larger than the core of the fibre cable to ensure that all the light passes through the cable with minimum loss. Figure 22.16(a) shows one of the most common coupling processes: direct coupling. Here, an optical epoxy can be used to bond the fibre cable to the light source. Application of the epoxy and curing process are very important during manufacturing to reduce losses. This method is the cheapest form of coupling because of the short alignment time and the simple assembly design that enables the alignment of the light source to the fibre cable. However, this method is less efficient than the other coupling methods.

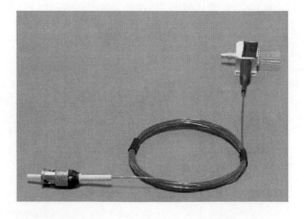

FIGURE 22.15 A light source coupled to a fibre cable.

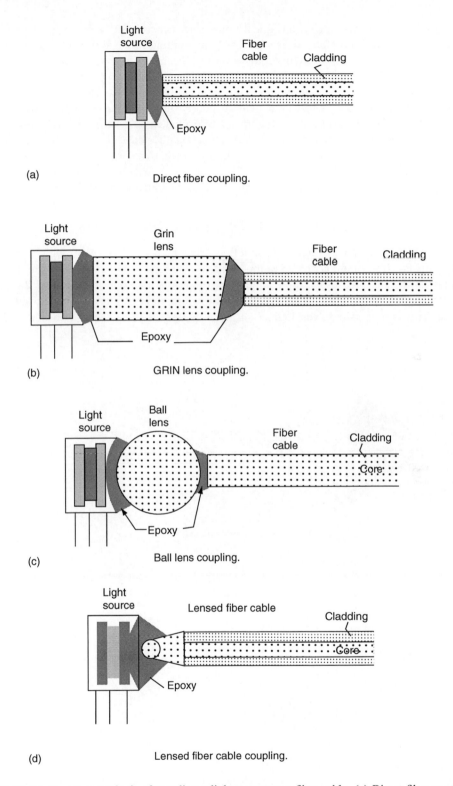

FIGURE 22.16 Common methods of coupling a light source to a fibre cable. (a) Direct fibre coupling. (b) GRIN lens coupling. (c) Ball lens coupling. (d) Lensed fibre cable coupling.

To improve the coupling process, an optical component such as a GRIN lens or ball lens, a grating or a prism can be used. A problem arises when the light-emitting area and the fibre cable are not the same size, creating losses. Using a lens that reduces the light beam size to match the fibre cable can solve this problem. Figure 22.16(b) illustrates the use of a GRIN lens placed between the light source and the cable. GRIN lenses are commonly used in building optical fibre devices, such as switches and polarization beamsplitters, because of easy alignment by a fairly skilled employee. A ball lens can also be used in the coupling process, as shown in Figure 22.16(c). Optical devices using ball lenses take more time for alignment to the light source. They are also more complicated to design and bigger in size than the devices using GRIN lenses. As well, light coupling suffers high loss because more epoxy is needed to bond the ball lens into position. Figure 22.16(d) illustrates the coupling of a lensed fibre cable to a light source. This arrangement has an easy coupling alignment with low loss and small packaging size.

22.15 LAUNCHING LIGHT CONDITIONS INTO FIBRE CABLES

The amount of light carried by each mode in a fibre cable is determined by the light input or light launch conditions. If the angular spread of the rays from the light source is greater than the angular spread that can be accepted by the fibre cable (i.e., the NA of the light input to the fibre cable is greater than the NA of the fibre), then the radius of the light input is greater than the radius of the fibre cable. This case is called overfilled, as shown in Figure 22.17(a). Here, a portion of the light

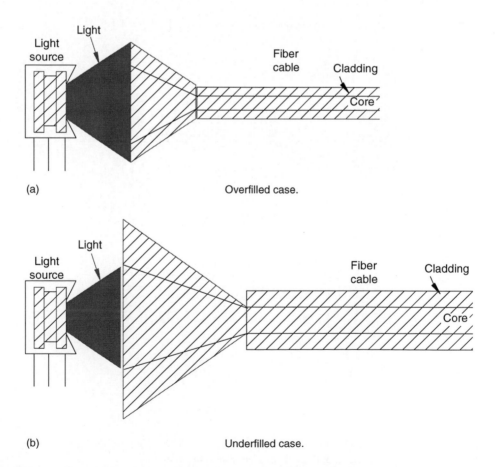

FIGURE 22.17 Launching light into a fibre cable. (a) Overfilled case. (b) Underfilled case.

source will be launched into the cladding and will be considered light loss. Conversely, when the input light NA is less than the NA of the fibre cable, i.e., the radius of the light input is smaller than the radius of the fibre cable, this is called underfilled, as shown in Figure 22.17(b). These two coupling conditions yield different attenuations, with the overfilled case having a higher loss than the underfilled case.

Other factors must also be taken into consideration in coupling: the condition the fibre cable end, the index of refraction of the epoxy, the epoxy curing process, the alignment method, the type of optical components, and the packaging. Similarly, it is necessary to consider some of these factors in any type of connection between two fibre cables. Common factors include the angular alignment, the air gap, the different numerical apertures or core diameters, and the cable end conditions. Fibre cable connections will be discussed in more detail in the fibre cable connections chapter.

22.16 FIBRE TUBE ASSEMBLY

The fibre tube assembly is typically the first step when manufacturing an optical device. It provides a means to handle, position, and glue the fibre optic cable to an optical component, such as a GRIN lens or beamsplitter. The fibre tube assembly is manufactured from single or double fibre optic cable and a glass tube, as shown in Figure 22.18. The fibre optic cable is attached to the tube using two types of epoxy: a hard type of epoxy fills the front part of the tube, while a soft type of epoxy fills the other end. Air bubbles in the epoxy or cracks in the tube will introduce insertion loss, especially under temperature variations. Fibre optic cables must also not be cracked or stressed, as this will increase losses. The fibre tube assembly end is polished at an angle of 2, 6, 8, or 12 degrees. The assembly end must be polished precisely and carefully to maintain a surface free of cracks and scratches. Figure 22.19 shows the fibre tube assemblies ready to be joined to a GRIN lens or an optical device. The fibre tube assembly has a polished end angle matching the polished end angle of the GRIN lens. An accurate optical alignment is achieved when the fibre tube assembly and the GRIN lens are facing each other in perfect alignment. The ends of the fibre tube assemblies are polished to the required angles using an industrial polishing machine for mass production.

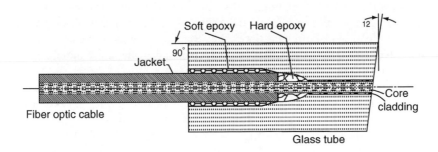

FIGURE 22.18 A fibre tube assembly/single fibre cable.

22.17 FIBRE OPTIC CABLES VERSUS COPPER CABLES

There is still a place for fibre optic and copper cables in communication systems, but the shrinking price gap, coupled with increasing bandwidth demands, makes fibre cables worth using in more situations than ever before. The customers of small and large communication providers are dispersed all over the world, and they need to send and receive lots of data. Customers require large amounts of bandwidth, and fibre cables are the only medium that can support this. Some of

FIGURE 22.19 Fibre tube assemblies.

those customers's bandwidth requirements have been growing exponentially since the beginning of the twenty-first century. In reality, all fibre networks have a lot of room for future growth. The ultimate choice is whether to use fibre optic cables. As mentioned above, the fibre optic technology is moving forward to create high-capacity fibres with low production and installation costs and increasing bandwidths. The overall cost difference between optical fibre and twisted-pair copper cabling has been reduced. Now the choice between optical fibre and twisted-pair copper cabling has shifted in favour of the fibre cable.

Desktop computers require very high bandwidths. One way to meet this need is to wire them with fibre optic cables. However, copper cable has continued to prove more capable than expected; every time that new, higher-speed network standards appear to be forcing a move to fibre cable, someone has found a way to pump more data through the old copper cables.

Still, fibre cables can be made even more economically attractive by rethinking the way the network is physically laid out. Because fibre cable can be run for longer distances than copper, the networks could be laid out without the wiring closets full of additional gear that are common in copper-based networks. Instead, fibre might run directly from the desktop back to the server or to the backbone connecting floors of a building, and the savings on the intermediate gear might more than cancel out the higher cost of installing the fibre cable. Running fibre cable to small enclosures close to users, and then covering the last short distances with copper, provides an economical alternative that minimizes the length of the twisted-pair cable used.

The increasing use of wireless networking also opens up a variation on the fibre cable network layout. Fibre can be run to a wireless access point that can then be used to serve a group. Thus, the copper cable can be eliminated altogether without actually taking fibre cable to every machine.

Many companies are removing existing copper cables and replacing them with fibre cables. When facilities are built or refurbished, fibre cables are installed. The choice between copper and fibre cables depends on several factors, including the applications being run on the network, the company's future plans, and the demands of costumers.

In particular, there are some specific situations in which fibre has advantages over copper cable. First, fibre cable is immune to electrical interference and tapping. It also carries high data capacity over long distances and is small and lightweight. When it comes to testing, fibre may still require some fairly sophisticated equipment—but the newer standards for copper cabling present the same issues.

In the end, the choice between fibre and copper cables comes down to the company's networking requirements, the needs of individual users, and the budget. Table 22.2 presents a side-by-side comparison of the important differences between fibre and copper cables.

TABLE 22.2
Fibre Optic Cables Versus Copper Cables

No.	Fibre Optic Cables	Copper Cables
1	Fibre-based systems are more expensive to buy and install	Copper-based systems are less expensive to buy and install
2	Fibre is clearly the superior technologically. Installing fibre ensures performance, as even higher speed networks will emerge in the future	Installing copper cable ensures performance for low-speed networks
3	Carry high data capacity over long distances	Carry low data capacity over short distances
4	Wide bandwidth	Limited bandwidth
5	Low loss per cable length	Conventional loss per cable length
6	Immune to electrical interference and tapping	Not immune to electrical interference and tapping
7	Small size and lightweight	Large size and heavyweight
8	New technology reduces installation time	Conventional technology keeps the same installation time
9	High safety	Low safety
10	Fast-developing technology	Steady-state developing technology

22.18 APPLICATIONS OF FIBRE OPTIC CABLES

Since the discovery of the laser, fibre optic cables and optical fibre devices have seen increased applications in every sector of industry. Light is a very important element in our lives, controlling and operating many types of devices, instruments, and systems. Fibre optics has emerged as a practical technology that is easy to work with. Fibre optic cables with other optic components are used in building optical fibre devices and systems. One of the large-scale applications of fibre optics is its use in communication systems. The small sizes and wide bandwidths, as well as their capacity to carry large amounts of information, make optical fibres very attractive for use in these systems. Later chapters in this book will explain in more detail communication systems that use optical fibre technology. Video, including broadcast television, is one of the main telecommunication applications. Other applications include cable television, high-speed Internet, wireless transition, remote monitoring, and surveillance. Fibre optic video transmission is successfully used around the world in surveillance and remote monitoring systems with many applications. Fibre optics applications in the military include communications, command-and-control links on ships and aircrafts, data links for satellite earth stations, and transmission lines for tactical command-post communications.

Fibre cables can be used throughout the communication network, including in the final link into the subscriber's home to wall outlets. This field has continued to develop since the discovery of optical amplifiers, dense wavelength division multiplexers (DWDM), fibre Bragg grating, and photonics crystal fibres.

One particular advantage of fibre optic cables is that they are immune to electromagnetic interference (EMI) from electricity. Therefore, optic cables can be placed near high-voltage power cables without any effect on data transmission. Similarly, the cables also can be laid along railway lines without suffering from EMI.

Optical fibres are applied in building night-vision viewing devices, scanning and sensing instruments, and vibration sensors, which are extensively used in military, medical, and other applications. Imaging techniques have been rapidly developed for a variety of medical applications, such as viewing inside human tissues and scanning microscopic particles.

Fibres are also used in monitoring and sensing technology. They are used as sensors to monitor the vibration in the structures of bridges and high buildings. They are also used as gas and DNA sensors.

The use of fibre optic cables in lighting systems can reduce energy consumption. Fibre optic lighting systems are developing quickly, with wide applications. Fibre optic lighting systems can be applied to the interior and exterior of commercial, retail, and residential buildings. New applications are being explored in landscaping, waterscaping, medical lighting instruments, and theme environments. More details will be discussed in the fibre optic lighting chapter.

Fibre optic cables can also be coated for special handling requirements and resistance to temperature, chemicals, or radiation. Radiation-resistant fibre is suitable for use in environments where electronics-based optical solutions are not viable, such as monitoring nuclear waste disposal in storage facilities. To make the fibres heat resistant, a chemical-resistant polyimide coating that can withstand temperatures of up to 300°C is applied. This is especially useful for manufacturers designing medical equipment for applications in which autoclave sterilization is necessary.

22.19 EXPERIMENTAL WORK

In this experiment the student will perform the following cases:

22.19.1 Case (a): Fibre Cable Inspection and Handling

After completing this lab experiment, students will be able to inspect, clean, and handle fibre optic cables. A full, detailed procedure is presented in the apparatus set-up section.

22.19.2 Case (b): Fibre Cable Ends Preparation

The students will also be able to prepare fibre optic cable ends. A full, detailed procedure is presented in the apparatus set-up section.

22.19.3 Case (c): NA and Acceptance Angles Calculation

The students will calculate the numerical aperture (NA) of a fibre optic cable. Figure 22.20 shows the laser beam emerging from a fibre optic cable core when the fibre output is projected onto a black/white card. The following formulas are used to calculate the half-acceptance (θ) and full-acceptance (β) angles, as well as the numerical aperture (NA) of the fibre optic cable used in this lab experiment. From the geometry of Figure 22.20, it is easy to calculate the half-

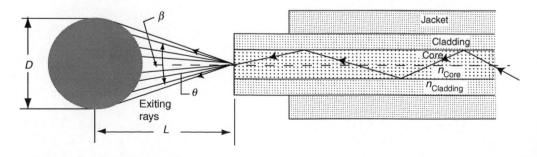

FIGURE 22.20 Numerical aperture (NA).

acceptance angle (θ) and the full-acceptance angle (β) by:

$$\tan \theta = \frac{D}{2L} \qquad (22.11)$$

Thus

$$\theta = \tan^{-1}\left(\frac{D}{2L}\right) \qquad (22.12)$$

$$\beta = 2\theta \qquad (22.13)$$

Assuming the refractive index (n_i) for air $= 1$, use Equation (22.2) to calculate the numerical aperture (NA) of a fibre optic cable:

$$NA = \sin \theta \qquad (22.14)$$

22.19.4 Case (d): Fibre Cable Power Output Intensity

Students will practise measuring the output intensity of a laser beam at a distance (L) from the fibre cable end, using a laser light sensor. The students will also understand how the intensity on the card varies with distance. They will determine the laser beam intensity on the card as a function of distance.

Light wave motion involves the propagation of energy. The rate of energy transfer is expressed in terms of intensity, which is the energy transported per unit time across a unit area. Energy/time is power, and we usually express intensity as power per unit area.

$$\text{Intensity}(I) = \frac{\text{Energy/Time}}{\text{Area}} = \frac{\text{Power}}{\text{Area}} \qquad (22.15)$$

The standard units of intensity (power/area) are watts per square metre (W/m^2).

The geometric shape of the laser beam, when emerging from the fibre cable end, is defined by the numerical aperture, as explained in Case (c). The intensity of the laser beam emerging from the fibre cable end and incident on a card is therefore inversely proportional to its distance from the fibre cable end. The geometric diagram of the laser beam emerging from the fibre cable and incident on the black/white card will be defined by the Equation (22.11) and Equation (22.12).

Again, as explained in Case (c), the laser beam spot has a diameter (D) incident on the black/white card at a distance (L) from the fibre cable end. Equation (22.11) and Equation (22.12) give the relation between (D) and (L). Therefore, (D) is directly proportional to ($2L \tan \theta$), as given by Equation (22.11). The area (A) of the laser beam spot is calculated by:

$$A = \frac{\pi D^2}{4} \qquad (22.16)$$

22.19.5 Technique And Apparatus

Appendix A presents details of the devices, components, tools, and parts. Figure 22.21 shows the fibre optic cable ends preparation and cleaning kit.

1. Fibre optic cable, 250 μm outside diameter, 500 m in length.
2. Safety goggles.
3. Clean room wipers.
4. Finger cots.
5. Rubber gloves.
6. Cleaver.

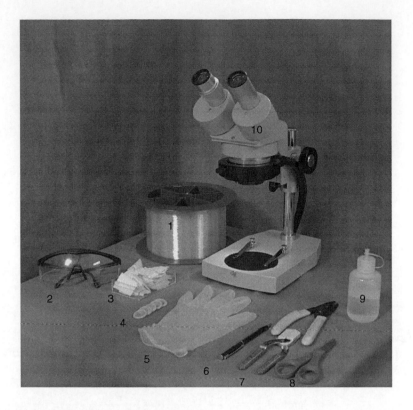

FIGURE 22.21 Fibre optic cable end preparation kit.

7. Fibre optic strippers.
8. Scissors.
9. Cleaner (Denatured Ethanol).
10. Microscope with a minimum of 10X magnification.
11. Thermal cleaver, as shown in Figure 22.26.
12. Mechanical cleaver, as shown in Figure 22.28.
13. 2×2 ft. optical breadboard.
14. HeNe laser light source and power supply.
15. Laser mount assembly.
16. Laser light sensor.
17. Laser light power meter.
18. 20X microscope objective lens. Use the 20X lens to focus the collimated laser beam onto the fibre optic cable core. The lens will increase the input power into the fibre cable core and consequently increase the output power.
19. Hardware assembly (clamps, posts, screw kits, screwdriver kits, lens/fibre cable holder/positioner, sundry positioners, brass fibre cable holders, fibre cable holder/positioner, etc.).
20. Lens/fibre cable holder/positioner assembly.
21. Fibre cable holder/positioner assembly.
22. Black/white card and cardholder.
23. Vernier.
24. Ruler.

22.19.6 PROCEDURE

Follow the laboratory procedures and instructions given by the professor and/or instructor.

22.19.7 SAFETY PROCEDURE

Follow all safety procedures and regulations regarding the use of fibre optic cables. You must wear safety glasses and finger cots or gloves when working with and handling fibre optic cables, optical components, and optical cleaning chemicals.

22.19.8 APPARATUS SET-UP

22.19.8.1 Case (a): Fibre Cable Inspection and Handling

Fibre Optic Cable Defect Types

The following are types of fibre optic cable defects:
1. Missing Jacket. The fibre optic cable jacket is removed, damaged, cracked, or shows signs of abrasion.
2. Pinch. The fibre optic cable has an indentation, but there are no breaks in the jacket.
3. Delamination. There is a separation between the fibre optic cable jacket and cladding or a change in the cladding colour caused by heat or chemicals, but there are no breaks in the jacket.
4. Contamination. There is a contaminate or any foreign material present on the outer surface of the fibre optic cable jacket.

Fibre Optic Cable Inspection

To find fibre optic cable defects, the following are the most common inspection procedures to use:
1. Visual Inspection. The jacket of the fibre optic cable should be inspected using an illuminated magnifying aid.
2. Touch Inspection. The jacket of the fibre optic cable should be inspected using your fingers. To perform touch inspection of the cable jacket, start from one end of the fibre optic cable. Gently squeeze the cable between the tips of the thumb and index finger, then slide the fibre optic cable, pulling by hand in one direction to the other end.
3. Combined Inspection. Fibre optic cable should be inspected using an illuminated magnifying aid and your fingers to perform the combined (visual and touch) inspection. Visual inspection procedure can be performed under a magnifying aid.
4. Inspection Under a Microscope. Fibre optic cables should be inspected under a powerful microscope (20X), as shown in Figure 22.22, to find any small crack or damage on the cable jacket that cannot be diagnosed using the above inspection procedures.
5. Test Inspection. This procedure is an expensive and precise inspection. The fibre optic cable should be tested by sending a signal from a light source and measuring the signal output power. This procedure determines the loss as well as the location of any light leakage from the fibre optic cable jacket.

Fibre Optic Cable Cleaning

The following steps are necessary in the fibre optic cable cleaning process:
1. The fibre optic cable should be inspected and kept clean during the manufacturing process.

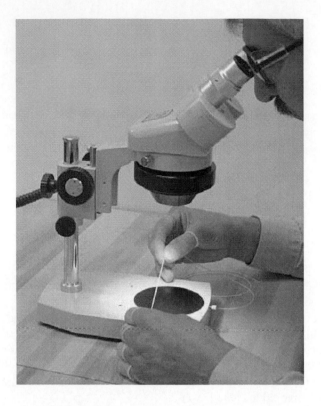

FIGURE 22.22 Fibre optic cable inspection.

2. If any type of contamination is found on the fibre optic cable, follow the recommended cleaning procedure. Each type of fibre optic cable has a cleaning procedure that uses a specific cleaning solvent. Check the fibre optic cable cleaning process before applying any cleaning solvent.

3. Use a swab or fibre optic cable cleaning pad dampened with solvent to remove the foreign material. Finger cots should be used when handling solvents. Gently rub the swab/pad along the fibre optic cable, in one direction only, to remove the contamination, as shown in Figure 22.23.

4. When cleaning do not use aggressive rubbing to remove the contamination. Do not use fingernails or any other hard-surfaced objects to scrape off the contamination.

5. If you are unable to remove the foreign material using the above methods, reject the fibre optic cable.

Fibre Optic Cable Handling

When handling fibre optic cable, observe the following rules:

1. Do not allow the fibre optic cable to rest on the floor.
2. Do not allow the fibre optic cable to be crushed or pinched.
3. Do not allow the fibre optic cable to drop directly over sharp edges, such as a table edge, tools, or handling trays.
4. Do not allow the fibre optic cable to come in contact with hot surfaces, such as a soldering iron, hot air pencil, or hot handling trays from an oven.
5. Do not allow anything to be placed on top of the fibre optic cable during the handling.

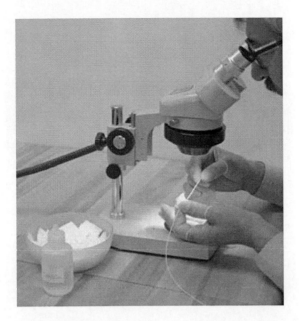

FIGURE 22.23 Fibre optic cable cleaning.

6. Do not permit any kind of macro bends to be present in a fibre optic cable. Bends increase attenuation and decrease the tensile strength of the fibre optic cable.
7. The fibre optic cable should be coiled, as shown in Figure 22.24(a). The coil diameter depends on the fibre optic cable diameter and type. The fibre cable can also be coiled around a rubber ring, as shown in Figure 22.24(b). The size of the rubber ring depends on the fibre optic cable diameter and type.
8. Always separate paper from the fibre optic cable. Put paper in a plastic bag when handling the fibre optic cable.

22.19.8.2 Case (b): Fibre Cable Ends Preparation

Manual Stripping Procedure

1. Hold and open the hand stripper with one hand ONLY.
2. Hold the fibre optic cable very tightly between the thumb and forefinger. Place the stripper on the fibre optic cable, making sure to insert the fibre optic cable through

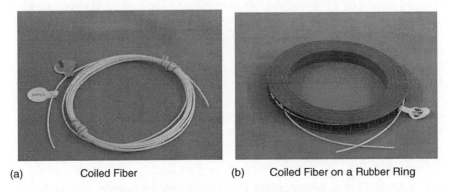

(a)   Coiled Fiber   (b)   Coiled Fiber on a Rubber Ring

FIGURE 22.24 Fibre optic cable handling. (a) Coiled fibre. (b) Coiled fibre on a rubber ring.

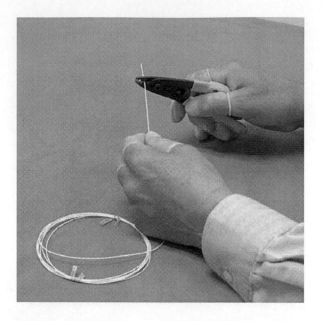

FIGURE 22.25 Manual stripping.

the "V's" in the heads, as shown in Figure 22.25. The stripper should be perpendicular to the fibre optic cable. Approximately one inch of the fibre optic cable should pass through to the other side of the stripper.
3. Gently squeeze the handles until the stripper closes completely. Keep the handles in this position.
4. While holding the fibre optic cable tightly, pull the stripper as straight as possible along the length of the fibre optic cable, toward the fibre optic cable end.
5. Always keep the hand stripper clean.

Thermal Stripping Procedure

In contrast to mechanical strippers, which create a mechanical stress and leave scratches on the surface of the cladding, thermal strippers are easy, quick, and safe tools used for the stripping of jackets. The following steps summarize the procedure:
1. Turn on the thermal stripper.
2. Hold and open the thermal stripper with one hand ONLY.
3. Hold the fibre optic cable very tightly between the thumb and forefinger. Insert the fibre cable into the stripper for the stripping length that is required, as shown in Figure 22.26. The stripper should be lined up with the fibre optic cable. Approximately one inch of the fibre optic cable should pass into the stripper.
4. Gently squeeze the handles until the stripper closes completely. Keep the handles in this position.
5. While holding the fibre optic cable tightly, pull the stripper as straight as possible along the length of the fibre optic cable, away from the fibre.
6. Always keep the thermal stripper clean. Follow the cleaning procedure of the device.

Cleaning Procedure After Stripping

1. Take a piece of cotton tissue dampened with denatured ethanol or any cleaning chemical suitable for the fibre.

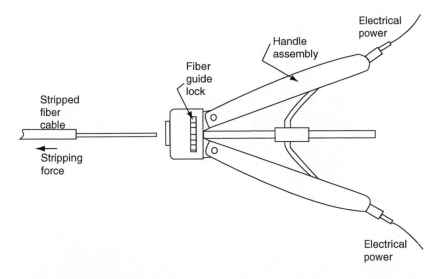

FIGURE 22.26 Thermal stripping.

2. Wrap it around the stripped end of the fibre optic cable.
3. Pull in one direction toward the end.
4. Clean more than once if necessary.

Manual Cleaving Procedure

1. Place the stripped length of fibre optic cable firmly on the inside part of your forefinger.
2. Hold the cleaver with your other hand perpendicular to the fibre optic cable. Make a gentle scratch across the cladding surface of the fibre optic cable at a distance of about a half of an inch from the end of the fibre optic cable, as shown in Figure 22.27.
3. Use the same tool to break off the fibre optic cable at the scribed mark.

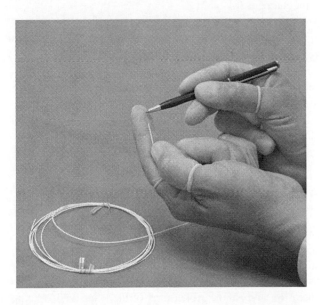

FIGURE 22.27 Cleaving.

FIGURE 22.28 Mechanical cleaving.

Mechanical Cleaving Procedure

A mechanical cleaver is used in most production lines. The mechanical cleaver gives a precise cut at a cleave angle of 90 degrees to the fibre optic cable end. Figure 22.28 shows a mechanical cleaver that cleaves the fibre optic cable end to a required length; this is typical of most manufacturing and test processes. To use the mechanical cleaver perform the following steps:

1. Get a mechanical cleaver ready.
2. Press the red lever (a), as shown in Figure 22.28.
3. Open the top mechanism (b).
4. Open the small cover (c).
5. Place the stripped fibre optic cable at the required cleave length on the disc blade on the block (d). This is where the cleaving will take place.
6. Close the small cover (c) and top mechanism (b).
7. Push block (d) in the direction of the arrow to cleave the fibre.
8. Press knob (e) to break the fibre.
9. Press the red lever (a), open the top mechanism (b), and open the small cover (c).
10. Remove the fibre optic cable.
11. The fibre optic cable is cleaved automatically to the required cleave length.

22.19.8.3 Case (c): NA and Acceptance Angles Calculation

1. Figure 22.29 shows the experimental apparatus set-up.
2. Bolt the laser short rod to the breadboard.
3. Bolt the laser mount to the clamp using bolts from the screw kit.
4. Put the clamp on the short rod.
5. Place the HeNe laser into the laser mount and tighten the screw. Turn on the laser device. Follow the operation and safety procedures of the laser device being used.
6. Check that the laser is aligned with the line of bolt holes and adjust if necessary.

FIGURE 22.29 Numerical aperture apparatus set-up.

7. Mount a lens/fibre cable holder/positioner to the breadboard so that the laser beam passes over the centre hole.
8. Add the 20X microscope objective lens to the lens/fibre cable holder/positioner.
9. Prepare a fibre cable with a good clean cleave at each end, as described in Cases (a) and (b).
10. Insert one end (input) of the fibre cable into the brass fibre cable holder and place it in the hole of the lens/fibre cable holder/positioner.
11. Extend the fibre cable so that the end is at the centre of the lens/fibre cable holder/positioner. This is a very important step for obtaining an accurate value of fibre cable NA.
12. Recheck the alignment of the light-launching arrangement by making sure that the input end of the fibre cable remains at the centre of the laser beam.
13. Mount the output end of the fibre cable into the brass fibre cable holder and place it in the hole of the fibre cable holder/positioner.
14. Place a black/white card on a cardholder in front of the output end of the fibre cable at a distance (L).
15. Turn off the lights of the lab and observe the circular red laser beam spot with the diameter (D) on the black/white card, as shown in Figure 22.30.
16. Measure the diameter (D) of the spot on the black/white card, and the distance (L) from the fibre cable end to the black/white card. Fill out Table 22.3.
17. Repeat this procedure five times for five different distances.
18. Turn on the lights of the lab.

22.19.8.4 Case (d) Fibre Cable Power Output Intensity

Repeat the set-up procedure explained in Case (c) of this experiment. Figure 22.31 shows the experimental apparatus set-up for fibre cable power output intensity.

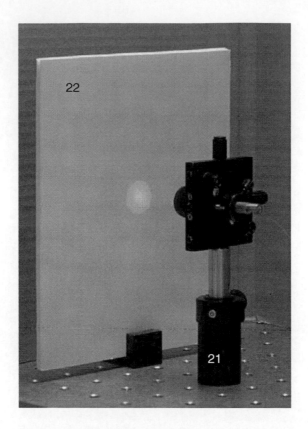

FIGURE 22.30 Red laser beam spot on the black/white card.

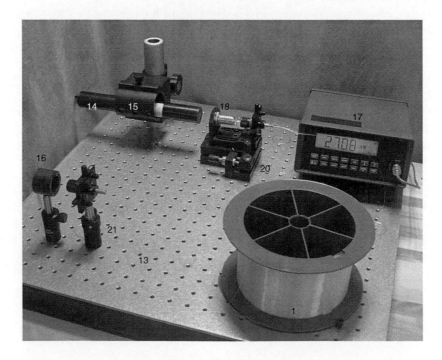

FIGURE 22.31 Fibre cable power output intensity apparatus set-up.

For Case (d), add the following steps:

1. Place the black/white card in front of the output end of the fibre cable at distance (L).
2. Measure the distance between the fibre cable end and the black/white card. This is the distance (L). Fill out Table 22.4.
3. Measure the diameter (D) of the spot on the black/white card from the fibre cable output end at distance (L), as shown in Figure 22.30. Fill out Table 22.4.
4. Place the laser sensor in front of the output end of the fibre cable at the same distance (L).
5. Measure the laser output power (P_{out}) at that distance (L). Fill out Table 22.4.
6. Repeat the measurement of the laser output power (P_{out}) and spot diameter for each of the five distances (L).

22.19.9 Data Collection

22.19.9.1 Case (a): Fibre Cable Inspection and Handling

No data collection is required for this case.

22.19.9.2 Case (b): Fibre Cable Ends Preparation

No data collection is required for this case.

22.19.9.3 Case (c): NA and Acceptance Angles Calculation

1. Measure the diameter (D) of the laser red spot at the distance (L) from the black/white card. Report these measurements for the five distances to get five diameters.
2. Fill out Table 22.3 with the measured data.

TABLE 22.3
Numerical Aperture Data Collection and Calculations

Measured Values		Calculated Values		
Distance (L)	Diameter (D)	Half-Acceptance Angle (θ)	Full-Acceptance Angle (β)	NA
(unit)	(unit)	(degrees)	(degrees)	
L_1 = ( )	D_1 = ( )			NA_1 =
L_2 = ( )	D_2 = ( )			NA_2 =
L_3 = ( )	D_3 = ( )			NA_3 =
L_4 = ( )	D_4 = ( )			NA_4 =
L_5 = ( )	D_5 = ( )			NA_5 =

22.19.9.4 Case (d): Fibre Cable Power Output Intensity

1. Measure the laser output power (P_{out}) at that distance (L).
2. Repeat the measurement of the laser output power (P_{out}) and spot diameter (D) for each of the five distances (L).
3. Fill out Table 22.4 with the measured data.

TABLE 22.4
Data Collection and Calculations

Measured Values			Calculated Values	
Distance (L) (unit)	Diameter (D) (unit)	Output Power (Pout) (unit)	Area (A) (unit^2)	Intensity (I) = (Pout/A) (unit)/(unit^2)
$L_1 =$ ( )	$D_1 =$ ( )	$P_{out1} =$ ( )	$A_1 =$ ( )	$I_1 =$ ( )
$L_2 =$ ( )	$D_2 =$ ( )	$P_{out2} =$ ( )	$A_2 =$ ( )	$I_2 =$ ( )
$L_3 =$ ( )	$D_3 =$ ( )	$P_{out3} =$ ( )	$A_3 =$ ( )	$I_3 =$ ( )
$L_4 =$ ( )	$D_4 =$ ( )	$P_{out4} =$ ( )	$A_4 =$ ( )	$I_4 =$ ( )
$L_5 =$ ( )	$D_5 =$ ( )	$P_{out5} =$ ( )	$A_5 =$ ( )	$I_5 =$ ( )

22.19.10 CALCULATIONS AND ANALYSIS

22.19.10.1 Case (A): Fibre Cable Inspection and Handling

No calculations and analysis are required for this case.

22.19.10.2 Case (b): Fibre Cable Ends Preparation

No calculations and analysis are required for this case.

22.19.10.3 Case (c): NA and Acceptance Angles Calculation

1. Plot the graph of (L) vs. (D).
2. Calculate the fibre cable half-acceptance angle (θ), full-acceptance angle(β), and numerical aperture (NA) using Equation (22.12) through Equation (22.14), respectively.
3. Fill out Table 22.3 with the calculated values.

22.19.10.4 Case (d): Fibre Cable Power Output Intensity

1. Calculate the area (A) of the laser beam spot and the output intensity of the laser beam using Equation (22.16) and Equation (22.15), respectively.
2. Plot the graph of the laser power output (P_{out}) as a function of the distance (L).
3. Plot the graph of the laser power intensity (I) as a function of the distance (L).
4. Fill out Table 22.4 with the calculated values.
5. Determine the relationship between the laser power intensity (I) and the distance (L).

22.19.11 RESULTS AND DISCUSSIONS

22.19.11.1 Case (A): Fibre Cable Inspection and Handling

Discuss fibre cable inspection and handling procedure and your observations.

22.19.11.2 Case (b): Fibre Cable Ends Preparation

Discuss the fibre cable inspection and handling procedures and your observations.

22.19.11.3 Case (c): NA and Acceptance Angles Calculation

1. Determine the numerical aperture (NA) value from the graph.
2. Report the numerical aperture (NA) value.
3. Compare the graphical and calculated values of the NA.
4. Compare the graphical and calculated values with the actual value of the numerical aperture (NA) for the fibre cable, which is provided by the manufacturer specifications.

22.19.11.4 Case (d): Fibre Cable Power Output Intensity

1. Present the plotted graph of the laser power output (P_{out}) as a function of the distance (L).
2. Present the plotted graph of the laser power intensity (I) as a function of the distance (L).
3. Report the relationship between the laser power intensity (I) as a function of distance (L).

22.19.12 CONCLUSION

Summarize the important observations and findings obtained in this lab experiment.

22.19.13 SUGGESTIONS FOR FUTURE LAB WORK

List any suggestions for improvements using different experimental equipment, procedures, and techniques for any future lab work. These suggestions should be theoretically justified and technically feasible.

22.20 LIST OF REFERENCES

List any references that were used in the report. Use one format in writing the references. Never mix reference formats in a report.

22.21 APPENDIX

List all of the materials and information that are too detailed to be included in the body of the report.

FURTHER READING

Agrawal, Govind P., *Fiber-Optic Communication Systems*, 2nd ed., Wiley, New York, 1997.

Beiser, Arthur, *Physics*, 5th ed., Addison-Wesley Publishing Company, Reading, MA, 1991.

Boyd, Waldo T., *Fiber Optics Communications, Experiments & Projects*, 1st ed., Howard W. Sams & Co., Washington, DC, 1987.

Buckler, Grant, Fiber versus copper, *CNS Mag.*, May/June, 10–14, 2004.

Chen, Kevin P., In-fiber light powers active fiber optical components, *Photonics Spectra*, April, 78–90, 2005.

Cole, Marion, *Telecommunications*, Prentice Hall, Englewood Cliffs, NJ, 1999.

Cornsweet, T. N., *Visual Perception*, Academic Press, New York, 1970.

Derickson, Dennis, *Fiber Optic Test and Measurement*, Prentice Hall PTR, Englewood Cliffs, NJ, 1998.

Dutton, Harry J. R. . . , *Understanding Optical communications*, IBM Prentice Hall, Inc., New Jersey, 1998.

Eggleton, B. J., Kerbage, C., Westbrook, P. S., Windeler, R. S., and Hale, A., Microstructured optical fiber devices, *Opt. Express Dec.*, 17, 698–713, 2001.

Goff, David R. and Hansen, Kimberley S., *Fiber Optic Reference Guide: A Practical Guide to the Technology*, 2nd ed., Butterworth-Heinemann, London, 1999.

Green, Paul E., *Fiber Optic Networks*, Prentice Hall, Englewood Cliffs, NJ, 1993.

Groth, David, Mcbee, Jim, and Barnett, David, *Cabling—The Complete Guide to Network Wiring*, 2nd ed., Sams & Company, Washington, DC, 2001.

Hecht, Jeff, *City of Light: The Story of Fiber Optics*, Oxford University Press, New 15York, 1999.

Hecht, Jeff, *Understanding Fiber Optics*, 3rd ed., Prentice Hall, Inc., Englewood Cliffs, NJ, 1999.

Hecht, Eugene, *Optics*, 4th ed., Addison-Wesley Longman, Inc., New York, 2002.

Horng, H. E., Chieh, J. J., Chao, Y. H., Yang, S. Y., Chin-Yih, Hong, and Yang, H. C., Designing optical-fiber modulators by using magnetic fluids, *Opt. Lett.*, 30.5, 543–545, 2005.

Hoss, Robert J., *Fiber Optic Communications—Design Handbook*, Prentice Hall, Englewood Cliffs, NJ, 1990.

Kao, Charles K., *Optical Fiber Systems: Technology, Design, and Applications*, McGraw-Hill, New York, 1982.

Kasap, S. O., *Optoelectronics and Photonics Principles and Practices*, Prentice Hall PTR, Englewood Cliffs, NJ, 2001.

Keise, Gerd, *Optical Communications Essentials*, 1st ed., McGraw-Hill, New York, 2003.

Keise, Gerd, *Optical Fiber Communications*, 3rd ed., McGraw-Hill, New York, 2000.

Kolimbiris, Harold, *Fiber Optics Communications*, Prentice Hall, Inc., Englewood Cliffs, NJ, 2004.

Lerner, Rita G. and Trigg, George L., *Encyclopedia of Physics*, 2nd ed., VCH Publishers, Inc, New York, 1991.

Malacara, Daniel, *Geometrical and Instrumental Optics*, Academic Press Company, Boston, 1988.

Mazzarese, David, Meeting the OM-3 challenge—Fabricating this new generation of multimode fiber requires innovative state of the art designs and testing processes, *Cabling Syst.*, December, 18–20, 2002.

Mynbaev, Djafar K. and Scheiner, Lowell L., *Fiber-Optic Communication Technology—Fiber Fabrication*, Prentice Hall PTR, Englewood Cliffs, NJ, 2005.

National Research Council, *Harnessing Light Optical Science and Engineering for the 21st Century*, National Material Advisory Board, National Academy Press, Washington, DC, 1998.

Newport Corporation, *Projects in Fiber Optics Applications Handbook*, Newport Corporation, Irvine, CA, 1986.

Newport Corporation, *Photonics 2000 Catalog*, Newport Corporation, Irvine, CA, 2000.

Palais, Joseph C., *Fiber Optic Communications*, 4th ed., Prentice Hall, Inc., Englewood Cliffs, NJ, 1998.

Romine, Gregory S., *Applied Physics Concepts into Practice*, Prentice Hall, Inc., Englewood Cliffs, NJ, 2001.

Salah, B. E. A. and Teich, M. C., *Fundamentals of Photonics*, Wiley, New York, 1991.

Senior, John M., *Optical Fiber Communications* 1986. Principle and Practice

Shamir, Joseph, *Optical Systems and Processes*, SPIE Optical Engineering Press, Washington, DC, 1999.

Shashidhar, Naga, Lensing technology, *Corning Incorporated Fiber Product News*, April, 14–15, 2004.

Shotwell, R. Allen, *An Introduction to Fiber Optics*, Prentice Hall, Englewood Cliffs, NJ, 1997.

Sterling, Donald J. Jr., *Technician's Guide to Fiber Optics*, 2nd ed., Delmar Publishers Inc., New York, 1993.

Sterling, Donald J. Jr., *Premises Cabling*, Delmar Publishers, New York, 1996.

Torrey, Scott, Gel-free outdoor loose tube and ribbon fiber optic cable, *Fiberoptic Product News*, August, 14–16, 2004.

Ungar, Serge, *Fiber Optics: Theory and Applications*, Wiley, New York, 1990.

Vacca, John, *The Cabling Handbook*, Prentice Hall, Englewood Cliffs, NJ, 1998.

Walker, Rob and Bessant, Neil, Image fiber delivers vision of the future, *Photonics Spectra*, March, 96–97, 2005.

Weisskopf, Victor F., How light interacts with matter, *Scientific American*, September, 60–71, 1968.

Yeh, Chai, *Handbook of Fiber Optics: Theory and Applications*, Academic Press, San Diego, CA, 1990.

Yeh, C., *Applied Photonics*, Academic Press, New York, 1994.

23 Advanced Fibre Optic Cables

23.1 INTRODUCTION

The scaling down of fibre light sources brings the benefits of this fibre technology to a wide range of applications. These developments have led to a surge of interest in advanced fibre cables for both industrial and military applications. Advanced fibre cables are also used for transmitting high volumes of data in communication systems over long distances for getting very clear images, and in building many sophisticated instruments for a variety of applications. By creating new core designs, adding dopants to the fibre core and cladding, and developing manufacturing processes, engineers achieve advanced fibre-optic cable technology.

For example, the core of the holey fibre consists of many air holes; each hole acts as a single fibre. This fibre enables a high data transmission rate and capacity, and, consequently, reduces the cost of the network.

As explained in the preceding chapter, light propagates throughout the fibres by the principle of total internal reflection. The advanced fibres also have the same characteristics and suffer the same kind of losses as the common types of fibres, but in general terms, they have very low loss.

This chapter proposes a few experiments, such as fibre cable end preparation, and the measurements of numerical aperture (NA) aperture, power output intensity, and calculations of loss. The experimental results from this chapter can be compared with the preceding chapter. Some facts surrounding the advanced fibres can be concluded.

23.2 ADVANCED TYPES OF FIBRE OPTIC CABLES

The advanced core and cladding designs of newer fibre cables attempt to achieve lower loss and low dispersion over a wide range of wavelengths, so the cable can operate at very high speeds over very long distances. These types are also called "holey" fibres because they have tubes or spaces in the core along the fibre's length. Some types of these advanced fibre-optic cables are presented below:

23.2.1 Dual-Core Fibre for High-Power Laser

Fibre lasers—also called fibre sources—are light sources made from fibre. They are built around fibres cores, which are doped with materials that can be stimulated to emit light. Light stimulation and amplification are generally explained by the laser principles.

Single-mode, rare-earth-doped fibre lasers and amplifiers are widely used in telecommunications and other applications requiring compact, rugged optical sources with high beam quality. Fibre sources provide high electrical-to-optical conversion efficiency. The glass host broadens the optical transitions in the rare-earth ion dopants, gives continuous tunability. The variety of possible rare-earth dopants and co-dopants, such as erbium (Er^{3+}), neodymium (Nd), ytterbium (Yb), aluminum (Al), and germanium (Ge), yields broad wavelength coverage in the near-IR spectral region.

The developments of power scaling of continuous wave (CW) and pulsed fibre sources bring the benefits of this technology to a wide range of applications. These developments have led to a surge of interest in fibre-based laser systems for both industrial and military applications.

High-power fibre sources incorporate double-clad fibre, as shown in Figure 23.1, in which the rare-earth-doped core is surrounded by a much larger and higher NA outer core. The inner core has a higher index of refraction than the outer core, and the outer core, in turn, has a higher index of refraction than the cladding. The cladding may be glass or polymer, always with an index lower than the outer core. The inner core is normally doped with a high light-emitting rare-earth element. Light from high-power multimode pump diodes can be launched efficiently into the outer core. The pump light is absorbed only in the inner core, where it operates in a single mode. When the pump light goes through the inner core many times, it can excite light-emitting atoms efficiently. This type of fibre is used in building a femtosecond switch in a dual-core-fibre nonlinear coupler and some devices that need higher laser power.

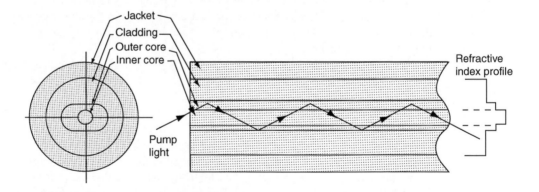

FIGURE 23.1 Dual-core fibre for high-power laser.

23.2.2 Fibre Bragg Gratings

Fibre Bragg grating (FBG) technology has gained favour in a wide range of applications because of its all-fibre configuration, great flexibility, and highly-efficient filtering functions. Fibre Bragg gratings are most commonly used for stabilizing pump lasers in optical amplifiers, wavelength division multiplexing, filtering, and chromatic dispersion compensation, because their efficiency reduces the cost of optical networking. Fibre Bragg gratings are made from a simple ordinary single-mode fibre. The grating is constructed by varying the index of refraction of the core along the

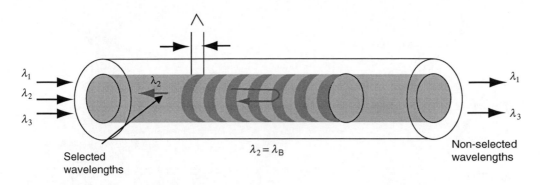

FIGURE 23.2 A fibre Bragg grating (FBG).

fibre's length. The index of refraction rises and falls along the fibre's length periodically in a pattern that would look like a uniform wave, as shown in Figure 23.2. All wavelengths are scattered by the regularly-spaced index variation, which causes the waves to interfere with each other in the fibre. Light at the wavelength that hits each index peak at the same phase experiences constructive interference and is selectively reflected back toward the source. These are called resonant waves. The spacing of the grating matches the resonant wavelength. Light scattered at other wavelengths by the grating is out of phase and cancels out by destructive interference. The waves transmitted though the fibre are called non-resonant waves. The FBG performance is influenced by the centre wavelength, bandwidth, and reflectance peak. The FBG has basic parameters that control the performance, such as the grating period, the grating length, the modulation depth, the index of refraction profile, and the wavelengths of the incoming light.

The grating selective reflected light at the Bragg wavelength (λ_B) is given by:

$$\lambda_B = 2n_{eff}\Lambda \tag{23.1}$$

where n_{eff} is the effective index of refraction of the core and Λ is the pitch of the grating in the core.

Fibre Bragg gratings are sensitive to temperature variation. Therefore, they must be kept at an operating temperature between -5 and $+70°C$ to operate accurately. When used in building WDM systems, temperature compensation techniques are necessary. Emerging manufacturing improvements—such as complex holographic phase masks—combined with automated processes will drive down costs and further improve the performance of FBG-based components.

23.2.2.1 Manufacturing Method

The simplest manufacturing method of FBGs is exposing a photosensitive fibre to an intensity pattern of ultraviolet radiation to write a FBG. The complete manufacturing process of FBG has four steps:

1. *Preparing the photosensitive optical fibre*. The preparation consists of stripping the coating (jacket) from the region to be exposed. Using chemicals for coating removal instead of mechanical stripping helps to avoid mechanical damage to the fibre.
2. *Recording of the grating*. Recording FBG inside a fibre usually involves interference between two coherent UV laser beams that precisely define the spacing between the grating's planes. For symmetric incidence and an angle (θ) between the UV beams, the pitch of the recorded grating is given as $\Lambda = \lambda/2 \sin(\theta/2)$, where λ is the illumination wavelength. The basic interferometric setup splits the laser beam to create an interference pattern on the fibre, as shown in Figure 23.3(a). Another interferometric

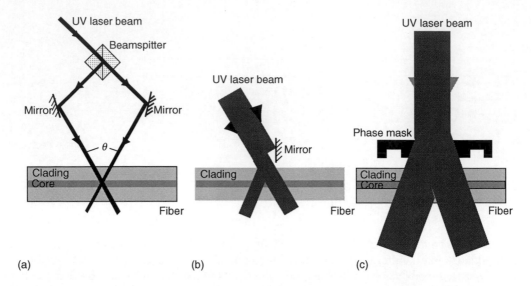

FIGURE 23.3 Several configurations can be used to make FBGs.

setup uses a mirror positioned perpendicularly to the fibre to fold half of the laser onto its other half, as shown in Figure 23.3(b). This process is adequate for short FBGs. A third method uses a phase mask to generate multiple beams. Two of the beams carry approximately 40% of the total energy and closely overlap the phase mask surface to create an interference pattern at the desired Bragg wavelength, as shown in Figure 23.3(c). The fibre-phase mask assembly can be illuminated either by a large beam to cover the full FBG length or by a small scanning beam. Because the fibre is usually close to the phase mask, the assembly is a very stable mechanical system, better than the process shown in (a).

3. *Thermal annealing.* Subjecting the FBG to thermal annealing at high temperatures ($>250°C$) helps ensure that their optical properties remain stable for at least 25 years. Annealing techniques must heat the grating locally without damaging the coating of the fibre.

4. *Packaging.* The final step of the manufacturing process consists of packaging the grating. Depending on the application, there are three packaging processes: recoating the fibre with coating similar to the fibre jacket, protecting the grating with a mechanical sleeve, and packaging the grating in a thermal module.

23.2.3 CHIRPED FIBRE BRAGG GRATINGS

Chirped fibre Bragg gratings (chirped FBGs) are suitable for compensating the chromatic dispersion that occurs along a fibre link. The chirped FBG is composed of an optical core that varies the period of the grating along the length of the grating, as shown in Figure 23.4. Chirped FBGs can be made by two methods: either by varying the period of the grating, or by varying the average index of refraction of the grating. Both methods have the effect of gradually changing the grating period.

The grating reflects light propagating through the fibre when its wavelength corresponds to the grating period. As the grating period varies along the axis, the different wavelengths are reflected by different portions of the grating and, accordingly, are delayed by different amounts of time. The net effect is a compression (or a broadening) of the input pulse that can be tailored to compensate for

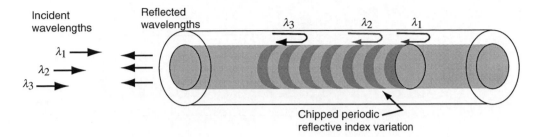

FIGURE 23.4 A chirped fibre Bragg grating (chirped FBG).

the chromatic dispersion accumulated along the fibre link. Although entering the grating at different times, the wavelength components of broadened pulses all return to the entrance at the same time.

23.2.3.1 Manufacturing Method

When fabricating a simple FBG, a standard phase mask is usually used, along with a fixed UV laser beam, as shown in Figure 23.5(a). This technique is similar to the one used for the fabrication of simple FBG devices, such as pump laser stabilizers. This recording technique is very efficient, but can be applied only to simple FBG filters. One of the most common techniques for fabricating an FBG filter that requires a complex profile, and that uses a standard phase mask, involves a complex recording scheme wherein the UV laser beam is scanned along the phase mask, as shown in Figure 23.5(b). A complex phase mask can be used for the fabrication of a complex FBG filter by transferring the pattern of the recording scheme directly into the phase mask, as shown in Figure 23.5(c).

Fibre Bragg grating technology is also well-suited to tunable dispersion compensation. Submitting the FBG to a temperature gradient allows the grating chirp to be changed, and, accordingly, the dispersion level to be tuned. Single gratings can be used for producing negative dispersion or for producing a similar positive dispersion range.

Fibre Bragg gratings are a mature technology that is very well-adapted to dispersion compensation in optical communications systems. With all of the aforementioned developments, FBG-

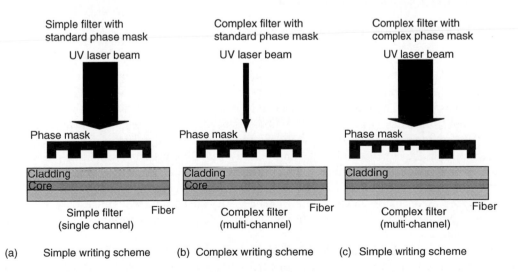

FIGURE 23.5 Several configurations can be used to make chirped FBGs.

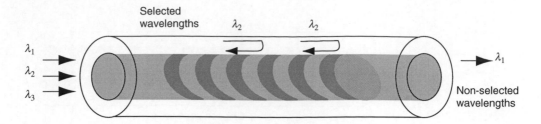

FIGURE 23.6 A blazed FBG. (a) Hollow core photonic band-gap fibre. (b) Endlessly photonic crystal fibre (PCF). (c) Photonic crystal fibre.

based technology can be considered a cost-effective alternative to competing technologies, while providing major advantages, such as small size, no nonlinear optical effects, and low insertion loss.

23.2.4 BLAZED FIBRE BRAGG GRATINGS

A blazed FBG is constructed when the grating is written at an oblique angle to the centre of the core, as shown in Figure 23.6. The selected wavelengths are reflected out of the fibre, and the non-selected wavelengths are passed through the fibre. In this case, a filter is constructed. Fibre Bragg gratings are used in erbium-doped fibre amplifiers (EDFAs). Manufacturing configurations are similar to the processes used in manufacturing the FBGs.

23.2.5 NONZERO-DISPERSION FIBRE-OPTIC CABLES

Nonzero-dispersion fibre was introduced to the market in 1993. This fibre type provides optimum performance for optically-amplified systems, over longer distances and with higher capacity. This cable can be used with optimum dispersion slope over the S-band (1480–1520 nm), C-band (1520–1570 nm), and L-band (1570–1620 nm).

The advanced core designs of newer single-mode fibres attempt to make lower loss and low dispersion coincide at the same wavelength, so the system can operate at very high speeds over very long distances. Dispersion-shifted fibres have a structure that changes the zero-dispersion wavelength, typically from 1300 to 1550 nm. Dispersion-flattened fibres have a structure that lowers dispersion over a wide range of wavelengths.

23.2.6 PHOTONIC CRYSTAL FIBRE CABLES

Recently, there has been a dramatic increase of interest in photonic crystal fibre (PCF), also called microstructure fibre (MOF) or holey fibre, which features an array of air holes running along the fibre length, as shown in Figure 23.7. Photonic crystal fibres use a micro-structured cladding region with air holes to guide light in a pure silica core, giving rise to novel functionalities. The PCFs are divided into two classes: photonic bandgap fibres associated with the photonic band gap structures, and index-guiding holey fibres based on the modified total internal reflection. The latter form their cores by filling one or several air holes in the array with fibre material. This fibre exhibits many unique properties of light guidance, such as single-mode operation over a wide range of wavelengths, highly adjustable effective mode area and nonlinearity, anomalous dispersion at visible and near-infrared wavelengths, and high birefringence. They are capable of high-capacity transmission over transoceanic distances, and extremely high-efficiency transmission over a wide range of wavelengths.

Photonic crystal fibre keeps light confined to a hollow core that is about five times thinner than a human hair. The fibre cross-section is an air core of about 15 µm in diameter surrounded by a web-like structure of glass and air. Photonic crystals work on the principle of refraction, or the bending

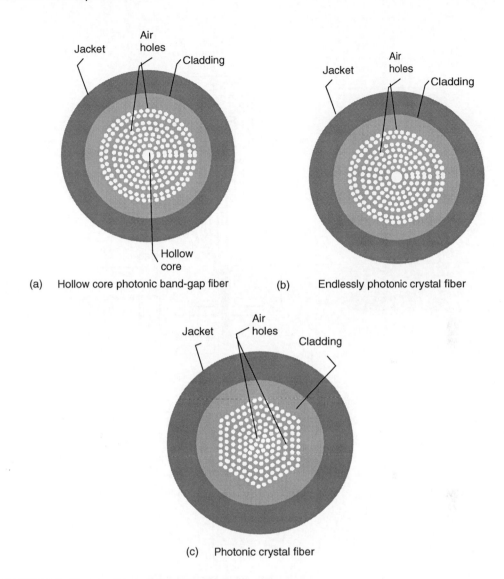

FIGURE 23.7 Cross-sections of the PCF cables.

of light as it passes from one material to another. Light travels through the core of the material just as it travels through the middle of hollow fibre-optic lines. However, photonic crystal is better than a reflective coating at keeping light from scattering or being absorbed, which keeps signals stronger over longer distances. Photonic crystal fibres are widely used in telecommunications—to filter light signals, to represent data in quantum communications, implementing fibre-based polarization splitters, transmitting exotic wavelengths in the IR or UV range, for applications in medical equipment and sensors, diode-pumped Nd^{3+} laser sources, and measurements in scientific experiments.

Figure 23.8 illustrates the stack-and-draw process, the most widely-used method for fabricating PCF. Silica capillary tubes are stacked by hand to make a preform. The preform has a diameter of 20–40 mm. The core is embedded by omitting several capillaries from the centre of the stack. Typically, several hundred capillaries are stacked in a close-packed array and inserted into a jacketing tube to create a fibre preform. This preform is drawn into fibre in two stages, adding

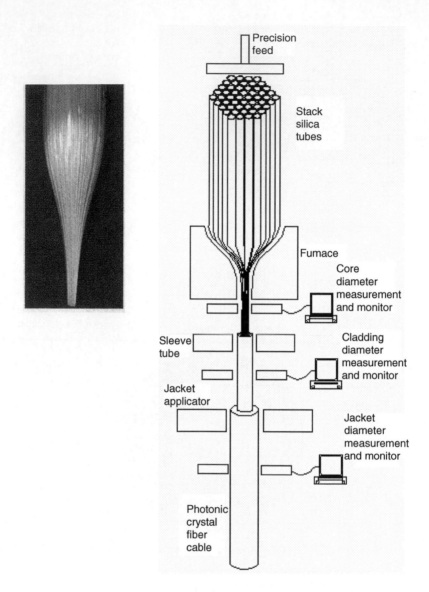

FIGURE 23.8 In the Stack-and-Draw process for fabricating PCFs.

a second jacketing tube before the final draw. The extra jacket enables the creation of fibres with a standard outer diameter while providing independent control over the photonic bandgap pitch, and hence, the operating wavelength. The preform is drawn into fibre, as shown in Figure 23.8. The preform is attached to a precision feed that moves it into the furnace at a proper speed to often and shrink the silica to a diameter of 2–4 mm. The drawing fibre passes through high-precision diameter control and is monitored by imaging or x-ray to detect any non-homogeneity or bubbles in the drawing fibre. It is then inserted into a silica sleeve tube and drawn down again to a fibre of typically 125 µm in diameter. After this stage, the diameter is continuously monitored by an accurate measurement and monitoring device. The fibre is then covered by a coloured jacket layer. Again, the jacket layer undergoes diameter control and monitoring. The end of the fibre cable is attached to a rotating spool, which turns steadily. The fibre cable is then tested for attenuation in dB per kilometer, for dispersion, and any other requirements specified by the customer. Drawn PCF lengths of a few kilometers are typical.

Photonic crystal fibres are available in different types, such as highly nonlinear, double-cladding high numerical aperture, large mode-field area, and polarization-maintaining. These types have different mode-field areas, numbers of holes, and working wavelengths. Some types of PCFs have inner and outer cladding layers that are used in highly-polarization-maintaining cables.

23.2.7 MICROSTRUCTURE FIBRE CABLES

Microstructure fibres can be used to create a supercontinuum (SC). This is a broadening of the narrow laser spectrum into a broad continuous spectrum with many of the properties of coherent light. Supercontinuums have applications in frequency metrology and optical coherence tomography. The addition of Bragg gratings to the SC light enhances the SC spectrum and alters the dispersion profile, thus making gratings an effective tool in tailoring SC sources.

A Grapefruit fibre is one type of MOF. In the Grapefruit MOF, the lower-order modes are confined by the airholes and are sensitive to surrounding media. Figure 23.9 illustrates a cross-section of the Grapefruit MOF. After the addition of Long Period Gratings (LPG) fibre, the higher-order modes interact with the holes and with the material with which they may be infused. The manufacturing process of this fibre is similar to the manufacturing process of the crystal fibres. This fibre has the same manufacturing process as that of holey fibres. These processes are explained in the above sections.

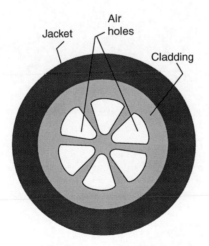

FIGURE 23.9 A cross-section of the Grapefruit microstructure fibre (MOF).

23.2.8 POLYMER HOLEY-FIBRE CABLES

The polymer holey-fibre has same cross-section as the photonics crystal fibres. A polymer fibre is fabricated by first drilling holes into a poly methylmethacrylate (PMMA) preform in which the holes drill down to an intermediate preform. The holes of this intermediate preform are filled with a solution of the dye rhodamine 6 G, which permeates the PMMA. The intermediate preform is heated to drive out the solvent molecules and to lock the dye molecules into the structure. Finally, the preform is drawn into a fibre with a 600-μm outer diameter and an 18-μm core diameter. Polymer fibres can be used in replacing the silica fibres in limited applications.

23.2.9 IMAGE FIBRE CABLES

Image fibre cables are used for remote or semi-flexible viewing systems, and ultra-thin image fibre delivers vision to previously inconceivable locations. Until recently, applications for this type of optical fibre have been in the medical industry, primarily in endoscopes. Because of the versatility of the image fibre, however, it is meeting a variety of specialized demands.

Figure 23.10 shows an image fibre made of silica glass that transmits a coherent picture from one end to the other. The technology behind ultra-thin image fibres is not new, but manufacturers and designers are increasingly seeking to expand into new imaging applications. Figure 23.11 shows the cross-section of an image fibre cable. The multiple cores are pure silica or silica doped with germanium, and they share a common cladding that is doped with fluorine. The cores, or pixels, are approximately three times thinner than the core of a single-mode telecom fibre; a 1600-pixel fibre has an outer diameter of approximately 210 μm, and a 10,000-pixel fibre has an outer diameter of 1700 μm. The fibre is drawn in a manner similar to that used in the production of communication fibre cables, then cut and bundled into a silica tube to make a preform. It can bend from 20 to 200 mm in radius. The versatility of image fibre is partially the result of its ability to function at wavelengths from the UV to the near-IR range.

With the customizable features of the image fibre cable, the fibre can be used to meet a variety of specific application requirements. One important area is in the field of biomolecular research. Scientists are now using the fibre in confocal microscopes for fluorescence imaging of cells and

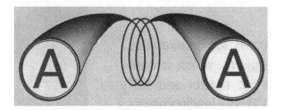

FIGURE 23.10 An image fibre transmits a picture from one end to the other.

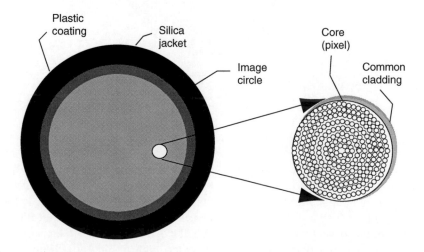

FIGURE 23.11 A cross-section of an image fibre cable.

proteins at the molecular level. In the automotive sector, the image fibre is used to perform quality checks inside an engine before initial testing. Previously, the manufacturers employed rigid rod-lens-based or video imaging systems. Similarly, in the testing and maintenance of car and jet engines and steam and gas turbines, industrial endoscopes can be used where access is limited or when there is no time to dismantle components. The advantage is that image fibrescopes are flexible, whereas rigid rod-lens-based systems are not. The flexibility helps them reach into pre-viously-inaccessible areas. There are other possible applications for image fibres; for instance, using infrared wavelengths, it could be employed to take snapshots of the blades inside gas or steam turbines, which would minimize, if not eliminate, the need to shut down power stations or refineries for such maintenance checks.

23.2.10 LIQUID CRYSTAL PHOTONIC BANDGAP FIBRE CABLES

As explained above, PCFs—usually made from silica—have a micro-structured cross section of airholes running along their length. Recently, much attention has been focused on the development and design of these fibres for various applications, including SC generation, nonlinear signal processing in high-speed optical time domain multiplexing communications systems, and high-power double-clad fibre lasers. Such applications use fibres with static properties.

In PCFs, the airholes provide access to the guided light in the core and allow liquids to be infiltrated into the capillaries. The optical properties of the liquids are usually easier to modify than those of the silica. This approach, therefore, paved the way to components based on tunable PCFs.

These fibres guide light using two different principles. The first principle is modified total internal reflection, which relies on guiding light in a high-index core surrounded by a low-index cladding material. The PCF guides light by the principle of modified total internal reflection when the holes are filled with air or with a material with a lower refractive index than that of silica. The second principle involves guiding light in a low-index core surrounded by a cladding arrangement of high-index rods. Infiltrating the airholes with a high-index oil or liquid crystal transforms the fibre into a photonic bandgap guiding type, which guides light in the low-index silica core.

Light coupled into the core can be guided through antiresonant reflection from the rods, which forms a sort of two-dimensional Fabry-Perot resonant cavity surrounding the core.

The optical modes of a single liquid-crystal-filled hole determine the spectral properties of a PCF infiltrated with a pneumatic liquid crystal. Because liquid crystals are highly anisotropic in molecular orientation, the optical modes are determined not only by the refractive indices of the liquid crystal, but also by the alignment within the hole. Therefore, changing the alignment of the liquid crystal and/or the refractive indices of the liquid crystal can modify the spectral properties of the fibre.

Pneumatic liquid crystals are known for exhibiting very large thermo-optic effects as a result of the existence of crystalline and isotropic liquid phases, which depend on temperature. When the liquid crystal is doped with a small amount of an absorbing dye, the high thermo-optic effect modifies its refractive index, allowing the tuning of spectral properties of the liquid crystal photonic bandgap fibre.

23.2.11 LENSED AND TAPERED FIBRE CABLES

Lensed fibres are used to couple light between optical fibres and lasers, semiconductor optical amplifiers, waveguides, and other fibre optical devices. Lensing technology enables highly-effec-tive coupling within a small space. In recent years, substantial changes in lensed fibre cable technology have increased the flexibility and extended the optical performance. Figure 23.12 shows lensed fibre variations. Two significant developments have occurred: the advancement of fibre-polishing technologies and the development of novel lens designs.

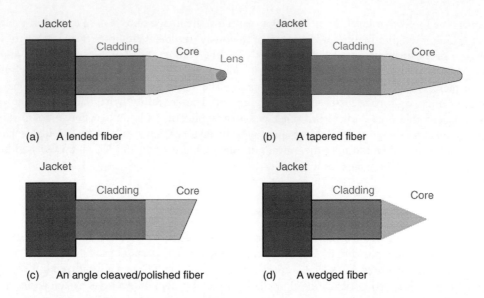

(a)   A lended fiber

(b)   A tapered fiber

(c)   An angle cleaved/polished fiber

(d)   A wedged fiber

FIGURE 23.12 Lensed/tapered fibre cable variations.

Advances in polishing technologies include process and equipment improvements and the use of laser polishing instead of mechanical polishing. The development of novel lens designs has significantly improved lens performance. A clear understanding of the optics of tapered lenses, their performance, and the various methods of manufacture and specifications will ensure that the most appropriate lensing technology is used to meet specific design requirements.

23.2.11.1 Advantages of Lensing Technology

A number of industry requirements are driving advancements in lensed fibre technology. These include:

- **Lower pricing**
    The price of lensed fibres has decreased because new advanced manufacturing methods are being developed to address the demand for lower pricing.
- **Higher consistency**
    Tapered lenses are made by polishing fibres into a conical shape with a spherical surface at the tip. These lenses, called conical lensed fibres, were inconsistent in quality, resulting in significant performance limitations. It was not uncommon, for example, for customers to have to test the lenses and select the ones that met optical criteria. Select rates for the lenses were as low as 60%.
- **Longer working distance**
    The working distance for lenses with a small spot size is very limited. A polished lensed fibre having a 3-μm mode-field diameter at the waist (MFDw) has a distance-to-beam waist (DBW) between 5 and 8 μm. Such a small DBW makes packaging difficult, increasing the risk of the lens hitting the facet of the device. This translates into longer process times and lower yields in component assembly.
- **Higher return loss**
    A polished lens with anti-reflective (AR) coating typically has a return loss of 30 dB. In some applications, where light is coupled into the waveguide, this could be a limitation, since the returned light acts as a feedback, increasing noise in the communication systems containing amplifiers or lasers.

- **Smaller beam diameter**

    Waveguides on materials, such as indium phosphide (InP) and gallium arsenide (GaAs), have very small MFDw. Using lenses with an MFDw that does not match the device decreases coupling efficiency, so efforts have been made to decrease the MFDw of the lens.

23.2.11.2 Manufacturing Technologies

There are a number of alternatives for manufacturing lensed fibres. Each alternative has advantages and disadvantages, and some are progressive improvements on earlier technologies. Like all technologies, not all techniques have matured or even sometimes progressed to commercial availability. These methods include mechanical polishing, chemical etching, thermally drawing the fibre to a taper, and lensing the fibre. These methods are explained briefly below.

Mechanical Polishing

Lensed fibres are made by mechanically polishing a fibre into a conical shape with the tip shaped into a spherical lens, as shown in Figure 23.13. The lens is spherical with a radius of curvature (Rc). A challenge for this type of lens is the misalignment between the centre of curvature of the spherical lens and the optical axis. The core-to-cladding concentricity for single-mode fibre (<0.5 μm) and the geometric repeatability of the polishing method defines the degree of misalignment. This misalignment can cause deviations from Gaussian beam profiles, resulting in decreased coupling efficiency. Additionally, significant aberrations can occur while trying to polish lensed fibres to the small centre of curvature required for lensed fibres with MFDw less than 3.5 μm.

    Advances in fibre-polishing equipment have increased the quality of the lensed fibres. Manufacturers can use laser ablation instead of mechanical polishing. This technique has the potential to better control the radius of curvature with mechanical polishing. It is also claimed that there is greater product consistency when a far-field monitoring feedback loop is added to the system.

Chemical Etching

Another manufacturing technique is chemical etching, wherein immersion in an acid solution chemically polishes the fibre. The cladding is dissolved away, leaving behind a bit of the core. Melting this stub forms a spherical lens with the desired radius of curvature. An advantage of this method is that several fibres can be shaped at a time, as opposed to other techniques where only one fibre can be shaped at a time. Although this technology has been reported upon, commercially-available lensed fibres based on this technology are rare.

Drawn Tapered Fibres

Tapered fibres are manufactured by heating the fibre cladding and core beyond the softening point of the glass, and then applying a force to make a taper. The tip is then melted back to create a spherical lens. This technique enables greater flexibility in MFDw, as well as improving the DBW compared to mechanical polishing. In this kind of tapered lens, the amount of heat and heating time

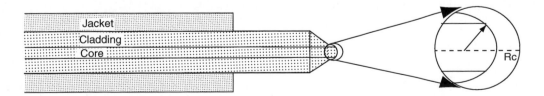

FIGURE 23.13 Lensed fibre-optic cable.

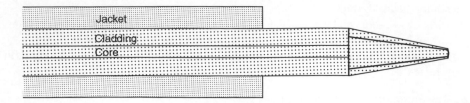

FIGURE 23.14 Lensed tapered fibre optic cable.

determine the Rc of the spherical lens. During manufacturing, the core of the fibre is also tapered, causing deviations from the expression relating MFDw to Rc.

In thermal methods, such as the drawn tapered fibre and lensed methods, the tapered fibre tip is made by melting the glass in such a way that the surface tension of molten glass forms a smooth surface. Since the tip is frozen liquid, the surface finish can be far smoother than polishing methods.

Lensed Tapers

Perhaps the most promising technology is tapered lensed fibres, also called lensed fibres. Figure 23.14 shows a lensed tapered fibre optic cable. These fibres are distinguished from tapered fibres because they are a two-element lens with a gradient index and a geometric (spherical/hyperbolic) lens at the tip. The combination of the two elements has enabled greater flexibility for meeting a variety of optical specifications.

In lensed tapers, a fibre with a gradient index profile is fused to a fibre pigtail. This spliced fibre is drawn into a taper with a thermally-formed spherical lens at the end. The advantage of this technology is that the gradient index tapered region directs the light to the spherical lens, giving the lensed fibre more consistency in the MFD. There are several factors that can be controlled in this kind of lens, including the gradient index profile, the radius of curvature, the length of the gradient index fibre, and the taper length. Since the gradient index fibre is fused to the pigtail, the lenses can be made on PM fibres.

Another advancement is the development of lensed tapers with a hyperbolic refractive lens. This technology has enabled a four-fold increase in the DBW, and has also increased the return loss. In this type of lensed fibre, a fibre with a gradient index is first spliced to the fibre. The gradient index fibre is then thermally shaped to make a hyperbolic shaped lens. The lensing action is a combination of the gradient index lens and the hyperbolic lens. This combination allows for a greater DBW.

Lensed fibre manufacturing has evolved significantly. Better methods for polishing and the development of more-sophisticated lens designs have increased flexibility, enabling lenses to meet a wide specification range for MFD and DBW. Each lensed fibre must be loaded individually into a fixture for anti-reflection coating. In addition to the MFDw specification, a DBW, and return loss (RL) specification may be added if important to an application.

23.2.12 BEND-INSENSITIVE FIBRE CABLES

Bend-insensitive fibre cables are designed for improved bend performance in reduced-radius applications. The fibre cables employ a moderately higher NA than standard single-mode telecommunication fibre cables, and offer improved bend performance for applications in the 1310- and 1550-nm range.

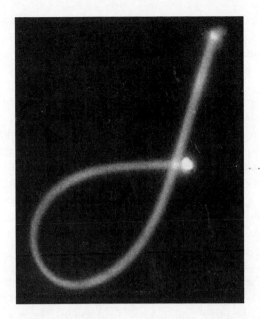

FIGURE 23.15 Nanoribbon fibre optic cables.

23.2.13 NANORIBBON FIBRE OPTIC CABLES

Crystalline oxide nanoribbon fibre cables are used as waveguides. They can be used with other nanowire optical components to create elementary systems that will suit the feasibility of building a nanowire-based photonic integrated circuitry. Photonic integrated circuits are suitable for applications as active and passive components in telecommunications, optical computing, and a wide variety of other sectors in photonics. Doping materials, such as SnO_2 and ZnO, are added in the core and cladding to enable a large amount of light guiding from end to end, and allow a small amount of light to escape from the fibre by scattering, as shown in Figure 23.15.

Nanoribbons with sizes on the order of 100–400 nm are suitable for waveguiding visible and ultraviolet radiation, although researchers found that synthesis process predominantly yielded ribbons for blue and green wavelengths. Measured losses depend on the area of a given ribbon in cross section and the presence of scattering centres.

To demonstrate the potential applications of waveguides in photonic circuitry, it is necessary to examine the approaches to coupling the ribbons and linking them to nanowires serving as input and output components. It has been found that staggering the ribbons for several microns with an air gap between them led to efficient evanescent tunneling and outperformed butt-coupling. As observed using a near-field scanning optical microscope, light could be injected into the ribbons from an optically pumped ZnO nanowire source and detected by a similar wire.

23.3 APPLICATIONS OF ADVANCED FIBRE CABLES

Fibre-optic cables are used in many applications, some of which are listed in the previous Fibre-Optic Cable Chapter. One of many applications for different fibre cable types is fibre-optic sensing. Fibre-optic sensing can be classed as intrinsic and extrinsic. In intrinsic sensors, the fibre simply conducts light from a sensing head to a detector; the interaction between light and the environment takes place outside the fibre cable. In intrinsic sensors, the interaction between the environment, the fibre, and the light itself generates information about a specific measurement. The key advantage of intrinsic fibre-optic sensors is the fundamental ability of a fibre to guide light around bends and over long distances. This enables the fibre to be confined within small physical volumes and magnifies the effects of very fine environment changes to a level that can be measured accurately and quantified.

Fibre Bragg gratings are commonly employed as passive temperature and pressure sensors. The ability to adjust these optical devices introduces a new dimension for the design of very fast, precise, and multifunctional fibre sensors without compromising their intrinsic advantages. As explained in the previous sections, an FBG consists of a short length of fibre-optic having a periodic modulation of refractive index along its length. This periodic structure causes the FBG to act as a narrowline filter with a peak reflectivity at a specific wavelength, called Bragg wavelength (λ_B), which is determined by the period length of the FBG and the refractive index of the fibre. The Bragg wavelength can be calculated by Equation (23.1). If the refractive index of the FBG is changed by a temperature shift, or the period is altered in some way (e.g., by compression or expansion), a shift in the peak reflected wavelength results. Detecting this shift is the basis of fibre sensing.

Fibre Bragg gratings are also used in many other applications, such as tunable FBGs. Tuning can be accomplished by several mechanisms, such as electric heating, the piezoelectric effect, mechanical stretching and bending, and acoustic modulation. These sensors are available in the market in different types, sizes, and wavelength ranges.

Photonics bandgap and photonics crystal fibres are both members of the family of MOFs, called holey fibres. Both types can be used in gas sensors. The sensors are made of a holey fibre for the 1550-nm spectral range. One end of the fibre is spliced to conventional single mode fibre from an optical source, such as a laser or an LED, and placed on the outer end in a V-groove in a vacuum chamber, as shown in Figure 23.16. A multi-mode fibre leading to a detector is installed 50 μm away from the end in the V-groove. The separation allows gas to flow into or out of the hollow core of the bandgap fibre, while still allowing for efficient optical coupling between the fibres. A 1-meter length of the fibre is filled with acetylene (C_2H_2) to a pressure of 10 m bar and illuminated with a tunable laser source. The expected spectral changes of acetylene are observed and displayed on a monitor.

Beyond pressure and temperature sensor technology, advanced fibre sensors types include chemical, strain, biomedical, electrical and magnetic, rotation, vibration, and displacement as major applications. One more application of the advanced fibre sensor technology is in the area of DNA analysis. Fibre-optic biosensors have the ability to detect the presence of short DNA sequences called oligonucleotides. This ability gives the diagnostic some excellent tools for early and accurate diagnosis. These fibres are efficiently constructed as waveguides with novel properties for communications and sensing applications.

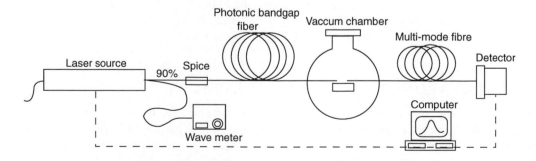

FIGURE 23.16 Bandgap fibre used in gas sensor.

23.4 EXPERIMENTAL WORK

This chapter presents delicate and expensive fibre cables. Testing these types of fibre cables in the lab is very difficult using low- or moderate-accuracy tools and measuring instruments. Therefore, students should test only the least-delicate fibre cables, such as nonzero-dispersion fibre-optic cables and the dual-core fibre for high-power laser. Students can perform the regular loss tests as explained in the Fibre-Optic Cables Chapter. A comparison between single- or multi-mode fibres can be conducted, for example, on one type of the advanced fibre cables. Finally, conclusions can be summarized for the performance of the two types that are used in the experiments and compared with the data provided by the manufacturers.

23.4.1 Conclusion

Summarize the important observations and findings obtained in this lab experiment.

23.4.2 Suggestions for Future Lab Work

List any suggestions for improvements using different experimental equipment, procedures, and techniques for any future lab work. These suggestions should be theoretically justified and technically feasible.

23.5 LIST OF REFERENCES

List any references that were used in the report. Use one format in writing the references. Never mix reference formats in a report.

23.6 APPENDIX

List all of the materials and information that are too detailed to be included in the body of the report.

FURTHER READING

Agrawal, G. P., *Nonlinear Fiber Optics. Optics and Photonics*, 2nd ed., Academic Press, New York, 1995.

Alkeskjold, T. T., Laegsgaard, J., Bjarklev, A., Hermann, D., Anawati, A., Broeng, J., Li, J., and Wu, S.-T., All optical modulation in dye-doped nematic liquid crystal photonic bandgap fibers, *Opt. Express*, 12 (24), 5857–5871, November 29, 2004.

Astle, T. B. et al., Optical components—The planar revolution, Merrill Lynch & Co., In-depth Report, May 17, 2000.

Bise, R. T., Windeler, R. S., Kranz, K. S., Kerbage, C., Eggleton, B. J., and Trevor, D. J., Tunable photonic bandgap fiber, *Proc. Opt. Fiber Commun. Conf. Exhibit*, 17, 466–468, March 17, 2002.

Blaze Photonics Limited, Photonic crystal fibers, Product summary, Blazephotonics, UK, 2003.

Burgess, D. S., Nanoribbon waveguides suited for photonics circuits. Photonics technology world, *Photonics Spectra*, 38 (11), 28, November, 2004.

Chen, K. P., In-fiber light powers active fiber optical components, *Photonics Spectra*, 39 (4), 78–90, April 2005.

Crystal, Fiber, *Photonic Crystal Fibers—Revolutionizing Optical Fiber Technology*, Crystal Fiber, Denmark, 2004.

Derickson, D., *Fiber Optic Test and Measurement*, Prentice Hall PTR, Englewood Cliffs, NJ, 1998.

Deveau, R. L., *Fiber Optic Lighting: A Guide for Specifiers*, 2nd ed., Prentice Hall PTR, Englewood Cliffs, NJ, 2000.

Eggleton, B. J., Kerbage, C., Westbrook, P., Windeler, R., and Hale, A., Photonics crystal fiber, honey-comb fiber, *Opt. Lett.*, 24, 1460, 1999.

Eggleton, B. J., Kerbage, C., Westbrook, P., Windeler, R., and Hale, A., Grapefruit fiber, *J. Lightwave Devices*, 18, 1084–1100, 2000.

Eggleton, B. J., Kerbage, C., Westbrook, P., Windeler, R., and Hale, A., Microstructured optical fiber devices, *Opt. Express*, 698–713, Dec 17, 2001.

Fedder, G. K., Santhanam, S., Reed, M. L., Eagle, S. C., Guillou, D. F., Lu, M. S.-C., and Carley, R., Laminated high-aspect-ratio microstructures in a conventional CMOS process, *Sens. Actuators A*, 57 (2), 103–110, 1996.

Goff, D. R. and Kimberly, S. H., *Fiber Optic Reference Guide: A Practical Guide to the Technology*, 2nd ed., Butterworth-Heinemann, London, 1999.

Guy, M. and François, T., Fiber Bragg gratings: better manufacturing—better performance, *Photonics Spectra*, 9(3), 106–110, March 2002.

Guy, M. and Yves, P., Fiber Bragg gratings: a versatile approach to dispersion compensation, 38(8), 96–101, August 2004.

Hitz, Breck, Holey fibers connect to conventional fibers with low loss, *Photonics Spectra*, 84–85, August, 2005.

Hood, D. C. and Finkelstein, M. A., Sensitivity to light, In *Handbook of Perception and Human Performance, Sensory Processes and Perception*, Boff, K. R., Kaufman, L., and Thomas, J. P., Eds., Vol. 1, Chapter 5, Wiley, New York, 1986.

Kao, Charles K., *Optical Fiber Systems: Technology, Design, and Applications*, McGraw-Hill, New York, 1982.

Kashyap, R., *Fiber Bragg Gratings*, Academic Press, New York, 1999.

Larsen, T., Bjarklev, A., Hermann, D., and Broeng, J., Optical devices based on liquid crystal photonic bandgap fibers, *Opt. Express*, 11 (20), 2589–2596, 2003.

Mynbaev, D. K. and Lowell, L. S., *Fiber-Optic Communication Technology—Fiber Fabrication*, Prentice Hall PTR, Englewood Cliffs, NJ, 2005.

OFS Leading Optical Innovations, Nonzero dispersion optical fiber, True Wave Reach Fiber, Product Catalog 2003, USA, 2003.

Opto-Canada, SPIE Regional Meeting on Optoelectronics, and Imaging, Vol. TD01, Technical Digest of SPIE—The International Society for Optical Engineering, Ottawa, Canada, pp. 9–10, May 2002.

Razavi, B., *Design of integrated circuits for optical communications*, McGraw-Hill, New York, 2003.

Sabert, H. and Jonathan, K., Hollow-core fibers seek the 'holey' grail. BlazePhotonics Ltd., *Photonics Spectra*, 37 (8), 92–94, 2003.

Sabert, H., Hollow-core fibers allow light to travel by air, *Laser Focus World*, 40 (5), 161–164, May 2004.

Savage, N., Into thin air: photonic bandgap fibers hold promise as new types of waveguides, *SPIE's oemagazine*, 4 (1), 44, January 2004.

Savage, N., Holy-y waveguides. Product trends, *SPIE's oemagazine*, 5 (5), 36, May 2005.

Shashidhar, N., Lensing technology, *Corning Incorporated, Fiber Product News*, 14–15, April 2004.

Shen, L. P., Huang, W. P., Chen, G. X., and Jian, S. S., Design and optimization of photonic crystal fibers for broad-band dispersion compensation, *IEEE Photonics Technol. Lett.*, 15 (4), 540–542, April 2003.

The 2004 Photonics Circle of Excellence Award Winners, Crystal fiber A/S, *Photonics Spectra*, 39 (1), 25, January 2005.

VonWeller, E. L., Photonics crystal fibers advances in fiber optics, *Appl. Opt*, March 1, 2005.

Walker, R. and Neil, B., Image fiber delivers vision of the future. Fujikura Europe Ltd., *Photonics Spectra*, 96–97, March 2005.

West, J. A., Venkataramam, N., Smith, C. M. Gallangher, M. T., Photonic crystal fibers, In *Proceedings of the 27th European Conference on Optical Communication (ECOC '01)*, Vol. 4, pp. 582–585, Corning Inc., USA, 2001.

Yeh, C., *Handbook of Fiber Optics: Theory and Applications*, Academic Press, San Francisco, CA, 1990.

Yeh, C., *Applied Photonics*, Academic Press, New York, 1994.

Zhang, L. and Changxi Y., Polarization selective coupling in the holey fiber with asymmetric dual cores, LEOS2003, Tucson, USA, paper ThP1, Oct 26–30, 2003.

24 Light Attenuation in Optical Components

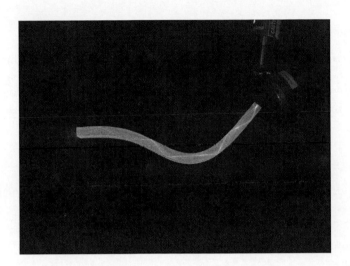

24.1 INTRODUCTION

Attenuation is the loss of power in a fibre-optic cable or any optical material, and can result from many causes. During transit, light pulses lose some of their energy. Light losses occur when the fibre-optic cables are subjected to any type of stress, temperature change, or other environmental effects. The most important source of loss is the bending that occurs in the fibre-optic cable during installation or in the manufacturing process.

Losses that occur in a fibre-optic cable, optical fibre devices, and systems of many devices can be calculated. The decibel (dB) is the standard unit to express the losses in optical fibre cables, devices, and systems. Attenuation for commercially-available fibre cables is specified in decibels per kilometre (dB/km). Attenuation also depends on the types and specifications of the fibre.

Attenuation in optical fibres varies with the wavelength of light. There are three low-loss windows of interest: 850, 1300, and 1550 nm. The 850-nm window is perhaps the most widely used, because 850-nm devices are inexpensive. The 1300-nm window offers lower loss, but at a modest increase in cost for light-emitting diodes, the main light sources. The 1550-nm window today is mainly of interest to long-distance communication applications.

The experimental work in this chapter will enable students to practise calculating the light loss in optical components, such as a glass slide, an epoxy layer, and a fibre cable. Light loss due to microscopic bending in a fibre cable can also be calculated.

24.2 LIGHT LOSSES IN AN OPTICAL MATERIAL

When light passes through an optical component, power is lost. The light loss in any optical component is dependent on the accumulative losses due to internal and external losses. Internal losses are caused by light reflection, refraction, absorption, dispersion, and scattering. External losses are caused by bending, stresses, temperature changes, and overall system losses. Losses due to refraction and reflection (such as Fresnel reflection, microscopic reflection, surface reflection, and back reflection) are generally explained by the laws of light. Common losses due to absorption, dispersion, and scattering mechanisms, as well as light losses in parallel optical surfaces, and in epoxy that occur in any optical material are explained below.

24.2.1 ABSORPTION

Every optical material absorbs some of the light energy. The amount of absorption depends on the wavelength of the light and on the optical material. Absorption loss depends on the physical characteristics of the optical material, such as transitivity and index of refraction. The wavelength of the light passing through an optical material is a function of the index of refraction of the material.

24.2.2 DISPERSION

Dispersion is caused by the expansion of light pulses as they travel through optical components. This occurs because the speed of light through the optical medium is dependent on the wavelength, the propagation mode, and the optical properties of the materials along the light path.

24.2.3 SCATTERING

Scattering losses occur in all optical materials. Atoms and other particles inevitably scatter some of the light that hits them. Rayleigh scattering is named after the British physicist Lord Rayleigh (1842–1919), who stated that such scattered light is not absorbed by the particles, but simply redirected. Light scattering in the core of the fibre-optic cable is a common example, as illustrated in Figure 24.1. The further the light travels through a material, the more likely scattering is to occur. Rayleigh scattering depends on the type of material and the size of the particles relative to the wavelength of the light. The amount of scattering increases quite rapidly as the wavelength decreases.

Scattering loss also occurs in optical material inhomogeneities introduced during glass preparation and the addition of dopants in the manufacturing process. Imperfect mixing and processing of chemicals and additives can cause inhomogeneities within the preparation of a preform. When the preform is used in the fibre-drawing method, rough areas will form in the core and thus increase the scattering of light in the fibre.

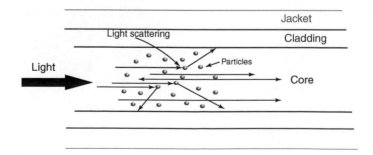

FIGURE 24.1 Light scattering in the fibre core.

24.2.4 Light Loss in Parallel Optical Surfaces

Loss of light due to reflection at a boundary between two parallel optical surfaces comprises a large portion of the total optical losses in a system. The simplest case of reflection loss occurs when an incident ray travels normal to the boundary, as shown in Figure 24.2. The reflection coefficient (ρ) is the ratio of the reflected electric field to the incident electric field. For a ray incident at the normal:

$$\rho = \frac{n_1 - n_2}{n_1 + n_2} \tag{24.1}$$

where n_1 is the refractive index of the incident medium and n_2 is the index of the transmitted medium.

If $n_2 > n_1$, then the reflection coefficient becomes negative. This indicates a $180°$ phase shift between the incident and reflected electric fields. The reflectance (R) is the ratio of the reflected ray intensity to the incident ray intensity. Because the intensity in an optical beam is proportional to the square of the beam's electric field, the reflectance is equal to the square of the reflection coefficient (ρ). The reflectance is calculated as:

$$R = \left[\frac{n_1 - n_2}{n_1 + n_2} \right]^2 \tag{24.2}$$

24.2.5 Light Loss in an Epoxy Layer

Adhesives are used in manufacturing optical devices and are a key technology in the fibre-optic communications market. In order to produce low-cost and highly reliable optical components and devices, an easy-to-use, durable adhesive is necessary. Requirements for optical adhesives are extremely dependent upon the specific application. A specialized group of companies develops and markets adhesives and adhesive resins for fibre-optic applications. These adhesives and resins are designed for a specific refractive index. They have high transmittance, precise curing time, heat-resistance, high elasticity, and permeability.

Epoxy adhesives come in several forms. The most commonly used types are one-part, two-part, and UV-curable systems. One-part systems typically require heat to cure the adhesive. Refrigeration of the liquid adhesive typically prolongs its shelf life. Two-part systems are based on a chemical reaction and thus must be used immediately after mixing. Setting times range from several minutes to several hours. UV-curable epoxy systems are one-part mixtures, which are activated by a UV-light source. As such, UV-systems do not require refrigeration. These can also be heat-treated to stabilize the cure.

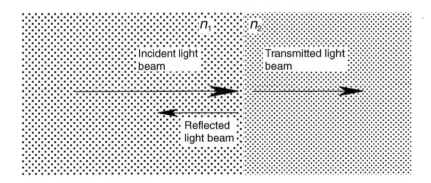

FIGURE 24.2 Light beam passing through two media.

Epoxy adhesives are a very important component in manufacturing fibre-optic devices. Each epoxy has a unique index of refraction and unique optical properties, like any other optical material. All epoxy adhesives affect the light passing through the epoxy. Light can be absorbed, dispersed, and scattered by the epoxy. The optical performance of an epoxy is dependent on the epoxy preparation procedure. Cleaning the optical contact surfaces where the epoxy is to be applied, epoxy curing time, epoxy curing temperature, and epoxy cooling rates all impact the optical epoxy performance. It is very important to understand an epoxy's physical properties and performance in relation to the signal wavelengths and the maximum signal power.

24.2.6 BENDING AND MICRO-BENDING

Any bending in a fibre-optic cable generates loss. Fibre-optic cable bending losses are caused by a variety of outside influences. These influences can change the physical characteristics of the cable and affect how the cable guides the light. Certain modes are affected and losses are accumulated over long distances. However, significant losses can arise from any kind of bending in a fibre cable. The cause of bending loss is easier to envisage using the ray model of light in a multimode fibre cable. When the fibre cable is straight, the ray falls within the confinement angle (θc) of the fibre cable. However, as shown in Figure 24.3, a bend will change the angle at which the ray hits the core-cladding interface. If the bend is sharp enough, the ray strikes the interface at an angle outside of the confinement angle, and the ray is refracted into the cladding and then to the outside as loss. These are referred to as leaky modes, whereby the ray leaks out and the attenuation is increased.

In another class of modes, called radiation mode, power from these modes radiates into the cladding and increases the attenuation. In radiation mode, the electromagnetic energy is distributed in the core and the cladding; however, the cladding carries no light.

When light is launched into a fibre cable, the power distribution varies as light propagates down the fibre cable. The power distribution decreases over long distances and eventually stabilizes. This characteristic of optical fibre is referred to as stable mode distribution. Stable mode distribution can be observed in short fibre cables by introducing mode-filtering devices. Mode filtering may be accomplished through the use of mode scrambling, which can be achieved by bending the fibre cable to form a corrugated path, as shown in Figure 24.4. The corrugated path introduces a coupling, which leads to the existence of both radiation and leaky modes. In high-power applications, stable mode distribution can be achieved because the effective portion of the signal that "leaks" is small in comparison to the full signal strength. Mode scrambling allows repeatable laboratory measurements of signal attenuation in fibre cables.

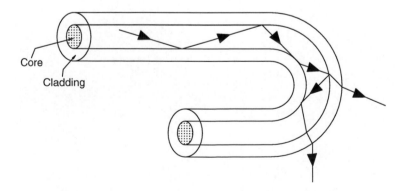

FIGURE 24.3 Light propagation around a bend in a fibre cable.

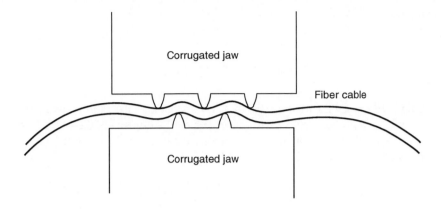

FIGURE 24.4 Mode scrambler for fibre cable micro-bends.

Figure 24.5 shows a microscopic bend in a fibre-optic cable. Micro-bends can be a significant source of loss. When the fibre cable is installed and pressed onto an irregular surface, tiny bends can be created in the fibre cable. Light is lost due to these irregularities.

24.3 ATTENUATION CALCULATIONS

Any incident light power passing through an optical component—such as a glass microscope slide, fibre-optic cable, and epoxy layer—is subjected to losses. Attenuation measures the reduction in light signal strength by comparing output power with input power. Measurements are made in decibels (dB). The decibel is an important unit of measure in fibre-optic components, devices, and systems loss calculations.

$$\text{Loss (dB)} = -10 \log_{10}\left(\frac{P_{\text{out}}}{P_{\text{in}}}\right) \tag{24.3}$$

Equation (24.3) is also used to calculate the loss between the input and output of the cable. In industry, fibre-optic cable loss is calculated as the loss per unit length. Therefore, the fibre-optic cable loss is calculated by dividing the cable loss (dB) by cable length (km), as given in the

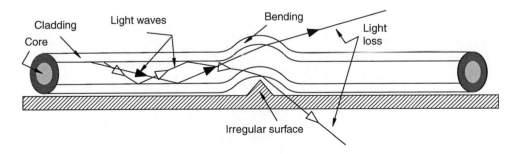

FIGURE 24.5 Fibre cable micro-bend.

following equation:

$$\text{Loss}_{\text{cable}}(\text{dB/km}) = \frac{-10 \log_{10}\left(\frac{P_{\text{out}}}{P_{\text{in}}}\right)}{\text{Cable Length}} \tag{24.4}$$

There is another type of loss, called excess loss. It is defined as the ratio between the sum of all power outputs (P_1 to P_n) and input signal power. Excess loss specifies the power lost within the system. It includes dispersion, scattering, absorption, and coupling loss. Excess loss is calculated using the following formula:

$$\text{Loss}(\text{dB}) = -10 \log_{10}\left(\frac{(P_1 + P_2 + P_3 + \dots + P_n)}{P_{\text{in}}}\right) \tag{24.5}$$

Figure 24.6 shows fibre-optic attenuation as a function of wavelength. In the 1970s, communication systems operated in the wavelength range of 800–900 nm. At that time, fibre-optic cables exhibited a local minimum in the attenuation curve, and optical sources and photo detectors operating at this range were available. In addition, the fibre-optic cables were faster than their counterpart, the copper cabling systems. This region is referred as the first window.

By reducing the concentration of hydroxyl ions and metallic impurities in the fibre core material, in the 1980s manufacturers were able to fabricate fibre-optic cables with very low loss in the 1100–1600-nm region. At the same time, the demand increased for high-speed data rate transmission over long distances. Thus, the second window was defined, concentrated around

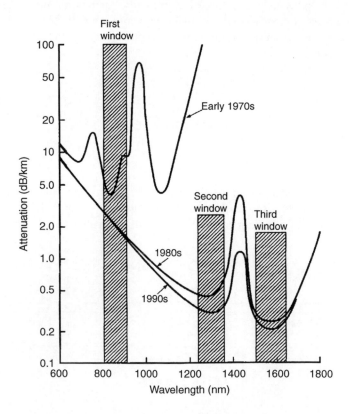

FIGURE 24.6 Fibre-optic attenuation as a function of wavelength.

1310 nm. This spectral band is referred to as the long-wavelength region. Due to further successful development in the fabrication of fibre-optic cables and optical amplifiers, the third window was defined around 1550 nm.

The most important aspect of the fibre optic communication link is that many different wavelengths can be sent along a fibre simultaneously, without interference, in the 1310–1625-nm spectrum. The technology of combining a number of wavelengths onto the same fibre cable is known as wavelength division multiplexing (WDM). Furthermore, this technology experienced advanced development by using the dense wavelength division multiplexing (DWDM). WDM technology is presented in detail in the wavelength division multiplexing/demultiplexing chapter.

24.4 EXPERIMENTAL WORK

This experiment is designed to determine the power loss of light through optical components, such as glass microscope slides, epoxy layer, and fibre-optic cables, in the following cases:

   a. To measure and analyze the power loss of laser light through a single and up to five glass microscope slides.
   b. To measure and analyze the power loss of laser light through a single microscope slide inclined at different angles from the normal.
   c. To measure and analyze the power loss of laser light through an epoxy layer between two glass microscope slides.
   d. To measure the power loss of laser light through a fibre-optic cable and power loss per length.
   e. To measure the power loss of laser light through a fibre-optic cable due to micro bending.
   f. To measure the power loss of laser light through a fibre cable with GRIN lens at the input and/or output.

24.4.1 TECHNIQUE AND APPARATUS

Appendix A presents the details of the devices, components, tools, and parts.
   1. 2×2 ft. optical breadboard.
   2. HeNe laser light source and power supply.
   3. Laser light sensor.
   4. Laser light power meter.
   5. Laser mount assembly.
   6. Hardware assembly (clamps, posts, screw kits, screwdriver kits, sundry positioners, etc.).
   7. Fibre-optic cable, 250 μm diameter, 500 m long.
   8. 20X microscope objective lens.
   9. Lens/fibre cable holder/positioner assembly.
   10. Fibre cable holder/positioner assembly.
   11. Standard glass microscope slides and slide holders/positioners.
   12. Rotation stage.
   13. Mode scrambler, as shown in Figure 24.7.
   14. GRIN lenses.
   15. GRIN lens/fibre cable holder/positioner assembly.
   16. Denatured ethanol.
   17. Tissue.
   18. Swabs.
   19. Epoxy.
   20. Micro-spatula.

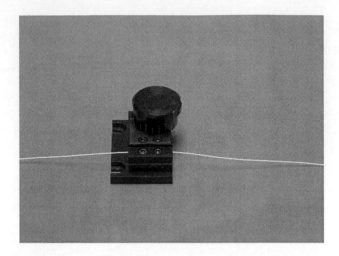

FIGURE 24.7 Mode scrambler.

21. Timer.
22. Microscope.
23. Black/white card and cardholder.
24. Ruler.

24.4.2 PROCEDURE

Follow the laboratory procedures and instructions given by the instructor.

24.4.3 SAFETY PROCEDURE

Follow all safety procedures and regulations regarding the use of optical components, electrical and optical devices, and test measurement instruments.

24.4.4 APPARATUS SETUP

24.4.4.1 Laser Light Power Loss through One to Five Microscope Slides

This part of the experiment measures the power of the laser beam using a laser light sensor that is located at a fixed distance from the laser source. The measured laser light input power is the same value for all other cases in this experiment. The laser light passes through one glass microscope slide, then more slides are continually added, up to five slides in total. The output power will be measured for each slide when added. As slides are added, the output power will decrease, since the light is being refracted and absorbed more and more with increasing glass thickness. The distances between the laser source, the slides, and the laser power sensor will also have an effect on the measured losses. Therefore, these distances should be kept constant throughout the experiments.

The following steps illustrate the experimental lab setup for Case (a):

1. Figure 24.8 shows the apparatus setup.
2. Bolt the laser short rod to the breadboard.
3. Bolt the laser mount to the clamp using bolts from the screw kit.
4. Put the clamp on the short rod.

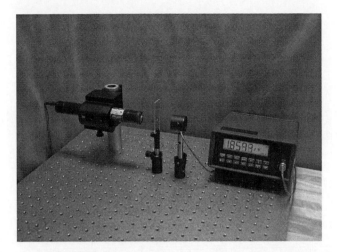

FIGURE 24.8 Laser light loss through one to five slides.

5. Place the HeNe laser into the laser mount and tighten the screw. Follow the operation and safety procedures of the laser device in use.

6. Check the laser alignment with the line of bolt holes on the breadboard and adjust when necessary.

7. Place the laser sensor in front of the laser source at a fixed distance for all cases.

8. Turn off the lights of the lab.

9. Turn on the laser power. Measure the laser input power (P_{in}). Fill out Table 24.1.

10. Prepare one standard glass microscope slide.

11. Clean the slide with a swab dampened with denatured ethanol. Follow the cleaning procedure for the optical components.

12. Mount a single slide into the slide holder/positioner.

13. Place the slide and slide holder/positioner in front of the laser source at a fixed distance between the laser source and the laser sensor, as shown in Figure 24.8. Make sure that the slide is perpendicular to the laser beam. Keep the same distance for all cases.

14. Measure the laser output power (P_{out}) from the slide. Fill out Table 24.1.

15. Add another slide, stacking the slides together. Place the slides and slide holder/positioner in front of the laser source at the same distance. Measure the laser output power from the slides. Fill out Table 24.1.

16. Repeat step 15 to add one slide at a time. Use up to five slides added together one at a time. Fill out Table 24.1.

17. Turn on the lights of the lab.

24.4.4.2 Laser Light Power Loss through a Single Slide Inclined at Different Angles

This case of the experiment deals with a single slide being rotated around the normal to the laser light source. The output power of the laser light depends on the angle that the slide is inclined. Most likely, there will be an angle for which the output power is the greatest; at all other angles, output power will be less. There is a specific inclination angle at which the laser beam will not pass through the slide. At this angle, the laser beam is reflected back to the incoming side of the laser beam.

The following steps illustrate the experimental lab set-up for Case (b):

1. Figure 24.9 shows the apparatus setup.

2. Repeat steps 2–13 in Case (a) for installing the laser source and measuring the input laser power.

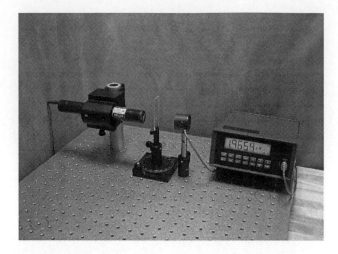

FIGURE 24.9 Laser light loss through a single slide inclined at different angles.

3. Measure the laser output power (P_{out}) from the slide. Fill out Table 24.2.
4. Mount the slide holder on the top of the rotation stage.
5. Rotate the rotation stage an angle of 10° from the normal of the light source. Measure the laser output power from the slide. Fill out Table 24.2.
6. Repeat step 5, rotating the rotation stage an angle increment of 10° each time. Measure the laser light output. Fill out Table 24.2.
7. Turn on the lights of the lab.

24.4.4.3 Laser Light Power Loss through an Epoxy Layer Between Two Slides

Having experienced the power loss of the laser light as it passes through a single slide in Case (a), add the following steps to measure the power loss of the laser light passing through two slides, as shown in Figure 24.10.

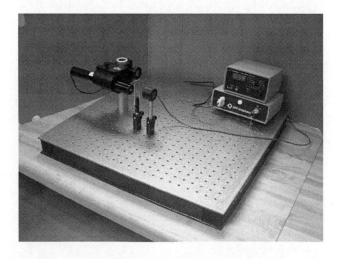

FIGURE 24.10 Laser light loss through air layer in a two-slide assembly.

1. Repeat steps 2–13 in Case (a) for installing the laser source.
2. Measure the input laser power. Fill out Table 24.3 for laser power output for a single slide.
3. Strip a piece of fibre cable and cut two short lengths of the cladding to be used as a spacer between the two slides.
4. Prepare two standard slides.
5. Clean the two standard slides with a swab dampened with denaturated ethanol. Follow the optical components cleaning procedure.
6. Arrange the two slides with the spacers between them.
7. Mount the two-slide assembly into the slide holder/positioner.
8. Position the two-slide assembly in front of the laser source at a fixed distance, as shown in Figure 24.10.
9. Measure the laser output power (P_{out}) from the two-slide assembly with an air gap between the slides Fill out Table 24.3.
10. Prepare the epoxy by mixing the two-epoxy materials together carefully.
11. Apply epoxy between the two slides. Use a micro-spatula to distribute the epoxy uniformly in the air gap between the slides. Follow the epoxy curing procedure for the epoxy that is given in the lab. Note that a number of epoxy types used in manufacturing optical fibre devices cures in an oven set at 120°C temperature for 12 min.
12. Assemble the spacers and the two slides.
13. Mount the assembly into the slide holder/positioner, as shown in Figure 24.11.
14. Inspect the condition of the cured epoxy using a microscope, as shown in Figure 24.11. During the inspection, air bubbles, inhomogeneous epoxy mixing, or the presence of uncured epoxy can be identified. Compare the colour of the cured epoxy with the manufacturer's specifications.
15. Position the assembly in front of the laser source at a fixed distance, as shown in Figure 24.12.
16. Measure the laser output power (P_{out}) from the assembly. Fill out Table 24.3.
17. Turn on the lights of the lab.

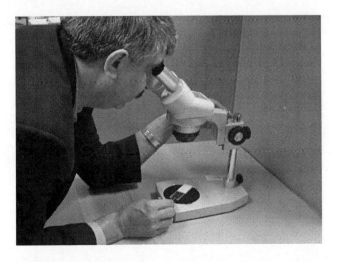

FIGURE 24.11 Epoxy inspection process.

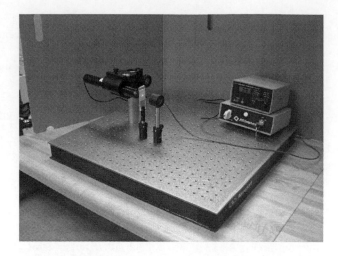

FIGURE 24.12 Laser light loss through the epoxy layer in the two-slide assembly.

24.4.4.4 Laser Light Power Loss through a Fibre Optic Cable

This case of the experiment deals with the loss per length in a fibre optic cable. Loss calculation in these optical components is similar to the above cases. The following steps illustrate the experimental lab setup for Case (d):

1. Figure 24.13 shows the apparatus set-up.
2. Bolt the laser short rod to the breadboard.
3. Bolt the laser mount to the clamp using bolts from the screw kit.
4. Put the clamp on the short rod.
5. Place the HeNe laser into the laser mount and tighten the screw. Turn on the laser device. Follow the operation and safety procedures of the laser device in use.
6. Check the laser alignment with the line of bolt holes of the breadboard and adjust when necessary.
7. Mount a lens/fibre cable holder/positioner to the breadboard so that the laser beam passes over its centre hole.

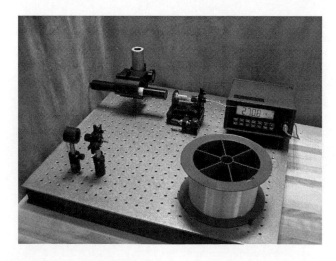

FIGURE 24.13 Fibre cable power loss apparatus set-up.

8. Add the 20X microscope objective lens to the lens/fibre cable holder/positioner.
9. Prepare a fibre cable with a good cleave at each end, as described in the end preparation section of the fibre-optic cable chapter.
10. Insert one end (input) of the fibre cable into the brass fibre cable holder and place it into the hole of the lens/fibre cable holder/positioner.
11. Mount the output end of the fibre cable into the brass fibre cable holder and place it in the hole of the fibre cable holder/positioner. Mount the assembly in front of the laser sensor.
12. Turn on the laser power.
13. Extend the end of the fibre cable until the fibre cable end is at the centre of the lens/fibre cable holder/positioner. This is a very important step for obtaining an accurate value of the laser beam intensity from the fibre cable output end.
14. Re-check the alignment of your light launching arrangement by making sure that the input end of the fibre cable remains at the centre of the laser beam.
15. Check to ensure that you have a circular output from the output end of the fibre cable. Point the output toward the centre of the laser sensor.
16. Turn off the lights of the lab.
17. Measure the laser input power (P_{in}) and output power (P_{out}). Fill out Table 24.4.
18. Find the net length of the fibre optic cable length that is being used in this experiment. Fill out Table 24.4.
19. Turn on the lights of the lab.

24.4.4.5 Laser Light Power Loss through a Fibre-Optic Cable Due to Micro-Bending

Continue the procedure as explained in Case (d) by adding a mode scrambler near the end of the fibre cable, as shown in Figure 24.14.

1. Place the mode scrambler at a convenient position on the fibre cable near the output end, as shown in Figure 24.14. Lay the fibre cable between the two corrugated surfaces of the mode scrambler, as shown in Figure 24.7. Rotate the knob clockwise until the corrugated surfaces just contact the fibre cable. Rotate the knob further clockwise and carefully observe the reduction in the power readout. Rotate the knob more until you break the fibre cable. In this case, no light will pass though the fibre cable.
2. Measure the output power while tightening the knob. Record the last power measurement before the fibre cable broke. Fill out Table 24.5.
3. Turn on the lights of the lab.

FIGURE 24.14 Mode scrambler apparatus setup.

24.4.4.6 Laser Light Power Loss through a Fibre-Optic Cable Coupled to a Grin Lens at the Input and/or Output

Coupling light into a fibre cable is accomplished using a quarter-pitch GRIN lens that focuses the collimated beam from the laser source onto a small spot on the core of the fibre cable. The GRIN lens is coupled to the input and output ends of a fibre-optic cable. When the GRIN lens focuses the power of the collimated laser beam onto the core of the fibre cable, the input power increases. Similarly, inserting a GRIN lens at the output of the fibre cable increases the output power.

Follow the instructions from Case (a) regarding the use of a 20X lens. Figure 24.15 shows the experimental setup for Case (f). Add the following steps:

1. Mount a GRIN lens/fibre cable holder/positioner to the breadboard at the input end of the fibre cable so that the laser beam passes through the centre hole.
2. Place a quarter-pitch GRIN lens into the groove of the GRIN lens/fibre cable holder/positioner, as shown in Figure 24.15.
3. Insert the input end of the fibre cable into the brass fibre cable holder and place it into the hole of the GRIN lens/fibre cable holder/positioner.
4. Extend the end of the fibre cable so that the fibre cable end is at the centre of the GRIN lens/fibre cable holder/positioner. This is a very important step for obtaining an accurate value of the laser beam intensity from the fibre cable output end.
5. Mount the output end of the fibre cable into the brass fibre cable holder and place it in the hole of the fibre cable holder/positioner. Mount the assembly in front of the laser sensor.
6. Verify the alignment of the input end of the fibre cable with the GRIN lens. Make sure that the focal point of the GRIN lens remains at the front of the input end of the fibre cable.
7. Check to ensure that you have a circular output from the output end of the fibre cable. Point the fibre cable output end toward the centre of the laser sensor.
8. Optimize the coupling of the laser beam by finely adjusting the fibre cable holder/positioner.
9. Measure the laser output power (P_{out}). Fill out Table 24.6.
10. Continue the experiment by adding a GRIN lens to the output cable end. Make a very fine adjustment of the GRIN lens to the fibre cable output end position.
11. Measure the laser output power (P_{out}) with the GRIN lens at the output. Fill out Table 24.6.
12. Turn on the lights of the lab.

FIGURE 24.15 Fibre cable power loss with a GRIN lens at the input and output ends.

24.4.5 DATA COLLECTION

24.4.5.1 Laser Light Power Loss through One to Five Microscope Slides

1. Record the input power (P_{in}) without a slide in front of the laser light.
2. Measure the output power (P_{out}) for one to five slides.
3. Determine the losses for the slides.
4. Fill out Table 24.1 for Case (a).

24.4.5.2 Laser Light Power Loss through a Single Slide Inclined at Different Angles

1. Record the input power (P_{in}) without a slide in front of the laser light.
2. Measure the output power (P_{out}) for each inclination angle.
3. Determine the losses of the slide for each inclination angle.
4. Fill out Table 24.2 for Case (b).

TABLE 24.1
Laser Power Loss Through a Single to Five Slides

Power Input P_{in} (unit)		
Number of Glass Microscope Slides	Power Output P_{out} (unit)	Loss (dB)
One glass microscope slide		
Two glass microscope slides		
Three glass microscope slides		
Four glass microscope slides		
Five glass microscope slides		

TABLE 24.2
Laser Power Loss Through a Single Slide Inclined at Different Angles

Power Input P_{in} (unit)		
Laser Beam Incident Angle with the Normal (degrees)	Power Output P_{out} (unit)	Loss (dB)
0		
10		
20		
30		
40		
50		
60		
70		
80		
90		

24.4.5.3 Laser Light Power Loss through an Epoxy Layer Between Two Slides

1. Record the input power (P_{in}) without a slide in front of the laser light.
2. Record the output power (P_{out}) of the laser light from the air gap filled between the two-slide.
3. Record the output power (P_{out}) of the laser light from the epoxy-filled two-slide.
4. Determine the loss for the slide and epoxy.
5. Fill out Table 24.3 for Case (c).

TABLE 24.3
Power Loss of the Laser Light Passing Through the Epoxy Layer in the Two-Slide Assembly

Power Input P_{in}(unit)			
	Case	Power Output P_{out}(unit)	Loss (dB)
(a)	Single glass microscope slide		
(b)	Air gap between two glass microscope slides		
(c)	Epoxy layer between two glass microscope slides		

24.4.5.4 Laser Light Power Loss through a Fibre-Optic Cable

1. Measure the laser input power (P_{in}) and output power (P_{out}).
2. Find the length of the fibre optic cable that is being used in this case.
3. Fill out Table 24.4 for fibre cable power loss measurements.

TABLE 24.4
Fibre Cable Power Loss Data

P_{in} (unit)	P_{out} (unit)	Loss (dB)	Cable Length (km)	Loss$_{cable}$ (dB/km)

24.4.5.5 Laser Light Power Loss through a Fibre-Optic Cable Due to Micro-Bending

1. Measure the laser output power (P_{out}) during the mode scrambler operation before the fibre cable breaks off. Measure power input (P_{in}) as in Case (d).
2. Fill out Table 24.5 for mode scrambler loss measurements.

TABLE 24.5
Mode Scrambler
Power Loss Data

P_{in} (unit)	P_{out} (unit)	Loss (dB)

24.4.5.6 Laser Light Power Loss through a Fibre-Optic Cable Coupled to a Grin Lens at the Input and/or Output

1. Measure the laser output power (P_{out}) with the quarter-pitch GRIN lenses placed at the input and output ends of the fibre cable.
2. Find the length of the fibre-optic cable being used in this lab.
3. Fill out Table 24.6 loss measurements.

TABLE 24.6
Laser Light Power Loss Through a Fibre-Optic Cable Coupled to a GRIN Lens at the Input or/and Output

Situation	P_{in} (unit)	P_{out} (unit)	Loss (dB)	Cable Length (km)	$Loss_{cable}$ (dB/km)
GRIN lens at the fiber cable input end					
GRIN lens at the fiber cable output end					

24.4.6 Calculations and Analysis

24.4.6.1 Laser Light Power Loss through One to Five Microscope Slides

1. Calculate the power loss (dB) of the laser light when it passes through a single and multiple of slides, using Equation 24.3.
2. Draw a curve for the power loss (dB) as a function of number of slides.

24.4.6.2 Laser Light Power Loss through a Single Slide Inclined at Different Angles

1. Calculate the power loss (dB) of the laser light when it passes through a single slide, using Equation 24.3.
2. Draw a curve for the power loss (dB) as a function of the angles with the normal.

24.4.6.3 Laser Light Power Loss through an Epoxy Layer Between Two Slides

1. Calculate the power loss (dB) of the laser light when it passes through a single, two slides with an air gap, and two slides with an epoxy layer using Equation 24.3. Fill out Table 24.3 for Case (c).

2. Compare power loss (dB) among the three situations in this case.
3. Calculate the reflection coefficient (ρ) and the reflectance (R) at the air–glass interface or glass–epoxy interface, using Equation 24.1 and Equation 24.2.

24.4.6.4 Laser Light Power Loss through a Fibre-Optic Cable

1. Calculate the cable power loss (dB) using Equation 24.3.
2. Calculate the cable power loss per unit length (dB/km) using Equation 24.4.
3. Fill out Table 24.4 for this case.

24.4.6.5 Laser Light Power Loss through a Fibre-Optic Cable Due To Micro-Bending

1. Calculate the cable power loss with the mode scrambler, using Equation 24.3.
2. Fill out Table 24.5 for this case.
3. Compare the losses with and without the mode scrambler.

24.4.6.6 Laser Light Power Loss through a Fibre-Optic Cable Coupled to a Grin Lens at the Input and/or Output

1. Calculate the cable power loss with the quarter-pitch GRIN lenses placed at the input and output ends of the fibre cable using Equation 24.3 and Equation 24.4.
2. Fill out Table 24.6 for this case.

24.4.7 Results and Discussions

24.4.7.1 Laser Light Power Loss through One to Five Microscope Slides

1. Report the power loss (dB) of the laser light.
2. Discuss the power loss as a function of number of slides.

24.4.7.2 Laser Light Power Loss through a Single Slide Inclined at Different Angles

1. Report the power loss (dB) of the laser light.
2. Discuss the power loss as a function of the angles with the normal.
3. Find the maximum and minimum losses that occur with the corresponding angles.

24.4.7.3 Laser Light Power Loss through an Epoxy Layer Between Two Slides

1. Report the power loss (dB) of the laser light.
2. Report the percentage of the light reflected and transmitted at the air–glass interface or glass–epoxy interface.
3. Compare and discuss the power loss in each situation. Compare the power loss when air between the two slides is replaced by an epoxy layer.

24.4.7.4 Laser Light Power Loss through a Fibre-Optic Cable

1. Report the cable power loss (dB) and cable power loss per unit length (dB/km).
2. Compare the measured power loss values with the actual value of the power loss for the fibre cable, which is provided by the manufacturer.

24.4.7.5 Laser Light Power Loss through a Fibre-Optic Cable Due to Micro-Bending

1. Report the calculated results for the fibre cable power loss (dB) with the mode scrambler.
2. Perform a comparison of the losses with and without the mode scrambler.

24.4.7.6 Laser Light Power Loss through a Fibre-Optic Cable Coupled to a Grin Lens at the Input and/or Output

1. Report the calculated result for the cable power loss (dB) for this case.
2. Compare and discuss the calculated losses when using quarter-pitch GRIN lenses in this case, and the losses in Case (a).

24.4.8 CONCLUSION

Summarize the important observations and findings obtained in this lab experiment.

24.4.9 SUGGESTIONS FOR FUTURE LAB WORK

List any suggestions for improvements using different experimental equipment, procedures, and techniques for any future lab work. These suggestions should be theoretically justified and technically feasible.

24.5 LIST OF REFERENCES

List any references that were used in the report. Use one format in writing the references. Never mix reference formats in a report.

24.6 APPENDICES

List all of the materials and information that are too detailed to be included in the body of the report.

FURTHER READING

Al-Azzawi, Abdul R. and Peter Casey, *Fiber Optics Principles and Practices*, Algonquin College Publishing Centre, ont., 2002.

Chen, Kevin P., In-fiber light powers active fiber optical components, *Photonics Spectra*, April, 78–90, 2005.

Derickson, Dennis, *Fiber Optic Test and Measurement*, Prentice Hall PTR, Englewood Cliffs, NJ, 1998.

Deveau, Russell L., *Fiber Optic Lighting: A Guide for Specifiers*, 2nd ed., Prentice Hall PTR, Englewood Cliffs, NJ, 2000.

Edmund Industrial Optics, *Optics and Optical Instruments Catalog, 2004*, Edmund Industrial Optics, Barrington, NJ, 2002.

Hecht, Jeff, *Understanding Fiber Optics*, 3rd ed., Prentice Hall, Inc., Englewood Cliffs, NJ, 1999.

Kao, Charles K., *Optical Fiber Systems: Technology, Design, and Applications*, McGraw-Hill, New York, 1982.

Keiser, Gerd, *Optical Fiber Communications*, 3rd ed., McGraw-Hill, New York, 2000.

Mynbaev, Djafar K. and Scheiner, Lowell L., *Fiber-Optic Communication Technology—Fiber Fabrication*, Prentice Hall PTR, Englewood Cliffs, NJ, 2005.

Okamoto, K., *Fundamentals of Optical Waveguides*, Academic Press, New York, 2000.

Palais, Joseph C., *Fiber Optic Communications*, 4th Ed., Prentice Hall, Inc., Englewood Cliffs, NJ, 1998.

Salah, B. E. A. and Teich, M. C., *Fundamentals of Photonics*, Wiley, New York, 1991.

Senior, John M., *Optical Fiber Communications*, 2nd ed., 1986. Principle and Practice

Shamir, Joseph., *Optical Systems and Processes*, SPIE Optical Engineering Press.

Shotwell, R. Allen, *An Introduction to Fiber Optics*, Prentice Hall, Englewood Cliffs, NJ, 1997.

Sterling, Donald J. Jr., Jr., *Technician's Guide to Fiber Optics*, 2nd ed., Delmar Publishers Inc., New York, 1993.

Ungar, Serge, *Fiber Optics: Theory and Applications*, Wiley, New York, 1990.

Yeh, Chai, *Handbook of Fiber Optics: Theory and Applications*, Academic Press, San Diego, CA, 1990.

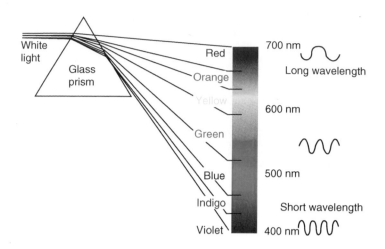

FIGURE 6.1 Dispersion of white light into the visible spectrum by a prism.

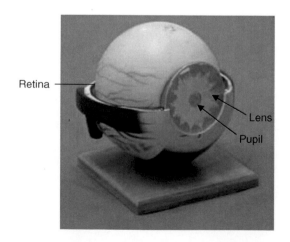

FIGURE 6.2 The major parts of the human eye.

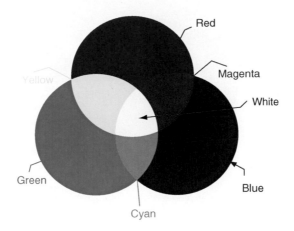

FIGURE 6.4 Additive colour mixing.

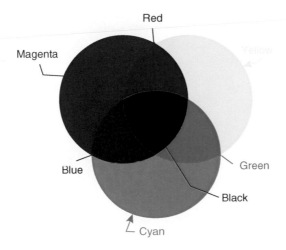

FIGURE 6.6 Subtractive colour mixing.

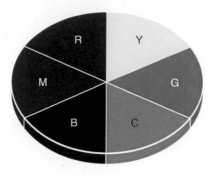

FIGURE 6.7 The Relationship between the additive and subtractive primary's complement.

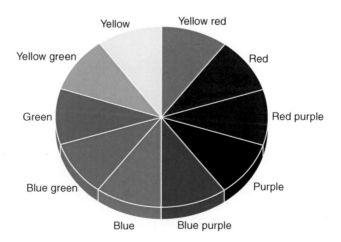

FIGURE 6.8 The colour wheel (the three primary additive colours sandwiched between variations of coloured light).

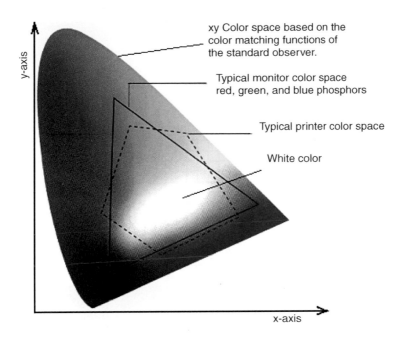

xy Color space based on the color matching functions of the standard observer.

Typical monitor color space red, green, and blue phosphors

Typical printer color space

White color

y-axis

x-axis

FIGURE 6.10 The C. I. E. Chromaticity diagram.

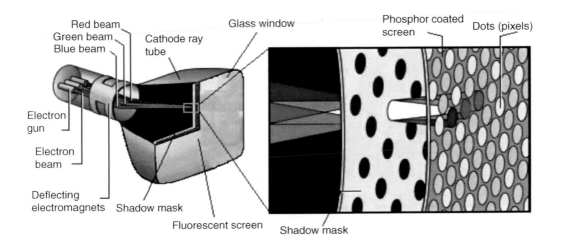

Red beam
Green beam
Blue beam

Cathode ray tube

Glass window

Phosphor coated screen

Dots (pixels)

Electron gun

Electron beam

Deflecting electromagnets

Shadow mask

Fluorescent screen

Shadow mask

FIGURE 6.11 The additive theory is used in colour television.

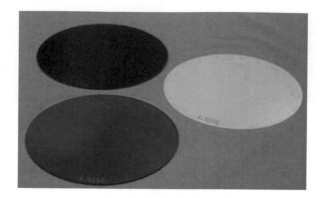

FIGURE 6.19 Subtractive colour filter set.

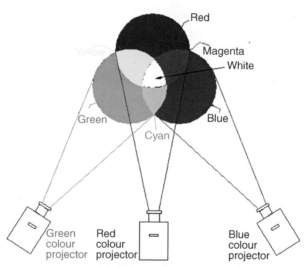

(a) Three slide projectors project lights onto the white cardboard.

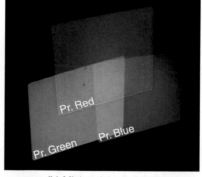

(b) Mixing three colours

FIGURE 6.22 Additive colour mixing. (a) Three slide projectors project lights onto the white cardboard. (b) Mixing three colours.

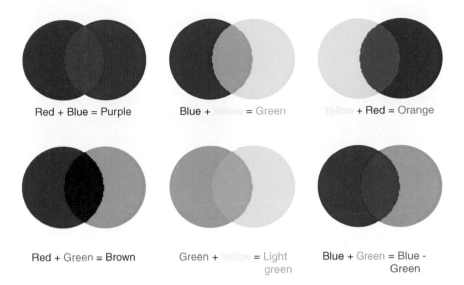

Red + Blue = Purple	Blue + Yellow = Green	Yellow + Red = Orange
Red + Green = Brown	Green + Yellow = Light green	Blue + Green = Blue - Green

FIGURE 6.23 Two colour additive mixing.

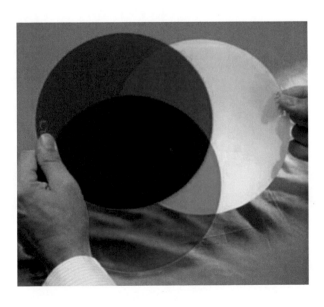

FIGURE 6.24 Subtractive colour mixing.

FIGURE 6.25 Newton's colour wheel apparatus set-up.

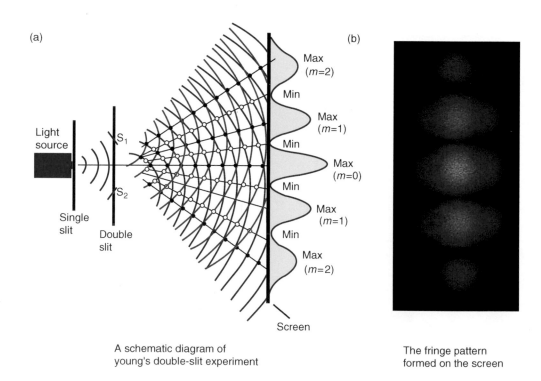

(a)

Max
(*m*=2)

Min

Max
(*m*=1)

Min

Max
(*m*=0)

Min

Max
(*m*=1)

Min

Max
(*m*=2)

Light
source

S₁

S₂

Single
slit

Double
slit

Screen

A schematic diagram of
young's double-slit experiment

(b)

The fringe pattern
formed on the screen

FIGURE 18.3 Young's double-slit experiment.

FIGURE 18.17 Diffraction patterns produced by a blade.

FIGURE 18.18 Diffraction patterns produced by a small disk.

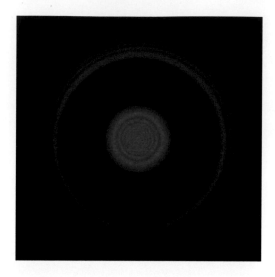

FIGURE 18.19 Diffraction patterns produced by a small washer.

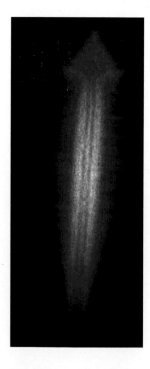

FIGURE 18.20 Diffraction patterns produced by an arrow shape.

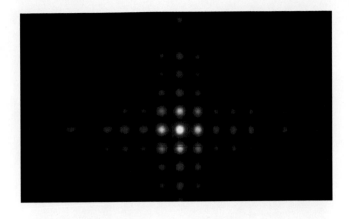

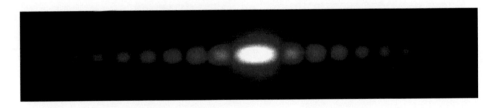

FIGURE 18.22 Diffraction patterns produced by a single-slit.

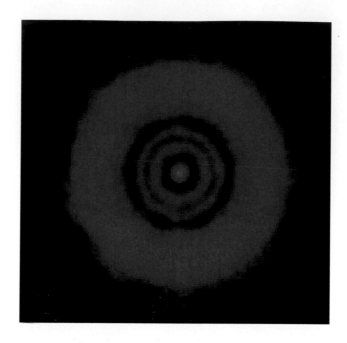

FIGURE 18.24 Diffraction patterns produced by a pinhole.

 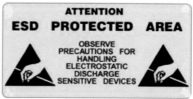

FIGURE 36.8 Electrostatic discharge warning symbols and signs.

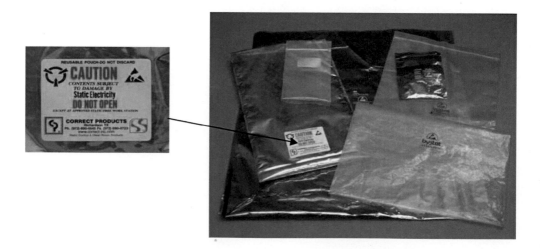

FIGURE 36.10 ESD bags.

ioerror
ING COMMAND: image

onary-

evel-

25 Fibre-Optic Cable Types and Installations

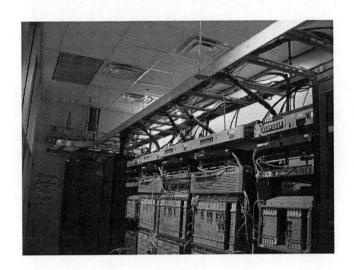

25.1 INTRODUCTION

Fibre-optic cables are significantly different from copper cables. Fibre-optic cables transmit data through very small cores over long distances at the speed of light. These cables come in a wide variety of configurations. Important considerations in any cable installation and operation are the bending radius; tensile strength; ruggedness; durability; flexibility; environmental conditions, such as temperature extremes; and even appearance. Due to fibre-optic cables' light weight and extreme flexibility, fibre-optic cables are more easily installed than their copper counterparts. They are easy to handle and they can be pulled through conduit and piping systems over long distances using various installation techniques. The minimum bend radius and maximum tensile loading allowed on a fibre-optic cable are critical during and after installation. A tensile load causes attenuation and may ultimately crack the fibre-optic core. The tensile loading allowed during installation is higher than the permissible loads during operation. The minimum bend radius allowed during installation is likewise larger than the bend radius allowed during operation.

This chapter presents the types of fibre-optic cables in current use and methods of installation. Testing fibre cables after installation is an important aspect of any communication system. Some kinds of hardware used in telecommunications systems are presented in detail. This chapter also introduces the student to fibre cable installation and the testing of fibre cable after the installation is completed.

25.2 FIBRE-OPTIC CABLE TYPES AND APPLICATIONS

There are many fibre-optic cable types and designs available. Fibre cables can be classified into the following categories: fibre-optic cables for indoor applications, those for outdoor applications, and for indoor/outdoor applications. The following sections explain each category in detail.

25.2.1 INDOOR FIBRE-OPTIC CABLE TYPES AND APPLICATIONS

Indoor cables (inside cable plant) are generally installed and operated in a controlled, stable environment. The cables must perform with minimal loss. Other factors that generate losses, such as environmental and mechanical stresses during and after installation, can cause failure. Outdoor cables (outside cable plant) have more factors that affect their performance. Indoor fibre cables are divided into the following types:

1. Simplex cables contain a single fibre cable, as shown in Figure 25.1. They are used for connections within equipment. They have a thicker outer jacket, which makes the cable easier to handle and adds mechanical protection.
2. Duplex cables contain two tight-buffered fibre cables inside a single jacket, as shown in Figure 25.2. They are used in equipment interconnections within workstations, test equipment, hubs and routers, etc.
3. Multi-fibre cables contain more than two fibre cables in one jacket, as shown in Figure 25.3. They have anywhere from three to several hundred fibre cables, all of which are colour-coded. They are used in vertical and horizontal fibre cable distributions between floors and telecommunication room.
4. Light cables, heavy cables, and plenum cables are application-dependent.
5. Breakout cables have several individual simplex cables.

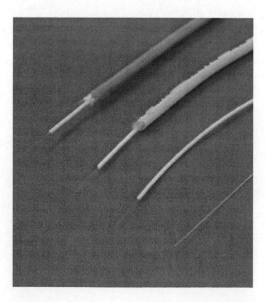

FIGURE 25.1 Simplex fibre cables.

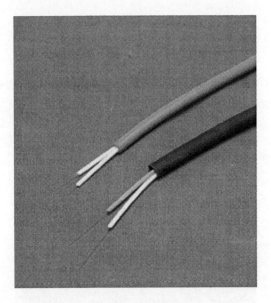

FIGURE 25.2 A duplex fibre cable.

25.2.2 OUTDOOR FIBRE-OPTIC CABLE TYPES AND APPLICATIONS

Outdoor cables (outside cable plant) must withstand a variety of environment and mechanical stresses during and after installation. The cables must perform with minimal losses over a wide range of temperature and humidity changes. The cable must also have waterproof capabilities, have strength to endure the difficult installation conditions, provide protection against ultraviolet radiation, and provide mechanical protection. There are many types and designs, depending on the manufacturer and application. Outdoor fibre cables are divided into the following types:

1. Overhead cables many be strung from telephone poles or along power lines.
2. Direct burial cables are placed directly in a trench dug in the ground and then covered by soil.

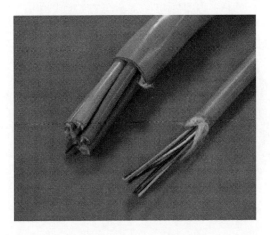

FIGURE 25.3 Multi-fibre cables.

3. Indirect burial cables are placed inside a duct or conduit system.
4. Submarine cables are laid underwater.

25.2.3 INDOOR/OUTDOOR FIBRE-OPTIC CABLE TYPES AND APPLICATIONS

There are many types of fibre cables that are used for indoor and outdoor applications. These cables use materials that enable them to pass the flame-retardant requirements of the indoor applications and provide reliable waterproof performance for outdoor applications. They are able to withstand difficult installation conditions and temperature variations. They are widely used within buildings and between buildings in campus applications.

FIGURE 25.4 A ribbon cable for indoor applications.

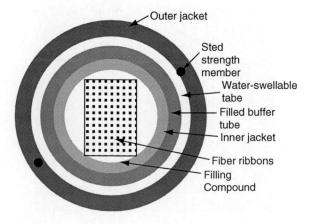

FIGURE 25.5 A cross-section of an outdoor ribbon cable.

25.2.4 Other Fibre-Optic Cable Types and Applications

One of the fibre cables in common use is the ribbon cable. Ribbon cables are made of many fibres, which are embedded in a plastic or PVC material in parallel, forming a flat, ribbonlike structure, as shown in Figure 25.4. The ribbon cable carries up to 12 fibre cables in a single ribbon. They are ideal for multi-fibre connector interconnect applications from equipment to a patch panel or as a patch cord.

Some ribbon cables consist up to 216 fibre cables, as shown in Figure 25.5. The cable consists of a single buffer tube that contains a stack of up to eighteen 12-fibre ribbons wrapped within a water-swellable foam tape and surrounded by a jacket. Dielectric or steel strength members located under the cable jacket provide tensile and mechanical strength. Some ribbon cables provide up to 864 fibre cables in a rugged, compact design to maximize the use of critical duct space. They are the ideal choice for maximizing usage of duct or pipe space and getting the service running faster than other cable types. They are ideal for outdoor applications and upgraded communication systems.

25.3 FIBRE-OPTIC CABLE INSTALLATION METHODS

This section describes some of the common fibre optic cable installation methods for inside and outside plants in local area networks, metropolitan area networks, and wide area networks.

25.3.1 Indoor Fibre-Optic Cable Installation

Generally, indoor cables are placed in conduits or trays. Since standard fibre-optic cables are electrically non-conductive, they may be placed in trays alongside high-voltage cables without the special insulation required by copper cabling. Plenum fibre-optic cables can be placed without special restrictions in any plenum area within a building.

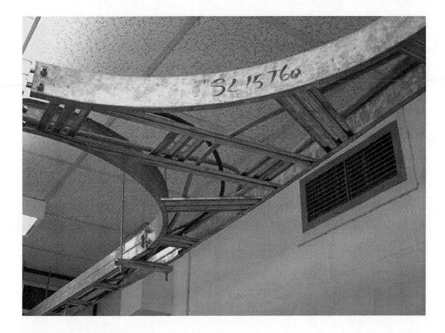

FIGURE 25.6 Fibre-optic cables in the tray system.

FIGURE 25.7 Heavy power cable crossing over the fibre-optic cable.

25.3.2 Cable Installation in Tray and Duct Systems

The primary consideration in selecting a route for fibre-optic cable (through trays and ducts) is the avoidance of potential cutting edges and sharp bends. Areas where particular caution must be taken include the corners and exit slots in the sides of the trays, as shown in Figure 25.6.

If a fibre-optic cable is in the same tray or duct as a very large and heavy electrical cable, care must be taken to avoid placing an excessive weight on the fibre-optic cable. Figure 25.7 illustrates such a case.

Cables in ducts and trays are not subjected to tensile forces. The tensile load must be considered when determining the minimum bend radius at the top of a vertical run. Long vertical runs must be clamped at intermediate points to prevent excessive tensile loading on the fibre cable. The clamping force should not exceed that which is necessary to prevent the possibility of slippage. Clamping forces are often determined experimentally, since the force is dependent on the type of clamping and jacket materials. The clamping force should be applied over as long a length of the fibre-optic cable as practical. Clamping surfaces should be made of a soft material, such as rubber or plastic. Tensile load during vertical installation is reduced by installing the fibre cable in a top–down manner.

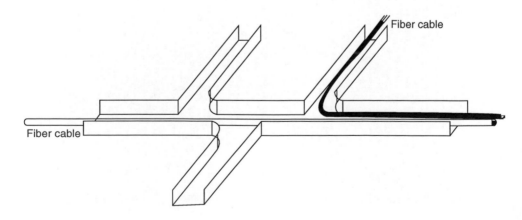

FIGURE 25.8 Wireway or tray with turn fittings.

25.3.3 Conduit Installation

Fibre-optic cables are pulled through a conduit (wireway or tray) by a wire or synthetic rope attached to the cable. Any pulling force must be applied to the cable strength member and not to the fibre-optic cable. In situations where the fibre-optic cable does not have connectors, the pull wire should be tied to the Kevlar strength member, or a pulling grip may be taped to the cable jacket or sheath. When determining the suitability of a conduit for a fibre-optic cable, the clearance between the conduit and any other cable is critical (described by a fill factor). Sufficient clearance must be provided, to allow the fibre-optic cable to be pulled through without excessive friction or binding. If the conduit must make a 90° turn, a fitting, such as that shown in Figure 25.8, must be used to allow the cable to be pulled in a straight line while avoiding any sharp bends in the cable.

25.3.4 Pulling Fibre-Optic Cable Installation

Fibre optic cables are pulled using many of the same tools and techniques that apply to wire cable installations. Departures from standard methods are required, since connectors are usually pre-installed on the fibre-optic cable, smaller pulling forces are allowed, and there are minimum bend radius requirements. The pull tape must be attached to the optical cable in such a way that the pulling forces are applied to the outer Kevlar layer. Connectors should be protected to prevent damage. The recommended method for attaching a pulling tape to a simplex cable is the Kellems grip, as shown in Figure 25.9. The connector should be wrapped in a thin layer of foam rubber and inserted in a plastic sleeve for protection. The fibre-optic cable grip should be stretched and then wrapped tightly with electrical tape to provide a firm grip on the fibre-optic cable. The duplex fibre-optic cable is supplied with Kevlar strength members extending beyond the outer jacket to provide a means of attaching the pulling tape. The Kevlar layer is attached with epoxy to the outer jacket and inner layers to prevent torsion while the cable is being pulled; the Kevlar is wrapped around the inner jacket in a helical pattern. The free ends of the Kevlar fibres are inserted into a loop at the end of the pulling tape and then epoxied back to themselves. The connectors are protected by foam rubber and a heat-shrink sleeving. The heat-shrink sleeving is clamped to the front of the steel ring in the pulling tape to prevent pushing the connectors along the rest of the fibre-optic cable. During an installation, the pulling force should be constantly monitored by a mechanical gauge. The pull tension can be monitored by a line tensiometer, a breakaway, or by using a dynamometer and pulley arrangement. If a power winch is used to assist the pulling, a power capstan with adjustable slip

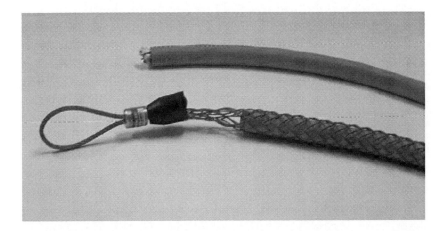

FIGURE 25.9 Kellems grip.

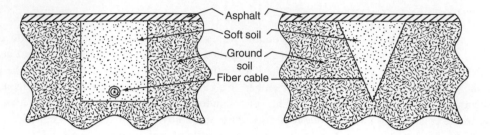

FIGURE 25.10 Fibre-optic cable direct burial methods.

clutch is recommended. The clutch, set for the maximum loading, will disengage if the set load is reached. The fibre optic cable should be continuously lubricated using gel or powder if necessary.

25.3.5 FIBRE-OPTIC CABLES DIRECT BURIAL INSTALLATION

Fibre-optic cables can be buried directly in the ground using either plowing or trenching methods, as shown in Figure 25.10. The plowing method uses a cable-laying plow, which opens the ground, lays the cable, and covers the cable in a single operation. In the trench method, a trench is dug with a machine (such as a backhoe), the cable is laid, and the trench is filled. The trench method is more appropriate for short-distance installations. Buried fibre-optic cables must be protected against frost, water seepage, attack by burrowing and gnawing animals, and mechanical stresses which

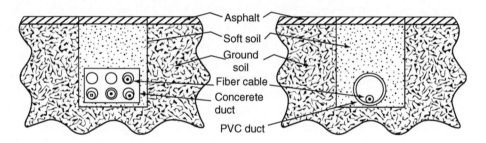

FIGURE 25.11 Fibre-optic cable indirect burial methods.

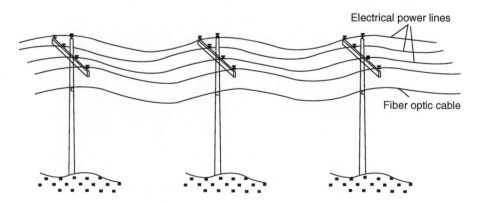

FIGURE 25.12 Fibre-optic cable aerial installation.

could result from earth movements. Armoured fibre-optic cables specially designed for direct and indirect burial are available. Cables should be buried so they are below the frost line. Other buried cables should be enclosed in sturdy concrete ducts and polyurethane or PVC pipes, as shown in Figure 25.11. The duct holes and pipes should have an inside diameter several times larger than the outside diameter of the fibre-optic cable to protect against earth movements. An excess length of fibre-optic cable in the duct or pipe prevents tensile loads from being placed on the fibre-optic cable.

25.3.6 FIBRE-OPTIC CABLE AERIAL INSTALLATION

Aerial installation includes stringing fibre-optic cables between telephone poles or along power lines, as shown in Figure 25.12. Unlike copper cables, fibre-optic cables may be run along power lines with no danger of inductive interference. Aerial fibre-optic cables must be able to withstand the forces of high winds, storms, ice loading, and birds. Self-supporting aerial cables can be strung directly from pole to pole. Other cables must be attached to a high-strength steel wire, which provides additional support. The use of a separate support structure by lashing is the preferred method.

25.3.7 AIR-BLOWN FIBRE CABLE INSTALLATION

There are many ways to upgrade or plan a new network system. Some network architects may choose a new and improved technology. The most recent technology is air-blown optical fibre invented by British Telecom (BT-London) and deployed in the early 1980s. This technology offers installation higher capacity, and better security over other technologies. The air-blown fibre process involves the deployment of tube cable in place of traditional inner ductwork, as shown in Figure 25.13. These tube cables contain several individual tube cells inside a protective outer jacket. The individual tube cells can be in different diameters. Once the tube cable is in place, fibre cable is blown through the tubes to various locations and terminated or interconnected. Fibre in different types (single or multimode) is blown into the network. The air-blown fibre system can work with any panels and connectors of traditional fibres. The technology can be used to provide pathways between buildings and cities. It also allows upgrade, growth, and change of high-speed voice, data, and video Local Area Networks.

25.3.8 OTHER FIBRE CABLE INSTALLATION METHODS

In 2001, Toronto became the first major city in North America to test-drive a system to lay fibre-optic cable in sewer pipes deep below city streets using a remote-controlled robot. With this new

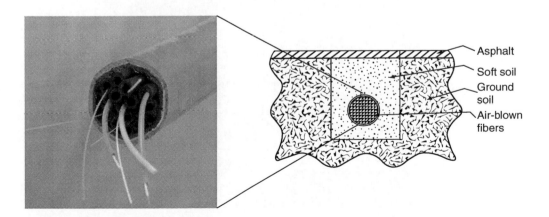

FIGURE 25.13 Air-blown fibre cable installation.

technology, the city of Toronto managed to upgrade and extend the communication networks without digging up the streets. The project uses air-blown cables installed by a robot. The robot navigated cables and performed all-at-once infrastructure installation. Connections between the points were achieved quickly and efficiently. This technology is planned for use in other cities in Canada.

Other cities caught on quickly, and now several cities worldwide enjoy the benefits of having their own fibre networks upgraded. Vienna and Berlin applied this technology to upgrade their network systems.

In 2003, the city of San Francisco, U.S.A., conducted a two-mile pilot project. The project enables San Francisco's Department of Telecommunications and Information Services to connect additional city facilities to E-Net. The project uses flexible cable, air-blown tubes, and robots to install the tubes through sewer clean-outs. Each tube houses up to 19 individual fibre tubes. Distribution boxes housing the individual fibre tube connections are strategically placed to avoid unnecessary splice points along the network. The project demonstrates the importance of using new technology to advance network systems.

25.4 STANDARD HARDWARE FOR FIBRE-OPTIC CABLES

This section describes some types of common fibre-optic hardware that are used in all telecommunication systems.

25.4.1 Fibre Splice Closures

There are different types of splice closures for fibre-optic cables. A splice closure is a standard piece of hardware in the telecommunication industry for protecting fibre-optic cable splices. Some small-closure types are for two to eight fibre mechanical splices and suitable for indoor/outdoor installation. Other closure types include splice trays and large splice closures for outdoor applications. Figure 25.14 shows a rack-mounted splice closure unit.

FIGURE 25.14 Rack-mounted splice closure unit.

FIGURE 25.15 Rack with panels.

25.4.2 RACK WITH PANELS

Consider building an application, such as a wiring centre, which will be used as a central distribution point, as shown in Figure 25.15. Fibre cables can be routed to their final destinations from the rack. An outdoor multi-fibre cable is brought (into the building) to the distribution rack, and then simplex or duplex fibre cables route the signals to different locations within the building.

25.4.3 CONNECTOR HOUSINGS

Connector housings are designed for main cross-connect, interconnect, intermediate cross-connects, for the local area network, and for data centre fibre distribution frames in telecommunication systems. Figure 25.15 shows a connector housing unit. They provide easy, open access to connectors for moves, adds and changes and for connector cleaning. They can easily be mounted on the racks or cabinets.

25.4.4 PATCH PANELS

Patch panels provide a convenient way to rearrange fibre cable connections and circuits, as shown in Figure 25.15. A simple patch panel is a metal frame containing bushings in which fibre-optic cable connectors plug in on either side. One side of the panel is usually fixed, meaning that the fibre cables are not intended to be disconnected. On the other side of the panel, fibre cables can be connected and disconnected to arrange the circuits as required. Patch panels are widely used in the telecommunication industry to connect circuits to transmission equipment. They can also be used in the telecommunication room of a building to rearrange connections. The splice organizer and the patch panel serve similar functions in distribution. The primary difference is that the organizer is intended for fixed connections, whereas a patch panel is used for flexible connections.

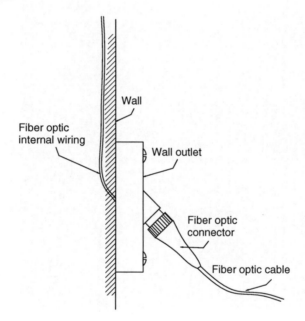

FIGURE 25.16 Wall outlet.

25.4.5 SPLICE HOUSINGS

Splice housings storage and protection of fibre splices in individually accessible trays. The splice trays are explained in the fibre optic connection chapter. They can easily be mounted on the racks for transition splicing between different cables at building entrance or pigtail splicing.

25.4.6 WALL OUTLETS

Fibre-optic wall outlets serve a similar function to electrical outlets, except that they allow connections to fibre cables carrying optical signals, as shown in Figure 25.16. In a building or office wired with fibre-optic cables, the outlet serves as a transition point between the horizontal cabling and the equipment. Wall outlets are designed to accommodate the fibre connectors at a 45° mating angle. This design avoids the fibre-optic cable bending at the back of the wall outlet. A short simplex or duplex cable, called a jumper or drop cable, typically runs from the wall outlet to the equipment being served.

25.4.7 FIBRE-OPTIC TESTING EQUIPMENT

Many types of fibre-optic testing equipment offer true multi-testing capability. They feature field-installable single-mode and multi-mode cables, optical time domain reflectometer modules, visual fault locator, optical return loss, fibre communicator, optical network simulator, built-in power meter, laser source, etc.

25.5 FIBRE-OPTIC CABLE TEST REQUIREMENTS

After the fibre cabling system is installed, it should be tested and certified that it meets performance specifications (TIA/EIA). Testing and certification have different implications when installing fibre-optic cables. Testing implies that certain values are measured, while certification, on the other hand, compares these measured values to standards or derived values to determine the values are within specified limits. Testing is a quantitative procedure, while certification is both quantitative and qualitative.

Attenuation tests at different operating wavelengths should be performed on all fibre-optic cables. The test results should be documented, complied with the standards, and should include the following information:

1. Fibre-optic cable ID.
2. Fibre-optic cable types.
3. Test equipment.
4. Testing and verification techniques.
5. Attenuation values.
6. Wavelengths.
7. Connector types.
8. Connector names.
9. Splice types and locations.
10. Reference setting at first wavelength.
11. Reference setting at second wavelength.
12. Building number and location.
13. Mapping and documentation requirements.
14. Relevant additional data.
15. Relevant additional comments.
16. Any special customer requirements.
17. Test results that must be supplied.
18. Date of test performed.
19. Contractors' names.
20. Technicians' names and signatures.
21. Any other requirements specified by the customer.

25.6 EXPERIMENTAL WORK

There are many types and applications of fibre-optic cable. The practical importance of the mechanical properties of fibre-optic cable is demonstrated here. The different types of common fibre-optic hardware, such as closures, organizers, rack boxes, distribution panels, and wall outlets, play an important role in telecommunication systems. The student will also study and practise using the

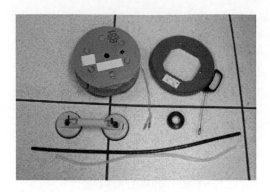

FIGURE 25.17 Fibre cable installation tools.

parts of the telecommunication system in the building. The student will become familiar with the correct manner of installing and testing a fibre-optic cable in a networking system.

25.6.1 TECHNIQUE AND APPARATUS

Appendix A presents the details of the devices, components, tools, and parts.

1. Figure 25.17 shows the tools that are required for this experiment.
2. Fibre-optic cable.
3. Fibre cable end preparation kit.
4. Fish wire.
5. Plastic tape/isolation tape.
6. Protective plastic spiral tube.
7. Vacuum handle for removing floor tiles.
8. Optical continuity checker.
9. Optical power meter and sensor.
10. Fibre cable measurement set.
11. Connector assembly kit.
12. Mechanical optical splice kit.
13. Fusion splice.

Note: Tools and parts listed in items 10–13 are listed in case they are needed to build a connector or splice as an adapter between the fibre-optic cable and the testing devices.

25.6.2 PROCEDURE

Follow the laboratory procedures and instructions given by the instructor.

25.6.3 SAFETY PROCEDURE

Follow all safety procedures and regulations regarding the use of fibre-optic cable, testing instrument, installation tools.

25.6.4 Apparatus Setup

25.6.4.1 Fibre-Optic Cable Installation

Choose the area where the cable is to be installed. Try to choose two types of cable installation procedures between two connections, such as an overhead cable tray system and an installation under raised floor tiles. The following steps will guide you throughout this experiment.

1. Locate the area and plan your fibre-optic cable installation in the Telecommunications Lab.
2. Determine the route that will be taken by the fibre-optic cable.
3. Ensure proper coiling of the fibre-optic cable. Arrange all of your cable in the correct entry position, using a spiral rubber tube to protect the fibre-optic cable from any sharp edges.
4. Remove every second tile from the floor, starting at the termination/distribution box along the fibre-optic cable raceways.
5. Properly attach the fibre-optic cable to the fish wire after passing it through the protective jacket in case of excessive bending.
6. Pull slowly and have someone assist with feeding the fibre-optic cable from the other end.
7. Proceed to pull the fibre-optic cable through the overhead cable tray using the fish wire.
8. Keep the fibre-optic cable neatly bundled, laying it on the floor away from the posts and other obstacles.
9. Carefully pull the fibre-optic cable through the tray and position it in the centre of the cable tray. Trim the excess fibre-optic cable, leaving sufficient service length of the cable for repair or maintenance.
10. Replace the tiles back into position, tidy up the site, and store the tools.
11. Terminate each cable end and make the required end connections.
12. Carry out the fibre-optic cable continuity test.
13. Conduct cable end connection tests.
14. Measure the fibre-optic cable power loss.

25.6.5 Data Collection

Record the power and losses measurements of the fibre cable.

25.6.6 Calculations and Analysis

Verify the power and loss measurements and compare to cable specifications.

25.6.7 Results and Discussions

Report the power and losses measurements with the cable specifications in charts or tables.

25.6.8 Conclusion

Summarize the important observations and findings obtained in this lab experiment.

25.6.9 SUGGESTIONS FOR FUTURE LAB WORK

List any suggestions for improvements using different experimental equipment, procedures, and techniques for any future lab work. These suggestions should be theoretically justified and technically feasible.

25.7 LIST OF REFERENCES

List any references that were used in the report. Use one format in writing the references. Never mix reference formats in a report.

25.8 APPENDIX

List all of the materials and information that are too detailed to be included in the body of the report.

FURTHER READING

Buckler, G., Air-blown fiber facts: Focus on engineering and design, *Cabling Systems*, 14–15, June/July 2003.

Boyd, W. T., *Fiber Optics Communications, Experiments & Projects*, 1st ed., Howard W. Sams & Co., Indianapolis, IN, 1987.

Chen, K. P., In-fiber light powers active fiber optical components, *Photonics Spectra*, 78–90, April 2005.

Cole, M., *Telecommunications*, Prentice Hall, Englewood Cliffs, NJ, 1999.

Derfler, F. J. Jr. and Freed, L., *How Networks Work*, Millennium edition, Que Corporation, IN, 2000.

Derickson, D., *Fiber Optic Test and Measurement*, Prentice Hall PTR, Upper Saddle River, NJ, 1998.

Dutton, H. J. R., *Understanding Optical Communications*, IBM, Prentice Hall, Englewood Cliffs, NJ, 1998.

Goff, D. R. and Hansen, K. S., *Fiber Optic Reference Guide: A Practical Guide to the Technology*, 2nd ed., Butterworth-Heinemann, London, 1999.

Golovchenko, E., Mamyshev, P. V., Pilipetskii, A. N., and Dianov, E. M., Mutual influence of the parametric effects and stimulated raman scattering in optical fibers., *IEEE Journal of Quantum Electronics*, 26, 1815–1820, 1990.

Groth, D., Barnett, D., and McBee, J., *Cabling—the Complete Guide to Network Wiring*, 2nd ed., Sams & Co., Indianapolis, IN, 2001.

Hecht, J., *Understanding Fiber Optics*, 3rd ed., Prentice Hall, Inc., Englewood Cliffs, NJ, 1999.

Herrick, C. N. and McKim, C. L., *Telecommunication Wiring*, 2nd ed., Prentice Hall, Inc., Englewood Cliffs, NJ, 1998.

Hioki, W., *Telecommunications*, 3rd ed., Prentice Hall, Englewood Cliffs, NJ, 1998.

Hitz, B., Photonic bandgap fiber eyed for telecommunications, *Photonics Spectra*, 122–124, April 2004.

Hoss, R. J., *Fiber Optic Communications—Design Handbook*, Prentice Hall Publishing Co., Englewood Cliffs, NJ, 1990.

Kao, C. K., *Optical Fiber Systems: Technology, Design, and Applications*, McGraw-Hill, New York, 1982.

Keiser, G., *Optical Fiber Communications*, 3rd ed., McGraw-Hill, New York, 2000.

Keiser, G., *Optical Communications Essentials*, 1st ed., McGraw-Hill, New York, 2003.

Kolimbiris, H., *Fiber Optics Communications*, Prentice Hall, Englewood Cliffs, NJ, 2004.

Mynbaev, D. K. and Scheiner, L. L., *Fiber-Optic Communication Technology—Fiber Fabrication*, Prentice Hall PTR, Upper Saddle River, NJ, 2005.

Palais, J. C., *Fiber Optic Communications*, 4th ed., Prentice Hall, Inc., Englewood Cliffs, NJ, 1998.

Pease, B., When it comes to optical fiber installations, some companies are really "blowing it", *Fiber-Optic Product News*, 12, May 2004.

Razavi, B., *Design of Integrated Circuits for Optical Communications*, McGraw-Hill, New York, 2003.

Shotwell, R. A., *An Introduction to Fiber Optics*, Prentice Hall, Englewood Cliffs, 2NJ, 1997.

Sterling, D. J. Jr., *Technician's Guide to Fiber Optics*, 2nd ed., Delmar Publishers Inc., Albany, NY, 1993.

Sterling, D. J. Jr., *Premises Cabling*, Delmar Publishers, Albany, NY, 1996.

Vacca, J., *The Cabling Handbook*, Prentice Hall, Inc., Englewood Cliffs, NJ, 1998.

Yeh, C., *Handbook of Fiber Optics: Theory and Applications*, Academic Press, San Diego, CA, 1990.

26 Fibre-Optic Connectors

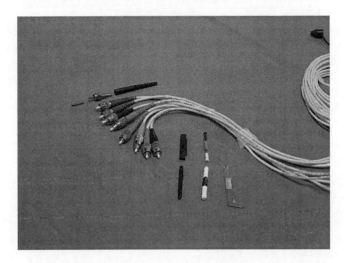

26.1 INTRODUCTION

The interconnection of optical components is a vital part of an optical system, having a major effect on performance. Interconnection between two fibre-optic cables is achieved by either connectors or splices which link the ends of the fibre cables optically and mechanically.

Connectors are devices used to connect a fibre-optic cable to an optical fibre device, such as a detector, optical amplifier, optical light power meter, or link to another fibre cable. They are designed to be easily and reliably connected and disconnected. The connectors create an intimate contact between the mated halves to minimize the power loss across the junction. They are appropriate for indoor applications. Splices are used to permanently connect one fibre-optic cable to another. Splices are suitable for outdoor and indoor applications. Some types of splices are used to temporarily connect for quick testing purposes.

The key to a fibre-optic interconnection is precise alignment of the mated fibre cable cores so that the light couples from one fibre, across the junction, into the other fibre. This precise alignment creates a challenge for designers.

There is a difference between a connection of two fibre cables and a coupling of a light source into a fibre cable. Coupling techniques are explained in the fibre-optic cables chapter. This chapter presents the operating principles of the connectors and splices, and describes their types, properties, and operations.

This chapter also presents four experimental cases for building a connection between two fibre-optic cables; linking two fibre cables by connectors and a fusion splice; and testing the connection for losses.

26.2 APPLICATIONS OF CONNECTORS AND SPLICES

Connectors and splices make optical and mechanical connections between two fibre cables. It is easy to connect and disconnect a cable with a connector from another cable or a device. There are many applications for fibre connectors and splices in fibre systems, such as:

- Connecting between a pair of fibre cables, using connectors or a splice, is an essential part of any fibre system.
- Interfacing devices to local area networks.
- Connecting and disconnecting fibre cables to patch panels where signals can be checked and routed in a fibre system.
- Connecting and splicing may be required on short fibre cables for wiring, testing devices, connecting instruments and devices, and at other intermediate points between transmitters and receivers.
- Dividing a fibre system into subsystems, which simplifies the selection, installation, testing, and maintenance of fibre systems.
- Temporarily connecting remote mobile systems and recording equipment in many fibre systems.

26.3 REQUIREMENTS OF CONNECTORS AND SPLICES

It can be very difficult to design a connector or a splice that meets all the requirements. A low-loss connector may be more expensive than a high-loss connector, or it may require relatively expensive application tooling. The lowest losses are desirable, but the other factors clearly influence the selection of the connector or splice as well.

The following is a list of the most desirable features for fibre connectors or splices required by customers and industry:

- Low loss (insertion and return)
    The connector or splice causes low loss of optical power across the junction between a pair of fibre cables.
- Easy installation and use
    The connector or splice should be easily and rapidly installed without the need for special tools or extensive training.
- Repeatability
    There should be no variation in power loss. Loss should be consistent whenever a connector is connected, disconnected and reconnected again, as many times as required.
- Economical
    The connector, splice, and special application tooling should be inexpensive.
- Compatibility with the environment
    The connector or splice should be waterproof and not affected by temperature variations.
- Mechanical properties
    The connector or splice should have high mechanical strength and durability to withstand the application and tension forces.
- Long life
    The connector or splice should be built with a material that has a long life in various applications.

26.4 FIBRE CONNECTORS

Fibre connectors are designed to be easily connected and disconnected. Fibre-optic cable can be easily connected to a transmitter, receiver, power meters, or another fibre cable. The key optical parameter for a fibre-optic connector is its attenuation. Signal attenuation in connectors is the sum of losses caused by several factors. The major factors are as follows:

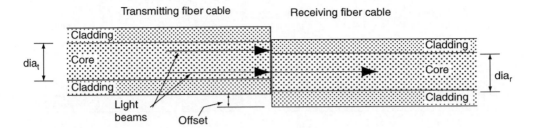

FIGURE 26.1 Overlap of fibre cable cores.

- Overlap of fibre cable cores (also called lateral displacement)
- Alignment of fibre axes
- Fibre cable numerical aperture
- Reflection at the fibre cable junction/interface
- Connector end polishing
- Fibre cable spacing
- Connector end face profiles
- Insertion loss

When the diameter of the transmitting fibre cable is greater than that of the receiving fibre cable, as shown in Figure 26.1, the diameter-mismatch loss (Loss $_{\text{dia}}$) is given by:

$$\text{Loss}_{\text{dia}} 10 \log_{10} \frac{(\text{dia}_t^2 - \text{dia}_r^2)}{\text{dia}_t^2} \tag{26.1}$$

When the numerical aperture NA of the transmitting fibre cable is greater than that of the receiving fibre cable, as shown in Figure 26.2, the NA-mismatch loss (Loss$_{\text{NA}}$) is given by:

$$\text{Loss}_{\text{NA}} = 10 \log_{10} \left(\frac{\text{NA}_r}{\text{NA}_t}\right)^2 \tag{26.2}$$

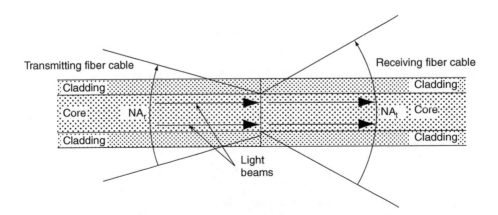

FIGURE 26.2 Numerical aperture (NA)-mismatch loss.

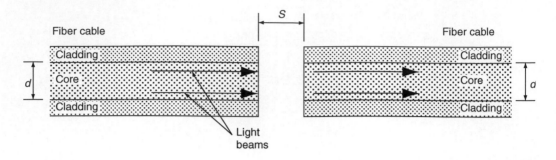

FIGURE 26.3 End separation loss.

The formula for the loss due to end separation (Loss$_{\text{Separation}}$) between two fibre-optic cables (at separation distance) is rather involved. Assume that the transmitting and receiving fibres are identical. Figure 26.3 illustrates the separation (some times called air gap) between a pair of fibre cables. The formula for end separation loss (Loss$_{\text{Separation}}$) is given by:

$$\text{Loss}_{\text{Separation}} = 10 \log_{10} \left(\frac{\frac{d}{2}}{\frac{d}{2} + S \tan(\arcsin)\left(\frac{\text{NA}}{n_0}\right)} \right) \tag{26.3}$$

where, d is core diameter, S is the fibre spacing, NA is the numerical aperture, n_0 is the refractive index of the material between the two fibre cables.

A material known as index-matching fluid or gel applied between the two fibre cables reduces the reflection loss between the surfaces of the fibre cable ends. This loss, called Fresnel reflection loss, generally occurs between parallel optical surfaces. Most mechanical splices also use an index-matching gel to fill the gap between the connected fibre cable ends. An antireflection coating can also be applied to reduce this loss.

Additional losses may be experienced when two different types of fibre cable connectors are connected using an adapter. You will find more details in the section dealing with adapters for different fibre connector types.

Insertion loss is a measure of the performance of a connector or splice. Insertion loss is calculated by:

$$\text{Loss(dB)} = 10 \log_{10} \frac{P_2}{P_1} \tag{26.4}$$

where P_1 is the initial power measured and P_2 is the power measured after the connector has been mated.

26.5 MECHANICAL CONSIDERATIONS

The optical characteristics of the fibre connectors are significant. However, the mechanical characteristics are also important, and in some cases critical. Virtually all fibre connectors are designed to remain in place under working conditions. Connectors must withstand physical stresses, such as forces encountered during mating and unmating, and sudden stress induced by bending and tension. Connectors must also prevent contamination caused by dirt and moisture in the fibre cable ends.

26.5.1 DURABILITY

Durability is a concern with any type of connector. Repeated mating and unmating of the fibre connectors can cause wear in the mechanical components. Allowing dirt into the optics and straining the fibre cable will damage the exposed fibre cable ends.

26.5.2 ENVIRONMENTAL CONSIDERATIONS

Fibre connectors designed for indoor applications must be protected from environmental extremes to avoid excessive connector loss and poor system performance. Special hermetically-sealed connectors are required for outdoor use.

26.5.3 COMPATIBILITY

Compatibility refers to the need for the connector to be compatible with other connectors or with specifications. Specifications describe the type of connector to be used in specific applications. Compatibility exists on several levels the most basic level being physical compatibility. The connector must meet certain dimensional requirements to allow it to mate with other connectors of the same style. The second level of compatibility involves connector performance, such as insertion loss, durability, operating temperature range, and other requirements specified by the customers.

26.6 FIBRE-OPTIC CONNECTOR TYPES

Figure 26.4 shows the most common types of fibre-optic connectors. Fibre connectors are unique in that they must make both optical and mechanical connections. They must also allow the fibre cables to be precisely aligned to ensure that a connection is robust. Fibre connectors use various methods to achieve solid connections. Some of the types of fibre-optic connectors currently in use are listed below:

- Subscriber Connectors (SC)
- SC/APC Connectors
- FC/PC Connectors

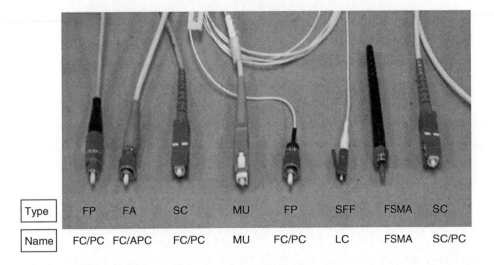

FIGURE 26.4 Fibre-optic connector types.

- FC/APC Connectors
- LC Connectors
- MU Connectors
- Straight-tip Connectors (ST)
- 5685C Connectors (duplex SC)
- FDDI Connectors (MIC)
- Biconic Connectors
- SMA Connectors
- Enterprise System Connection (ESCON)
- Duplex Connectors (ST)
- Polarizing Connectors
- MT Multifibre Connectors
- MT-RJ Connectors
- D4-style Connectors
- Biconic Connectors
- MFS/MPO Connectors
- Plastic-Fibre Connectors
- E-2000 Diamond
- Fibre-Optic Connectors Self-Latch in Push/Pull System
- Special Connectors

26.7 ADAPTERS FOR DIFFERENT FIBRE-OPTIC CONNECTOR TYPES

An adapter is a passive device used to join two different types of connectors together. The type of adapter is identified by a nomenclature, such as SC, FC, ST, or 568SC. Hybrid adapters join dissimilar connectors together, such as SC to FA. Figure 26.5 shows examples of some adapters.

FIGURE 26.5 Fibre-optic adapter types.

26.8 FIBRE-OPTIC CONNECTOR STRUCTURES

Most fibre-optic connectors are built from a ferrule, a connector body, an epoxy material, and a strain relief boot. Most connectors use a ferrule to hold the fibre and provide alignment. The most popular ferrule size is a 2.5-mm diameter, which is standard. Manufacturers offer a few types of ferrules made from different materials, such as ceramic, plastic, and stainless steel.

Connectors may be attached to a device–outlet box or adapter—by direct connection, by coupling a threaded nut, or by twisting a spring-loaded bayonet socket. The connector body is made from steel, ceramic, or plastic. Epoxy is usually applied to secure the fibre cable end in the connector body. A strain relief boot made from plastic or rubber is used at the junction between the connector body and the fibre cable.

26.9 FIBRE-OPTIC CONNECTOR ASSEMBLY TECHNIQUES

The following sections present common assembly techniques that are used in building fibre-optic connectors.

26.9.1 COMMON FIBRE CONNECTOR ASSEMBLY

The most common fibre connector assembly techniques use a fibre cable and a suitable connector. The fibre cable is most often epoxied into the connector. Epoxy provides good tensile strength to the connector to prevent the fibre cable from moving within the connector body, maintaining a good alignment. After the epoxy cures, the ferrule end is polished to a smooth finish by one of the many available procedures. Then the connector undergoes many inspections and test procedures to issue a data sheet for the customer.

26.9.2 HOT-MELT CONNECTOR

Hot-melt connectors are widely used in North American telecommunication systems. The hot-melt connectors use preloaded epoxy so that external mixing is not required. The prepared end of the fibre-optic cable is inserted into the connector ferrule, as shown in Figure 26.6. The cable (with the connector inserted) is loaded onto the connector holder and placed in an oven for a few minutes, which softens the epoxy around the fibre cable and cures the epoxy at the same time. The curing time is dependent on the type of epoxy. The end of the connector is then polished to a smooth finish. The polishing can be done by hand or by an industrial polishing machine. When such connectors are assembled in the field, a portable hand polisher is used.

26.9.3 EPOXYLESS CONNECTOR

Epoxyless connectors, also called crimp connectors, have been widely used for quick cable connections in telecommunication systems. When the connector is crimped, an insert compresses around the fibre cable. A front clamp on the bare fibre cable and a rear compression clamp add a higher clamping force on the fibre cable buffer coating to provide the necessary tensile strength. Special gripping tools are used in the assembly of the epoxyless connectors. The end of the fibre connector is polished to a smooth finish using a portable hand polisher before the connector is assembled in

FIGURE 26.6 Hot-melt connectors assembly.

the field. The main advantage of an epoxyless connector is the speed of assembly. Some customers will tolerate a slightly higher loss to achieve a fast, easy termination. The epoxyless approach is a technology that is not limited to one connector type.

26.9.4 AUTOMATED POLISHING

All fibre-optic connector-polishing machines are designed for accuracy, easy setup, and production efficiency. The polishing pressure, speed, and duration can be adjusted to meet exact requirements. These machines precisely polish the ends of fibre-optic connectors in a repeatable and reliable manner. Polishing machines are available for dry or wet polishing process.

26.9.5 FLUID JET POLISHING

Fluid jet polishing (FJP) is another technique for shaping and polishing small surface areas of complicated optics made of brittle materials. This technique uses a fluid jet system to guide pre-mixed slurry, at low pressures, onto the optical surface being machined. The surface is altered by the erosive effect of the abrasive particles in the stream.

26.9.6 FIBRE-OPTIC CONNECTOR CLEANING

Contamination of connector ends can occur from something as simple as dust particles or fingerprints that can reduce light propagation through the fibre cable. This will degrade device performance, causing data error and loss. To avoid this, it is common practise to clean fibre connectors prior to assembly and testing.

There are three major components of the fibre-optic connector system that users must consider when cleaning, mating, and testing fibre-optic connectors: the adapter split sleeve, the outer diameter of the ferrule, and the tip of the ferrule. There are many techniques for cleaning connectors, either wet or dry, by hand with recommended cleaning chemicals, or with automated machines. Follow the cleaning procedure for each fibre connector type. Do not use the same procedure for other types of connectors. Cleaning standards for fibre-optic connectors promise savings in time and cost.

26.9.7 CONNECTOR TESTING

There are many testing instruments available for testing connectors. Testing instruments range from a simple view scope to a sophisticated system. The condition of the end of the ferrule after the polishing process is usually inspected using simple instruments, as shown in Figure 26.7. This procedure is adequate for inspecting a connector built and polished in the field.

Handheld devices can measure the losses, optical powers, light sources, etc. The basic test measures the attenuation of the fibre cable with connectors by comparing the power through the fibre cable to that of a known reference fibre cable. The power through the fibre cable under test is measured in absolute units. The power through the reference fibre cable is also measured. Figure 26.8 illustrates a connector test set-up.

Using sophisticated systems for testing connectors saves time and cost in industrial production. These systems are very accurate. Each connector type refers to a standard test. The preferred test methods are compliant with the commercial building telecommunications cabling standard, TIA/EIA-568-B.1. The ANSI/TIA/EIA standards group developed intermateability standards for several connector types to ensure compatibility among manufacturers.

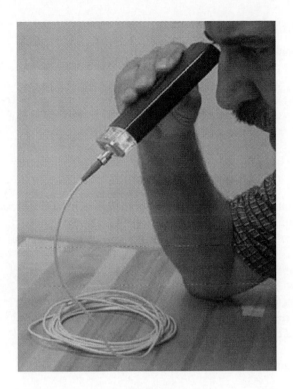

FIGURE 26.7 Connector end inspection with a view scope.

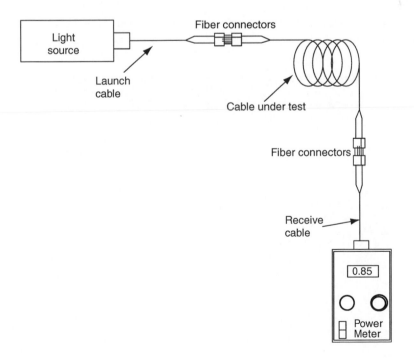

FIGURE 26.8 Connector testing set-up.

26.10 FIBRE SPLICING

The splicing process joins fibre-optic cable ends permanently. In general, a splice has a lower loss than a connector. Splices are typically used to join lengths of cable for outside applications. Splices may be incorporated into lengths of fibre-optic cable or housed in indoor/outdoor splice boxes, whereas connectors are typically found in patch panels or attached to equipment at fibre cable interfaces. The sources of loss, described in the section on fibre cable connectors, are also applicable to splices. There are two types of splices: mechanical and fusion.

26.10.1 MECHANICAL SPLICING

Mechanical splices join two fibre cable ends together both optically and mechanically by clamping them within a common structure. In general, mechanical splicing requires less expensive equipment; however, higher consumable costs are experienced. Figure 26.9 shows the most common types of fibre-optic mechanical splices. A few important types of mechanical splices are listed below:

- Table-type splices
- Key lock slices
- Fibre lock splices
- Twist lock splices
- Fastomeric splices
- Capillary splices
- Rotary or polished-ferrule splices
- V-groove splices
- Elastomeric splices
- Finger splices
- Inner lock splices

Many other types are available.

Some mechanical and fusion splices are used with one type of splice closure. Figure 26.10 shows small and large splice closures. Splice closures are standard pieces of hardware in the

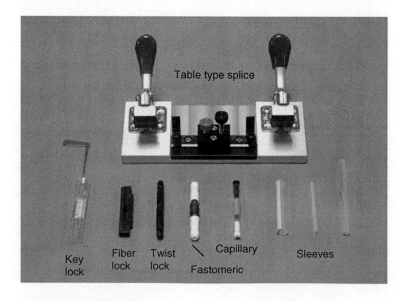

FIGURE 26.9 Some fibre-optic splice types.

(a) A small closure

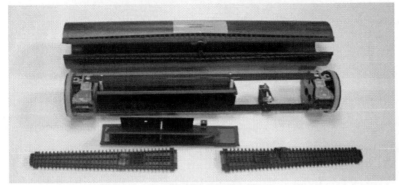

(b) A large closure

FIGURE 26.10 Splice closures: (a) a small closure, (b) a large closure.

telecommunication industry for protecting fibre-optic cable splices. Splices are protected mechanically and environmentally within the sealed closure. Splice closures are waterproof. Water is kept out by using non-flowing gel under permanent compression. They are suitable for indoor, outdoor, and underground cable system installations. There are small and large closures available for different applications.

Splice trays are designed to hold fusion and mechanical splices, as shown in Figure 26.11. They are available in different sizes. Fusion and mechanical splices are held in a specially-designed splice organizer and splice holder, respectively. They are not sealed off environmentally. These trays are installed in a wall-mounted fibre splice cabinet in a communication system.

The following sections explain temporary mechanical splices that are used in different cases of the testing.

FIGURE 26.11 Splice trays.

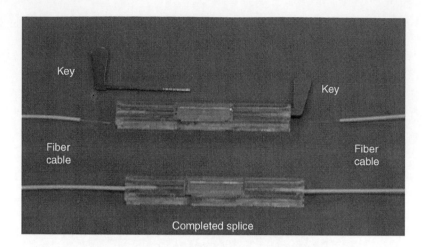

FIGURE 26.12 Key-lock mechanical fibre-optic splice.

26.10.1.1 Key-Lock Mechanical Fibre-Optic Splices

Figure 26.12 shows the key-lock mechanical fibre-optic splice, commonly used to quickly mate and unmate fibre optic cables. It is made from a U-shaped metal part covered by a transparent plastic body with two holes on each end. The prepared ends of the fibre cables are made longer than half of the length of the metal part. The fibre cable is inserted in the centre hole. When the key is inserted in the second hole towards the edge of the splice and turned by 90°, the metal part opens and one fibre cable end can be easily inserted. This operation can be repeated on the other side to insert the second fibre cable. This type of splice provides a quick and easy way of joining two fibre cables with low signal loss. It may be used to temporarily or permanently connect fibre cables, wavelength division multiplexing components, and other fibre-optic elements.

26.10.1.2 Table-Type Mechanical Fibre-Optic Splices

Figure 26.13 shows a custom-made mechanical splice, used for quick mating and unmating of connections. This splice works like any other mechanical splice. The fibre cable ends are prepared and inserted into the mid-point of the block assembly. Screws are tightened to align the fibre cables

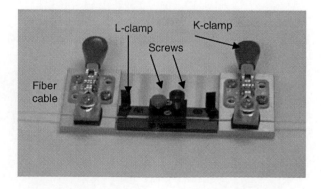

FIGURE 26.13 Table-type mechanical fibre-optic splice.

on both sides. L-clamps and K-clamps are placed in position to secure the fibre cables on both sides. Most fibre-optic companies use this kind of mechanical splice for quickly mating and unmating during manufacturing and testing processes. The splice loss associated with these instruments is acceptable by industry standards.

26.11 FUSION SPLICES

Fusion splicing is performed by placing the tips of two fibre cables together and heating them by a fast electrical fusion process so that they melt into one piece. Fusion splices automatically align the two fibre cable ends and apply a spark across the tips to fuse them. They also include instrumentation to test the splice quality and display optical parameters pertaining to the join. A fusion splice is shown in Figure 26.20. When the fusion splice is completed, a cylindrical fusion protector is placed over the splice location. Fibre fusion protectors are made from metal or polymer, and they are applied to ensure mechanical strength and environmental protection. Some types of fusion splice protectors (sleeves), as shown in Figure 26.9, are designed for use in place of the heat shrink method for fast, easy, and reliable permanent installations. Fusion splices provide lower loss than mechanical splices.

26.11.1 SPLICE TESTING

Attenuation can be measured as the splice is being performed. Many fusion splicers come with instrumentation that measure the loss as the fibre cables are being aligned, and test the loss when the splice is completed.

26.12 CONNECTORS VERSUS SPLICES

There are definite differences between connectors and splices. Most companies make connectors and splices to satisfy customer requirements for smaller size and lower loss. As mentioned in this chapter, fibre-optic technology is moving forward to create high-durability connectors and splices with small sizes and low cost in production and installation. Table 26.1 compares general important factors between the fibre connectors and splices.

TABLE 26.1
Connectors versus Splices

Connectors	Splices
Provide temporary connections	Provide permanent connections
Higher loss	Lower loss
Larger sizes	Smaller sizes
Immune, or not immune, to environmental effects (depends on the connector type)	Immune to environmental effects
It takes a long time to build a connector	It takes a very short time to build a splice
Diverse applications	Connection between a pair of fibre cables
Many types	Few types
New technology reduces installation time	Conventional technology keeps the same installation time
Building reasonable mechanical stability at the connection points	Building better mechanical stability at the connection points

26.13 EXPERIMENTAL WORK

The purpose of this experiment is to build a connector on a fibre cable end. Two fibre cables can be connected by two connectors with a sleeve or by a mechanical splice or by a fusion splice. The student will terminate the fibre-optic cable using an FSMA connector. The student will also be required to conduct mechanical and fusion splices and perform tests on the cable connectors and splices. Losses within the connectors, mechanical splices, and fusion splices will be measured.

In this experiment the student will perform the following cases:

a. Building a connector
b. Connecting using two connectors
c. Connecting by a mechanical splice
d. Connecting by a fusion splice
e. Testing connection loss in two connectors
f. Testing connection loss in a mechanical splice
g. Testing connection loss in a fusion splice

26.13.1 TECHNIQUE AND APPARATUS

Appendix A presents the details of the devices, components, tools, and parts.

1. 2×2 ft. optical breadboard.
2. HeNe laser light source and power supply.
3. Laser light sensor.
4. Laser light power meter.
5. Laser mount assembly.
6. $20\times$ microscope objective lens.
7. Lens/fibre cable holder/positioner assembly.
8. Fibre cable holder/positioner assembly.
9. Hardware assembly (clamps, posts, screw kits, screwdriver kits, lens/fibre cable holder/-positioner, sundry positioners, brass fibre cable holders, fibre cable holder/positioner, etc.).
10. Fibre cable end preparation procedure and kit, and cleaning kit, as explained in the end preparation section of the fibre-optic cables chapter.
11. Connector holder/positioner assembly.
12. Black/white card and cardholder.
13. Water spray bottle.
14. 50%/50% mixing epoxy.
15. Fibre optic cable, 900 micrometers diameter, 500 meters long.
16. Polishing disk.
17. FSMA connector kit, as shown in Figure 26.15.
18. Polishing pads, as shown in Figure 26.15.
    Size 63.0 µm GREY Colour,
    Size 9.0 µm BLUE Colour,
    Size 1.0 µm VIOLET Colour, and
    Size 0.3 µm WHITE Colour.
19. Key-lock mechanical fibre-optic splicing unit, as shown in Figure 26.9 and Figure 26.12.
20. Table-type mechanical fibre-optic splice unit, as shown in Figure 26.9 and Figure 26.13.
21. Fusion splicing machine, as shown in Figure 26.20.
22. Ruler.

26.13.2 Procedure

Follow the laboratory procedures and instructions given by the instructor.

26.13.3 Safety Procedure

Follow all safety procedures and regulations regarding the use of fibre-optic cable. You must wear safety glasses and finger cots or gloves when working with and handling fibre-optic cables, optical components, and optical cleaning chemicals.

26.13.4 Apparatus Setup

26.13.4.1 Case (a): Building FSMA Connectors

The students will assemble a FSMA connector in this experiment. The structure of this connector is shown in Figure 26.14.

Connector Body

The connector body is a continuation of the ferrule, as shown in Figure 26.14. The connector body accommodates the strain relief boot. The connector is attached to an adapter or a device by a threaded coupling nut.

Epoxy and Polish

A fibre cable is often epoxied to a connector. Because of its curing time, epoxy is generally considered to be an undesirable but necessary process in fibre-optic termination. After the epoxy cures, the ferrule end is polished to a smooth finish.

Strain Relief Boot

A black polymer strain relief boot shields the junction of the connector body and the fibre cable.
Figure 26.15 shows the connector assembly kit. Building and testing a connector involves the following steps:

1. Cut two four-foot lengths of fibre cables from the spool.
2. Strip, cleave, and clean the fibre cable ends so that there is about 1 to $1\frac{1}{4}$ inch of bare fibre extending beyond the jackets. Follow the fibre cable end preparation procedure as explained in the fibre-optic cables chapter.
3. Prepare a connector for the assembly process (the connector consists of two parts: ferrule-connector body and strain relief boot).

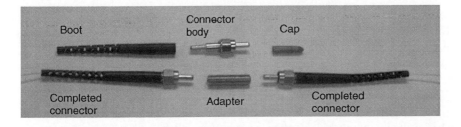

FIGURE 26.14 FSMA connectors.

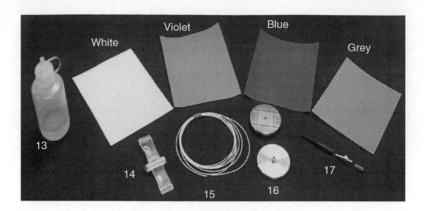

FIGURE 26.15 Connector assembly kit.

4. Insert the prepared end of the fibre cable into the connector until the fibre cable stops and the bare fibre emerges from the ferrule.
5. Prepare the epoxy according to the epoxy manufacturer's preparation procedure.
6. Pull the fibre cable ½ inch backwards; then, with a toothpick or small probe, apply the epoxy onto the fibre cable. Push and pull the cable (back and forth), until the epoxy adequately fills the gap between the fibre cable and the ferrule. The quantity of epoxy must be sufficient to support the fibre cable inside the connector body.
7. Insert the fibre cable into the ferrule completely.
8. Using a small probe, place a very small drop of epoxy onto the ferrule face where the fibre cable exits. This will seal off the space between the ferrule face hole and the fibre cable. The size of this drop is important; too large an epoxy drop will extend the polishing time.
9. Set the connector aside for the specified time to allow the epoxy to fully cure. Curing time depends on the type of epoxy.
10. Insert the fibre cable through the strain relief boot, positioning the boot on the connector body.
11. Cleave the protruding fibre flush with the ferrule face. Remove the protruding piece of fibre and dispose of it in the designated container.
12. Place the size 63.0 μm GREY colour polishing pad on a flat and clean surface, such as a piece of glass sheet.
13. Screw the connector gently onto the polishing disk until finger-tight. Do not over-tighten.
14. Place several small drops of water on the polishing pad. Begin the polishing process by moving the polishing disk with your fingers in a figure 8 motion, as shown in Figure 26.16.
15. Polish until the epoxy bead and excess fibre cladding are removed and the fibre cable is flush with the ferrule face surface.
16. Apply only light pressure on the polishing disk, using enough water to keep the polishing pad and disk clean; apply constant motion.
17. About 20–30 motions should be sufficient to complete the rough polishing.
18. Rinse the polishing disk with water.
19. Inspect the ferrule face using an inspection microscope. Look at the quality of the polished ferrule face.
20. Repeat steps 14–19 using polishing pads of size 9.0 μm (BLUE colour), size 1.0 μm (VIOLET colour), and size 0.3 μm (WHITE colour).
21. About 20–30 motions should be enough to achieve a good quality finish on the ferrule face when using the white pad. Caution: do not over-polish.
22. Inspect the finished connector using the inspection microscope.

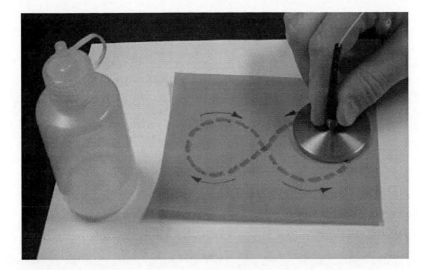

FIGURE 26.16 Polishing connector by hand.

23. Once the two connectors are completed, test the connectors to measure the connection loss, which is calculated using Equation (26.4).

26.13.4.2 Case (b): Testing Connection Loss in Two Connectors

1. Figure 26.17 shows the experimental setup for testing connection loss in two connectors.
2. Follow the instruction of the fibre-optic cable loss measurements and calculations explained in the fibre-optic cables chapter.
3. Measure the laser input power into the fibre cable. Fill out Table 26.2.
4. Couple the laser beam output to the fibre cable input.
5. Carefully align the laser with the lens/fibre cable holder/positioner so that the maximum

FIGURE 26.17 Mated connectors set-up test.

amount of the laser beam is entering the core of the fibre cable.

6. Check to ensure that you have a circular output from the first optic connector. Direct the output towards the centre of the laser sensor.
7. Measure the laser output power from the first optic connector. Fill out Table 26.2.
8. Screw the two optic connectors into the in-line adapter; with one connector on each side, as shown in Figure 26.17.
9. Measure the laser output power from the second fibre cable end, as shown in Figure 26.17. Fill out Table 26.2.
10. Determine the connection assembly loss from the data. Fill out Table 26.2.
11. Mate and unmate the connectors three times to check the repeatability of the connection loss figures.

26.13.4.3 Case (c): Testing Connection Loss in a Mechanical Splice

Testing Connection Loss Using a Key-Lock Mechanical Splice

1. Figure 26.18 shows the experimental setup for testing connection loss in a mechanical splice. Prepare a key-lock mechanical splice for the fibre cable connection process.
2. Repeat steps 2–7 from Case (b).
3. Insert the key fully into the edge hole and turn the key by 90° to the open position.
4. Insert the prepared first fibre cable output end into the centre hole on the side of the key-lock mechanical splice. The fibre cable end should be at the midpoint of the metal part of the mechanical splice. Then turn the key to the closed position to secure the fibre cable in the mechanical splice. Remove the key.
5. Repeat step 4 and gently insert the second fibre cable end into the other side of the mechanical splice. The two fibre cables should meet approximately at the centre of the metal part. Make sure that the fibre cable ends are as close (face-to-face) as possible without an air gap between them. If the mechanical splice shines during the test, the light is escaping because the two ends are not close enough.

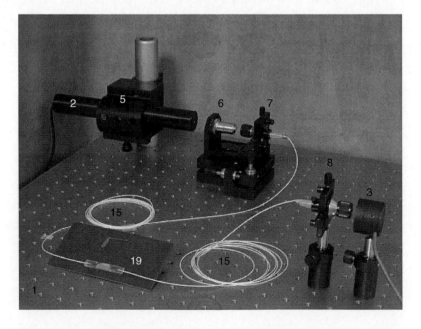

FIGURE 26.18 Key-lock mechanical splicing set-up test.

6. Measure the laser output power from the second fibre cable, as shown in Figure 26.18. Fill out Table 26.2.
7. Determine the key-lock mechanical splicing loss from the collected data. Fill out Table 26.2.
8. Repeat the mechanical splicing process three times to check the repeatability of the connection loss figures.

Testing Connection Loss in a Table-Type Mechanical Splice

1. Figure 26.19 shows the experimental setup for testing connection loss in a table-type mechanical splice.
2. Prepare the table-type mechanical splice by cleaning and oiling it to be ready for the connection process.
3. Repeat the fibre cable ends as explained above.
4. Insert the first prepared fibre cable output end into the table-type mechanical splice to the midpoint of the black block assembly on one side.
5. Tighten the screw on the side where the first fibre cable was inserted to secure the first fibre cable in the splice.
6. Rotate the L-clamp and clamp down the K-clamp into position to secure the first fibre cable.
7. Repeat steps 4–6 to insert the second fibre cable into the other side of the splicer.
8. Measure the laser power at the second fibre cable output. Fill out Table 26.2.
9. Determine the connection loss from the data. Fill out Table 26.2.
10. Repeat the mechanical splicing process three times to check the repeatability of the connection loss figures.

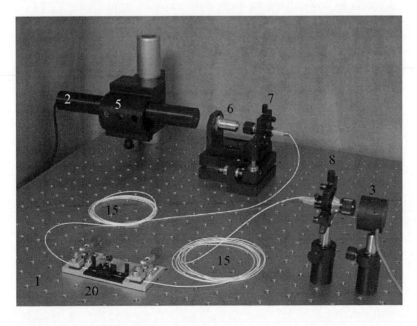

FIGURE 26.19 Table-type mechanical splicing set-up test.

26.13.4.4 Case (d): Testing Connection Loss in a Fusion Splice

1. Figure 26.20 shows the experimental setup for testing connection loss in a fusion splicer machine.
2. Prepare a fusion splicer machine for the fibre cables connection process.
3. Repeat the steps in the mechanical splice for the fibre cable connection process.
4. Turn on the fusion splicer.
5. Insert one prepared end of each fibre cable into the bonder.
6. Make sure that the fibre cable ends are lined up straight, to ensure proper bonding.
7. Look through the window of the fusion splicer to inspect the positioning of the two fibre cables; ensure that they are lined up and that they are flush. Perform a fine alignment for the two fibre cable ends.
8. Push the fusion key to start the fusion process.
9. The fusion splicer connects the two fibre cables permanently. When the splicing is completed, remove the fibre cable.
10. Add a fibre-optic fusion protector (sleeve) to the fusion location on the fibre cable.
11. The fusion splicer has a display panel and can display the splice test results and other optical parameters, as shown in Figure 26.20.
12. Set up the fusion splice test, as shown in Figure 26.20.
13. Measure the laser power at the second fibre cable output. Fill out Table 26.2.
14. Determine the fusion splice connection loss from the data. Fill out Table 26.2.
15. Find the connection loss from the data that is provided by the fusion splicer machine.

26.13.5 DATA COLLECTION

26.13.5.1 Case (a): Building a Connector

No data collection is required for this case.

26.13.5.2 Case (b): Testing Connection Loss in Two Connectors

1. Measure the laser power at the first fibre cable input and output.
2. Measure the laser power from the second fibre cable output.
3. Fill out Table 26.2 for Case (b).

FIGURE 26.20 Fusion splice set-up test.

26.13.5.3 Case (c): Testing Connection Loss in a Mechanical Splice

Data collection is similar to Case (b). Fill out Table 26.2 for Case (c).

26.13.5.4 Case (d): Testing Connection Loss in a Fusion Splice

Data collection is similar to Case (b). Fill out Table 26.2 for Case (d).

TABLE 26.2
Connectors and splices Tests

		Cable 1			Cable 2	Cables-Connectors/
		P_{in}	$P_{out\,1}$	Loss	$P_{out\,2}$	Splices Loss
		(unit)	(unit)	(dB)	(unit)	(dB)
Case (b)						
Case (c)	(1)					
	(2)					
Case (d)						

26.13.6 CALCULATIONS AND ANALYSIS

26.13.6.1 Case (a): Building a Connector

No calculations or analysis are required for this case.

26.13.6.2 Case (b): Testing Connection Loss in Two Connectors

1. Calculate the power loss in the connectors, using Equation (26.4).
2. Fill out Table 26.2 for Case (b).

26.13.6.3 Case (c): Testing Connection Loss in a Mechanical Splice

1. Calculate the power loss in the mechanical splicer, using Equation (26.4).
2. Calculate the power loss in the table type mechanical splice, using Equation (26.4).
3. Fill out Table 26.2 for Case (c).

26.13.6.4 Case (d): Testing Connection Loss in a Fusion Splice

1. Calculate the power loss in the fusion splicer using Equation (26.4). Read the fusion splice loss value from the fusion splicer once the fusion splice is complete.
2. Fill out Table 26.2 for Case (d).

26.13.7 RESULTS AND DISCUSSIONS

26.13.7.1 Case (a): Building a Connector

1. Report your observations of the connector assembly and of the connector inspection for the quality of polishing the end of the ferrule.
2. Compare the quality of the polished end of the ferrule with a similar industrial pre-assembled connector.

26.13.7.2 Case (b): Testing Connection Loss in Two Connectors

Report the calculated results for the power loss (dB) in the connector.

26.13.7.3 Case (c): Testing Connection Loss in a Mechanical Splice

1. Report the calculated results for the power loss (dB) in the mechanical splice.
2. Report the calculated results for the fibre cable power loss (dB) in the table type mechanical splice.

26.13.7.4 Case (d): Testing Connection Loss in a Fusion Splice

1. Report the calculated results for the power loss (dB) in the fusion splicer.
2. Compare the results of the loss calculations in the connector, key lock mechanical splice, table type mechanical splice, and fusion splice.

26.13.8 CONCLUSION

Summarize the important observations and findings obtained in this lab experiment.

26.13.9 SUGGESTIONS FOR FUTURE LAB WORK

List any suggestions for improvements using different experimental equipment, procedures, and techniques for any future lab work. These suggestions should be theoretically justified and technically feasible.

26.14 LIST OF REFERENCES

List any references that were used in the report. Use one format in writing the references. Never mix reference formats in a report.

26.15 APPENDIX

List all of the materials and information that are too detailed to be included in the body of the report.

FURTHER READING

Al-Azzawi, A. and Casey, R. P., *Fiber Optics Principles and Practices*, Algonquin College Publishing Centre, Ottawa, ON, Canada, 2002.

Agrawal, G. P., *Fiber-Optic Communication Systems*, 2nd ed., Wiley, New York, 1997.

Boyd, W. T., *Fiber Optics Communications, Experiments and Projects*, 1st ed., Howard W. Sams and Co., Indianapolis, IN, 1987.

Camperi-Ginestet, C., Kim, Y. W., Wilkinson, S., Allen, M., and Jokerst, N. M., Micro-opto-mechanical devices and systems using epitaxial lift off, In *JPL, Proceedings of the Workshop on Microtechnologies and Applications to Space Systems*, pp. 305–316, (SEE N94-29767 08-31), Category Solid-State Physics, Georgia Inst. of Tech., Atlanta, U.S.A., June, 1993.

Cole, M., *Telecommunications*, Prentice Hall, Englewood Cliffs, NJ, 1999.

Derickson, D., *Fiber Optic Test and Measurement*, Prentice Hall PTR, Upper Saddle River, NJ, 1998.

Dutton, H. J. R., *Understanding Optical Communications*, IBM/Prentice Hall, Inc., Research Triangle Park, NC/Englewood Cliffs, NJ, 1998.

Golovchenko, E., Mamyshev, P. V., Pilipetskii, A. N., and Dianov, E. M., Mutual influence of the parametric effects and stimulated Raman scattering in optical fibers, *IEEE J. Quantum Electron.*, 26, 1815–1820, 1990.

Green, P. E., *Fiber Optic Networks*, Prentice Hall Publishing Co., Englewood Cliffs, NJ, 1993.

Hecht, J., *Understanding Fiber Optics*, 3rd ed., Prentice Hall, Inc., Englewood Cliffs, NJ, 1999.

Herrick, C. N. and McKim, C. L., *Telecommunication Wiring*, 2nd ed., Prentice Hall, Inc., Englewood Cliffs, NJ, 1998.

Hoss, R. J., *Fiber Optic Communications—Design Handbook*, Prentice Hall Publishing Co., Englewood Cliffs, NJ, 1990.

Kao, C. K., *Optical Fiber Systems: Technology, Design, and Applications*, McGraw-Hill, New York, 1982.

Keiser, G., *Optical Fiber Communications*, 3rd ed., McGraw-Hill, New York, 2000.

Keiser, G., *Optical Communications Essentials*, 1st ed., McGraw-Hill Pub., New York, 2003.

Kolimbiris, H., *Fiber Optics Communications*, Prentice Hall Inc., Englewood Cliffs, NJ, 2004.

L-com Connectivity Products, *Fiber Optic Connectors*, Master Catalog 2006, L-com Connectivity Products, U.S.A., June, 2006.

Mazzarese, D., Meeting the OM-3 challenge—fabricating this new generation of multimode fiber requires innovative state of the art designs and testing processes, *Cabling Syst.*, 18–20, December 2002.

Mynbaev, D. K. and Scheiner, L. L., *Fiber-Optic Communication Technology—Fiber Fabrication*, Prentice Hall PTR, Englewood Cliffs, NJ, 2005.

Newport Corporation, *Projects in Fiber Optics Applications Handbook*, Newport Corporation, Fountain Valley, CA, 1986.

OZ Optics Limited, *Fiber Optic Components for Optoelectronic Packaging*, Catalog 2006, OZ Optics, Ottawa, ON, Canada, 2006.

Pease, B., When it comes to optical fiber installations, some companies are really "blowing it", *Fiberoptic Product News*, 12, May 2004.

Salah, B. E. A. and Teich, M. C., *Fundamentals of Photonics*, Wiley, New York, 1991.

Senior, J. M., *Optical Fiber Communications*, 2nd ed., Principle and Practice, 1986.

Shotwell, R. A., *An Introduction to Fiber Optics*, Prentice Hall, Englewood Cliffs NJ, 1997.

Sterling, D. J. Jr., *Technician's Guide to Fiber Optics*, 2nd ed., Delmar Publishers Inc., Albany, NY, 1993.

Sterling, D. J. Jr., *Premises Cabling*, Delmar Publishers, Albany, NY, 1996.

Ungar, S., *Fiber Optics: Theory and Applications*, Wiley, New York, 1990.

Vacca, J., *The Cabling Handbook*, Prentice Hall, Inc., Englewood Cliffs, NJ, 1998.

Yariv, A., Universal relations for coupling of optical power between microresonators and dielectric waveguides, *Electron. Lett.*, 36, 321–322, 2000.

Yeh, C., *Handbook of Fiber Optics: Theory and Applications*, Academic Press, San Diego, CA, 1990.

Zhang, L. and Yang, C., Polarization selective coupling in the holey fiber with asymmetric dual cores, Paper ThP1, presented at LEOS2003, Tucson, AZ, October 26–30, 2003.

27 Passive Fibre Optic Devices

27.1 INTRODUCTION

A variety of passive fibre optic devices are used in optical fibre communication systems to perform specific tasks. Passive devices work without using an external power supply, while active devices need external power to work. Passive devices split, redirect, or combine light waves. This chapter describes in detail some common passive fibre optic devices.

The simplest passive fibre optic devices are couplers. Couplers direct multiple input light waves to multiple outputs. Normally, couplers split input signals into two or more outputs, or combine two or more inputs into one output. Couplers can have more than two inputs or outputs. Couplers work as power splitters, power taps, and wavelength selectors.

Other passive devices, such as wavelength division multiplexers and de-multiplexers, filters, circulators, and isolators, are also explained in detail in this chapter and the next chapters. One section of this chapter allows the student to practise designing, manufacturing, and testing a 50/50 Y-coupler in the lab. The student can plan the manufacturing process, and determine the process and handling times required to build one Y-coupler. The student will manufacture and test a Y-coupler in the lab, and test a 3 dB coupler, a 1×4 3 dB coupler, a Y-coupler, a 1×4 coupler, and a proximity sensor.

27.2 2×2 COUPLERS

In an optical network, there are many situations where it is necessary to combine and/or split light signals. Directional couplers form the basis of many data distribution networks. Figure 27.1 shows a four-port directional coupler with two inputs and two outputs. The arrows indicate the directions of power flow through the coupler. Power P_1 is incident on the input port, Port 1, of the coupler; the signal is divided between output ports, Ports 2 and 3, based on a set splitting ratio. Ideally, no power will reach Port 4, called the isolated port. By convention, the power P_2 emerging from Port 2

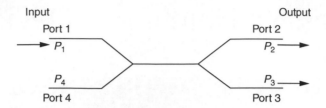

FIGURE 27.1 A four-port directional coupler.

is equal to or greater than the power P_3 emerging from Port 3, depending on the designed splitting ratio of the coupler. The splitting ratio is denoted by $P_2:P_3$ or else percentage of power e.g. 1:1 same as 50/50% splits power in half.

The main parameters of couplers are optical power losses. There are several types of coupler losses, given in decibels (dB):

1. Throughput loss (L_{THP}) specifies the transmission loss between the input power P_1 at Port 1 and transmission power P_2 at Port 2. This loss is calculated by:

$$L_{THP} = -10 \log_{10}\left(\frac{P_2}{P_1}\right) \qquad (27.1)$$

2. Tap loss (L_{TAP}) specifies the transmission loss between the input power P_1 at Port 1 and the tap power P_3 at Port 3. This loss is calculated by:

$$L_{TAP} = -10 \log_{10}\left(\frac{P_3}{P_1}\right) \qquad (27.2)$$

3. Directional loss (L_D) specifies the loss between the input power P_1 at Port 1 and the isolated power P_4 at Port 4. Ideally, the isolated power loss is zero. This loss is calculated by:

$$L_D = -10 \log_{10}\left(\frac{P_4}{P_1}\right) \qquad (27.3)$$

4. Excess loss (L_E) specifies the power lost within the coupler. It includes radiation, scattering, absorption, and coupling to the isolated port. This loss is calculated by:

$$L_E = -10 \log_{10}\left(\frac{P_2 + P_3}{P_1}\right) \qquad (27.4)$$

Table 27.1 lists splitting ratio values, throughput loss, and tap loss for several common couplers.

TABLE 27.1
Characteristics of Several Common Couplers

Coupler Description (dB)	Splitting ratio	L_{THP}(dB)	L_{TAP}(dB)
3	1:1	3	3
6	3:1	1.25	6
10	9:1	0.46	10
12	15:1	0.28	12

27.3 3 DB COUPLERS

A simple four-port coupler is often called a 3 dB coupler, if the input light splits into two equal portions at the output ports. The signal is split in half (3 dB = half). The 3 dB comes from the power loss formula: $[-10 \times \log_{10}(P_2/P_1) = -10 \times \log_{10}(0.5/1.0) = -10 \times (-0.3) = 3 \text{ dB}]$. Half of the light entering at Port 1 will exit at Port 2 and half at Port 3, as shown in Figure 27.2.

It is often useful to cascade many 3 dB couplers, as shown in Figure 27.3. The configuration shown is called a splitter. This splitter divides a single input into four equal outputs and is denoted as a 1×4 coupler. As might be expected, if the device is perfect, each output port will contain one fourth of the input power.

Figure 27.4 shows how a 1×8 coupler can be constructed by cascading several 3 dB couplers in a tree configuration. The signal input power will divide into eight equal outputs. Each output port will contain one eighth of the input power.

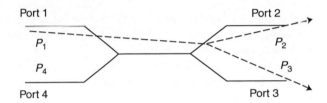

FIGURE 27.2 3 dB coupler.

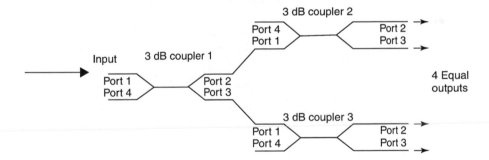

FIGURE 27.3 Cascaded 3 dB couplers to produce a 1×4 coupler.

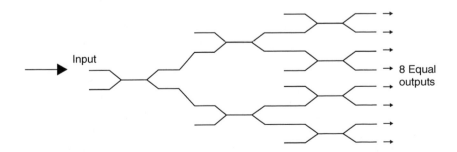

FIGURE 27.4 Cascaded 3 dB couplers to produce a 1×8 coupler.

27.4 Y-COUPLERS

Y-couplers or splitters, sometimes called 3 dB couplers, split the light equally. Y-couplers are 3 dB couplers in which Port 4 is not used. In the Y-coupler (splitter), as shown in Figure 27.5, the light entering Port 1 will be split equally between Ports 2 and 3 with almost no loss. They are extremely efficient at splitting light with little loss. Y-couplers are difficult to construct in fibre optics, but they are easy to construct in planar waveguide systems. The power loss in the Y-coupler system can be calculated by:

$$\text{Loss (dB)} = -10 \log_{10}\left(\frac{P_{\text{out}}}{P_{\text{in}}}\right) \tag{27.5}$$

Y-couplers are very seldom built as separate planar devices; instead they are manufactured on the same substrate as other devices. Connecting Y-couplers to a fibre optic cable is expensive; and significant loss is experienced in the connections. However, Y-couplers of this kind are used extensively in complex planar devices.

It is often useful to cascade Y-couplers, as shown in Figure 27.6. The splitter configuration shown divides a single input into four equal outputs. If the device is perfect, each output port will contain one fourth of the input power. Figure 27.6 shows how a 1×4 Y-coupler can be constructed by cascading three 1×2 Y-couplers in a tree arrangement.

Figure 27.7 shows how a 1×8 Y-coupler can be constructed by cascading seven Y-couplers in a tree configuration. The power of the input signal will divide into eight equal outputs. Each output port will contain one eighth of the input power.

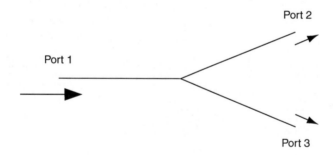

FIGURE 27.5 Y-coupler (splitter).

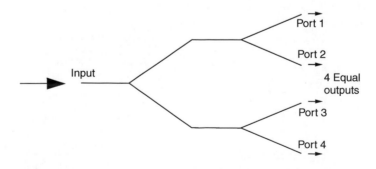

FIGURE 27.6 Cascaded 1×2 Y-couplers to produce a 1×4 coupler.

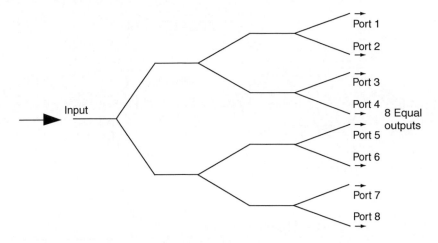

FIGURE 27.7 Cascaded Y-couplers to produce a 1×8 Y-coupler.

27.5 STAR COUPLERS

A star coupler is simply a multiple output coupler in which each input signal is made available on every output fibre. There are two star coupler designs, as shown in Figure 27.8. Figure 27.8(a) shows an 8×8 coupler. This coupler distributes the power from any input port to all of the output ports, splitting equally among the output ports. This type of coupler is called a transmission star coupler. Figure 27.8(b) shows a reflection star coupler; any input is split equally and is reflected back among all fibres. Star couplers are typically used in local area networks (LAN) and metropolitan area networks (MAN).

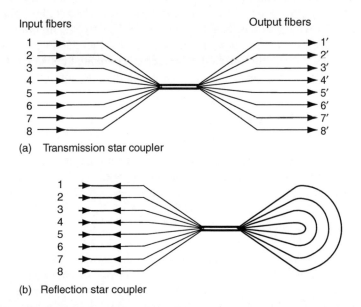

FIGURE 27.8 Star couplers.

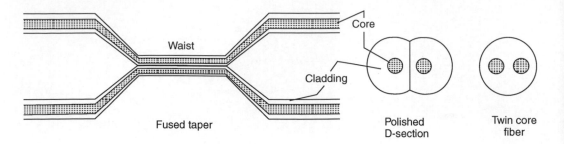

FIGURE 27.9 Some coupler configurations.

27.6 COUPLER CONSTRUCTION

In practise, many techniques can be used to construct couplers. Figure 27.9 shows the most common designs of manufactured couplers.

27.6.1 FUSED TAPER COUPLERS

Fused taper couplers consist of two regular single-mode fibre optic cables which are in contact with one another, as shown in Figure 27.9. In this design, two fibre cables are heated, and then tension is applied to the ends of the fibre cables. Both the claddings and the cores of the fibre cables are drawn, and thus become thinner. During this process, the fibre cables fuse together, forming the waist. Sometimes the fibre cables are twisted tightly together before heating and stretching. Fused taper couplers are very common commercial devices.

27.6.2 POLISHING D-SECTION COUPLERS

The design of polishing D-Section couplers involves embedding fibre cables into a solid material such as plastic. The cladding surface is polished along the length of the fibre cables until one side of the cladding is flat and removed to within 4 μm of the core. A D-shaped fibre cable section results, as shown in Figure 27.9. The plastic is then dissolved away. The two sections are joined along their flat surfaces, using index matching epoxy resin. Sometimes, the cladding on the fibre cables is thinned-out by etching with hydrofluoric acid, to reduce the amount of polishing. This is a relatively precise technique. However, it is more expensive to make than the fused taper method.

27.6.3 TWIN CORE FIBRE COUPLERS

This design uses twin cores having a very small separation, as shown in Figure 27.9. The manufacturing cost of the twin core fibre cable is very low compared to the cost of other designs. However, it is more difficult to connect twin core couplers to regular fibre cables than it is for other coupler designs.

27.7 THE PRINCIPLE OF RECIPROCITY

Reciprocity is a principle that applies to optical couplers and similar optical devices. The principle states that couplers work in both directions (forward and reverse) symmetrically. In Figure 27.10, a resonant coupler is set up to split input from Port 1 equally between Ports 2 and 3. The fused tapered directional coupler is designed to provide low-loss signal coupling with a range of splitting ratios. The construction of this coupler is described in Figure 27.9.

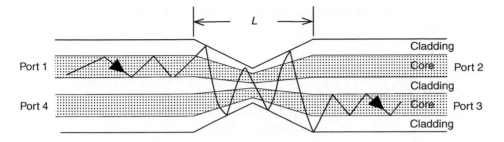

FIGURE 27.10 Fused biconically tapered directional coupler.

In multimode fibre cables, coupling occurs because higher-ordered modes no longer strike the core-cladding interface beyond the critical angle in the tapered regions. These modes are trapped by total reflection at the outer surface of the cladding; they have been converted into cladding modes. Rays from lower-ordered modes do not travel near the critical angle and are not converted. Power associated with these modes remains in the originating fibre. Because the fused waveguides in Figure 27.10 share the same cladding, power from higher-ordered input modes is common to both fibre cables. The output tapers convert the cladding modes back into core-guided waves. The splitting ratio depends on the length of the taper (L) and the cladding thickness.

27.8 PROXIMITY SENSOR

Fibre optic cables can be used as sensors. An example of a proximity sensor that uses a bifurcated (Y-branched) fibre bundle is shown in Figure 27.11. Optical power from a light source is launched into one arm of the fibre bundle. It is reflected from a surface (reflecting mirror, white surface, black surface, mat colour surface, etc.) at the output end of the fibre cable bundle. A portion of the reflected light enters the second arm of the bundle where it is routed to the sensor. The amount of light sensed by the sensor depends on the distance between the end of the bundle and the

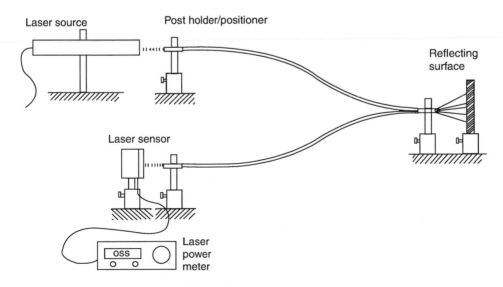

FIGURE 27.11 Proximity sensor using a fibre optic cable bundle.

reflecting surface. A fibre optic bundle can be manufactured, tested, and used in the experiment as a proximity sensor.

27.9 MACH–ZEHNDER INTERFEROMETER

Mach–Zehnder interferometers (MZIs) are used in a wide variety of applications within optics and fibre communication systems. Wavelength division multiplexers, opto-mechanical switches, and modulators can be made using MZI techniques. Mach–Zehnder interferometers can be either passive or active devices.

The basic structure of the MZI has the balanced configuration of a splitter and combiner, connected by a pair of matched waveguides or couplers. The MZI consists of three parts: an input Y-coupler, which splits the signal into two beams; the arms; and an output Y-coupler, which recombines the two components of the signal to regenerate the input signal. MZI operations and applications are generally presented in opto-mechanical switches.

27.10 OPTICAL ISOLATORS

Optical isolators are used in a wide variety of applications within fibre communication systems. Connectors, splices, and other optical components are generally presented in fibre optic connectors. Return lights are generated from the various parallel optical surfaces and the end faces of these devices. Return lights are known as back reflections. Back reflections have a destabilizing effect on oscillation of the laser source and on the operation of fibre optic amplifiers, thus resulting in poor transmission performance in the data lines. Optical isolators are used to block the back reflections.

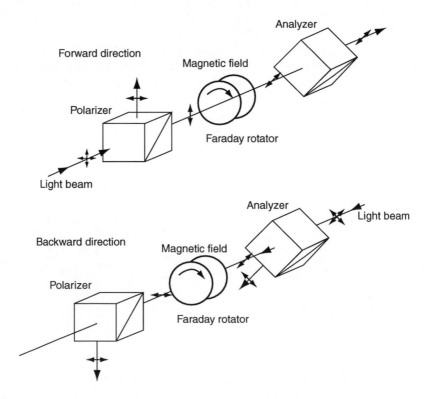

FIGURE 27.12 Isolator operation.

Figure 27.12 illustrates the isolator operation. An optical isolator is composed of a magnetic garnet crystal acting as a Faraday rotator; a permanent magnet for applying a designated magnetic field; and polarizing elements that permit only forward light to pass, while shutting out backward light. For this reason, optical isolators are indispensable devices for eliminating the adverse effect of back reflection in fibre optic systems. The optical isolator consists of two polarization elements and a 45° Faraday rotator placed between the polarization elements. The polarizer and the analyzer have a 45° difference in the direction of their light transmission axes. Forward light passing the optical isolator undergoes the following:

- When passing through the polarizer, the incident light is transformed into linearly polarized light.
- When passing through the Faraday rotator, the polarization plane of the linearly polarized light is rotated 45°.
- This light passes through the analyzer without loss, since the light polarization plane is now in the same direction as the light transmission axis of the analyzer, which is tilted 45° from the polarizer in the direction of Faraday rotation.

On the contrary, backward light undergoes a slightly different process:

- When passing through the analyzer, the backward light is transformed into linearly polarized light with a 45° tilt in the transmission axis.
- When passing through the Faraday rotator, the polarization plane of the backward light is rotated 45° in the same direction as the initial tilt.
- This light is completely shut out by the polarizer because its polarization plane is now tilted 90° from the light transmission axis of the polarizer.

27.11 OPTICAL CIRCULATORS

Optical circulators are used in a wide variety of applications within fibre communication systems. In advanced optical communication systems, optical circulators are used for bi-directional transmissions, wavelength division multiplexing (WDM) networks, fibre amplifier systems, optical time domain reflectometers (OTDR), etc. Optical circulators are nonreciprocal devices that redirect a signal from port to port sequentially, in only one direction. The operation of an optical circulator is similar to that of an optical isolator; however, its construction is more complex. Figure 27.13(a) shows a three-port optical circulator. An input signal (λ_1) at Port 1 exits at Port 2, an input signal

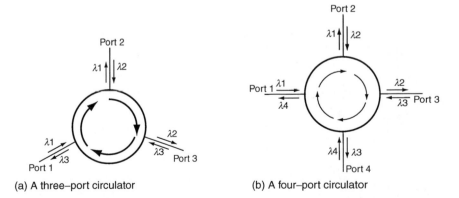

(a) A three–port circulator   (b) A four–port circulator

FIGURE 27.13 Circulator principle.

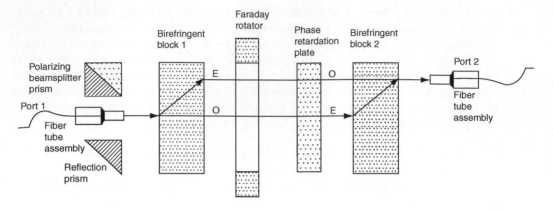

FIGURE 27.14 Optical circulator directs light beam from Port 1 to Port 2.

(λ_2) at Port 2 exits at Port 3, and an input signal (λ_2) at Port 3 exits at Port 1. Similarly, in a four-port optical circulator, as shown in Figure 27.13(b), one could ideally have four inputs and four outputs. In practise, many applications do not need four inputs and four outputs. Therefore, in a four-port circulator, it is common to have three input ports and three output ports. This is done by making Port 1 an input-only port, Ports 2 and 3 input and output ports, and Port 4 an output-only port. A typical application of the isolators is demonstrated in Figure 21.13. Input signals of different wavelengths are circulated to the next port in a clockwise direction.

There are many different circulator designs available for industrial applications. Circulators are commonly made of an assembly of the following optical components: polarizing beamsplitter, reflector prism, birefringent blocks, Faraday rotator, and retardation plate.

Figure 27.14 shows the basic operation of such a three-port circulator for a light beam entering at Port 1 and exiting at Port 2. The light input at Port 1 is split into two separate orthogonal polarization

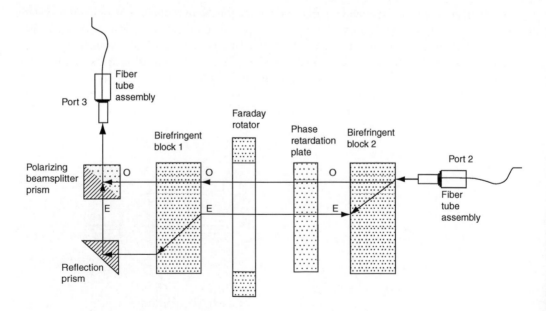

FIGURE 27.15 Optical circulator directs light beam from Port 2 to Port 3.

rays, which are called the ordinary (O) and extraordinary (E) components. The ordinary component passes through birefringent block 1 without refraction. The extraordinary component is refracted when it passes through birefringent block 1. Both ordinary and extraordinary components pass through the Faraday rotator and phase retardation plates. As a result, both components are rotated 90° clockwise. The component that was the ordinary in block 1 becomes the extraordinary component. The extraordinary component in block 1 becomes the ordinary component. The two components meet birefringent block 2, which is similar to block 1. Then, the two components re-combine to form the original light ray when exiting block 2, which enters Port 2, as shown in Figure 27.14.

On the other hand, a light beam entering at Port 2 will exit from Port 3, as shown in Figure 27.15. The light entering Port 2 is split by birefringent block 2 into two separate rays, similar to the light passing through birefringent block 1. The ordinary component passes through birefringent block 2 without refraction. The extraordinary component is refracted when it passes through birefringent block 2. The two components meet birefringent block 1, the ordinary component passes through to the polarizing beam splitter, and the extraordinary component refracts towards the reflector prism. The two components recombine at the polarizing beamsplitter prism to form the original light ray, which exits from Port 3. Using the same principles, light can be traced when it enters at Port 3 and exits at Port 1.

27.12 OPTICAL FILTERS

Optical filters are used in a wide variety of applications within optics and optical fibre devices. They are also used in a wide variety of optical applications in the fields of microscopy, photometry, radiometry, imaging, instrumentation, displays, charge-coupled devices (CCDs), astronomy, aerospace, etc. Scientific, electronic, analytical, imaging, and medical instrument companies are designing the next evolution of their products using a range of selected optical filters. There are many devices that are not called filters but have the same characteristics as a filter. Such devices are switches, modulators, array waveguide gratings, grating diffractions, grating multiplexers, etc. These devices are presented in detail in different chapters throughout the book.

Optical filters are devices that allow specific wavelengths to pass while rejecting all other wavelengths. Optical filters can be divided into two categories: fixed filters and tunable filters.

27.12.1 FIXED OPTICAL FILTERS

Fixed optical filters are commonly made from coloured glass (silica), thin metallic films, or thin dielectric films, as shown in Figure 27.16. Some metallic films, such as inconel, chromium, and nickel, are particularly insensitive to wavelength for absorption. On the other hand, the amount of absorption by coloured glass can vary as much as several orders of magnitude over only tens of nanometres of wavelength.

There is an extensive range of optical filter types. The following filters can be found in the market: visible filters, combination filters, infrared radiation cut-off filters, narrow bandpass filters, calibration filters, laser application filters, ultraviolet transmitting filters, infrared transmitting filters, light balancing filters, skylight filters, sharp cut filters, contrast filters, colour temperature conversion filters, special application filters, etc.

Optical filter selection depends on a multitude of factors, including wavelength selection, shape and passband width, blocking outside the passband, transmittance colour matching, material and thickness, vibration and shock resistance, ordinary and advanced anti-reflection coatings, heat absorption, and long-term stability.

FIGURE 27.16 Different types of filters.

27.12.2 TUNABLE OPTICAL FILTERS

Tunable optical filters are versatile devices that are used in many photonic applications. They are essential in wavelength-flexible WDM systems, and they can also play a key role in wavelength-tunable lasers for WDMs.

There is an extensive range of optical tunable filter types. The following filters can be found in the market:

- Micron optic fibre Fabry-Perot (FFP) tunable filters.
- Digitally tunable optical filters based on dense wavelength division multiplexer (DWDM) thin film filters and semiconductor optical amplifiers.
- Narrowband tunable optical filters using fibre Bragg gratings.
- High-speed tunable optical filters using a semiconductor double-ring resonator.
- Wide-bandpass tunable optical filters.
- Micromachined in-plane tunable filters using the thermo-optic effect of crystalline silicon.
- Acousto-optic tunable filters (AOTFs).
- Liquid crystal tunable filters (LCTFs).
- Others not listed here.

Some common types of tunable filters will be explained in detail in the following section.

Optical tunable filter selection depends on various factors, such as fast tuning speed, simple control mechanism, being scalable without additional insertion loss, and long-term operation temperature stability.

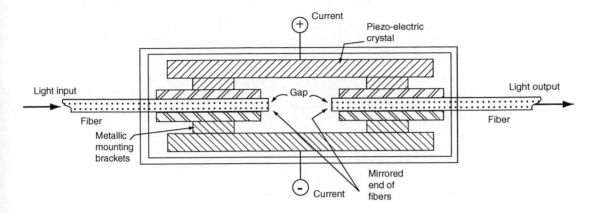

FIGURE 27.17 Fibre Fabry-Perot (FFP) tunable filter.

27.12.2.1 Fibre Fabry-Perot Tunable Filters

The FFP tunable filter principle is based on Fabry-Perot etalon technology. An FFP tunable filter passes wavelengths that are equal to integer fractions of the cavity (etalon) length; all other wavelengths are attenuated or reflected back. The key to the design of the FFP tunable filter is its lensless fibre construction. There are no collimating optics or lenses; thus, the FFP tunable filter achieves high precision, maintains low loss, and good transmission profile. Figure 27.17 shows a cross-section of an FFP tunable filter design. The design has two pieces of fibre, the ends of which are polished and silvered, so that each end acts like a mirror. The ends are placed precisely opposite one another with a specific gap between them. The fibre assemblies are mounted on two piezo-electric crystals and packaged in a box. By applying a voltage across the crystals, the distance between the fibre ends changes, thus changing the resonant cavity length, and therefore, a change in the wavelength selection.

This design of tunable filter eliminates the pitfalls of other Fabry-Perot component technologies, including misalignment and environmental sensitivity. Fibre Fabry-Perot tunable filters have low loss, high isolation, long-term alignment stability, high reliability, and accurate power or wavelength measurements. Fibre Fabry-Perot tunable filters are used in optical performance monitoring, tunable optical noise filtering, dropping of a tunable channel for ultra dense WDM, etc.

The design of FFP tunable filter can be modified by putting a liquid crystal material into the gap between the ends of the two fibres. The index of refraction of the liquid crystal can be changed very quickly, by passing current through the liquid crystal. By changing the index of refraction of the crystal, a change in the wavelengths passing through the crystal can be achieved, thus eliminating unwanted wavelengths.

27.12.2.2 Mach-Zehnder Interferometer Tunable Filters

The Mach–Zehnder interferometer tunable filter has a ladderlike structure, in which each section resembles a Mach–Zehnder interferometer, as shown in Figure 27.18. The output waveguide (across the top) and the input waveguide (across the bottom) are joined by regularly spaced linking waveguides, each longer than the previous one by ΔS, as illustrated in Figure 27.18. For constructive interference to occur at each coupler in the output waveguide, ΔS must be equal to an integral number of wavelengths

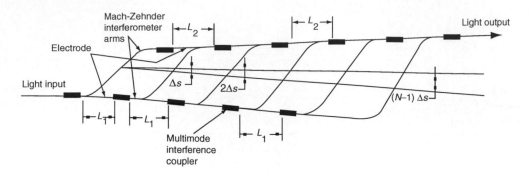

FIGURE 27.18 Mach–Zehnder interferometer tunable filter.

of the input light. However, similar to a Mach–Zehnder that can be tuned by adjusting the refractive index of one or both arms, this filter can be tuned by adjusting the refractive index of the arms with an injected current. More details are presented on the Mach–Zehnder interferometer in Section 27.9 opto-mechanical switches.

27.12.2.3 Fibre Grating Tunable Filters

Fibre grating tunable filters are used in a wide variety of applications within optics and fibre communication systems. They are an important element in wavelength division multiplexer systems for combining and separating individual wavelengths. Fibre Bragg Grating (FBG) transmits one wavelength and reflects all others. Basically, a grating is a periodic structure within an optical material. This variation in the structure of the optical material reflects or transmits light in a certain direction depending on the light wavelength. Therefore, gratings can be categorized as either transmitting or reflecting.

Figure 27.19 shows a design of an in-fibre Bragg grating tunable filter. The filter contains two Bragg gratings and a four-port circulator. The gratings have high reflectance in wavelength bands at specified wavelengths. The Bragg grating fibre is glued to the piezoelectric crystal contacts. Current can be applied on one or two Bragg gratings. The current deforms the crystal, stretching the gratings to match the wavelength of the signals. The wavelength filtered depends on the current level. This type of filter is used in aircraft or spaceborne differential absorption systems that measure water vapour in Earth's atmosphere. It is also used for a unique optical receiver that couples a laser radar signal from a telescope to the in-fibre Bragg grating filter.

27.12.2.4 Liquid Crystal Tunable Filters

Liquid crystal tunable filters (LCTFs) use electrically controlled liquid crystal elements, which select a specific visible wavelength of light for transmission through the filter. A typical wavelength-selective LCTFs is constructed from a stack of fixed filters that consist of interwoven birefringent crystal/liquid-crystal combinations and linear polarizers. The spectral region transmitted through an LCTF depends upon the choice of polarizers, optical coatings, and the liquid crystal characteristics (nematic, cholesteric, smectic, etc.). In general, visible-wavelength devices of this type usually perform in the range of 400–700 nm. This type of filter is ideal for use with electronic imaging devices, such as CCDs, because it offers excellent imaging quality with a simple optical pathway.

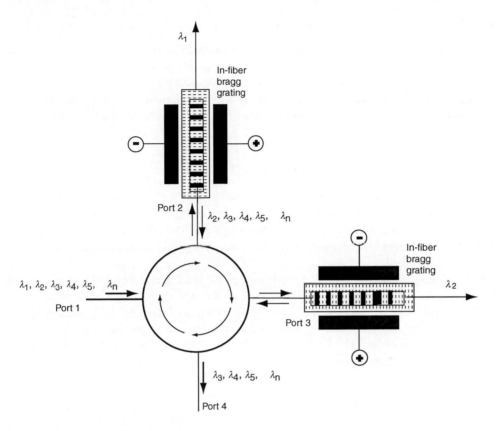

FIGURE 27.19 Fibre grating tunable filter.

27.12.2.5 Acousto-Optic Tunable Filters

Acousto-optic tunable filters (ATOFs) apply the same technology used in acousto-mechanical switches. An incident beam of light impacts a dioxide crystal of an AOTF. The dioxide crystal is sandwiched between an acoustic transducer and absorber that can be regulated by the acoustic power and acoustic frequency sliders. Upon encountering the standing wave in the dioxide crystal, a portion of the incident light beam is diffracted into the output port, while the remainder of the beam passes through the crystal and is absorbed by a beam stop. As the slider moves, the amplitudes of the waves passing through the AOTF are increased or decreased. Wavelength selection is controlled by the acoustic frequency slide. Acousto-optic tunable filters are employed to modulate the wavelength and amplitude of incident laser light.

27.12.2.6 Thermo-Optic Tunable Filters

Thermo-optic tunable filters apply the thermo-optic effect on an optical material, such as crystalline silicon. Changing the temperature results in a change in the index of refraction of the material. For example, current applied to a resistive element creates heat which increases the temperature of the filter. Thus, by applying current, wavelength selection can be achieved. This type of filter is used for spectroscopy and in optical communication systems.

27.12.2.7 Other Types of Tunable Filters

There are many other types of tunable filters used in building various optical devices and systems. Many tunable filters perform specific functions. Tunable filters are also available in the market, such as wide-band-pass filters, gas tunable filters, active optical filters, volume holographic grating-based filters, digitally tunable filters based on dense wavelength division multiplexer (DWDM) thin film filters, and high-speed tunable filters using a semiconductor double-ring resonator.

27.13 OPTICAL FIBRE RING RESONATORS

Optical fibre ring resonators are popular in communication systems. These devices are used in wavelength filtering, routing, switching, modulation, and multiplexing/demultiplexing applications. They are sometimes called filters based on optical ring resonators. There are many arrangements for coupling ring resonators to achieve the desired function. Ring resonators are either passive or active (when power is applied).

Optical ring resonators consist of a waveguide, such as a fibre, in a closed loop coupled to one or more input/output directional couplers for in and outcoupling, as shown in Figure 27.20. When light of the appropriate wavelength is coupled into the loop via the input waveguide, the light builds up in intensity over multiple round-trips due to constructive interference at the coupler. Since only specific wavelengths resonate in the loop, it functions as a filter. Wavelength selection depends on the ring loop length being an integral number of wavelengths.

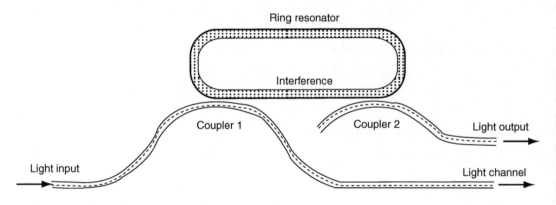

FIGURE 27.20 Optical fibre ring resonator.

27.14 OPTICAL MODULATORS

There are many devices that are not called optical modulators but which have the characteristics of modulators. Such devices include switches, filters, array waveguide gratings, grating diffractions, and grating multiplexers. These devices are presented in detail in different chapters throughout the book.

Optical modulators also use materials that change their optical properties under the influence of an electric, acoustic, or magnetic field; electro-absorption; vibration; light; or heat. Optical modulators also use the DWDM and in-fibre Bragg grating techniques.

An optical fibre modulator, which operates by using magnetic fluids, is a common example. The modulator consists of a bare fibre core surrounded by magnetic fluids, instead of a SiO_2 cladding layer. When applying a magnetic field, the index of refraction of the magnetic fluid changes. When light propagates along the fibre, controlling the occurrence of total reflection at the interface between the fibre core and the magnetic fluid is very rapid. As a result, the intensity of the output light is modulated by varying the magnetic field strength. This allows only specific wavelengths to pass through the fibre, reflecting the rest.

27.15 OPTICAL ATTENUATORS

Optical attenuators are devices that decrease the optical power in a fibre by a fixed or adjustable amount. Optical attenuators can be divided into two categories: fixed and variable. They are widely used in fibre optic telecommunication systems. The basic principle of operation of a common attenuator is controlled by the offset of the input and output fibres, or an air gap between the input and output fibres. Other techniques are used in the operation of advanced optical attenuators; they use temperature, twisting, tension, or current acting on some type of optical material. Some of these techniques are explained throughout the book.

Optical attenuator selection depends on a multitude of factors, including wavelength, attenuation levels, temperature stability, etc.

27.15.1 FIXED ATTENUATORS

Once the desired level of attenuation is determined, fixed attenuators can be set to deliver a precise power output. Fixed attenuators are used to reduce the optical power transmitted through fibre optic cables. The most common uses include equalizing power between fibres and multi-fibre systems and reducing receiver saturation. They are available in plug and inline styles.

27.15.2 VARIABLE ATTENUATORS

Variable attenuators allow a range of adjustability, delivering a precise power output at multiple decible loss levels. Variable attenuators produce precise levels of attenuation, with flexible adjustment. By using simple adjustment controls, the attenuation can be easily modified to any level, as shown in Figure 27.21. They are available with other connector and inline styles.

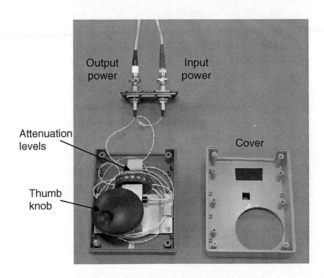

FIGURE 27.21 Inside a variable attenuator.

27.16 OTHER TYPES OF OPTICAL FIBRE DEVICES

There are many other types of fibre optic devices developed for research and development in the photonics industry. Most of the new devices are small in size, tunable to a wide range of wavelengths, multi-functional, and sometimes low in cost.

27.17 EXPERIMENTAL WORK

In this experiment, the student will measure power levels at the input and output ports, calculate the power losses, and determine the splitting ratio for optical couplers in the following cases:

a. Testing a 3 dB coupler
b. Testing a 1×4 3 dB coupler
c. Manufacturing a Y-coupler
d. Testing a Y-coupler
e. Testing a 1×4 coupler
f. Testing a proximity sensor

27.17.1 TECHNIQUE AND APPARATUS

Appendix A presents the details of the devices, components, tools, and parts.

1. 2×2 ft. optical breadboard.
2. HeNe laser light source and power supply.
3. Laser light sensor.
4. Laser mount assembly.
5. 20× microscope objective lens.
6. Lens/fibre cable holder/positioner.
7. Hardware assembly (clamps, posts, screw kits, screwdriver kits, sundry positioners, etc.)
8. Fibre optic cable.
9. Fibre cable end preparation procedure and kit, and cleaning kit, as explained in the Fibre Optic Cables Chapter, end preparation section.
10. Fibre cable holder/positioner assembly.
11. 3 dB coupler, as shown in Figure 27.22.

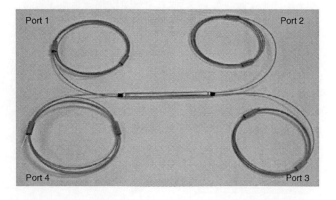

FIGURE 27.22 3 dB coupler.

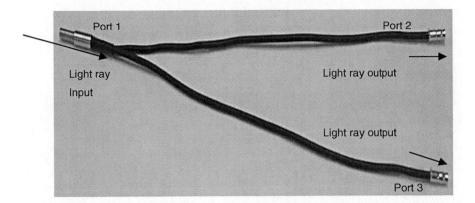

FIGURE 27.23 Y-coupler.

12. Y-Coupler, as shown in Figure 27.23. [Note: for Case (f), a Y-coupler (splitter) is used as a proximity sensor, as shown in Figure 27.28].
13. Reflecting surfaces (mirror, white card, and black card).
14. Ruler.

27.17.2 PROCEDURE

Follow the laboratory procedures and instructions given by the professor and/or instructor.

27.17.3 SAFETY PROCEDURE

Follow all safety procedures and regulations regarding the use of fibre optic cable. You must wear safety glasses and finger cots or gloves when working and handling fibre optic cables, optical components, epoxy, and optical cleaning chemicals.

27.17.4 APPARATUS SET-UP

27.17.4.1 Testing a 3 dB Coupler

1. Figure 27.24 shows the experimental set-up.
2. Bolt the laser short rod to the breadboard.
3. Bolt the laser mount to the clamp using bolts from the screw kit.
4. Put the clamp on the short rod.
5. Place the HeNe laser into the laser mount and tighten the screw. Turn on the laser device. Follow the operation and safety procedures of the laser device in use.
6. Check the laser alignment with the line of bolt holes and adjust when necessary.
7. Mount a lens/fibre cable holder/positioner to the breadboard, so that the laser beam passes through its centre hole.
8. Add the 20× microscope objective lens to the lens/fibre cable holder/positioner. It is better to use a 20× microscope objective lens to focus the collimated laser beam onto the input port.
9. Prepare the end of each port with a good cleave, as described in the end preparation procedure of the Fibre Optic Cables Chapter.

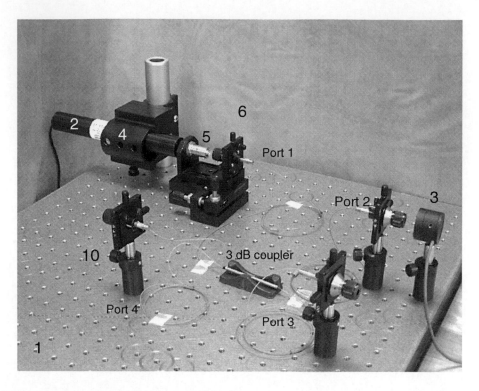

FIGURE 27.24 Apparatus set-up for 3 dB coupler.

10. Mount Port 1 into the lens/fibre cable holder/positioner, as explained in the Fibre Optic Cables Chapter.
11. Mount Ports 2–4 of the 3 dB coupler into a fibre cable holder/positioner, one for each port.
12. Align the laser source with Port 1 of the 3 dB coupler.
13. Verify the alignment of your laser beam launching arrangement, by making sure that the 3 dB coupler input port remains at the centre of the laser beam.
14. Measure the input power of the laser before it enters Port 1 of the 3 dB coupler.
15. Place the laser sensor in front of Ports 2, 3, or 4 of the 3 dB coupler.
16. Measure the output power from Ports 2–4 of the 3 dB coupler. Fill out Table 27.2.

27.17.4.2 Testing a 1×4 3 dB Coupler

A 1×4 coupler can be built by cascading three 3 dB couplers, as shown in Figure 27.3 and Figure 27.25. To connect the 3 dB coupler ports, a mechanical splice can be used, as shown in Figure 27.25. Fibre connectors and splices technique is presented in the Fibre Connectors Chapter. Continue the procedure explained in Section 27.21.1, adding the following steps:

1. Figure 27.25 shows the experimental set-up.
2. Add two fibre cable holder/positioners to the breadboard.
3. Mount the output ports of 3 dB coupler 2 into a fibre cable holder/positioner one for each port.

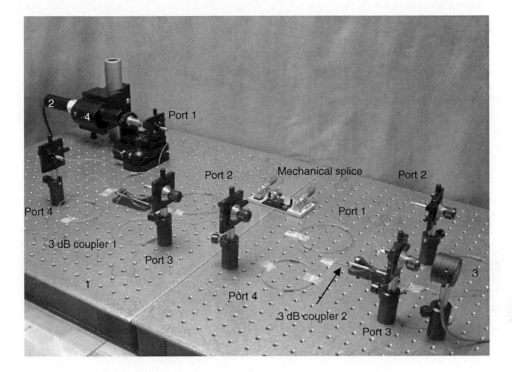

FIGURE 27.25 Apparatus set-up for 1×4 coupler.

4. Use mechanical splices to connect Port 2 of 3 dB coupler 1 to Port 1 of 3 dB coupler 2, as shown in Figure 27.25.
5. Place the laser sensor in front of Ports 2, 3, or 4 of 3 dB coupler 2.
6. Measure the power at each output port of cascaded 3 dB coupler 2. Fill out Table 27.3.
7. Repeat steps 4 to 6, by adding 3 dB coupler 3 to the set-up, to form a 1×4 coupler. This can be constructed by connecting Port 3 of 3 dB coupler 1 to Port 1 of 3 dB coupler 3. Fill out Table 27.3.

27.17.4.3 Manufacturing a Y-Coupler in the Lab

This case enables the student to practise designing, manufacturing, and testing a 50/50 Y-coupler in the lab. The student can plan the manufacturing process and determine the process and handling times required to build one Y-coupler. The student can also list tools, instruments, and parts requirements and design a floor plan for a production line for such a product. The manufactured Y-coupler can be used in the lab both as a self-sufficient product and a testing tool. The following steps explain the manufacturing process for a Y-coupler:

1. Cut ten lengths of fibre optic cable, 80 cm long. Make sure that all lengths are equal.
2. Cut two lengths (50 cm) and one length (25 cm) of heat shrink tubing.
3. Insert the ten fibre cables through the 25 cm long heat shrink tube. Make sure that all the fibre cables are even at the ends, and that they are slightly sticking out of the heat shrink tube.
4. Apply heat, using an air heat gun, to shrink the tube tidily around the fibre cables.

5. At the end of the 25 cm heat shrink tube, insert five fibre cables through one 50 cm heat shrink tube. And insert the other five fibre cables into the other 50 cm heat shrink tube. Leave the fibre cables extending from the heat shrink tubes.
6. Apply heat, using an air heat gun, to shrink the tube tidily around the fibre cables.
7. Add a metal tube on each end of the Y-coupler. Fix the metal tubes using an epoxy. Follow the epoxy instructions for the curing time.
8. Polish each end of the Y-coupler. Follow the polishing process, as explained in the Fibre Optic Connectors Chapter.
9. Use the microscope to inspect the polishing quality of the ends. The polishing process may need repeating, to get a mirror-finished surface on each end.

27.17.4.4 Testing a Y-Coupler

As explained in Section 27.21.1 and Section 27.21.2, a Y-coupler can be tested, as shown in Figure 27.26.

1. Figure 27.26 shows the apparatus set-up.
2. Install the laser source as explained Case (a).
3. Mount one post holder/positioner in front of the laser source. Check the laser alignment, so that the laser beam passes over the post's centre hole.
4. Insert input Port 1 of the Y-coupler into the post holder/positioner.

FIGURE 27.26 Apparatus set-up for Y-coupler test.

5. Mount the two output ports of the Y-coupler into the posts.
6. Verify the alignment of the laser light launching arrangement, by making sure that input Port 1 of the Y-coupler remains at the centre of the laser beam.
7. Measure the laser input power before it enters the Y-coupler. Fill out Table 27.4.
8. Place the laser sensor in front of the output ports of the Y-coupler.
9. Measure the laser output power from the output ports of the Y-coupler. Fill out Table 27.4.

27.17.4.5 Testing a 1×4 Y-Coupler

Continue the procedure explained in Case (d), adding the following steps, to create a 1×4 coupler.

1. Figure 27.27 shows the experimental set-up.
2. Add four post holder/positioners to the breadboard.
3. Couple two new Y-coupler inputs to the first Y-coupler outputs. Be sure that the coupling is good, with minimum light loss.
4. Mount the two output ports of the two new Y-couplers into the posts, as shown in Figure 27.27.
5. Measure the input power of the laser beam before it enters the cascaded Y-coupler. Fill out Table 27.5.
6. Measure the power at each output port of the cascaded Y-coupler. Fill out Table 27.5.

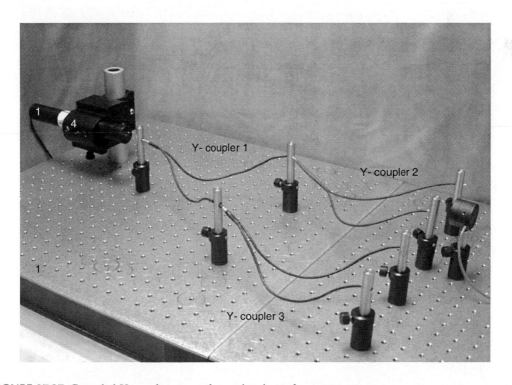

FIGURE 27.27 Cascaded Y-couplers to produce a 1× 4 couple apparatus set-up.

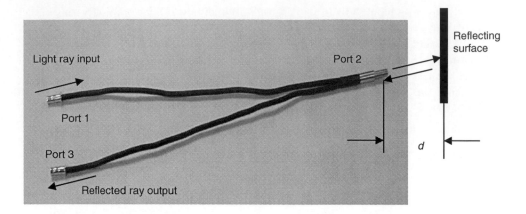

FIGURE 27.28 A proximity sensor.

27.17.4.6 Testing a Proximity Sensor

For the purpose of the lab experiment set-up, a proximity sensor can be made from a Y-coupler (splitter). Continue the procedure explained in Case (d), adding the following steps, to use a proximity sensor.

1. Rotate the Y-coupler 180° so that it works as a proximity sensor, as shown in Figure 27.28.
2. Figure 27.29 shows the experimental set-up.
3. Mount Port 1 of the Y-coupler into the post, so that Port 1 is facing the laser beam.
4. Align the laser beam with Port 1 of the proximity sensor.
5. Mount Ports 2 and 3 of the proximity sensor into a post for each port, as shown in Figure 27.29.
6. Measure the input power of the laser beam before it enters Port 1 of the proximity sensor.

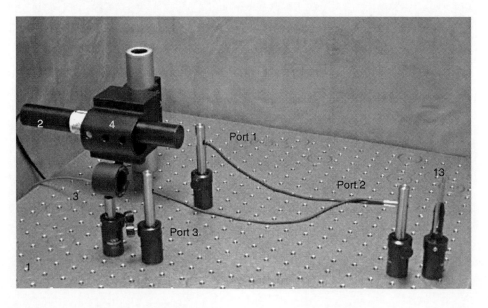

FIGURE 27.29 Proximity sensor apparatus set-up.

7. Place a mirror and mirror holder at a distance (d) from Port 2 of the proximity sensor, such that the mirror reflects the light back into the proximity sensor.
8. Measure the output power at Port 3. Fill out Table 27.6.
9. Repeat steps 7 and 8 for at least five distances, measuring the output power for each distance.
10. Repeat steps 7–9 using white and black reflecting surfaces.

27.17.5 DATA COLLECTION

27.17.5.1 Testing a 3 dB Coupler

1. Measure the input power of the laser beam before it enters the 3 dB coupler. Then measure the output power at Ports 2–4 of the 3 dB coupler.
2. Fill out Table 27.2 with the collected data.

TABLE 27.2
3 dB Coupler Test

	Input Power (unit)	Output Power (unit)			Calculated Loss (dB)			
	Port 1	Port 2	Port 3	Port 4	L_{THP}	L_{TAP}	L_D	L_E
Splitting Ratio								

27.17.5.2 Testing a 1×4 3 dB Coupler

1. Measure the input power and output power of the laser beam at each port of the cascaded 3 dB couplers that produced the 1×4 3 dB coupler.
2. Fill out Table 27.3 with the collected data.

TABLE 27.3
1×4 3 dB Coupler Test

	3 dB Coupler 1							
	Input Power (unit)	Output Power (unit)			Calculated Loss (dB)			
	Port 1	Port 2	Port 3	Port 4	L_{THP}	L_{TAP}	L_D	L_E
Splitting Ratio								

(continued)

TABLE 27.3   *(Continued)*

	3 dB Coupler 2						
	Input Power (unit)	Output Power (unit)			Calculated Loss (dB)		
	Port 1	Port 2	Port 3	Port 4	L_{THP}	L_{TAP}	L_D	L_E
Splitting Ratio							

	3 dB Coupler 3						
	Input Power (unit)	Output Power (unit)			Calculated Loss (dB)		
	Port 1	Port 2	Port 3	Port 4	L_{THP}	L_{TAP}	L_D	L_E
Splitting Ratio							

27.17.5.3 Manufacturing a Y-Coupler in the Lab

1. List the tools, instruments, parts that are required. Design a floor plan for a production line to manufacture a Y-coupler in the lab.
2. Estimate the manufacturing, testing, and handling times required to manufacture a Y-coupler in the lab.

27.17.5.4 Testing a Y-Coupler

1. Measure the input power of the laser beam before it enters the Y-coupler. Measure the output power at Ports 2 and 3 of the Y-coupler.
2. Fill out Table 27.4 with the collected data.

TABLE 27.4
Y-Coupler Test

	Input Power (unit)	Output Power (unit)		Calculated Loss (dB)	
	Port 1	Port 2	Port 3	$Loss_{1-2}$	$Loss_{1-3}$
Splitting Ratio					

27.17.5.5 Testing a 1×4 Y-Coupler

1. Measure the input power of the laser beam before it enters the cascaded Y-couplers. Measure the output power at each port of the cascaded Y-couplers.
2. Fill out Table 27.5 with the collected data.

TABLE 27.5
1×4 Y-Coupler Test

	Y-Coupler 1				
	Input Power (unit)	Output Power (unit)		Calculated Loss (dB)	
	Port 1	Port 2	Port 3	Loss$_{1\text{-}2}$	Loss$_{1\text{-}3}$
Splitting Ratio					

	Y-Coupler 2				
	Input Power (unit)	Output Power (unit)		Calculated Loss (dB)	
	Port 1	Port 2	Port 3	Loss$_{1\text{-}2}$	Loss$_{1\text{-}3}$
Splitting Ratio					

	Y-Coupler 3				
	Input Power (unit)	Output Power (unit)		Calculated Loss (dB)	
	Port 1	Port 2	Port 3	Loss$_{1\text{-}2}$	Loss$_{1\text{-}3}$
Splitting Ratio					

27.17.5.6 Testing a Proximity Sensor

1. Measure the input power of the laser beam before it enters the sensor.
2. Measure the output, which is caused by backscattering reflection from the mirror, white surface, or black surface. Repeat this for five distances, measuring the output power.
3. Fill out Table 27.6(I) for the mirror, (II) the white surface, and (III) the black surface with the collected data.

TABLE 27.6
Proximity Sensor Test

(I) Mirror

Input Power Port 1 (unit) =

Output Power Port 2 (unit) =

Calculated Loss Loss$_{1\text{-}2}$ (dB) =

Distance (d) (unit)	Output Power (unit) Port 3	Calculated Loss (dB) Loss$_{1\text{-}3}$

(II) White Surface

Input Power Port 1 (unit) =

Output Power Port 2 (unit) =

Calculated Loss Loss$_{1\text{-}2}$ (dB) =

Distance (d) (unit)	Output Power (unit) Port 3	Calculated Loss (dB) Loss$_{1\text{-}3}$

(continued)

TABLE 27.6 *(Continued)*

(III) Black Surface

Input Power Port 1 (unit) =

Output Power Port 2 (unit) =

Calculated Loss Loss$_{1-2}$ (dB) =

Distance (d) (unit)	Output Power (unit) Port 3	Calculated Loss (dB) Loss$_{1-3}$

27.17.6 CALCULATIONS AND ANALYSIS

27.17.6.1 Testing a 3 dB Coupler

1. Calculate the power loss in each output port of the 3 dB coupler, using Equation (27.1) through Equation (27.4).
2. Calculate the splitting ratio of each output port.
3. Fill out Table 27.2.

27.17.6.2 Testing a 1×4 3 dB Coupler

1. Calculate the power loss in each output port of the 3 dB couplers, using Equation (27.1) through Equation (27.4).
2. Calculate the splitting ratio of each output port.
3. Fill out Table 27.3.

27.17.6.3 Manufacturing a Y-Coupler in the Lab

Estimate the manufacturing, testing, and handling times that are required to manufacture a Y-coupler in the lab.

27.17.6.4 Testing a Y-Coupler

1. Calculate the power loss in each output port of the Y-coupler, using Equation (27.5).
2. Calculate the splitting ratio of each output port.
3. Fill out Table 27.4.

27.17.6.5 Testing a 1×4 Y-Coupler

1. Calculate the power loss in each output port of the Y-coupler, using Equation (27.5).
2. Calculate the splitting ratio of each output port.
3. Fill out Table 27.5.

27.17.6.6 Testing a proximity sensor

1. Calculate the power loss in each output port of the proximity sensor, using Equation (27.5).
2. Fill out Table 27.6 (I)–(III).
3. Plot the output power as a function of $(1/d^2)$, when using the mirror, white surface, and black surface.
4. Find the point where a linear $(1/d^2)$ dependence begins. This is the distance at which the finite diameter of the fibre core is no longer significant in determining the amount of reflected light accepted by the proximity sensor. Explain why the amount of reflected power accepted by the proximity sensor drops from a maximum when approaching the reflecting surface being used (mirror, white surface, and black surface).

27.17.7 RESULTS AND DISCUSSIONS

27.17.7.1 Testing a 3 dB Coupler

1. Report the measurements of the input power into Port 1 and output power from Ports 2–4 of the 3 dB coupler.
2. Report the loss calculation for the 3 dB coupler.
3. Determine the splitting ratio of the 3 dB coupler.
4. Compare the results to the manufacturer's specifications.

27.17.7.2 Tesing a 1×4 3 dB Coupler

1. Report the measurements of the input power of Port 1 and the output power from Ports 2–4 of the 3 dB couplers.
2. Report the loss calculation results for the 3 dB couplers.
3. Report the splitting ratio of the 1×4 coupler.
4. Compare the results to the manufacturer's specifications.

27.17.7.3 Manufacturing a Y-Coupler in the Lab

1. Report the manufacturing and testing process, and floor plan. Propose any other manufacturing process, which would reduce manufacturing time and manpower and using different materials.
2. Report the manufacturing, testing, and handling times required to manufacture a Y-coupler in the lab.

27.17.7.4 Testing a Y-Coupler

1. Report the input power and output power of the Y-coupler.
2. Report the loss calculation for each output port.
3. Report the splitting ratio of the Y-coupler.
4. Compare the results to the theoretical ratio.

27.17.7.5 Testing a 1×4 Y-Coupler

1. Report the input power and output power of the cascaded Y-couplers.
2. Report the loss calculation for each output port.
3. Report the splitting ratio of the 1×4 Y-coupler.
4. Compare the results to the theoretical ratio.

27.17.7.6 Testing a Proximity Sensor

1. Report the input power and output power of the proximity sensor.
2. Report the loss calculation for each output port.
3. Report the calculation results for the output power, when using the mirror, white surface, and black surface.
4. Compare the results of using the three different reflective surfaces.

27.17.8 CONCLUSION

Summarize the important observations and findings obtained in this lab experiment.

27.17.9 SUGGESTIONS FOR FUTURE LAB WORK

List any suggestions for improvements using different experimental equipment, procedures, and techniques for any future lab work. These suggestions should be theoretically justified and technically feasible.

27.18 LIST OF REFERENCES

List any references that were used in the report. Use one format in writing the references. Never mix reference formats in a report.

27.19 APPENDIX

List all of the materials and information that are too detailed to be included in the body of the report.

FURTHER READING

Agrawal, G. P., *Fiber-Optic Communication Systems*, 2nd ed., Wiley, New York, 1997.

Bise, R. T., Windeler, R. S., Kranz, K. S., Kerbage, C., Eggleton, B. J., and Trevor, D.J., Tunable photonic bandgap fiber, *Proceedings of Optical Fiber Communication Conference and Exhibit*, pp. 466–468, OFC, U.S.A., 2002.

Camperi-Ginestet, C., Kim, Y. W., Wilkinson, S., Allen, M., and Jokerst, N. M., Micro-opto-mechanical devices and systems using epitaxial lift off, *JPL Proceedings of the Workshop on Microtechnologies and Applications to Space Systems*, pp. 305–316, (SEE N94-29767 08-31), Category Solid-State Physics, Georgia Inst. of Tech., Atlanta, U.S.A., June 1993.

Cantore, F. and Della Corte, F. G., 1.55-µm silicon-based reflection-type waveguide-integrated thermo-optic switch, *Opt. Eng.*, 42 (10), 2835–2840, 2003.

Dutton, H. J. R., *Understanding Optical Communications*, IBM/Prentice Hall, Inc., Upper Saddle River, New Jersey, U.S.A., 1998.

Eggleton, B. J., Kerbage, C., Westbrook, P., Windeler, R., and Hale, A., Microstructured optical fiber devices, *Opt. Express*, 698–713, December 2001.

Fedder, G. K., Santhanam, S., Reed, M. L., Eagle, S. C., Guillou, D. F., Lu, M. S.-C., and Carley, L. R., Laminated high-aspect-ratio microstructures in a conventional CMOS process, *Sens. Actuators, A*, 57 (2), 103–110, 1996.

Fedder, G. K. and Howe, R. T., Multimode digital control of a suspended polysilicon microstructure, *IEEE J. MEMS*, 5 (4), 283–297, 1996.

Fedder, G. K., Iyer, S., and Mukherjee, T, Automated optimal synthesis of microresonators, *Technical Digest of the IEEE International Conference on Solid-State Sensors and Actuators (Transducers '97)*, Chicago, IL, June 16–19, Vol. 2, pp. 1109–1112, 1997.

Francon, M., *Optical Interferometry*, Academic Press, New York, pp. 97–99, 1966.

Gebhard, B. and Knowles, C. P., Design and adjustment of a 20 cm Mach–Zehnder interferometer, *Rev. Sci. Instrum.*, 37 (1), 12–15, 1996.

Goff, D. R. and Hansen, K. S., *Fiber Optic Reference Guide: A Practical Guide to the Technology*, 2nd ed., Butterworth-Heinemann, Boston, MA, 1999.

Hariharan, P., Modified Mach–Zehnder interferometer, *Appl. Opt.*, 8 (9), 1925–1926, 1969.

Hecht, J., *Understanding Fiber Optics*, 3rd ed., Prentice Hall, Inc., Englewood Cliffs, NJ, 1999.

Heidrich, H., Albrecht, P., Hamacher, M., Nolting, H.-P., Schroeter-Jannsen, H., and Weinert, C. M., Passive mode converter with a periodically tilted InP/GaInAsP rib waveguide, *IEEE Photon. Technol. Lett.*, 4, 34–36, 1992.

Hibino, K., Error-compensating phase measuring algorithms in a Fizeau interferometer. Mechanical Engineering Laboratory, AIST, 1–2, Namiki, Tsukuba, Ibaraki, 305-8564, Japan, *Opt. Rev.*, 6 (6), 529–538, 1999.

Him, F. C., Effects of passivation of GaN transistors, Paper presented to Epitaxy Group of the IMS at NRC, Ottawa, Canada, 2005.

Hitz, B., A new tunable optical filter, *Photonics Spectra*, Technology World, Vol. 10, no. 9, pp. 32, October 2003.

Ishida, O., Takahashi, H., and Noue, Y. I., Digitally tunable optical filters using arrayed-waveguide grating (AWG) multiplexers and optical switches, *J. Lightwave Technol.*, 15, 321–327, 1997.

Jackel, J., Goodman, M. S., Baran, J. E., Tomlinson, W. J., Chang, G.-K., Iqbal, M. Z., Song, G. H. et al., Acousto-optic tunable filters (AOTF's) for multiwavelength optical cross-connects: crosstalk considerations, *J. Lightwave Technol.*, 14, 1056–1066, 1996.

Kao, C. K., *Optical Fiber Systems: Technology, Design, and Applications*, McGraw-Hill, New York, 1982.

Koga, M., Compact quartzless optical quasi-circulator, *Electron. Lett.*, 30, 1438–1440, 1994.

Kolimbiris, H., *Fiber Optics Communications*, Prentice Hall, Inc., Englewood Cliffs, NJ, 2004.

Lequine, M., Parmentier, R., Lemarchand, F., and Amra, C., Toward tunable thin-film filters for wavelength division multiplexing applications, *Appl. Opt.*, 41 (16), 3277–3284, 2002.

Li, X., Chen, J., Wu, G., and Ye, A., Digitally tunable optical filter based on DWDM thin film filters and semiconductor optical amplifiers, *Opt. Express*, 13 (4), 1346–1350, 2005.

Litchinitser, N., Dunn, S., Steinvurzel, P., Eggleton, B., White, T., McPhedran, R., and de Sterke, C., Application of an arrow model for designing tunable photonic devices, *Opt. Express*, 12 (8), 1540–1550, 2004.

Loreggia, D., Gardiol, D., Gai, M., Lattanzi, M. G., and Busonero, D., Fizeau interferometer for global astrometry in space, *Appl. Opt.*, 43 (4), 721–728, 2004.

Mynbaev, D. K. and Scheiner, L. L., *Fiber-Optic Communication Technology—Fiber Fabrication*, Prentice Hall PTR, Englewood Cliffs, NJ, 2005.

OFS Leading Optical Innovations, Nonzero dispersion optical fiber, *True Wave Reach Fiber Product Catalog*, OFS Laboratories, Somerset, New Jersey, U.S.A., 2003.

Okamoto, K., *Fundamentals of Optical Waveguides*, Academic Press, San Diego, CA, 2000.

OZ Optics Limited, *Fiber Optic Components for Optoelectronic Packaging*, Catalog 2006, OZ Optics, Ottawa, ON, Canada, 2006.

Page, D. and Routledge, I., Using interferometer of quality monitoring, *Photonics Spectra*, 147–153, November 2001.

Palais, J. C., *Fiber Optic Communications*, 4th ed., Prentice Hall, Inc., Englewood Cliffs, NJ, 1998.

Sadot, D. and Boimovich, E., Tunable optical filters for dense WDM networks, *IEEE Commun. Mag.*, 36, 50–55, 1998.

Shamir, J., *Optical Systems and Processes*, SPIE Optical Engineering Press, Bellingham, WA, 1999.

Stone, J. and Stulz, L. W., Pigtailed high finesse tunable FP interferometer with large, medium and small FSR, *Electron. Lett.*, 23, 781–783, 1987.

Sugimoto, N., Shintaku, T., Tate, A., Kubota, E., Terui, H., Shimokozono, M., Ishii, M., and Inoue, Y., Waveguide polarization-independent optical circulator, *IEEE Photon. Technol. Lett.*, 11, 355–357, 1999.

Tolansky, S., *An Introduction to Interferometry*, Longmans/Green and Co., London, pp. 115–116, 1955.

Ungar, S., *Fiber Optics: Theory and Applications*, Wiley, New York, 1990.

Vail, E., Wu, M. S., Li, G. S., Eng, L., and Chang-Hasnain, C. J., GaAs micromachined widely tunable Fabry–Perot filters, *Electron. Lett.*, 31 (3), 228–229, 1995.

Yeh, C., *Applied Photonics*, Academic Press, San Diego, CA, 1994.

Yun, S.-S. and Lee, J.-H., A micromachined in-phase tunable optical filter using the thermo-optic effect of crystalline silicon, *J. Micromech. Microeng.*, 13, 721–725, 2003.

28 Wavelength Division Multiplexer

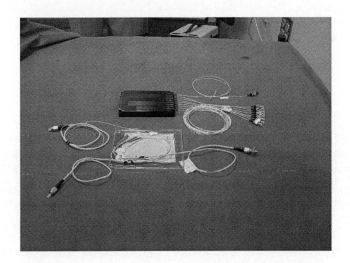

28.1 INTRODUCTION

A powerful aspect of fibre communication systems is that signals of many different wavelengths can be sent along a single optical fibre cable simultaneously, without signal interference. The technology that combines a number of wavelengths into the same fibre cable is known as wavelength-division multiplexing (WDM). Similarly, separating and distributing the signals is known as de-multiplexing. Multiplexing and de-multiplexing techniques are used in fibre communication systems to increase link capacity, flexibility, and speed, and to reduce cost.

This chapter presents the operating principles of WDM and describes the components needed for its utilization in systems. These components range from simple devices to sophisticated tunable optical sources and wavelength filters.

Multiplexer and de-multiplexer devices can be either passive or active in design. Passive designs are based on the use of a prism, diffraction grating, or filter. Active designs combine passive devices with tunable filters.

This chapter presents two experiments in both multiplexing/de-multiplexing of light signals from many light sources, using different techniques.

28.2 WAVELENGTH DIVISION MULTIPLEXING

Wavelength division multiplexing (WDM) is an optical technology that permits several wavelengths to be coupled into the same fibre cable, effectively increasing the aggregation bandwidth per fibre cable. Figure 28.1 illustrates the components of a basic communication system. The de-multiplexer (de-mux) decouples what the multiplexer has coupled. The de-mux separates several wavelengths in a single fibre cable, and directs them individually onto many fibre cables, which are connected to the receiver channels.

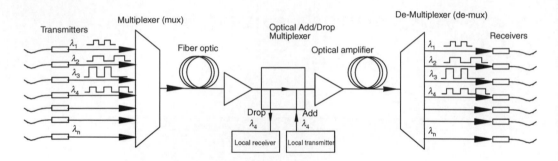

FIGURE 28.1 A schematic diagram of basic communication system.

WDM systems are based on the ability of a fibre cable to carry many different wavelengths without mutual interference. Each wavelength represents an optical channel within the fibre cable. Several optical methods are available to combine individual channels within a fibre cable, and to extract them at appropriate points along a network. WDM technology has evolved to the point that channel wavelength separations can be as small as a few nanometres, giving rise to dense wavelength division multiplexing (DWDM) systems.

The output of each laser transmitter in a WDM system is set to one of the channel frequencies. These signals of various frequencies must be then be multiplexed (superimposed or combined) and then inserted into a single fibre optic cable. These signals then travel through the cable from the multiplexer to an add-drop multiplexer. The add-drop multiplexer routes one wavelength, λ_4, to a point and picks up another signal at the same wavelength, also λ_4. Note that this is a different signal, as shown in Figure 28.1. A de-multiplexer is used to extract the multiplexed channels at the receiver end. Multiplexing and de-multiplexing devices employ narrowband filters. Multiplexer and de-multiplexer devices can be cascaded and combined to achieve the desired results. Several devices exist to perform such filtering, including thin-film fibre Bragg gratings, optic gratings, tapered fibres, liquid crystal filters, and integrated optical devices. These multiplexing technologies are also used for other optical fibre applications. These applications include telephone and data communications, SONET/SDH networks, inter-exchange networks, and in links for trunk exchange and local exchange hubs.

28.3 TIME-DIVISION MULTIPLEXING

Time-division multiplexing (TDM) is a technique for transmitting multiple digitized data, voice, and video signals simultaneously over one fibre cable. This is accomplished by interleaving pulses representing bits from different channels or time slots. The public-switched telephony network (PSTN) is based on the TDM technologies and is often called a TDM access network.

The time-division multiplexer is a device that uses TDM techniques to combine several slower speed data streams into a single high speed data stream, as shown in Figure 28.2. Data from multiple sources is broken into portions (bits or bit groups); these portions are transmitted in a defined sequence. The transmission order must be maintained so that the input streams can be reassembled at the destination. Typically, using the same TDM techniques, the same device can also perform the reverse process: de-compose the high-speed data streams into multiple low speed data streams, a process called de-multiplexing. Therefore, a Time Division Multiplexer and De-multiplexer are very often packaged in the same box.

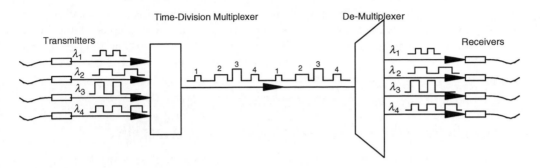

FIGURE 28.2 A schematic diagram of time-division multiplexing.

28.4 FREQUENCY-DIVISION MULTIPLEXING

Frequency-division multiplexing (FDM) is a scheme in which numerous analogue signals are combined for transmission on a single communications line or channel. Each signal is assigned a different frequency (sub-channel) within the main channel. This technology is used in broadcast radio, television, and cable television. Home local area networks (LAN) use this technology to ensure compatibility between the different services sharing the same telephone wire, specifically voice, and the home network. To eliminate interference, each service has a frequency spectrum that is different from the others. Traditionally, frequency-division multiplexing is used for analogue signals, but it also can be used for digital signals.

When FDM is used in a communication network, each input signal is sent and received at maximum speed at all times. However, if many signals must be sent along a single long-distance line, the necessary bandwidth is large, and careful design is required to ensure that the system will perform properly. In some systems, time-division multiplexing is used instead.

28.5 DENSE WAVELENGTH DIVISION MULTIPLEXING

DWDM is an acronym for dense wavelength division multiplexing, an optical technology used to increase bandwidth over existing fibre optic backbones. DWDM is a fibre-optic transmission technique that employs light wavelengths to transmit data as parallel bits or a serial string of characters. Using DWDM, up to 80 (and theoretically more) separate wavelengths or channels of data can be multiplexed into a single light stream, and then transmitted on a single fibre optic cable. Each channel carries a time division multiplexed (TDM) signal. In a system with each channel carrying data, billions of bits per second,can be delivered by the fibre optic cable. DWDM is also sometimes called wave division multiplexing (WDM). Since each channel is de-multiplexed at the end of the transmission back into the original source, different data formats can be transmitted together, at different rates. Specifically, internet data, synchronous optical network (SONET) data, and asynchronous transfer mode (ATM) data can all be transmitted at the same time within the same optical fibre. Utilizing DWDM technology is a suitable solution for high-speed data transmission, without the addition of more fibre cables.

28.6 COARSE WAVELENGTH DIVISION MULTIPLEXING

CWDM is an acronym for coarse wavelength division multiplexing, which is a technology that combines up to 16 wavelengths onto a single fibre. When there are just a few channels (up to 16 channels) and they are spaced more widely (10 nm or more) apart, the system is called CWDM. The coarse wavelength division multiplexer and de-multiplexer (CWDM) are designed to multiplex and de-multiplex wavelength signals in metropolitan, access and enterprise networks, and for cable television applications. They are a low-cost approach for systems that use un-cooled laser sources, and are an alternative to more expensive DWDM components based on 100 or 200 GHz channel spacing. CWDMs are used to isolate a specific wavelength channel, whereas CWDM channel splitters are used to isolate a band of channels.

28.7 TECHNIQUES FOR MULTIPLEXING AND DE-MULTIPLEXING

The following presents some of the techniques that are used in the multiplexing and de-multiplexing of many wavelengths.

28.7.1 MULTIPLEXING AND DE-MULTIPLEXING USING A PRISM

A simple way of multiplexing or de-multiplexing wavelengths can be done using a prism. Figure 28.3 shows the de-multiplexing of multiple wavelengths exiting from the fibre cable. The first lens makes the diverging light beam become parallel and incident on the prism surface. Each component of the light is refracted differently when it exits from the prism. This spreading of the wavelengths produces the rainbow effect. Each wavelength is refracted by a different angle from the next wavelength. This angle is called the angle of refraction. The angle of refraction depends on the wavelength, the apex angle, and the refraction index of the prism. The second lens focuses each wavelength to the designated output receiver via a fibre optic assembly. The same components can be used in reverse to multiplex different wavelengths onto one fibre tube assembly. Therefore, this device is bi-directional.

28.7.2 MULTIPLEXING AND DE-MULTIPLEXING USING A DIFFRACTION GRATING

Another technology based on the principles of diffraction uses a diffraction grating. When a light source is incident on a diffraction grating, each wavelength is diffracted at a different angle, and therefore, to a different point in space. It is necessary to use a lens to focus the wavelengths onto individual fibres, as shown in Figure 28.4. Separate wavelengths can be combined onto the same output port, or a single mixed input may be split into multiple outputs, one per wavelength. This device is bi-directional.

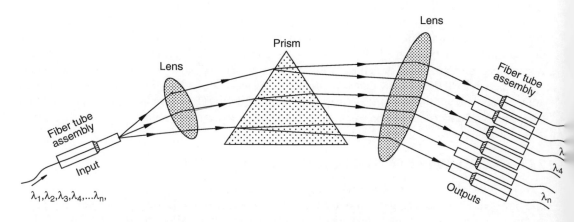

FIGURE 28.3 Multiplexing and de-multiplexing of wavelengths using a prism.

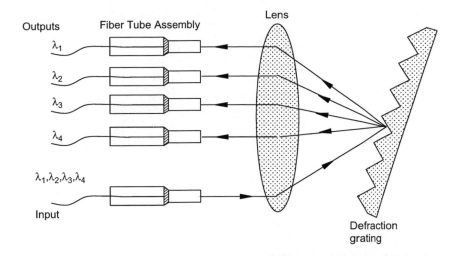

FIGURE 28.4 Multiplexing and de-multiplexing of wavelengths using a diffraction grating.

28.7.3 Optical Add/Drop Multiplexers/De-Multiplexers

Figure 28.5 illustrates a schematic representation of a design for an optical add/drop multiplexer/ de-multiplexer (OADM), which is widely used in communication systems. Between multiplexing and de-multiplexing points in a DWDM system, there is a span where multiple wavelengths exist. An OADM removes or inserts one or more wavelengths at some point along this span. Rather than combining or separating all wavelengths, the OADM can remove some, while passing others on. OADMs are a key part of moving toward the goal of all-optical networks. The design shown in Figure 28.5 includes both pre- and post-amplification components that may or may not be present in an OADM design.

There are two general types of OADMs. The first generation is a fixed device, physically configured to drop specific preset wavelengths while adding others. The second generation is reconfigurable and capable of dynamically selecting which wavelengths are added and dropped. Thin-film filters are used for OADMs in metropolitan DWDM systems, because of their low loss, low cost, and high stability. The new third generation of OADMs involves other technologies, such as tunable fibre gratings, fibre Bragg gratings, and circulators.

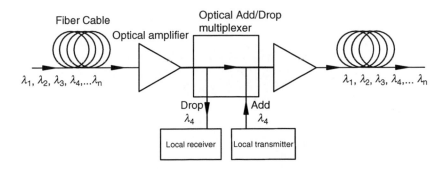

FIGURE 28.5 An optical add/drop multiplexer.

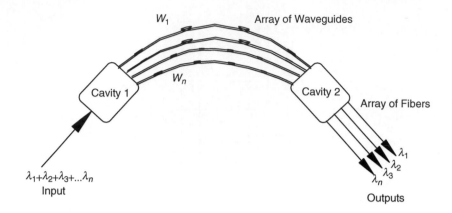

FIGURE 28.6 An arrayed waveguide grating device.

28.7.4 ARRAYED WAVEGUIDE GRATINGS

Arrayed waveguide gratings (AWGs) are also based on the principles of diffraction. An AWG device is sometimes called an optical waveguide, a waveguide grating router, a phase array, or a phasar. An AWG device consists of an array of curved-channel waveguides (W_1, W_2, W_3,...W_n) with a fixed difference in the length of optical path between the adjacent channels, as shown in Figure 28.6. The waveguides are connected to cavities at the input (S_1) and output (S_2). When light enters the input cavity, it is diffracted and enters the waveguide array. There the optical path length difference of each waveguide creates phase delays in the output cavity, where an array of fibres is coupled. The process results in different wavelengths having constructive interference at different locations, where the output ports are aligned.

28.7.5 FIBRE BRAGG GRATING

A Bragg grating is made of a small section of fibre cable, which is modified by exposure to ultraviolet radiation to create periodic changes in the refractive index of the core of the fibre cable. Figure 28.7 shows a fibre Bragg grating fibre cable. The Bragg grating reflects some of the light waves when travelling through it. The reflected waves usually occur at one particular wavelength. The reflected wavelength, known as the Bragg resonance wavelength, depends on the change in refractive index that is applied to the Bragg grating fibre. This also depends on the basic parameters of the grating (the grating period, the grating length, and the modulation depth).

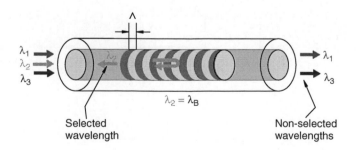

FIGURE 28.7 A fibre Bragg grating.

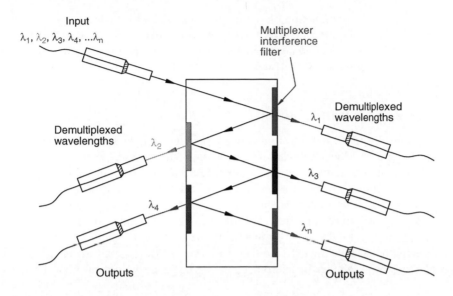

FIGURE 28.8 Multilayer interference filters.

28.7.6 THIN FILM FILTERS OR MULTILAYER INTERFERENCE FILTERS

Figure 28.8 shows one multiplexing technique that uses interference filters in devices called thin film filters, or multilayer interference filters. By positioning the thin filters in the optical path, wavelengths can be distributed. The property of each filter is such that it transmits one wavelength while reflecting others. By arranging the thin filters in a device, a de-multiplexer is created, and many wavelengths can be de-multiplexed.

There are several designs that use spectral filters positioned in the optical path to sort out wavelengths. These designs can be used as de-multiplexers.

28.7.7 PERIODIC FILTERS, FREQUENCY SLICERS, INTERLEAVERS MULTIPLEXING

Figure 28.9 is a schematic diagram of periodic filters, frequency slicers, and interleaver components that share the same functions, and are usually used together to make a multiplexer device. Stage 1 is a type of periodic filter, called an AWG. Stage 2 represents a frequency slicer. In this instance, stage 2 is

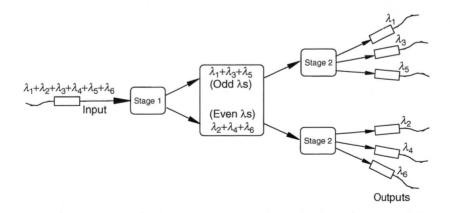

FIGURE 28.9 Periodic filters, frequency slicers, and interleavers multiplexing.

another AWG; an interleaver function on the output is provided by six Bragg gratings. Six wavelengths are received at the input to the AWG at stage 1, which then breaks the wavelengths down into odd and even wavelengths. The odd and even wavelengths go to their respective stage 2 frequency slicer, and then are delivered by the interleaver in the form of six discrete, interference-free optical channels.

28.7.8 Mach–Zehnder Interferometer

Interferometers are based on the interferometric properties of light and the principles of Mach–Zehnder interferometer. Mach–Zehnder interferometers can be used to direct a specific wavelength to a specific output port, as shown in Figure 28.10. The interferometer consists of parallel titanium-diffused waveguides on a lithium niobate substrate. The incoming signal is split evenly along the arms of the Mach–Zehnder interferometer, and then recombined at the output. Electrodes are fixed on the arms of the Mach–Zehnder interferometer, and two couplers are connected to the interferometer. While voltage is applied to the electrodes, a specific wavelength can be directed to either port 2 or 3. More detail about the principle and operation of the Mach–Zehnder interferometer is presented in the optical fibre devices chapter.

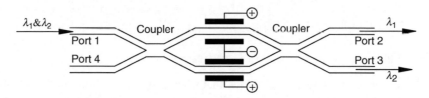

FIGURE 28.10 Mach–Zehnder interferometer.

28.8 WAVELENGTH DIVISION MULTIPLEXERS AND DE-MULTIPLEXERS

There are many types of wavelength division multiplexers and de-multiplexers available in the market to suit the requirements of any communication system. The following sections present the most widely used devices.

28.8.1 2-Channel WDM Devices

Figure 28.11 shows single mode wavelength division multiplexer devices (1X2 WDM 1310/1550, 1480/1550 nanometre). These two examples combine or separate light at different wavelengths and from different directions. The devices work to multiplex or de-multiplex optical signals within the 1310 and 1550 nanometre windows. These devices utilize a micro-optic filter-based technology to provide high performance, low insertion loss, low polarization dependence, high isolation, and environmental stability. These devices are extensively used in EDFA, CATV, WDM networks, and in fibre optics instrumentation.

28.8.2 8-Channel WDM Devices

Figure 28.12 shows a wavelength division multiplexer device, which is used for increasing fibre optic network signal capacity. The capacity is increased by enabling the simultaneous transmission of several wavelengths over the same common fibre. This device combines or separates light of eight different wavelengths. The optical channel wavelengths cover the 1480–1625 nm range, with 20 nm channel spacing. These devices utilize a micro-optic filter based technology to provide high performance, low insertion loss, low polarization dependence, high isolation and environmental stability. These devices are extensively used in EDFA, CATV fibre links, long-haul loops, subscriber loops, WDM networks, and fibre optics instrumentation.

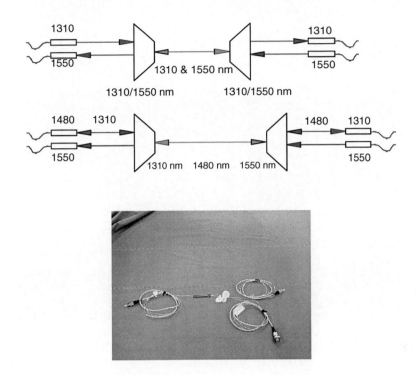

FIGURE 28.11 2-Channel WDM device.

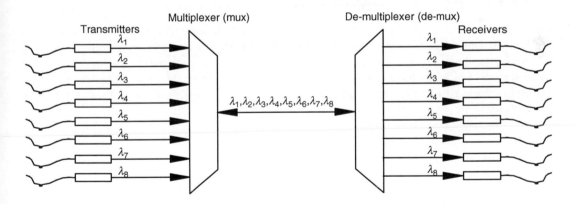

FIGURE 28.12. 8-Channel WDM device.

28.9 EXPERIMENTAL WORK

This experiment is designed to demonstrate the concept and practicality of wavelength division multiplexing (WDM) and wavelength division de-multiplexing (de-mux) devices. These devices are the key components in optical telecommunications systems. WDM is a process that combines "n" different signals for simultaneous transmission over the same fibre cable. Similarly, de-mux is a process that separates "n" different signals from the same fibre cable. WDM and de-mux devices use the concept of passive optical diffraction to combine and separate signals.

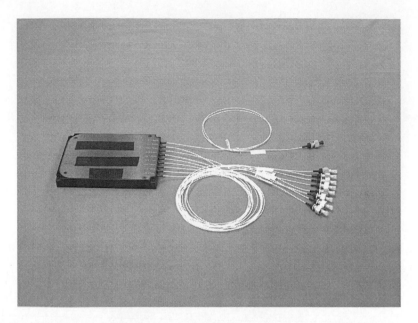

FIGURE 28.12 (*continued*)

Students will measure the power, wavelength, and insertion loss of the individual signals from these devices. The following cases will be examined:

28.9.1 WAVELENGTH DIVISION MULTIPLEXER

Figure 28.13 shows a schematic diagram of the set-up used in this experiment. Three laser beams are launched into a fibre cable through a prism. The prism acts as part of a diffraction grating, combining the beams onto a fibre cable. The power, wavelength, and losses of the laser beams can be measured at different locations of the experimental set-up. The index of refraction of the prism can be calculated, by measuring the angles of the prism, and the angles of refraction and deviation of the beam.

28.9.2 WAVELENGTH DIVISION DE-MULTIPLEXER

Figure 28.14 shows a schematic diagram of white light dispersed into a coloured spectrum by a glass prism. The wavelengths of such a spectrum are examined in this experiment. White light is launched into a lens, and later through a prism. The prism acts as part of a diffraction grating that separates the light into a coloured spectrum, and directs each colour onto a fibre optic cable. Each colour of the spectrum has a specific wavelength that can be measured. The index of refraction of the prism can be calculated by measuring the angles of the prism, and the angles of refraction and deviation of the beams of colour.

28.9.3 TECHNIQUE AND APPARATUS

Appendix A presents the details of the devices, components, tools, and parts.
1. 2×2 ft. optical breadboard
2. Three laser light sources and power supplies
3. Three laser clamps

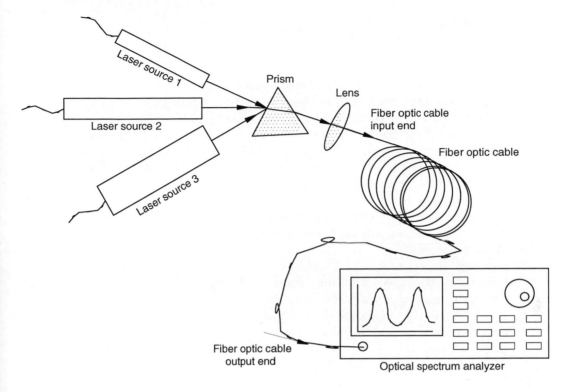

FIGURE 28.13 A schematic diagram of the WDM set-up.

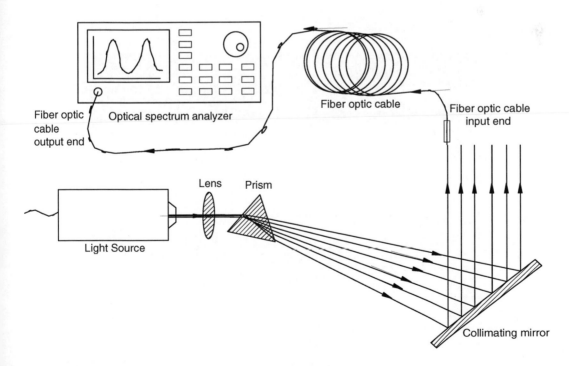

FIGURE 28.14 A schematic diagram of the de-mux set-up.

4. Laser light power detector
5. Laser light power meter
6. Light source and jack
7. Optical spectrum analyzer
8. Multi-axis translation stages, as shown in Figure 28.15
9. Magnification lens, lens holder, and rotation stages, in Figure 28.16
10. Collimating mirror and mirror holder, as shown Figure 28.17
11. White card and cardholder, as shown in Figure 28.18
12. Hardware assembly (clamps, posts, screw kits, screwdriver kits, lens/fibre cable holder/positioner, positioners, post holders, brass fibre cable holders, fibre cable holder/positioner, laser holder/clamp, etc.)
13. Prism and prism holder
14. Fibre optic cable end preparation procedure and kit, and cleaning kit, as presented in the fibre optic cables chapter, end preparation section
15. Connector assembly
16. Building connectors kit
17. Fibre optic cable, 250 micrometres diameter, 500 metres long
18. Fibre cable end holder/positioner assembly
19. 20X, 30X, or 40X microscope objective lens
20. Lens/fibre cable holder/positioner
21. Rotation stages
22. Black/white card and cardholder
23. Protractor
24. Ruler

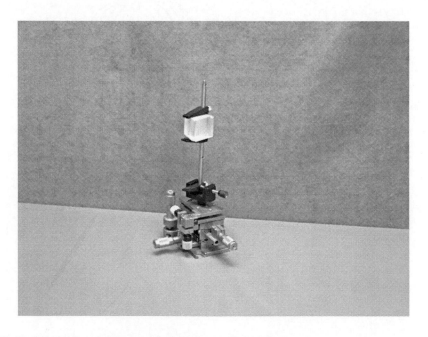

FIGURE 28.15 Prism, prism holder, and multi-axis translation stage.

FIGURE 28.16 Magnification lens and lens holder.

FIGURE 28.17 Collimating mirror and mirror holder.

FIGURE 28.18 White card and cardholder.

28.9.4 Procedure

Follow the laboratory procedures and instructions given by the professor and/or instructor.

28.9.5 Safety Procedure

Follow all safety procedures and regulations regarding the use of electric and optical devices, optical components, fibre optic cable, laser devices, measurement instruments, and optical cleaning chemicals.

28.9.6 Apparatus Set-Up

28.9.6.1 Wavelength Division Multiplexer

1. Figure 28.13 and Figure 28.19 show the apparatus set-up.
2. Mount three multi-axis translation stages on the breadboard.
3. Place laser source into the laser mount clamp and tighten the screw.
4. Bolt three laser sources to the multi-axis translation stages, one to each stage.
5. Align each laser to point at a central location on the breadboard.
6. Turn the lasers on individually. Follow the operation and safety procedures of the laser device in use. Use the optical spectrum analyzer (or the laser power detector for quick power out results) to measure the power and wavelength of each laser source. Ask the supervisor/instructor how to operate the optical spectrum analyzer and reference the laser. Refer to manufacturer's operation manual for the optical spectrum analyzer in use.
7. Measure the laser power and wavelength of each laser beam, after it exits from the laser sources. Figure 28.20 shows a sample of data collected by the optical spectrum analyzer and printed on a paper chart. Fill out Table 28.1.

FIGURE 28.19 Apparatus set-up for a WDM.

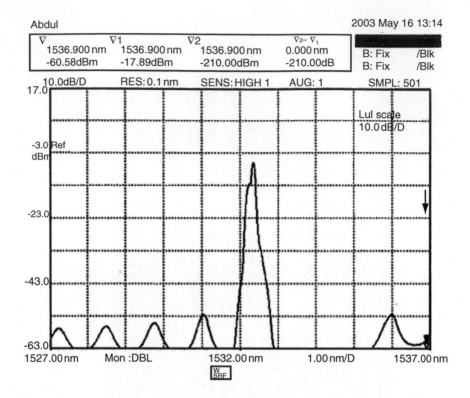

FIGURE 28.20 A sample of spectral data printed on a chart.

8. Mount a multi-axis translation stage at a central location on the breadboard.
9. Mount and clamp a prism on top of the stage platform, and ensure that one side of the prism is facing the three laser beams.
10. Make a rough alignment of the three laser beams that pass through the prism and exit from one spot on the other side of the prism, as shown in Figure 28.21. Check for output from all laser beams, separately, with a piece of black/white card.
11. Finely align the three laser beams passing through the prism. Ensure the output of all laser beams are all on one spot on the black/white card, as shown in Figure 28.21.
12. Figure 28.22 shows another way to align beams, using a glass rod. The beams exit from the glass rod in concentric rings. One beam hits a point on the black/white card. This point will be the centre for the second and third beams, as shown in Figure 28.22. The glass rod can only be used for quick beam alignment purposes. When the alignment is completed as explained above, use the first laser beam to continue the lab set-up.
13. Measure the laser power, wavelength, and insertion loss of each laser beam after exiting from the prism. Figure 28.20 shows a sample of collected data printed on a chart. Fill out Table 28.1.
14. Mount a lens/fibre cable holder/positioner to the breadboard, so that the laser beam passes over the centre hole.
15. Add the 20X, 30X, or 40X microscope objective lens to the lens/fibre cable holder/positioner.
16. Measure the laser power, wavelength, and insertion loss of each laser source, after it exits from the lens. Fill out Table 28.1.
17. Prepare a fibre cable to have a good cleave at each end.
18. Insert one end (input) of the fibre cable into the brass fibre cable holder, and put it into the hole of the lens/fibre cable holder/positioner.

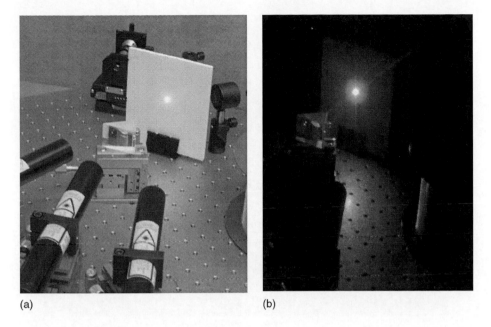

(a)   (b)

FIGURE 28.21 Red laser beam spot exiting from the prism.

19. Locate the input of the fibre cable at the focal point of the microscope objective lens.
20. Mount the output end of the fibre cable into the brass fibre cable holder, and place it in the hole of the fibre cable holder/positioner.
21. Finely align the input end of the fibre cable holder, so that the input end is at the centre of the lens/fibre cable holder/positioner, and is positioned at the focal point of the microscope

FIGURE 28.22 Laser beams exiting from the glass tube falling on the black/white card.

FIGURE 28.23 Red laser beam spot on the black/white card.

objective lens. This is a very important step for obtaining an accurate value of the laser beam power from the fibre cable output end.

22. Check to make sure that you have a proper circular spot from the output end of the fibre cable, as shown in Figure 28.23.

23. Repeat steps 21 and 22 to check the alignment of the second and third laser beams through the fibre cable.

24. Connect the output end of the fibre cable to the optical spectrum analyzer. There are many ways to achieve the connection, either by using a fibre cable holder/connector to connect the fibre cable directly to the input of the optical spectrum analyzer, or by building a connector on the output end of the fibre cable to suit the optical spectrum analyzer input connection point.

25. Measure the laser power, wavelength, and insertion loss of each laser beam after exiting from the fibre cable. Fill out Table 28.1.

26. Find the prism angles. Fill out Table 28.2.

27. Use a protractor (or ruler and geometry) to measure the angle of incidence of each laser beam incident on the prism surface, and the refracted angle of the laser beams exiting from the prism. Fill out Table 28.2.

28. Now, turn on the three laser sources at the same time. Measure the laser power outputs, wavelengths, and insertion loss for two laser sources exiting from the prism, the lens, and the fibre cable. Fill out Table 28.1.

29. Turn on two laser sources at the same time, by blocking either laser source # 1, # 2, or #3. Measure the laser power outputs, wavelengths, and insertion loss for the other laser beams exiting from the prism, the lens, and the fibre cable. Fill out Table 28.3.

28.9.6.2 Wavelength Division De-Multiplexer

1. Figure 28.14 and Figure 28.24 show the apparatus set-up.
2. Mount a light source on a jack and plug the light power supply into a 110V wall outlet.

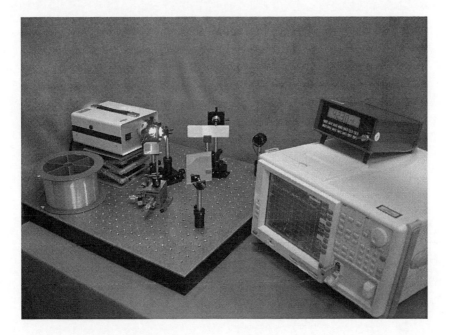

FIGURE 28.24 Apparatus set-up for a de-mux.

3. Place the input end of the fibre cable into the brass fibre cable holder, and place it in the hole of the fibre cable holder/positioner assembly.
4. Locate the input end of the fibre cable in front the light source.
5. Connect the output end of the fibre optic cable to the optical spectrum analyzer. There are many ways to achieve the connection to the optical spectrum analyzer. Either use a fibre cable holder/connector to connect the fibre cable directly to the input of the optical spectrum analyzer, or build a connector on the output end of the fibre cable to suit the optical spectrum analyzer input connection point.
6. Measure the power and wavelength of the light source using the optical spectrum analyzer (or the light power detector for quick power out results). Ask the supervisor/-instructor how to operate the optical spectrum analyzer and reference the laser. Refer to manufacturer's operation manual for the optical spectrum analyzer in use.
7. Figure 28.20 shows a sample of data collected by the optical spectrum analyzer printed on a paper chart. Fill out Table 28.4.
8. Mount a magnification lens into a lens holder/positioner and onto the breadboard, so that the light beam passes over the centre of the lens.
9. Mount a multi-axis translation stage on the breadboard at a location facing the lens.
10. Mount and clamp a prism on top of the stage platform, and ensure that one side of the prism is facing the light beam exiting from the lens.
11. Use the prism multi-axis translation stage to make a rough alignment of the light beam passing through the prism and exiting from the other side of the prism.
12. Mount the white/black card and cardholder to intercept the light exiting the prism. This light will disperse into a spectrum (rainbow), as shown in Figure 28.25.
13. Use the prism multi-axis translation stage to finely align the spectrum exiting the prism. A clear coloured spectrum (rainbow) can be captured on the white/black card.
14. Replace the white/black card and cardholder with the collimating mirror and mirror holder, to intercept the spectrum (rainbow) of light exiting the prism.

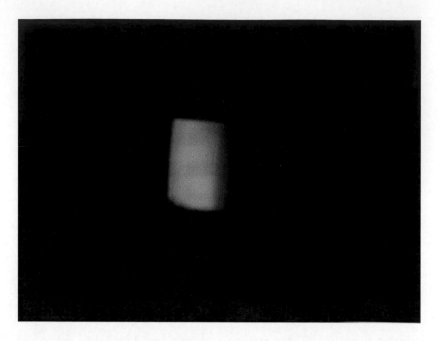

FIGURE 28.25 Spectrum (rainbow) of the light from the prism falling on the black/white card.

15. Mount the white card and cardholder, to intercept the spectrum (rainbow) of the light that reflects from the mirror.
16. Locate the fibre cable input end holder/positioner assembly behind the white card, to capture the first colour of the spectrum.
17. Measure the power and wavelength of each colour of the spectrum. Fill out Table 28.4.
18. Find the prism angles. Fill out Table 28.5.
19. Use a protractor (or ruler and geometry) to measure the angle of incidence of the light source incident on the prism surface, and the angles of refraction and deviation of each colour of the spectrum. Fill out Table 28.5.

28.9.7 Data Collection

28.9.7.1 Wavelength Division Multiplexer

1. For each laser source, measure the power and wavelength of the beams at the laser sources. Measure the power, wavelength, and insertion loss of the laser beams, after exiting the prism, the lens, and the fibre cable. Fill out Table 28.1
2. Find the prism angles. Measure the angle of incidence of each laser beam incident on the prism, and the refracted angle of the laser beams leaving the prism. Find the deviation angle of the laser beams. Fill out Table 28.2.
3. Measure the power, wavelength, and insertion loss of the laser beams when all laser sources are turned on together. Fill out Table 28.1
4. Measure the power, wavelength, and insertion loss of the laser beams, after exiting the prism, the lens, and the fibre cable, when two laser sources (#1 and #2, #1 and #3, and #2 and #3) are turned on. Fill out Table 28.3.

TABLE 28.1
Laser Powers and Wavelengths Measurements

	At Laser Source		Exiting from Prism			Exiting from Lens			Exiting from Fiber Cable		
	Laser Power P_{laser} (unit)	Wavelength λ_{laser} (unit)	Laser Power P_{prism} (unit)	Wavelength λ_{prism} (unit)	Prism Insertion Loss (unit)	Laser Power P_{lens} (unit)	Wavelength λ_{lens} (unit)	Prism + Lens Insertion Loss (unit)	Laser Power P_{cable} (unit)	Wavelength λ_{cable} (unit)	Prism + Lens + Cable Insertion Loss (unit)
Laser Source # 1											
Laser Source # 2											
Laser Source # 3											
Laser Sources Turnd on at the Same Time											

TABLE 28.2
Prism Angles, Incidence, and Reflected Angles of the Laser Beams

Prism Angles Apex Angle (°), Side Angle (°), and Side Angle (°)			
	Laser Beam Incidence Angle on Prism (°)	Laser Beam Refracted Angle from Prism (°)	Deviation Angle (°)
Laser Source # 1			
Laser Source # 2			
Laser Source # 3			

TABLE 28.3
Laser Powers and Wavelengths of Two Laser Sources Turned on at the Same Time

	Exiting from Prism			Exiting from Lens			Exiting from Fiber Cable		
	Laser Power P_{prism} (unit)	Wavelength λ_{prism} (unit)	Prism Insertion Loss (unit)	Laser Power P_{lens} (unit)	Wavelength λ_{lens} (unit)	Prism + Lens Insertion Loss (unit)	Laser Power P_{cable} (unit)	Wavelength λ_{cable} (unit)	Prism + Lens + Cable Insertion Loss (unit)
Laser Source # 1 + Laser Source # 2									
Laser Source # 1 + Laser Source # 3									
Laser Source # 2 + Laser Source # 3									

28.9.7.2 Wavelength Division De-Multiplexer

1. Measure the power and wavelength of the light source input to the prism. Fill out Table 28.4.
2. Measure the power and wavelength of each colour of the spectrum exiting the prism. Fill out Table 28.4.
3. Find the prism angles. Measure the angle of incidence of light on the prism, and the angles of refraction and deviation for each colour of the spectrum exiting the prism. Fill out Table 28.5.

TABLE 28.4
Light Source and Colours Powers and Wavelengths

	Light Source Input to Prism	Spectrum Colors						
		Color 1	Color 2	Color 3	Color 4	Color 5	Color 6	Color 7
Power (unit)								
Wavelength (unit)								

TABLE 28.5
Prism, Light Source, and Colors Angles

Prism Angles Apex Angle (°), Side Angle (°), and Side Angle (°) Light Source Incidence Angle on Prism (°) Colors	Angle of Refraction of the Colors Exiting from Prism (°)	Angle of Deviation of the Colors (°)
Color 1		
Color 2		
Color 3		
Color 4		
Color 5		
Color 6		
Color 7		

28.9.8 CALCULATIONS AND ANALYSIS

28.9.8.1 Wavelength Division Multiplexer

1. Report the power, wavelength, and insertion loss of the laser beams exiting from the prism, the lens, and the fibre cable. Report this for the laser beams turned on individually, three together, or two at a time.
2. Calculate the index of refraction (n) of the prism, using Equation 28.1.

$$n = \frac{\sin(\frac{A+\delta}{2})}{\sin\frac{A}{2}} \qquad (28.1)$$

where

A is the apex angle of the prism, and
δ is the deviation angle of the light exiting from the prism.

28.9.8.2 Wavelength Division De-Multiplexer

1. Report the power, wavelength, and insertion loss of the laser beams exiting from the prism, the lens, and the fibre cable.
2. Calculate the index of refraction (n) of the prism, using Equation 28.1.

28.9.9 RESULTS AND DISCUSSIONS

28.9.9.1 Wavelength Division Multiplexer

1. Compare the power and wavelength of the laser beams from the laser sources (input) to the laser beams exiting from the fibre cable (output).

2. Compare the powers, wavelengths, and insertion loss, when the laser sources are turned on individually, three together, or two at a time.
3. Compare the insertion loss at the prism, the lens, and the fibre cable, among the above set-ups.

28.9.9.2 Wavelength Division De-Multiplexer

1. Report the power and wavelength of the light source, and each colour of the spectrum.
2. Report the angles of refraction and deviation of each colour of the spectrum.
3. Compare the power and wavelength of the light source, and each colour of the spectrum.

28.9.10 CONCLUSION

Summarize the important observations and findings obtained in this lab experiment.

28.9.11 SUGGESTIONS FOR FUTURE LAB WORK

List any suggestions for improvements using different experimental equipment, procedures, and techniques for any future lab work. These suggestions should be theoretically justified and technically feasible.

28.10 LIST OF REFERENCES

List any references that were used in the report. Use one format in writing the references. Never mix reference formats in a report.

28.11 APPENDIX

List all of the materials and information that are too detailed to be included in the body of the report.

FURTHER READING

Agrawal, G. P., *Fiber-Optic Communication Systems*, 2nd ed., John Wiley & Sons, Inc., New York, 1997.
Cole, M., *Telecommunications*, Prentice Hall, New Jersey, 1999.
Derfler, F. J., Jr., Freed, L., *How Networks Work,* Millennium Edition, Que Corporation, IN, U.S.A., 2000.
Dutton, H. J. R., *Understanding Optical Communications*, IBM, Prentice Hall, Inc., Engelwood Cliffs, NJ, 1998.
EXFO, *Guide to WDM Technology and Testing*, EXFO Electro-Optical Engineering, Inc., Quebec City, Canada, 2000.
Glance, B., Wavelength-tunable add/drop optical filter, *IEEE Photon. Techn. Lett.*, 8, 245–247, 1996.
Goff, D. R. and Hansen, K. S., *Fiber Optic Reference Guide: A Practical Guide to the Technology*, 2nd ed., Butterworth-Heinemann, London, 1999.
Hecht, J., *Understanding Fiber Optics*, 3rd ed., Prentice Hall, Inc., Englewood Cliffs, 1999.
Hioki, W., *Telecommunications*, 3rd ed., Prentice Hall, New Jersey, 1998.
Jackel, J., Goodman, M. S., Baran, J. E., Tomlinson, W. J., Chang, G.-K., Iqbal, M. Z., Song, G. H. et al., Acousto-optic tunable filters (AOTF's) for multiwavelength optical cross-connects: Crosstalk considerations, *J. Lightwave Techn.*, 14, 1056–1066, 1996.
Kao, C. K., *Optical Fiber Systems: Technology, Design, and Applications*, McGraw-Hill, New York, 1982.
Kartalopoulos, S. V., *Introduction to DWDM Technology—Data in a Rainbow*, SPIE Optical Engineering Press, Bellingham, WA, 1999.
Keiser, G., *Optical Communications Essentials*, 1st ed., McGraw-Hill, New York, 2003.

Keiser, G., *Optical Fiber Communications*, 3rd ed., McGraw-Hill, New York, 2000.

Kolimbiris, H., *Fiber Optics Communications*, Prentice Hall, Inc., New Jersey, 1998.

Laude, J. P., *DWDM Fundamentals, Components, and Applications*, Artech House, Boston, 2002.

Li, X., Chen, J., Wu, G., and Ye, A., Digitally tunable optical filter based on DWDM thin film filters and semiconductor optical amplifiers, *Opt. Express.*, 13 (4), 1346–1350, 2005.

Malacara, D., *Geometrical and Instrumental Optics*, Academic Press, Boston, 1988.

Mazzarese, D., Meeting the OM-3 challenge—Fabricating this new generation of multimode fiber requires innovative state of the art designs and testing processes, *Cabling Systems* (December 2002), pp. 18–20, 2002.

Palais, J. C., *Fiber Optic Communications*, 4th ed., Prentice Hall, Inc., Englewood Cliffs, 1998.

Razavi, B., *Design of Integrated Circuits for Optical Communications*, McGraw-Hill, New York, 2003.

Senior, J. M., *Optical Fiber Communications Principle and Practice,* 2nd ed., Prentice Hall, U.S.A., 1992.

Shamir, J., *Optical Systems and Processes*, SPIE Optical Engineering Press, Bellingham, WA, 1999.

Watanabe, T., Inoue, Y., Kaneko, A., Ooba, N., and Kurihara, T., Polymeric arrayed-waveguide grating multiplexer with wide tuning range, *Electron. Lett.*, 33 (18), 1547–1548, 1997.

Yeh, C., *Handbook of Fiber Optics: Theory and Applications*, Academic Press, San Diego, 1990.

29 Optical Amplifiers

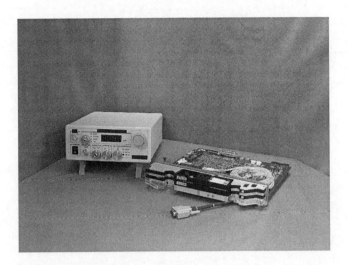

29.1 INTRODUCTION

The development of the optical amplifier (OA) in the late 1980s revolutionized communication systems. Optical amplifiers had an important impact similar to the invention of the laser in early 1960s. Both devices contributed to the development of communication systems and other applications, such as lower pump optical power, single pixel multicolour displays, and light emitting devices. An optical amplifier is a device that amplifies the optical signal directly, without converting it to an electrical signal and then to an optical signal again. OAs are used for amplifying a weak optical signal in order to increase the distance the signal can be transmitted down the transmission lines. In comparison, repeaters and generators convert the signal to electrical form, regenerate or amplify the signal, and then convert it to optical form again. The conversion of the signal from one form to another is a complex process, subject to high losses, slow speed, and more costly than simpler optical amplifiers.

This chapter presents the fundamentals of OAs, their characteristics, and classifications of common ones; it also describes their applications. The three basic technologies are: Erbium-doped fibre optical amplifiers (EDFAs), semiconductor optical amplifiers (SOAs), and Raman optical amplifiers. These technologies and applications are presented in detail in this chapter.

29.2 BASIC APPLICATIONS OF OPTICAL AMPLIFIERS

Optical amplifiers are used to boost signals transmitted over long distances in network systems. Figure 29.1 is a schematic diagram illustrating the components of a basic communication system.

The characteristics and advantages of the OAs have led to many applications. OAs can be used to boost signal power after multiplexing, or before demultiplexing, or at any point in modern optical networks. OAs are ideal for Merto and Long-Haul dense wavelength multiplexing (DWDM) as well as single wavelength applications. The optical design, coupled with sophisticated control circuitry, allows these OAs to provide constant gain even with signals being added to the network, such as λ_4

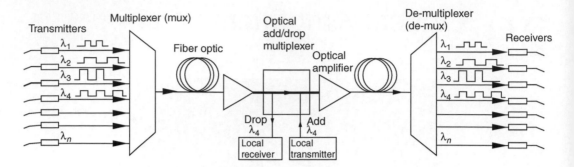

FIGURE 29.1 A schematic diagram of basic communication system.

in Figure 29.1. OA operation is independent of the signal data rate. OAs have different designs and packaging shapes. The following sections present the general applications of OAs in optical networks:

29.2.1 IN-LINE OPTICAL AMPLIFIERS

An optical amplifier is used as a in-line amplifier allowing signals to be amplified within the optical signal path, as shown in Figure 29.2. Optical amplifiers can compensate for transmission loss and thus increase the distance between the transmitters and receivers. This application enables the signal to travel through lines hundreds of kilometres long.

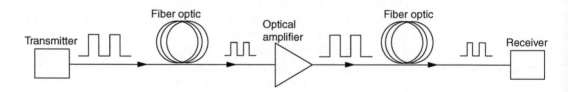

FIGURE 29.2 In-line optical amplifier.

29.2.2 POSTAMPLIFIER

An optical amplifier is used as a postamplifier when placed immediately after the transmitter, as shown in Figure 29.3. An optical amplifier will boost the light signal to the required power level at the beginning of a fibre line. This arrangement enables the signal to travel hundreds of kilometres down the fibre cable. A common application of the postamplification technique, together with an

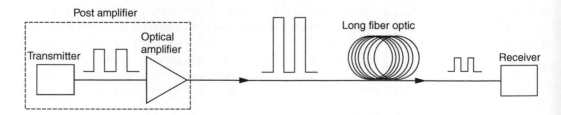

FIGURE 29.3 Postamplifier application.

optical preamplifier at the receiving end, enables continuous underwater transmission distances up to a few kilometres.

29.2.3 PREAMPLIFIER

An optical amplifier is used as a preamplifier when placed immediately before the receiver, as shown in Figure 29.4. An OA will boost the light signal to the required power level before being received by an optical receiver. The preamplifier enables the signal to be processed directly by the receiver. The preamplification technique is commonly used before optical herodyne detectors or avalanche photodiodes.

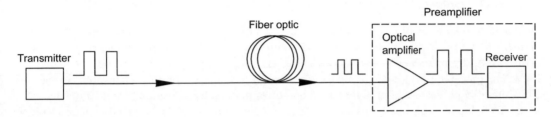

FIGURE 29.4 Preamplifier application.

29.2.4 IN LOCAL AREA NETWORKS

Optical amplifiers are used to boost signals in local area networks, when placed in sub centres within the transmission lines and branches, as shown in Figure 29.5. This type of arrangement also enables optical amplifiers to have the characteristics suitable for analogue transmission. The video

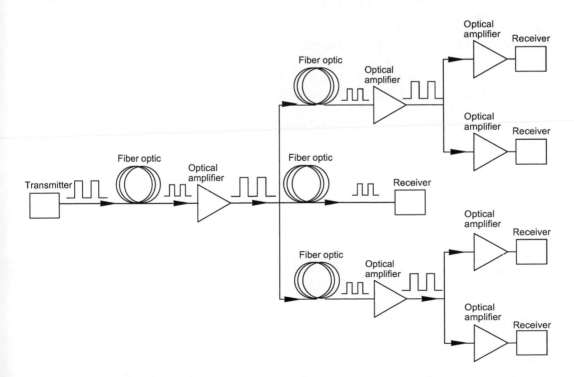

FIGURE 29.5 In local area network application.

content multiplexed by head-end equipment is converted from optical/electrical transmission into the 1550 nm band-wavelength optical video signal that is suitable for optical amplification. The optical video signals are optically amplified to compensate the losses of splitting and transmission. The signals repeatedly undergo amplification, splitting, and transmission.

29.3 TYPES OF OPTICAL AMPLIFIERS

Fundamental operation of each type of OA will be presented in the classifications of the following sections.

29.3.1 DOPED FIBRE OPTICAL AMPLIFIERS

Doped fibre optical amplifiers are made from optical fibre whose core is doped with atoms of an element that can be excited by an external pump light to a state where stimulated emission can occur. Pump light from an external laser source is steadily pumped in one end or both ends of the fibre. The pump light is guided along the fibre length where it excites the doped atoms of the core. The core also guides the input light signal and the amplified light resulting from stimulated emission. The doped material types and doping concentrations in the fibre core depend on the wavelengths of light to be amplified. This is a general description of the operation of the doped fibre optical amplifiers. The following sections present details of each type.

29.3.1.1 Erbium-Doped Fibre Optical Amplifiers

In the late 1980s, a group of researchers at the University of Southampton in the United Kingdom successfully developed the Erbium doped fibre amplifier (EDFA). The EDFA then became the dominant type of optical amplifier. EDFAs combined with wavelength division multiplexing technology are widely used in long-distance optical communications, networks, and signal modulation. EDFAs operate in the 1540–1570 nm range, called the C-band by convention. Erbium ions have quantum levels that allow them to be stimulated to emit light in the C-band, which is the wavelength band having the least power loss in most silica-based fibre, where high-quality amplifiers are most needed.

Figure 29.6 shows the basic operation of the EDFA. An Erbium doped fibre amplifier consists of a 10–30 metre length of optical fibre, the core of which is doped with a rare-earth element of Erbium (Er^{3+}). The fibre is pumped with a laser light at 980 or 1480 nm to raise the Erbium ions to a high energy state. When an Erbium ion is in a high energy state, an incident photon of input light signal can stimulate the Erbium ions to give up some of its energy in the form of light and return to a lower energy state, which is more stable. This operation is called stimulated emission, and is generally presented in light production and laser theory.

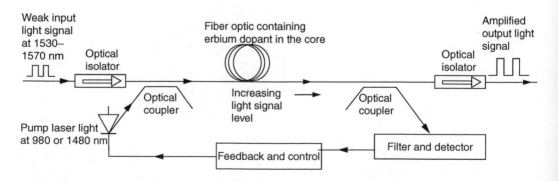

FIGURE 29.6 Erbium-doped fibre optical amplifier basic operation.

The pump laser power supplies the optical energy for the amplifier. The pumped laser light is mixed with the input light signal via a coupler at the input fibre cable. The mixed light is guided into the fibre section with Erbium ions included in the core. A photon of the laser excites the Erbium ion to its higher-energy state. When the photon of the input light signal meets the exited Erbium atom, the Erbium atom gives up some energy in the form of a photon and returns to its lower-energy state. The new photon is in exactly the same phase and direction as the light signal that is being amplified. Thus, the light signal is amplified along the fibre core in a forward direction of travel only. Figure 29.6 also shows the need to have a pump laser beam along the length of a fibre to provide the energy for EDFAs. This design requires power and optics, such as couplers and filters.

EDFAs also have gain that varies with a signal's wavelength, which creates problems in many WDM applications. This can be solved by using special optical passive filters that are designed to compensate for the gain variation of the EDFA.

Pumping power can be applied in a forward direction, as shown in Figure 29.6, backward from the output end, or in both directions. Optical isolators are commonly used at the output end or both ends of the EDFA, to prevent the pump power signal and light signal from returning back down the fibre, or unwanted reflections that may affect laser stability.

As explained above, the pump laser light is supplied using a coupler at the inlet of the fibre cable, as shown in Figure 29.6. Pump laser light can be pumped in the direction and/or opposite direction of the light signal. The pump signal can be coupled in various locations in the amplification system, as shown in Figure 29.7; Figure 29.7 also shows that the pump power can be (a) at the end of the fibre cable using a coupler, or (b) coupled on both ends. Figure 29.7(c) shows a design for the remote pumping of the power, used where the pump laser is a long distance from the amplifier, such as in undersea systems.

EDFAs have a number of main technical characteristics, such as efficient pumping, wavelength selection, minimal polarization sensitivity, low insertion loss, low distortion and interchannel crosstalk, high power output, low noise, very high sensitivity, low power consumption, and low cost.

29.3.1.2 Praseodymium-Doped Fluoride Optical Amplifiers

EDFAs have shifted the optical telecommunication emphasis towards the third transmission window, called long wavelength band in the 1510–1600 nm range. There is still great interest in 1300 nm in O-band amplifiers. This is mainly because a substantial part of the fibre optic network worldwide is designed for operation in the second transmission window of about 1310 nm. Praseodymium-doped fluoride fibre amplifiers (PDFFAs) can provide substantial gain in this region. However, to compete with EDFAs, the quantum efficiency of the 1310 nm transition of Pr3+ should be increased. Low-phonon-energy glass hosts are needed for this purpose. Other alternatives are directed towards Gallium-Lanthanum-Sulfide (GLS) and Gallium-Sulfide-Iodide (GSI) glasses.

29.3.1.3 Neodymium-Doped Optical Amplifiers

Neodymium (Nd)-doped optical amplifiers amplify in the 1310 nm band. Nd will amplify over the 1310–1360 nm range when doped into Fluorozirconate (ZBLAN) glass and over the 1360–1400 nm range when doped into silica. The most efficient pump wavelengths are at 795 and 810 nm.

29.3.1.4 Telluride-Based, Erbium-Doped Fibre Optical Amplifiers

Telluride-based, Erbium-doped optical amplifiers offer the potential optical bandwidth of over 76 nm in the 1532–1608 nm band, thus increasing the potential bandwidth of an Erbium doped optical amplifier from 30 to over 110 nm.

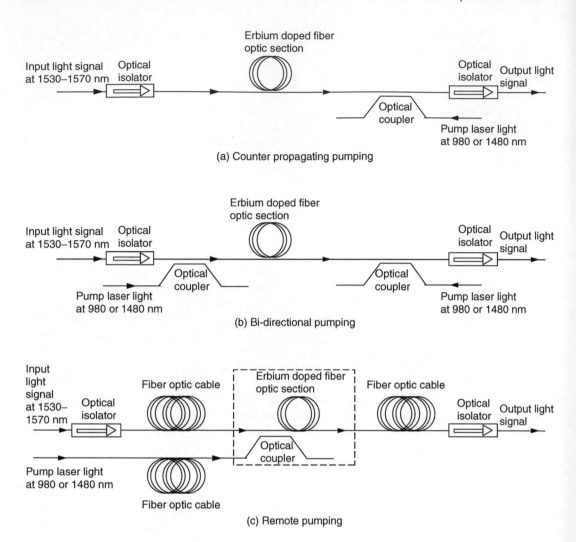

FIGURE 29.7 Erbium-doped fibre optical amplifier basic operation.

29.3.1.5 Thulium-Doped Optical Amplifiers

Thulium-doped optical amplifiers amplify between about 1450–1500 nm in the S-band.

29.3.1.6 Other Doped Fibre Optical Amplifiers

Other fibre optical amplifiers use doping materials, such as Ytterbium (Yb). The host fibre material can be silica, a fluoride-based glass, or a multi-component glass. Some plastic fibre amplifiers are under research and development. Modern plastics have characteristics similar to doped glass.

29.3.2 Semiconductor Optical Amplifiers

The functional applications of Semiconductor optical amplifiers (SOAs) were first studied in the early 1990s. Since then, the diversity and scope of such applications have been steadily growing. SOAs are another common type of in-line amplifiers that are developed to support dense wavelength division multiplexing (DWDM) and to expand to the other wavelength bands supported by fibre optics. They have many applications in optical fibre communication, switching, and signal processing systems.

SOAs are based on the same technology as basic semiconductor Fabry–Perot laser diodes, but they have anti-reflection (AR) coating at the endfaces. Fabry–Perot laser diodes are generally presented in the laser theory. The structure of the SOAs is much the same as the diode, with two stacked slabs of specially designed semiconductor material, with another material between them that forms the active layer, as shown in Figure 29.8. An electrical current is passed through the device in order to excite electrons to high-state level. The electrons then fall back to the non-excited ground state, emitting photons by stimulated emission. An incoming optical signal stimulates emission of photons at its own wavelength. This is accomplished by blocking the cavity reflectors using an antireflection (AR) coating on both end faces. Fibre optic cables are attached to both ends. As explained in EDFAs, optical isolators are commonly used at both ends of the SOA to prevent light signals from returning back. Depending on the material of the active layer, they operate from 1310 to 1550 nm in telecommunication systems.

SOAs are typically constructed in a small package. In addition, they transmit bi-directionally, making the reduced size of the device an advantage over EDFAs. They can be integrated with optical devices, such as semiconductor lasers, modulators, and DWDM. But the actual performance is still not compatible with EDFAs. They have high noise, less gain, medium polarization dependence, and high optical gain nonlinearity with fast transient time. High nonlinearity makes the SOAs attractive for optical signal processing, such as all-optic switching, wavelength conversion and regeneration, time demultiplexing, clock recovery, and pattern recognition. A number of SOA chips can be integrated on the same substrate to create high-density switching matrices.

SOAs are classified into two groups: Fabry–Perot Cavity Amplifiers (FPA) and Travelling Wave Amplifiers (TWA). The difference depends on the efficiency of the reflection value of the antireflection coating material used.

29.3.3 RAMAN FIBRE OPTICAL AMPLIFIERS

Raman optical amplifiers differ in principle from EDFAs. They utilize stimulated Raman scattering (SRS) to create optical gain. Stimulated Raman scattering occurs when light waves interact with molecular vibrations in a material having a solid lattice structure. In Raman scattering, the molecule absorbs the light, then quickly re-emits a photon with energy equal to the original photon, plus or

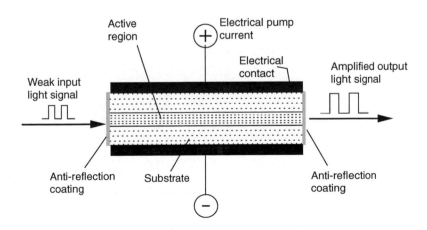

FIGURE 29.8 Semiconductor optical amplifier.

minus the energy of a molecular vibration mode. This has the effect of both scattering light and shifting its wavelength.

When a fibre transmits two suitably spaced wavelengths, stimulated Raman scattering can transfer energy from one wavelength to the other. In this case, one wavelength excites the molecular vibration; then light of the second wavelength stimulates the molecule to emit energy at the second wavelength.

Figure 29.9 shows the topology of a typical Raman optical amplifier. The pump laser and circulator comprise the two key elements of the Raman optical amplifier. The pump laser, in this case, has a wavelength of 1535 nm. Raman amplifiers work in the 1550 nm window. The circulator provides a convenient means of injecting light backwards into the transmission fibre with minimal optical loss. The pump laser is coupled into the transmission fibre either in the same direction as the transmission signal, which is called "co-directional pumping", or is coupled into the transmission fibre in the opposite direction, which is called "contra-directional pumping." Contra-directional pumping is more common because co-directional pumping has the problem of optical nonlinearity (nonlinear amplification). In contra-directional pumping, the attenuation of the pump light is so small that it travels a great distance, several kilometres along the transmission fibre. It also keeps pump photons from reaching the receiver, where they could interfere with reception of the desired signal.

Optical isolators are commonly used at both ends of the Raman amplifier to prevent pump power and light signal from returning back down the fibre or unwanted reflections that may affect laser stability. Raman amplifiers are used as pre-amplifiers, power amplifiers, and distributed amplifiers in digital and analogue transmissions in communication systems.

Figure 29.10 shows another technique for pumping light into Raman amplifiers. The amplification bandwidth can be extended, by using multiple pump light sources along with a wavelength division multiplexer (WDM). This can be done by using more than two pump light sources, producing a broadband amplifier over bands of more than 100 nm, for example in the range of 1500–1600 nm.

Raman amplifiers have a number of main technical characteristics, such as efficient pumping, simplicity, wider wavelength coverage, minimal polarization sensitivity, low insertion loss, high gain, low noise, fast reaction to changes of the pump power, low power consumption, and low cost. These amplifiers are used in long-haul and ultra-haul DWDM transmission systems

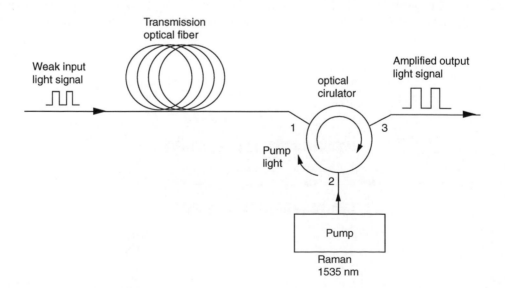

FIGURE 29.9 Raman optical amplifier.

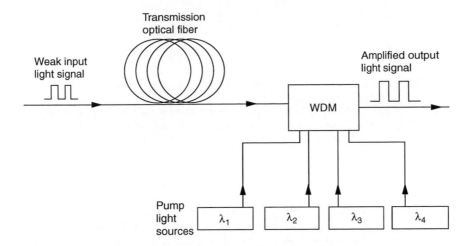

FIGURE 29.10 Raman optical amplifier with multiple pump light sources.

29.3.4 PLANER WAVEGUIDE OPTICAL AMPLIFIERS

Rare-earth-doped planar waveguide optical amplifiers are becoming increasingly important and provide compact and inexpensive alternatives to fibre amplifiers. In addition, planar technology is quite suitable for optical integration and will be essential to the development of fully integrated advanced optical devices.

29.3.5 LINEAR OPTICAL AMPLIFIERS

The design of the linear optical amplifier is similar to that of the semiconductor optical amplifier. The device has an active waveguide gain region; the input and output fibres are aligned to this waveguide. Unlike a semiconductor optical amplifier, the linear optical amplifier also features an integrated orthogonal laser that shares the gain region of the waveguide, as shown in Figure 29.11. This laser makes the amplifier gain linear. It also acts as an ultra-fast optical feedback circuit that responds to changes in the network. During the operation, the multiple wavelength signals to be

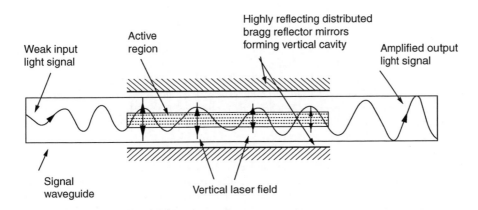

FIGURE 29.11 Linear optical amplifier.

amplified pass horizontally through the device, directly through the path of the laser, which is pumping photons of light vertically in the same device. The linear optical amplifiers are designed to be small and low cost. They are used in high data rate telecommunication systems.

29.4 OTHER TYPES OF OPTICAL AMPLIFIERS

There are many other types of optical amplifiers under research and development. The main objective is to increase the wavelength range and to increase the bit rates in communication systems along with improving the performance and decreasing the cost.

FURTHER READING

Agrawal, Govind P., *Fiber-Optic Communication Systems*, 2nd ed., Wiley, New York, 1997.

Bar-Lev, A., *Semiconductors and Electronic Devices*, 2nd ed., Prentice Hall International, Inc., New York, 1984.

Becker, Philippe M. et al., Erbium-doped fiber amplifiers: Fundamentals and technology, *Elsevier*, 1999.

Becker, P. C., Olsson, N.A, and Simpson, J. R., *Erbium-Doped Fiber Amplifiers: Fundamentals and Technology*, Academic Press, San Diego, 1999, chap. 6, See also pp. 66–75, p. 184, pp. 161–197.

Bise, R.T., Tunable photonic bandgap fiber, Presented at Proc, Optical Fiber Communication Conference and Exhibit,17March, pp. 466–468, 2002.

Cantore, Francesca and Della Corte, Francesca G., 1.55-μm silicon-Based reflection-type waveguide-integrated thermo-optic switch, SPIE: The International Society for Optical Engineering, *Opt. Eng.*, 42 (10), 2003.

Chee, J. K. and Liu, J. M., Polarization-dependent parametric and Raman processes in a birefringent optical fiber, *IEEE J. Quantum Electron.*, 26, 541–549, 1990.

Chen, Kevin P., In-fiber light powers active fiber optical components, *Photonics Spectra*, 78–90, April 2005.

Cole, Marion, *Telecommunications*, Prentice Hall, New Jersey, 1999.

Connelly, Michael J., *Semiconductor Optical Amplifiers*, 1st ed., Kluwer Academic Publishers, U.S.A., 2002.

Derfler, Frank J. Jr., Freed, Les, *How Networks Work*. Millennium Edition, Que Corporation, 2000.

Derickson, Dennis, *Fiber Optic Test and Measurement*, Prentice Hall PTR, New Jersey, 1998.

Digonnet, Michael J.F., *Rare-Earth Fiber Lasers and Amplifiers*, Marcel Dekker, New York, 2001.

Dutta, Niloy K. and Wang, Qiang, *Semiconductor Optical Amplifiers*, 1st ed., World Scientific, Singapore, 2005.

Dutton, Harry J.R., *Understanding Optical Communications, IBM*, Prentice Hall, Inc., Upper Saddle River, NJ, 1998.

Eggleton, B. J. et al., Microstructured optical fiber devices, *Opt. Express*, 17, 698–713, 2001.

Fedder, G. K. and Howe, R. T., Multimode digital control of a suspended polysilicon microstructure, *IEEE J. MEMS*, 5 (4), 283–297, 1996.

Fedder, G. K., Iyer, S., and Mukherjee, T., Automated optimal synthesis of microresonators. Presented at the IEEE Int. Conf. on Solid-State Sensors and Actuators (Transducers '97), 16–19 June, in Chicago, IL, *Tech. Dig.*, 2, 1109–1112, U.S.A., 1997.

Fludger, C. R. S., Handerek, B., and Mears, R. J., Pump to signal RIN transfer in Raman fiber amplifiers, *Lightwave Tech.*, 19, 1140–1148, 2001.

Green, Paul E., *Fiber Optic Networks*, Prentice Hall Pub Co., Englewood Cliffs, NJ, 1993.

Goff, David R. and Hansen, Kimberly S., *Fiber Optic Reference Guide: A Practical Guide to the Technology*, 2nd ed., Butterworth-Heinemann, Science and Technology Books Home, North America, 1999.

Headley, Clifford and Agrawal, Govind, *Raman Amplification in Fiber Optical Communication Systems*, Academic Press, San Diego, 2005.

Hecht, Jeff, *Understanding Fiber Optics*, 3rd ed., Prentice Hall, Inc., New Jersey.

Hioki, Warren, *Telecommunications*, 3rd ed., Prentice Hall, New Jersey, 1998.

Ho, M. C., Uesaka, K., Marhic, M., Akasaka, Y., and Kazovsky, L. G., 200-nm-bandwidth fiber optical amplifier combining parametric and Raman gain, *J. Lightwave Tech.*, 19, 977–981, U.S.A., 2001.

Hoss, Robert J., *Fiber Optic Communications—Design Handbook*, Prentice Hall Pub. Co., New Jersey, 1990.

IGI Consulting, Inc., *Optical Amplifiers: Technology and Systems*, Global Information, Inc., U.S.A., 1999.

Jeong, Y. et al., Ytterbium-doped double-clad large-core fibers for pulsed and CW lasers and amplifiers, *SPIE*. 140–150, 2004.

Kao, Charles K, *Optical Fiber Systems: Technology, Design, and Applications*, McGraw-Hill, New York, 1982.

Kashyap, R., *Fiber Bragg Gratings. 458*, Academic Press, New York, 1999.

Keiser, Gerd, *Optical Fiber Communications*, 3rd ed., McGraw-Hill Pub., New York, 2000.

Keiser, Gerd, *Optical Communications Essentials*, 1st ed., McGraw-Hill Pub., New York, 2003.

Kolimbiris, Harold, *Fiber Optics Communications*, Prentice Hall, Inc., New jersey, 2004.

Li, Xinwan et al., Digitally tunable optical filter based on DWDM thin film filters and semiconductor optical amplifiers, *Opt. Express*, 13 (4), 2005.

Litchinitser, N. M. et al., Application of an ARROW model for designing tunable photonic devices, *Opt. Express*, 19, 1540–1550, 2004.

Nortel, Networks, Products, Services, and Solutions Change Notification, Nortel Equipment, Nortel Products Manual, Canada, 2005.

Palais, John C., *Fiber Optic Communications*, 4th ed., Prentice Hall, Inc., New Jersey, 1998.

Ralston, J. M. and Chang, R. K., Spontaneous-Raman-scattering efficiency and stimulated scattering in silicon, *Phys. Rev. B.*, 2, 1858–1862, 1970.

Saini, S., Michel, J., and Kimerling, L. C., Index scaling for optical amplifiers, *IEEE J. Lightwave Tech.*, 21 (10), 2368–2376, 2003.

Salah, B. E. A. and Teich, M. C., *Fundamentals of Photonics*, Wiley, New York, 1991.

Senior, John M., *Optical Fiber Communications,* 2nd ed., Prentice Hall, U.S.A., 1993.

Shamir, J., *Optical Systems and Processes*, SPIE-The International Society for Optical Engineering, SPIE Press, U.S.A., 1999.

Shotwell, R. A., *An Introduction to Fiber Optics*, Prentice Hall, Englewood Cliffs, NJ, 1997.

Tran, A. V. et al., A bidirectional optical add-drop multiplexer with gain using multiport circulators. Fiber Bragg gratings, and a single unidirectional optical amplifier, *IEEE Photon. Technol. Lett.*, 15, 975–977, 2003.

Uskov, A. V., Mork, J., Tromborg, B., Berg, T. W., Magnusdottir, I., and O'Reilly, E. P., On high-speed cross-gain modulation without pattern effects in quantum dot semiconductor optical amplifiers, *Opt. Commun.*, 227 (4–6), 363–369, 2003.

Uskov, A. V., Berg, T. W., and Mrk, J., Theory of pulse train amplification without patterning effects in quantum dot semiconductor optical amplifiers, *IEEE J. Quantum Electron*, 40 (3), 306–320, 2004.

Uskov, A. V., O'Reilly, E. P., Manning, R. J., Webb, R. P., Cotter, D., Laemmlin, M., Ledentsov, N. N., and Bimberg, D., On ultra-fast optical switching based on quantum dot semiconductor optical amplifiers in nonlinear interferometers, *IEEE Photon. Technol. Lett.*, 16 (5), 1265–1267, 2004.

Vail, E., Wu, M. S., Li, G. S., Eng, L., and Chang-Hasnain, C. J., GaAs micromachined widely tunable fabry-perot filters, *Electron. Lett.*, 31 (3), 228–229, 1995.

Yariv, A., *Optical Electronics*, Wiley, New York, 1997.

Yariv, A., Universal relations for coupling of optical power between microresonators and dielectric waveguides, *Electron. Lett.*, 36, 321–322, 2000.

Yeh, C., *Handbook of Fiber Optics: Theory and Applications*, Academic Press, San Diego, CA, 1990.

Yeh, C., *Applied Photonics*, Academic Press, San Diego, CA, 1994.

Zirngibl, M., Joyner, C. H., and Glance, B., Digitally tunable channel dropping filter/equalizer based on waveguide grating router and optical amplifier integration, *IEEE Photon. Technol. Lett.*, 6, 513–515, 1994.

30 Optical Receivers

30.1 INTRODUCTION

Optical receivers are an essential part of a communication system. An optical receiver converts an optical signal, transmitted through an optical fibre cable into an electrical signal suitable for a receiving device installed at the other end of the communication system. The conversion process in the receiver is performed by two essential parts: a detector and an electronic signal processor. The detector converts the optical signal into an electrical signal. The electronic signal processor converts the raw detector signal into a form decipherable by the receiving device, such as a telephone, camera, or scanner.

This chapter presents two experimental cases: measuring light power using two types of photodetectors, and measuring the power output and calculating the efficiency of a solar cell with/without a filter and/or a lens.

30.2 FIBRE OPTIC RECEIVERS

The function of the optical receiver is to pick up an optical signal and convent it into an electrical signal suitable for a receiving device at the end of the communication line. Optical signals can be data, video, or audio. Optical detectors perform this conversion of an optical signal into an electrical signal. Therefore, optical receivers are sometimes called optical detectors. Optical detectors perform the opposite function of optical transmitters, such as light-emitting diodes and semiconductor lasers.

One application of an optical receiver is the conversion of an optical signal into a digital form, as shown in Figure 30.1. The incoming optical signal is converted to an electronic signal using a photodetector, such as a *p-i-n* photodiode or an avalanche photodiode. The signal is then amplified by a preamplifier, and passed through a bandpass filter that removes unwanted wavelengths. Further amplification, with a gain feedback control circuit, provides stable signal levels for the rest of the process. This control circuit controls the bias current, and thus the sensitivity of the photodiode.

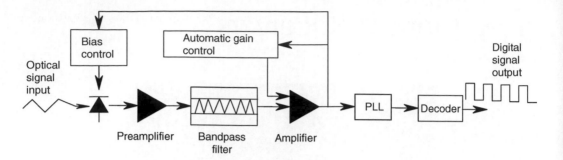

FIGURE 30.1 Optical signal conversion process.

A phase-locked loop (PLL) recovers the data bit stream and the timing information. The stream of bits needs to be decoded, from the coding used on the line, into its data format coding. This decoding process varies, depending on the encoding, and is occasionally integrated with the PLL, depending on the code in use.

The most common type of optical detector is the semiconductor photodiode, which produces current in response to incident light. Detectors operate based on the principle of the semiconductor diode (p–n junction). An incident photon striking the diode gives an electron in the valence band of the atom. If the photon has sufficient energy to move the electron to the conduction band, this creates a free-moving electron and a hole. If the creation of these carriers occurs in a depleted region, the charge carriers (electrons and holes) will quickly separate and create a current. As they reach the edge of the depleted area, the electrical forces diminish and current ceases.

30.3 PRINCIPLES OF SEMICONDUCTORS

The following section describes the fundamental principles of semiconductors. In semiconductors, "doping" means adding a small amount of material to the semiconductor. Doping changes the conductivity by creating an excess of electrons (n doped) or holes (p doped). This creates a p-region or n-region. The junction between n and p forms a diode. Superscripts of $+$ and $-$ indicate light and heavy doping. The intrinsic region has no doping.

30.3.1 P–N JUNCTION AND DEPLETION REGION

The descriptions of the n-region and p-region are generally presented in the semiconductor theory. A bias voltage V is applied to the p–n junction. Electrons diffuse away from the n-region into the p-region, leaving behind positively-charged ionized atoms (called "donors"), as shown in

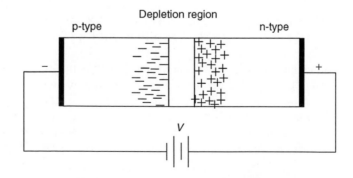

FIGURE 30.2 Junctions and depletion region with a reverse voltage bias.

Figure 30.2. In the p-region, these electrons recombine with the abundant holes. Similarly, holes diffuse away from the p-region, leaving behind negatively-charged ionized atoms (called "acceptors"). In the n-region, the holes recombine with the abundant mobile electrons.

This diffusion process cannot continue indefinitely, however, because it causes a disruption of the charge balance in the two regions. As a result, a narrow region on both sides of the junction becomes almost totally depleted of mobile charge carriers (electrons and holes). This region is called the depletion region. This region has a built-in electric field E_o, due to the accepters and donors beyond its edges. This occurs even without an applied voltage.

30.3.2 THE FUNDAMENTALS OF PHOTODETECTION

The fundamental principle behind the photodetection process is optical absorption. If the energy ($h\nu$) of an incident photon exceeds the energy of the band gap (between the conduction and valence bands) and is absorbed in the depletion region, then an electron moves up to the conduction band leaving a hole in the valence band. Thus an electron-hole pair is generated each time a photon is absorbed by the semiconductor. Under the influence of the electric field, electrons and holes are swept across the semiconductor in opposite directions. This flow of carriers results in the flow of electrical current called "generated photocurrent" when connected to an electric circuit. An applied voltage serves to speed up the carrier movement, increasing the current.

The fraction of light absorbed by the photodiode depends on:

1. Wavelength (λ) of the light, determined by the photon energy ($\varepsilon = h\nu = hc/\lambda$).
2. The thickness of the absorption material (depletion region width or depletion layer thickness).

30.3.3 LEAKAGE CURRENT

The current resulting from absorbed incident light is called "generated photocurrent." The current passing through the detector in the absence of light is called "dark current" or "leakage current." Low leakage current is an important measure of device quality. If the dark current is high, the generated photocurrent needs to be larger, in order to provide a good signal. Otherwise, the leakage current will dominate the detector current. Therefore, it is important to control the dark current below a certain value.

30.3.4 SOURCES OF LEAKAGE CURRENT

The three fundamental sources of leakage current are:

1. Generation-recombination (g–r) current: Arises from the generation and recombination of electron-hole pairs in the diode depletion region. The g–r current dominates the leakage current at low temperature.
2. Diffusion current: Arises from the diffusion of the minority carriers, toward or away from the junction, in the diode neutral region. In the case of P^+-n junction with the intrinsic region width larger than the hole diffusion length, the intrinsic region alone may be considered. Then diffusion current dominates the leakage current at high temperature.
3. Tunneling current: Refers to the band-to-band tunneling in the presence of high electric fields. A high field reduces the effective band gap barrier, allowing carriers to cross the band gap.

30.4 PROPERTIES OF SEMICONDUCTOR PHOTODETECTORS

The following sections describe the most common properties of semiconductor photodiodes.

30.4.1 QUANTUM EFFICIENCY

The quantum efficiency (QE) of a photodetector is a measure of how effectively the detector converts light into electrical current. QE denoted by η, is defined as the ratio of the flux of generated electron-hole pairs (EHPs) that contribute to the detector current, to the flux of the incident photons. The quantum efficiency of a detector is the ratio of the number of photons actually detected, to the number of incident photons. The QE range is $0 \leq \eta \gtreqless 1$. The quantum efficiency of the photodiode is defined as:

$$\eta = \frac{\text{Number of Free EHP Generated and Collected}}{\text{Number of Incident Photons}} \qquad (30.1)$$

Since QE is a function of photon energy, the QE is calculated at a particular photon energy. The measured photocurrent (I_{ph}) in the external circuit is due to the flow of electrons to the terminals of the photodiode. The number of electrons collected at the terminals per second is I_{ph}/e, where e is the charge of an electron. For incident optical power P_o, the number of incident photons arriving per second is $P_o/h\nu$. Thus, the QE η can also be defined as:

$$\eta = \frac{I_{ph}/e}{P_o/h\nu} \qquad (30.2)$$

30.4.2 RESPONSIVITY

The responsivity (R) of a photodetector is defined as the ratio of the photocurrent (I_{ph}) flowing in the device, to the incident optical power (P_o). Responsivity is measured in amps per watt. Thus:

$$R = \frac{\text{Photocurrent}}{\text{Incident Optical Power}} = \frac{I_{ph}}{P_o} \qquad (30.3)$$

Since R is a function of photon wavelength, R is calculated at a particular wavelength λ. Substituting Equation (30.2) into Equation (30.3) gives:

$$R = \eta \frac{e}{h\nu} = \eta \frac{e\lambda}{hc} \qquad (30.4)$$

As given in Equation 30.2 and Equation 30.4, the efficiency and responsivity depend on wavelength. R is also called the spectral responsivity or radiant sensitivity. The R vs. λ characteristics represent the spectral response of the photodiode. Spectral response curves are generally provided by the manufacturer. The spectral response characteristics for various quantum efficiencies are shown in Figure 30.3, and can be calculated using Equation 30.4. The outer area of the detector has a higher responsivity than the centre area, which can cause problems when aligning the fibre cable to the detector.

30.4.3 RESPONSE TIME

Response time is defined as the time needed for the photodiode to respond to an optical input by producing photocurrent. When light incident on the photodiode generates an electron-hole pair in a

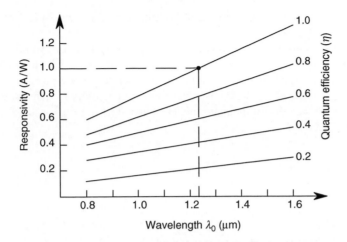

FIGURE 30.3 Responsivity vs. wavelength for various quantum efficiencies.

photodetector material, an electrical charge is generated in an external circuit, as shown in Figure 30.4. This electrical charge, due to the electron and hole, equals 2e (e is the charge of an electron).

The charge delivered to the external circuit, by the movement of carriers in the photodetector material, is not provided instantaneously. The charge is delivered over an extended period. It is as if the motion of the charged carriers in the material draws charge slowly away from the wire on one side of the device and then pushes it slowly into the wire at the other side. In this way, each charge passing through the external circuit is spread out in time. This phenomenon is

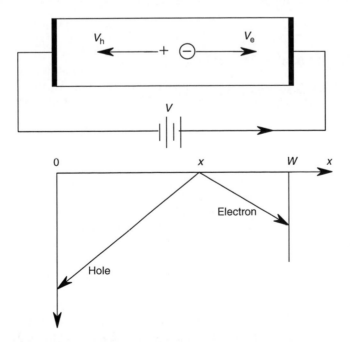

FIGURE 30.4 Generated electron-hole pair at position x.

called transit-time spread. It is an important limiting factor for the speed of operation of all semiconductor photodetectors.

Consider an electron-hole pair generated (by photon absorption, for example) at position x in a semiconductor material. The semiconductor material has width W, to which a voltage V is applied, as shown in Figure 30.4. When an electron-hole pair is generated at position x, the hole moves to the left with velocity v_h, and the electron moves to the right with velocity v_e. This movement terminates when the carriers reach the edge of the material. The current (i) in the external circuit, generated by this movement, is given by:

$$i(t) = -\frac{Q}{W} v(t) \tag{30.5}$$

where

t is the time
Q is the total charge of the photo-generated electrons.

If the voltage is increased, the electron velocity increases, and thus current increases. This means that for an input light pulse, the output current pulse will have a faster response time, for a higher applied voltage.

Response time can be affected by dark current, noise, responsivity linearity, back-reflection, and detector edge effect. Edge effect occurs because detectors only provide a fast response in their centre area.

30.4.4 SENSITIVITY

A photodetector is a device that converts photon energy into an electrical signal. A photodetector usually detects the energy of some photons better than others. Detection sensitivity is a function of the photon's energy being detected. The sensitivity is usually given as a function of the wavelength and expressed as the quantum efficiency. A high sensitivity allows a low level of light to be detected.

30.5 TYPES OF OPTICAL DETECTORS

There are many types of optical detectors including the phototransistor, photovoltaic, metal-semiconductor-metal (MSM), *pin* photodiode, and avalanche photodiode (APD). These detectors are explained in the following sections.

30.5.1 PHOTOTRANSISTORS

Phototransistors are the simplest type of photodetector. The basic operating principle of a phototransistor is shown in the cross-section in Figure 30.5. The device consists of an n–p–n junction, in which n is the emitter, p is the base, and the other n is the collector. The base terminal is normally open, and there is a voltage applied between the collector and emitter terminals (just as in the normal operation of common bipolar junction transistor (BJT)).

A large space charge layer (SCL) forms between the base and collector. The SCL region is called the absorption region. The operation of this device begins when an incident photon is absorbed in the SCL, and generates an electron-hole pair. The electrical field E_o drifts the electron and hole in opposite directions. Phototransistors operate as a photodetector that amplifies the photocurrent. An applied voltage V will increase E to become E_o plus V. When the drifting electron reaches the collector, it gets collected (and thereby neutralized) by the power supply (applied voltage). On the other hand, when the hole enters the neutral base region, it can only be neutralized

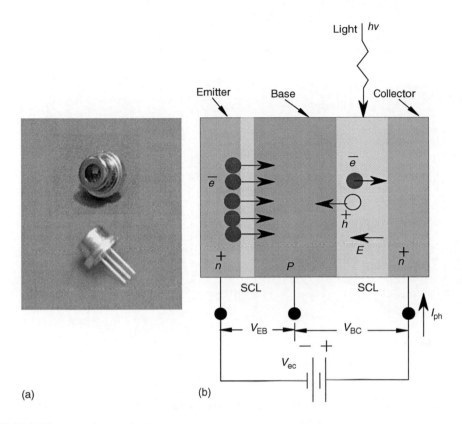

FIGURE 30.5 Phototransistor.

by injecting a large number of electrons into the base. It forces a large number of electrons to be injected from the emitter.

Normally, the electron recombination time in the base is very long, compared with the time it takes for electrons to diffuse across the base. This means that only a small fraction of electrons injected from the emitter can recombine with holes in the base. Thus, the emitter has to inject a large number of electrons to neutralize this extra hole in the base. These electrons diffuse across the base and reach the collector, and thereby create a photocurrent, which is amplified compared to the original electron. Thus, phototransistors have photocurrent gain.

30.5.2 PHOTOVOLTAICS

Photovoltaic panels, or solar cells, convert the incident solar radiation, through the photovoltaic effect, into electrical current. The basic principle behind this effect relies on the small energy gap between the valence and conduction bands of the photovoltaic material. When light photons incident on a photovoltaic have enough energy to excite electrons from the valence to the conduction band, the resulting accumulation of charge leads to a flow of current.

Figure 30.6 shows a typical solar panel and its cross-section. Consider a p–n junction with a very narrow and more heavily doped n-region. Solar radiation is incident on the thin n-side. The electrodes attached to the n-side must allow illumination to enter the cell and at the same time have a small series resistance. The electrodes are deposited on the n-side to form an array of finger electrodes on the surface. A thin antireflection coating on the surface reduces reflections and allows more light to enter the cells.

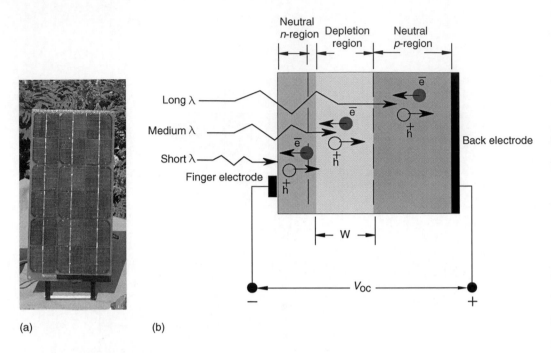

(a)   (b)

FIGURE 30.6 Photovoltaic panel.

The width (W) of the depletion region or the space charge layer (SCL) extends primarily into the p-side. Most photons are absorbed within the n-region and depletion region. Thus, short and medium wavelengths are absorbed. The generated electron-hole pairs are swept away by the built-in field E_o in the depletion layer. This creates an open circuit voltage V_{oc} between the electrodes. If an external load is connected, a photocurrent results.

The efficiency of a solar cell is one of its most important characteristics; it allows the device to be assessed economically, in comparison to other energy conversion devices. The solar cell efficiency refers to the fraction of incident light energy converted to electrical energy. This conversion efficiency depends on the semiconductor material properties, the device structure, and the incident light wavelength spectrum, which is mostly solar radiation. The efficiency of a solar cell decreases with increasing temperature. Therefore, the temperature of solar cells must be controlled for maximum efficiency.

Most solar cells are silicon based; silicon based semiconductor fabrication is a very developed technology, enabling cost effective devices for energy production in remote applications. A solar cell fabricated by making a p–n junction in the same crystal is called "homojunction." A silicon homojunction solar cell is called a "single crystal passivated emitter rear locally diffused" (PERL) cell. It has higher efficiency than other types of semiconductor solar cells.

30.5.3 METAL-SEMICONDUCTOR-METAL DETECTORS

Metal-semiconductor-metal (MSM) detectors are probably the fastest and simplest optical detector to fabricate. The basic idea is to create a Shottky barrier, which forces the material at the surface to be depleted. This barrier is created by contacting a metal to the semiconductor surface. Figure 30.7 shows the cross-section of a metal area. The barriers are often in the form of inter-digitated metal fingers separated by a small distance, typically on the order of microns. The metal is usually opaque to the incoming light; the remainder of the surface area absorbs the light. All the depletion layers are

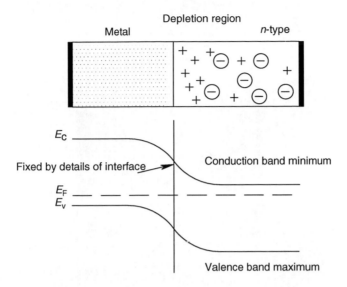

FIGURE 30.7 Shottky barrier and energy diagram.

connected together. Any absorbed light generates electron-hole pairs, which are quickly swept out to the contacts. Full-width at half maximum (FWHM) pulses are measured in picoseconds for such structures, since the response time is so quick.

When using the MSM for the detection of 1300–1550 nm ranges, the MSM suffers from two serious drawbacks:

1. Shottky barriers on indium phosphate (InP) tend to have high dark current, and therefore low receiver sensitivity.
2. Low quantum efficiency results, because the metal fingers prevent some of the incoming light from reaching the absorption layer.

30.5.4 THE P–I–N PHOTODIODES

In p–i–n photodiodes, the conversion of light into electrical current is achieved by the creation of free electron hole pairs by the absorption of photons. This absorption process creates electrons in the conduction band and holes in the valence band. Figure 30.8 shows the simplified structure of a typical p–i–n junction photodiode. The structure of the photodiode is the p^+-intrinsic-n^+ junction. The intrinsic silicon (i-Si) layer has much less doping than both the p^+ and n^+ regions, and it is much wider than these regions. At long wavelengths, where penetration depth is large, the photons can be absorbed in the wide depletion region. Thus, the width depends on the particular wavelength used in the application. In contrast, a p–n junction has a narrow depletion region and fewer photons are absorbed.

When the structure is formed, holes diffuse from the p^+ side, and electrons from the n^+ side, into the i-Si layer, also called the depletion region. In this region, they recombine (with other holes and electrons) and disappear. This leaves behind a thin layer of exposed, negatively-charged acceptor ions in the p^+ side, and a thin layer of exposed, positively-charged donor ions in the n^+ side. The two charges are separated and create the built-in electric field E in the i-Si layer. An exterior voltage increases E, which increases the response speed. While the photogenerated carriers

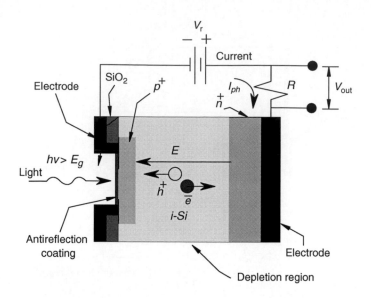

FIGURE 30.8 A p–i–n junction photodiode.

are drifting through the *i*-Si layer, they create an external photocurrent, when a voltage is applied. This photocurrent can be detected as the voltage across a small external resistor *R*, as shown in Figure 30.8. A larger thickness of the i-Si layer increases QE, but slows the response time since carriers have further to travel.

In some photodiodes, such as pyroelectric detectors, the energy conversion generates heat, which increases the temperature of the detector. The temperature increase changes the polarization and relative permittivity of the photodiode.

The p–i–n photodetectors offer high bandwidth, high quantum efficiency, and low dark current. High bandwidth and low dark current are important characteristics for good receiver sensitivity. However, the device has no gain, which places a lower limit on the sensitivity achievable, before dark current becomes significant. The p–i–n photodetectors are small devices with small capacitance, thus can detect high-speed signals with high sensitivity. The most important applications of the photodiodes are in optical communications.

30.5.5 AVALANCHE PHOTODIODES

Avalanche photodiodes (APDs) are high performance devices and are widely used in many applications, such as optical communication, due to their high speed and internal gain. Although not as fast as p–i–n photodiodes, the devices offer superior receiver sensitivity in their own bandwidth range. The device bandwidth at high gain is limited by the gain-bandwidth product (GBW), for InGaAs-InP APDs. However, there are difficulties in fabricating the device; this process requires stringent process control. For this reason, commercial high performance APDs cost more than similar photodetectors.

Figure 30.9 shows the cross-section of the structure of an InGaAs-InP avalanche photodiode with separate absorption and multiplication (SAM) regions. The InP multiplication (avalanche) layer has a wider bandgap than InGaAs absorption layer. The p-type and n-type doping of InP is indicated by capital letters, P and N.

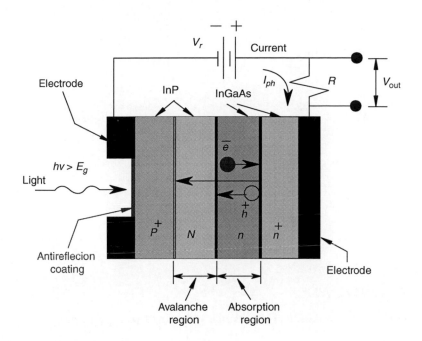

FIGURE 30.9 Avalanche photodiode.

The main depletion layer forms between the P⁺-InP and the N-InP layers and is within the N-InP. The electric field is greatest in this N-InP layer; this is therefore where avalanche multiplication takes place. With sufficient reverse voltage bias, the depletion layer in the n-InGaAs extends into the N-InP layer.

The electric field in the n-InGaAs depletion layer is not as great as that in the N-InP. Although long wavelength photons are incident on the InP side, they are not absorbed by InP, since the photon energy is less than the bandgap energy of InP (E_g = 1.35 eV). Long wavelength photons pass through the InP layer and are absorbed in the n-InGaAs layers.

The electric field E in the n-InGaAs layer drifts the holes to the multiplication region, where impact ionization multiplies the carriers. The impact ionization, from the physics point of view, is the mechanism that creates the internal current gain. Primary electrons and/or holes (carriers) are generated through the absorption of photons. Carriers can acquire large amounts of energy from a high E field, when the device has a strong reverse-voltage bias. This can be translated into high-speed motion. When a collision between a carrier and the lattice occurs, the energy from the carrier can be transferred to the lattice. Sufficient energy can be absorbed by the lattice for an electron to be promoted from the valance to the conduction band, creating an electron-hole pair. This process is called "impact ionization." These new carriers are swept out by E and can acquire high energy, causing further electron-hole pairs to be created. The entire process in which many carriers are created from one initial carrier is called "avalanche multiplication."

30.6 COMPARISON OF PHOTODETECTORS

The most common candidates for photodetectors are p–i–n photodiodes and APDs. They both have the same basic operating principle, where light is absorbed and converted to photocurrent. Table 30.1 highlights the major differences between p–i–n and APD detectors.

TABLE 30.1
Comparison of Photodetectors

P–I–N Detectors	APD Detectors
Fast	Not as fast
High bandwidth, up to 40 GHz at quantum efficiency >80%	Significantly less bandwidth
Low dark current	High dark current
No gain, which leads to lower sensitivity	Built-in gain, which extends the sensitivity to lower levels of received light
Conversion efficiency (responsivity) from 0.5 to 1.0 Amps/Watt	Conversion efficiency (responsivity) from 0.5 to 100.0 Amps/Watt
Less expensive	More expensive

30.7 EXPERIMENTAL WORK

The experiments on optical detectors include the following two experiments:

30.7.1 MEASURING LIGHT POWER USING TWO PHOTODETECTOR TYPES

Students will measure the power of a light source using two types of photodetectors, and compare among them with data sheets provided by the manufacturers. Figure 30.10 illustrates the experimental set-up.

30.7.2 PHOTOVOLTAIC PANEL TESTS

The purpose of this experiment is to measure the intensity of the solar radiation using a solar radiation meter (black and white pyranometer). Students will determine the converted power and efficiency of a photovoltaic panel subjected to solar radiation. The students will also determine the efficiency of the photovoltaic panel using different set-ups described by the following cases:

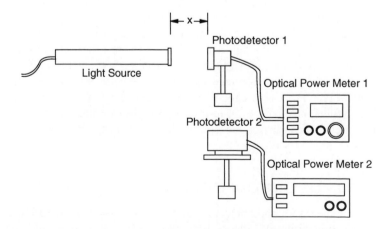

FIGURE 30.10 Measuring light power using a photodetector.

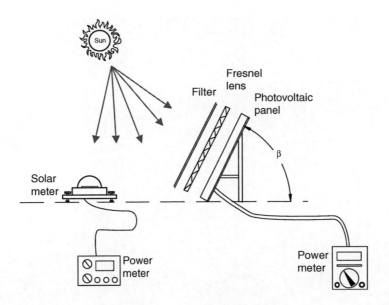

FIGURE 30.11 Photovoltaic panel with a lens and filter combination.

a. A photovoltaic panel producing electricity without using a filter or lens. Figure 30.11 (without a filter or lens) shows the experimental set-up for a photovoltaic panel producing electricity.

b. A photovoltaic panel producing electricity using a lens. Figure 30.11 (without a filter) shows the experimental set-up for a photovoltaic panel producing electricity using a lens to concentrate solar radiation.

c. A photovoltaic panel producing electricity using four filters (one at a time). Figure 30.11 (without a lens) shows the experimental set-up for a photovoltaic panel producing electricity using four types of filters. These filters are used to filter different wavelengths of the solar radiation.

d. A photovoltaic panel producing electricity using a combination between a lens and four filters (one at a time). Figure 30.11 shows the experimental set-up for a photovoltaic panel producing electricity using a combination of a lens and one filter. These filters are used to filter different wavelengths of the solar radiation.

30.7.3 TECHNIQUE AND APPARATUS

Appendix A presents the details of the devices, components, tools, and parts.

1. Fresnel lens and lens holder, as shown in Figure 30.12.
2. Four colour filters and filter holder, as shown in Figure 30.13.
3. Light source, as shown in Figure 30.14.
4. Photodetector, as shown in Figure 30.14.
5. Power light meter, as shown in Figure 30.15.
6. Photovoltaic panel and stand, as shown in Figure 30.15.
7. Solar radiation sensor (black and white pyranometer), as shown in Figure 30.15.
8. Solar radiation meter, as shown in Figure 30.15.
9. Hardware assembly (clamps, etc.).
10. Ruler.

FIGURE 30.12 A Fresnel lens.

FIGURE 30.13 Four colour filters.

FIGURE 30.14 Measuring light power using a photodetector.

FIGURE 30.15 Photovoltaic panel set-up.

30.7.4 Procedure

Follow the laboratory procedures and instructions given by the professor and/or instructor.

30.7.5 Safety Procedure

Follow all safety procedures and regulations regarding the use of pyranometer, photovoltaic panel, lens, filters, and power meters.

30.7.6 Apparatus Set-Up

30.7.6.1 Measuring Light Power Using Two Photodetector Types

1. Figure 30.14 shows the experimental apparatus set-up.
2. Prepare a light source on the table and connect it to the power supply.
3. Mount a photodetector, connected to an optical power meter, at a distance of 2 cm facing the light source.
4. Align the photodetector and light source to face each other.
5. Turn on the power supply, to provide electrical power to light source.
6. Turn off the lights of the lab.
7. Measure light power emitted by the light source, using the photodetector. Fill out Table 30.2.
8. Repeat steps 3–7 for the second type of the photodetector.
9. Turn on the lights of the lab.
10. Illustrate the location of the light sources and the photodetector head in a diagram.

30.7.6.2 Photovoltaic Panel Tests

Photovoltaic Panel

1. Figure 30.15 (without a filter or lens) shows the experimental apparatus set-up.
2. Mount a photovoltaic panel on the stand. Place the photovoltaic panel and stand on a flat surface. Place the photovoltaic panel facing the sun at the south orientation. Normally, the photovoltaic panel is positioned inclined at an angle equal to the geographic location of the site, to collect maximum solar energy without tracking the sun movement during the day.
3. Measure the electrical power produced by the photovoltaic panel. Fill out Table 30.3.
4. Place a solar radiation meter on the flat surface near the photovoltaic panel.
5. Measure the solar radiation intensity, using the solar radiation meter. Fill out Table 30.3.
6. Illustrate the locations of the photovoltaic panel and solar radiation meter in a diagram.

Photovoltaic Panel with a Lens

1. Repeat the procedure explained in Case (a) of this experiment, but use a Fresnel lens to concentrate solar radiation.
2. Mount the Fresnel lens on the photovoltaic panel. Keep a distance between the lens and the photovoltaic panel. The distance should be less than the focal length of the lens.
3. Measure the electrical power produced by the photovoltaic panel. Fill out Table 30.4.
4. Measure the solar radiation intensity, using the solar radiation meter. Fill out Table 30.4.
5. Illustrate the locations of the photovoltaic panel, Fresnel lens, and solar radiation meter in a diagram.

Photovoltaic Panel with a Filter

1. Repeat the procedure explained in Case (a) of this experiment, but use a filter without a lens. Using one colour filter, filter out some wavelengths of the solar radiation spectrum.
2. Measure the electrical power produced by the photovoltaic panel. Fill out Table 30.5.
3. Measure the solar radiation intensity, using the solar radiation meter. Fill out Table 30.5.
4. Repeat steps 2–3 using the three other colour filters.
5. Illustrate the locations of the photovoltaic panel, filter, and solar radiation meter in a diagram.

Photovoltaic Panel with a Lens and Filter Combination

1. Repeat the procedure explained in Case (a) of this experiment, but use a Fresnel lens and one filter.
2. Mount the Fresnel lens and one colour filter on the photovoltaic panel.
3. Measure the electrical power produced by the photovoltaic panel. Fill out Table 30.6.
4. Measure the solar radiation intensity, using the solar radiation meter. Fill out Table 30.6.
5. Repeat steps 3–4 using the three other colour filters.
6. Illustrate the locations of the photovoltaic panel, Fresnel lens, filter, and solar radiation meter in a diagram.

30.7.7 DATA COLLECTION

30.7.7.1 Measuring Light Power Using Two Photodetector Types

Measure the light power emitted by the light source, using two types of photodetectors. Fill out Table 30.2.

TABLE 30.2
Measuring Light Power Using Two Photodetector Types

Light Source Power (unit)	Optical Power Measurements by	
	Photodetector # 1 (unit)	Photodetector # 2 (unit)

30.7.7.2 Photovoltaic Panel Tests

30.7.7.2.1 Photovoltaic Panel

1. Measure the solar radiation intensity, using the solar radiation meter.
2. Measure the electrical power converted by the photovoltaic panel.
3. Fill out Table 30.3.

TABLE 30.3
Photovoltaic Panel Set-Up

Solar Radiation Intensity (unit)	Electrical Power (unit)

30.7.7.2.2 Photovoltaic Panel with a Lens

1. Measure the solar radiation intensity, using the solar radiation meter.
2. Measure the electrical power converted by the photovoltaic panel.
3. Fill out Table 30.4.

TABLE 30.4
Photovoltaic Panel with a Lens Set-Up

Solar Radiation Intensity (unit)	Electrical Power (unit)

30.7.7.2.3 Photovoltaic Panel with a Filter

1. Measure the solar radiation intensity, using the solar radiation meter.
2. Measure the electrical power converted by the photovoltaic panel.
3. Fill out Table 30.5.

TABLE 30.5
Photovoltaic Panel with a Filter Set-Up

Filter Types	Solar Radiation Intensity (unit)	Electrical Power (unit)
Red		
Yellow		
Green		
Blue		

30.7.7.2.4 Photovoltaic Panel with a Lens and Filter Combination

1. Measure the solar radiation intensity, using the solar radiation meter.
2. Measure the electrical power converted by the photovoltaic panel.
3. Fill out Table 30.6.

TABLE 30.6
Photovoltaic Panel with a Lens and
Filter Combination Set-Up

Filter Types	Solar Radiation Intensity (unit)	Electrical Power (unit)
Red		
Yellow		
Green		
Blue		

30.7.8 CALCULATIONS AND ANALYSIS

30.7.8.1 Measuring Light Power Using Two Photodetector Types

No calculations or analysis are required for this experiment.

30.7.8.2 Photovoltaic Panel Tests

No calculations or analysis are required for Cases (a), (b), (c), and (d).

30.7.9 RESULTS AND DISCUSSIONS

30.7.9.1 Measuring Light Power Using Two Photodetector Types

1. Study the specifications of the light source and the photodetectors.
2. Discuss the measurement and accuracy of the two types of photodetectors.
3. Compare the two types of photodetectors.
4. Verify the optical power measurements with the data provided by the manufacturer.

30.7.9.2 Photovoltaic Panel Tests

30.7.9.2.1 Photovoltaic Panel

1. Discuss the measurements and weather conditions.
2. Illustrate the locations of the photovoltaic panel and solar radiation meter in a diagram.

30.7.9.2.2 Photovoltaic Panel with a Lens

1. Discuss the measurements and how the power output is affected by the weather conditions. Verify the measurements obtained in this case with Case (a).
2. Illustrate the locations of the photovoltaic panel, lens, and solar radiation meter in a diagram.

30.7.9.2.3 Photovoltaic Panel with a Filter

1. Discuss the measurements and how the power output is affected by the weather conditions. Verify the measurements obtained in this case with Cases (a) and (b).
2. Illustrate the locations of the photovoltaic panel, filter, and solar radiation meter in a diagram.

30.7.9.2.4 Photovoltaic Panel with a Lens and Filter Combination

1. Discuss the measurements and how the power output is affected by the weather conditions. Verify the measurements obtained in this case with Cases (a), (b), and (c).
2. Illustrate the locations of the photovoltaic panel, lens, filter, and solar radiation meter in a diagram.

30.7.10 CONCLUSION

Summarize the important observations and findings obtained in this lab experiment.

30.7.11 SUGGESTIONS FOR FUTURE LAB WORK

List any suggestions for improvements using different experimental equipment, procedures, and techniques for any future lab work. These suggestions should be theoretically justified and technically feasible.

30.8 LIST OF REFERENCES

List any references that were used in the report. Use one format in writing the references. Never mix reference formats in a report.

30.9 APPENDICES

List all of the materials and information that are too detailed to be included in the body of the report.

FURTHER READING

Agrawal, G. P. and Dutta, N. K., *Semiconductor Lasers*, 2nd ed., Van Nostrand, New York, 1993.

Bar-Lev, A., *Semiconductors and Electronic Devices*, 2nd ed., Prentice Hall International, Inc, Upper Saddle river, NJ, 1984.

Bean, J., Optical wireless: Secure high-capacity bridging, *Fiber Opt. Technol.*, 10–13, January 2005.

Derfler, F. J. Jr. and Freed, Les, *How Networks Work, Millenium Edition*, Que Corporation, 2000.

Donati, S., *Photodetectors*, Prentice Hall PTR, Upper Saddle river, NJ, 2000.

Fedder, G. K. and Howe, R. T., Multimode digital control of a suspended polysilicon microstructure, *IEEE J. MEMS 5*, 4, 283–297, 1996.

Fedder, G. K., Santhanam, S., Rud, M. L., Eagle, S., Guillon, D. F., Lu, M., and Carliy, L. R., Laminated high-aspect-ratio microstructures in a conventional CMOS process, *Sens. Actuators A 57*, 2, 103–110, 1996.

Fedder, G. K., Iyer, S., and Mukherjee, T., Automated optimal synthesis of microresonators, Paper presented at IEEE Internationl Conference on solid-state sensors and actuators (Transducers '97), 16–19 June, in Chicago, IL, 1997.

He, S. and Mrad, R. B, A vertical bi-directional electrostatic comb driver for optical MEMS devices. Paper presented at Opto-Canada, SPIE Regional Meeting on Optoelectronics, Photonics and Imaging. SPIE, vol. TD01, 2002.

Heidrich, H., Albruht, P., Hamachr, H. P., Notting, H., Schroiter-Janssery, ??., and Uneinert, C.M, Passive mode converter with a periodically tilted InP/GaInAsP rib waveguide, *IEEE Photon. Technol. Lett.*, 4 (34), 36, 1992.

Him, F. C., Effects of passivation of GaN transistors, Paper presented to Epitaxy Group of the IMS at NRC in Ottawa, Canada, 2005.

Hioki, W., *Telecommunications*, 3rd ed., Prentice Hall, Upper Saddle river, NJ, 1998.

Horng, H. H., Designing optical-fiber modulators by using magnetic fluids, *Opt. Lett.*, 30 (5), 543–545, 2005.

Hoss, R. J., *Fiber Optic Communications—Design Handbook*, Prentice Hall Pub. Co, Englewood Cliffs, NJ, 1990.

Kaiser, R., Trommer, D., Heidrich, H., Fidorra, F., and Hamachr, M., Heterodyne receiver PICs as the first monolithically integrated tunable receivers, *Opt. Quantum Electron.*, 28, 565–573, 1996.

Kasap, S. O., *Optoelectronics and Photonics Principles and Practices*, Prentice Hall PTR, Upper Saddle river, NJ, 2001.

Litchinitser, N. M., Dunn, S. C., Steinvurzel, P. E., Eggletan, B. S., White, T. P., Mophedran, R. C., and de Sterky, C. M., Application of an ARROW model for designing tunable photonic devices., *Opt. Express*, 19 April, 1540–1550, 2004.

Mouthaan, T., *Semiconductor Devices Explained Using Active Simulation*, Wiley, New York, 1999.

Ralston, J. M. and Chang, R. K., Spontaneous-Raman-scattering efficiency and stimulated scattering in silicon, *Phys. Rev. B*, 2, 1858–1862, 1970.

Razavi, B., *Design of Integrated Circuits for Optical Communications*, McGraw-Hill Higher Education, U.S.A., 2003.

Salah, B. E. A. and Teich, M. C., *Fundamentals of Photonics*, Wiley, New York, 1991.

SCIENCETECH, Modular Optical Spectroscopy, *Designers and Manufacturers of Scientific Instruments Catalog*, SCIENCETECH, London, Ontario, Canada, 2005.

Senior, John M., *Optical Fiber Communications*, 2nd ed., Principle and Practice, Prentice Hall Europe, 1986.

Setian, Leo, *Applications in Electro-Optics*, Prentice Hall PTR, Upper Saddle River, NJ, 2002.

Simin, G., ELCT882: Basics of heterostructures. Online course taught at University of South Carlina, 2005.

Singh, J., *Semiconductor Devices: Basic Principles*, Wiley, 2001.

Sze, S. M., *Physics of Semiconductor Devices*, 2nd ed., Wiley, New York, 1981.

Watanabe, T., Inoue, Y., Kaneko, A., Ooba, N., and Kurinara, T., Polymeric arrayed-waveguide grating multiplexer with wide tuning range, *Electron. Lett.*, 33 (18), 1547–1548, 1997.

Yariv, A., *Quantum Electronics*, 3rd ed., John Willey & Sons, Inc., New York, U.S.A., 1989.

Yariv, A., *Optical Electronics*, Wiley, New York, 1997.

31 Lasers

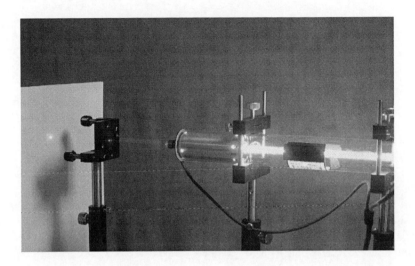

31.1 INTRODUCTION

The term *laser* is an acronym for light amplification by stimulated emission of radiation. Laser technology is one of the most rapidly developing areas in modern technology. When the first laser was developed by T.H. Maiman in the United States of America in 1960, it was called 'a solution in search of many problems' because there were no applications for the laser at that time. Since then, the laser has become the solution for many diverse applications, from a simple laser pointer to a sophisticated satellite tracking laser system.

To explain how the laser can be applied in such diverse areas, we need to understand the basic physical principles of the operation of a laser. Basically, the laser is a device that transforms various forms of energy into electromagnetic radiation. The energy put into the laser can be in many forms, such as electromagnetic radiation, electrical energy, or chemical energy. Energy is always emitted from the laser as electromagnetic radiation. Laser devices produce a narrow, intense beam of monochromatic coherent light. Laser beams are used to produce signals for fibre optic telecommunication systems and also for industrial, medical, and military applications, micromachining, etc.

This chapter has three experimental sections. One section involves the building a laser from individual components, learning how to align the components of a laser, and studying the effects of alignment on the production of various transverse modes. In the second section, the student will practise some alignment techniques for a laser beam. In the third section, the student will study laser beam expansion by building a Galilean telescope and a Keplerian telescope.

31.2 LIGHT EMISSION

Light can be produced through the rapid change of state of an electron from a state of relatively high energy to a lower energy or ground state. The energy of the electron has to leave the atom, usually in the form of a photon of light. A series of rapid energy state transitions will produce a stream of

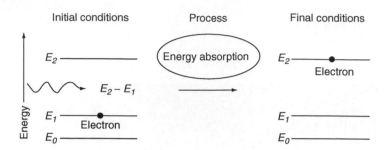

FIGURE 31.1 Absorption process.

photons, which can provide illumination. This principle of light emission is used in the operation of light bulbs, light emitting diodes, and lasers.

Artificial light can be produced by either spontaneous emission or a stimulated emission process. Absorption of energy occurs before either emission process.

When an atom at a low energy state E_1 absorbs energy, it will be elevated to a higher energy state E_2, as shown in Figure 31.1. The absorbed energy can come from many energy sources, such as electrical, thermal, chemical, optical, or nuclear energy sources.

The atom will eventually fall to a lower energy state; photons will be emitted, with photon energy being the difference between the two atomic energy states. This is the process of spontaneous emission. The energy of the photon is given by:

$$E_{photon} = E_2 - E_1 = hf \tag{31.1}$$

where h is Planck's constant (6.6261×10^{-34} J s) and f is the frequency of light (Hz).

The spontaneous emission of light serves as the basis of most lighting systems. For example, by passing an electric current through a metal wire (filament), the filament will begin to glow. The type and temperature of the wire will determine the wavelength range of the light being generated. This process occurs when an atom is initially in the higher energy level; it may drop spontaneously to the lower energy level and release its energy in the form of a photon, as shown in Figure 31.2. As such, light that is generated by this method is comprised of many different wavelengths and distributed in all directions.

The stimulated emission of light occurs when an incoming photon of energy hf stimulates the emission process by suddenly inducing the electron at E_2 to transit down to E_1. The emitted photon is in phase with, is in the same direction as, and has the same energy as the incoming photon, as shown in Figure 31.3. These two stimulated photons will stimulate more transitions. If conditions are appropriate, this process continues and results in monochromatic light being amplified to produce laser light. Again, *laser* is an acronym for light amplification by stimulated emission of

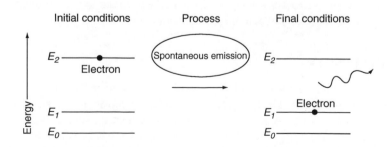

FIGURE 31.2 Spontaneous light emission.

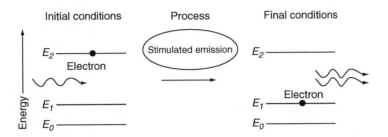

FIGURE 31.3 Stimulated light emission.

radiation. Stimulated emission can be achieved by using either light or electricity (or both). Lasers are the most common example of stimulated radiation emission.

31.3 PRINCIPLE OF THE LASER

The following is a brief introduction to the development of the laser. The laser was conceived after the maser, another acronym, which stands for Microwave Amplification by Stimulated Emission of Radiation. The first maser, using vibrations of ammonia molecules (NH_3), was developed by James Gordon, H.J.D. Zeiger, and Charles Townes in 1954. The ammonia molecules, when pumped to an excited state, radiated electromagnetic waves (microwaves) at a precise frequency (the same frequency used in the first atomic clock and the basis of extremely sensitive microwave amplifiers used by astronomers). Development of the maser was connected with radar development for military research. The first high power laser was developed by T.H. Maiman in the United States of America in 1960. It was a pink ruby rod with its ends silvered and placed inside a helical shaped flashlamp. This laser was only capable of pulsed operation. The laser is still widely used in many applications. Some of these applications will be presented later in this chapter.

The emitted laser beam is a nearly perfect plane wave. In contrast, an ordinary light wave emits light in all directions, and the emitted light is incoherent. The underlying aspects of laser production are based on quantum theory.

Electrons exist at specific energy levels, or states, characteristic of the particular atom or molecule. The energy levels can be imagined as rings or orbits around a nucleus. Electrons in the outer rings are at higher energy levels than those in the inner rings. When external optical or electrical energy is applied to a material, the atoms of the material get excited and are bumped up to higher energy levels. An atom that is already in the excited state may fall spontaneously to the lower state, resulting in the emission of energy in the form of a photon. However, if a photon with this same energy encounters the excited atom, the incident photon can stimulate the electron to make the transition to the lower state sooner.

Figure 31.4 shows dots that represent the energy state of one atom. In a normal situation, most of the electrons are found in a lower, or ground state, energy level of the atoms, as in Figure 31.4(a). If an external energy source is applied to excite many electrons into a higher energy level, a population inversion can be created in which more electrons are in the higher level than in the lower level, as shown in Figure 31.4(b). With the normal population in thermal equilibrium, as shown in Figure 31.4(a), some stimulated emission still occurs. But only in the non-equilibrium case, as shown in Figure 31.4(b), will light amplification occur due to stimulated emission. In stimulated emission, the original photon and a second one of the same frequency exist. These two photons are exactly in phase and move in the same direction, producing coherent light as a laser beam.

For further explanation, assume that the atoms have been excited to a higher state, as shown in Figure 31.5. Some of the excited atoms drop down spontaneously soon after being excited. If an emitted photon strikes another atom in the excited state, it stimulates this atom to emit a photon of

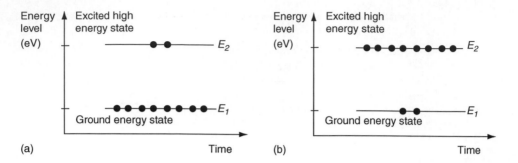

FIGURE 31.4 Two-Energy levels for a collection of atoms. (a) Normal population, (b) Inverted population.

the same frequency, moving in the same direction and phase. These two photons then move on to strike other atoms, causing more stimulated emissions. As the process continues, the number of photons multiplies. A laser device has two mirrors at each end. When the photons strike the end mirrors, most are reflected back. As they move in the opposite direction, the photons continue to stimulate more atoms to emit photons. As the photons move back and forth between the mirrors, a small percentage pass through the partially transparent mirror at one end, creating a narrow coherent laser beam.

Depending on the particular lasing material, specific wavelengths of light are absorbed and specific wavelengths are emitted. As a result, laser beams are very different from normal light. Laser beams have the following properties:

1. Laser light is monochromatic. This means that it contains one specific wavelength of light. The wavelength of laser light is determined by the amount of energy released when the electron drops to the lower orbit.

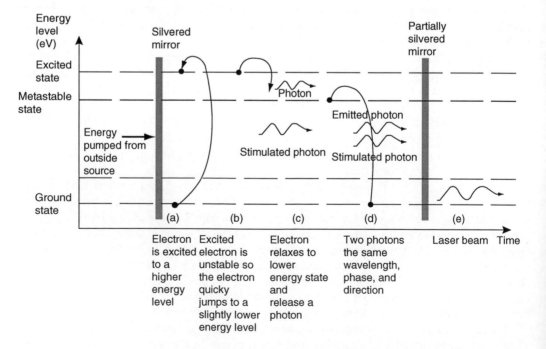

FIGURE 31.5 Energy level diagram showing excited atoms stimulated to emit laser light.

2. Laser light is coherent. This means that it is organized with each photon moving in step with the others. All of the photons' wave fronts are in unison.
3. Laser light is highly directional and has a tight concentrated beam. An ordinary light bulb, on the other hand, releases light in many directions, and the light is relatively weak and diffuse.

31.4 THERMAL EQUILIBRIUM AND POPULATION INVERSION

It is only possible to achieve gain in a laser, as described above, if the population of electrons in the laser's upper energy state is greater than the population in the lower state. This condition is called a population inversion. In thermal equilibrium, which is the typical state of normal matter (such as the air around us), the population ratio between two states is governed by the Maxwell–Boltzmann equation, which shows that higher energy levels always have lower populations than lower energy levels:

$$\frac{N_2}{N_1} = \exp\left(\frac{-(E_2 - E_1)}{k_B T}\right) = \exp\left[-\frac{h\nu}{k_B T}\right] \tag{31.2}$$

where N_2 is the population in the upper state, N_1 is the population in the lower state, k_B is Boltzman's constant $= 8.61738 \times 10^{-5}$ (eV/K), and T is the temperature.

For example, consider two energy states, $E_2 = 1.30$ eV and $E_1 = 1.00$ eV. Assume that there are 2×10^{15} electrons/cm^3 in E_1. At a temperature of 600°K, the number of electrons in state E_2 is calculated using Equation 31.2 as follows:

$$N_2 = 2.0 \times 10^{15} \times \exp\left(\frac{-(1.30 - 1.00)}{8.61738 \times 10^{-5} \times 600}\right) = 6.067 \times 10^{12} \text{ (electron/cm}^3\text{)}$$

The negative sign in the exponent suggests that a population inversion is only permitted under the conditions of negative temperature. This result was very disturbing to early laser researchers, as negative temperatures are not physically possible.

However, the Boltzman equation only describes conditions of thermal equilibrium. Lasers are not operated in thermal equilibrium. Instead, the upper state is populated by pumping it via some non-equilibrium process. A pulse of light, an electrical spark, or a chemical reaction can all be used to populate the upper laser state.

In order to determine when lasing will occur, the rate of change in population between the two energy levels in an atom needs to be estimated. The approach for this was initially done by Einstein for his new vision of the thermal radiation law. A few of his analysis steps are outlined below, using an approach very similar to his original paper.

Essentially, Einstein was rethinking the Blackbody thermal radiation law of Planck when he produced his seminal paper. Start with the concept of a flux of radiation u_ν impinging on a medium. The medium is absorbing radiation to a higher state and re-emitting radiation when moving to a lower state. For simplicity, only consider two states, a higher energy state E_2 and a lower state E_1. The rate of change, called the transition rate, of the number of atoms is proportional to the number of atoms in that state (N_1). The more atoms there are in a state, the more atoms that can leave this state per second.

When the transition occurs due to absorption, the process is driven by the photon flux. This is called stimulated absorption. The transition rate of absorption (from state 1 to state 2) is given by:

$$\left(\frac{dN_1}{dt}\right) = -B_{12} N_1 u_\nu \tag{31.3}$$

where B is a constant of proportionality between the two states. The minus is required because N_1 is decreasing.

Similarly, The transition rate of stimulated emission (from state 2 to state 1) is given as:

$$\left(\frac{dN_2}{dt}\right) = -B_{21}N_2 u_\nu \tag{31.4}$$

In the case of spontaneous emission, the rate is independent of the radiation field and is given as:

$$\left(\frac{dN_2}{dt}\right) = -A_{21}N_2 \tag{31.5}$$

The A and B constants are called the Einstein coefficients. Einstein assumed that there is thermodynamic equilibrium between the radiation field and the atoms, that the energy density has blackbody characteristics, and that the electron densities of the two states are governed by the Maxwell–Boltzmann distribution.

In equilibrium, the down transitions (emission) must equal the up transitions (absorption). In other words, the rate of absorption equals the rate of stimulated emission plus the rate of spontaneous emission. Therefore:

$$B_{12}N_1 u_\nu = B_{21}N_2 u_\nu + A_{21}N_2 \tag{31.6}$$

Dividing both sides by N_1 and rearranging gives:

$$\frac{N_2}{N_1} = \frac{B_{12}u_\nu}{A_{21} + B_{21}u_\nu} \tag{31.7}$$

Note that for this two-level systems N_2 can never be greater than N_1. A three or four-level system is needed for this to occur.

Using the Maxwell–Boltzmann Equation (31.2), the following equation can be obtained:

$$e^{-h\nu/k_B T} = \frac{B_{12}u_\nu}{A_{21} + B_{21}u_\nu} \tag{31.8}$$

Solving for u_ν and simplifying (assuming that B_{12} equals B_{21}, without proof) gives:

$$u_\nu = \frac{A}{B}\left[\frac{1}{e^{h\nu/k_B T} - 1}\right] \tag{31.9}$$

Planck's law of black body radiation describes radiant energy as a function of frequency. This law produces the following equations to generate the radiation curve for a blackbody as a function of frequency:

$$U_\nu d\nu = \frac{8h\nu^3}{c^3}\left[\frac{d\nu}{\exp\left(\frac{h\nu}{kT}\right) - 1}\right] \tag{31.10}$$

where ν is the frequency, $U_\nu\, d\nu$ is the energy density of radiation between the frequencies ν and $\nu + \delta\nu$, h is Planck's constant, 6.63×10^{-34} J s, c is the speed of light, 3.0×10^8 m/s, and k is Boltzmann's constant, 1.38×10^{-23} J.

Using Planck's law and comparing Equations (31.9) and (31.10), the following equation can be derived:

$$\frac{A}{B} = \frac{8h\nu^3}{c^3} \tag{31.11}$$

Therefore, there is a way of finding A in terms of B. If A or B can be measured, the rates for the system of interest can be found. Then the pump rate (the rate of energy pumped into a laser) needed to make a system laze can be estimated. However, to obtain conditions for population inversion (in a radiation field), a similar, but more complex, analysis of three or four-level systems is required.

Also note that the spontaneous emission rate increases as ν^3. This is why it is much more difficult to build short wavelength lasers (for example, UV and x-ray) than long wavelength (for example, infrared) lasers. At short wavelengths, electrons jump to a lower energy state spontaneously at a higher rate; therefore, fewer are available to emit lasing light by stimulation.

31.5 TRANSVERSE AND LONGITUDINAL MODES

The output spot of the laser beam, observed on a screen, is called the transverse electromagnetic mode (TEM). This intensity curve is a round mode with a Gaussian (bell shaped) profile in cross-section. However, it is possible to operate a laser having a wide variety, or combination, of other transverse modes. In these cases, the output beam may have a strange shape. Figure 31.6(a) shows that an off-axis transverse mode is able to self-replicate (resonate) after one round trip along the optical cavity. Figure 31.6(b) shows wavefronts in a self-replicating wave. Figure 31.6(c) shows four possible low order transverse cavity modes. Figure 31.6(d) shows the intensity patterns of

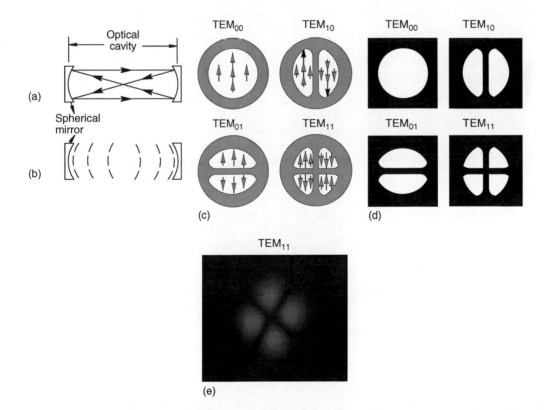

FIGURE 31.6 Laser cavity and transverse modes.

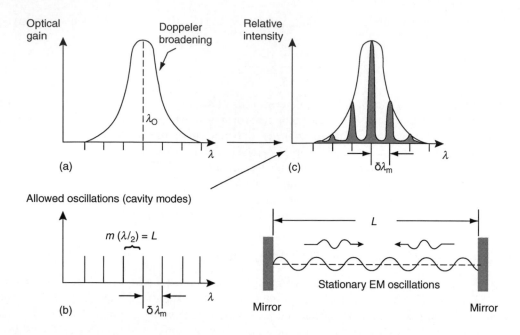

FIGURE 31.7 (a) Optical gain curve; (b) optical cavity modes; (c) output spectrum.

these modes. Most commercial lasers produce the transverse electromagnetic TEM_{00} mode. Figure 31.6(e) shows a photograph of the TEM_{11} mode.

A laser can only laze, however, at those wavelengths for which an integral number of half wavelengths fit into the laser (Fabry–Perot) cavity. The set of possible integral multiples of the cavity length is called the longitudinal electromagnetic modes of the cavity. The frequencies of these modes are given by:

$$\nu = \frac{mc}{2nL} \tag{31.12}$$

where m is the mode number, an integer as 1, 2, 3, and so on, n is the refractive index in the cavity, and L is the length of the cavity (resonator).

Notice that m will be a very large number in a typical laser system.

The spacing between longitudinal modes is frequently of interest and is given as:

$$\delta\nu = \frac{c}{2nL} \tag{31.13}$$

The spacing between modes is independent of wavelength. The longitudinal mode spacing can be expressed in wavelength, denoted by $\delta\lambda_m$ in Figure 31.7(b) and Figure 31.7(c).

31.6 GAIN

A given laser does not have gain at all frequencies. The function that describes the frequency dependence of the gain, g, is called the gain profile $g(\nu)$, or in terms of wavelength λ, $g(\lambda)$. Figure 31.7(a) shows an optical gain curve of the gain versus wavelength; this curve is characteristic of the lasing medium. Figure 31.7(b) shows the allowed modes and their wavelengths due to

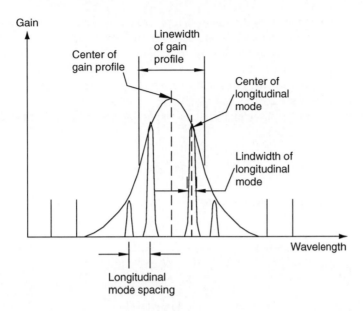

FIGURE 31.8 Optical gain curve of the lasing medium.

stationary (resonating) electromagnetic waves within the optical cavity. There are an infinite number of integral multiples (modes) of the cavity length. However, only a finite number will fit into the gain profile of the laser gain material. The actual output of the laser is the intersection of the set of possible longitudinal modes with the gain profile, as shown in Figure 31.7(c) and Figure 31.8. These figures show that the output spectrum (relative intensity versus wavelength) is determined by satisfying (a) and (b) simultaneously, assuming no cavity losses. Figure 31.8 shows that the centre longitudinal mode is not necessarily the same as the centre of the gain curve.

31.7 THRESHOLD CONDITION

As noted above, the laser consists of a pair of mirrors, between which is some active lasing medium made into an amplifier in a limited frequency range.

One of the reflectors is usually made slightly transparent at the wavelength of interest, and it has a reflectivity that can be denoted by r. At each reflection, a fraction $(1 - r)$ of the intensity is not returned to the resonator. The reflection coefficients are r_1 and r_2 for the pair of mirrors. Then the wave energy diminishes at each set of reflections by $r_1 r_2$.

In most lasers, there are other losses, but for the sake of simplicity consider only the reflective losses in this case. The fraction of intensity remaining after a full round trip passage through the laser cavity is:

$$e^{-2\gamma} = r_1 r_2 \tag{31.14}$$

The value of γ is positive and represents the amount of loss in a single passage. Solving for this loss gives:

$$\gamma = -\frac{1}{2} \log r_1 r_2 \tag{31.15}$$

If the medium is able to amplify the signal as it passes through and is able to compensate for the losses in the cavity then the intensity is increased by $e^{\alpha L}$ by virtue of the amplification in the gain material. Therefore, by taking the losses into consideration, the intensity changes from 1 to $F = e^{(\alpha L - \gamma)}$, in which F is the intensity factor. If F is greater than 1, the oscillations will grow; if less than 1, they will die out. At the threshold, F is equal to1; thus, threshold is clearly defined as $\alpha L = \gamma$. Lasing occurs when $\alpha L \geq \gamma$.

The amplification α in the laser material, as a function of frequency and relative population inversion n, can be shown to be:

$$\alpha = k(\nu)_0 n \tag{31.16}$$

where $k(\nu)_0$ is the absorption in the unexcited laser material. The relative population inversion is given as:

$$n = \frac{1}{N_0}\left(\frac{g_1}{g_2}N_2 - N_1\right) \tag{31.17}$$

where g is the degeneracy or multiplicity of the laser level.

A laser of a given length and mirror reflectivity will operate only if the population inversion is large enough to ensure that:

$$\alpha_m = nk_0 \geq \frac{\gamma}{L} \tag{31.18}$$

When the absorption coefficient k_0 has been determined experimentally, this equation can be used with a substitution of the relative population inversion to determine the threshold population inversion:

$$\left(\frac{g_1}{g_2}N_2 - N_1\right) = \frac{N_0\gamma}{k_0 L} \tag{31.19}$$

31.8 POWER AND ENERGY

Lasers can operate in either a continuous wave (CW) or a pulsed mode. Pulsed operation is occasionally used to reduce the heating of the laser (common for semiconductor diode lasers). However, in most cases, pulsed operation is combined with techniques such as Q-switching (which concentrates the laser energy into the pulse) and mode-locking (which shortens the width of the pulse in time). Q-switched and mode-locked lasers can concentrate very high peak power densities due to the relatively short length of the pulse.

The excitation of the atoms in a laser can be done continuously or in pulses. In a pulsed laser, the atoms are excited by periodic inputs of energy. The multiplication of photons continues until all the atoms have been stimulated to jump down to the lower energy state; this process is repeated with each input pulse. In a continuous laser, the energy input is continuous so that as atoms are stimulated to jump down to the lower level. They are soon excited back up to the upper energy level.

Some care must be taken in discussing the properties of pulsed versus CW lasers because the use of watts describing peak power can easily be confused with the use of watts describing average power. The following definitions will be used:

1. In a CW laser, the average power is the power (watts).
2. For pulsed lasers, the average power equals the energy per pulse (joules) times the pulse repetition rate (Hz):

$$P_{avg} = E_{pulse}R_{reprate} \tag{31.20}$$

3. The peak power is the energy per pulse (joules) divided by the temporal length of the pulse (seconds), given as:

$$P_{peak} = \frac{E_{pulse}}{t_{pulse}} \tag{31.21}$$

where P_{avg} is the average power density, in watts per unit area (cm^2 or m^2), and P_{peak} is the peak power density, in watts per unit area (cm^2 or m^2).

31.9 THREE AND FOUR LEVEL LASER SYSTEMS

Lasers are classified by the number of energy levels involved in the actual lasing process. Lasers classified as three or four energy level lasers are shown in Figure 31.9. As noted previously, it is not possible to achieve population inversion with a two-level system. These lasers are the most typical classes found in commercial systems.

To produce the required population inversion for laser activity, atoms or molecules must be excited to specific energy levels. Either light or electricity can provide the energy necessary to excite atoms to higher energy levels.

As previously discussed, the amount of time spent by an atom in an excited state is critical in determining whether it will be stimulated to emission or lose its energy through spontaneous emission. Excited states commonly have lifetimes of only nanoseconds before they release their energy by spontaneous emission, a period that is not long enough to likely undergo stimulation by another photon. A critical requirement for laser action, therefore, is an upper energy state that is long-lived. Such states do exist for certain materials and are referred to as metastable states. The average lifetime before spontaneous emission occurs for a metastable state is on the order of a microsecond to a millisecond, quite a lengthy period of time on the atomic timescale. With lifetimes this long, excited atoms can produce significant amounts of stimulated emission. Laser action is only possible if population builds up faster than it decays in the upper energy level, thus maintaining a population larger than that of the lower level. The longer the spontaneous emission lifetime, the more suitable the atom is for laser applications.

Figure 31.9(a) illustrates the simplest functional energy-level structure for laser operation, the three-level system. In this system, the ground state is the lower lasing level, and a population inversion is created between the ground level and a higher-energy metastable state. Most of the atoms are initially excited to a short-lived high-energy state that is higher than the metastable level. From this state they quickly decay to the intermediate metastable level, which has a much longer lifetime than the higher state (often on the order of 1000 times longer). Because each atom's residence time in the metastable state is relatively long, the population tends to increase and leads to a population inversion between the metastable state and the lower ground state (which is being depopulated to the highest level). Stimulated emission occurs because more atoms are available in the upper excited (metastable) state than in the lower state where absorption of light would most likely occur. Then, laser emission occurs between the metastable level and the ground state.

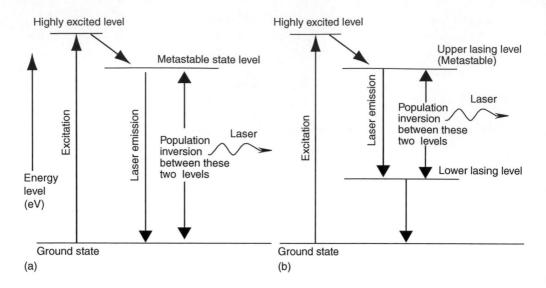

FIGURE 31.9 Three and four level laser systems: (a) three-level laser; (b) four-level laser.

Although the three-level laser system works for all practical purposes, as exemplified by the first laser (Ruby), a number of problems limit the effectiveness of this approach. The central problem is that the laser has difficulties operating efficiently. This occurs because the lower lasing level is the ground level, which is the normal state for most atoms or molecules. In order to produce the population inversion, a majority of ground state electrons must be promoted to the excited energy level, requiring a significant input of external energy. In addition, the population inversion is difficult to sustain for an appreciable time, and, therefore, three-level lasers generally must be operated in pulsed mode rather than continuously.

Figure 31.9(b) illustrates a four-level structure. The energy level structure is similar to that of the three-level system. The difference occurs after the atoms drop from the highest level to the metastable upper state; they do not drop all the way to the ground state in a single step. Because the population inversion is not created between the ground state and the upper level, the number of atoms that must be elevated is dramatically reduced in this design. In a typical four-level laser system, if only one or two percent of the atoms or molecules reside in the lower laser level (which is above the ground state) then exciting only two to four percent of the total to the higher level will achieve the required population inversion. Another advantage of separating the lower laser level from the ground level is that the lower level atoms will naturally fall to the ground state. If the lower laser level has a lifetime that is much shorter than the upper level, atoms will decay to the ground level at a rate sufficient to avoid accumulation in the lower laser level. Then, laser emission occurs between the metastable level and the lower level. Many of the lasers designed under these constraints can be operated in a continuous mode to produce an uninterrupted laser beam.

31.10 TYPES OF LASER

Most lasers are constructed of three elements: an active medium, a pumping source, and a resonant cavity. The active medium is a collection of atoms or molecules, which can be excited into a population inversion situation and can release electromagnetic radiation by the stimulated emission process. The active medium can be in any of the four states of matter: solid, liquid, gas, or plasma. The wavelengths of the emitted light are determined by the specific transitions between the laser energy levels in the material. The basic physics of the laser is similar for all types of lasers, and the term active medium will be used. It is assumed that the active medium is composed of atoms. In reality, the active medium can be atoms, molecules, and ions, according to the laser type.

The pumping source provides the energy required to pump atoms to higher energy levels so that stimulated emission can occur. Lasers can be optically pumped, electrically pumped (also by using an electric discharge), or pumped by a chemical reaction. There are three sources for optical pumping:

1. Flash lamps, which are built from a quartz tube filled with gas (such as mercury vapour) at low pressure;
2. Noble gas discharge tubes, which usually use xenon gas (but sometimes when higher energy is required, other noble gases with lower atomic weights, such as krypton or helium, are used); and
3. Another laser or any other light source, such as the light from the sun.

The resonant cavity provides a regenerative path for the photons. In essence, the functions of the resonant cavity are to shorten the laser and to construct the electromagnetic mode. Although the resonant cavity is a key part of most lasers, there are lasers for which the resonant cavity is not essential. It is certainly possible to make a laser long enough so that a reasonable intensity emerges without a resonant cavity. Nitrogen laser and most x-ray lasers are made this way. However, such lasers tend to have poor output-beam quality.

31.10.1 Gas Lasers

The gas laser is probably the most frequently encountered type of laser. The red–orange, green, and blue beam of the HeNe, Ar^+, and He–Cd gas lasers, respectively, are common for many applications.

The Kr^+ laser readily produces hundreds of milliwatts of optical power at wavelengths ranging from 350 nm in the ultraviolet to 647 nm in the red. It can be operated simultaneously on a number of lines to produce white laser light. Gas lasers are classified into two categories: pulsed gas lasers, such as N_2, Excimer, and TEA CO_2 and CW gas lasers, such as Copper Vapour, CO_2, Argon Ion, and Helium–Neon.

31.10.1.1 Helium–Neon (HeNe) Laser

Ali Javan, William Bennet, and Donald Herriot made the first gas laser using helium and neon. The helium–neon (HeNe) laser is used for classroom demonstrations, laboratory experiments, and was initially used in optical scanning systems at supermarket checkouts. In the HeNe laser, the lasing material is a gas, a mixture of about 15 percent He and 85 percent Ne. The low pressure gas mixture is placed in a glass tube that has two parallel mirrors one at each end. Figure 31.10 shows a HeNe laser beam.

Essentially, the helium is used for energizing and the neon for amplification. Atoms in the gas mixture are excited by applying a high voltage to the tube so that an electric discharge takes place within the gas mixture. In the process, some of the He atoms are raised to the metastable state E_3, as shown in Figure 31.11, which corresponds to a jump of 20.61 eV, almost exactly equal to an excited state in neon, 20.66 eV. The He atoms do not quickly return to the ground state by spontaneous emission but instead often give their excess energy to the Ne atoms when they collide. In such a collision, the He drops to the ground state and the Ne atom is excited to the state E_3' (the prime refers to neon states). The slight difference in energy $(20.66 - 20.61 = 0.05$ eV$)$ is supplied by the kinetic energy of the moving molecules. In this manner, the E_3' state in Ne, which is metastable, becomes more populated than the E_2' level. This inverted population between E_3' and E_2' is what is needed to produce the laser beam. The HeNe laser produces a characteristic reddish-pink sidelight. The most

FIGURE 31.10 Helium–Neon (HeNe) laser beam.

common HeNe lasing wavelength is red, at 632.8 nm. The HeNe laser will also laze at other wavelengths (notably, green light).

31.10.1.2 Argon Ion (Ar$^+$) Laser

The Argon laser was invented in 1964 by William Bridges in the United State of America. The argon ion (Ar$^+$) laser provides powerful, CW visible coherent light of several watts. The laser operation starts when the argon atoms are ionized by electron collisions in a high current electrical discharge. Further multiple collisions with electrons excite the argon ion to an energy level, about 35 eV above the atomic ground state, as shown in Figure 31.12. The atoms cannot stay long at the high energy level. Thus a population inversion forms between the high energy and the next energy level, about 33.5 eV above the argon ground level. Consequently, the stimulated radiation between the two energy levels contains a series of wavelengths ranging from 351.1 to 528.7 nm. Maximum light power is concentrated in the 488 and 514.5 nm emissions. The Ar$^+$ ion at 33.5 eV, the lower laser level, returns to its neutral atomic ground state via a radiative decay to the Ar$^+$ ion ground

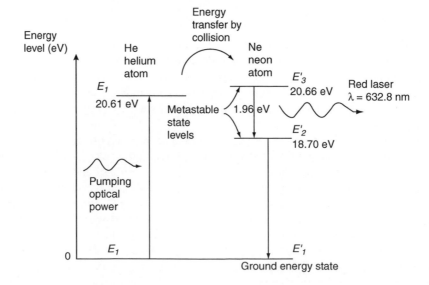

FIGURE 31.11 Energy levels for the HeNe laser.

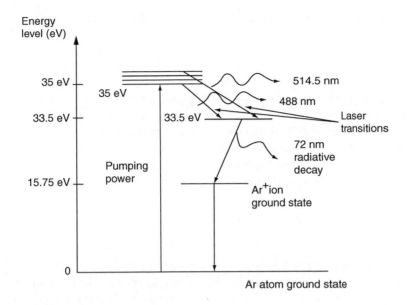

FIGURE 31.12 Energy levels for the Ar-ion laser.

state. The Ar^+ ion then recombines with an electron to form the neutral atom, Ar. The Ar atom is then ready for pumping again. The operation cycle continues as long as pumping power is available.

31.10.1.3 Carbon Dioxide Gas Laser

Carbon dioxide (CO_2) gas lasers are capable of high power output in the infrared region, tunable between 9 and 11 microns. Although the high and medium power CO_2 laser has many industrial applications, such as cutting and welding, there are also many medical uses. CO_2 laser therapy removes tattoo pigment after vapourization of the epidermis and superficial dermis and is also used for skin resurfacing. Although effective, this laser has a high incidence of scarring and pigmentary disturbances in inexperienced hands. CO_2 lasers can operate in continuous or pulsed modes.

31.10.2 SOLID STATE LASERS

There are several types of solid state lasers, such as Ruby, Ti Sapphire, Alexandrite, and rare-earth lasers, such as Neodymium-Yag (Nd^{3+}-YAG), Neodymium-glass, and Erbium-silica (Er^{3+}-Si) fibre. When placed in an optical resonator that provides feedback, all of these materials behave as laser oscillators and have their own set of advantages and disadvantages. The following explains the principles and operation of the Ruby laser.

31.10.2.1 Ruby Laser

The first successful optical maser or laser was developed in 1960 by Theodore Maiman using a ruby crystal. Ruby is an aluminum oxide crystal (Al_2O_3) in which some of the aluminum atoms have been replaced with chromium atoms. Chromium gives ruby its characteristic red colour and is responsible for the lasing behaviour of the crystal. Chromium atoms absorb green and blue light and emit red light. Figure 31.13 shows the components of the first Ruby laser device. A cylindrical crystal of ruby is used. A fully reflecting mirror is placed on one end, and a partially reflecting mirror on the other. A high-intensity lamp is spiraled around the ruby cylinder, to provide a flash of white light that triggers the laser action. The green and blue wavelengths in the flash light excite electrons in the chromium atoms to a higher energy level. Upon returning to their normal state, the electrons emit their characteristic ruby red light. The mirrors reflect some of this light back and forth inside the ruby crystal, stimulating other excited chromium atoms to produce more red light, until the light pulse builds up to high power and drains the energy stored in the crystal.

As shown in Figure 31.14, the atoms are excited from state E_1 to state E_3. This process is called optical pumping. The atoms quickly decay either back to E_1 or to the intermediate state E_2, which is metastable with a lifetime of about 3×10^{-3} s (compared to 10^{-8} s for ordinary levels). With strong pumping, more atoms can be forced into the E_2 state than are in the E_1 state. Thus, we have the inverted population needed for lasing. As soon as a few atoms in the E_2 state jump down to E_1, they emit photons that produce stimulated emission of the other atoms, and the lasing action begins. A Ruby laser thus emits a beam whose photons have 1.8 eV of energy and a wavelength of 694.3 nm (ruby-red light).

The lasing operation process of the Ruby laser is typical for most laser types, and is illustrated below:

1. Figure 31.15 shows the Ruby laser in its non-lasing state.
2. The flash tube fires and injects external light into ruby crystal. The chromium ions (Cr^{3+}) in a ruby crystal are excited by light from the flash tube when high-voltage electricity is applied to the this external light source, as shown in Figure 31.16. This light source must have photons of the right frequency to raise to the atoms to excited states.
3. The Cr^{3+} ions are raised in the pumping process to energy level E_3, from which they decay to the metastable level E_2 by losing energy to other atoms in the crystal. Some of these atoms emit photons, as shown in Figure 31.17.
4. A few Cr^{3+} ions in the E_2 level then spontaneously fall to the E_1 level. Some of these photons propagate parallel to the axis of the ruby crystal, so they bounce back and forth between the reflecting end mirrors of the ruby rod, as shown in Figure 31.18. This initiates the continuing process of stimulated emission and amplification.
5. The presence of light of the right frequency now stimulates the other Cr^{3+} ions in the E_2 level to radiate; the result is an avalanche of photons that produces a large pulse of red light. Some of the photons leave through the partially silvered mirror at one end. The red light is the laser light, as shown in Figure 31.19.

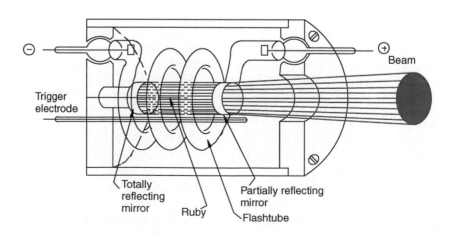

FIGURE 31.13 The first ruby laser produces pulses of light.

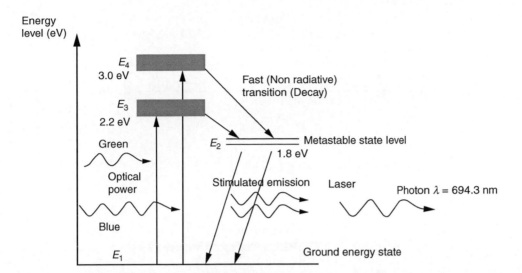

FIGURE 31.14 Energy level diagram for the ruby laser.

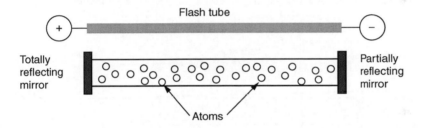

FIGURE 31.15 The ruby laser in its non-lasing state.

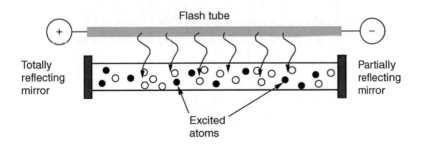

FIGURE 31.16 The flash tube injects external light into the ruby rod.

31.10.2.2 Neodymium-YAG Laser

Another example of a solid-state laser, the neodymium-YAG, is shown in Figure 31.20. The Neodymium ion (Nd^{3+}) is doped into the yttrium–aluminum–garnet (YAG) host crystal. Neodymium-YAG lasers have become very important because they can be used to produce high-power laser light. The energy diagram of this laser is shown in Figure 31.21. Such lasers have been constructed to produce over a kilowatt of continuous laser power at 1,064 nm and can achieve

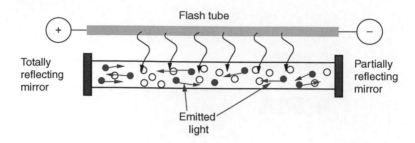

FIGURE 31.17 Some atoms emit photons spontaneously.

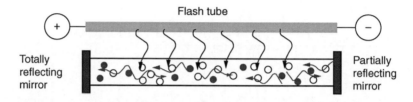

FIGURE 31.18 Some photons traverse the ruby's axis and reflect between the mirrors.

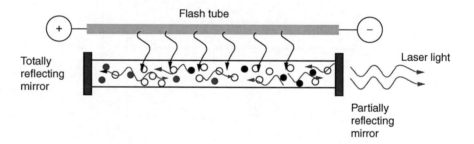

FIGURE 31.19 Laser light leaves the ruby through the partially silvered mirror.

extremely high power in a pulsed mode. Neodymium-YAG lasers are used in pulse mode in laser oscillators to produce short pulses of several nanoseconds. Furthermore, neodymium laser light can be passed through a second harmonic generating crystal that doubles its frequency, thereby providing a strong source of green light at 532 nm. It can also be quadrupled, providing ultraviolet radiation at 266 nm.

Aside from Ruby, Nd^{3+}, and Er^{3+}, other commonly encountered optically pumped solid-state laser amplifiers and oscillators include:

1. Alexandrite (Cr^{3+}:Al_2BeO_4), which has a tunable output in the wavelength range between 700 nm and 800 nm;
2. Ti^{3+}:Al_2O_3 (Ti: sapphire), which is tunable over an even broader range, from 660 nm to 1180 nm; and
3. Er^{3+}:YAG, which is often operated at 1660 nm. Ti: sapphire is often chosen for ultra-short pulse generation because of its broad bandwidth because pulse length is inversely proportional to bandwidth.

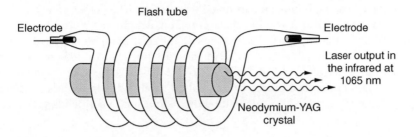

FIGURE 31.20 Neodymium-YAG laser.

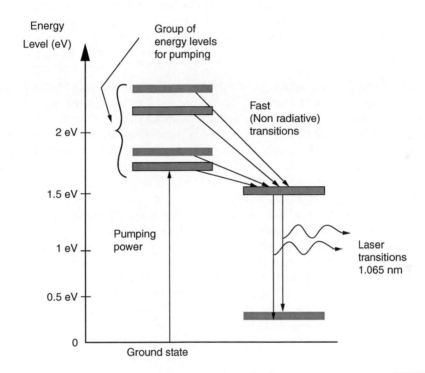

FIGURE 31.21 Energy level diagram for the Neodymium-YAG laser.

31.10.2.3 Alexandrite Laser

The Alexandrite laser is a solid-state laser in which the crystal ($BeAl_2O_4$) is doped with 0.01%–0.40% of Chromium ions (Cr^{+3}). The Alexandrite laser has an energy level structure similar to that of the Ruby laser. The Alexandrite laser operates as a four level laser, which can be tuned over a range of wavelengths: 720–800 nm. It was the first tunable solid-state laser to reach the market.

31.10.3 DYE LASERS

The Dye laser was first demonstrated in 1965 at IBM laboratories in the United States of America by Peter Sorokin and J. Lankard. They discovered the Dye laser action during research into fluorescence of organic dye molecules that were excited by a Ruby laser. In 1967, scientists discovered the possibility of tuning the emitted wavelength using a grating at the end of the optical cavity. A Dye laser can be considered as a special device to convert electromagnetic radiation from one wavelength into another tuned wavelength. The output of a Dye laser is always coherent radiation tunable over a specific spectral region, determined by the dye material.

The wavelength of the Dye laser can be tuned for specific applications. For example, the wavelength range of one tunable medical Dye laser is from 577 to 585 nm. This wavelength range increases the penetration depth from 0.5 to 1.2 mm in skin, when scanning skin tissue with light. By using microsecond pulses from a pulsed tunable dye laser, it is possible to confine thermal injury to the target blood vessels without heat diffusion to surrounding tissue.

31.10.4 SEMICONDUCTOR LASERS

All semiconductor lasers, sometimes called laser diodes, are built from semiconductor materials. The first semiconductor laser was created by Robert Hall in 1961 in the United States of America. In 1975, the first semiconductor laser capable of operating continuously at room temperature was introduced. This development has led to the use of semiconductor lasers in the CD-ROM, DVD-ROM, laser pointer, and many other useful devices. Figure 31.22 shows a semiconductor laser device.

Researchers succeeded in porducing coherent electromagnetic radiation from a forward biased diode (p–n junction) made from semiconductor materials. The characteristics of the laser beam depend on the semiconductor materials used in building the device. The majority of semiconductor materials are based on a combination of elements in the third group of the Periodic Table (such as Al, Ga, and In) and the fifth group (such as N, P, As, and Sb), hence referred to as the III–V compounds. Examples include GaAs, AlGaAs, InGaAs, and InGaAsP alloys. The CW laser emission wavelengths are normally within 630–1600 nm, but InGaN semiconductor lasers were found to generate CW at 410 nm (blue light) at room temperature. Semiconductor lasers can generate blue–green light using materials that are the combination of elements of the second group (such as Cd and Zn) and the sixth group (S and Se), referred to as the II–VI compounds.

The conductivity of a semiconductor increases with temperature, up to a certain temperature level. The efficiency of a semiconductor laser decreases when the temperature increases.

The Fabry–Perot laser structure is the simplest example of semiconductor lasers. Fabry–Perot single-spatial-mode laser diodes are available that operate from 730 to 770 nm, with tens of mW of optical power. Applications of the Fabry–Perot laser include medicine, spectroscopy, gas detection, sensors, measurement instruments, and high-power laser seeding.

31.10.4.1 Energy Bands

As explained in the principle of light-emitting diodes, when a diode is forward biased, holes from the p-region are injected into the n-region, and electrons from the n-region are injected into the p-region, as shown in Figure 31.23. If electrons and holes are present in the same region, they may radiatively recombine; that is, the electron falls into the hole and emits a photon with the energy of the band gap. This is called spontaneous emission and is the main source of light in a light-emitting diode. There is a spread in emitted wavelength.

Under suitable conditions, the electron and the hole may coexist in the same area for quite some time (on the order of microseconds) before they recombine. If a photon of exactly the correct frequency happens along within this time period, a recombination (of hole and electron) may be stimulated by the photon. This causes another photon of the same frequency to be emitted with exactly the same direction, polarization, and phase as the first photon.

In a laser diode, the semiconductor crystal is fashioned into three layers very thin in thickness and rectangular in the other two dimensions. An optical waveguide is made in the middle layer such that the light is confined to a relatively narrow line. This is called the active layer. The top of the crystal is p-doped, and the bottom is n-doped, resulting in a p–n junction, which forms the diode. The two ends of the crystal are cleaved so as to form perfectly smooth, parallel edges like mirrors;

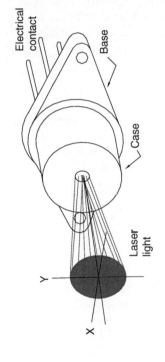

FIGURE 31.22 A packaged semiconductor laser.

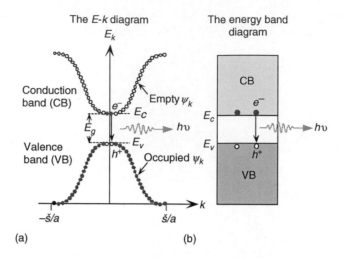

(a)   (b)

FIGURE 31.23 *E-k* and energy band diagrams for a diode: (a) the *E-k* diagram; (b) the energy band diagram.

these two reflective parallel edges form a resonator called a Fabry–Perot cavity, as shown in Figure 31.24.

Photons emitted in precisely the right direction will travell along the cavity and be reflected several times from each end before they exit the cavity. Each time photons pass through the cavity, the light is amplified due to stimulated emission. Hence, if there is more amplification than loss, the diode begins to laze. The wavelength emitted is a function of the band-gap between the energy levels of the p and n regions. No photons with energy higher than the band-gap will be emitted.

31.10.4.2 Types of Lasers

There are essentially two ways that laser diodes can direct laser light that can be suitable for optical communications, as shown in Figure 31.25. If the emitted light emerges from an area on the edge of the device, i.e., from an area on a crystal face perpendicular to the active layer, as shown in

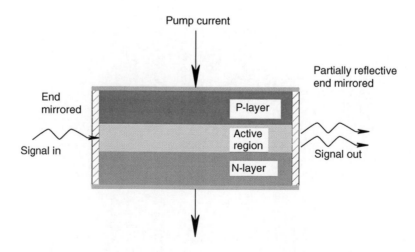

FIGURE 31.24 Laser diode.

Figure 31.25(a), then the device is called an edge emitting laser diode. If the emitted light emerges from an area on the plane of the active layer, as shown in Figure 31.25(b), then the device is called a surface emitting laser diode.

There are many types of laser diodes on the market, varying in device structure, wavelength of emitted light, and application. The basic types of laser diodes are listed in the following sections.

31.10.4.3 Heterojunction Laser Diodes

The type of laser diode described above is called a heterojunction laser diode, as shown Figure 31.26. The heterojunction is a junction between two different semiconductors with different bandgap energies. Unfortunately, heterojunction laser diodes are extremely inefficient. They require so much power that they can only be operated in short pulses.

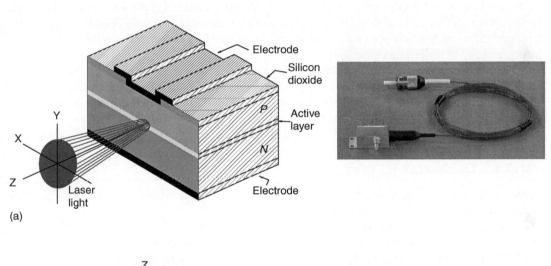

(a)

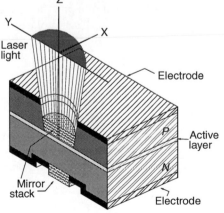

(b)

FIGURE 31.25 Emitting laser diodes: (a) edge emitting; (b) surface emitting.

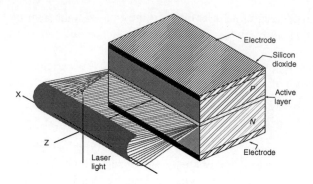
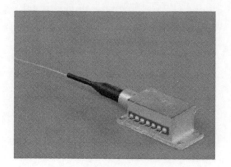

FIGURE 31.26 A heterostructure laser.

31.10.4.4 Double Heterostructure Laser Diodes

In double heterostructure devices, a layer of low bandgap material is sandwiched between two high bandgap layers. One commonly used pair of materials is GaAs with Al GaAs. Each of the junctions between different bandgap materials is called a heterostructure. Hence, the name double hetero-structure laser describes this type of laser.

The advantage of a double heterostructure laser is that the region where free electrons and holes exist simultaneously, the active region, is confined to the thin middle layer. As a result, many more of the electron-hole pairs can contribute to amplification, and fewer are left out in the poorly amplifying periphery. In addition, light is reflected from the heterojunctions. Hence, the light is confined to the active region where the amplification takes place.

31.10.4.5 Quantum Well Lasers

If the middle layer of a laser diode is made thin enough, it starts behaving like a quantum well. This means that in the vertical direction electron energy is quantized.

The quantum layers are usually stacked one on top of the other. The energy difference between quantum well levels can be used to produce the laser action instead of the bandgap. This is very useful because the wavelength of light emitted can be tuned by simply altering the thickness of the layer. The efficiency of a quantum well laser is greater than that of a bulk laser due to a tailoring of the distribution of electrons and holes that are involved in the stimulated emission (light producing) process.

31.10.4.6 Separate Confinement Heterostructure Lasers

The problem with heterostructure laser devices is that the thin layer is simply too small to effectively confine the light. To compensate, another two layers are added on outside the first three. These layers have a lower refractive index than the centre layers and, hence, confine the light effectively. Such a design is called a separate confinement heterostructure (SCH) laser diode. Almost all commercial laser diodes since the 1990s have been SCH quantum well diodes.

31.10.4.7 Distributed Feedback Lasers

Distributed feedback lasers (DFB) are the most common laser type used to transmit signals in dense wave division multiplexer systems (DWDM). To create a precise lasing wavelength, a diffraction grating is etched close to the p–n junction of the diode. This grating acts like an optical filter, causing only a single wavelength to be fed back to the active region where the lasing is amplified. At least one face of a DFB is anti-reflection coated. The DFB laser has a stable wavelength that is set during manufacturing by the pitch of the grating and can only be tuned slightly with temperature.

31.10.4.8 Tunable Laser Diodes

As the number of wavelengths required in networks increases, tunable lasers play an increasingly important role in the development of dynamic network systems. With the assistance of electronic controls, a tunable laser can be tuned to emit different wavelengths. This reduces the need for spare wavelength-specific line cards and thus lowers a network's overall cost. There are several different technologies for tunable lasers. Figure 31.27 shows the basic principle of a tunable laser.

The tuning can be described by the following equations.

The wavelength of the mode is given by:

$$\frac{m\lambda}{2} = \bar{n}L \tag{31.22}$$

where m is the mode number, λ is the wavelength, $\bar{n}$ is the effective refractive index, and L is the effective cavity length.

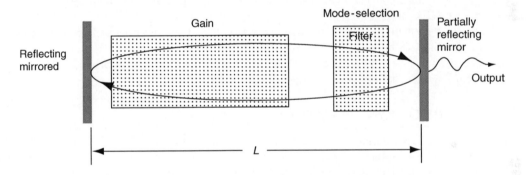

FIGURE 31.27 Common principle of a tunable laser.

From the Equation 31.23, the relative change in wavelength is given as:

$$\frac{\Delta\lambda}{\lambda} = \frac{\Delta n}{n} + \frac{\Delta L}{L} - \frac{\Delta m}{m} \tag{31.23}$$

where $\Delta n/n$ is the tuning portion due to the change in the net cavity index, $\Delta L/L$ is the tuning portion due to the change in the physical length of the crystal, and $\Delta m/m$ is the tuning portion due to the mode-selection filtering (via index or grating angle).

31.10.5 OTHER TYPES OF LASERS

Some other types of lasers include:

1. Chemical lasers, in which the energy input comes from the chemical reaction of highly reactive gases;
2. Ultrafast, Gas Dynamic, x-ray, and Bragg Reflector (DBR) lasers, such as Distributed, Tunable, and External Cavity;

3. Vertical Cavity Surface Emitting Lasers (VCSELs);
4. In-Fibre lasers;
5. Free Electron lasers; and
6. Excimer lasers.

31.11 COMPARISON OF SEMICONDUCTOR AND CONVENTIONAL LASERS

Semiconductor lasers are quite different from conventional lasers. In particular:

1. The gain of the semiconductor laser material is very high and is generated by a population inversion between the conduction and valence bands of the semiconductors. In some sense, a semiconductor laser is a two-state laser system.
2. Since the longitudinal mode is on the order of the size of the laser device, the transverse mode of the semiconductor laser is quite different from that of a conventional laser. In particular, the beam is not Gaussian, the beam profile tends to be elliptical, and the beam divergence tends to be large.
3. The gain spectrum is quite large (many THz or hundreds of angstroms).
4. The short cavity length (several hundred microns) means that the longitudinal mode spacing is much larger than that of a conventional gas or solid-state laser (on the order of GHz or angstroms).
5. Due to the small size of semiconductor lasers, they have the potential for mass production and can be easily integrated on PC boards.
6. The properties of the semiconductor lasers are being rapidly improved. They are becoming increasingly powerful and efficient laser sources.

31.12 CLASSIFICATION OF LASERS

Full details on the laser classifications and safety issues are presented in the photonics laboratory safety chapter. This chapter also covers laser devices, safety, operation, and utilization as light sources in lab experiments. Lasers are classified into four basic laser classes, which include Classes 1, 2, 2a, 3a, 3b, and 4. Higher class numbers reflect an increased potential to harm users. Operating wavelength and mode (CW and pulse) are the criteria used to classify the hazard level of lasers. A summary of the laser classifications and applications is listed bellow:

1. Class 1 lasers output light in the visible range ($450\,\text{nm} < \lambda < 500\,\text{nm}$), and have very low output power (between 0.04 and 0.40 mW). Some examples of Class 1 laser devices include: CD players, scanners, laser pointers, and small measurement equipment.
2. Class 2 lasers output light in the visible range ($400\,\text{nm} < \lambda < 700\,\text{nm}$) and have low output power (must be less than 1 mW of visible CW). Some examples of Class 2 lasers are classroom demonstration laser sources and laser source devices for testing and telecommunications.
3. Class 2a lasers are low output power devices, which are considered to be visible light lasers. A supermarket bar-code scanner is a typical example of a Class 2a laser device. A Class 2a laser beam should never be viewed directly.
4. Class 3 lasers are divided into two subgroups (Class 3a and Class 3b lasers), so there is no plain Class 3.
5. Class 3a lasers are intermediate power devices. Class 3a lasers are allowed to exceed the output power limit of Class 2 lasers by no more than a factor of 5 or produce visible light power less than 5 mW. They are considered to be CW lasers. Some examples of Class 3a laser devices are: laser scanners, laser printers, and laser source devices for testing and telecommunications.

6. Class 3b lasers are intermediate power devices; they output between 5 and 500 mW of CW, or else pulsed 10 J/cm^2 power. They are considered to be CW lasers. Some examples of Class 3b lasers are laser source devices for testing.

7. Class 4 lasers are high power devices; they either output more than 500 mW of CW power or pulsed 10 J/cm^2 power. They are considered to be very high power lasers. Some applications of Class 4 laser devices include: surgery, drilling, cutting, welding, and micromachining.

31.13 LASER BEAM ALIGNMENT

Researchers and users of commercial laser based systems need to align the laser beam source to an optical axis of an optical component. In the laboratory, operational alignment flexibility is key when set-ups are modified and reconfigured routinely. In commercial products, the critical aspects are quick alignment during assembly and simple realignment to accommodate efficient replacement or repair of the laser or optics.

The advantage of using mirrors for alignment is that they provide uniform performance over multiple laser wavelengths. Metallic mirrors offer the broadest wavelength performance. Dielectric mirrors offer a narrower reflectivity range but can provide higher peak reflectivity.

To define an optical axis, it is necessary to fix the $X–Y$ location of the beam at some arbitrary point along the Z-axis and to set the angular direction of the beam at this location (Ø, ß). The $X–Y$ plane is perpendicular to the direction of beam propagation, which is the Z-axis. There are two common ways of establishing such a reference:

1. The casy way: Pass the laser beam through two apertures, adjusting the beam angle and position until it passes through the centre of each. The first aperture sets the $X–Y–Z$ coordinates, and the second determines the angle of propagation. If using adjustable irises rather than pinholes, open the apertures during initial alignment to save considerable effort finding the axis.

2. The accurate way: Replace the apertures with position-sensitive detectors. Quantitative lateral and angular alignment is achieved by centring the beam on each detector or, in the case of a quad cell, by balancing the signals. Because these detectors are not transmissive, the first detector must be mounted on a removable base or placed off-axis in conjunction with an on-axis beamsplitter.

There are several techniques to accurately perform the positional and angular alignment of a laser beam. They can be divided into two categories. In one category, the laser source device is aligned using a one, two, or multi-axis stages. In the second category, the laser beam exiting from the laser source device is aligned directly to an optical application. Some of these techniques are listed below.

31.13.1 ALIGNMENT USING ONE, TWO, AND MULTI-AXIS POSITIONERS

Using stages or positioners, as shown in Figure 31.28, is a simple, easy, and low-cost way of aligning a small laser device like a HeNe laser tube. Alignment performance parameters, such as specifications for positioner accuracy, precision, and resolution are the most quoted. These parameters refer to how well a component performs along the desired axis of travell and are determined primarily by the drive mechanism.

For the one-axis stage, performance is additionally measured by the deviation from the desired axis of travell. For the single axis translation stage, as shown in Figure 31.28(a), one coordinate of translation is designed to vary. The position of any optical mount can be defined uniquely in terms of six independent coordinates: three translational and three rotational, with respect to some

(a)   (b)

(c)   (d)

FIGURE 31.28 Laser holder/positioners: (a) one-axis laser holder/stage; (b) two-axis laser holder/stage; (c) multi directions laser holder/positioner; and (d) multi-axis laser holder/stage.

arbitrary fixed coordinate system. The error in motion, then, is defined in terms of the remaining five independent coordinates, which ideally stay fixed in space. Similarly, the two-axis translation stage, as shown in Figure 31.28(b), provides linear motion in two directions. Three or multi-axis stages overcome the limitations of one-and two-stages. Some of the multi-axis positioners and stages are shown in Figure 31.28(c) and Figure 31.28(d).

Figure 31.29(a) illustrates a schematic diagram for a set-up to align a HeNe laser tube so that the laser beam is reflected by mirror M1 towards the target card. Figure 31.29(b) illustrates a set-up to align a HeNe laser tube so that the laser beam bounces between two mirrors M1 and M2 and is captured on a target card.

31.13.2 ALIGNMENT USING TWO MIRROR MOUNTS

The most common method of beam steering is to use flat mirrors on tip/tilt adjustable multi-axis mounts. At least two mirrors are needed to deliver the four required degrees of freedom for the alignment steps.

Figure 31.30 shows aligning a laser beam to the two irises. The first step is to roughly position the mirrors so that the beam falls on Iris 1. Then the beam on Iris 1 is centred, using fine adjustments on mirror M_1 only. This step is verified by reducing the aperture of Iris 1 until it is just slightly smaller than the laser beam. When the beam is properly centred, only a halo of light is seen around the aperture. Next, Iris 1 is reopen and the beam is centred on Iris 2 by adjusting mirror M_2 so that only a halo of light is seen on Iris 2.

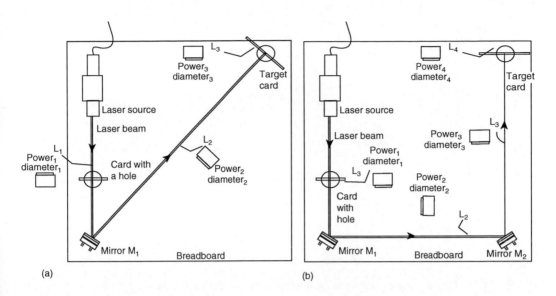

FIGURE 31.29 A schematic diagram for laser beam alignment: (a) using one mirror; (b) using two mirrors.

Iris 1 is rechecked for beam centration using the halo effect. Iterative adjustment of mirror M_1 on Iris 1 and mirror M_2 on Iris 2 will quickly result in precise alignment. However, if final alignment leaves the beam near the edge of one of the mirrors then the mirror should be grossly repositioned accordingly and the alignment process repeated. Placing mirror M_2 as close as possible to Iris 1 reduces the number of iterations necessary to reach alignment by reducing crosstalk between the degrees of freedom. As the separation approaches zero, adjustments to mirror M_2 have no effect on the position of the beam at Iris 1.

There are several reasons for the popularity of this technique. First, it uses simple, inexpensive components available in most laboratories. But just as important, it is a quickly converging and straightforward technique. The intuitive feel is further enhanced by the fact that the beam's angular displacement is essentially linear over the entire adjustment range of the mounts.

By using closely spaced, large mirrors (compared to the beam diameter), this approach offers fairly large translational and angular adjustment ranges. One potential limitation of a two-mirror mount is that it produces a lateral offset. In many applications, the overall optical layout can accommodate this offset. In others, a zero offset can be achieved by using three or more mirrors.

31.13.3 Alignment Using Three and Four Mirror Mounts

A zero offset alignment can be achieved by using three or more mirrors, as shown in Figure 31.31. In a zero-offset layout, the mirrors are roughly aligned, and then the final alignment is performed by adjusting only the first mirror (for Iris 1) and last mirror (for Iris 2). The three-mirror system is simple but yields incident angles greater than 45°. The four-mirror system, as shown in Figure 31.32, produces incidence angle close to 45°, which matches well with standard mirror coatings and leads to higher reflectivity.

The choice of mirror mount and actuator depends on the exact dimensions of the set-up and the angular tolerance of the application. Increasing mirror separation produces larger X–Y displacements at the second mirror, for a given actuator adjustment. This can increase the lateral adjustment range of the system but also amplifies the effects of drift or vibration. Also, as mirror separation increases, maintaining a given linear adjustment resolution at the second mirror requires greater angular resolution in the actuator of the first mirror.

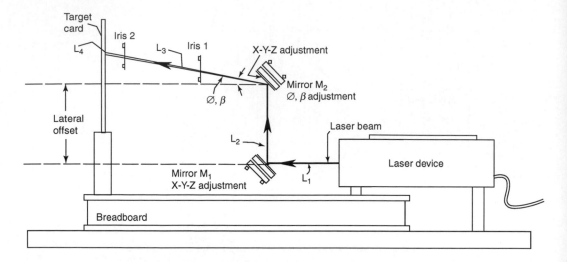

FIGURE 31.30 Two adjustable mirrors aligned to the optical axis of two irises.

31.13.4 Alignment Using a Risley Prism Pair

An alternative alignment technique is based on the Risley prism pair. The Risley prism pair consist of two round, paired flats with identical wedge angles. A pair of Risley prisms is wedged and circular windows mounted in independent rotation stages. Each window acts as a thin, refractive prism. The window rotation tilts the beam, as shown in Figure 31.33. By adjusting the two matched prisms, the beam can be deflected at any angle (Ø, ß) within a cone (set by the wedge angle and index of refraction of the prisms). As with mirrors, two sets of the prisms are needed for adjustments in four degrees of freedom (X, Y, Ø, ß). All surfaces in a Risley prism pair system need antireflective coating and/or the air gap to be filled in with a refractive-index-matching gel.

An advantage of Risley prisms over two-mirror systems is that they do not introduce a significant lateral beam offset. Also, they offer excellent mechanical stability because of their independent rigid mountings. In a typical rotation stage, the locking mechanism acts orthogonal to the adjustment mechanism, allowing rigid locking without affecting the precise alignment. In contrast, spring tension maintains mirror mount alignment, making the mounts more susceptible to

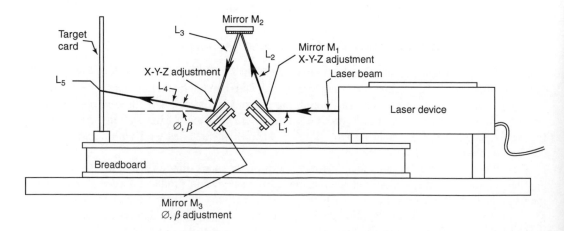

FIGURE 31.31 Three mirrors provide beam alignment with no lateral offset.

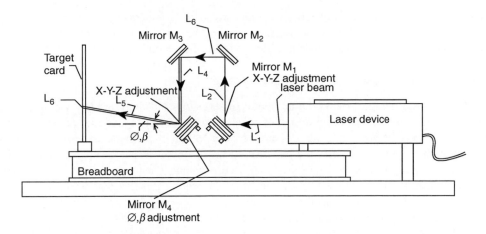

FIGURE 31.32 Four mirrors provide beam alignment with no lateral offset.

vibration and long-term drift. Consequently, Risley prisms are often found in commercial and aerospace applications where vibration and shock are common.

One significant limitation of the Risley prisms is their lack of orthogonal adjustment. Indeed, their operation is nonintuitive. Rotating them produces tip/tilt redirection. Furthermore, the mechanical adjustment is nonlinear. The result is that it may take more time to adjust precisely. Other drawbacks are the increased optical/mechanical complexity and cost.

In addition, the prisms operate by refraction, with dispersion effects making beam deflection angle a function of wavelength. Unlike mirrors, this can preclude their use in multi-wavelength applications. Prisms also have an inherent trade-off between adjustment range and resolution

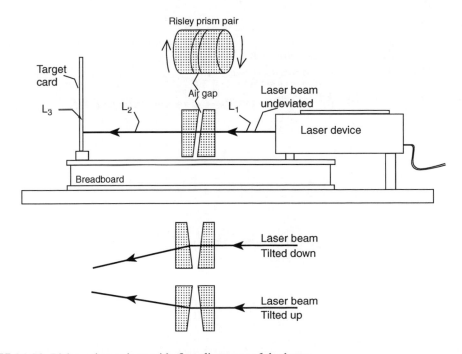

FIGURE 31.33 Risley prism pair provide fine alignment of the beam.

because the maximum possible deflection depends on the prism wedge angle. Increasing the wedge angle increases the adjustment range but reduces the minimum resolution provided by a given rotation mount. The Risley prism pair is better suited to fine-tuning than gross alignment.

31.13.5 ALIGNMENT USING AN ADJUSTABLE WEDGE

A simple way of aligning a beam with no lateral offset is to use a single wedge with a continuous angle adjustment, as shown in Figure 31.34. A matching lens pair is made by mating the curved surfaces of a plano-convex lens with a plano-concave lens that has the same radius of curvature. Sliding the convex lens relative to the concave surface adjusts the wedge angle. Like the Risley prisms, the adjustable wedge is better suited to fine adjustments than to gross alignment. Two sets of optics are required to provide the full four degrees of freedom. Besides zero offset, adjustment of the variable wedge is intuitive; pushing the plano-convex lens in one direction moves the beam in the same direction. Also, for small adjustments beam deflection is fairly linear.

The thin gap between any lenses can contain air or a refraction-index-matching fluid. The advantage of an air gap is a higher damage threshold. But beam-focusing errors can arise if a constant, minimal air gap is not maintained. To maintain throughput efficiency, air/glass surfaces must be anti-reflection coated, whereas surfaces that contact index-matching fluid do not require coating. Although it lowers the damage threshold, an index-matching fluid can have two advantages over an air gap. This fluid reduces the number of air/glass interfaces, which reduces reflection between parallel surfaces of the lenses, and the fluid lubricates the interface while holding the lenses together with surface tension.

Regardless of which alignment method is used, some practical tips should be kept in mind:

1. The height of the optical axis and laser beam above the mounting surface should be minimized. Using low-profile mounts and posts limits the effects of vibration and thermal drift. If vibration is a problem, using damped mounting posts should be considered.

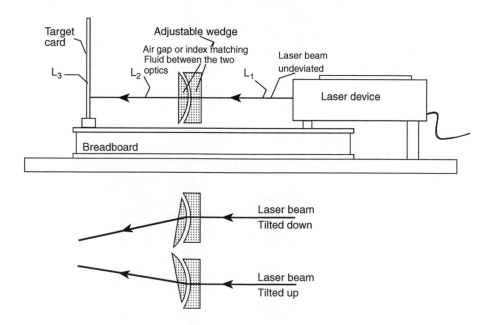

FIGURE 31.34 Matched lenses provide fine alignment of the beam.

2. Each mirror has a thickness when the back is silver coated. The thickness of the mirror refracts the laser beam and increases laser beam losses. It is better to use other types of mirrors, such as metallic or dielectric mirrors. They are ideal for sensitive applications, such as imaging with no chromatic aberration.

3. For each lens there are two reflections, one from each surface. When the centres of the two reflections are at the height of the laser beam, the height of the lens is properly adjusted. When they are overlapping, the beam is at the centre of the lens. And when they are centred about the laser output, the lens is not tilted with respect to the beam.

4. For each optical component there are many types of losses. The main losses are caused by absorption, dispersion, scattering, and reflection of the light.

5. A plan should be devised before attempting to align the system. All of the angular tolerances required and the long-term stability and vibration sensitivity of the system should be considdered, then components should be specified accordingly. As with any optical set-up, too much performance adds unnecessary cost while too little presents the risk that the system will not work.

31.14 LASER BEAM EXPANSION

Often when a laser is used in an optical system, there is a requirement for either a larger beam or a small beam divergence. In some cases, the size of the beam becomes critical. For example, when measuring the distance from the Earth to the Moon, a beam one metre in diameter travelling to the Moon expands to several hundred metres in diameter and returns to the Earth's surface with a diameter of several kilometres. The signal returned from this expansion is millions of times smaller than the original signal, so the divergence of a laser beam needs to be reduced to produce strong and detectable signals. Reduction of the beam divergence is called beam collimation. Even in earth-bound experiments, higher beam collimation is required for many applications.

As stated earlier, light from a laser source is very different from that of most other light sources. After a description of the simplest type of beam, the transverse mode TEM_{00} Gaussian beam and its parameters will be discussed below. This beam type should be understood before examining ways of collimating the beam.

31.14.1 CHARACTERISTICS OF A GAUSSIAN BEAM

For a laser beam, the term Gaussian describes the variation in the irradiance along a line perpendicular to the direction of propagation and through the centre of the beam, as illustrated in Figure 31.35.

The irradiance, I, is symmetric about the beam axis and varies radially outward from the propagation axis with the form given by:

$$I(r) = I_0 e^{-2r^2/r_1^2} \tag{31.24}$$

This equation is illustrated in Figure 31.36, and expressed in terms of a beam diameter as:

$$I(d) = I_0 e^{-2d^2/d_1^2} \tag{31.25}$$

where r_1 is the radius of the beam and d_1 is the diameter of the beam.

By definition, r_1 and d_1 occur when I equals $1/e^2$ of I_0.

Figure 31.36 assumes a beam of parallel rays. In reality, a Gaussian beam either diverges from a region where the beam is smallest, called the beam waist, or converges, as illustrated in Figure 31.37. The amount of divergence or convergence is measured by the full angle beam

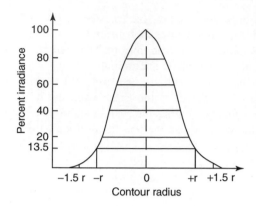

FIGURE 31.35 Gaussian beam profile.

divergence θ, which is the angle subtended by the $1/e^2$ diameter points for distances far from the beam waist, as illustrated in Figure 31.37. In some laser references, the half angle is measured from the beam axis to the $1/e^2$ asymptote. However, it is the full angle divergence, as defined here, that is usually given in the laser specifications. Because of symmetry on either side of the beam waist, the convergence angle is equal to the divergence angle.

According to the laws of geometrical optics, a Gaussian beam converging at an angle of θ should collapse to a point. Because of diffraction, this does not occur. However, at the intersection of the asymptotes that define θ (radians), the beam does reach a minimum beam waist diameter d_0. It can be shown that for a TEM_{00} mode beam, d_0 depends on the beam divergence angle as:

$$d_0 = \frac{4\lambda}{\pi\theta} \tag{31.26}$$

where λ is the wavelength of the beam.

Note that for a Gaussian beam of a particular wavelength, the product $d_0\theta$ is constant. Therefore, for a very small beam waist, the divergence must be large; for a highly collimated

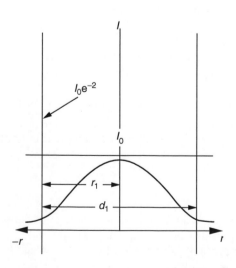

FIGURE 31.36 Gaussian beam profile.

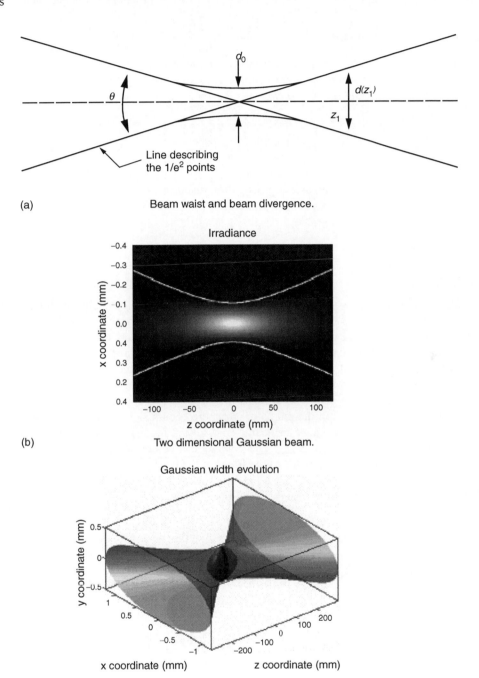

FIGURE 31.37 Variation of Gaussian beam diameter in the vicinity of the beam waist: (a) beam waist and beam divergence; (b) two dimensional gaussian beam; (c) three dimensional gaussian beam.

beam (small θ), the beam waist must be large. The variation of the beam diameter in the vicinity of the beam waist is illustrated in Figure 31.37 and given as:

$$d^2 = d_0^2 + \theta^2 z^2 \tag{31.27}$$

where d is the diameter at a distance $\pm z$ from the waist along the beam axis.

31.14.2 The Rayleigh Range

It is useful to characterize the extent of the beam waist region with a parameter called the Rayleigh range. In other descriptions of Gaussian beams, this extent is sometimes characterized by the confocal beam parameter and is equal to twice the Rayleigh range. Rewriting Equation 31.28 gives:

$$d = d_0 \sqrt{1 + \left(\frac{\theta z}{d_0}\right)^2}$$

(31.28)

The Rayleigh range is defined as the propagation distance from the beam waist to where the diameter has increased to $d_0\sqrt{2}$. Obviously, this occurs when the second term under the radical is unity, when:

$$z = z_R = \frac{d_0}{\theta}$$

(31.29)

Although the definition of z_R might seem rather arbitrary, this particular choice offers more than just convenience. Figure 31.38 illustrates a plot of the radius of curvature of the wavefronts in a Gaussian beam as a function of z. For large distances from the beam waist, the wavefronts are nearly planar, and the radius tends toward infinity. At the beam waist, the wavefronts are also planar. Therefore, the absolute value of the radius of curvature of the wavefronts must go from infinity at large distances, through a minimum, and return to infinity at the beam waist. This is also true on the other side of the beam waist but with the opposite sign. It can be shown that the minimum in the absolute value of the radius of curvature occurs at $z = \pm z_R$, that is, at a distance of one Rayleigh range on either side of the beam waist. From Figure 31.38, the collimated region of Gaussian beam waist can be taken as $2z_R$. The point z_R can be considered as the transition of the beam from the near field to the far field. At the focus of the beam using a lens, the Rayleigh range transforms to the depth of the focus.

The Rayleigh range can be expressed in a number of ways:

$$z_R = \frac{d_0}{\theta} = \frac{4\lambda}{\pi\theta^2} = \frac{\pi d_0^2}{4\lambda}$$

(31.30)

This shows that all three characteristics of a Gaussian beam are dependent on each other. Given any one of the three quantities, d_0, θ, z_R, and the wavelength of the radiation, the behaviour of the beam is completely described.

Figure 31.39 illustrates an example of a helium–neon laser ($\lambda = 632.5$ nm) transverse mode TEM$_{00}$, with a beam waist diameter of 1 mm. The beam divergence angle θ from Equation 31.31 is:

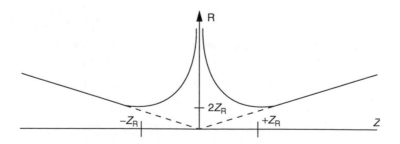

FIGURE 31.38 Radius of curvature (R) versus distance (z) from the beam waist.

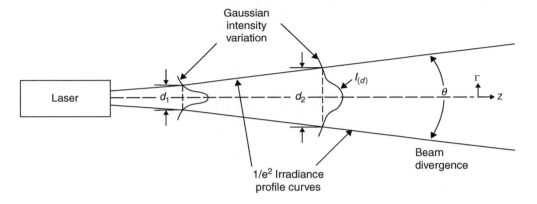

FIGURE 31.39 A Gaussian TEM$_{00}$ laser beam.

$$\theta = \frac{4\lambda}{\pi d_0} = \frac{4 \times 632.5 \times 10^{-9}}{3.141 \times 1 \times 10^{-3}} = 0.805 \times 10^{-3} \text{ rad}$$

The Rayleigh range calculated using Equation 31.30 is:

$$z_R = \frac{d_0}{\theta} = \frac{1 \times 10^{-3}}{0.805 \times 10^{-3}} = 1.242 \text{ m}$$

31.14.3 Expansion and Collimation of A Laser Beam

Through the use of lenses, the divergence beam waist and Rayleigh range of the Gaussian beam can be changed. However, from the above discussion it is clear that the relations between the various beam parameters cannot be changed. Thus, to increase the collimation of a beam by reducing the divergence requires that the beam waist diameter be increased, since the beam waist diameter-divergence product is constant.

There are two ways of collimating a Gaussian beam. One method uses a Galilean telescope, which consists of a negative eyepiece lens of short focal length and a positive objective lens of long focal length, as shown in Figure 31.40. Collimation can be done by first creating a beam with a strong divergence and small beam waist, using the negative lens. Then by putting the beam waist at the focal point of the long focal length lens. This amounts to putting the beam through a telescope backwards. The laser beam goes in the eyepiece lens and comes out the objective lens. Thus, the diverging beam is collimated with a large beam waist and small divergence. It can be shown that the decrease in the divergence is equal to the original divergence divided by the magnification of the telescope. The magnification of the telescope is equal to the ratio of the focal length of the objective and the focal length of the eyepiece.

The second method uses a Keplerian telescope, as illustrated in Figure 31.41. The eyepiece lens is a positive lens of short focal length. The beam comes to a focus and then diverges to be collimated by the objective lens of long focal length.

Each telescope has distinct advantages for beam expansion. The advantage of the Galilean type of beam expander is well utilized in high power or pulsed laser systems. Since the beam does not come to a focus anywhere inside the beam expander's optical path, the power density of the beam decreases. Thus, if the lenses and environment can survive the initial beam, they can survive the beam anywhere in the optical path. Although the Keplerian beam expander can give similar ratios of beam expansion, the power density at the focus of the first lens is very large. In fact, when using a high-power pulsed laser it is possible to cause a breakdown of the air in the space between the

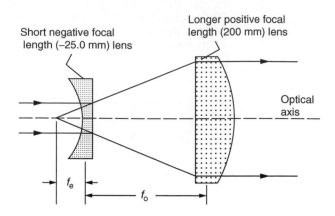

FIGURE 31.40 Gaussian beam collimation by Galilean telescope.

lenses. This breakdown is caused by the strong electrical field that results from focusing the beam to a small diameter, creating miniature lightning bolts.

31.15 LASER APPLICATIONS

The unique feature of light from a laser, as mentioned before, is that it is a coherent narrow beam of a single frequency (or several distinct frequencies). Because of this feature, the laser has found many applications. The main applications are as follows:

1. Industrial applications: Industry accepted the laser as a tool soon after the laser was invented in 1960. At first, the laser was used for alignment and measurements, but with time, applications using high power laser beams became more common. The intense heat produced in a small area by a laser beam is also used for welding and machining metals and for drilling tiny holes in hard materials. The beam of a laser is narrow in itself (typically a few mm). But because the beam is coherent, monochromatic, and essentially parallel and narrow, lenses can be used to focus the light into incredibly small areas without the usual aberration problems. The limiting factor thus becomes diffraction. Also, the energy intensity can be very large. The precise straightness of a laser beam is also useful to surveyors for lining up equipment precisely, especially in inaccessible places.

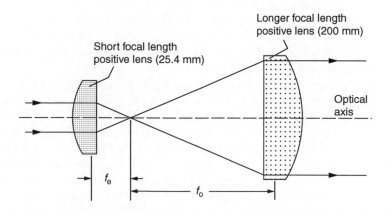

FIGURE 31.41 Gaussian beam collimation by Keplerian telescope.

2. Applications in chemistry: There are a variety of applications of lasers in chemistry, such as excitation of molecules to specific levels, examination of the emitted radiation, measurements of the relaxation time of specific excited levels of molecules, disruption of chemical bonds in molecules at specific region, Raman spectroscopy, and Raman scattering, which is a process of inelastic scattering of a photon by a molecule.

3. Medical applications: Lasers are useful surgical tools. The narrow intense beam can be used to destroy tissue in a localized area or to break up gallstones and kidney stones. Because of the heat produced, a laser beam can be used to weld broken tissue, such as a detached retina. For some types of internal surgery, the laser beam can be carried by an optical fibre to the surgical point and sometimes has an additional fibre-optic path on an endoscope for tissue treatment. An example is the removal of plaque that clogs human arteries. Tiny organelles within a living cell have been destroyed using lasers to study how the absence of that organelle affects the behaviour of the cell. Laser beams have been used to destroy cancerous and precancerous cells; at the same time, the heat seals off capillaries and lymph vessels, thus cauterizing the wound in the process. Single and multi-coloured tattoos can be removed by Q-switched lasers.

4. Military applications: Since the invention of the laser, its potential military uses have been exploited. A large number of projects on lasers were done in secret laboratories, and many new types of equipment were put into service. Lasers are used in range finders, detecting laser signals and laser weapons, tracking systems, etc. Directed energy weapons (DEW) include lasers as well as microwaves, and practical beams are still being actively developed by the U.S. military.

5. Daily applications: Lasers are used as bar-code readers, compact disc (CD) players, and CD-ROMs. The laser beam reflects off the stripes and spaces of a bar-code and off the tiny pits of a CD. Lasers are used in remote controls and laser printers.

6. Laboratory applications: Lasers are used in spectroscopy, advanced microscopes, laser fusion, and laser cooling of atoms.

7. Imaging applications: One of the most interesting applications of laser light is the production of three-dimensional images called holograms. In an ordinary photograph, the film simply records the intensity of light reaching it at each point. When, the photograph or transparency is viewed, light reflecting or passing through it gives us a two-dimensional picture. In holography, the images are formed by interference. When a laser hologram is made on film, a broadened laser beam is split into two parts by a half-silvered mirror, as shown in Figure 31.42. One part goes directly to the film; the rest passes to the object to be photographed, from which it is reflected to the film. Light from every point on the object reaches each point on the film, and the interference of the two beams allows the film to record both the intensity and relative phase of the light at each point. It is crucial that the light be coherent, which is why a laser is used. After the film is developed, it is placed again in a laser beam and a three-dimensional image of the object is created. When viewed from different sides, the hologram image looks like the original object; but when the image is touched, there is nothing material there.

8. Special Applications: The number and variety of laser applications are expanding, and many new special applications have been discovered in fields such as energy transmission from space to Earth, optical gyroscopes, and fibre-lasers.

31.16 EXPERIMENTAL WORK

The experimental works are conducted for the following cases:

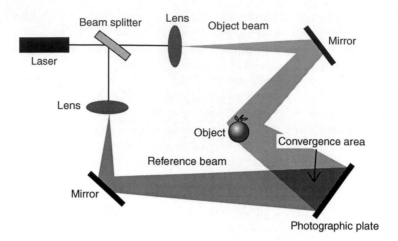

FIGURE 31.42 Production of three-dimensional images by holography.

31.16.1 Experiment One Summary: Laser Alignment, Transverse Modes, and Power Gain/Loss

Build and study the various aspects of a Helium–Neon laser by performing the following steps:

Building a laser from individual components, which involves aligning the laser bore with the optics. Called the autocollimator method, this is only one method of aligning laser optics. Autocollimation is the best method for small-bore laser tubes like the HeNe laser used in this experiment.
1. Varying the cavity length of the laser while studying the effect on transverse mode structure
2. Placing a loss element into the laser cavity and estimating the power gain/loss of the laser by adjusting the intra-cavity loss

Figure 31.43 illustrates all the loss elements that a beam of photons encounters as it completes a round trip (two passes) through the laser cavity. The purpose of adding the slide/loss element is to introduce just enough reflective loss so that the laser switches off. At the point where the laser switches off (lasing light disappears), the gain equals the loss. Therefore, if all the other cavity loss elements are known, a reasonably accurately estimate of the gain of the laser can be made.

To calculate the single pass gain, first add up all the losses to get the overall loss, then divide the losses by the number of passes (two passes in this case). For the p polarization component, the reflection loss at the mirrors is calculated using the appropriate Fresnel equation as given below:

$$r_p = \left(\frac{E_{0r}}{E_{0i}}\right)_p = +\frac{\tan(\theta_i - \theta_t)}{\tan(\theta_i + \theta_t)} \tag{31.31}$$

where r_p is the reflection coefficient.

Calculation of r_p requires a calculation of the refraction angle first. As an example:

Suppose the laser output decreases to zero at 51° angle of incidence and also at 59°. There will be a maximum output with the loss plate (slide) in the cavity at Brewster's angle.

For the 51° case:

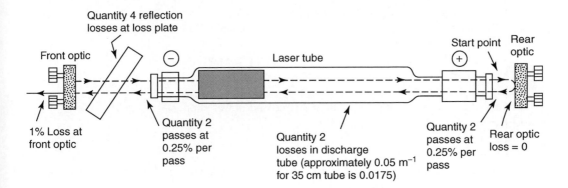

FIGURE 31.43 Loss elements in the laser.

The refraction angle, θ_t, is calculated using Snell's law, assuming a refractive index of 1.5 for the glass:

$\sin 51°/\sin \theta_t = 1.5$. Therefore, $\theta_t = 31.2°$

Using the formulae given above, the reflection coefficient (r_p) at one surface can be calculated as:

$r_p = \tan 19.8°/\tan 82.2° = 4.9\%$.

The reflection (R_p) is the reflection coefficient squared. Therefore, the reflection is $= 0.25\%$. Two surfaces of the slide add up to 0.5% loss.

For a round trip, as noted in Figure 31.43, the total (for two passes) is a 1% loss.

31.16.2 EXPERIMENT TWO SUMMARY: LASER BEAM ALIGNMENT

Students will practise some alignment techniques for a laser beam using optical components and measure the power and diameter of a laser beam at several distances from the output of the laser source in the following experimental cases:

1. Alignment Using One, Two, and Multi-Axis Positioners
2. Alignment Using Two Mirror Mounts
3. Alignment Using Three and Four Mirror Mounts
4. Alignment Using a Risley Prism Pair
5. Alignment Using an Adjustable Wedge

31.16.3 EXPERIMENT THREE SUMMARY: LASER BEAM EXPANSION

This experiment demonstrates the design of two types of laser beam expanders: the Galilean telescope and the Keplerian telescope. From the cases of this experiment, the student will gain experience in aligning, expanding, and measuring the power and diameter of the laser beam. The student will also calculate the diameter of a converging Gaussian laser beam, at various distances from the output of the laser.

31.16.4 TECHNIQUE AND APPARATUS

Appendix A presents the details of the devices, components, tools, and parts.

1. 2×2 ft. optical breadboard
2. Experimental Helium Neon (HeNe) laser kit, as shown in Figure 31.44

  3. High voltage power supply (approximately 1.5 kV)
  4. Alignment light source: small tungsten filament lamp shown in Figure 31.45, with a variable power supply
  5. Alignment beamsplitter: thick (~ 4 mm) glass plate and holder
  6. Loss-element: rotation stage assembly with microscope glass slide, as shown in Figure 31.46
  7. Laser light source (HeNe laser light source and power supply)
  8. Laser clamp/laser mount assembly
  9. Laser light detector
 10. Laser light power meter
 11. Hardware assembly (clamps, posts, screw kits, screwdriver kits, positioners, post holders, laser holder/clamp, etc.)
 12. Two or four mirrors, as shown in Figure 31.47
 13. Mirror holders/positioners, as shown in Figure 31.47
 14. Four mirror holder/rotator/stage assemblies, as shown in Figure 31.48
 15. Lenses (200 mm, -25 mm, and 25.4 mm focal lens), as shown in Figure 31.49
 16. Two lens holders/positioners, as shown in Figure 31.49
 17. Card with a hole and card holder, as shown in Figure 31.50
 18. Risley prism pair and prism holder/rotator assemblies, as shown in Figure 31.51
 19. Adjustable wedge and wedge holder/rotator assemblies, as shown in Figure 31.52
 20. Target card and cardholder, as shown in Figure 31.53
 21. Black/white card and cardholder
 22. Ruler/tape measure

31.16.5 Procedure

Follow the laboratory procedures and instructions given by the professor and/or instructor.

31.16.6 Safety Procedure

Follow all safety procedures and regulations regarding the use of optical instruments and measurements, and light source devices.

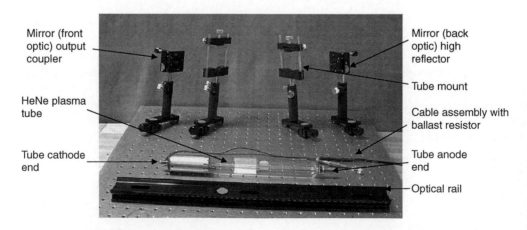

FIGURE 31.44 Helium–Neon (HeNe) laser kit.

FIGURE 31.45 Alignment light source.

FIGURE 31.46 Rotation stage assembly with microscope slide.

FIGURE 31.47 Mirror and mirror holder/positioners.

FIGURE 31.48 Mirror and mirror holder/rotating stage assembly.

FIGURE 31.49 Two lenses and holders/positioners.

FIGURE 31.50 Card with a hole and card holder.

FIGURE 31.51 Risley prism pair and holder.

FIGURE 31.52 Adjustable wedge and holder.

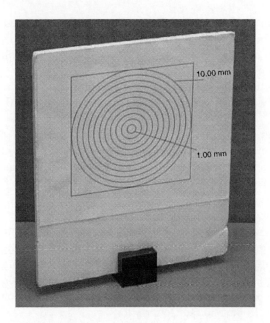

FIGURE 31.53 Target card and cardholder.

31.16.7 APPARATUS SET-UP

31.16.8 EXPERIMENT ONE: LASER ALIGNMENT, TRANSVERSE MODES, AND POWER GAIN/LOSS

31.16.8.1 Building a Laser from Individual Components

1. Mount the Helium–Neon (HeNe) laser tube horizontally on the optical rail using the two tube mounts.
2. Ensure that the cable assembly is not connected to the high voltage power supply. Connect the appropriate leads of the cable assembly to the cathode ($-$) and anode ($+$) of the tube. (This is to avoid misaligning everything when trying to connect the leads to the tube later).
3. Place the tungsten filament lamp at the cathode end of the tube.
4. Connect the tungsten filament lamp to the variable power supply. Mount the filament to be oriented vertically, for easy alignment. This will generate a thin vertical filament line, which is visible through the bore of the tube.
5. Line up the tungsten filament with the bore of the laser tube, while viewing through the tube, as shown in Figure 31.54(a). The image of the filament, when sighted by eye through the laser bore, is shown in Figure 31.55. Adjust the lamp voltage so that the filament is clearly visible. Adjust the bulb position, until it is clearly visible and centred in the tube bore.
6. Mount the beamsplitter approximately 45° to the tube axis, at a reasonable distance from the AR coated front window, as shown in the Figure 31.54(b). Reflected light passes back down the tube, and can be observed via the beamsplitter.

7. The filament lamp will now appear to be out of position, if viewed from the rear through the tube bore. (This is because the thick glass plate displaces the filament image.) The lamp will need to be moved laterally so that it again appears centred and vertical in the bore of the discharge tube.

8. Identify which optic is the front and which is the rear optic by using another laser beam. The laser beam will pass through the front optic, while it will not pass through the rear optic.

9. As shown in Figure 31.54(c), insert the rear optic into the mount and place it at approximately one centimetre away from the AR coated rear end of the laser tube.

10. The filament lamp needs to be viewed in the rear optic via the beamsplitter, as shown in Figure 31.54(c). The rear optic must now be precisely aligned to be perpendicular to the laser bore. Adjust the two knobs on the rear optic, while looking through the beamsplitter, until an image of the filament is seen. If two images are visible, align only one image. This is done by adjusting the two knobs on the rear optic until the filament is clearly visible and centred in the optic. The rear optic is now aligned.

11. Remove the beamsplitter and mount the front optic at a reasonable distance from the AR coated front end, as shown in Figure 31.56. Adjust the front optic to be in line with the bore, so that you can see straight through both optics and the bore.

12. Place the target card beyond the front optic.

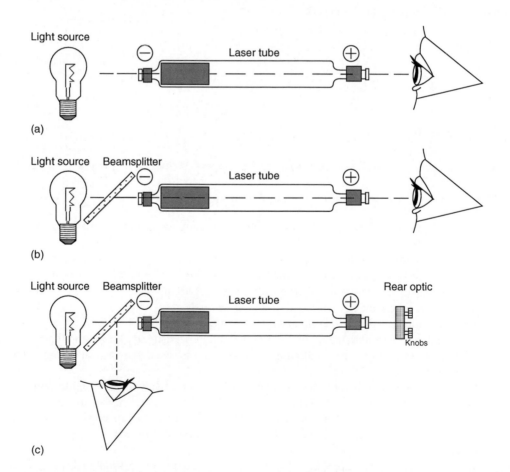

(a)

(b)

(c)

FIGURE 31.54 Alignment of the filament to the bore: (a) without beamsplitter; (b) with the beamsplitter; (c) with the beamsplitter and the rear optic.

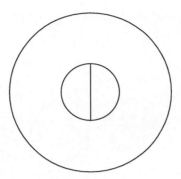

FIGURE 31.55 Image of the filament as seen though the bore.

13. Do not proceed to start the discharge tube without the presence of the professor/instructor for safety issues. Connect the cable assemble to the high voltage power supply. Then turn the power on to start the tube discharge. Keep away from the high voltage power.
14. Observe the discharge tube glowing.
15. Roughly align the front optic, by adjusting its knobs, until the image reflects from the front optic onto the laser bore entrance.
16. Precisely align the front optic, by making fine systematic adjustments of the tilt and lateral knobs, until the lasing beam is constantly viewed on the card, as shown in Figure 31.57. The laser now is lasing.
17. To increase the image size and make the modes more recognizable, move the card a distance (several metres) away from the front optic. It is also possible to use a diverging lens to expand the laser beam.
18. A wide variety of modes can be obtained with an aligned and lasing tube. The number and types of modes will vary with the alignment of the front and back optics. Some of the lasing modes that can be obtained are shown in Figure 31.58.
19. Gently adjust the back and front optics, one knob at a time, to get different TEM modes.

31.16.8.2 Varying the Laser Cavity Length while Observing Transverse Modes

1. The laser tube assembly, as shown in Figure 31.54, must be aligned and lasing as explained in Section 31.16.6.1.
2. The laser cavity length is the distance between the front and back optics. Gradually expand the length of the laser cavity by moving the back and front optics, one at a time, to get different TEM modes. Again, some of the modes that can be obtained are shown in Figure 31.55. The types of modes may vary from those modes that were observed in Section 31.16.6.1.

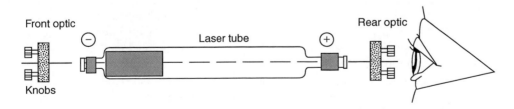

FIGURE 31.56 Alignment of the front optic.

FIGURE 31.57 Laser light beam lasing from the laser kit.

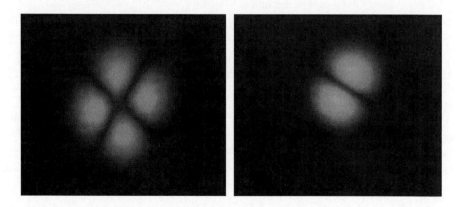

FIGURE 31.58 Transverse electro-magnetic modes.

31.16.8.3 Estimating the Power Gain/Loss from a Loss Element Placed in the Cavity

1. The laser tube assembly, as shown in Figure 31.54, must be aligned and lasing as explained in Section 31.16.6.1.
2. Maximize the beam power output before inserting the loss element. Measure the beam power emitted from the front and back ends without the loss element. Measure the length of the plasma (inner) tube; this is the gain element. Fill out Table 31.1
3. Mount an intra-cavity loss element (clean glass microscope slide) onto a rotation stage assembly. Place the assembly into the lasing cavity, between the front optic and the end of the plasma tube, as shown in Figure 31.59.
4. Find the equilibrium point, by rotating the slide gradually until the laser just switches off. Note the angle where the laser gain equals the loss; there will be two such angles.
5. Record all the other cavity elements and their losses; these losses will be needed to accurately estimate the gain of the laser. Fill out Table 31.2.

FIGURE 31.59 Lasing plasma tube with intra-cavity loss element.

6. Place the laser power meter in the path of the transmitted laser beam, as shown in Figure 31.59. Rotate the slide gradually, while measuring the transmitted power, for several angles at which the laser has a measurable light output. Fill out Table 31.3.

31.16.9 EXPERIMENT TWO: LASER BEAM ALIGNMENT

The following cases are designed for aligning the laser beam towards many directions in space:

31.16.9.1 Alignment Using One, Two, and Multi-Axis Positioners

This case has two set-ups: Section Laser Beam Alignment Using One Mirror describes the use of one mirror to align the laser beam from the laser source, and Section Laser Beam Alignment Using Two Mirrors describes the use of two mirrors to direct the laser beam in a two stage alignment process.

Laser Beam Alignment Using One Mirror

1. Figure 31.60 shows the experimental apparatus set-up.
2. Bolt the laser short rod to the breadboard.
3. Bolt the laser mount to the clamp using bolts from the screw kit.
4. Put the clamp on the short rod.
5. Place the HeNe laser into the laser mount and tighten the screw. Turn on the laser device. Follow the operation and safety procedures of the laser device in use.
6. Align the laser beam to be parallel to the edge of the breadboard.
7. Mount a mirror and mirror holder assembly (M_1) to the breadboard at the corner facing the laser beam.
8. Mount a card with a hole and card holder assembly between the laser assembly and mirror M_1. Make sure that the hole diameter is slightly larger than the laser beam diameter (about 2 mm) so that the laser beam will pass through and back reflections from the mirrors can be easily seen.
9. Adjust the position of the laser assembly such that the laser beam passes through the hole and is parallel to the edge of the breadboard.

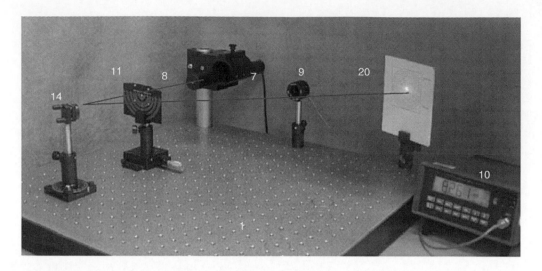

FIGURE 31.60 Laser beam aligned using one mirror.

10. Adjust the height of mirror M_1 until the laser beam intersects the centre of mirror M_1. Finely rotate mirror M_1 post to make sure the laser beam is perpendicular to mirror M_1. In this case, the laser beam reflects back from the mirror, through the hole, to the laser source. Rotate mirror M_1 an angle of 30° away from the laser beam towards the right corner of the breadboard.
11. Place the target card at the corner of the breadboard that is facing mirror M_1. Adjust the position of the target card so that the laser beam is incident on the centre of the target card, as shown in Figure 31.61.
12. Measure the laser beam power and diameter at several locations between each optical component along the beam path. Fill out Table 31.4 for the listed locations.

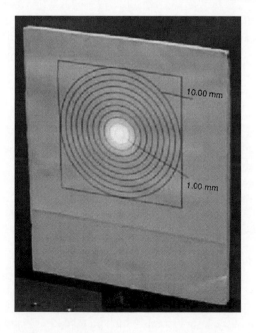

FIGURE 31.61 Incident laser beam on a target card.

Laser Beam Alignment Using Two Mirrors

Using the same technique as in Section Laser Beam Alignment Using One Mirror, the laser beam can be aligned between two mirrors, as illustrated in Figure 31.29 and Figure 31.30. Figure 31.62 shows the two-mirror experimental setup. Start with the following steps to perform alignment of the laser beam using two mirrors:

1. As explained in Section 31.16.7.1.1, align the laser beam. The laser beam returning back to mirror M_1 should go back into the laser source.
2. Rotate mirror M_1 an angle of 45° away from the laser beam, towards the right corner of the breadboard.
3. Place a second mirror and mirror holder assembly (M_2) on the breadboard at the adjacent corner.
4. Rotate mirror M_2 an angle of 45° away from mirror M_1, towards the right corner of the breadboard. After this step, the laser beam is successfully aligned, from the laser source to mirror M_1 and mirror M_2. The laser beam is now the same height and is parallel to the breadboard, as shown in Figure 31.62.
5. Place the target card at the corner of the breadboard that is facing mirror M_2. Adjust the position of the target card so that the laser beam is incident on the centre of the target card, as shown in Figure 31.62.
6. Measure the laser beam power and diameter at several locations between each optical component along the beam path. Fill out Table 31.5 for the listed locations.

31.16.9.2 Alignment Using Two Mirror Mounts

Using the same technique as in Section 31.16.7.1, the laser beam can be aligned using two mirror mounts, as illustrated in Figure 31.30. Figure 31.63 shows the experimental setup for Section 31.16.7.2. Start with the following steps:

1. Mount a laser device to the middle of the edge of the breadboard, as shown in Figure 31.63.
2. Mount two mirror assemblies, as shown in Figure 31.48 and Figure 31.65, facing each other in the middle of the breadboard in front of the laser device.

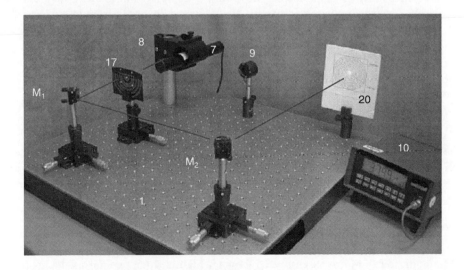

FIGURE 31.62 Laser beam aligned using two mirrors.

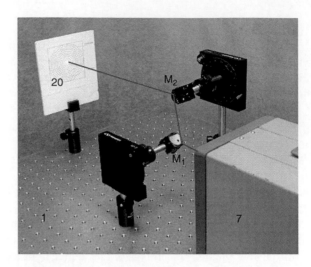

FIGURE 31.63 Laser beam alignment using two mirror mounts.

3. Align mirror M_1 to reflect the laser beam towards mirror M_2. Adjust M_2 to reflect the laser beam towards the target card, as shown in Figure 31.63.
4. Measure the laser beam power and diameter at several locations between each optical component along the beam path. Fill out Table 31.6 for the listed locations.

31.16.9.3 Alignment Using Three and Four Mirror Mounts

Using the same technique as in Section 31.16.7.2, the laser beam can be aligned using three and four mirror mounts, as illustrated in Figure 31.31 and Figure 31.32. Figure 31.64 and Figure 31.65 show the three and four mirror mounts experimental setups for Section 31.16.7.3. Start with the following steps to perform alignment of the laser beam using the three mirror mounts:

Laser Beam Alignment Using Three Mirror Mounts

1. Mount a laser device near the middle edge of the breadboard, as shown in Figure 31.64.
2. Mount three mirror assemblies, as shown in Figure 31.48 and Figure 31.64, one mirror facing the other two mirrors, in the middle of the breadboard in front of the laser device.
3. Align mirror M_1 to reflect the laser beam towards mirror M_2. Do likewise from mirror M_2 towards mirror M_3. Adjust M_3 to reflect the laser beam towards the target card, as shown in Figure 31.64.
4. Measure the laser beam power and diameter at several locations between each optical component along the beam path. Fill out Table 31.7 for the listed locations.

Laser Beam Alignment Using Four Mirror Mounts

Using the same technique as in Laser Beam Alignment Using Three Mirror Mounts, he laser beam can be aligned using four mirror mounts, as illustrated in Figure 31.32. Figure 31.65 shows the four-mirror mounts experimental setup. Add the following steps to perform alignment of the laser beam using the four mirror mounts:

1. Continue the procedure as explained in the three-mirror mounts in Section Laser Beam Alignment Using Three Mirror Mounts.
2. Mount the fourth mirror assembly, as shown in Figure 31.65, two mirrors facing the other two mirrors, at a location in the middle of the breadboard in front of the laser device.

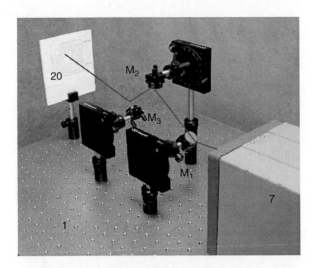

FIGURE 31.64 Laser beam alignment using three mirror mounts.

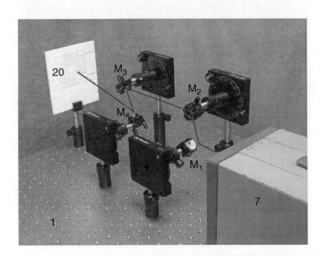

FIGURE 31.65 Laser beam alignment using four mirror mounts.

3. Align mirror M_1 to reflect the laser beam towards mirror M_2. Do likewise from mirror M_2 towards mirror M_3, then towards mirror M_4. Adjust M_4 to reflect the laser beam towards the target card, as shown in Figure 31.65.
4. Measure the laser beam power and diameter at several locations between each optical component along the beam path. Fill out Table 31.8 for the listed locations.

31.16.9.4 Alignment Using a Risley Prism Pair

As explained in the theory, the laser beam can be aligned using a Risley prism pair and mount assemblies, as illustrated in Figure 31.33. Figure 31.66 shows the experimental set-up for Section 31.16.7.4.

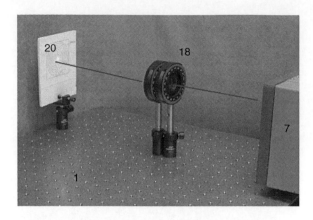

FIGURE 31.66 Laser beam alignment using a Risley prism pair.

1. Mount a laser device near the middle edge of the breadboard, as shown in Figure 31.66.
2. Mount the Risley prism pair and mount assemblies, as shown in Figure 31.48 and Figure 31.66, in the middle of the breadboard in front of the laser device.
3. Try to adjust the Risley prism pair to reflect the laser beam towards different spots on the target card, as shown in Figure 31.66.
4. Align the Risley prism pair to reflect laser beam towards the target card.
5. Measure the laser beam power and diameter at different locations: between the laser device and the Risley prism pair, between the Risley prism pair and the target card, and at the target card. Fill out Table 31.9.

31.16.9.5 Alignment Using an Adjustable Wedge

As explained in the theory, the laser beam can be aligned using an adjustable wedge and mount assembly, as illustrated in Figure 31.34. Figure 31.67 shows the experimental setup for Section 31.16.7.5.

1. Mount a laser device near the middle side of the breadboard, as shown in Figure 31.67.
2. Mount the adjustable wedge and mount assembly, as shown in Figure 31.49 and Figure 31.67, in the middle of the breadboard in front of the laser device.
3. Try to adjust the adjustable wedge to reflect the laser beam towards different spots on the target card, as shown in Figure 31.67.
4. Align the adjustable wedge to reflect the laser beam towards the target card.
5. Measure the laser beam power and diameter at different locations: between the laser device and the adjustable wedge, between the adjustable wedge and the target card, and at the target card. Fill out Table 31.10.

31.16.10 EXPERIMENT THREE: LASER BEAM EXPANSION

The following cases are designed to study laser beam expansion.

31.16.10.1 Laser Beam Alignment

The laser beam needs to be aligned, as explained in Experiment Two, Laser Beam Alignment. Laser beam alignments are shown in Figure 31.60 and Figure 31.62.

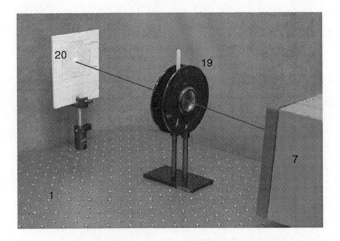

FIGURE 31.67 Laser beam alignment using an adjustable wedge.

31.16.10.2 Laser Beam Expansion by Galilean Telescope

Once the laser beam is aligned as in Section 31.16.8.1, the laser beam can be expanded by a Galilean telescope arrangement, illustrated in Figure 31.40, by adding the following steps. Figure 31.68 shows the experimental setup for this case.

1. Insert a short focal length (-25.0 mm) negative lens (Lens 1) into a lens holder/positioner assembly and mount it 127 mm from mirror M_1. Align the lens height in the lens holder and slide the lens holder/positioner so that the diverging beam is centred on mirror M_2.
2. Insert a longer focal length (200 mm) positive lens (Lens 2) into a holder/positioner and place it about 175 mm (the sum of the focal lengths of the two lenses, remembering that the first lens is a negative lens) from the first lens in the diverging laser beam path.
3. Align lens (Lens 2) height in the lens holder and slide the lens holder/positioner so that the diverging beam is centred on mirror M_2.

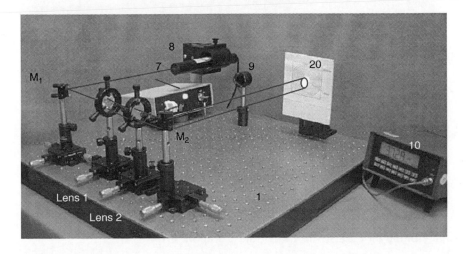

FIGURE 31.68 Laser beam expansion by a Galilean telescope.

4. Carefully adjust the position of Lens 2, by moving it back and forth along the beam, until the laser beam is expanded and incident on the target card, as shown in Figure 31.68.
5. Measure the power and diameter of the laser beam close to the laser source. Fill out Table 31.12.
6. Measure the power and diameter of the laser beam at several distances, two metres apart, from two to ten metres away from mirror M_2. Fill out Table 31.13.

31.16.10.3 Laser Beam Expansion by Keplerian Telescope

Once the laser beam is aligned as in Section 31.16.8.1, the laser beam can be expanded by a Keplerian telescope arrangement, illustrated in Figure 31.41, by adding the following steps. Figure 31.69 shows the experimental set-up for this case.

1. Replace the negative lens with a short focal length positive lens (25.4 mm) and use the same adjustments to centre the beams in the lenses (Lens 1 and Lens 2). Adjust the distance between the two lenses to be the sum of their focal lengths.
2. Carefully adjust the position of Lens 1 and Lens 2, by moving them back and forth along the beam, until the laser beam is expanded and incident on the target card, as shown in Figure 31.69.
3. Measure the power and diameter of the laser beam close to the laser source. Fill out Table 31.14.
4. Measure the power and diameter of the laser beam at several distances, two metres apart, from two to ten metres away from mirror M_2. Fill out Table 31.15.

FIGURE 31.69 Laser beam expansion by a Keplerian telescope.

31.16.11 Data Collection

31.16.12 Experiment One: Laser Alignment, Transverse Modes, and Power Gain/Loss

31.16.12.1 Building a Laser from Individual Components

No data collection is required for this case.

31.16.12.2 Varying the Laser Cavity Length while Observing Transverse Modes

Sketch or photograph each of the observed modes.

31.16.12.3 Estimating the Power Gain/Loss from a Loss Element Placed in the Cavity

1. Measure the beam power emitted from the front and back ends without the loss element. Fill out Table 31.1.
2. Measure the length of the plasma tube to calculate the gain element. Using the assumed numbers in Figure 31.43, record the losses for the front/rear optics/coated. Fill out Table 31.2.
3. Measure the transmitted component of light power at various angles at which the laser has a measurable light output. Fill out Table 31.3.
4. Record the equilibrium angles

TABLE 31.1
Cavity Characteristics

Plasma Tube Length (x)

Beam Power

Front (unit)	Back (unit)

TABLE 31.2
Cavity Losses

Element	Reflectance	Number of Passes/ Surfaces	Loss (%)
Front Optic			
Rear Optic			
Tube			
Front Window			
Rear Window			
Slide at Equilibrium Angles			
Total			

TABLE 31.3
Transmitted and Reflected Power
Components

Angle (°)		Power
Incident θ_i	Refracted θ_r	Transmitted (unit)

31.16.13 EXPERIMENT TWO: LASER BEAM ALIGNMENT

31.16.13.1 Alignment Using One, Two, and Multi-Axis Positioners

Laser Beam Alignment Using One Mirror

Measure the laser beam power and diameter at several locations between each optical component along the beam path. Fill out Table 31.4.

TABLE 31.4
Laser Beam Power and Diameter Measurements Using One
Mirror Mount

Location	Laser Beam Power P (unit)	Laser Beam Diameter d (unit)
Between Laser Source and M_1 L_1		
Between M_1 and Target Card L_2		

Laser Beam Alignment Using Two Mirrors

Measure the laser beam power and diameter at several locations between each optical component along the beam path. Fill out Table 31.5.

TABLE 31.5
Laser Beam Power and Diameter Measurements for Laser Beam Alignment Using Two Mirrors

Location	Laser Beam Power P (unit)	Laser Beam Diameter d (unit)
Between Laser Source and M_1 L_1		
Between M_1 and M_2 L_2		
Between M_2 and Target Card L_3		
At Target Card L_4		

31.16.13.2 Alignment Using Two Mirror Mounts

Measure the laser beam power and diameter at several locations between each optical component along the beam path. Fill out Table 31.6.

TABLE 31.6
Laser Beam Power and Diameter Measurements for Laser Beam Alignment Using Two Mirror Mounts

Location	Laser Beam Power P (unit)	Laser Beam Diameter d (unit)
Between Laser Source and M_1 L_1		
Between M_1 and M_2 L_2		
Between M_2 and Target Card L_3		
At Target Card L_4		

31.16.13.3 Alignment Using Three and Four Mirror Mounts

Laser Beam Alignment Using Three Mirror Mounts

Measure the laser beam power and diameter at several locations between each optical component along the beam path. Fill out Table 31.7.

TABLE 31.7
Laser Beam Power and Diameter Measurements for Laser Beam Alignment Using Three Mirror Mounts

Location	Laser Beam Power P (unit)	Laser Beam Diameter d (unit)
Between Laser Source and M_1 L_1		
Between M_1 and M_2 L_2		
Between M_2 and M_3 L_3		
Between M_3 and Target Card L_4		
At Target Card L_5		

Laser Beam Alignment Using Four Mirror Mounts

Measure the laser beam power and diameter at several locations between each optical component along the beam path. Fill out Table 31.8.

TABLE 31.8
Laser Beam Power and Diameter Measurements for Laser Beam Alignment Using Four Mirror Mounts

Location	Laser Beam Power P (unit)	Laser Beam Diameter d (unit)
Between Laser Source and M_1 L_1		
Between M_1 and M_2 L_2		
Between M_2 and M_3 L_3		
Between M_3 and M_4 L_4		
Between M_4 and Target Card L_5		
At Target Card L_6		

31.16.13.4 Alignment Using a Risley Prism Pair

Measure the laser beam power and diameter at several locations: between the laser device and the Risley prism pair, between the Risley prism pair and the target card, and at the target card. Fill out Table 31.9.

TABLE 31.9
Laser Beam Power and Diameter Measurements Using Risley Prism

Location	Laser Beam Power P (unit)	Laser Beam Diameter d (unit)
Between Laser Device and Risley Prism Pair L_1		
Between Risley Prism Pair and Target Card L_2		
At Target Card L_3		

31.16.13.5 Alignment Using an Adjustable Wedge

Measure the laser beam power and diameter at several locations: between the laser device and the adjustable wedge, between the adjustable wedge and the target card, and at the target card. Fill out Table 31.10.

TABLE 31.10
Laser Beam Power and Diameter Measurements Using Adjustable Wedge

Location	Laser Beam Power P (unit)	Laser Beam Diameter d (unit)
Between Laser Device and Adjustable Wedge L_1		
Between Adjustable Wedge and Target Card L_2		
At Target Card L_3		

31.16.14 EXPERIMENT THREE: LASER BEAM EXPANSION

31.16.14.1 Laser Beam Alignment

1. Measure the laser beam power (P) and diameter (d) at several locations: between the laser source and mirror M_1, between mirror M_1 and mirror M_2, between mirror M_2 and the target card, and at the target card.
2. Fill out Table 31.11.

TABLE 31.11
Laser Beam Power and Diameter Measurements

Location	Laser Beam Power P (unit)	Laser Beam Diameter d (unit)
Between Laser Source and M_1 L_1		
Between M_1 and M_2 L_2		
Between M_2 and Target Card L_3		
At Target Card L_4		

31.16.14.2 Laser Beam Expansion by Galilean Telescope

1. Measure the laser beam power (P) and diameter (d_0) close to the laser source.
2. Record the measured data in Table 31.12.

TABLE 31.12
Galilean Telescope Laser Beam Measurements
Close to the Laser Source

Laser Beam Power P (unit)	Laser Beam Diameter d_0 (unit)	Wavelength λ (unit)	Divergence Angle θ (unit)	Rayleigh Range z_R (unit)

TABLE 31.13
Galilean Telescope Laser Beam Measurements at Various
Distances from M_2

At Distance z (unit)	Laser Beam Power P (unit)	Laser Beam Diameter d (unit)	Divergence Angle θ (unit)	Laser Beam Diameter at z dz (unit)	Rayleigh Range z_R (unit)
z_1					
z_2					
z_3					
z_4					
z_5					

3. Measure the laser beam power (P) and diameter (d) at several distances, two metres apart, from two to ten metres away from mirror M_2.
4. Record the measured data in Table 31.13.

31.16.14.3 Laser Beam Expansion by Keplerian Telescope

1. Measure the laser beam power (P) and diameter (d_0) close to the laser source.
2. Record the measured data in Table 31.14.
3. Measure the laser beam power (P) and diameter (d) at several distances two metres apart, from two to ten metres away from mirror M_2.
4. Record the measured data in Table 31.15.

TABLE 31.14
Keplerian Telescope Laser Beam Measurements Close to the Laser Source

Laser Beam Power P (unit)	Laser Beam Diameter d_o (unit)	Wavelength λ (unit)	Divergence Angle θ (unit)	Rayleigh Range z_R (unit)

TABLE 31.15
Keplerian Telescope Laser Beam Measurements at Various Distances from M_2

At Distance z (unit)	Laser Beam Power P (unit)	Laser Beam Diameter d (unit)	Divergence Angle θ (unit)	Laser Beam Diameter at z d_z (unit)	Rayleigh Range z_R (unit)
z_1					
z_2					
z_3					
z_4					
z_5					

31.16.15 CALCULATIONS AND ANALYSIS

31.16.16 EXPERIMENT ONE: LASER ALIGNMENT, TRANSVERSE MODES, AND POWER GAIN/LOSS

31.16.16.1 Building a Laser from Individual Components

No calculations or analysis are required for this case.

31.16.16.2 Varying the Laser Cavity Length while Observing Transverse Modes

No calculations or analysis are required for this case.

31.16.16.3 Estimating the Power Gain/Loss from a Loss Element Placed in the Cavity

1. Plot the angles versus power output for the transmitted power.
2. Calculate the refraction angle for the two equilibrium angles using Snell's law (assume $n_g = 1.5$).
3. Calculate r_p for the equilibrium points using Fresnel equations for Reflectance: $R_p = [(n \cos \theta_i - \cos \theta_r)/(n \cos \theta_i + \cos \theta_r)]^2$. Note that r_p equals zero at the polarizing angle.
4. Based on the assumed numbers in Figure 31.43 and the measured tube length, calculate the loss for the tube.
5. Add the losses in Table 31.2 to calculate the round trip loss with the loss element.
6. Then divide the round trip loss by two to get the single pass gain.
7. Accurately estimate the gain of the laser.

31.16.17 EXPERIMENT TWO: LASER BEAM ALIGNMENT

31.16.17.1 Alignment Using One, Two, and Multi-Axis Positioners

1. There are no calculations and analysis required for this case.
2. Plot laser beam diameter along the path between the laser source and the target card.
3. Plot laser beam power along the path between the laser source and the target card.

31.16.17.2 Alignment Using Two Mirror Mounts

Repeat the steps as explained in Section 31.18.2.1.

31.16.17.3 Alignment Using Three and Four Mirror Mounts

Repeat the steps as explained in Section 31.18.2.1.

31.16.17.4 Alignment Using a Risley Prism Pair

Repeat the steps as explained in Section 31.18.2.1.

31.16.17.5 Alignment Using an Adjustable Wedge

Repeat the steps as explained in Section 31.18.2.1.

31.16.18 EXPERIMENT THREE: LASER BEAM EXPANSION

31.16.18.1 Laser Beam Alignment

1. Calculations and analysis are not required for this part.
2. Plot a graph of the power (P) of the laser beam versus distance at various locations (L).
3. Plot a graph of the diameter (d) of the laser beam versus distance at various locations (L).

31.16.18.2 Laser Beam Expansion by Galilean Telescope

1. Calculate the divergence angle (θ) and Rayleigh range (z_R) for the laser beam close to the laser source.
2. Fill out Table 31.12 with the calculated data.
3. Calculate the divergence angle (θ), diameter (d_z), and Rayleigh range (z_R) for the laser

   beam at various distances (z) away from the mirror M_2.
4. Fill out Table 31.13 with the calculated data.
5. Plot a graph of the power $= (P)$ of the laser beam vs. distance at various distances (z).
6. Plot a graph of the measured (d) and calculated (d_z) diameters of the laser beam vs. various distances (z).
7. Plot a graph of the calculated divergence angle (θ) vs. various distances (z).
8. Divergence of the beam diameter along the principle axis of the laser beam can be called the beam profile. To map the beam profile, plot a graph of the beam radius ($\pm d_c/2$) vs. distance (z) along the axis. The plot of the laser profile can be mapped from the mirror M_2, up to ten metres away.

31.16.18.3 Laser Beam Expansion by Keplerian Telescope

Repeat the steps as explained in Section 31.18.3.2. Fill out Table 31.14 and 31.15.

31.16.19 RESULTS AND DISCUSSIONS

31.16.20 EXPERIMENT ONE: LASER ALIGNMENT, TRANSVERSE MODES, AND POWER GAIN/LOSS

31.16.20.1 Building a Laser from Individual Components

Discuss the alignment procedure used in this case. Propose any improvement to the alignment process.

31.16.20.2 Varying the Laser Cavity Length while Observing Transverse Modes

1. Compare the observed modes to the standard modes for a circular discharge beam, and identify the observed modes.
2. There are a wide variety of modes that can be obtained with a laser. The number of modes will vary with the alignment, cavity configuration (optic separation), as well as the gain of the individual system used.

31.16.20.3 Estimating the Power Gain/Loss from a Loss Element Placed in the Cavity

Referring to Figure 31.43, discuss the gain of the laser tube.

31.16.21 EXPERIMENT TWO: LASER BEAM ALIGNMENT

31.16.21.1 Alignment Using One, Two, and Multi-Axis Positioners

1. Report the graph for the laser beam diameter along the path between the laser device and the target card.
2. Report the graph for the laser beam power along the path between the laser device and the target card.
3. Discuss the graphs for the laser beam diameter and power along the path between the laser device and the target card.

31.16.21.2 Alignment Using Two Mirror Mounts

Repeat the steps as explained in Section 31.19.2.1.

31.16.21.3 Alignment Using Three and Four Mirror Mounts

Repeat the steps as explained in Section 31.19.2.1.

31.16.21.4 Alignment Using a Risley Prism Pair

Repeat the steps as explained in Section 31.19.2.1.

31.16.21.5 Alignment Using an Adjustable Wedge

Repeat the steps as explained in Section 31.19.2.1.

31.16.22 EXPERIMENT THREE: LASER BEAM EXPANSION

31.16.22.1 Laser Beam Alignment

1. Report the graph for the laser beam diameter along the path between the laser device and the target card.
2. Report the graph for the laser beam power along the path between the laser device and the target card.
3. Discuss the graphs for the laser beam diameter and power along the path between the laser device and the target card.

31.16.22.2 Laser Beam Expansion by Galilean Telescope

Repeat the steps as explained in Section 31.19.3.1.

31.16.22.3 Laser Beam Expansion by Keplerian Telescope

Repeat the steps as explained in Section 31.19.3.1.

31.16.23 CONCLUSION

Summarize the important observations and findings obtained in this lab experiment.

31.16.24 SUGGESTIONS FOR FUTURE LAB WORK

List any suggestions for improvements using different experimental equipment, procedures, and techniques for any future lab work. These suggestions should be theoretically justified and technically feasible.

31.17 LIST OF REFERENCES

List any references that were used in the report. Use one format in writing the references. Never mix reference formats in a report.

31.18 APPENDICES

List all of the materials and information that are too detailed to be included in the body of the report.

FURTHER READING

Agrawal, G. P. and Dutta, N. K., *Long Wavelength Semiconductor Lasers*, Van Nostrand, New York, 1986.
Agrawal, G. P. and Dutta, N. K., *Semiconductor Lasers*, 2nd ed., Van Nostrand, New York, 1993.
Agrawal, G. P., *Fiber-Optic Communication Systems*, 2nd ed., Wiley, New York, 1997.
Alda, J., Laser and Gaussian beam propagation and transformation, *Encyclopedia of optical engineering*, Barry Johnson, R., et al., Eds., Marcel Dekker, Inc., New York, 2002.

Black, E., *An Introduction to Pound-Drever-Hall Laser Frequency Stabilization*, California Institute of Technology & Massachusetts Institute of Technology, LIGO, 2000.

Buus, J., *Tunable Laser Diodes and Related Optical Sources*, 2nd ed., Wiley, Santa Barbara, 2005.

Charschan, S., *Lasers in Industry*, Van Nostrand, New York, 1972.

Csele, M., *Fundamentals of Light Sources and Lasers*, Wiley-Interscience, New York, 2004.

Chee, J. K. and Liu, J. M., Polarization-dependent parametric and raman processes in a birefringent optical fiber, *IEEE J. Quantum Elect.*, 26, 541–549, 1990.

Chen, J.-H., Su, D.-C., and Su, J.-C., Holographic spatial walk-off polarizer and its application to a 4-port polarization independent optical circulator, *Opt. Express*, 11, 2001–2006, 2003.

Davis, C. C., *Lasers and Electro-Optics, Fundamental and Engineering*, Cambridge University Press, New York, 1996.

Derickson, D., *Fiber Optic Test and Measurement*, Prentice Hall PTR, New Jersey, 1998.

Duarte, F. J. and Piper, J. A., Dispersion theory of multiple-prisms beam expanders for pulsed dye lasers, *Opt. Commun.*, 43, 303–307, 1982.

Duarte, F. J. and Piper, J. A., Narrow-linewidth, high prf copper laser-pumped dye laser oscillators, *Appl. Opt.*, 23, 1391–1394, 1984.

Duarte, F. J. and Piper, J. A., Multi-pass dispersion theory of prismatic pulsed dye lasers, *Optica Acta.*, 33, 331–335, 1984.

Duarte, F. J., Note on achromatic multi-prism beam expanders, *Opt. Commun.*, 53, 259–262, 1985.

Duarte, F. J., *Narrow-Lindwidth Pulse Dye Laser Oscillators, in Dye Laser Principles*, F.J. Duarte and L.W. Hillman, Academic, New York, 1990.

Duarte, F. J., *Tunable Lasers Handbook*, Elsevier, New York, 1999.

Dutton, H. J. R., *Understanding Optical Communications*, IBM, Prentice Hall, Inc., New Jersey, 1998.

Fedder, G.K., Iyer, S., and Mukherjee, T., Automated optimal synthesis of microresonators, In Technical Digest of the IEEE 9th International Conference on Solid-State Sensors and Actuators (Transducers '97), Chicago, IL, U.S.A., 2:1109–1112, 1997.

Griffel, G., Abeles, J. H., Menna, R. J., Braun, A. M., Connolly, J. C., and King, M., Low-threshold InGaAsP ring lasers fabricated using bi-level dryetching, *IEEE Photon. Technol. Lett.*, 12, 146–148, 2000.

Hecht, J., *Understanding Fiber Optics*, 3rd ed., Prentice Hall, Inc., New Jersey, 1999.

Hine, T. J., Cook, M., and Rogers, G. T., An illusion of relative motion dependent upon spatial frequency and orientation, *Vision Res.*, 35, 3093–3102, 1995.

Hoss, R. J., *Fiber Optic Communications—Design Handbook*, Prentice Hall Pub. Co., New Jersey, 1990.

Iovine, J., *Homemade Holograms—The Complete Guide to Inexpensive, Do-It-Yourself Holography*, Tab Books, Division of McGraw-Hill, Inc., Pennsylvania, 1990.

Jackson, R. A., The laser as a light source for the mach-zehnder interferometer, *J. Sci. Instrum.*, 42, 282–283, 1965.

Kao, C. K., *Optical Fiber Systems: Technology, Design, and Applications*, McGraw-Hill, New York, 1982.

Kasap, S. O., *Optoelectronics and Photonics Principles and Practices*, Prentice Hall PTR, Indiana, 2001.

Keiser, G., *Optical Fiber Communications*, 3rd ed., McGraw-Hill Pub, New York, 2000.

Keiser, G., *Optical Communications Essentials*, 1st ed., McGraw-Hill Pub, New York, 2003.

Kolimbiris, H., *Fiber Optics Communications*, Prentice Hall, Inc., New Jersey, 2004.

Kuhn, K., *Laser Engineering*, Prentice Hall, Inc., New Jersey, 1998.

Lengyel, B., *Lasers*, Wiley, New York, 1971.

Litchinitser, N. M., Dunn, S., Steinvurzel, P., Eggleton, B., White, T., McPhedran, R., and de Sterke, C., Application of an arrow model for designing tunable photonic devices, *Opt. Express*, 12 (8), 1540–1550, 2004.

McComb, G., *The Laser Cookbook – 88 Practical Projects*, Tab Book, Division of McGraw-Hill, Inc., Pennsylvania, 1988.

Mouthaan, T., *Semiconductor Devices Explained Using Active Simulation*, Wiley, New Jersey, 1999.

Nanni, C. A. and Alster, T. S., Laser-assisted hair removal: side effects of Q-switched Nd:YAG, long-pulsed ruby, and alexandrite lasers, *J. Am. Acad. Dermatol.*, 41 (2:1), 165–171, 1999.

Ralston, J. M. and Chang, R. K., Spontaneous-raman-scattering efficiency and stimulated scattering in silicon, *Phys. Rev. B*, 2, 1858–1862, 1970.

Razavi, B., *Design of Integrated Circuits for Optical Communications*, McGraw-Hill, Ohio, 2003.

Senior, J. M., *Optical Fiber Communications: Principle and Practice*, 2nd ed., Prentice Hall, Inc., New Jersey, 1986.

Shashidhar, N., Lensing technology, corning incorporated, *Fiber Prod. News*, 14–15, 2004.

Simin, G., High Speed Semiconductor Devices, Basics of heterostructures, University of South Carolina, U.S.A., Online Course 02/ELCT882, 2005.

Thompson, G. H. B., *Physics of Semiconductor Laser Device*, Wiley, Chichester, 1980.

Topping, A., Linge, C., Gault, D., Grobbelaar, A., and Sanders, R., A review of the ruby laser with reference to hair depilation, *Ann. Plastic Surg.*, 44 (6), 668–674, 2000.

Venkataramanan, V., *Introduction to Laser Safety*, Photonics Research, Ontario, 2002.

Yeh, C., *Handbook of Fiber Optics. Theory and Applications*, Academic Press, San Diego, 1990.

32 Optical Switches

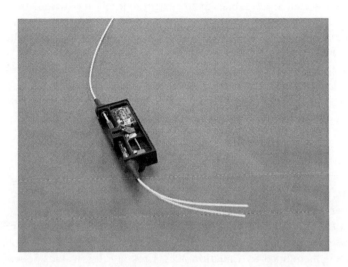

32.1 INTRODUCTION

Many optical networks integrate optical switches into their design. Opto-mechanical switches redirect optical signals from one port to another by moving a fibre tube assembly or an optical component, such as a mirror or prism. There are many different types of optical switches incorporated into networks. In practice, most optical switches are still operated mechanically and controlled by an electronic control circuit. Speed is a crucial parameter in network applications, since a high-speed data transmission of tenths of milliseconds is required. In the near future, dynamic optical routing will require much faster switching speeds. More technology exists for optical switches than any other functional component within the optical network. Researchers are developing optical switches to increase the number of outputs, and to reduce size, cost, and switching time. Presently, optical switches include many types, for example: opto-mechanical switches, thermo-optic switches, electro-optic switches, micro-electro-mechanical switches (MEMS), and micro-opto-mechanical switches (MOMS). New types of optical switches are in the research and development stages.

This chapter illustrates a few switch designs, which are manufactured for use in communication systems, and other applications. Opto-mechanical and electro-mechanical switches are the oldest type of optical switch and the most widely deployed at this time. These devices achieve switching by moving fibre or other optical components by means of stepper motors or relay arms. This causes them to have relatively slow switching time; however, their reliability is excellent, and they offer low insertion and crosstalk losses.

This chapter presents four cases in building opto-mechanical switches using a movable mirror or prism to switch between the input and output ports.

32.2 OPTO-MECHANICAL SWITCHES

Figure 32.1 illustrates common switch configurations. The input signal comes through the input fibre cable on the left side of the switch. A mechanical slider moves that fibre up and down, latching into one of the two output fibre cables on the right side of the switch. In OFF/ON positions, the switch directs light from the input fibre into one of the two outputs. This arrangement is called "1×2 switch configuration." As input at port 1, the signal can be switched to either port 2 or port 3.

For the following definitions, assume the switch is configured to couple to port 2. The insertion loss L_{IL} (in decibels) is defined by Equation 32.1. Insertion loss depends on fibre cable alignment at the input and output ports. Low insertion loss value can be obtained on switches with good mechanical alignment. A good switch provides similar values of insertion loss for all switch positions.

$$L_{IL} = -10 \times \log_{10} \frac{P_2}{P_1}$$

(32.1)

where,

P_1 is the power going into port 1 and P_2 is the power exiting from port 2.

Crosstalk loss L_{CT} is one of the important losses, which should be considered in opto-mechanical switches. Crosstalk loss is a measure of how well the uncoupled port is isolated. The crosstalk loss L_{CT} (in decibels) is defined by Equation 32.2. Crosstalk loss values depend on the particular design of the switch.

$$L_{CT} = -10 \times \log_{10} \frac{P_3}{P_1}$$

(32.2)

where,

P_1 is the power going into port 1 and P_3 is the power exiting from port 3.

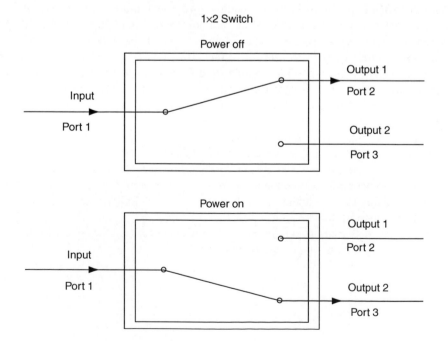

FIGURE 32.1 A typical 1×2 switch configuration.

There are other important optical parameters that need to be specified for each switch type. These parameters include: polarization dependent loss (PDL), return loss (RL), and the Etalon effect. The PDL is defined as the maximum difference in insertion loss between any two-polarization states. It is caused by mechanical stress and temperature variation on optical components or fibre cables. This causes changes in the birefringence and a gradient of index of refraction (n) of the optical material. The RL is defined as the light reflected back into the input path. It is caused by scattering and reflection from optical surfaces like mirrors, lenses, and connectors or from defects, such as cracks and scratches. The back reflection is equal to the RL with a negative quantity. Elaton effect is defined as light resonance (ripple) at a certain wavelength. It is caused by reflection of light from parallel optical surfaces and interference between the signals. All the above losses are measured in decibels (dB). Special optical parameters can be specified by the customers.

Another important parameter of the optical switches is the repeatability—achieving the same insertion loss each time the switch is returned to the same position. Switching speed is also another important specification of a switch. The switching speed is defined as how fast the switch can change the signal from one port to the other. It is an important factor in some switch applications in communication systems.

Figure 32.2 shows a schematic diagram of a mechanical switch configuration with two inputs and two outputs. The inputs are located on the one side and the outputs on the other side of the switch. This configuration is called a 2×2 switch. The signal enters port 1 and port 4, and exits from port 2 and port 3, respectively. This case is called the bypass state, in the OFF position. When the latching mechanism changes position between port 2 and port 3, signal enters port 1 and exits port 3 and from port 4 to port 2. This case is called the operate state, in the ON position.

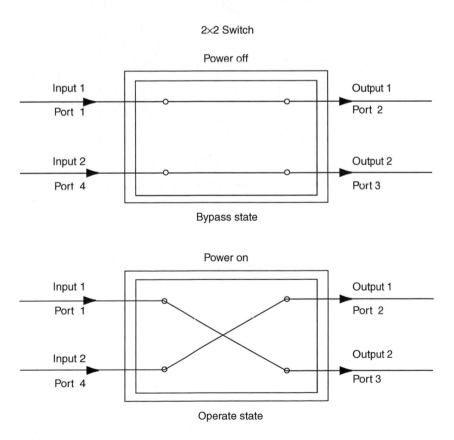

FIGURE 32.2 A typical 2×2 switch configuration.

Opto-mechanical switches collimate the optical beam from each input and output fibre and move these collimated beams around inside the switch. This creates low optical loss, and allows distance between the input and output fibre. These switches have more bulky components compared to newer alternatives, such as the micro-opto-mechanical switches.

Figure 32.3 shows a schematic diagram of a two-position switch. The switch consists of a sliding prism and quarter pitch graded index (GRIN) lenses at the input and output ports. The components are assembled in a packaging base and sealed with a lid. Each GRIN lens is connected to the fibre tube assembly using an epoxy. Figure 32.3 illustrates the OFF/ON positions of a 1×2 switch. As explained above, the GRIN lens collimates the divergence beam exiting from the input fibre. The right angle prism deflects the light by total internal reflection (TIR) at its two slanting surfaces. The GRIN lens refocuses the collimated beam onto a fibre cable at one of the output ports. To direct the signal from port 1 to port 3, the prism slides to a new position, as shown in Figure 32.3 in the OFF position. Figure 32.3 also shows the signal directed between port 1 and port 2, in the ON position, when the prism changes position.

Opto-mechanical switches drive optical fibre networks mechanically. They can switch between light paths at high speed and with low insertion loss. They are widely used in rapidly developing areas of the fibre-optic field, such as optical cross connection and wavelength multiplexing.

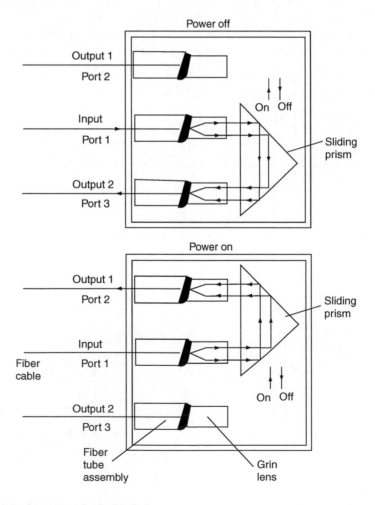

FIGURE 32.3 A 1×2 opto-mechanical switch.

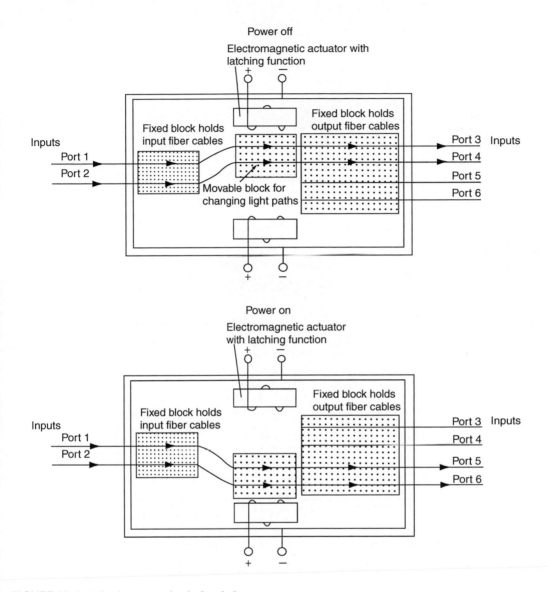

FIGURE 32.4 A 2×4 opto-mechanical switch.

Figure 32.4 shows the design of a 2×4 opto-mechanical switch. The switch uses an electromagnetic actuator with a latching function to drive a movable block to change the light path between the ports. Figure 32.4 shows the switch in the OFF position. The light passes through from port 1 and port 2 to port 3 and port 4, respectively. When the power is turned ON, an electromagnetic actuator (with latching function) drives a movable block to change light path from port 1 and port 2 to port 5 and port 6, respectively, as shown in Figure 32.4. The optical and mechanical components of a switch are assembled in a packaging box with minimal alignment work. There are three configurations of this switch: 1×2, 2×2, and 2×4.

A practical electro-magnetic bypass switch is illustrated in Figure 32.5. The switch contains a quarter pitch GRIN Lens connected to fibre tube assembly at the input and output ports, a relay, and an iron bar with mirror end faces. The components are assembled in a packaging base, which is sealed with a lid. When the power is turned OFF, a spring pulls the iron bar out of the signal path,

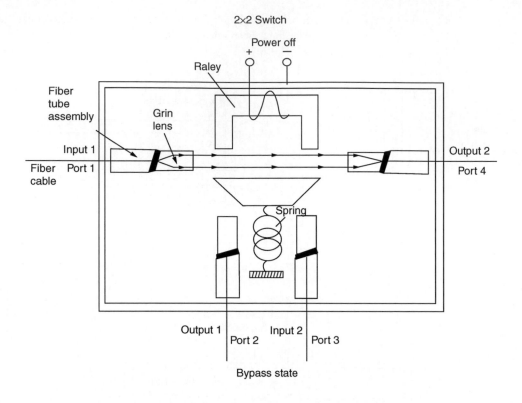

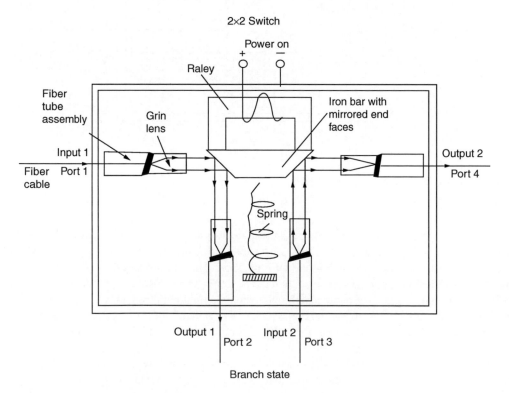

FIGURE 32.5 An electro-magnetic bypass switch.

returning the switch to the bypass condition. This is called the bypass state. In the bypass state, the signal passes directly from port 1 and port 4. When the power is ON, the electromagnet is activated and the iron bar is raised. This is called the branch state. In the branch state, mirrors direct the signal between port 1 and port 2, and between port 3 and port 4.

Another type of bypass switch is also used in communication networks. Figure 32.6 illustrates the function of this type of bypass switch. When the power is in the OFF position, the input signal comes through the input fibre cable on port 1 on the left side and leaves through the output port 4 on the right side of the switch. This is called the bypass state. When the power is ON, a mechanical slider moves two fibre connections to the up position, latching into two output fibre cables at port 2 and port 3 on the side of the switch. In this position, the input signals from port 1 and port 4 are launched into port 2 and port 3, respectively. This position is called branch state. As input at port 1, the signal can be directed to either port 4 or port 2. Also, an input signal at port 4 can be directed to output port 3.

This section presents new switch designs of a 1×8 latching switch configuration using prisms. These switches are commercially available in the market. There are two types of models: the linear and triangular models. The linear model directs the signal from the input to the outputs by arranging the prisms linearly, as shown in Figure 32.7 and Figure 32.8. The triangular model directs the input to the outputs by arranging the prisms triangularly, as shown in Figure 32.9 and Figure 32.10.

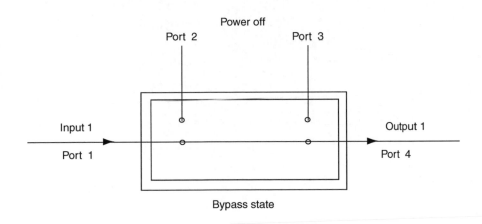

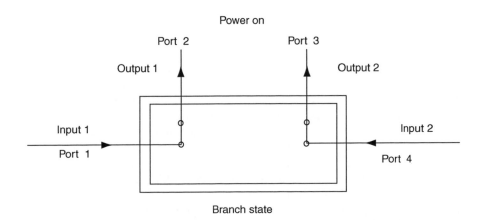

FIGURE 32.6 A bypass switch.

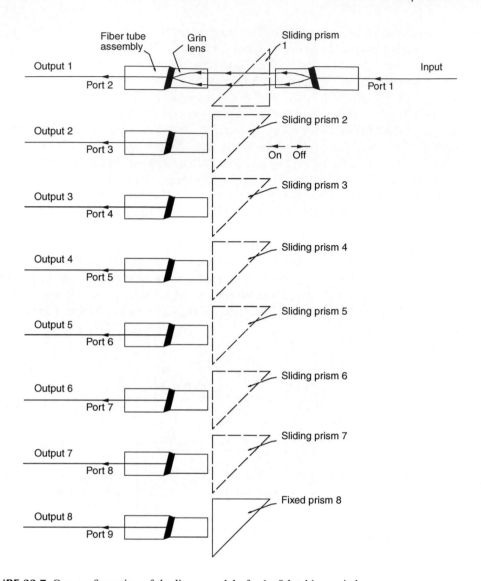

FIGURE 32.7 One configuration of the linear model of a 1×8 latching switch.

These models have come into wide use because they are simple, offer 8 outputs, and are cost effective. They are also used in back-up systems to re-route signals around broken fibre optic cable and in fibre optical instruments.

Figure 32.7 illustrates a schematic diagram for one configuration of a linear model of a 1×8 latching switch. The common element of this type of opto-mechanical switch is that their operation involves mechanical sliding motion of prisms in OFF/ON positions to direct the signal from one port to another. Figure 32.7 shows the input located on the one side and the outputs on the other side of the switch. This configuration is called a 1×8 switch in the linear model. Light enters port 1 and exits from port 2 when sliding prism 1 is in the OFF position. When the latching mechanism places the sliding prisms 1 and 2 in position, the light enters port 1 and exits from port 3, as shown in Figure 32.8. Similarly, light enters port 1 and exits from port 4, when the sliding prism 2 is in the OFF position and sliding prism 3 is in the ON position. This switch configuration is more complicated than the second switch configuration because the input is located on one side and the outputs

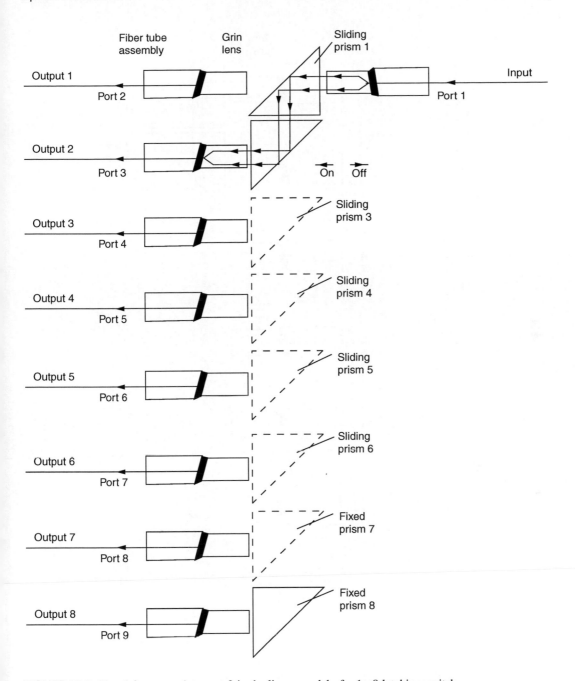

FIGURE 32.8 Signal from port 1 to port 3 in the linear model of a 1×8 latching switch.

on the other side of the switch. This configuration includes a complex mechanism, controls, seven sliding prisms, and one fixed prism.

Figure 32.9 illustrates a schematic diagram of the second configuration of the linear model of a 1×8 latching switch. This configuration is different because the input and the outputs are located on the same side of the switch, as shown in Figure 32.9. Prisms are also used in the operation of this type of switch configuration. Light enters port 1 and exits from port 2 when fixed prism 1 and sliding prism 2 are in position. When the latching mechanism places the sliding prism 2 in the OFF

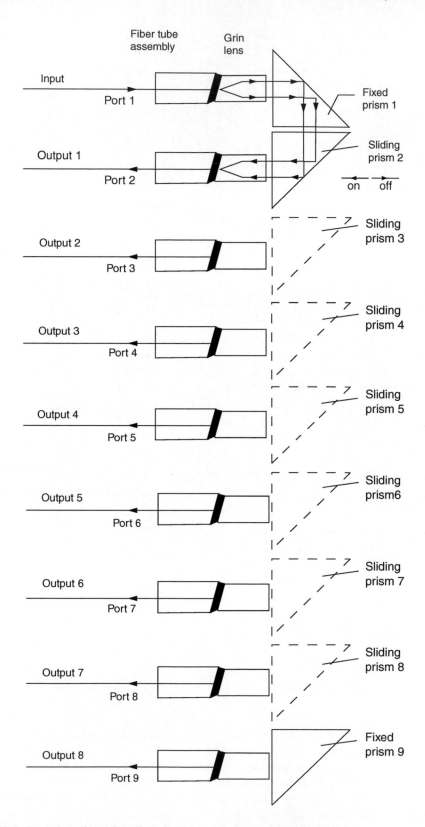

FIGURE 32.9 Second configuration of the linear model of a 1×8 latching switch.

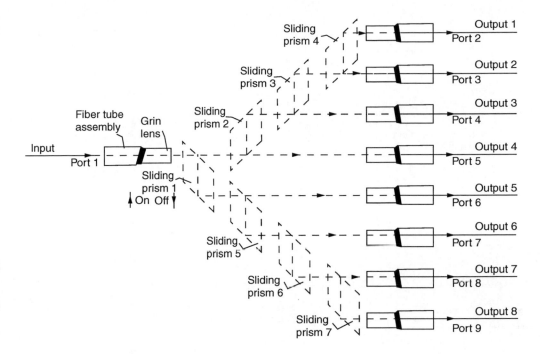

FIGURE 32.10 The triangular model of a 1×8 latching switch.

position and sliding prism 3 in the ON position, the light enters port 1 and exits from port 3. Similarly, light enters port 1 and exits from port 4 when the sliding prism 3 is in the OFF position and sliding prism 4 is in the ON position. The same procedure is used for the signal exiting from other ports. This switch configuration is simpler than the first configuration because the input and outputs are located on the same side of the switch. This configuration includes less complex mechanism, controls, seven sliding prisms, and two fixed prisms.

Figure 32.10 illustrates a schematic diagram of a configuration of the triangular model of a 1×8 latching switch. The common element of this type of opto-mechanical switch is that the operation involves mechanical sliding motion of parallelogram prisms in OFF/ON positions to direct the signal from one port to another. Figure 32.10 shows the input located on one side and the outputs on the other side of the switch. This configuration is called a 1×8 switch in the triangular model. Light enters port 1 and exits from one of the outputs. When the latching mechanism places the sliding parallelogram prism 1 in the OFF position and sliding prism 2 into position, the light enters port 1 and exits from port 4, as shown in Figure 32.10. Similarly, light enters port 1 and exits from port 3 when the sliding prisms 2 and 3 are in the ON position. Figure 32.11 shows light entering port 1 and exiting from port 5, when sliding prism 1 is in the OFF position. This switch configuration is more complicated than the linear model because the input is located on one side, and the outputs on the other side of the switch. Seven sliding parallelogram prisms with additional mechanisms and controls form this configuration. Both models have difficulty achieving precise alignment and low losses during the manufacturing processes.

Many other modern opto-mechanical switches are used in telecommunication networks management, monitoring, restoration, and protection. They have excellent optical performance and the high reliability necessary for network applications. They feature low insertion loss, high RL and channel isolation, excellent repeatability, and fast switching speeds. The switches are available in single-mode and multi-mode, and cover wide wavelength ranges. They are available

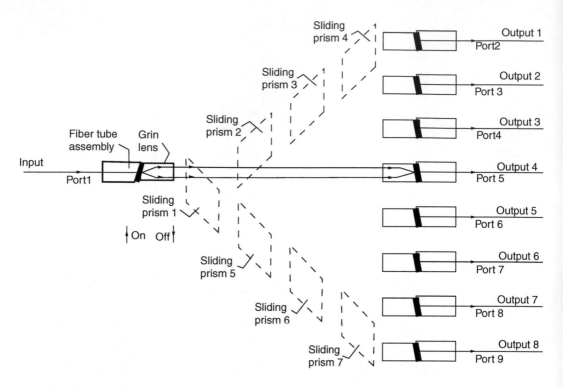

FIGURE 32.11 Signal from port 1 to port 5 in a 1×8 latching switch.

in 1×1, 1×2, and 2×2 configurations. The switching mechanism is latching and remains in its selected state following a loss of power. The switch consists of a quarter pitch GRIN lens glued to a double bore fibre tube assembly, relay, and mirror mounted on a shaft. The components are assembled inside a packaging base box and covered by a lid. Figure 32.12 illustrates a schematic diagram of a 2×2 switch. This figure illustrates the switch in the OFF/ON positions. When the power is off, light transmits from port 1 to port 2 and port 4 to port 3. This configuration is called the transmission state. When the power is ON, the mirror is in position, the light is reflected by the mirror, light exits port 1 and reflects to port 4 and similarly, port 2 reflects to port 3. This state is called the reflection state.

32.3 ELECTRO-OPTIC SWITCHES

Switches with no moving parts can be built by using some of the passive devices, such as Mach-Zehnder Interferometers (MZIs) and couplers. Some optical materials, such as lithium niobate crystal (LiNiO$_3$), Avalanche Photo Diode (APD) (NH$_4$H$_2$PO$_4$), and KDP (KH$_2$PO$_4$) exhibit an electro-optic effect. The index of refraction (RI) of the optical material changes in the presence of an electric field. These optical materials are used in building devices, such as the MZI, APD, and KDP. An electric field applied across the lithium niobate crystal causes a variation in the RI. This changes the transit time, creating a phase shift of the optical signal passing through the lithium niobate crystal.

Mach-Zehnder Interferometers are used in building optical devices, which are used in a wide variety of applications in optical communication systems. The basic requirement of the Mach-Zehnder Interferometer is to have a balanced configuration of a splitter and a combiner connected

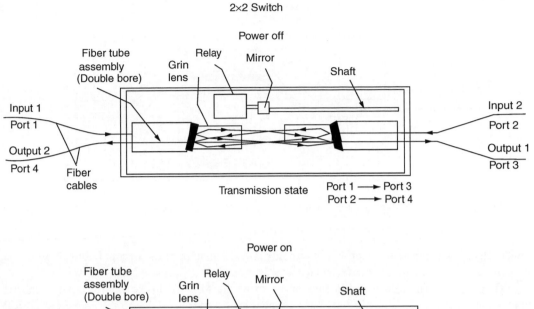

FIGURE 32.12 A 2×2 opto-mechanical switch.

by a pair of optically matched waveguides, as shown in Figure 32.13. The optical signal entering the Mach-Zehnder Interferometer input port is split through a "Y" splitter section into two equal components. Each component goes to one of the two arms of the Y splitter. When there is no phase change in signal components after passing through both arms of the interferometer, the signal components is recombined at the "Y" coupler immediately before the optical signal exits the Mach-Zehnder Interferometer. The recombination of the two signal components takes place as

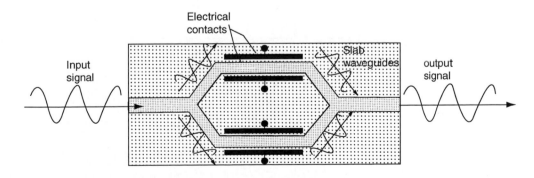

FIGURE 32.13 Mach-Zehnder Interferometer acts as a passive device.

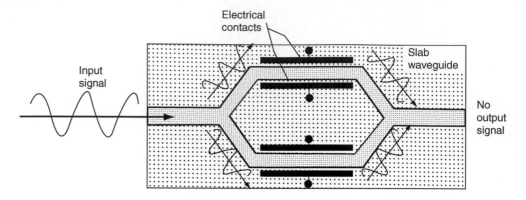

FIGURE 32.14 An electro-optic switch using a Mach-Zehnder Interferometer.

constructive interference between two components and regenerates the original optical signal. In this case the Mach-Zehnder Interferometer acts as a passive device.

When an electric field is applied to one arm of the Mach-Zehnder Interferometer, the RI changes and causes 180° shift in the phase of the signal component, due to the change in optical path length of this arm. As shown in Figure 32.14, when there is a difference in phase at the destination "Y" coupler, the signal components will be out of phase with one another. The signal components re-combination will be lost because the components will cancel each other in destructive interference. If the phase difference is a full 180°, then the output will be zero. In other words, applying the electric field to one of the arms of the Mach-Zehnder Interferometer will make the phase shift one of the signal components. The Mach-Zehnder Interferometer acts as an active device, when an external electric voltage is applied causing the switching.

Using the same principle as discussed above, one can build an electro-optic switch using two branching waveguides arranged like a 3 dB coupler to switch one input between two outputs. You can replace the one input with two parallel outputs coupled to the pair of switching waveguides by a combining coupler, as shown in Figure 32.15. An electric field applied to one arm of the waveguide causes a 180° shift in the phase of the signal component. The electrical voltage is raised or lowered to shift the delay between waveguides by 180°. This directs the output from one waveguide on the right side to the other output. Because signal interference depends on phase shift, it is possible to further increase the voltage to switch the signal back to the other output. Table 32.1 presents the possible outcomes achieved by applying different voltages across the waveguide arms.

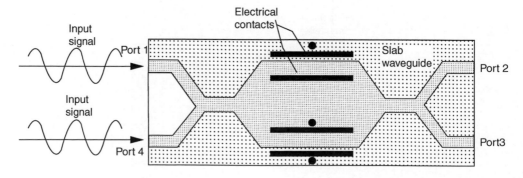

FIGURE 32.15 A 2×2 electro-optic switch.

TABLE 32.1
Input and Output Signals Connections

Voltage	Connections
V_1	Port 1 to Port 2 and Port 4 to port 3
V_2	Port 1 to Port 3 and Port 4 to Port 2

32.4 THERMO-OPTIC SWITCHES

A novel rib waveguide-integrated thermo-optic switch has appeared recently. The device is based on the TIR phenomenon and the thermo-optic effect (TOE) in hydrogenated amorphous silicon (a-Si:H) and crystalline silicon (c-Si). It takes advantage of a bandgap-engineered a-Si:H layer to explore the properties of an optical interface between materials showing similar refractive indexes but different thermo-optic coefficients. In particular, the modern plasma-enhanced chemical vapour deposition techniques, the refractive index of the amorphous film can be properly tailored to match that of c-Si at a given temperature. TIR may be achieved at the interface by acting on the temperature, because the two materials have different thermo-optic coefficients. The switch is integrated in a 4-pm-wide and 3-μm-thick single-mode rib waveguide, as shown in Figure 32.16. The substrate is a silicon-on-insulator wafer with an oxide thickness of 500 nm. The active middle region has an optimal length of 282 μm. The device performance is analysed at a wavelength of 1.55 μm.

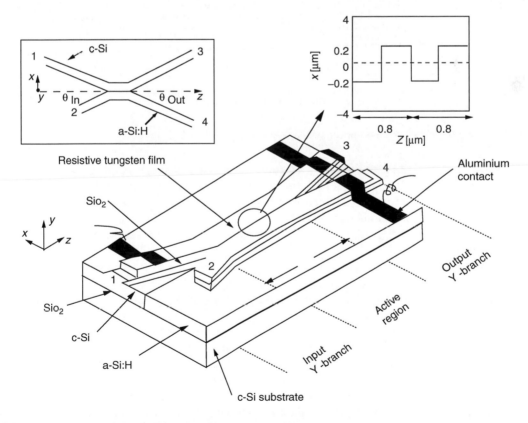

FIGURE 32.16 A Waveguide-integrated thermo-optic switch.

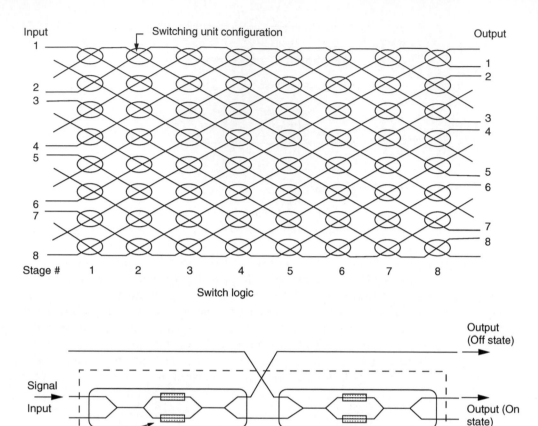

FIGURE 32.17 An 8×8 thermo-optical switch using Mach-Zehnder Interferometer.

As shown in Figure 32.12, the optical waveguide-integrated switch consists of a 2×2 waveguide structure with an input Y branch and an output Y branch. They are joined by a middle active region, although in this work, only the 1×2 switch operation will be considered. The device is composed of rib channel waveguides, which guarantee both an effective optical confinement and low propagation losses. When properly designed, single mode operation can be achieved in the input and output of the Y branches.

As shown in the top-left inset of Figure 32.16, the device structure is symmetric with respect to the (yz) plane. It consists of a core layer of c-Si in the upper half, and a core layer of a-Si:H in the lower half, both laying on a SiO_2 layer grown on a highly doped crystalline silicon substrate. The thickness of the two guiding layers together is 3 µm. Due to the refractive index of Sia2 ($n_{SiO_2} = 1.48$), a 500-nm-thick under cladding layer ensures the optical confinement for both waveguides, as suggested by electromagnetic field propagation simulations. In the top-right inset of Figure 32.16, a detail of the interface between the a-Si:H waveguide and the crystalline Si waveguide in the active region is also shown. The irregular profile at the TIR interface takes into account surface roughness that may result from the fabrication process.

We can exploit the TOE in a-Si:H and c-Si by changing the refractive index of the core layers and thereby switching the light beam at the output of the structure. A 300-nm-thick tungsten heating film is introduced on top of the stacked structures. It is separated from the active region by a 100-nm-thick SiO$_2$ film. This reduces the optical absorption by the electrodes due to the evanescent field of the optical mode. Finally, the heating structure is completed by aluminum bonding pads.

The operating principle of the device is the TIR, which can be activated or dropped by exploiting the different thermo-optic coefficients in a-Si:H and c-Si. In particular, by choosing a proper gas phase composition during the deposition process of a-Si:H, the two materials develop the same refractive index at a given temperature. By changing the device operating temperature, a refractive index discontinuity is created at the (yz) interface, producing the desired optical switching. At room temperature, an incident channel-guided light beam coming from port 1 will encounter a refractive index discontinuity between c-Si and a-Si:H, and the reflection (straight state) exists. Under these conditions TIR will occur and the incident light beam at port 1 will be reflected to port 3.

Another type of thermo-optical switch uses the Mach-Zehnder interferometer for the switching process. This type of thermo-optical switch is used in communication systems. Figure 32.17 illustrates the logic of an 8×8 thermo-optical switch. The 8×8 optical matrix switch employs Mach-Zehnder interferometer with a thermo-optic phase shifter as the switching mechanism. The small switch offers low loss, low crosswalk, low return loss, excellent stability, high reliability, and low power consumption. Applications of such switches include: space division switching systems (with analog and/or digital signals), wavelength routing (such as cross-connect and add-drop), protective switching, video switching, and inter-module connection.

32.4.1 Switch Logic

32.4.1.1 Switching Unit Configuration

A high-speed all-optical switch using a fibre optic coupler and a light-sensitive variable-index material is illustrated in Figure 32.18. Refractive index variation with light is the principle of the switch operation. Evanescent-wave coupling between two mono-mode fibres is extremely sensitive to the RI of the material surrounding the coupling region. Two ground and polished fibres, producing an evanescent field, can be brought in close proximity so that light in one fibre will couple into the other fibre in any desired ratio. Such polished couplers have been constructed to produce very low losses.

A similar coupler may be made by timed etching of the fibres. Hydrofluoric acid may be used to remove as much cladding as desired; this exposes the core and produces an evanescent field. Within the etched regions, fibre-to-fibre optical coupling will occur for fibres placed in close proximity.

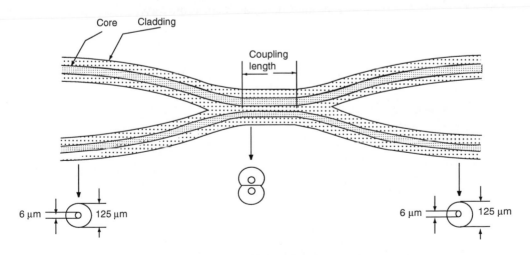

FIGURE 32.18 A high-speed optical switch using a fibre optic coupler.

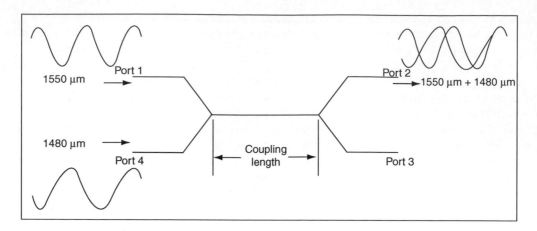

FIGURE 32.19 Two wavelengths entering at port 1 and port 2.

The coupling efficiency will depend on the RI of the surrounding medium, the core-to-core separation, the length of the interaction, and the amount of etching.

Another evanescent-field coupler is based on a non-etched, fused, and drawn coupler. This type of coupler is drawn to such an extent that the core is essentially lost and the cladding reaches a diameter near that of the undrawn core. The claddings become the core, and the evanescent field is forced outside the new core (waist region) into the air. Twisting two fibres together and fusing by using heat makes fused bi-conical tapered couplers.

The contribution of charge carriers (electrons) to the RI provides a simple way to modulate the index by the introduction or withdrawal of such carriers in the material. This can easily be done by the creation of electron-hole pairs if the material is also photoconductive (optical modulation).

One of the important parameters of the couplers is the coupling length. The coupling length is wavelength dependent. Thus the shifting of power between the two parallel waveguides will take place at different places along the coupler for different wavelengths. Figure 32.19 shows two wavelengths entering at port 1 and port 2. When coupler length is made exactly to match the wavelength of the signal, the coupler works to combine wavelengths. Combined signals exit from port 2.

Figure 32.20 shows the reverse process where two different wavelengths arrive on the same input fibre at port 1. At a particular location along the coupler, the wavelengths will be in different

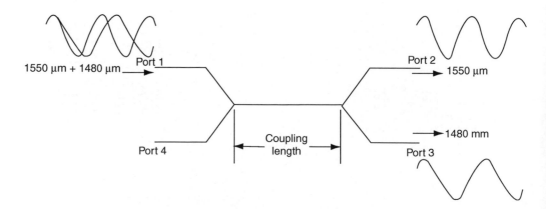

FIGURE 32.20 Two wavelengths entering at port 1.

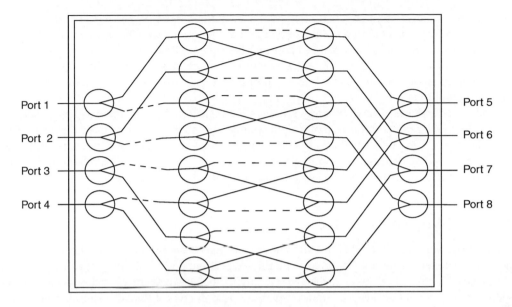

FIGURE 32.21 A 4×4 optical space-division switch.

waveguides. Then the wavelengths separate exactly and each wavelength exits from a different port. In this case, one wavelength exits from port 2 and the other from port 3. The processes described in Figure 32.19 and Figure 32.20 are performed in the same coupler. This process is bi-directional. The coupler in Figure 32.19 works as a splitter; the same coupler in Figure 32.20 also works as a combiner.

There are other types of switches that use the same elements of either the micro-opto-mechanical switch or micro-electro-optical switch. They employ couplers for switching between the inputs and outputs. These types of optical switches are used in communication systems. Figure 32.21 shows the configuration of a 4×4 optical space-division switch. The switch is designed to connect

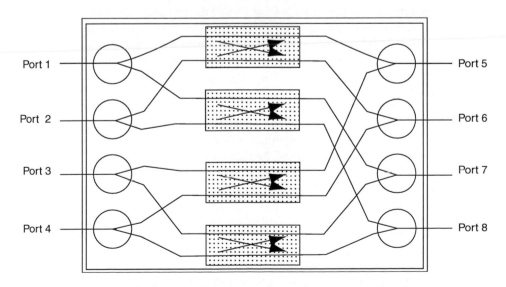

FIGURE 32.22 A cross connect switch.

any input port to any output port as desired by the user. Any input may be switched to any output; however, two inputs may not go to the same output at the same time. The device is bi-directional such that once a connection has been established between an input port and an output port, that particular connection may be used in either one or both directions. The switches have no moving parts, are very stable, and reliable while exhibiting very low loss; thereby reducing the need for expensive amplifiers.

Figure 32.22 shows the cross connect switch, which selects outputs by optical cross connecting. This results in a significant reduction of overall complexity and number of required elements. These switches are used in protection/backup switching, optical cross-connecting, network testing and monitoring, optical routing, and optical burst switching. A 4×4 switch configuration can be cascaded to build 16×16–256×256 switch configurations.

32.5 ACOUSTO-OPTIC SWITCHES

Sound waves are generated when a material is in mechanical vibration mode. They can also be generated by acoustic transducers. Like any light wave, a sound wave is a moving wave, which has a frequency. Light waves travel at the speed of light; sound waves travel at the speed of sound, which is slower than light waves. Sound waves are used to control light transmission in acousto-optic switches and modulators. The refractive index of some optical material is altered by the presence of sound waves. The sound wave causes regular zones of compression and tension within the optical material. This creates a regular pattern of changes in index of refraction n of the optical material; this is called a Bragg diffraction grating. Within the optical material, there is interference between sound and light waves. The power of the deflected light is controlled by the intensity of the sound wave. The angle of deflection is controlled by the frequency of the sound wave. Figure 32.23 illustrates a design of an acoustic-optic switch. The figure shows that an incident light can be controlled by the frequency of the sound wave. The incident light exits the switch from one or more selected output ports depending on the sound wave intensity.

32.6 MICRO-ELECTRO-MECHANICAL SYSTEMS

Micro-electro-mechanical systems (MEMS), is a rapidly growing technology for the fabrication of miniature devices using processes similar to those used in the integrated circuit industry. MEMS are widely used in optical switching in telecom networks. The appeal of MEMS goes beyond just switching applications in defense, aerospace, and medical industries. MEMS technology provides a way to integrate electrical, electronic, mechanical, optical, material, chemical, and fluids engineering on very small devices ranging in size from a few microns to one millimetre. MEMS devices have many important advantages over conventional opto-mechanical switches. First, like integrated circuits, they can be fabricated in large numbers, so that cost of production can be reduced substantially. Second, they can be directly incorporated into integrated circuits, so that far more complicated systems can be made than with other technologies. Third, MEMS have small size, low cost, and high reliability and stability. Fourth, MEMS have the important capability of high-density digital transmission communication with different bandwidths.

There are two categories of MEMS switches: MEMS 2D and 3D. They are typically fabricated onto a substrate that may also contain electronics needed to drive the MEMS switching element. MEMS-based optical switches route light from one fibre to another to enable equipment to switch traffic completely in the optical domain without requiring any optical-to-electrical conversion. At the core of MEMS 2D matrix switches is an array of micro mirrors capable of redirecting light either in free space or within a waveguide framework. The 2D switch architecture shown in Figure 32.24 employs one mirror for every possible switched node in a matrix switch, and thus requires N^2 mirrors for an $N \times N$ array. 2D mirror arrays are characterized by two-state mirror

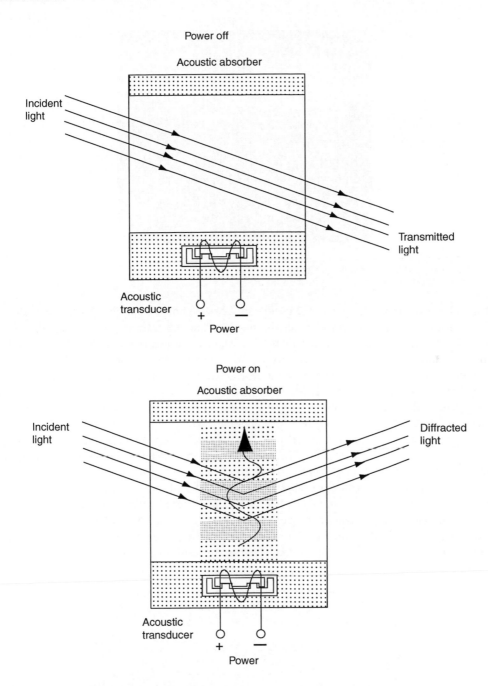

FIGURE 32.23 A design of an acoustic-optic switch.

positioning. One state is inactive and requires only that the mirror can be parked out of the optical path. During the switching state, the mirror redirects the light path. Mirror positioning accuracy, repeatability, and stability are critical in determining switch performance. Unlike 3D switch architectures that require servo positioning of individual mirrors, a 2D switch can rely on passive positioning control of the switched mirror, simplifying the control scheme. But a successful MEMS 2D approach must provide means for actuating the mirror into a highly predictable, stable state and hold it there indefinitely.

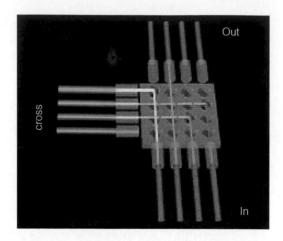

FIGURE 32.24 MEMS 2D switch architecture.

MEMS use an array of pop-up MEMS mirrors fabricated on the surface of a silicon wafer. The mirror is hinged to allow its rotation off the plane of the substrate to an angle of 90° where it redirects a light channel from the through to cross state, as shown in Figure 32.25. An addressing scheme is required to select individual mirrors for actuation into the popped-up state and also for positioning them with sufficient accuracy for efficient coupling into the switched channel. The 2D MEMS array described here, called MagO×C, which stands for "magnetically optical actuated cross-connect," uses a combination of magnetic and electrostatic actuation to rotate the mirrors and to select and deselect individual mirrors for clamping into the up or down state.

To rotate unclamped mirrors into the up state, magnetic actuation is implemented globally by applying an external field generated with a small electromagnet. The magnetic signal only needs to be applied momentarily. Using a global field avoids the need to fabricate individual magnetic actuators for each chip. Mirrors are fabricated with a layer of attached nickel to produce torque on the mirror hinge in response to the applied field. The nickel plate aligns with the magnetic field lines and generates a magnetostatic torque on the mirror. This lifts the mirror off the substrate and orients it near the desired vertical position, where electrostatic force can take over in setting and holding the final desired mirror position. An electrostatic field is applied mirror by mirror, either to

FIGURE 32.25 Pop-up mirror array with passive mirror alignment.

hold the mirror down against the torque produced by the magnet or to hold the mirror in the up position against the restoring force of the elastic hinge. Since the magnetic field is applied globally, all mirrors will attempt to rotate when the field is turned on. Only the mirrors to be rotated into the up position are unclamped; all others are held down electrostatically. Similarly, mirrors clamped in the up state remain so until the electrostatic signal is removed; the magnet is no longer needed to hold the mirror up. The combination of magnetic and electrostatic actuation provides an effective means for configuring a mirror array without resorting to complex individual actuators for each mirror. Since electrostatic clamping of the mirrors requires virtually no current flow, the switch array consumes very little steady state power. Power is consumed only during transitions when the magnet is activated. The components of the switch are packed in a packaging base and lid.

32.7 3D MEMS BASED OPTICAL SWITCHES

3D MEMS based optical switch routes light from any of 80 input fibres to any of 80 output fibres. Designed for fibre-based test and measurement, 10-Gbit/s Ethernet, high-definition video, and telecom applications, this all-optical micro-photonic subsystem fits in the palm of a hand. The switch design is based on 3D MEMS mirror arrays, which is called Reflexion. It can switch signals within 10 ms, which is well within the telecommunications requirements for communications applications. 3D designs have switching elements accommodating hundreds, even thousands, of ports. Figure 32.26 shows a 3D MEMS optical switch. The design of the 3D MEMS is simple,

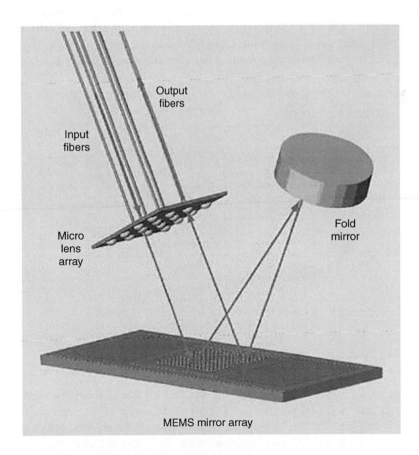

FIGURE 32.26 A 3D MEMS optical switch architecture.

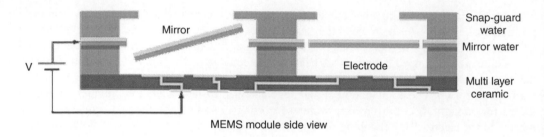

FIGURE 32.27 A 3D MEMS optical switch design.

solves mechanical and optical issues, is easy to fabricate, and achieves manufacturing tolerances that are accepted by the telecommunications industry.

A 3D MEMS design is shown in Figure 32.27. A micro-mirror array rests atop a single piece of silicon on a ceramic substrate. No bonding pads or other integrated electronics exist on the chip. All routing to the rest of the actuation electronics is done on the back end of the ceramic substrate. Additional electronics is located on a photodetector card, along with constant-delay two-pole Bessel filters, and an analog-to-digital converter with a conversion-phasing time. Parallel-plate electrostatic actuation of the mirrors, with potentials of about 200 V, is provided by high-voltage linear amplifiers.

Figure 32.28 shows a torsional micro mirror, which is driven by a vertical bi-directional comb driver (micro electrostatic actuator). Underneath the mirror plate is the substrate electrode. During operation, a voltage is applied to the electrode in order to generate an electrostatic attractive force on the mirror plate. The mirror plate rotates around the supporting axis in two dimensions. Such tilting mirror position can direct light from many distinct input ports to any of many distinct output ports.

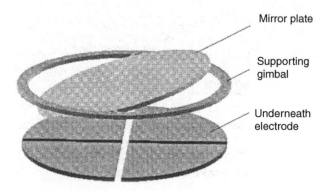

FIGURE 32.28 A torsional micro mirror.

32.8 MICRO-OPTO-MECHANICAL SYSTEMS

On-chip integrated MOMS were developed for a variety of applications for optical telecommunications. It is a new technology that allows for the integration of multiple passive and active components at the chip level. The technology is an extension of integrated optics and is based on suspended waveguides fabricated on chips, which are integrated with other optical components. It is used for a variety of optical solutions, including optical switching and cross-connect, signal dispersion correction, configurable optical add/drop multiplexers, and signal intensity equalizers. The technology bases its main technological concept on integrating optics at the chip level. The technology explores low-cost, high-performance, planar optical waveguide switches. Waveguides are used to channel light, rather than allowing it to propagate in free-space. The use of wave-guides allows a degree of freedom in reducing size, while at the same time operating in a controlled physical environment. Switching time and losses are very low. By using waveguides, photons can be channeled in a controlled fashion, making it possible to lay out a photonic network within the chip with low losses. Since silicon is used as the propagating material, wavelength transparency for the 1300–1600 nm is utilized. The use of silicon for the waveguide material creates a tight confinement of light that allows the use of very small curvature radii, enabling a significant reduction of footprint chip size.

32.9 EXPERIMENTAL WORK

The purpose of this lab is to build and test an opto-mechanical switch. The following cases illustrate the experimental set-up for the different types of switch arrangements that can be performed in an experimental laboratory. Different optical components (mirror, double side silvered mirror, and prism) may be used with one of the experimental set-ups to create different opto-mechanical switch configurations as follows:

  a. A 1×2 switch with one laser source.
  b. Two 1×2 switches with two laser sources.
  c. A 2×2 switch using a movable mirror.
  d. A 1×2 switch using a prism.

32.9.1 A 1×2 SWITCH WITH ONE LASER SOURCE

Figure 32.29 illustrates the design of a 1×2 switch using fixed and movable mirrors to redirect the reflected light onto one of the output ports. The input light comes through the input fibre cable on one side of the switch and leaves on one of the two output ports on the same side. A movable mirror assembly activated by a relay moves the mirror up and down in the OFF and ON positions. When the movable mirror is in the OFF position, the incident light on the fixed mirror reflects onto the output port 1. When the switch is in the ON position, the movable mirror reflects the incident light onto the second output port 2, as shown in Figure 32.29. This arrangement is called a 1×2 switch configuration.

32.9.2 TWO 1×2 SWITCHES WITH TWO LASER SOURCES

We can use the same arrangement as explained in this section to create a two-1×2 switch, as shown in Figure 32.30. This set-up has two laser light sources at port 1 and 2. Each laser source has two independent outputs. The outputs are at port 3 and 4. The switch operates as a two-1×2 switch. Each laser light input gives two outputs by the movable mirror in OFF/ON positions.

32.9.3 A 2×2 SWITCH USING A MOVABLE MIRROR

As explained in Section 32.9.1 for 1×2 switch, we can use a movable mirror to create a 2×2 switch, as shown in Figure 32.31. We need two laser sources, two outputs, and a mirror mounted on a movable arm assembly arrangement to build a 2×2 switch. Figure 32.31 shows a schematic diagram of a two-position switch that can be built in the laboratory. The following experimental

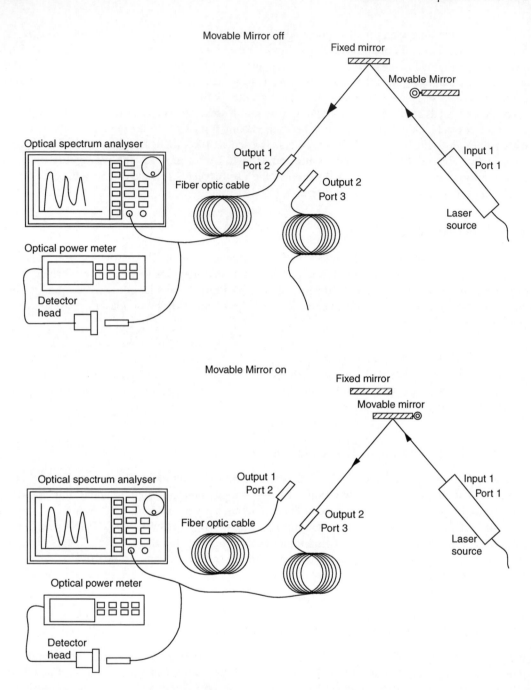

FIGURE 32.29 A 1×2 switch configuration.

set-up has two laser light sources at ports 1 and 2, and two outputs at ports 3 and 4. Each laser source has two independent outputs. The figure shows the switch in the OFF/ON states. When the power is OFF, light is transmitted from port 1 to port 3 and from port 2 to port 4. This case is called the transmission state. When the power is ON and the mirror moves into position, the light is reflected by the mirror, light from laser source 1 is reflected to port 4, and light from laser source 2 is reflected to port 3. This case is called the reflection state.

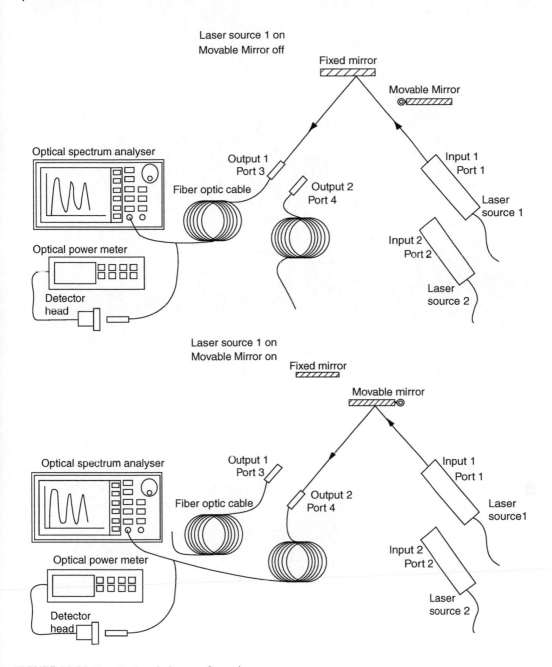

FIGURE 32.30 Two 1×2 switches configuration.

32.9.4 A 1×2 Switch Using a Prism

As explained in Section 32.9.1, instead of a movable mirror, we can use a movable prism to create a 1×2 switch, as shown in Figure 32.32. We need one laser source, one fixed mirror, two outputs, and a prism mounted on a movable arm assembly to build a 1×2 switch. Figure 32.32 shows a schematic diagram of a two-position switch that can be built in the laboratory. This experiment set-up has one laser light source as a port 1 and two outputs at ports 2 and 3. The figure shows the switch in the OFF/ON positions. When the power is OFF, light from port 1 is reflected to port 2.

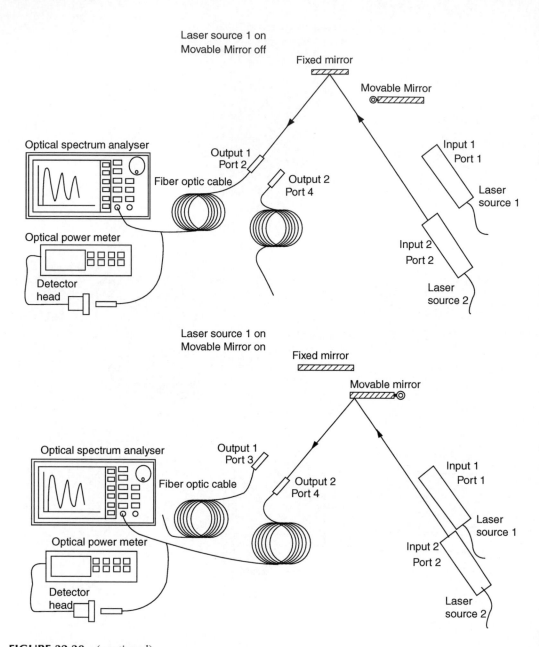

FIGURE 32.30 (*continued*)

When the power is ON, the prism moves into position. The light refracts through the prism and is reflected by the mirror onto output port 3.

32.9.5 TECHNIQUE AND APPARATUS

Appendix A presents the details of the devices, components, tools, and parts:

1. 2×2 ft. Optical breadboard.
2. Two laser light sources.

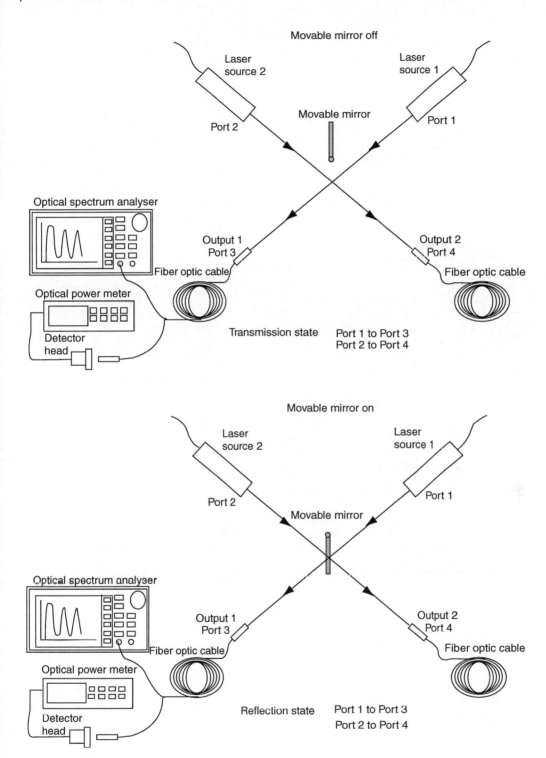

FIGURE 32.31 A 2×2 switch configuration.

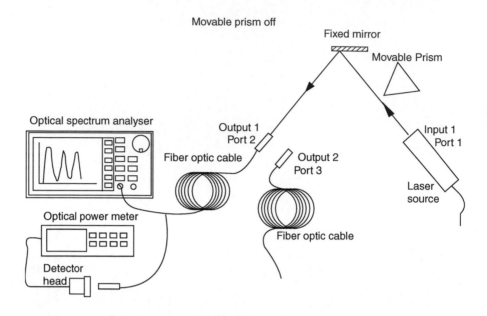

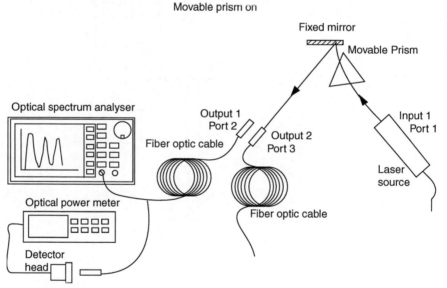

FIGURE 32.32 A 1×2 switch using a prism.

3. Two laser power supplies.
4. Two laser mount assemblies.
5. Laser light detector.
6. Laser light power meter.
7. Optical spectrum analyser.
8. Four multi-axis translation stages.
9. Hardware assembly (clamps, posts, screw kits, screwdriver kits, fibre cable holder/positioner, positioners, brass fibre cable holders, laser holder/clamp, etc.).
10. Two mirrors and mirror holder assemblies (large size mirrors can be used because of the size of the multi-axis translation stage and the space between the components).

11. Double side-silvered mirror.
12. Prism and prism holder assembly.
13. Fibre optic cable end preparation kit.
14. Fibre cable end holder.
15. Two fibre optic cables, 250 micrometres diameter, 500 metres long.
16. Fibre cable holder/positioner assembly.
17. Movable arm assembly for mirror and prism.
18. Black/white card and cardholder.
19. Ruler.

32.9.6 Procedure

Follow the laboratory procedures and instructions given by the professor and/or instructor.

32.9.7 Safety Procedure

Follow all safety procedures and regulations regarding the use of fibre optic cables, fibre optic devices and instruments, optical components and instruments, light source devices, and optical cleaning chemicals.

32.9.8 Apparatus Set-up

32.9.8.1 A 1×2 Switch with One Laser Source

1. Figure 32.33 shows the experimental apparatus set-up. (Note: You could arrange this experiment set-up to fit with the measurement instruments that you have in the lab).

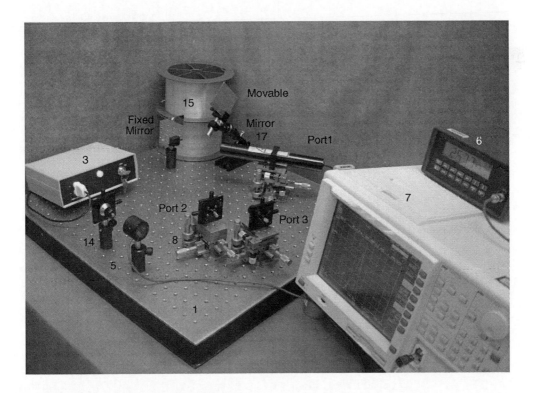

FIGURE 32.33 A 1×2 opto-mechanical switch.

2. Mount three multi-axis translation stages on the breadboard. Multi-axis translation stages are for port 1 (laser source), port 2, and port 3.
3. Place a laser source into the laser mount clamp and tighten the screw.
4. Bolt the laser source to the multi-axis translation stage.
5. Place a fibre cable holder/positioner assembly on the multi-axis translation stages to hold ports 2 and 3.
6. Mount a fixed mirror and mirror holder facing the laser source at an angle with the normal line of the mirror.
7. Turn the laser on. Follow the operation and safety procedures of the laser device in use.
8. Turn off the laboratory lights before taking measurements.
9. Align the laser beam reflected off the mirror to fibre cable at port 2.
10. Use the optical spectrum analyser to measure the laser power and wavelength of the laser source (port 1). Figure 32.34 shows a sample of collected data by the optical spectrum analyser printed on a paper chart. Ask your supervisor/instructor for the optical spectrum analyser calibration procedure and referencing the source. Otherwise use the optical power meter for quick power out results. When using the optical power meter, you need to calibrate it; ask your supervisor/instructor for the calibration procedure. Record your results for port 1 in Table 32.2.
11. Measure the laser power and wavelength of the laser light at port 2. Record your results for port 2 in Table 32.2.
12. Mount the movable mirror assembly between the laser source and the fixed mirror. Locate the movable mirror parallel to the fixed mirror.
13. Move down the movable mirror, as shown in Figure 32.35.

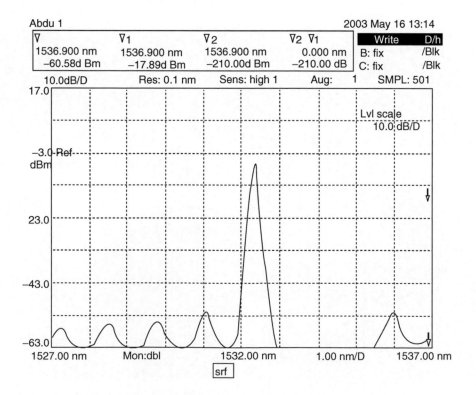

FIGURE 32.34 A sample of collected data printed on a chart.

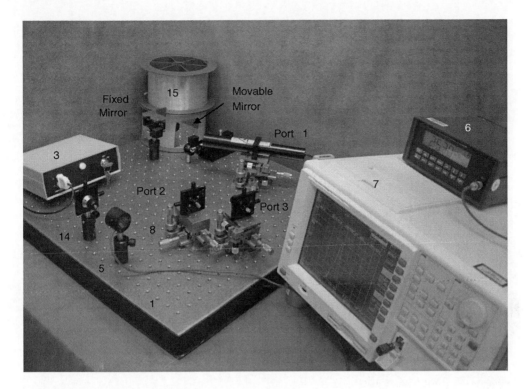

FIGURE 32.35 Movable mirror is in the ON position.

14. Align the fibre cable at port 3 to receive the reflected laser beam from the movable mirror.
15. Measure the laser power and wavelength of the laser light at port 3. Record your results for port 3 in Table 32.2.

32.9.8.2 Two 1×2 Switches with Two Laser Sources

Continue the procedure as explained in Section 32.9.1 of this experiment. Figure 32.36 shows the experimental apparatus set-up for Section 32.9.2. Add the following steps:

1. Mount the forth multi-axis translation stage on the breadboard.
2. Place a second laser source into the laser mount clamp and tighten the screw.
3. Bolt the laser source to the multi-axis translation stage.
4. Align each laser beam to point at a fibre cable.
5. As explained in Section 32.9.1, measure the first laser power and wavelength of the laser light at port 1. Record your results in Table 32.3.
6. Measure the laser power and wavelength of the laser light at port 3. Record your results for port 3 in Table 32.3.
7. Put movable mirror in position, as shown in Figure 32.37.
8. Measure the laser power and wavelength of the laser light at port 4. Record your results in Table 32.3.
9. Turn off the first laser and the second laser on. Swing the movable mirror upwards.
10. Align laser source 2 to reflect onto port 3.

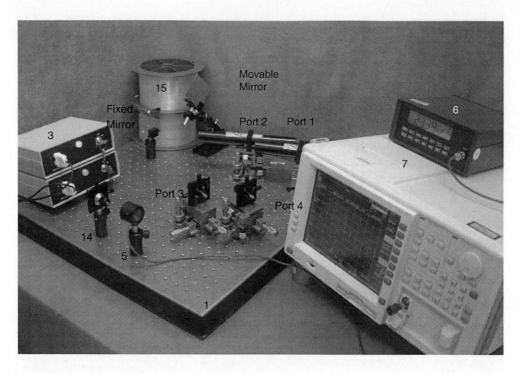

FIGURE 32.36 Two 1×2 switches.

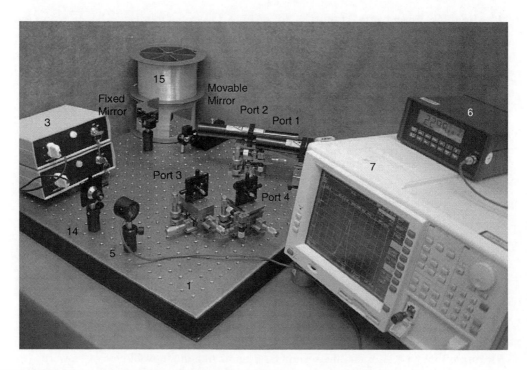

FIGURE 32.37 A movable mirror is in the ON position.

11. Measure the second laser power and wavelength of the laser light at port 2. Record your results in Table 32.3.
12. Measure the laser power and wavelength of the laser light at port 3. Record your results for port 3 in Table 32.3
13. Put movable mirror in position, as shown in Figure 32.37.
14. Measure the laser power and wavelength of the laser light at port 4. Record your results in Table 32.3.

32.9.8.3 A 2×2 Switch Using a Movable Mirror

You can build a 2×2 opto-mechanical switch using a double sided mirror mounted on a movable arm mechanism, to direct the light from one light source onto two outputs in two modes. Figure 32.38 shows the experimental apparatus set-up for Section 32.9.3.

1. Mount four multi-axis translation stages on diagonals of the breadboard close to the corners.
2. Place two laser sources into the laser mount clamps and tighten the screw. Mount each laser source on one multi-axis translation stage. Bolt each multi-axis translation stage on the same side of the breadboard pointing to the centre.
3. Mount a movable mirror and mirror holder assembly in a position where the two laser beams intersect.
4. Place a fibre cable holder/positioner assembly on the remaining two multi-axis translation stages.

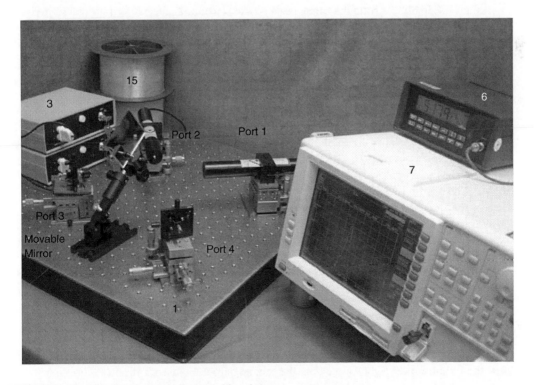

FIGURE 32.38 A 2×2 switch using a movable mirror.

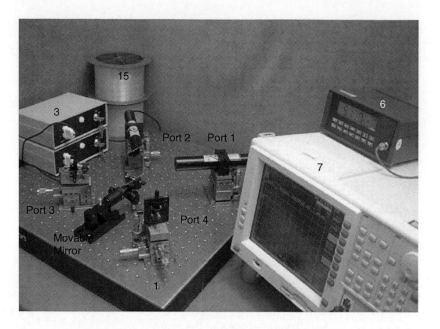

FIGURE 32.39 A 2×2 switch when the movable mirror is in the ON position.

5. Bolt each multi-axis translation stage on the same side of the breadboard on the opposite side of the laser sources (i.e., each laser source is facing one fibre cable holder assembly). Arrange so that laser source 1 (port 1) faces one fibre cable holder assembly (port 3) and laser source 2 (port 2) faces the second fibre cable holder assembly (port 4).

6. Turn on the two laser sources. Follow the operation and safety procedures of the laser device in use.

7. Turn off the laboratory lights before taking measurements.

8. As explained in Section 32.9.1, measure the laser power and wavelength of the laser light at port 1 and 2. Record your results in Table 32.4.

9. Align the laser beam 1 to point at fibre cable on port 3; align the laser beam 2 to point at fibre cable on port 4.

10. Measure the laser power and wavelength of the laser sources at port 3 and port 4. Record your results in Table 32.4.

11. Put the movable mirror in position, as shown in Figure 32.39.

12. Carefully align the position of the mirror assembly so the laser beam from laser source 1 reflects by one side of the movable mirror to port 4 and the laser beam from laser source 2 reflects by other side of the movable mirror to port 3, as shown in Figure 32.39.

13. Measure the laser power and wavelength of the laser beam at port 3 and port 4. Record your results in Table 32.4.

32.9.8.4 A 1×2 Switch Using a Prism

You can build a 1×2 opto-mechanical switch using a prism mounted on a movable arm mechanism to direct the light from the light source onto the second output. Figure 32.40 shows the experimental apparatus set-up for the experiment described in Section 32.9.4.

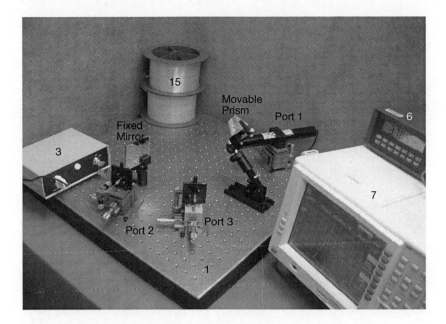

FIGURE 32.40 A 1×2 switch using a prism.

1. Mount three multi-axis translation stages on the breadboard.
2. Place a laser source into the laser mount clamps and tighten the screw. Mount the laser source on one multi-axis translation stage. Bolt the multi-axis translation stage to the breadboard.
3. Turn the laser on. Follow the operation and safety procedures of the laser device in use.
4. Turn off the laboratory lights before taking measurements.
5. As explained in Section 32.9.1, measure laser power and wavelength of the laser light at port 1. Record your results in Table 32.5.
6. Place a fibre cable holder/positioner assembly on each multi-axis translation stage for ports 2 and 3.
7. Mount a fixed mirror and mirror holder facing the laser source at an angle with the normal line of the mirror.
8. Bolt each multi-axis translation stage to the breadboard, as shown in Figure 32.40 (i.e., the laser source is facing one fibre cable holder assembly). Arrange the laser source (port 1) to face the fixed mirror by an angle with the normal. The light from port 1 is reflected by the fixed mirror onto port 2.
9. Measure the laser power and wavelength of the laser source at port 2. Record your results in Table 32.5.
10. Mount a prism and prism holder mechanism into a position such that the laser beam refracts from the prism, is incident on the fixed mirror, and reflects onto the second output at port 3.
11. Put the prism in position, as shown in Figure 32.41.
12. Carefully align the position of the prism assembly so the laser beam from the laser source refracts through the prism onto the fixed mirror and reflects by the fixed mirror onto port 3, as shown in Figure 32.41.

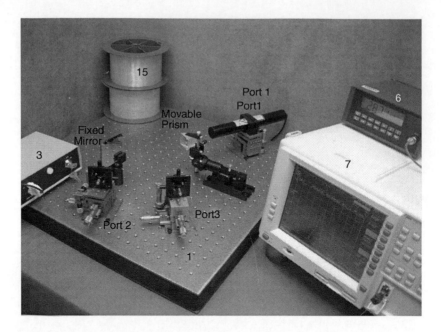

FIGURE 32.41 A 1×2 switch when the movable prism is in the ON position.

13. Measure the laser power and wavelength of the laser source at port 3. Record your results in Table 32.5.
14. Turn on the laboratory lights.

32.9.9 Data Collection

32.9.9.1 A 1×2 Switch with One Laser Source

1. Measure the power and wavelength of the laser source beam at port 1.
2. Measure the power and wavelength of the laser beam at ports 2 and 3.
3. Fill out the Table 32.2 with the collected data.

TABLE 32.2
A 1×2 Switch Data Collection

Port 1		Port 2			Port 3		
Power (unit)	**Wavelength** (unit)	**Power** (unit)	**Wavelength** (unit)	**Loss** (dB)	**Power** (unit)	**Wavelength** (unit)	**Loss** (dB)

32.9.9.2 Two 1×2 Switches with Two Laser Sources

1. Measure the power and wavelength of the laser source beams at ports 1 and 2.
2. Measure the laser power and wavelength of the laser beams at ports 3 and 4.
3. Fill out the Table 32.3 with the collected data.

TABLE 32.3
Two 1×2 Switches Data Collection

Port 1		Port 3			Port 4		
Power (unit)	Wavelength (unit)	Power (unit)	Wavelength (unit)	Loss (dB)	Power (unit)	Wavelength (unit)	Loss (dB)

Port 2		Port 3			Port 4		
Power (unit)	Wavelength (unit)	Power (unit)	Wavelength (unit)	Loss (dB)	Power (unit)	Wavelength (unit)	Loss (dB)

32.9.9.3 A 2×2 Switch Using a Movable Mirror

1. Measure the power and wavelength of the laser source beams at ports 1 and 2.
2. Measure the power and wavelength of the laser beams at ports 3 and 4.
3. Fill out the Table 32.4 with the collected data.

TABLE 32.4
2×2 Switch Using a Movable Mirror Data Collection

Port 1		Port 3			Port 4		
Power (unit)	Wavelength (unit)	Power (unit)	Wavelength (unit)	Loss (dB)	Power (unit)	Wavelength (unit)	Loss (dB)

Port 2		Port 3			Port 4		
Power (unit)	Wavelength (unit)	Power (unit)	Wavelength (unit)	Loss (dB)	Power (unit)	Wavelength (unit)	Loss (dB)

32.9.9.4 A 1×2 Switch Using a Prism

1. Measure the power and wavelength of the laser source beam at port 1.
2. Measure the power and wavelength of the laser beams at ports 2 and 3.
3. Fill out the Table 32.5 with the collected data.

TABLE 32.5
1×2 Switch Using a Prism Data Collection

Port 1		Port 2			Port 3		
Power (unit)	Wavelength (unit)	Power (unit)	Wavelength (unit)	Loss (dB)	Power (unit)	Wavelength (unit)	Loss (dB)

32.9.10 CALCULATIONS AND ANALYSIS

To calculate the power output loss at each port, use the light power loss formula given by Equation 32.3.

$$\text{Loss(dB)} = -10 \times \log_{10} \frac{P_{\text{out}}}{P_{\text{in}}} \tag{32.3}$$

32.9.10.1 A 1×2 Switch with One Laser Source

1. Calculate the loss at each port using the light power loss formula Equation 32.3.
2. Fill out the Table 32.2 with the calculated data.

32.9.10.2 Two 1×2 Switches with Two Laser Sources

1. Calculate the loss at each port using the light power loss formula Equation 32.3.
2. Fill out the Table 32.3 with the calculated data.

32.9.10.3 A 2×2 Switch Using a Movable Mirror

1. Calculate the loss at each port using the light power loss formula Equation 32.3.
2. Fill out the Table 32.4 with the calculated data.

32.9.10.4 A 1×2 Switch Using a Prism

1. Calculate the loss at each port using the light power loss formula Equation 32.3.
2. Fill out the Table 32.5 with the calculated data.

32.9.11 RESULTS AND DISCUSSIONS

32.9.11.1 A 1×2 Switch with One Laser Source

1. Determine the power and wavelength of the laser source beam at each port.
2. Compare the power and wavelength of the laser source beam at the input and output ports.

32.9.11.2 Two 1×2 Switches with Two Laser Sources

1. Determine the power and wavelength of the laser source beams at each port.
2. Compare the power and wavelength of the laser source beams at the input and output ports.

32.9.11.3 A 2×2 Switch Using a Movable Mirror

1. Determine the power and wavelength of the laser source beam at each port.
2. Compare the power and wavelength of the laser source beam at the input and output ports.

32.9.11.4 A 1×2 Switch Using a Prism

1. Determine the power and wavelength of the laser source beam at each port.
2. Compare the power and wavelength of the laser source beam at the input and output ports.

32.9.12 Conclusion

Summarize the important observations and findings obtained in this lab experiment.

32.9.13 Suggestions for Future Lab Work

List any suggestions for improvements using different experimental equipment, procedures, and techniques for any future lab work. These suggestions should be theoretically justified and technically feasible.

32.10 LIST OF REFERENCES

List any references that were used in the report. Use one format in writing the references. Never mix reference formats in a report.

32.11 APPENDICES

List all of the materials and information that are too detailed to be included in the body of the report.

FURTHER READING

Allan, R., 3D MEMS-based optical switch handles 80-by-80 fibers, Electronic Design, Infineon Technologies, 2002.

Bar-Lev, A., *Semiconductors and Electronic Devices*, 2nd ed., Prentice Hall International, Inc., Englewood Cliffs, NJ, 1984.

Berlin, A. and Gabriel, K. J., Distributed MEMS: New challenges for computation, *IEEE Comput. Sci. Eng. J.*, 4 (1), 12–16, 1997.

Boyd, W. T., *Fiber Optics Communications, experiments and Projects*, 1st ed., Howard W. Sams & Co. Indian Polis, In, U.S.A., October 1982.

Brower, D. L., Ding, W. X., and Deng, B. H., Fizeau interferometer for measurement of plasma electron current, Electrical Engineering Department, University of California, Los Angeles, U.S.A., Vol. 75, issue 10, pp. 3399–3401, October 2004.

Camperi-Ginestet, C., Kim, Y. W., Wilkinson, S., Allen, M., and Jokerst, N. M., Micro-opto-mechanical devices and systems using epitaxial lift off. *JPL, Proceedings of the Workshop on Microtechnologies and Applications to Space Systems*, pp. 305–316, June 1993.

Derickson, D., *Fiber Optic Test and Measurement*, Prentice Hall PTR, Upper Saddle River, NJ, 1998.

Dutton, H. J. R., *Understanding Optical Communications*, Prentice Hall, Inc., Englewood Cliffs, NJ, 1998.

Eggleton, B. J., Kerbage, C., Westbrook, P. S., Windeler, R. S., and Hale, A., December microstructured optical fiber devices, *Opt. Express*, 17, 698–713, December 2001.

Fedder, G. K., Santhanam, S. et al., Laminated high-aspect-ratio microstructures in a conventional CMOS process, *Sen. Actuator, A*, 57 (2), 103–110, 1996.

Francon, M., *Optical Interferometry*, Academic Press, New York, 1966 pp. 97–99

Gabriel, K. J., Tabata, O., and Sugiyama, S., Surface-normal electrostatic pneumatic actuators, *Technical Proceedings of MEMS '92, Travemunde, Germany*, 110–114, 1992.

Gebhard, B. and Knowles, C. P., Design and adjustment of a 20cm Mach-Zehnder interferometer, *Rev. Sci. Instrum.*, 37 (1), 12–15, 1966.

Goff, D. R. and Hansen, K. S., *Fiber Optic Reference Guide: A Practical Guide to the Technology*, 2nd ed., Butterworth-Heinemann, London, UK, 1999.

Hariharan, P., Modified Mach-Zehnder interferometer, *Appl. Optics*, 8 (9), 1925–1926, 1969.

He, S., Mrad, R. B., A vertical bi-directional electrostatic comb driver for optical MEMS devices, Paper presented at Opto-Canada, SPIE Regional Meeting on Optoelectronics, Photonics and Imaging, *SPIE TD01*, 2002.

Hecht, J., *Understanding Fiber Optics*, 3rd ed., Prentice Hall, Inc., Englewood Cliffs, NJ, 1999.

Hibino, K., Error-compensating phase measuring algorithms in a Fizeau interferometer Mechanical Engineering Laboratory, AIST, 1–2, Namiki, Tsukuba, Ibaraki, 305-8564, Japan, *Opt. Rev.*, 6 (6), 529–538, 1999.

Hoss, R. J., *Fiber Optic Communications-Design Handbook*, Prentice Hall Pub. Co., Englewood Cliffs, NJ, 1990.

Jackel, J., Goodman, M. S., Baran, J. E., Tomlinson, W. J., Chang, G.-K., Iqbal, M. Z., and Song, G. H., Acousto-optic tunable filters (AOTF's) for multiwavelength optical cross-connects: Crosstalk considerations, *J. Lightwave Technol.*, 14, 1056–1066, 1996.

Jackson, R. A., The laser as a light source for the Mach-Zehnder interferometer, *J. Sci. Instrum.*, 42, 282–283, 1965.

Jacobs-Cook, A. J., MEMS versus MOMS from a systems point of view, *J. Micromech. Microeng.*, 6, 148–156, 1996.

JDS Uniphase Corporation, Switch, SN Series, 2003.

Johnstone, R. D. M. and Smith, W., A design for a 6 in. Mach-Zehnder interferometer, *J. Sci. Instrum.* 42, 231–235, 1965.

Kao, C. K., *Optical Fiber Systems: Technology, Design, and Applications*, McGraw-Hill, New York, 1982.

Kolimbiris, H., *Fiber Optics Communications*, Prentice Hall, Inc., Englewood Cliffs, NJ, 2004.

Kranz, M. S. and Fedder, G. K., Micromechanical vibratory rate gyroscopes fabricated in conventional CMOS, Paper presented at the Symposium Gyro Technology, 16–17 September, in Stuttgart, Germany, 1997.

Li, J., Kahrizi, M., and Landsberger, L. M., In-plane electro-thermal actuated optical switches, Paper presented at Opto Canada, SPIE Regional Meeting on Optoelectronics, Photonics and Imaging. *SPIE TD01*, 2002.

Loreggia, D., Gardiol, D., and Gai, M., Fizeau interferometer for global astrometry in space, *Appl. Optics*, 43 (4), 721–728, 2004.

Marrakchi, A., Ed, *Photonic Switching and Interconnects*, Marcel Dekker, New York, 1994.

Mehregany, M. and Roy, S., *Introduction to MEMS*, Aerospace Press, AIAA, Inc., El Segundo, CA, 1999.

NEL NTT Electronics Corporation, *Thermo-Optic Switches* Catalog, Tokyo, Japan, 2004.

Page, D. and Routledge, I., November using interferometer of quality monitoring, *Photon. Spectra*, 147–153, November 2001.

Palais, J. C., *Fiber Optic Communications*, 4th ed., Prentice Hall, Inc., Englewood Cliffs, NJ, 1998.

Panepucci, R. R., Integrated micro opto-mechanical systems (MOMS), *Abstracts 2002/Mechanical & MEMS Devices*, 74, 2002.

Salah, B. E. A. and Teich, M. C., *Fundamentals of Photonics*, Wiley, New York, 1991.

Senior, J. M., Optical fiber communications Principle and Practice. 2nd ed. Prentice Hall, NJ, U.S.A., 1992.

Setian, L., *Applications in Electro-Optics*, Prentice Hall PTR., Upper Saddle River, NJ, 2002.

Shamir, J., *Optical Systems and Processes*, SPIE - The International Society for Optical Engineering, SPIE Press, U.S.A., Optical Engineering Press, WA, 1999.

Uskov, A. V., O'Reilly, E. P., Manning, R. J., Webb, R. P., Cotter, D., Laemmlin, M., Ledentsov, N. N., and Bimberg, D., On ultra-fast optical switching based on quantum dot semiconductor optical amplifiers in nonlinear interferometers, *IEEE Photon. Technol. Lett.*, 16 (5), 1265–1267, 2004.

Vandemeer, J. E., Kranz, M. S., and Fedder, G. K., Nodal simulation of suspended MEMS with multiple degrees of freedom, Paper presented at The Winter Annual Meeting of ASME in the 8th Symposium on MEMS, Dallas, TX, pp. 16–21 November 1997.

Wood, R. L., Mahadevan, R., and Hill, E., MEMS 2-D matrix switch, MEMS Business Unit, JDS Uniphase Corporation, 3026 Cornwallis Road, Research Triangle Prak, NC 27709, 2000.

Yariv, A., Universal relations for coupling of optical power between microresonators and dielectric waveguides, *Electron. Lett.*, 36, 321–322, 2000.

33 Optical Fibre Communications

33.1 INTRODUCTION

The availability of information depends on the transmission speed of data, voice, and multimedia across telecommunication networks. Despite new technologies that enable legacy copper telephone lines to carry information more efficiently, optical networks remain the most ideal medium for high-bandwidth communications. There are two distinct modes of optical communications: (1) fibre optics (fibre optic cable), and (2) optical wireless, based on free-space optics technology. For long-distance network deployments, nothing is better than fibre. When coupled with the wavelength division multiplexing technologies, fibre optic cables are capable of carrying information more densely across the globe. Optical fibre communications technology became available to bring the connection to the residential areas.

The advantages of optical communications systems for many applications are well recognized: expanded data handling capacity, immunity to electrical noise, electrical isolation, enhanced safety, and data security. Many large telecommunication systems are using fibre optic components, a trend certain to continue for more applications and innovative fields.

Fibre optic cost in communication systems is often high, the process long, and the investment irreversible, but it is profitable over a long period. Optical wireless comes into play because it complements fibre optics in metro area networks and local area networks, with considerably less expense and faster deployment.

33.2 THE EVOLUTION OF COMMUNICATION SYSTEMS

Light has always surrounded us since the creation of the universe. Earth cannot live without light. Communications using light occurred when human beings first communicated by hand signals for short distance under sunlight. This is one of the early methods of communication between people, and it is actually a form of optic communication. In daylight, the sun is the source of light that enables a person to see another's hand signals. The information is carried from the sender to the receiver by the sun's light. Hand motion acts as signal generator or modulator. The eye is the message detector device; the brain processes this message and converts it to an action understood by the receiver. Information transfer for such a system is slow. The transmission distance is limited and dependent on light, and the chance of error is great.

Another optical communication system was later developed which used smoke to signal across long distances. The message was sent by varying the pattern and colour of smoke from a fire. The pattern and colour were carried by light. This system required a coding method known to both the sender and the receiver of the message and has several advantages. It is useful for long distances and is accurate, but it also depends on light. This is comparable to modern digital systems that use pulse codes.

Before the 19th century, communication was slow and used either optical or acoustic means of sending information. The telegraph was introduced in 1838 by Samuel Morse; this was the start of electrical communications. In 1880, the Canadian scientist Alexander Graham Bell (1847–1922) invented the telephone. The telephone was the first basic system that used cable networks for telecommunications.

The development of light bulbs allowed the construction of a simple communication system. This type of system is widely used in ship-to ship and ship-to-shore links.

In the 1900s, radio, television, and microwave used electronic communications by superimposing data onto a carrier, using frequency modulation (FM) or amplitude modulation (AM).

All these communication systems have low rates of information transmission. In 1960, the invention of the laser was a major breakthrough that led to the rapid development of communication systems. The laser provided a coherent and narrow bandwidth beam at specific wavelengths which opened the door to optical communication. Optical communication uses lasers to superimpose electronic data onto an optical carrier using AM. The development of laser diodes along with the advancement of the manufacturability of fibre optic cables paved the way for the mass deployment of optical telecommunication systems. In the late 1980s, the erbium doped fibre amplifier (EDFA) was developed and used in long-distance optical communication networks and signal modulation.

33.3 ELECTROMAGNETIC SPECTRUM OVERVIEW

We are surrounded by waves of visible and invisible energy. These waves are created in many different ways. Waves are generated from natural sources, such as sunlight and plants, and from man-made waves, such as telecommunication and radio/television signals, microwaves, remote control infrared rays, and x-rays. All these types of energy waves are collectively known as the electromagnetic spectrum.

In order to understand the spectrum and radiation, it is necessary to study two concepts. Wavelength refers to the length of the energy waves (between its peaks). Frequency refers to the number of times, or cycles, per second that a wave cycle occurs; it is measured in hertz (Hz). Radio frequency (RF) is the generic name given to electromagnetic waves that can be used for communications. RFs are at the bottom of the electromagnetic spectrum, having the lowest

TABLE 33.1
Working Bands in Communication Systems

Band	Wavelength (nm)
Original band (O-band)	1260–1360
Extended band (E-band)	1360–1460
Short band (S-band)	1460–1530
Conventional band (C-band)	1530–1565
Long band (L-band)	1565–1625
Ultalong band (U-band)	1625–1675

frequencies, and commonly range from 9 kHz to 30 GHz. A higher frequency wave has a shorter range of travel, and vice versa. Activities with RF of higher frequencies (shorter wavelengths) in electrical communications led to the birth of radio, television, radar, and microwave links.

In optical communications, it is customary to specify the band in terms of wavelength instead of frequency. However, due to the high speed of multiple wavelength systems in the middle of the 1990s, the optical source started to be specified in terms of optical frequency.

The optical spectrum ranges from about 50 to about 100 mm, and from ultraviolet to far infrared waves, respectively. In optical fibre communication systems, the focus will be on the wavelengths ranging from 800 to 1600 nm. The electromagnetic spectrum covers a wide range of wavelengths and photon energies. Light used to see an object must have a wavelength about the same size as or smaller than the object. For example, a source that generates light in the far ultraviolet and soft x-ray regions spans the wavelengths suited to studying molecules and atoms. Table 33.1 presents the wavelength bands used in optical communication systems. Figure 33.1 shows the applications of the electromagnetic spectrum regions.

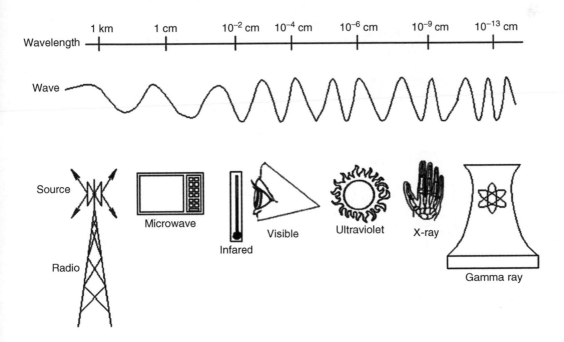

FIGURE 33.1 Applications of the electromagnetic spectrum regions.

33.4 THE EVOLUTION OF FIBRE OPTIC SYSTEMS

Figure 33.2 shows the evolution of operating ranges of optical fibre systems. The system contains the major key link components: light sources (transmitters), optical fibres, optical amplifiers (and other optical devices), and photodetectors (receivers). The vertical dashed lines indicate the three main generations of operating windows for optical fibre systems (850, 1310, and 1550 nm). Optical transmission rates are expressed in bits per second (b/s). More details on the three operating windows follow.

The 1st generation. This generation encompasses fibre links operated around 850 nm (early silica fibres). These links used gallium–aresenide (GaAs-based) optical sources, silicon photodetectors, and multimode fibres. Intercity transmission ranged from 45 to 140 Mb/s, with repeater spacing of around 10 km.

The 2nd generation. The development of the optical sources and photodetectors allowed the shift from 850 to 1310 nm. Bit rates for long-distance links range are between 155 Mb/s Optical Carrier (OC-3) and 622 Mb/s (OC-12). In some cases transmission is up to 2.5 Gb/s (OC-48) over repeater spacing of 40 km on a single mode fibre. Bit rates for local area network (LAN) range from 10 to 100 Mb/s, over a distance of 0.5–10 km.

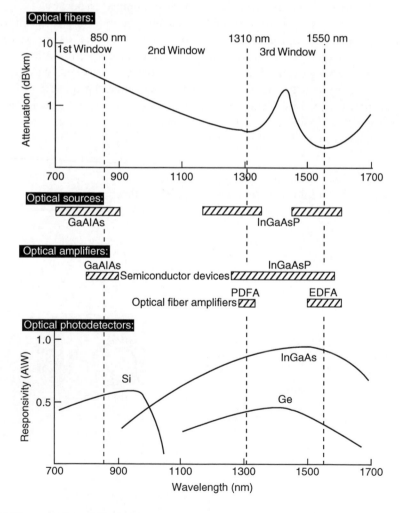

FIGURE 33.2 The evolution of fibre optic systems.

The 3rd generation. Systems operating at 1550 nm are used for high-capacity long-span terrestrial and undersea transmission links, which carry traffic over a 90 km repeaterless distance at 2.5 Gb/s (OC-48) and 10 Gb/s (OC-192). The introduction of optical amplifiers gave a major boost to fibre transmission capacity. Common optical amplifiers include GaAlAs-based optical amplifiers, praseodymium-doped fibre amplifiers (PDFAs) operating at 1310 nm, and EDFAs operating at 1550 nm.

33.5 UNDERSEA DWDM CABLE NETWORK (SEA-ME-WE-3)

The use of the wavelength-division multiplexing (WDM) offers a further boost in transmission capacity. The basic principle of WDM is to use multiple sources operating at slightly different wavelengths to send several independent packets of information data over the same fibre cable. For example, one of many of the world's WDM optical networks is the SEA-ME-WE 3-cable system, as shown in Figure 33.3. This undersea network runs from Germany to Singapore, connecting many countries in between; hence the name SEA-ME-WE, which refers to Southeast Asia (SEA), the Middle East (ME), and the West Europe (WE). The network has two pairs of undersea fibres with a capacity of eight STM-16 wavelengths per fibre (equivalent to eight OC-48 which is 8×2.5 Gb/s).

Another submarine cable system, the undersea network (see Figure 33.4), connects many countries spanning between:

1 Portugal (Sesimbra),
2 Spain (Chipiona),
3 Spain (Altavista),
4 Senegal (Dakar),
5 Côte d'Ivoire (Abidjan),

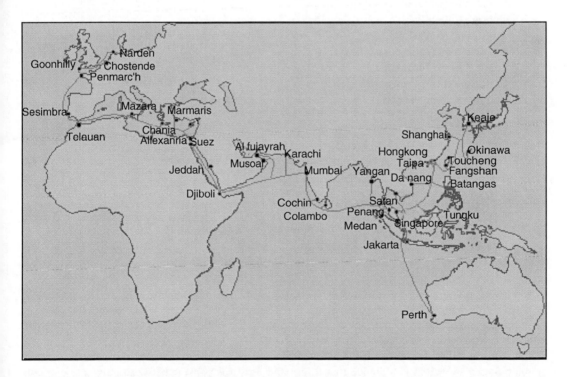

FIGURE 33.3 The SEA-ME-WE WDM cable network connects countries between Germany and Singapore.

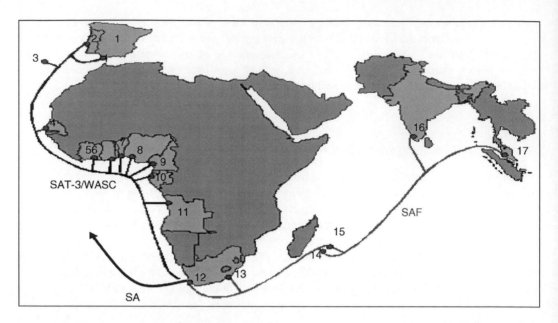

FIGURE 33.4 The SAT-3/WASC-SAF cable network connects countries between Portugal and Malaysia.

   6 Ghana (Accra),
   7 Benin (Cotonou),
   8 Nigeria (Lagos),
   9 Cameroon (Douala),
  10 Gabon (Libreville),
  11 Angola (Cacuaco),
  12 South Africa (Melkbosstrand),
  13 South Africa (Mtunzini),
  14 La Reunion (St. Paul),
  15 Mauritius (Baie Jacotet),
  16 India (Cochin), and
  17 Malaysia (Penang).

The route distance between Sesimbra and Penang is 23,455 km.

33.6 BASIC COMMUNICATION SYSTEMS

The following are basic definitions for the networks used in communication systems:

Station is a collection of devices with which users wish to communicate, such as computers, terminals, telephones, and videos. Stations are also called data terminal equipment (DTE) in network systems, and they can be connected directly to a transmission line.

Network is a group of two or more stations linked or interconnected by a transmission medium, such as a fibre optic cable or coaxial cable.

Topology is the logical manner or structure in which nodes/devices are linked together via information-transmission channels to form a network.

Switching is the transfer of information from source to destination through a series of intermediate nodes. A switch is a device that filters and forwards packets between network segments. Switches operate at the data link layer (layer 2) and sometimes at the network layer (layer 3) of the open system interconnection (OSI) reference model. A description of the layers is presented in Section 33.10. Switches support any packet protocol. More details on optical mechanical switches are presented in this book.

Routing is the process of moving a packet of data from the source to the destination by the selection of a suitable path through a network. Routing is usually performed by a dedicated device called a router.

Protocol is an agreed-upon format for transmitting data between two devices. The protocol determines the following items:

- Type of error checking to be used;
- Data compression method, if any; and
- Hand shaking, how the device indicates when finished sending a message.

A popular protocol used in optical LANs is the fibre distributed data interface (FDDI) protocol; SONET/SDH protocols are used in optical networks in metro or wider areas. Logical topologies are bound to the network protocols that direct how the data moves across a network. The Ethernet protocol is a common logical bus topology protocol.

33.7 TYPES OF TOPOLOGIES

Topology refers to the shape of a network, or the network's layout. There are different nodes in a network which communicate by the network's topology. Topologies are either physical or logical. Connections between the nodes are via optical couplers. The five most common network topologies are (1) bus topology, (2) ring topology, (3) star topology, (4) mesh topology, and (5) tree topology.

33.7.1 BUS TOPOLOGY

In a bus topology, all stations are connected to a central cable which is called the bus or backbone, as shown in Figure 33.5.

33.7.2 RING TOPOLOGY

In a ring topology, all stations are connected to one another in the shape of a closed loop, so that each station is connected directly to two other stations on either side, as shown in Figure 33.6.

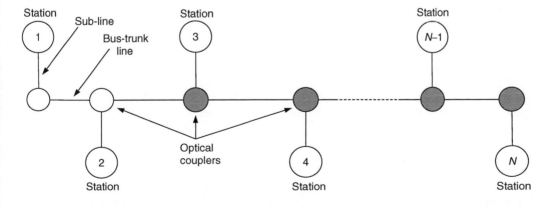

FIGURE 33.5 Bus topology.

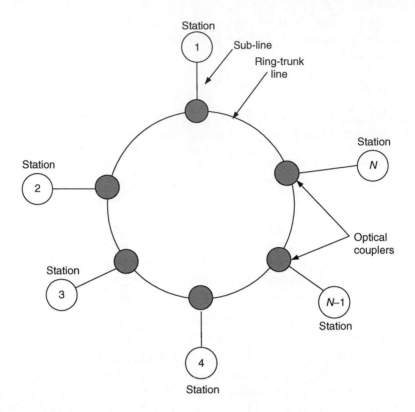

FIGURE 33.6 Ring topology.

33.7.3 STAR TOPOLOGY

In a star topology, all stations are connected to a central hub, which is an optical star coupler. Stations communicate across the network by passing data through the hub, as shown in Figure 33.7.

33.7.4 MESH TOPOLOGY

In a mesh topology, all stations are connected via many redundant interconnections, as shown in Figure 33.8. In a true mesh topology every station has a connection to every other station in the network.

33.7.5 TREE TOPOLOGY

In a tree (hybrid) topology, all stations are connected by various topologies, as shown in Figure 33.9. Groups of star-configured networks are connected to a linear bus backbone.

33.8 TYPES OF NETWORKS

Networks are divided into five types based on the size of the area that the network covers: (1) home-area networks (HANs), (2) local-area networks (LANs), (3) campus-area networks (CANs), (4) metropolitan-area networks (MANs), and (5) wide-area networks (WANs)

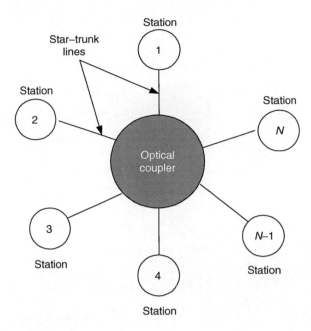

FIGURE 33.7 Star topology.

33.8.1 Home-Area Networks

As shown in Figure 33.10(a), (b), and (c), a HAN is a network that connects the digital devices to computers, which are contained within an individual user's home.

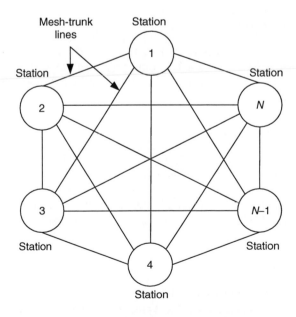

FIGURE 33.8 Mesh topology.

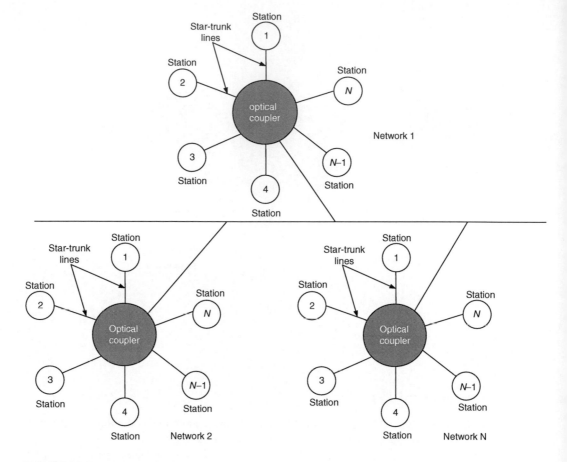

FIGURE 33.9 Tree topology.

33.8.2 LOCAL-AREA NETWORKS

A LAN interconnects users in a localized area such as a department, a small group of buildings, or a factory complex. Figure 33.11 shows a LAN connected to the Internet (through a firewall). This figure shows how workstations are connected to a hub and router. The router provides connections to file servers, and printers.

33.8.3 CAMPUS-AREA NETWORKS

As shown in Figure 33.12, a CAN is a computer network made up of an interconnection of LANs which are located within a limited geographical area, such as a university or college campus. In the case of a university campus-based CAN, the network is likely to link a variety of campus buildings, including academic departments, the university library, and student halls of residence. A CAN is larger than a LAN, but smaller than a MAN. In addition, CAN stands for corporate-area network.

33.8.4 METROPOLITAN-AREA NETWORKS

As shown in Figure 33.13(a) and (b), a MAN provides interconnections within a city or in a metropolitan area surrounding a city. MANs typically use wireless infrastructure or optical fibre connections to link their sites (nodes).

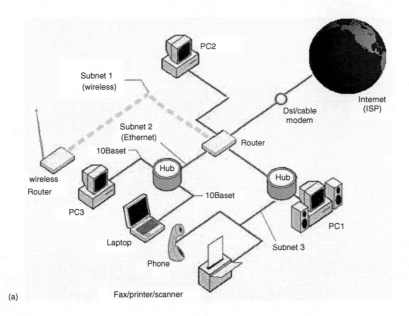

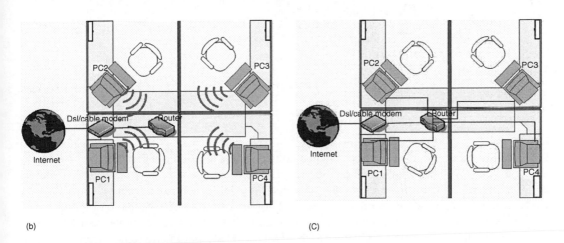

FIGURE 33.10 Home-Area Networks (HANs): (Small and Home Offices SOHO).

For instance, a university or college may have a MAN that joins together many of their LANs, which are situated in a site whose area is less than a square kilometre. Beyond that, the MAN could have several WAN links to other universities or the Internet.

Some of the technologies used for this purpose are asynchronous transfer mode (ATM) and fibre distributed data interface (FDDI). These older technologies are in the process of being displaced by Ethernet-based MANs (e.g., Metro Ethernet) in most areas. MAN links between LANs have been built without cables, by using microwave, radio, or infra-red free-space optical communication links.

33.8.5 Wide-Area Networks

As shown in Figure 33.14(a) and (b), a WAN covers a large geographical area connecting cities, countries, and continents. The best example of a WAN is the Internet. WANs are used to connect

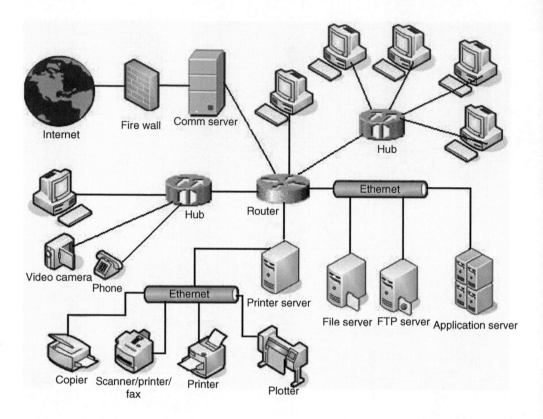

FIGURE 33.11 Local-Area Network (LAN).

LANs or MANs together, so that users and computers in one location can communicate with users and computers in other locations. Many WANs are built for one particular organization and are private. Others, built by Internet service providers, provide connections from an organization's LAN to the Internet. A router connects to the LAN on one side of the line, and a hub within the WAN on the other. Network protocols, including TCP/IP, deliver transport and addressing functions. Protocols include Packet over SONET/SDH, ATM, and Frame Relay. These protocols are often used by service providers to deliver the links used in WANs. X.25 was an important early WAN protocol, and is often considered to be the grandfather of Frame Relay, as many of the underlying protocols and functions of X.25 are still in use today (with upgrades) by Frame Relay.

33.9 SUBMARINE CABLES

Oceans cover over 70% of our planet, separating the continents and people. People rely on submarine cable networks for voice, data and Internet communication. Extreme demands between continents for network reliability and high capacity is achieved by using submarine networks that are reliable and well designed. Figure 33.15 shows the map of submarine cable systems across the globe. Submarine cables are sometimes known as underwater cables. The world-leading supplier of submarine networks connected every continent from Europe to Japan, the length of the Americas, and across the Pacific. For high capacity, dense wavelength division multiplexer (DWDM) technology is used in telecommunication systems.

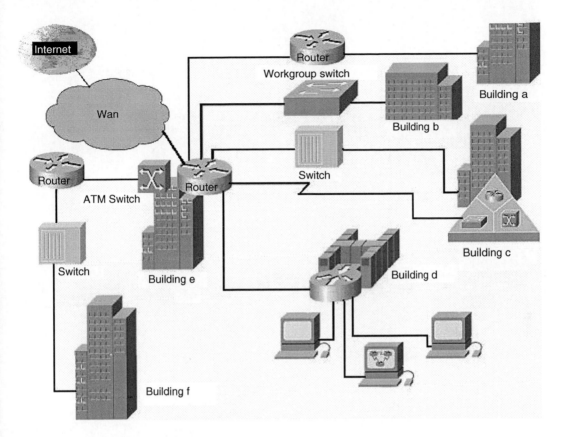

FIGURE 33.12 Campus-Area Network (CAN).

33.10 OPEN SYSTEM INTERCONNECTION

An OSI is a model that defines a networking framework used for implementing protocols in seven layers of communications. Control is passed from one layer to the next, starting at the application layer in one station, and proceeding through to the optical bottom layer. Control passes over the channel (in optical pulses) to the next station and back up the hierarchy. Figure 33.16(a) shows the seven layers of OSI and data process, while Figure 33.16(b) shows the host and media layers located in the OSI Model.

33.10.1 PHYSICAL (LAYER 1)

This layer conveys the bit stream (electrical impulse, light, or radio signal) through the network at the electrical and mechanical level. This is transmission of raw data over a communication medium. It provides the hardware means of sending and receiving data on a carrier which includes defining cables, cards, and physical aspects. SONET/SDH, Fast Ethernet, RS232, and ATM are protocols with physical layer components.

33.10.2 DATA LINK (LAYER 2)

Layer 2 includes transfer of data frames/packets, addressing, and error correction. At this layer, data packets are encoded and decoded into bits. Layer 2 furnishes transmission protocol knowledge and management, and handles errors in the physical layer, flow control, and frame synchronization. The society of electrical and electronic engineers formed the 802 committee which was responsible for

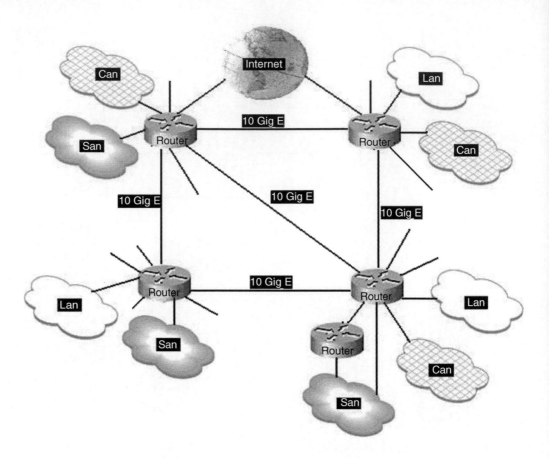

FIGURE 33.13. Metropolitan-Area Network (MAN).

dividing the data link layer into two sublayers: the media access control (MAC) layer and the 802.2 logical link control (LLC) layer. The MAC sublayer controls how a computer on the network gains access to the data and permission to transmit it. The LLC layer controls frame synchronization, flow control, and error checking.

33.10.3 NETWORK (LAYER 3)

This layer provides switching and routing functions, creating logical paths, known as virtual circuits (e.g., X.25 connection), for transmitting data from node to node. Other functions of this layer are routing and forwarding, as well as addressing, internetworking, error handling, congestion control, and packet sequencing. Routing of data packets across networks provides software interface between the physical and data link layers.

33.10.4 TRANSPORT (LAYER 4)

This layer provides transparent transfer of data between end systems, or hosts, and is responsible for end-to-end error recovery and flow control. It ensures complete data transfer.

33.10.5 SESSION (LAYER 5)

This layer establishes, manages, and terminates internode connections between applications and uses standards to move data between the applications. The session layer sets up, coordinates, and

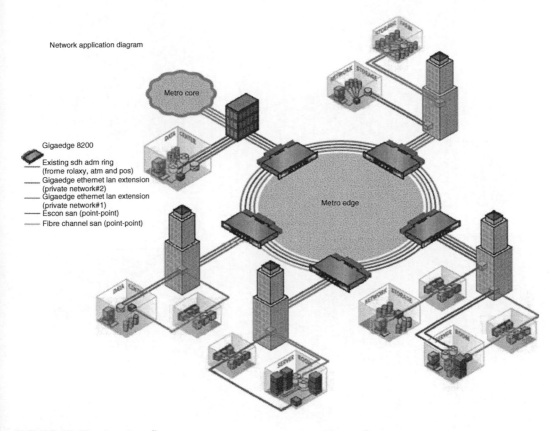

FIGURE 33.13   (*continued*)

terminates conversations, exchanges, and dialogues between the applications at each end. It deals with session and connection coordination.

33.10.6 PRESENTATION (LAYER 6)

This layer involves data formatting, character conversion, security, and coding. It provides independence from differences in data representation (e.g., encryption) by translating from application to network format, and vice versa. The presentation layer transforms data into the form that the application layer can accept. This layer formats and encrypts data to be sent across a network, providing freedom from compatibility problems. It is sometimes called the syntax layer.

33.10.7 APPLICATION (LAYER 7)

This layer supports application and end-user processes. Communication partners and quality of service are identified, user authentication and privacy are considered, and any constraints on data syntax are recognized. Everything at this layer is application-specific. This layer provides application services for file transfers, e-mail, network operating systems, application programs, and other network software services. Telnet and FTP are applications that exist entirely in the application level. Tiered application architectures are part of this layer.

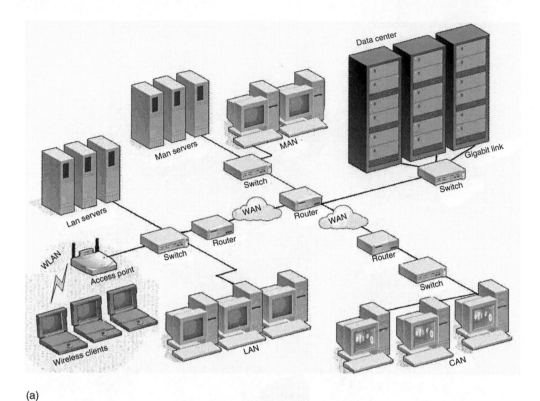

(a)

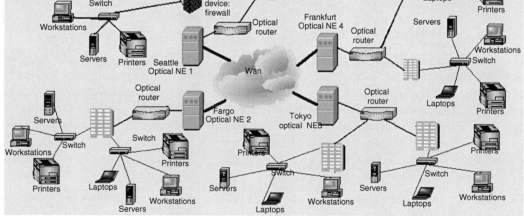

(b)

FIGURE 33.14 Wide-Area Network (WAN).

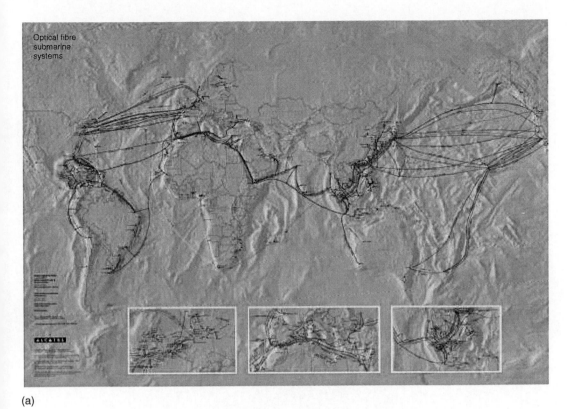

(a)

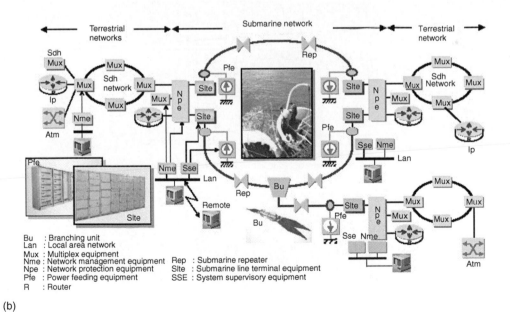

Bu   : Branching unit
Lan : Local area network
Mux : Multiplex equipment
Nme : Network management equipment   Rep : Submarine repeater
Npe : Network protection equipment   Slte : Submarine line terminal equipment
Pfe : Power feeding equipment   SSE : System supervisory equipment
R   : Router

(b)

FIGURE 33.15 Map of submarine cable systems. (a) Map of submarine cable systems across the globe (Courtesy of Alcatel). (b) Map of submarine configuration (Courtesy of Fujitsu).

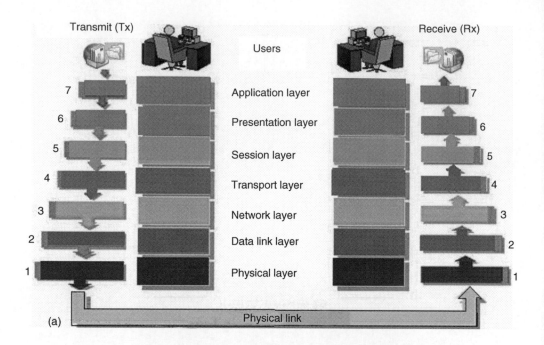

(a)

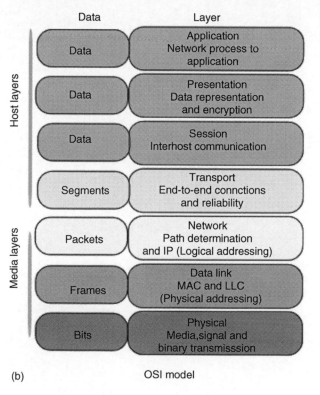

(b)   OSI model

FIGURE 33.16 Open System Interconnection (OSI). (a) OSI and data process. (b) OSI Model.

33.11 PERFORMANCE OF PASSIVE LINEAR OPTICAL NETWORKS

To evaluate the performance of passive linear networks, consider the fraction of optical power (F_C) lost at a particular interface or component along the transmission path, as shown in Figure 33.17. The power ratio (A_0) over an optical fibre of length (x) will be:

$$A_0 = \frac{P_{(x)}}{P_{(0)}} = 10^{-\alpha x/10} \tag{33.1}$$

where $P_{(x)}$ is the power received, $P_{(0)}$ is power transmitted, and α is the fibre attenuation (dB/km).

If F_C is lost at each port of the coupler, then the connecting loss (L_C) will be:

$$L_C = 10 \log(1 - F_C) \tag{33.2}$$

For example, if $F_C = 20\%$, then $L_C = -0.9691$ dB. The optical power gets reduced by the L_C of 1 dB at any connection junction.

The power extracted from the bus is called tap loss (L_{tap}), and is given by:

$$L_{tap} = 10 \log C_T \tag{33.3}$$

where C_T is the fraction power removed from the bus and delivered to the detected port.

Then the throughput coupling loss (L_{thru}) is given by:

$$L_{thru} = 10 \log(1 - C_T)^2 = 20 \log(1 - C_T) \tag{33.4}$$

In addition to the losses L_C and L_{tap}, there is an intrinsic loss (L_i) associated with each bus coupler. If the fraction of power lost in the coupler is F_i, then:

$$L_i = 10 \log(1 - F_i) \tag{33.5}$$

All losses are measured in decibels (dB).

A linear bus configuration, as shown in Figure 33.18, consisted of a number of stations (N) separated by a various lengths. For simplicity, assume a constant distance L.

The fibre attenuation between two adjacent stations is given by:

$$L_{fibre} = 10 \log A_0 = \alpha L \tag{33.6}$$

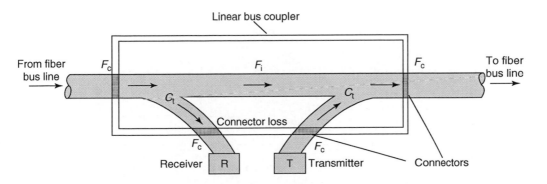

FIGURE 33.17 Losses in a passive linear-bus coupler consisting of two cascaded directional couplers.

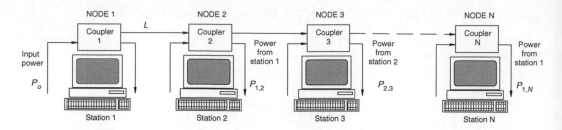

FIGURE 33.18 Topology of a simplex linear bus consisting of N uniformly spaced stations.

33.11.1 POWER BUDGET CALCULATION

To calculate the power budget of a fibre link consisting of N stations, as shown in Figure 33.18, the fractional power losses F_C should be examined first.

33.11.2 NEAREST-DISTANCE POWER BUDGET

If P_0 is the optical power launched from the optical source at station 1, then the optical power detected $P_{1,2}$ at the station 2 is given by:

$$P_{1,2} = A_0 C_T^2 (1 - F_C)^4 (1 - F_i)^2 P_0 \tag{33.7}$$

Then the power budget (considering all losses) between stations 1 and 2 is:

$$10 \log \left(\frac{P_0}{P_{1,2}} \right) = \alpha L + 2L_{\text{tap}} + 4L_c + 2L_i \tag{33.8}$$

33.11.3 LARGEST-DISTANCE POWER BUDGET

If P_0 is the optical power launched from the optical source at station 1, then the optical power detected $P_{1,N}$ at the station N is given by:

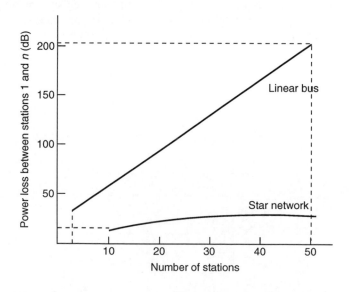

FIGURE 33.19 Total loss versus number of station in linear-bus and star networks.

$$P_{1,N} = A_0^{N-1}(1-C_T)^{2(N-2)}C_T^2(1-F_C)^{2N}(1-F_i)^N P_0 \qquad (33.9)$$

Then the power budget (considering all losses) between stations 1 and N is:

$$10\log\left(\frac{P_0}{P_{1,N}}\right) = N(\alpha L + 2L_c + L_{thru} + L_i) - \alpha L - 2L_{thru} + 2L_{tap}$$

$$= [\text{fibre} + \text{connector} + \text{coupler throughput} + \text{ingress/egress}$$

$$+ \text{coupler intrinsic}] \text{ losses} \qquad (33.10)$$

As shown in Figure 33.19, the losses (in dB) of a linear bus configuration in Figure 33.18 increase linearly with the number of stations N.

33.12 PERFORMANCE OF STAR OPTICAL NETWORKS

The optical input power is evenly divided among the output ports in an ideal star fibre coupler, as shown in Figure 33.20.

The total optical power loss of the coupler consists of its splitting loss and the excess loss in each path through the star configuration. The splitting loss (L_{split}) among the N stations is given by:

$$L_{split} = 10\log\left(\frac{1}{N}\right) = 10\log N \qquad (33.11)$$

The star fibre coupler excess loss (L_{excess}) is defined as the ratio of the single input power (P_{in}) to the total output power ($P_{out,i}$) of the N stations ($i = 1...N$).

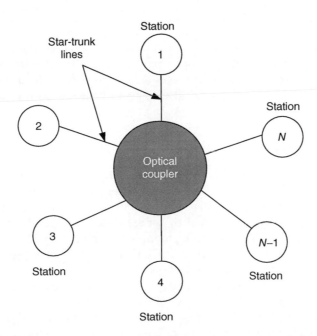

FIGURE 33.20 A star coupler.

$$L_{\text{excess}} = 10 \log \left(\frac{P_{\text{in}}}{\sum\limits_{i=1}^{N} P_{\text{out},i}} \right) \tag{33.12}$$

The total loss within the fibre star coupler is given by:

$$\text{Total Loss} = L_{\text{Spilt}} + L_{\text{excess}} \tag{33.13}$$

The optical power balance equation between any two stations in a star network, include all losses, and is defined as:

$$P_S - P_R = L_{\text{excess}} + \alpha(2L) + 2L_c + L_{\text{split}} = L_{\text{excess}} + \alpha(2L) + 2L_c + 10 \log N \tag{33.14}$$

where P_S is the fibre coupled output power from source in dBm; P_R is the minimum optical power required at the receiver; L_{excess} is the star fibre coupler excess loss; L_{split} is the splitting loss; Lc is the connector loss; L is the distance from the star coupler, assuming all stations at the same distance; and α is the fibre attenuation.

As more stations are added, the loss in a star network increases much slower than the loss in a linear bus network, as shown in Figure 33.19.

33.13 TRANSMISSION LINKS

Transmission links in an optical communication system are the fibre optic cables that carry light signals between the senders and receivers, and between the communication system's components and peripherals. Signals sent by sender sources over a transmission link or channel can be classified into two formats: analogue and digital signals. Each format has its advantages and disadvantages.

33.13.1 ANALOGUE SIGNALS

An analogue signal carries information through a continuous vibration (waves) of particles in time. Waves can be optical, electrical, or acoustical, and have intensities and frequencies, as shown in Figure 33.21. Audio (sound) and video messages are examples of analogue signals. Due to enormous capabilities of computers, the trend in communications has been a fast conversion from analogue to digital transmission format.

33.13.2 DIGITAL SIGNALS

A digital transmission system transmits signals that are in digital form. A digital signal is an ordered sequence of discrete symbols selected from a finite set of elements. The digital signals typically represent information that is alphabetical, numerical, and other symbols, such as @, #, or %. These discrete symbols are normally represented by unique patterns of pulses of electric voltages or optical intensity, which can take on two or more levels. Figure 33.22 shows common digital signal configurations in the binary waveform. A binary waveform is represented by a sequence of two types of pulses of known shape. The information contained in a digital signal is given by the particular sequence of a presence or high (a binary one, or simply either one or 1), and an absence or low (a binary of zero or simply either zero or 0) of these pulses. These pulses are commonly known as bits. The bit is derived from binary digits. Since digital logic is used in the generation and processing of 1 and 0 bits, these bits often are referred to as a logic one (or logic 1) and a logic zero (or logic 0), respectively. Sometimes two voltages are used as high or low level.

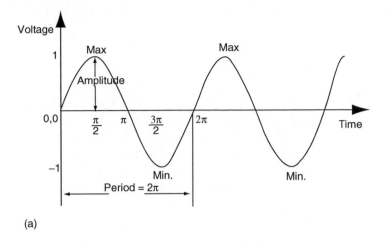

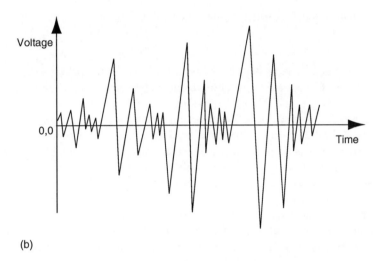

FIGURE 33.21 Analogue signals. (a) Sine wave ($y = \sin \theta$).(b) Analogue voice wave.

The time slot T in which a bit occurs is called the bit interval, bit period, or bit time. (Note that this T is different from T used for designating the period of a wave). The bit intervals are regularly spaced, and occur every $1/R$ seconds (s) at a rate of R bits per second, where R is the rate that data bits are sent, called the bit rate.

Digital signals are easy to process with electronics and optics. It is simple to design a circuit to detect whether a signal is high or low level (on or off).

Fibre optic cables work very well with digital signals, and were initially used mainly for digital systems. Fibre cables have the high transmission capacity needed for digital transmission with a wide range of wavelengths.

33.13.3 CONVERTING ANALOGUE SIGNAL TO DIGITAL SIGNAL

The term digital is used frequently because digital circuits are becoming so widely used in computation, robotics, medical science and technology, communications, transportation, etc. Digital electronics developed from the principle that the circuitry of a transistor could be designed and easily fabricated to have an output of one or two voltage levels, based on its input voltage.

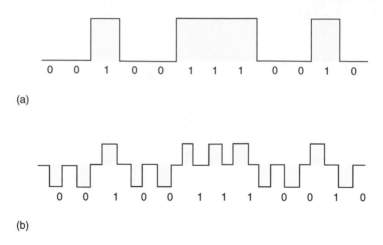

(a)

(b)

FIGURE 33.22 Common digital signal configurations. (a) Digital pulse stream. (b) Dipolar return-to-zero pulse.

Figure 33.23 shows an analogue-to-digital converter (ADC) used to convert an analogue signal into a digital signal. The ADC can be a single chip, or can be one circuit within a chip. The two voltage levels are usually 5 V (high) and 0 V (low); the levels can be represented by 1 and 0.

The binary numbering system (base-2 numbering system) is the main numbering system used in digital electronics. A digital value is represented by a combination of on and off voltage levels, and expressed as a string of 1s and 0s. Signals that have a theoretically infinite number of levels are converted into signals that have two defined levels.

To convert an analogue signal to a digital form, one starts by taking instantaneous measures of the height of the analogue signal wave at regular intervals, this is called sampling the signal. One way to convert these analogue samples to a digital format is to simply divide the amplitude of the analogue signal into N equally spaced levels, which are designated by integers, and to assign values to one of these integers. This process is called quantization. Since the signal varies continuously in time, this process generates a sequence of real numbers.

Figure 33.24 shows the equally spaced levels that are the simplest method of quantization produced by a uniform quantizer. If the digitization samples are taken frequently enough relative to the rate at which the signal varies, then a good approximation of the signal can be recovered from the samples by drawing a straight line between the sample points. The resemblance of the reproduced signal to the original signal depends on the fineness of the quantizing process, and on the effect of noise and distortion added into the transmission system. If the sampling rate is at least two times the highest frequency, then the receiving device can easily reconstruct the analogue signal. Thus, if a signal is limited to bandwidth of B Hz, then the signal can be reproduced without

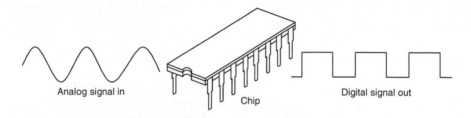

FIGURE 33.23 A chip used to convert an analogue to digital signal.

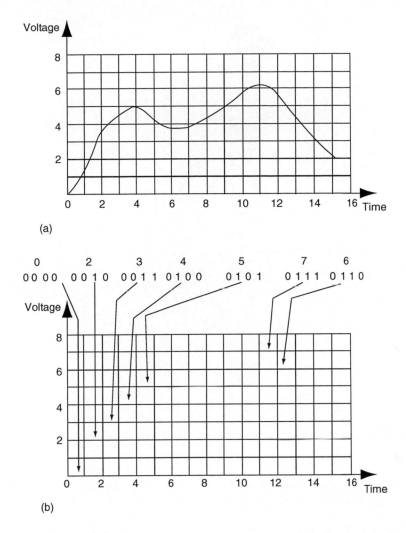

FIGURE 33.24 Digitization of analogue waveforms. (a) Analogue signal varying between 0 and V volts.(b) Quantized and sampled digital signal.

distortion if it is sampled at a rate of $2B$ times per second. These data samples are represented by a binary code.

As shown in Figure 33.24, eight quantized levels having upper bounds, $V_1, V_2, V_3, \ldots V$, can be described by 4 binary digits (for example 8 in binary becomes $2^3 = 2^3 \, 2^2 \, 2^1 \, 2^0 = 1\,0\,0\,0$). More digits can be used to give finer sampling levels. Thus, if n binary digits represent each sample, then one can have 2^n quantization levels. Figure 33.25 shows a conversion process of an analog signal to digital form, and then to bar code.

33.13.4 Bit Error Rate (BER)

The performance of a digital communication system is measured by the probability of an error occurring in a data bit; a bit error rate (BER) equal to 0 is ideal. Define p_1 as the probability of misinterpreting a 1 bit as a 0, and p_0 as the probability of misinterpreting 0 as a 1. If the 0 or 1 bits

FIGURE 33.25 Conversion process of an analogue signal to digital and to bar code.

are equally likely to be transmitted, then BER $= 1/2\, p_1 + 1/2\, p_0$. In telecommunication systems, a typical acceptable BER is 10^{-9} (i.e., an average of one error every 10^9 bit).

33.13.5 FIBRE OPTIC TELECOMMUNICATION EQUIPMENT AND DEVICES

Optical fibre devices used in any communication system, such as light sources (transmitters), photodetectors (receivers), fibre optic cables, multiplexers, demultiplexers, optical amplifiers, isolators, circulators, and optical switches are explained in detail throughout the book.

33.14 SONET/SDH

33.14.1 DEFINITION OF SONET AND SDH

With the advent and development of fibre optic cables and optical fibre amplifiers, the next important evolution of the digital time-division multiplexing (TDM) scheme was a standard signal format. This format is called Synchronous Optical NETwork (SONET) in North America, and synchronous digital hierarchy (SDH) in other parts of the world. SONET and SDH are both optical interface standards that allow internetworking of services from multiple service providers. SONET and SDH are almost identical standards dedicated to transporting data, voice imaging, and video over optical networks. Figure 33.26 illustrates a standard optical interface between SONET and SDH network elements (NE) from vendors A, B, and C.

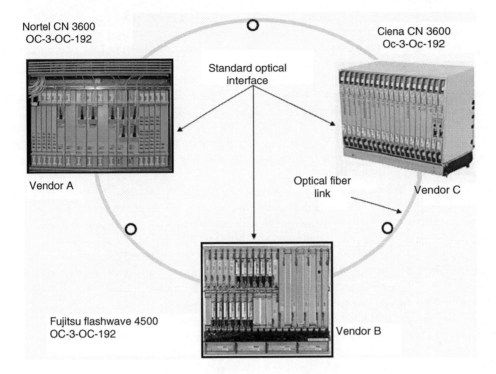

FIGURE 33.26 SONET/SDH.

SONET was specified primarily by Bellcore (now Telcordia) in the late 1980s. It was submitted to the international standards bodies as a proposed international standard. After some negotiation, SDH emerged as the international standard, of which SONET can be considered a complete subset. The standards define a hierarchy of high-speed transmission rates, which currently range from 51.84 Mb/s (SDH starts at 155 Mb/s) up to 40 Gb/s.

The standard specifies the physical interfaces, such as the optical wavelengths, pulse shapes, and link budgets. For some of the lower rates in the hierarchy, electrical interfaces are specified as well, including line coding and electrical pulse shapes. The organization of the digital information crossing the interface is defined by specifications for frame structures, payload mappings, and overhead assignments. In addition, special signals are specified which allow communication between the two optical network elements for the purpose of operations, administration, mainten-ance, and provisioning (OAM&P). The OAM&P of SONET/SDH networks rely on the Telecommunication Management Network architecture (TMN) as defined by the ITU. SONET/SDH network elements are managed by using translation language 1 (TL1) protocol.

33.14.2 SONET/SDH Purposes and Features

There are several purposes for implementing SONET or SDH. Some of the purposes and features are further explained in the following descriptions.

Multi-vendor networks. The primary motivation for developing the SONET standard was to allow the deployment of multi-vendor optical networks. Prior to SONET, fibre optic transport systems could only be deployed in point-to-point configurations, resulting in a lot of unnecessary equipment to terminate the proprietary interfaces. With SONET, a completely optical transport network is envisaged whereby high capacity fibres interconnect network elements that provide access to the transport network and manage the fibre bandwidth.

Cost reduction. It was also assumed that the resulting network would be cheaper due to the consolidation of network functions, elimination of unnecessary functions, and increased competition between vendors. Network providers feel that they no longer have to lock themselves into a single vendor's proprietary (non-standard) solution.

Survivability and availability. With the increasing concern over network reliability and robustness, SONET also provides the ability to build survivable networks; networks that can restore traffic within 50 ms, in the event of fibre cuts and equipment failure.

New high-speed services. The demand for higher bandwidth pipes continues to increase. SONET provided the opportunity to define an interface, and therefore a network that was capable of carrying a variety of payload types with differing bandwidth requirements, including broadband payloads beyond 50 Mb/s.

Bandwidth management. SONET manages bandwidth by introducing the concept of the payload envelope, into which all payloads are mapped as they enter the SONET network. This allows the infrastructure to be concerned only with transporting the envelopes, regardless of their contents. A limited number of envelopes are defined, with payload capacities ranging from 1.5 Mb/s to 10 GB/s. In addition, different envelopes may be combined on the same fibre.

Network management/Single-ended operations. Each rate in the SONET/SDH hierarchy is an integer multiple of a basic rate. For SONET, this basic rate is 51.84 Mb/s; for SDH, it is 155.52 Mb/s. The signal format at each level is created by synchronously multiplexing the basic format. This format is called the synchronous transport signal-level 1 (STS-1) in SONET and the synchronous transport module-level1 (STM-1) in SDH. Synchronous multiplexing simplifies bandwidth management, allowing access to individual tributaries within the fibre signal without having to completely de-multiplex the fibre signal. The creation of an all-optical network enables management of the network bandwidth using automated techniques. Spare bandwidth on one route may be reallocated to another route, and new connections between end offices can be created quickly when required. SONET includes overhead allocations for a variety of OAM&P functionality. Examples include a data communications channel to allow network elements to be monitored from a central operations system, and integrity checks to allow single-ended performance monitoring of fibre systems and end-to-end networks.

33.15 MULTIPLEXING TERMINOLOGY AND SIGNALING HIERARCHY

33.15.1 Existing Multiplexing Terminology and Digital Signaling Hierarchy

In order to understand the role that the SONET standard plays, it is first necessary to understand what interface standards existed previously. In North America, the standard pre-SONET digital hierarchy consisted mainly of digital signals of several levels. The digital signal-level 1 (DS1), is capable of transmitting and receiving data at a bit rate of 1.544 Mb/s (150 Mb/s) and the DS3 has the capability of 44.736 Mb/s, as shown in Figure 33.27. The DS2 really only exists as an intermediate step in the DS1–DS3 multiplex. The DS1 is the main interface to digital voice switches and channel banks, whereas the DS3 is the main interface to fibre optic transmission systems. The M13 multiplexer links the two together, and is named for its ability to multiplex DS1s into a DS3. This hierarchy is based on time division multiplexing (TDM).

A similar hierarchy, called electrical signal E, exists in most of the rest of the world. The 2.048 Mb/s interface is the key digital switch interface, whereas most pre-standard fibre optic transmission systems carry the 139 Mb/s signal. 2.048 Mb/s is called E1, and the hierarchy is based on multiples of 4 E1s:

E2 = 4×E1 = 8 Mb/s
E3 = 4×E2 = 34 Mb/s

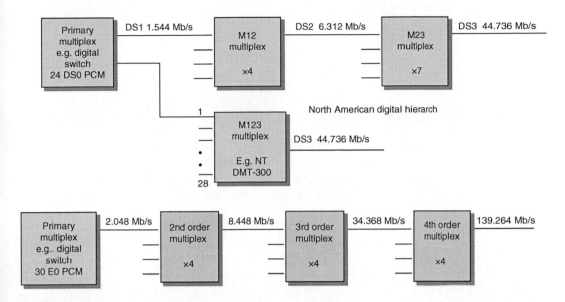

FIGURE 33.27 Existing digital/electrical hierarchies.

$$E4 = 4 \times E3 = 140 \text{ Mb/s}$$
$$E5 = 4 \times E4 = 565 \text{ Mb/s}$$

The E3 tributaries are faster than the E2 tributaries, while the E2 tributaries are faster than the E1 tributaries, and so forth. To synchronize with other tributaries, extra bits, called justification bits, are added. These tell the multiplexers which bits are data and which are spare. Multiplexers on the same level of the hierarchy remove the spare bits, and are synchronized with each other at that level only. Multiplexers on one level operate on a different timing from multiplexers on another level. For instance, the timing between primary rate muxes (which combine 30×64 Kb/s channels into 2.048 Mb/s E1) will be different from the timing between 8 Mbit muxes (which combine up to 4×2 Mb/s into 8 Mb/s).

DS1 is sometimes called transport level-1 (T1). T1 is the optimal rate for accessing low-level devices. It is a type of telephone service capable of transporting the equivalent of 24 conventional telephone lines, using only two pairs of wires. T1 uses two pairs of copper wires (four individual wires) to carry up to 24 simultaneous conversations (channels) that would normally need one pair of wires each. Each 64 Kbit/s channel can be configured to carry voice or data traffic. Most telephone companies allow customers to buy just some of these individual channels, a service called fractional T1. Typically, fractional T1 lines are sold in increments of 56 Kbps (the extra 8 Kbps per channel are used for administration purposes). One of the most common uses of a T1 line is an Internet T1. This connection is used to provide Internet access to businesses of all sizes, assisting these businesses to meet the challenges of e-commerce. A T1 line can transmit large amounts of data at speeds of 256 Kbit/s, 512 Kbit/s, 1.544 Mbit/s, and sometimes 3 Mbit/s.

33.15.2 SONET MULTIPLEXING TERMINOLOGY AND OPTICAL SIGNALING HIERARCHY

Table 33.2 presents the optically transmitted SONET signal which is referred to as an optical carrier—level N (OC-N). The OC-N is essentially the optical equivalent of the STS-N; however, the STS-N terminology is used when referring to the SONET format. As shown in Figure 33.28, the STS-N consists of a synchronous multiplex of N STS-1s. The STS-1 has a bit rate of 51.84 Mb/s, therefore the STS-N and the OC-N have a bit rate of N times 51.84 Mb/s.

TABLE 33.2
SONET Terminology

Optical carrier-level N (OC-N)	Optical SONET signal at N times the basic rate of 51.84 Mb/s
Synchronous transport signal-level N (STS-N)	The electrical SONET signal, or SONET format, at N times the basic rate of 51.84 Mb/s, consists of a multiplex of N STS-1s
Synchronous transport signal-level 1 (STS-1)	Electrical SONET signal at 51.84 Mb/s, also used to refer to the SONET format
Synchronous payload envelope (SPE)	In SONET, all payloads are mapped into several types: the VT SPEs carry 1.5–Mb/s payloads, the STS-1 SPE carries 50 Mb/s payloads, and the STS-Nc SPE carries 150 Mb/s and higher payloads
Virtual tributary group (VTG)	A logical grouping of VTs prior to multiplexing into the STS-1 SPE
Virtual tributary (VT)	The unit into which the STS-1 SPE can be subdivided to carry payloads that require much less than 51.84 Mb/s

In SONET, all payloads are mapped into synchronous payload envelopes (SPE) at the edge of the SONET network, as shown in Figure 33.27. The core of the SONET network transports the envelopes. Carried within the STS-1 is the STS-1 SPE, which has a payload capacity of approximately 51.84 Mb/s. N STS-1s may be concatenated to carry an STS-Nc SPE, which has a payload capacity of N×51.84 Mb/s, as shown in Table 33.2.

In Figure 33.28, the STS-1 SPE can be subdivided into VTs. There are four sizes of VT: VT1.5, VT2, VT3, and VT6. The VT1.5 carries a VT1.5 SPE that has a payload capacity of approximately 1.5 Mb/s (Note: VT1.5 bit rate is equivalent to the bit rate of DS1 or T1). Similarly, there are VT2, VT3, and VT6 SPEs with capacities of about 2, 3, and 6 Mb/s, respectively. Within the STS-1 SPE, the VTs are grouped by type into VT Groups. Within a VT Group, there may be four VT1.5s, three VT2s, two VT3s, or one VT6. There are seven VT Groups within an STS-1 SPE, and different groups may contain different VT sizes within the same STS-1 SPE.

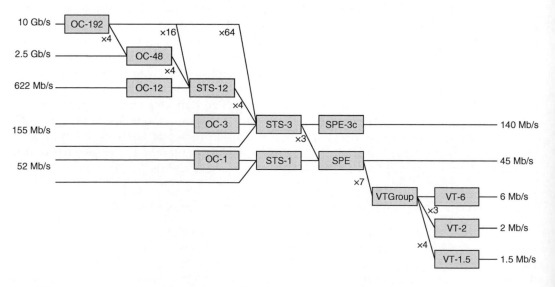

FIGURE 33.28 SONET multiplexing hierarchies.

TABLE 33.3
SDH Terminology

Synchronous transport module-level *N* (STM-N)	A synchronous multiplex of *N* STM-1s
Synchronous transport module-level 1 (STM-1)	The basic rate (155.52 Mb/s) and format of the SDH hierarchy; also refers to the optical signal
Administrative unit group (AUG)	A logical grouping of like AUs
Administrative unit (AU)	Similar to the TU; consists of a higher order VC and a payload pointer
Tributary unit group (TUG)	A logical grouping of like TUs
Tributary unit (TU)	A logical element consisting of a lower order VC and a payload pointer
Virtual container (VC)	The SDH structure into which all payloads are mapped

33.15.3 SDH Multiplexing Terminology and Optical Signaling Hierarchy

As discussed previously, a European standard, SDH, was developed parallel to the SONET standard. SDH uses a different terminology, as shown in Table 33.3 and Figure 33.29. Aside from the terminology, most other features of SONET can be extended to SDH.

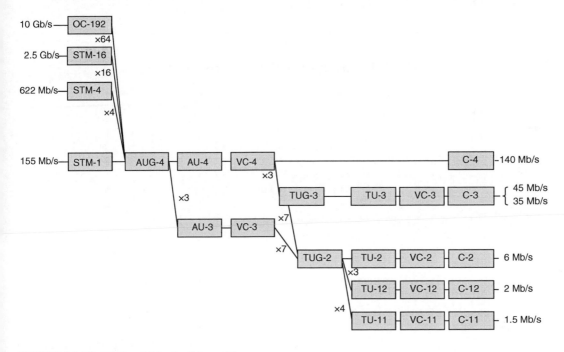

FIGURE 33.29 SDH multiplexing hierarchies.

33.16 SONET AND SDH TRANSMISSION RATES

Although the SONET multiplexing scheme would in theory allow any multiple of STS-1s, only certain rates are defined as standard transmission rates. Physical (photonic) interfaces are specified for these rates. Table 33.4 lists the most commonly supported SONET rates with their SDH equivalents.

TABLE 33.4
SONET/SDH Rates

SONET	SDH	Rate (Mb/s)
OC-1	STM-0	51.84
OC-3	STM-1	155.52
OC-12	STM-4	622.08
OC-48	STM-16	2488.32
OC-192	STM-64	9953.28
OC-768	STM-256	39813.12

33.17 NORTH AMERICAN OPTICAL AND DIGITAL SIGNAL DESIGNATION

Table 33.5 shows parts of the optical signal and carrier designations, as well as the complete digital signal and carrier designations which are used in North America. The T designation refers to the bit rate and the copper transmission system, and the DS designation refers to the bit format and framing. However, often the terms are used interchangeably. A single 64 kbps channel is called a DS0. The T1 rate of 1.544 Mbps for 24 channels of 64 kbps each is referred to as a DS1. The T3 rate of 44.736 Mbps for 28 T1s is referred to as a DS3.

TABLE 33.5
Optical and Digital Signal Designation in North American

Digital Signal Designation	Transmission Rate	Carrier Designation	Number of Channels
DS-0	64 Kbps	—	1
DS-1	1.544 Mbps	T1	24
DS-2	6.312 Mbps	T2	96
DS-3	44.736 Mbps	T3	672
DS-4	274.186 Mbps	T4	4032
OC-1	51.840 Mbps	STS-1	—
OC-3	155.52 Mbps	STS-3	—
OC-12	622.08 Mbps	STS-12	—

The optical carrier (OC) is the fundamental unit used in SONET. OC indicates an optical signal and the number following OC represents increments of 51.84 Mbps, the minimum transmission rate. The standard SONET frame format for 51.84 Mbps is called STS-1; the equivalent optical transmission rate is called OC-1. SONET standardizes higher transmission bit rates, OC-N, as OC-3, OC-12, OC-48, and OC-192, which are exact multiples of OC-1 (N×51.84 Mbps). SONET also standardizes the overhead formats and other details of optical transmission to implement mid-span links between different vendors' equipment.

33.18 SONET SYSTEMS

SONET network elements are combined to create systems that are classified based on the mechanism used to provide traffic protection and survivability. The two main types of protection provided in SONET are linear and ring.

FIGURE 33.30 Linear SONET system.

Linear systems. Linear systems transport traffic along a single route that may consist of one or more working fibres. A one plus one $(1+1)$ system has one protection fibre and one working fibre, as shown in Figure 33.30. The traffic is permanently bridged onto both fibres, so that the receiving end can autonomously choose the fibre that is operating better.

Ring systems. Ring systems transport traffic around a ring, allowing traffic to be added and dropped anywhere along the ring, as shown in Figure 33.31. Spare capacity is allocated around the ring so that when a failure occurs at any one point, the affected traffic can be restored using the spare capacity. In a unidirectional path switched ring (UPSR), traffic added to the ring is bridged onto both directions, such that the drop node can autonomously select the better path. In a bi-directional line switched ring (BLSR), the traffic affected by a failure at any point on the ring is rerouted the other way around the ring. A protocol operating around the ring coordinates this action.

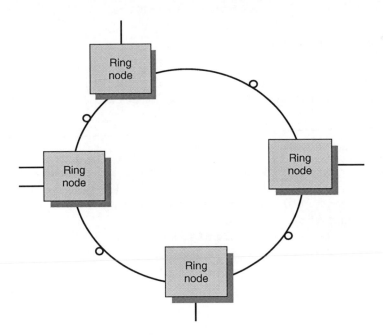

FIGURE 33.31 SONET ring system.

33.19 STS-1 FRAME STRUCTURE

The 810 bytes of an STS-1 frame are most conveniently represented as a matrix consisting of 9 rows and 90 columns, as shown in Figure 33.32. The intersection of a row and a column is one byte. The order of transmission is from left to right, and from top to bottom. The first three columns carry transport overhead, which consists of section overhead and line overhead. The remaining 87 columns carry the STS-1 synchronous payload envelope (SPE). The STS-1 SPE has its own frame consisting of 9 rows and 87 columns. Although the SPE fits within the STS-1 envelope capacity, the first byte of the SPE does not necessarily occupy the first byte position within the

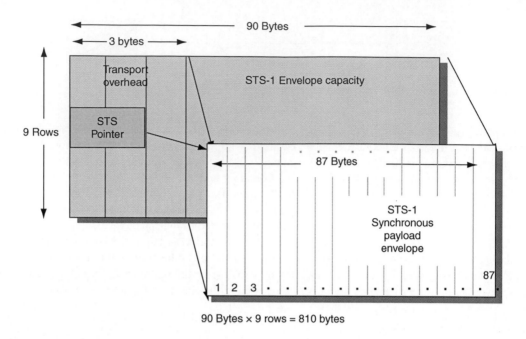

FIGURE 33.32 STS-1 frame structure.

envelope capacity. Thus the SPE is represented as being offset from the STS-1 frame. The STS payload pointer is a line overhead function that locates the start of the SPE within the envelope capacity.

33.19.1 SERIAL TRANSMISSION

A SONET signal is transmitted as a continuous serial bit stream, as shown in Figure 33.33. The bit stream is organized into bytes, and the bytes are grouped in 125 μs blocks called frames. At the basic STS-1 rate, there are 810 bytes per frame, which results in a basic rate of 51.84 Mb/s. All STS-N rates are N times the basic rate.

Currently, the common supported rates are OC-1, OC-3, OC-12, OC-48, OC-192, and OC-768.

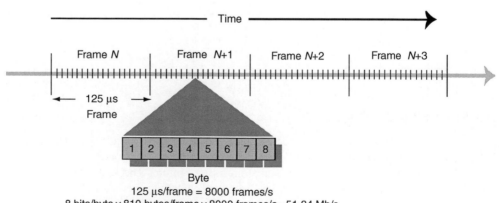

FIGURE 33.33 Serial transmission.

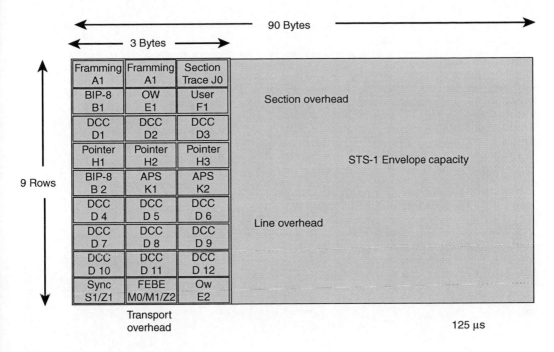

FIGURE 33.34 Transport overhead.

33.19.2 TRANSPORT OVERHEAD

Transport overhead is divided into two main types-section overhead and line overhead-as shown in Figure 33.34. Section overhead is allocated to the first three rows of the transport overhead. Subsequent slides describe each section overhead byte. Line overhead is allocated to the remaining six rows of the transport overhead. In a detailed treatment of SONET theory, each of the bytes and their functions would be described in detail; this information can be found in many of the SONET references.

33.19.3 STS-1 SPE PATH OVERHEAD

Path overhead consists of the first column of the STS synchronous payload envelope, as shown in Figure 33.35. STS path overhead allows for integrity verification of the end-to-end STS path. In general, it is not modified at line terminations.

33.19.4 MULTIPLEXING METHOD

When multiplexing STS-1s, the individual STS-1 frames must first be aligned. The STS payload pointer is used to accomplish alignment. The STS-1s are byte interleaved, as shown in Figure 33.36. In other words, a byte is transmitted from the first STS-1, then a byte from the second STS-1, and so on, until the first byte has been transmitted from every STS-1. Then the second byte is transmitted, and so on.

33.20 METRO AND LONG-HAUL OPTICAL NETWORKS

Metro optical networks can be thought of as consisting of core networks and access networks, as shown in Figure 33.37. SONET/SDH metro and long-haul core networks are typically configured

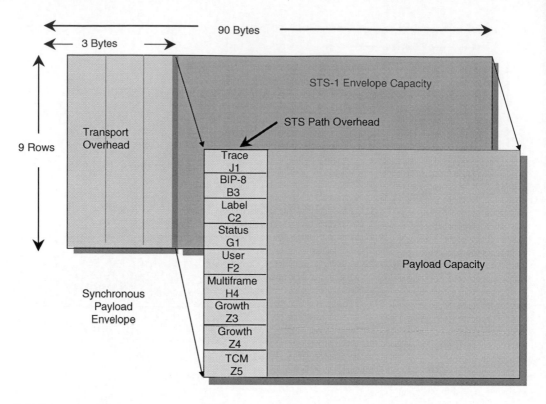

FIGURE 33.35 STS-1 SPE path overhead.

as point-to-point or ring connections that are spaced in tens to thousands of kilometres apart. The metro optical network consists of optical links between the end users and a central office (CO). The ring configurations shown in Figure 33.37 contain three or four nodes. Optical add/drop multiplexers provide the capability to add or drop multiple wavelengths to other locations or networks.

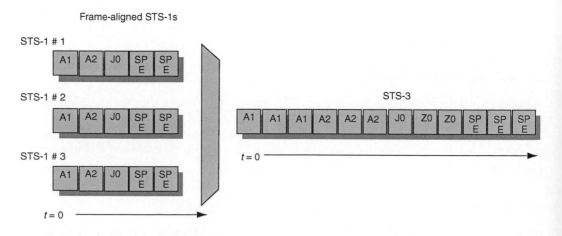

FIGURE 33.36 Multiplexing method.

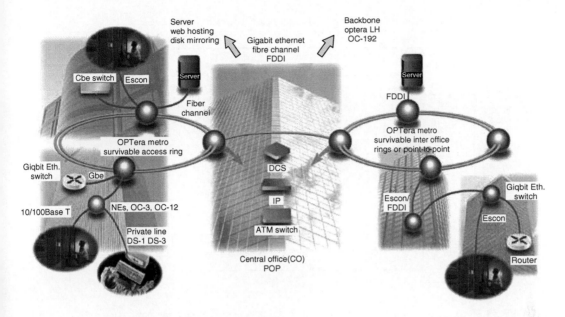

FIGURE 33.37 Metro and long-haul optical networks.

33.21 NETWORK CONFIGURATION

Network configuration is explained in the following text.

33.21.1 AUTOMATIC PROTECTION SWITCHING (APS)

Ring automatic protection switching (APS) provides an increased level of survivability for SONET/SDH networks by allowing as much traffic as possible to be restored, even in the event of a cable cut or node failure.

33.21.2 SONET/SDH RING CONFIGURATIONS

SONET/SDH has two main types of ring configurations: (1) a UPSR, and (2) a bidirectional line switched ring (BLSR).

A UPSR consists of two fibres on each span transmitting in opposite directions between adjacent nodes.

There are two types of bidirectional line switched ring (BLSR): two-fibre BLSR and four-fibre BLSR. In the two-fibre BLSR, there are two fibres transmitting in opposite directions between adjacent nodes. In the four-fibre BLSR, there are four fibres between adjacent nodes, with two fibres transmitting in one direction and other two fibres in the opposite direction. The four-fibre BLSR supports more traffic.

For a UPSR network, the selection of data path is made on a per path basis using the path layer integrity information. Thus, the ring is called path switched. In the case of a BLSR network, the decision to switch to the other path is made by the nodes adjacent to the failure using line layer integrity information. Thus, the ring is called line switched.

33.21.2.1 Two-Fibre UPSR Configuration

Figure 33.38 shows a two-fibre UPSR network. By convention, in a unidirectional ring, the normal working traffic travels clockwise around the ring on the primary (working) path. For example, the

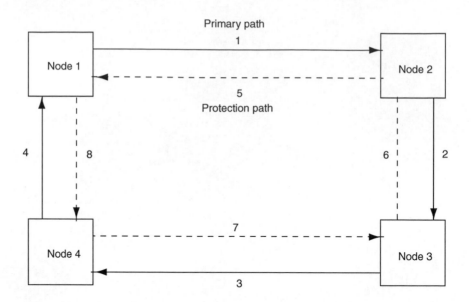

FIGURE 33.38 A 2-fibre unidirectional ring with a counter-rotating protection path.

connection from the node 1 to node 3 uses links 1 and 2, whereas the traffic from the node 3 to node 1 traverses links 3 and 4. In a UPSR ring, the counterclockwise path is used as an alternate route for protection against link or node failures. This protection path (links 5–8) is indicated by dashed lines. The signal from a transmitting node is dual-fed into both the primary and protection fibres. This establishes a designated protection path on which traffic flows counterclockwise; namely, from node 1 to node 3 via protection links 8 and 7.

Two-fibre UPSR configuration (Traffic flow). In Figure 33.39, two signals from node 1 arrive at their destination at node 3 from opposite directions. The receiver at node 3 selects the signal from the primary (working) path. However, if the quality of received signal on the primary path is poor, then it selects the signal from the protection path. In case of any failure in the node 2 equipment or on the primary path 2, node 3 will switch to the protection path via node 4 to receive the signal from node 1.

33.21.2.2 Four-Fibre BLSR Configuration

In Figure 33.40, two primary fibre loops are used for normal bidirectional communication while the other two secondary fibre loops are standby links for protection purposes. The two primary fibre loops have fibre segments labeled 1p through 8p, which provides for an arrangement of 1p, 2p, 3p, and 4p in one primary loop, and 8p, 7p, 6p, and 5p in the second primary loop. The two secondary fibre loops have fibre segments labeled 1s through 8s, grouping 1s, 2s, 3s, and 4s in one secondary loop, and 8s, 7s, 6s, and 5s in the second secondary loop.

Consider the connection from node 3 to node 1. The traffic from node 1 to node 3 flows in a clockwise direction along the links 1p and 2p. The traffic in the return path flows counterclockwise from node 3 to node 1, along links 6p and 5p. Thus, the information between node 1 and node 3 does not tie up any of the primary channel bandwidth in the other half of the ring.

Four-fibre BLSR reconfiguration (Failure 1). Consider the scenario shown in Figure 33.41, where a transmitter or receiver circuit card used on the primary ring fails (in either node 3 or 4, in this case). In this case the affected nodes detect a lose-of signal (LOS) condition and switch both primary fibres, connecting them to the secondary protection pair. The protection between these nodes (3 or 4, in this case) now becomes part of the primary bidirectional loop.

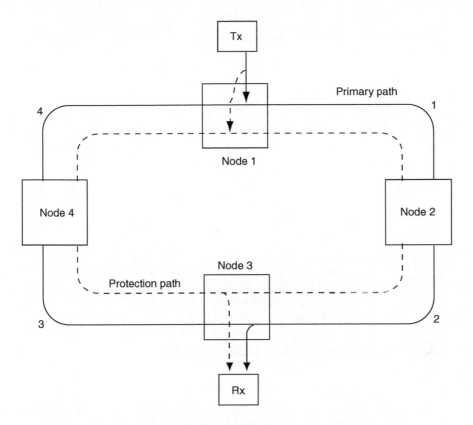

FIGURE 33.39 Flow of primary and protection traffic from Node 1 to Node 3.

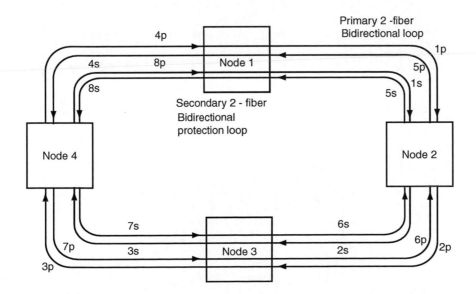

FIGURE 33.40 Four-fibre bi-directional line switched ring network.

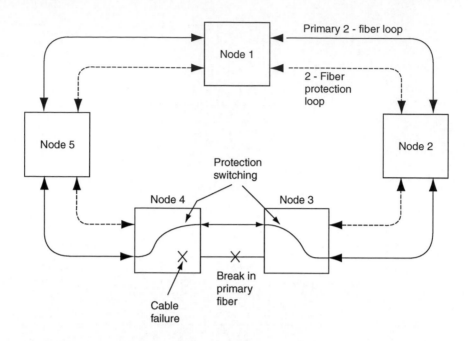

FIGURE 33.41 Reconfiguration under transceiver card or line failure.

Four-fibre BLSR reconfiguration (Failure 2). The exact same reconfiguration scenario as in failure 1 will occur when the primary fibre connecting two nodes (in this case, nodes 3 and 4) breaks, as in Figure 33.41. Note that in any case, the other links remain unaffected.

Four-fibre BLSR reconfiguration (Failure 3). In Figure 33.42, consider the scenario where an entire node fails (in this case node 3), or both the primary and the protection fibres in a given span are severed, which could happen if they are in the same cable duct between 2 nodes (in this case

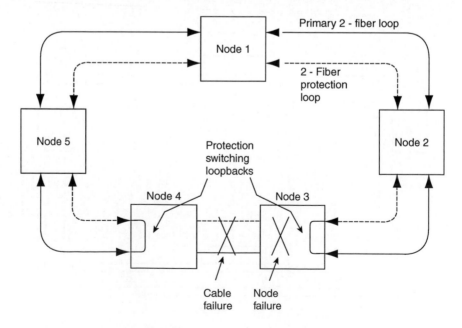

FIGURE 33.42 Reconfiguration under node or fibre cable failure.

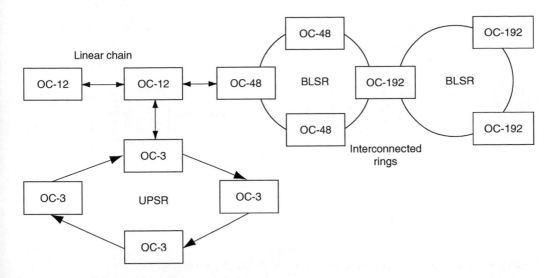

FIGURE 33.43 Generic configuration of large SONET network consisting of various types of interconnected systems.

nodes 3 and 4). In this scenario, the nodes on either side of the failed internal span will internally switch the primary path connection from their receivers and transmitters to the protection fibres, in order to loop traffic back to the previous node.

33.21.3 GENERIC SONET NETWORK

SONET/SDH architecture allows the interconnections and interoperability of a variety of network configurations, as shown in Figure 33.43. One can build point-to-point links, linear chains, UPSR, bidirectional link switched rings (BLSR), and interconnected rings. Each of the individual configurations has their own failure-recovery and protection mechanism, and SONET/SDH network management procedures.

33.21.4 SONET ADM

One of the important features in SONET/SDH architecture is the add/drop multiplexer (ADM), as shown in Figure 33.44. Various pieces of equipment are fully synchronized, and a byte-oriented

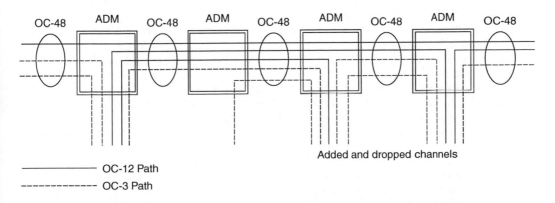

FIGURE 33.44 Functional concept of an Add/Drop Multiplexer (ADM) for SONET applications.

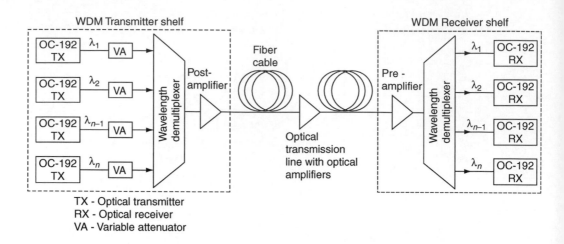

FIGURE 33.45 DWD deployment of n wavelengths in an OC-192 trunk ring.

multiplexer is used to add and drop sub-channels within an OC-N stream. Here, various OC-12s and OC-3s are multiplexed into OC-48 stream. Upon entering an ADM, these multiplexed sub-channels can be individually dropped by the ADM, and others can be added.

33.21.5 DENSE WDM DEPLOYMENT

SONET/SDH architectures can also be implemented with wavelengths. Figure 33.45 shows an example of dense wavelength division multiplexer (DWDM) deployment one an OC-192 trunk ring for n wavelengths (e.g., one could have $n = 16$). The different wavelength outputs from each OC-192 transmitter are first passed through the variable attenuator (VA) to equalize the out powers. These are then fed into a wavelength multiplexer, possibly amplified by a post optical amplifier, and sent out over the transmission fibre. Additional optical amplifiers might be located at intermediate points or at the receiver end.

FURTHER READING

Agrawal, G. P., *Fiber-optic Communication Systems*, 2nd ed., Wiley, New York, 1997.

Bean, J., Optical Wireless: Secure High-Capacity Bridging (Cover story). *Fiber Optic Technology*, pp. 10–13, U.S.A., January 2005.

Black, U. and Sharleen, W., *SONET and T1 Architectures for Digital Transport Networks*, Prentice Hall, Upper Saddle River, NJ, 1997.

Boyd, W. T., *Fiber Optics Communications, Experiments & Projects*, 1st ed., Howard W. Sams & Co., Washington, DC, 1987.

Cole, M., *Telecommunications*, Prentice Hall, Englewood Cliffs, NJ, 1999.

Derfler, F. J. Jr. and Les, F., *How Networks Work*, Millennium Edition, Que Corporation, 2000.

Derickson, D., *Fiber Optic Test and Measurement*, Prentice Hall PTR, Upper Saddle River, NJ, 1998.

Dutton, H. J. R., *Understanding Optical Communications*, IBM, Prentice Hall, Inc., Englewood Cliffs, NJ, 1998.

Goff, D. R. and Hansen, K. S., *Fiber Optic Reference Guide: A Practical Guide to the Technology*, 2nd ed., Butterworth-Heinemann, London, UK, 1999.

Goralski, W. J., *SONET*, 2nd ed., McGraw-Hill, New York, 2000.

Green, P. E., *Fiber Optic Networks*, Prentice Hall Pub Co., Englewood Cliffs, NJ, 1993.

Hecht, J., *Understanding Fiber Optics*, 3rd ed., Prentice Hall, Inc., Englewood Cliffs, NJ, 1999.

Hioki, W., *Telecommunications*, 3rd ed., Prentice Hall, Englewood, Chffs. NJ, 1998.

Hoss, R. J., *Fiber Optic Communications—Design Handbook*, Prentice Hall Pub. Co., Englewood Cliffs, NJ, 1990.

Kao, C. K., *Optical Fiber Systems: Technology, Design, and Applications*, McGraw-Hill, New York, 1982.

Keiser, G., *Optical Fiber Communications*, 3rd ed., McGraw-Hill Pub, Boston, 2000.

Keiser, G., *Optical Communications Essentials*, 1st ed., McGraw-Hill Pub, New York, 2003.

Kolimbiris, H., *Fiber Optics Communications*, Prentice Hall, Inc., Englewood Cliffs, NJ, 2004.

Nortel Networks, *Products, Services, and Solutions Change Notification, Nortel Equipment, Nortel Products*, Manual, Ottawa, ON, Canada, 2005.

Palais, J. C., *Fiber Optic Communications*, 4th ed., Prentice Hall, Inc., Englewood Cliffs, NJ, 1998.

Razavi, B., *Design of Integrated Circuits for Optical Communications*, McGraw-Hill, New York, 2003.

Senior, J. M., *Optical Fiber Communications: Principle and Practice*, 2nd ed., Prentice-Hall, Englewood Cliffs, NJ, 1986.

Shamir, J., *Optical Systems and Processes*, SPIE Optical Engineering Press, Bellingham, 1999.

Ungar, S., *Fiber Optics: Theory and Applications*, Wiley, New York, 1990.

Yeh, C., *Handbook of Fiber Optics: Theory and Applications*, Academic Press, San Diego, CA, 1990.

Yeh, C., *Applied Photonics*, Academic Press, San Diego, CA, 1994.

34 Fibre Optic Lighting

34.1 INTRODUCTION

This chapter will explain the basic principles of fibre optic lighting using fibre optic cables to transfer light from the source to applications. Fibre optic lighting is used to provide the light power required for residential, commercial, and office buildings, sport centres, theatres, retirement homes, swimming pools, backyards, walkways, stairways, show cases, warning signs, and advertisement panels. In medical applications, fibre optic lighting is important for delivering illumination to remote parts of the body and for carrying coherent images.

Fibre optic lighting transfers light from an electrical or solar light source through a bundle of fibres in a cable to the locations where the light is needed. In this way, light from the outside can be transferred to indoor spaces during the daytime. A fibre optic cable is considered a tool used to transfer light between two points. The use of fibre optic cable in lighting systems is one of the simple and easy ways to achieve green building conditions, help the environment, and reduce the cost of energy consumption in the building sector. This chapter will present multiple types of fibre optic lighting systems. Particular emphasis will be given to the study of the design of a lighting system and application. Also in this chapter, along with the theoretical presentation, three experimental cases will demonstrate the principles of the fibre optic lighting system.

34.2 LIGHT

Light waves have a wide range of wavelengths. Lighting applications use wavelengths in the visible region. Light frequently behaves as a particle, and at other times it behaves as a wave. Light makes things visible. Light transmits from a source to a receiver in the form of either pulses or waves. A light source can be an electrical bulb, a laser, or solar rays.

Fibre optic cables are used as a light carrier. The types of fibre optic cables are explained in the fibre optic cables principles. If the diameter of the fibre optic cable is large, then it is suitable for use in fibre optic lighting system designs.

34.3 ELECTRICAL ENERGY CONSUMPTION BY LIGHTING

With energy rates rising, high operating costs are a concern for building owners. Maintenance costs for items such as light bulbs, which require frequent replacement, must also be considered. Society consumes large amounts of electricity, much of which is used for indoor and outdoor lighting. In the average household, 18%–22% of total energy consumption is used for lighting, as illustrated in Figure 34.1.

One objective of saving energy is to demonstrate that buildings can be attractively and effectively lit while using only a fraction of normal electrical demand. Four lighting strategies can be employed to meet this objective: natural daylight, energy-efficient light fixtures, task lighting, and fibre optic lighting using solar rays.

Natural daylight is the most efficient source of building lighting; however, its full potential has yet to be demonstrated in buildings. Day lighting provides two very different challenges to designers: (1) perimeter spaces in buildings can suffer from over lighting and glare, and (2) interior spaces without windows are too dark for normal use. Solutions to these problems depend on the building design. Window size and placement can provide sufficient daylight for a building without increasing heating and cooling loads. Translucent fabric window coverings such as roller-blinds and horizontal blinds can diffuse light and reduce glare.

Rapid advances in lighting and appliance technologies have made significant energy reductions possible. Research on typical apartment buildings suggests that savings on lighting systems can average 75% through the use of fluorescent technologies. Even greater savings are possible using induction lamp technologies. Artificially-lit, energy efficient light fixtures are used in buildings to cut energy consumption. Most modern light fixtures have electronic dimmable ballasts in indirect/direct lighting fixtures. These lights use 35% less electricity than 40 W tubes with magnetic ballasts, while providing the same lighting level.

Electric lights are controlled by a modulating dimming system, in order to maintain desired light levels. Lights will dim on bright days, and brighten on dark days. Motion sensors and timers

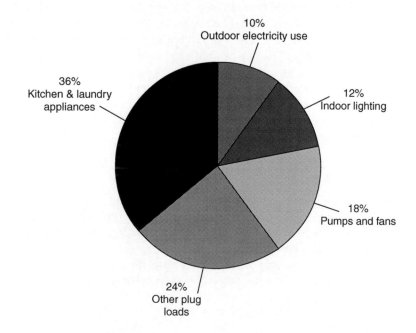

FIGURE 34.1 Distribution of electrical energy consumption in the advanced houses in Canada.

ensure that lights are on only when required. In addition, the lighting design emphasizes task lighting. Task lighting does not illuminate an entire room; it focuses on a small area, such as a desk top, where illumination is required. Compact fluorescents can be used for task and accent lighting while reserving halogen spotlights with parabolic reflectors for very small task lit areas. The use of automated controls, occupancy sensors, and energy management systems can produce additional savings. More efficient lighting systems also reduce building cooling requirements.

Solar technology utilizes free energy from the sun. The simplest form of this technology occurs when sunlight passing through windows heats a building and provides light. There are many passive solar heating and lighting designs that provide at least about one-third of the annual space heating and lighting for buildings. In a fibre optic lighting system, a fibre optic cable bundle is used to transmit solar rays for indoor lighting. The system can be located on the south-facing roof of the building, driveways, or empty spaces to facilitate the collection of solar rays. The system consists of solar collectors, fibre optic cable bundles, and diffusers. One design of a fibre optic light system is presented in detail in section 34.4 of this chapter.

Fibre optic lighting systems can also utilize a hybrid design, in order to provide light during both day and night. At night, an efficient electric light bulb can provide the required lighting. Even during cloudy days, the hybrid system provides lighting while electric light bulbs are used to supplement the daylight. The hybrid system eliminates the need for storage batteries and the resulting ventilation, maintenance, and disposal problems associated with some other solar photovoltaic technologies.

Each year, hundreds of dollars are spent to provide lighting in a typical home. A fibre optic lighting system would provide the same comfort for less energy, which is especially helpful as energy prices go up. It's good for the environment, too. Whenever combustion fuels are burned in a home or in generating stations to produce electricity, carbon dioxide, nitrogen oxides, and other emissions are released. In conclusion, by utilizing lighting and appliances equipment, and fibre optic lighting systems, homes will consume less energy and make less of an impact on the environment. Using a fibre optic lighting system is good for both budget and comfort.

34.4 LIGHT MEASUREMENT

34.4.1 LUMINOUS FLUX OR LIGHT OUTPUT

The study of the measurement of light is called photometry. Two important measurable quantities in photometry are the luminous flux (I), or the light output, and the illumination of a surface (E). Luminous flux measures the brightness of a light source, and it is defined as the total quantity of light emitted per second by a light source. The quantity of light emitted varies with the wavelength, reaching a maximum at the wavelength of 555 nm. The unit for luminous flux (I) is the candle or candela (cd). The early use of certain candles for standards of intensity led to the name of this unit. Currently, a platinum source is used at a specific temperature as the standard for comparison. Another unit, the lumen (lm), is often used to measure the flux of a light source; one candle produces 4π lumens. The lumen is defined as the luminous flux associated with a radiant flux of 1683 W at the wavelength of 555 nm in air. Lumens are equal to the quantity of light emitted by a lamp.

34.4.2 LUMINOUS EFFICACY

The luminous efficacy or efficiency of a light source is defined as the ratio of the light output (lm), to the energy input (watts). The effectiveness of a lighting system is measured in lumens per watt (lm/W), and can vary dramatically—from less than 10 to more than 200 lm/W.

34.4.3 Luminous Flux Density of Lighting Level

The luminous flux is a density expression that relates the amount of luminous radiative flux to a specific surface area. The luminous flux density is also known as the illuminance, or the quantity of light on a surface, or the lighting level. It is assumed that the surface being illuminated is perpendicular to the light source. The unit of the luminous flux density or lighting level is the lux (lx). The relation is given as:

$$\text{Illuminance flux} = \frac{\text{Luminous flux (lumens)}}{\text{Area}} \qquad (34.1)$$

where:

lux (lx) = lm/m^2,
footcandles (fc) = lm/ft.^2, and
1 fc = 10.76 lux.

In specific technical terms:

$$E_{ave} = \frac{\Phi_S}{A_S} \qquad (34.2)$$

where:

E_{ave} the average surface illuminance (lx or fc),
Φ_S the total luminous flux, or light (lumens), that falls onto the total surface area, and
A_S the total surface area (m^2 or ft.^2).

The lighting level is measured by a photometer, as shown in Figure 34.2. The minimum required lighting levels for different tasks are given in Table 34.1.

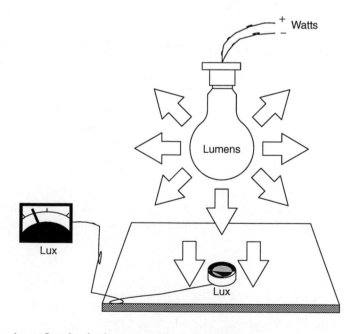

FIGURE 34.2 Luminous flux density is measured by a photometer.

TABLE 34.1
Lighting Levels by Building Area and Task

Building Area and Task	Lighting Level		Comments
	fc	lux	
Auditoriums	15	150	Include provision for higher levels
Banks – Tellers' stations	75	750	
Barber shops	75	750	
Bathrooms	30	300	
Building Entrances	5	50	
Cashiers	30	300	
Cleaning	15	150	
Conference Rooms	30	300	Plus task lighting
Corridors	15	150	
Dance Halls	7.5	75	
Drafting - High contrast	75	750	
Drafting - Low contrast	150	1500	
Elevators	15	150	
Exhibition Halls	15	150	Include provision for higher levels
Floodlighting - Bright surroundings	30	300	Less for light surfaces - more for dark
Floodlighting - Dark surroundings	15	150	Less for light surfaces - more for dark
Hospital - Examination rooms	75	750	High color rendition Variable (dimming
Hospital - Operation rooms	150	1500	or switching)
Kitchen	75	750	
Laundry	30	300	
Lobbies	30	300	
Office - General	75	750	
Parking Areas - Covered	5	50	
Parking Areas - Open	2	20	
Reading/Writing	75	750	
Restaurant - Dining	7.5	75	
Restaurant - Food display	75	750	
Stairways	15	150	
Stores - Sales area	75	750	
Street lighting - Highways	1.5	15	
Street lighting - Roadways	0.5	5	
Utility rooms	30	300	
Video display terminals	7.5	75	

34.5 ELECTRICAL LIGHTING SYSTEM

Figure 34.3 illustrates basic types of lights in an electrical lighting system, which consists of the following main elements:

1. A *lighting unit*, which serves as a light source. Light source can be any type of lamp(s).
2. A *ballast*, which is a device used with a gas discharge lamp and provides the lamp with the necessary starting and operating electrical conditions.
3. *End fixtures*, which produce the desired quality of light for the task. The end fixtures can be any design of light diffusers that diffuse light uniformly over the illuminated areas. The fixtures are available in different designs. Some fixtures have diffusers or reflectors, or they are covered with a lens to distribute light uniformly over the illuminated areas.
4. *Copper wire*, which carries electrical power from the main power supply to the fixture through the ballast (in the case of high intensity discharge lamp) and to the lamp(s).

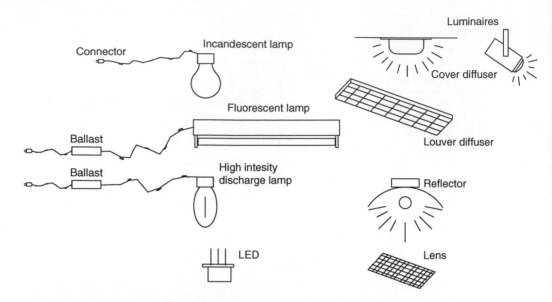

FIGURE 34.3 A basic design of an electrical lighting system.

34.6 FIBRE OPTIC LIGHTING SYSTEM

This technology is based on light being transferred from a source to several destination outlets through fibre optic cables. Since no heat or electricity is spent within the system it is more efficient. This technology is not limited in its application around water or in situations involving adverse environments, such as outdoor lighting.

Figure 34.4 illustrates a basic design of a fibre optic lighting system, which consists of the following main elements:

1. An *illuminator* (or engine), which serves as a light source (or provider or supplier). Light sources can be electric light bulbs, solar ray collectors/concentrators, and hybrids (a combination of electric light bulbs and solar collectors/concentrators).
2. A *fibre optic bundle* carries the light from the illuminator (or engine or solar collector/-concentrator) to the task. Fibre optic bundles can be classified as follows:
   a. Glass fibre bundles
   b. Small core diameter plastic fibre cables
   c. Large core diameter plastic fibre cables
   d. Hollow core fibre tubing
   e. Liquid core tubing
   f. Free air optics
3. The *end fixtures* produce the quality of light for the task. They can be any design of light diffusers that diffuse light uniformly over the illuminated areas. Unlike the standard electric lighting, fibre optic lighting enables changing of the light colour without changing the light fixture. This can be done by placing a small rotator that carries a wheel with coloured filters. The wheel rotates and changes the colour of the light exiting from the fixture. The user can control the light intensity and select the colour.

Standardization of fibre optic lighting components and modifications of the building codes will enable the usage of fibre optic lighting systems in both old and new buildings. These systems are one of many ways to reduce energy consumption in lighting. The systems are commercially

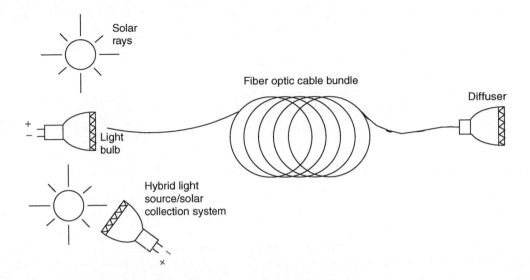

FIGURE 34.4 A basic design of a fibre optic lighting system.

available for common lighting applications. Currently, fibre optic lighting systems are widely used in a variety of applications in buildings around the world.

34.7 ADVANTAGES OF FIBRE OPTIC LIGHTING

Fibre optic lighting can significantly reduce electrical lighting requirements and energy costs in many residential, commercial, and industrial buildings as well as schools, libraries, and hospitals. Using fibre optic lighting in buildings may also reduce air-conditioning costs, because fibre optic lighting produces no heat.

Fibre optic lighting systems using solar rays capture the visible light while excluding the ultraviolet and infrared light, which are separated from the visible wavelengths by the focusing action of lenses and filters. Fibre optic lighting increases fire and electromagnetic safety by transporting light rather than electricity. This technology eliminates heat transfer and structural support hardware, simplifying light fixtures, installation, and architectural requirements. It produces more pleasing light with colour changes and special effect options. It centralizes maintenance and cleaning, and reduces maintenance and labour costs.

Some of the advantages of the fibre optic lighting systems are:

1. Their small size and light weight,
2. The elimination of hazardous area lighting concerns,
3. The ability to select a range of visible colour/specific wavelengths from the light source,
4. Their electrical safety,
5. Their low maintenance,
6. The maintenance cost reductions,
7. The elimination of running cost,
8. The simplicity of the system,
9. Their virtually unbreakable equipment
10. The elimination of vibration,
11. The elimination of noise,
12. The elimination of heat,
13. The elimination of ultraviolet rays,

14. Their flexibility and durability,
15. The elimination of consumables and replacements,
16. Their long-life running systems,
17. The variety of system designs and selections,
18. Their sound design for the environment.
19. The reduction of air-conditioning loads in buildings, and
20. An increase in human inspiration and comfort.

34.8 FIBRE OPTIC LIGHTING APPLICATIONS

Fibre optic lighting systems can be applied to the interior and exterior of commercial, retail, and residential buildings. New applications are being explored in landscapes, waterscapes, medical lighting instruments, and theme parks.

Electric lighting represents a large portion of the total energy consumption in buildings. When power plants generate electrical energy, they produce gas emissions. Such emissions proportionally increase with increased demands for lighting. Fibre optic lighting using a solar collector is one way to replace conventional electric lighting. Fibre optic lighting is a tool to be considered in renewable energy applications. It can be coupled with renewable energy sources like solar energy. Figure 34.5 shows a lighting system design using fibre optic bundles to guide sunlight from a solar concentrator inside a building. This system might be suitable for indoor, outdoor, and emergency lighting applications.

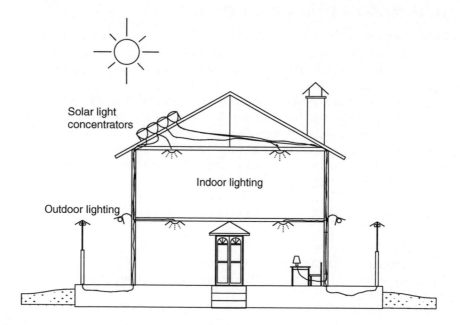

FIGURE 34.5 Proposed fibre optic lighting system.

34.9 EXPERIMENTAL WORK

This experiment is designed to launch light into a bundle of fibre optic cables. Figure 34.6 illustrates a schematic diagram for the experimental cases. The experiment can also be conducted using focusing and diverging lenses that are located at the input of the fibre optic cable bundle and at

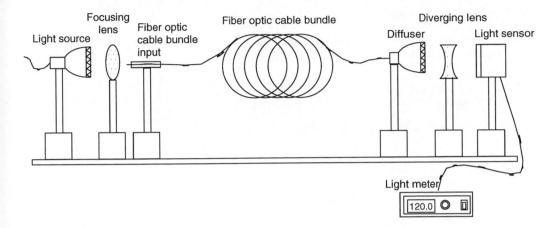

FIGURE 34.6 A fibre optic lighting system.

the diffuser side, respectively. The lenses are used to improve the collection and the diffuser efficiencies. Other types of lenses may also be used in this design. In the fibre optic lighting system experiment, the student will perform the following cases:

a. Fibre optic lighting with diffuser (light source, fibre optic bundle, and diffuser).
b. Fibre optic lighting with lens and diffuser (light source with focusing lens, fibre optic bundlc, and diffuser).
c. Fibre optic lighting with lenses and diffuser (light source with a focusing lens, fibre optic bundle, and diffuser with a diverging lens).

34.9.1 TECHNIQUE AND APPARATUS

Appendix A presents the details of the devices, components, tools, and parts.

1. 2×2 ft. optical breadboard
2. Light source
3. Fibre optic cable
4. Fibre cable holder/positioner assembly
5. Diffuser
6. Hardware assembly (clamps, posts, screw kits, screwdriver kits, etc.)
7. Fibre cable end preparation kit
8. Focusing lens
9. Diverging lens
10. Lens holder/positioner assembly
11. Light sensor
12. Light meter
13. Post holder/positioner assembly
14. Black/white card and cardholder
15. Ruler

34.9.2 PROCEDURE

Follow the laboratory procedures and instructions given by the professor and/or instructor.

34.9.3 SAFETY PROCEDURE

Follow all safety procedures and regulations regarding the use of optical components, electrical and optical devices, and optical cleaning chemicals.

34.9.4 APPARATUS SET-UP

34.9.4.1 Fibre Optic Lighting with Diffuser

1. Figure 34.7 (without lenses) shows the apparatus set-up.
2. Mount a light source on the breadboard.
3. Make a bundle of 10 fibre optic cables packed together, six feet long.
4. Prepare the fibre optic cable ends with a good cleave at each end, as described in fibre optic cable end preparation procedure in the Fibre Optic Chapter.
5. Pack the fibre cables together so that the ends meet at the same position.
6. Insert each end of the fibre optic cable into a 10 mm long metal tube with a diameter of 2.5 mm.
7. Dip a needle into clear epoxy and apply to each end of the fibre optic cable bundle filling the space in the metal tube and in between the fibre optic cables. Place the tube metal about 3 mm from the fibre optic cable bundle ends. Allow the epoxy to cure according the epoxy manufacturer's instructions.
8. Insert the fibre optic cable bundle into a heat shrink tube for mechanical protection of the fibre optic cables.
9. Hold the bundle end upright, and polish the ends, following the polishing procedure for a fibre optic connector described in the fibre optic connectors chapter.
10. Inspect the bundle ends under the microscope to see the condition of the polished end surfaces.
11. Turn on the light source.

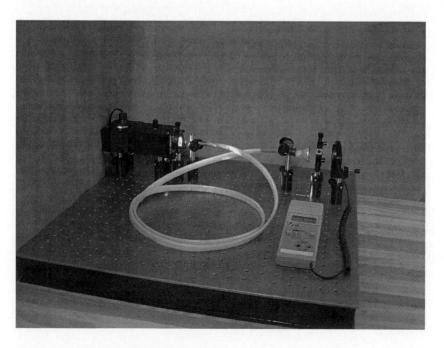

FIGURE 34.7 Fibre optic lighting apparatus set-ups.

12. Mount a fibre cable holder/positioner onto the breadboard so that the light from the light source passes over the centre hole.
13. Insert one end (input) of the bundle into the fibre cable holder/positioner.
14. Mount a second fibre cable holder/positioner on the other side of the breadboard facing the first fibre cable holder/positioner.
15. Insert the second end (output) of the bundle into the fibre cable holder/positioner.
16. Extend the end of the bundle so that the bundle input end is at the centre of the fibre cable holder/positioner. This is a very important step for obtaining maximum light from the light source to the output end.
17. Verify the alignment of your light launching arrangement by ensuring that the input end of the bundle remains at the centre of the light source.
18. Ensure that you have light output from the output end of the bundle. Point the output towards the centre of the light sensor head.
19. Place the light sensor head in front of the output end of the bundle at a distance (L). Remember to keep the same distance in following experimental cases.
20. Turn off the lights of the lab.
21. Measure the light intensity at the input and output ends. Record data in Table 34.2.
22. Turn on the lights of the lab.

34.9.4.2 Fibre Optic Lighting with Lens and Diffuser

In addition to what you have learned in Case (a), you can add a focusing lens at the input end of the bundle to focus the light from the light source into a small spot onto the bundle input end.

1. Figure 34.7 (without the diverging lens) shows the apparatus set-up.
2. Mount a lens holder/positioner to the breadboard at the input end of the bundle so that the light passes through the lens.
3. Place a focusing lens into the lens holder/positioner.
4. Verify the alignment of the input end of the bundle with the focusing lens. Ensure that the focal point of the focusing lens remains at the front of the input end of the bundle.
5. Turn off the lights of the lab.
6. Measure the light intensity at the input and output ends. Record date in Table 34.2.
7. Turn on the lights of the lab.

34.9.4.3 Fibre Optic Lighting with Lenses and Diffuser

In addition to what you have learned in Case (b), you can add a diverging lens at the output end of the bundle to diverge light from the output end of the bundle.

TABLE 34.2
Light Level Adjustment

Factor	Reduce Lighting Level by 30%	Increase Lighting Level by 30%
Reflectance of task background	Greater than 70%	Less than 70%
Speed of accuracy	Not important	Critical
Workers' age (average)	Under 40 years	Over 55 years

1. Figure 34.7 shows the apparatus set-up.
2. Mount a lens holder/positioner to the breadboard at the output end of the bundle so that the light passes over the centre of the diverging lens.
3. Place the diverging lens into the lens holder/positioner.
4. Verify the alignment of the output end of the bundle with the diverging lens.
5. Point the light output from the diverging lens towards the light sensor head. Keep the same distance (L) between the diverging lens and the light sensor head as in Case (a).
6. Turn off the lights of the lab.
7. Measure the light intensity at the input and output of the bundle ends. Record data in Table 34.3.
8. Turn on the lights of the lab.

34.9.5 Data Collection

34.9.5.1 Fibre Optic Lighting with Diffuser

1. Measure the light intensity at the input and output.
2. Record the collected data in Table 34.3.

TABLE 34.3
Light Intensity Measurements

Case	Light Intensity		Calculated	L
	At Input (unit)	At Output (unit)	Efficiency	(unit)
(a)				
(b)				
(c)				

34.9.5.2 Fibre Optic Lighting with Lens and Diffuser

1. Measure the light intensity at the input and output.
2. Record the collected data in Table 34.3.

34.9.5.3 Fibre Optic Lighting with Lenses and Diffuser

1. Measure the light intensity at the input and output.
2. Record the collected data in Table 34.3.

34.9.6 Calculations and Analysis

34.9.6.1 Fibre Optic Lighting with Diffuser

1. Calculate the efficiency of the fibre optic lighting system in Case (a).
2. Record the calculated result in Table 34.3.
3. Consider this case to be the basic case for comparison with other cases.

34.9.6.2 Fibre Optic Lighting with Lens and Diffuser

1. Calculate the efficiency of the fibre optic lighting system in Case (b).
2. Record the calculated result in Table 34.3.
3. Compare the efficiency of the fibre optic lighting system against Case (a).

34.9.6.3 Fibre Optic Lighting with Lenses and Diffuser

1. Calculate the efficiency of the fibre optic lighting system in Case (c).
2. Record the calculated result in Table 34.3.
3. Compare the efficiency of the fibre optic lighting system against Cases (a) and (b).

34.9.7 RESULTS AND DISCUSSIONS

34.9.7.1 Fibre Optic Lighting with Diffuser

1. Report the light intensity.
2. Consider this case to be the basic case for comparison with other cases.

34.9.7.2 Fibre Optic Lighting with Lens and Diffuser

1. Report the light intensity.
2. Compare the light intensity in this case with Case (a).
3. Discuss the effect of the presence of the focusing lens on the light intensity when the focusing lens is placed at the fibre cable bundle input.

34.9.7.3 Fibre Optic Lighting with Lenses and Diffuser

1. Report the light intensity.
2. Compare the light intensity with Cases (a) and (b).
3. Discuss the effect of the presence of the focusing and diverging lenses on the light intensity when the focusing and diverging lenses are placed at the fibre cable bundle input and output.

34.9.8 CONCLUSION

Summarize the important observations and findings obtained in this lab experiment.

34.9.9 SUGGESTIONS FOR FUTURE LAB WORK

List any suggestions for improvements in using different experimental equipment, procedures, and techniques for any future lab work. These suggestions should be theoretically justified and technically feasible.

34.10 LIST OF REFERENCES

List any references that were used in the report. Use one format in writing the references. Never mix reference formats in a report.

34.11 APPENDIX

List all of the materials and information that are too detailed to be included in the body of the report.

FURTHER READING

Al-Azzawi, A. R. and Casey, P., *Fiber Optics Principles and Practices*, Algonquin College, Algonquin Publishing Centre, Ont., Canada, 2002.
Beiser, A., *Physics*, 5th ed., Addison-Wesley, Reading, MA, 1991.
Blaze Photonics Limited, Product Summary of *Photonic Crystal Fibers Catalog, Blazephotonics*, UK, September 2003.
Derickson, D., *Fiber Optic Test and Measurement*, Prentice Hall PTR, Upper Saddle River, New Jersey, 1998.
Duffie, J. A. and Beckman, W. A., *Solar Energy Thermal Processes*, Wiley-Interscience Publication, New York, 1974.
Falk, D., Brill, D., and Stork, D., *Seeing the Light Optics in Nature, Photography, Color, Vision, and Holography*, Wiley, New York, 1986.
Hood, D. C. and Finkelstein, M. A., In *Sensitivity to Light Handbook of Perception and Human Performance. Sensory Processes and Perception*, Boff, K. R., Kaufman, L., and Thomas, J. P., Eds., Vol. 1, Wiley, Toronto, 1986.
IESNA ED-150.5A, *IESNA Lighting Education—Intermediate Level*, Illuminating Engineering Society of North America, U.S.A., 1993.
Jameson, D. and Hurvich, L. M., Theory of brightness and color contrast in human vision, *Vision Research Journal*, Vol. 4, no.1, pp. 135–154, May 1964.
Kao, C. K., *Optical Fiber Systems: Technology, Design, and Applications*, McGraw-Hill, New York, 1982.
Key, G., *Fiber Optics in Architectural Lighting Methods, Design, and Applications*, McGraw-Hill Book Company, New York, 2000.
Kreith, F. and Kreider, J. K., *Principles of Solar Engineering*, McGraw-Hill Book Company, New York, 1980.
Lerner, R. G. and Trigg, G. L., *Encyclopedia of Physics*, 2nd ed., VCH Publishers, Inc., New York, 1991.
Malacara, D., *Geometrical and Instrumental Optics*, Academic Press Co., Boston, MA, 1988.
McDaniels, D. K., *The Sun: Our Future Energy Source*, 2nd ed., Wiley, New York, 1984.
National Resources Canada, Advanced houses, *Testing New Ideas for Energy-Efficient Environmentally Responsible Homes*, CANMET's Buildings Group, National Resources Canada, Ottawa, ON, Canada,1993.
National Resources Canada, Household lighting, *Consumer's Guide to Buying and using Energy Efficient Lighting Products, Energy Guide*, National Resources Canada, Ottawa, ON, Canada, October 1993.
Nolan, P. J., *Fundamentals of College Physics*, Wm. C. Brown Publishers, Dubuque, IA, 1993.
Ocean Optics, Inc, *Product Catalog*, Ocean Optics, Inc., Florida, 2003.
Overheim, D. R. and Wagner, D. L., *Light and Color*, Wiley, New York, 1982.
Pedrotti, F. L. and Pedrotti, L. S., *Introduction to Optics*, 2nd ed., Prentice Hall, Inc., Englewood Cliffs, NJ, 1993.
Pritchard, D. C., *Environmental Physics: Lighting*, Longmans, Green, London, 1969.
Salah, B. E. A. and Teich, M. C., *Fundamentals of Photonics*, Wiley, London, 1991.
SEPA, *Solar Power*, Solar Electric Power Association, USA, 2002.
Serway, R. A., *Physics for Scientists and Engineers*, Saunders Golden Sunburst Series, London, 1990.
Warren, M. L., *Introduction to Physics*, W.H. Freeman and Company, San Francisco, CA, 1979.
Weisskopf, V. F., How light interacts with matter, *Scientific American*, Vol. 219, no. 3, pp. 60-71, U.S.A., September 1968.
Wilson, J. D. and Buffa, A. J., *College Physics*, 5th ed., Prentice Hall, Inc., Englewood, Chffs. NJ, 2000.

Section V

Testing

35 Fibre Optic Testing

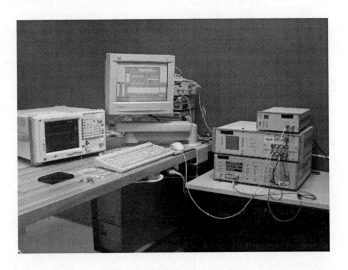

35.1 INTRODUCTION

This chapter describes the basic techniques employed to test optical fibre devices, using optical switches as an example. It presents common testing methods required by industry standards. In these tests, losses and other parameters, such as wavelength, will be measured. Students will become familiar with types of losses and customer requirements. Students will also learn how to use some optical testing instruments, such as an optical spectrum analyser (OSA). This chapter also presents three experimental cases to demonstrate the principles of testing a fibre optic device.

35.2 TESTING PHOTONICS COMPONENTS

The testing of optical fibre devices is based on two different measurements aspects of light output: power and frequency. Power measurements refer to light intensity, as measured by photosensitive devices, usually called photodetectors (e.g., photodiodes and photosensors.). The photodetectors generate an electrical signal proportional to the intensity (or power) of the incident light, thus allowing measurement of the light power. Photosensitive devices will not, however, react differently to different light frequencies. They will be more sensitive to certain frequencies while not being able to detect others. Because of this frequency dependence, power meters are calibrated for predetermined frequencies only. To ensure accurate readings, it is very important to verify the calibration of the power meter, before making any power measurements.

Photodetectors also depend on the polarization of the incident light. Different polarizations will generate different current intensities in the photosensitive devices, thus affecting the power readings during measurement. Polarization effects can be seen when measuring light power for a fibre optic device while moving the fibres or optical components. Movements will change the polarization of the light passing into the device, causing the power readings to fluctuate slightly. For this reason, fibres and/or optical components should not be moved during test measurements.

Polarization measurements require more complex calculations than power measurement alone. Many methods have been developed over the years to calculate the polarization dependent loss (PDL) of optical fibre devices. For each state of polarization, a device will absorb varying amounts of light. Thus, in order to determine the PDL of a device, the device needs to be illuminated with all possible polarization states while measuring the loss for each. Then the maximum difference in loss for all the polarization states can be calculated. This process is tedious and time-consuming. However, another calculation method, known as the Mueller matrix, allows the calculation of the PDL with enough accuracy using only four basic polarization states (linear 0, 90, 45° and circular polarizations). Using this matrix calculation method significantly reduces the time required to measure the PDL of any device. This algorithm is used in most PDL meters today to perform measurements.

The frequency of light refers to the colour of the light. Colours cannot be determined simply with the use of a photosensitive cell alone. Instead, comparison methods are used to determine the colour of light. A reference beam of light (a laser with a known and fixed single wavelength, such as a HeNe laser or another gas laser) is combined with the beam of light to be tested. From the interference of the two beams, the frequency of the incident beam can be determined. This method requires a more complex test circuit and more components (e.g., reference laser) than power measurements. This is why wavelength meters are more expensive and bigger than power meters. Furthermore, the alignment between the test beam and the reference beam is a critical factor in analysing the interference of the two beams. Thus, it is crucial to exercise caution when handling wavelength meters, to avoid misalignment of the internal optical components of the wavelength meter.

Calibration is always an issue in measurement practices for optical devices. The calibration of a power meter involves knowledge of the electrical properties of photodetectors. From this knowledge, calculations can establish the relationship between the current generated by the photodetector and the intensity of the test beam measured at a specific light wavelength. Lasers with adjustable light intensity are used to calibrate power meters with respect to a master power meter. The power measured by the power meter has to be equal to the power detected by the master power meter. Any deviation is recorded, and the values (constants) are adjusted in the calculation algorithms of the power meter. Power meters need to be calibrated on a regular basis, because the properties of the photodetectors are known to change over time. Power meters can be calibrated in a very short time using software.

Wavelength meters also need to be calibrated for power measurements. They too use photo-detectors, which measure the power of the interfering beam of light in order to determine the light colour. Improper power calibration could result in incorrect readings. Frequency measurements do not, however, only require power calibration in terms of seconds; they also require proper optical alignment. Thus, full calibration of this equipment requires opening the device box and realigning the internal optical components. Further calibration of a wavelength meter requires the skills of a qualified technician for aligning the optical parts.

As explained earlier, the optical fibre measurements can be divided into two types:

1. Optical power measurements (intensity I in Watts)
2. Optical frequency measurements (wavelength λ in metres)

35.3 OPTICAL POWER MEASUREMENTS (INTENSITY)

The following sections describe the parameters for optical power measurements.

35.3.1 Optical Power Measurement Units

Three common units are used in optical power measurements:

1. Linear (mW): The milliwatt is the standard unit of measurement. For example, typical communications lasers used are in the range of 1–10 mW.
2. Logarithmic (absolute) (dBm): In the absolute scale, 0 dBm = 1 mW.

$$P_{dBm} = 10 \times \log_{10}(P_{mW}) \tag{35.1}$$

3. Logarithmic (relative) (dB): Indicates a change in power level independent of the absolute power at the input,

$$L_{dB} = P_{in}(dBm) - P_{out}(dBm) \tag{35.2}$$

35.3.2 Optical Power Loss Measurements

The following sections describe common optical power loss measurements.

35.3.2.1 Insertion Loss

Insehrtion loss (IL) occurs in all optical fibre devices, such as fibre optic cables, optical passband filters, and optical switches, which are discussed in more detail throughout the text.

Consider an optical switch as an example. An optical switch redirects the optical signal in one or another direction or path, with low attenuation (loss) of the light. Thus, a very important parameter to test for a switch is the IL. Internal components of a switch (such as lenses, mirrors, or prisms) will always attenuate the light intensity, because of either absorption, diffraction, diffuse reflection, or scattering. Thus, the power transmitted out of the switch is less than the power incident on the switch. Thus, IL is never zero. For best switch performance, the IL should be as low as possible for all the paths of the switch.

A common procedure is used in testing optical fibre devices for IL. Using a single-wavelength laser source, the IL can be measured with a power meter calibrated for that wavelength. All optical testing procedures described below are based on this IL test. The procedure remains more or less the same; only the test parameters will change.

The intensity of light at one of the output ports, specified in dB, is attenuated, relative to the input signal power. The general light-loss formula is used to calculate the insertion loss, as given in the following equation:

$$IL(dB) = -10 \log_{10} \frac{P_{out}}{P_{in}} \tag{35.3}$$

A perfect optical device would have no internal losses and would transmit 100% of the incident light ($P_{in} = P_{out}$). In other words, IL = 0 dB in an ideal device.

35.3.2.2 Crosstalk

The purpose of an optical switch, as mentioned above, is to redirect the light in one, and only one, direction (channel). Ideally, when activating a switch, all the incoming light will go through one specific output port (active port), and no light at all will go through any other output port (inactive port). Therefore, in addition to measuring the IL for the active path, the loss for all the inactive paths must be measured, in order to discover if light leaks to other channels. In theory, all light is transmitted through the active channel, and absolutely no light goes through the inactive channels. In practice, however, this is not always the case. Diffuse reflection, diffraction or backreflection of the incident light on an internal component of a switch will redirect the light elsewhere inside the switch. This light can escape from the package by any other input or output port of the switch, depending on where the light is headed. This leakage of light into another channel can be measured in a manner similar to that used for the IL; the IL for the inactive channel (through which no light should be traveling) can be measured. The same procedure as used for the IL test is employed here, except that the optical path chosen is not the active path. The leak towards a wrong or inactive output port is calculated with the same formula used to calculate the IL of the active port, but P_{out} is measured at an inactive output port.

Crosstalk (XT) is the leakage from the active channel into the inactive channel. Crosstalk between channels (ports) occurs in optical devices, such as filters, wavelength separators, and switches. Crosstalk is critically detrimental in wavelength division multiplexer systems. When signals from one channel arrive in another undesired channel, they become noise in the other channel. This can have serious effects on the signal-to-noise ratio, and, hence, on the error rate of the communication system.

A perfect device would have no crosstalk (P_{out} of inactive channel $= 0$ mW). In other words, $XT = -\infty$ in a perfect device.

35.3.2.3 Polarization Dependent Loss

The PDL is another important parameter for optical fibre devices, such as an optical switch. An optical switch should not affect or distort the optical signal, only re-route it. Therefore, the PDL should be as low as possible for the active path. The loss in the switch should not vary when the polarization of the incident light is changed. Any change in loss due to polarization is called PDL.

PDL can be measured in many ways. One way is to illuminate the device under test (DUT) with laser light in which the polarization state is changing. Then, the loss for all possible states of polarization can be measured. However, as mentioned earlier, this method consumes too much time and too many resources to be cost-effective for companies to use routinely.

In order to measure the PDL property of a device, one of the most popular techniques used in manufacturing is the Mueller method. Mueller discovered that, by using the four basic light polarization states of the light (linear 0, 45, 90°, and circular polarization), the PDL can be calculated with great accuracy in a few seconds. Quite complex, the calculations take the form of the Mueller matrix. Most PDL meters use this matrix to calculate the PDL of a device.

PDL meters have a built-in laser source of known polarization, combined with a power detector, and also include polarizing filters, which can be moved in and out of the laser path to vary the polarization.

Because the components included in the PDL meter are expensive and their alignment needs to be precise, the cost of a PDL meter is much higher than that of a power meter. Also, because the PDL meter has to adjust the polarization of the light to take four measurements and then perform a calculation, it takes approximately 1 s before the PDL value is refreshed on the display. Furthermore, any change in the light path can modify the PDL value, so it is very important that the components and fibres do not move during the measurement. If the fibre optic cables of the device are moving during testing, the PDL values will change on the display unit and will not be as accurate.

There are three other polarization effects: polarization dependent gain (PDG) in optical amplifiers; polarization mode dispersion in optical fibres; and polarization dependent modulation in electro-optic modulators. These polarization effects occur in almost all other optical fibre devices transmitting polarized light. Such devices are polarizing beam-splitter devices and erbium doped fibre amplifiers.

35.3.2.4 Return Loss or Backreflection

The return loss (RL) of the DUT should be as high as possible, depending on the definition of loss. This is because having a large amount of incident light reflecting back into the system is not desired. Light reflected back towards the light source can damage the source, especially when laser light is employed.

A small portion of incident light will always be reflected due to a change in the refractive index between two adjacent surfaces. This reflection can be reduced when an antireflective coating is applied to one of the surfaces, or by using angled surfaces along the path of light. It is difficult, however, to eliminate any undesired reflection completely. Furthermore, a microbend in a fibre optic cable, glue joint, or anywhere else along the path of light, will create backreflection.

In order to measure the backreflection, the path of light must be terminated. A situation must be created in which all the transmitted light in the device will dissipate or diffract in the fibre and escape the device. To achieve this situation, the fibre needs to be coiled into fibre loops of a very small diameter (usually around 1–2 cm). This is the only situation in which it is acceptable to coil the fibres in very small loops. Careful coiling of the fibres is important so as not to coil part of the fibre onto another fibre loop. This coiling could crush the fibre underneath and create damage.

35.3.2.5 Temperature Dependent Loss

In most cases, an optical device, such as an optical switch, is used inside a module containing many other devices. These can include optical devices (e.g., couplers and filters) and electrical devices (e.g., switch controller, power supply, and fan). Because of the electronic components creating heat inside the module, the temperature can rise and/or fluctuate, so that the switch will almost never operate at ambient temperature. Therefore, the temperature properties of the DUT need to be measured.

The temperature dependent losses (TDL) are measured through the use of a test called the thermal gradient stability test, which consists of cycling the temperature of the DUT and measuring the optical losses of the DUT over the temperature range.

Usually, the operating temperature range of an optical module is small, varying from -10 to $-40°C$, to less than $+100°C$. Under these small temperature changes, the properties of the materials composing the switch will only change slightly (on the order of microns), and the effect on the optical properties will be small. Such a small loss, usually less than 1 dB, is significant for the IL and PDL, but are not significant for XT or RL. This is why usually only the IL and sometimes PDL are measured over the temperature range for a switch.

However, only the steady state losses are measured to calculate the TDL; the transient losses are not considered. Thus, the measurements are performed when the temperature has been achieved and the losses stabilized. Then a simple subtraction (maximum loss value achieved over the temperature range minus the minimum loss value achieved over the temperature range) is calculated.

The parameters most affected by temperature are the IL and PDL, as mentioned above. The variation of IL over temperature is called TDL. However, the PDL difference over the temperature range is not measured; instead, the maximum PDL value achieved over the temperature range is given, and it is called Max PDL over temperature.

35.3.2.6 Wavelength Dependent Loss

Because of the properties of the antireflecting (AR) coating on the lens and other optical components of an optical fibre device, the IL can vary depending on the wavelength of the incident light. The variation of IL over a wavelength range is called the wavelength dependent loss (WDL). It is also sometimes called the wavelength flatness or transmission curve. The WDL is a direct consequence of any coating or physical properties of the materials composing the optical fibre device.

In an optical switch, the AR coating must be able to transmit all the wavelengths with the same efficiency and have a WDL as low as possible. Thus, the difference between the IL measured at 1310, 1480 or 1550 nm should be near zero.

35.3.2.7 Chromatic Dispersion

All forms of dispersion degrade a light wave signal, reducing the data-carrying capacity through pulse broadening. Chromatic dispersion results from a wavelength-dependent variation in propagation delay and is affected by materials and dimensions of the waveguide of an optical fibre device or the fibre. Chromatic dispersion measurements characterize the way in which the velocity of propagation changes with wavelength, while traveling down the length of the fibre or through the waveguide of optical components.

The concept of optical phase should be considered in a discussion of chromatic dispersion. A mathematical relationship exists between optical phase and chromatic dispersion or group delay. It is important to mention optical phase before any explanations of chromatic dispersion or group delay. Group delay is defined as the first derivative of optical phase with respect to optical frequency. Chromatic dispersion is the second derivative of optical phase with respect to optical frequency. These quantities are represented as follows:

$$\text{Group Delay} = \frac{\partial \varphi}{\partial f} \tag{35.4}$$

$$\text{Chromatic Dispersion} = \frac{\partial^2 \varphi}{\partial f^2} \tag{35.5}$$

where Optical phase φ is the measured modulation in degrees. Optical frequency f is measured in Hz or THz. For example, a typical communications wavelength is 1550 nm = 193.4 THz.

Both of these phenomena occur because all optical signals have a finite spectral width, and different spectral components propagate at different speeds along the length of the fibre. One cause of this velocity difference is that the index of refraction of the fibre core is slightly different for different wavelengths. This condition is called material dispersion; it is the dominant source of chromatic dispersion in single-mode fibres.

Another cause of dispersion is the wavelength dependence of the cross-sectional distribution of light within the fibre. Shorter wavelengths are more completely confined to the fibre core, while a larger portion of the optical power at longer wavelengths propagates in the cladding. Since the index of the core is greater than the index of the cladding, the wavelengths in the core travel slightly more slowly. Thus, this difference in spatial distribution causes a change in propagation velocity among the wavelengths. This phenomenon is known as waveguide dispersion, which is relatively small compared to material dispersion.

Chromatic dispersion can also cause bit errors in digital communications, distortion, and a higher noise in analogue communications. These outcomes can pose a serious problem in high-bit-rate systems, if it is dispersion is not measured accurately, and if some form of dispersion compensation is not employed.

35.4 OPTICAL FREQUENCY MEASUREMENTS

The relation between light wavelength λ and frequency f is given by the following equation:

$$\lambda = \frac{c}{nf} \tag{35.6}$$

where λ is the light wavelength in nm, c is the seed of light in a vacuum in m/s, f is the light frequency in Hz or THz, and n is the index of refraction of an optical material.

Optical measurements are typically calibrated to the light wavelength in a vacuum ($n = 1$). Therefore, when $n = 1$, Equation 35.3 can be rewritten as:

$$\lambda = \frac{c}{f} \tag{35.7}$$

35.5 TESTING OPTICAL FIBRE SWITCHES

The following are the most common tests on optical fibre switches required by industry standards and consumers. These tests simulate the conditions that may be present during the switch's lifetime operation in the field.

There are a few reasons why testing of the optical performance is not the only important factor in testing an optical device. Mechanical, electrical, and environmental tests of devices are as important as optical testing, before delivering the devices to the market.

35.5.1 MECHANICAL TESTS

Mechanical tests are carried out to test the durability of the moving mechanical parts inside optical devices, such as switches. Inside an optical switch, there usually is a moving part (either a moving mirror, moving prism, or moving lens) that is moved in and out of the optical path; this changes the direction of the optical path (e.g., if the mirror is out of the optical path, the light will pass straight through the device, unaffected, to one output port; if the mirror is placed in the optical path, the light will be reflected by the mirror into a different output port).

Like all other devices, optical components wear out with time. For example in an optical switch, the physical joint between the mirror or lens, and the moving arm holding them, are affected by mechanical stresses. These stresses are induced by movement, change in ambient temperature, and humidity level, and they are due to the different intrinsic physical properties (e.g., thermal coefficient, stiffness, and elasticity) of the epoxy, mirror, and moving arm. Of course, manufacturers try to choose an epoxy closely matching the physical properties of the parts, but in practice, there is always a slight difference. Because of this difference, a stress situation can occur in which temperature increases will cause the epoxy to expand more (or less) than the mirror. This expansion causes the mirror angle to change, and thus the beam direction will be deviated and no longer be optimized. Therefore, the loss properties of the optical device will change due to mechanical stress.

Another situation can arise if there are internal moving parts in an optical device. The physical joint between moving parts, for example, between a mirror and the arm to which it is attached, can become weak after many movement cycles. This weakness causes the mirror to move a little with each cycle and induces instability in the beam deviation, and this weakness will of course affect the optical properties of the device.

Still another factor that needs to be taken into consideration concerns the device's electrical properties. Although it is not obvious, when dealing with the mechanical properties of a device, electrical signals are often linked to the observed mechanical defects. As an example, take an actuator with a coil creating a magnetic field used to move an arm up and down, depending on the orientation of the electric signal. A mirror is glued to the tip of the moving arm. The mirror can be moved either in or out of the optical path of a beam of light, forming an electro-optical switch. If the amplitude of the electrical current changes, the magnetic field will change accordingly, and thus will make the arm move more quickly or more slowly. This will affect the speed at which the mirror moves in and out of the optical path. Such factors can seem insignificant, but in some applications where the speed of the transmission of a signal is important (and customers always want faster components), this variation in time can become a real problem.

35.5.2 Environmental Tests

Industry standards require that a representative number of samples of a product be subjected to a programme of environmental challenges including high and low temperature storage, temperature cycling, and humidity. An environmental testing procedure might consist of three to six stages of temperature and humidity tests on selected devices. The characteristics of a DUT must be measured before and after each test stage. In some cases, continuous or interval testing is conducted during each test stage.

Environmental test systems are usually integrated in an automated test system, which is intended for long-term reliability testing of optical components under environmental stress conditions. Such stress conditions are listed in Telcordia specifications GR-326-CORE, GR1209-CORE, and GR-1221-CORE. The Telcordia GR-xxx-CORE standards are quality standards used in the fibre optic industry, and they are more complete and restrictive than the ISO quality standards.

Devices under test are subjected to a range of environmental stress conditions in a test chamber, usually over a period of many weeks. Chamber conditions are recorded at specified time intervals during the environmental tests, and the required parameters for each DUT are measured. User selected parameters are calculated from these responses and recorded along with the time and environmental data.

Switches combined with the appropriate source and monitoring hardware along with software can create fully automated measurement systems. An optional polarization controller is installed when PDL measurements are required. A computer is used to set up the tests, control the measurements, and monitor the test conditions and results.

For measuring IL and PDL, the optical component environmental test system (OCETS) uses a combination of up to three internal Fabry Perot lasers, and a broadband source (BBS) with a filter or an external source. The light from any of these sources can be routed to either end of each DUT. The power meter measures either the insertion loss through the DUT, in either direction, or the back-reflection from either end. A polarization controller enables PDL to be measured. Second and subsequent tests can be added for additional sets of DUTs, up to the switch capacity limit, while the first test is running. In this way, an environmental chamber running a long-term test (on a first prototype, for example) could evaluate the performance of product improvements, by installing later devices in the chamber and configuring a second test to run with the same conditions and measurements.

35.5.3 Repeatability Test

If there are moving parts in the optical path, the optical signal may change slightly each time the parts move from one position to another. This change in optical signal, or loss variation at each cycle, is a measure of repeatability. The difference in performance between test cycles is due to the physical properties of the components in the optical device. Minimal difference between the cycles indicates good repeatability. For example, the repeatability test for an optical switch measures the maximum IL variation, for a given optical path, over number of cycles, at a given constant temperature. The results of repeatability tests are often displayed as a graph. This graph enables a quick view of the readings stability, and it allows determination of the occurrence of spikes or non-regular variations.

The repeatability test is usually very sensitive to external vibrations (such as someone walking near the test station or banging on the table) because such vibrations induce additional movement to the internal parts of the DUT. This test can also be affected by the orientation of the DUT (due to gravitational forces on the DUT's internal moving parts). It is also important to keep the temperature constant during this test, because the physical properties of the materials composing the DUT change with temperature. This induces additional measurement variations, which are not due to the moving parts themselves.

35.5.4 Speed Test

Moving parts do not move instantly; they need an amount of time to react to the signal that controls them in order to move. The time elapsed between the application of the control pulse and the moment when the optical signal achieves its steady state is called the switching time for optical switches (also often called speed time for switches).

For an optical switch, the switching time is usually defined as the time between the application of the activation pulse and the time when the optical signal reaches 90% of its steady state, at a given constant temperature. In order to measure the switching time, a very fast optical power meter (OPM) is required. The sampling speed of the OPM should be at least ten times faster than the switching speed to be measured. For example, if a switching time in the order of milliseconds is to be measured, the sampling speed of the OPM should be at least 0.1 ms faster.

The results of this test are often displayed as a graph to see if the steady state of the optical signal is flat and high enough, and if the dynamic range between light-on signal and light-off signal is good.

This test is very sensitive to fluctuations of the laser light source. For example, a variation in optical signal could be perceived as unwanted movement inside of a switch. Also, when the OPM is set in the fast-reading mode, it is very sensitive to the power level of the laser because the response time of the OPM detector varies with the power of light: the lower the power, the slower the response time. Poor OPM performance results in additional noise in the measurements. However, if the laser power is too high, the detector will saturate and give inaccurate readings. Thus, it is important to find the acceptable power range for optimum response time of the OPM detector for this test.

Temperature is also an important factor for this test, because the properties of the materials will change with temperature (electrical current moves faster at higher temperatures, for example). So the temperature should be kept constant during the speed test.

The orientation of the DUT is also a factor to be considered. If the moving parts of the DUT are heavy, then gravitational forces will not be negligible and will have an effect on the speed. Thus, it is important to perform some tests to determine if the orientation of the DUT will have an effect on the speed test measurements.

35.6 LIGHT WAVELENGTH MEASUREMENTS

Wavelength measurements provide various information about most optical fibre devices. In these measurements, optical properties, such as IL, are measured over a wide range of wavelengths. For example, WDL is the variation in insertion loss over a specified wavelength range shown in a spectral plot.

From the spectral plot (IL versus λ), many device parameters can be measured:

1. Flatness (WDL)
2. Passband width (bandwidth, BW)
3. Full width half maximum (FWHM) (-3 dB BW)
4. Cross-channel isolation to reduce crosstalk
5. Maximum/minimum loss in passband
6. Free spectral range (FSR)

Light wavelength measurements can be achieved using either a tunable laser source with fixed detector or a broadband light source with an OSA. In addition, advanced wavelength measurement systems conduct different measurements, including different types of losses.

35.7 DEVICE POWER HANDLING TESTS

There are two categories of power-handling failures that occur during the power-handling test of an optical fibre device:

1. Increasing the input optical power until a device failure occurs: this happens in an epoxy joint, when the epoxy ruptures and the surface coating burns: in this case, a graph of loss versus power shows the results for each DUT.
2. Applying a higher than normal input optical power over a prolonged period of time, until a device failure occurs: in this case, a graph of losses versus time for different input powers shows the results for each DUT.

35.8 TROUBLESHOOTING

The main sources of problems that can arise during any of the preceding tests are:

1. Unwanted reflections occur when the refractive index along the light path changes, for example, at a glass-air interface, dirty and/or damaged connectors, and terminated reflections from unused fibre ends. Reflection between two parallel optical surfaces can be prevented, by using non-perpendicular surface angles.
2. Unstable laser source occurs if laser temperature is fluctuating, light is reflected back into the laser source (which can be solved by installing an isolator device after the laser source). It also occurs because of dirty and/or damaged connectors, damaged jumpers, or a sharply bent or too tightly coiled cable. It can also occur if the power meter is in fast-sampling mode.
3. Inaccurate power measurements occur when the power meter needs calibration, or are caused by stray light (low power measurement can be affected by room light; therefore, connections must be properly shielded). It also occurs due to dirty and/or damaged connectors or detectors or poor switch repeatability. It can also occur because of the connection to the detector differs between the reference and measurement.

When a problem arises during testing and troubleshooting is required, one must first ensure that these above-mentioned problems are not present at the test station before calling the DUT a failure. Even if none of these situations occurs during testing, there are other sources of errors, intrinsic to any test system, that must be taken into account.

35.9 SOURCES OF ERROR DURING FIBRE OPTIC MEASUREMENTS

The following are the main categories for sources of error in a test system:

35.9.1 RESOLUTION

Resolution is defined as the size of the smallest increment or unit in which the instrument can read. The resolution of the system should be $\leq 1/10$ of the total tolerance width. The resolution of equipment is an intrinsic property of the equipment's internal components, which the person performing the test has no control over (unless, of course, that person decides to buy another piece of equipment with a better resolution). Consider a ruler: the smallest divisions are millimeters. The human eye can differentiate, without the use of a magnifier, details as small as one tenth of a millimeter. However, when measuring the length of a line, it would be difficult to say whether the measurement is 11.2 or 11.3 mm. Therefore, the resolution of the human-eye-and-ruler system is ± 0.5 mm. Another example is a digital meter: the resolution of the digital meter is $\pm$ the last digit; if the meter gives a reading of 0.67 dB, then the resolution of the measurement is ± 0.01 dB.

35.9.2 Accuracy

Accuracy is defined as the difference between the observed average for a series of measurements and the true value being measured. The true value is measured using a reference measurement system, a master power meter. A master power meter has been referenced and calibrated with the world's master power meter located at the National Institute of Standards and Technology,and used to calibrate all the master power meters in the world (not that many). This reference ensures that any power meter in any facility worldwide, when calibrated with a master meter, will give the same measurement result for a given device. The accuracy of a power meter therefore usually depends on how long ago the meter was calibrated. Likewise, when measuring wavelength, the reference is a master wavelength meter.

35.9.3 Stability (Drift)

Stability is defined as the change in measurements over time. Measurement readings will vary over time, because the equipment wears out with time, dust can accumulate, and temperature and humidity, along with other factors, can affect the readings. If a device remains connected to a power meter for a long time and measurements are repeatedly performed at predetermined time intervals, the reading will vary over time. Most of the time, the readings variation follows a cycle, such as a sinusoid curve or a series of abrupt changes (steps). This pattern can be caused by instability in, for example, electrical components and heating components. In fact, the instability is closely related to inaccuracy. Stability can be considered a measure of the variation in accuracy over time. The more time that passes, the greater the likelihood that a drift will occur in the readings. This is why periodic calibration has to be performed on any measurement equipment

35.9.4 Linearity

Linearity is defined as the difference in accuracy values over the expected operating range. Some electrical and optical components will not have linear response depending on the intensity of electric current or incident light. Linearity is the linear correlation of an input power or signal to the corresponding optical power output. A mathematical formula is used to calculate the linearity value of a device. The correlation between electric current and light power, for an optical diode, is not perfectly linear; the correlation graph will usually be a curve. The more the curve bends, the worse the linearity is. This means that for a power value near the middle of the measurement range of the apparatus, the readings will be closer to the real value than when the power value is close to one end of the curve. The linearity and the correlation coefficients are adjusted during calibration of the meter or equipment.

35.9.5 Repeatability Error

Repeatability error is defined as the consistency of a given system, making repeated measurements of the same part, using the same measurement instrument. It is a measure of the inherent variation in the system. Basically, to determine the repeatability error of a measurement system (a complex automated system or a simple power meter and an operator), a number of devices (between five and thirty, but usually around ten to fifteen) have to be selected and tested. The selected devices should have measured values covering the entire range of probable results. These devices should include those that have both passed and failed testing. All of the devices are tested, and the test results are recorded as the first set of results. The order of the device testing is changed. Then, a new test is performed, and the results are recorded. This procedure can be repeated a number of times (between two times and five times, usually three to five times), and each time the testing order of the devices is changed. After each test, the results are correlated to identify all of the devices throughout the tests. Once all the series of tests are performed, the results can be recorded in a table, wherein each row corresponds to a device and the columns contain the test results for each measurement performed for each device. From the difference in test results for each device, analysis can be performed to determine the repeatability error of the test system.

35.9.6 REPRODUCIBILITY

Reproducibility is defined as consistency of different operators measuring the same part using the same measurement instrument. To determine the reproducibility error of a measurement system, a number of devices (between five and thirty, usually around ten to fifteen) have to be selected and tested. The same devices used for the repeatability-error test can be used for this. Selected devices should have measured values covering the entire range of probable results, and the devices should include those that passed and failed testing. A number of people are then selected (between two and five) to perform the tests. The devices are identified with a code, and one person is asked to test all of the devices and record the test results. This is the first set of results. Then the ID codes of the devices are changed, and another person tests the devices again and records the results. The same procedure continues until all the selected persons have tested the devices. The results can be recorded in a table, wherein each row corresponds to a device, and the columns correspond to the test results obtained for each person. From the difference between person-to-person results for each device, analysis can be performed to determine the reproducibility error of the test system.

The following points should be considered, when examining the sources of measurement errors.

1. The errors stem mostly from the measurement system and not only from the persons performing the test. If the test system were perfect, it would always give exactly the same results regardless of test operator or the number of tests. But systems are never perfect, and there will always be variations in the test results or errors. It is possible to minimize error occurrence by using experienced persons. System error itself can be reduced with more frequent calibration and a more-stable or better-designed system.

2. The repeatability and reproducibility errors will almost always be the major source of error in any test system. It is important to characterize these two sources of error, in order to determine if the results obtained from the test system are reliable or not. The measurement tool is sometimes called gauge. The most common way to determine the repeatability and reproducibility error factors of a system is to perform a gauge capability study (GCS) of the system, also known as gauge repeatability and reproducibility study (Gauge R&R Study). This GCS study includes simultaneously performing repeatability and reproducibility studies.

3. If the gauge R&R results show a systematic error in the measurements (results are always lower/higher than the real value), then this can be reduced by a proper calibration and increasing the frequency of calibrations.

4. If the gauge R&R results show random errors in the measurement results, then it is better to add a guardband to the test results. For example, if the test specification for the IL value is less than 1.0 dB, and the system gives a random error of ± 0.05 dB, then the test must fail any device that has an IL higher than 0.95 dB.

5. The resolution, accuracy, stability, and linearity sources of error are quoted in technical equipment specifications. These values can be improved with the calibration of the equipment.

35.10 EXPERIMENTAL WORK

In this experiment, students will practise testing of optical fibre devices. Students will work with the following cases:

35.10.1 Testing A Fibre Optic Device Using an Optical Spectrum Analyser

Case One familiarizes students with the use of an OSA for testing optical components. Refer to the manufacturer's operation manual for the OSA in use. As its name implies, an OSA analyses the optical spectrum of a DUT. In other words, an OSA measures the optical power of the DUT relative to a band of frequencies (or wavelengths) selected by the user. Because an OSA is quite big, heavy, very expensive, and more complicated to use than a power meter, its use is usually restricted to R&D laboratories. It is also used for testing fibre optic devices, which could not be fully tested using a power meter, such as pass band filters, WDM devices, and BBNS.

To measure the wavelength of the light coupled to its input port, the OSA uses a free-space optics interferometer. An adjustable power meter measures the optical power of the light for each wavelength in the range specified by the user. The data is stored in memory and displayed as a graph along with the analysed parameters. The user can then see the power distribution of a device for a determined wavelength or frequency range.

Students will learn how to configure the proper settings of an OSA for testing optical devices, and familiarize themselves with its basic functions.

35.10.2 Testing Mechanical Properties of Fibre Optic Devices

This experiment familiarizes students with tests performed on fibre optic devices, which require mechanical and environmental testing in addition to optical testing. Mechanical testing is required for any device containing moving parts (e.g., switches, tuneable filters, and variable attenuators). Environmental testing is usually performed during the R&D stage of a new product, to ensure that the product will perform within normal specifications over time, over different stress conditions, and over variable temperature and humidity levels. The students will learn about the most common mechanical tests performed in industry for optical devices, as well as develop intuition in determining the required tests for new devices.

35.10.3 Testing A Fibre Optic Cable Using an Optical Spectrum Analyser

This experiment introduces students to measuring the losses in a multimode fibre cable. Students will also verify the test results with the technical specifications provided by the manufacturer. An OSA is used to obtain accurate measurements for the losses in the fibre optic cable. Students will also practise alignment, connection, and set-up of optical tests, as shown in Figure 35.1.

This experiment can be repeated using a fibre optic cable with a connector on each end, and employing a laser source with output connectors matching the cable connectors. In this case, the

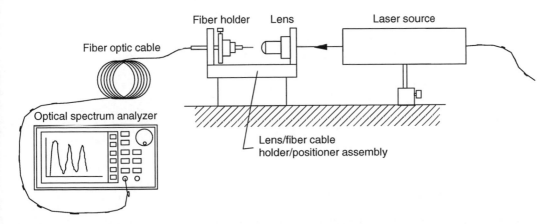

FIGURE 35.1 Fibre optic cable testing.

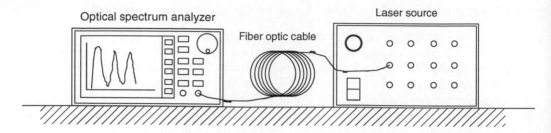

FIGURE 35.2 Fibre optic cable testing by direct connection.

cable can be connected directly to the laser source, as shown in Figure 35.2. This experimental set-up gives more precise measurements, because the direct connection between the laser source and the OSA reduces the losses caused by outside light interference, lens coupling, and air gap between the connectors.

35.10.4 TECHNIQUE AND APPARATUS

Appendix A presents the details of devices, components, tools, and parts.

1. C-band BBNS or white light source
2. Optical spectrum analyser (OSA)
3. Single mode fibre (SMF) cable with an angled connector FC/APC at one end, to connect to the BBNS
4. Single mode fibre (SMF) cable with a flat connector FC/PC at one end to connect to the OSA
5. Multi mode fibre (MMF) cable with an angled connector FC/APC at one end, to connect to the BBNS
6. Devices for testing:
   An SMF cable
   An MMF optic cable
   A pass band filter
   A WDM device
7. 2×2 ft. optical breadboard
8. HeNe laser light source and power supply
9. Laser mount assembly
10. Lens/fibre cable holder/positioner assembly
11. Multi-axis translation stage
12. Hardware assembly (e.g., clamps, posts, screw kits, screwdriver kit, positioners, and holder)
13. Optical assembly
14. Objective lens
15. Black/white card and cardholder

35.10.5 PROCEDURE

Follow the laboratory procedures and instructions given by the professor and/or instructor.

35.10.6 SAFETY PROCEDURE

Follow all safety procedures and regulations regarding the use of optical components, electrical and optical devices, light sources, measurement instruments, and optical cleaning chemicals.

35.10.7 APPARATUS SET-UP

35.10.7.1 Testing A Fibre Optic Device Using an Optical Spectrum Analyser

1. Connect the SMF cable with the FC/APC connector to the BBNS output port. Turn the BBNS on. The light source usually needs up to 30 min to warm-up and stabilize. In the meantime, gather all the material and equipment and prepare the set-up for the measurements. If the DUT does not have connectors on input and/or output ports, make a connection to attach the input port to the BBNS light source and the output port to the OSA. Gather all the necessary tools and prepare the fibre tips for connectors. Clean the tips of the fibres properly and set them aside. If the DUTs have connectors on their input/output ports, clean the connectors properly. Find the appropriate mating sleeve to connect the DUT's input port to the BBNS source fibre.

2. Turn on the OSA. The initialization and warm-up process takes from a few seconds up to a minute.

3. During this time, for non-connectorized DUTs, connect the SMF cable with the FC/PC connector to the input port of the OSA. Connect the BBNS source fibre to the measurement fibre of the OSA.

4. For connectorized DUTs, connect the BBNS source fibre directly to the OSA input port.

The following is the set-up procedure to prepare the OSA for optical measurements. Figure 35.3 shows the settings for the OSA.

5. Set up the scanning parameters on the OSA to take optical measurements as follows:
   Press the SETUP key.
   Press the RESOLN soft key.
   Turn the rotary knob, until a resolution of 20 nm is displayed.

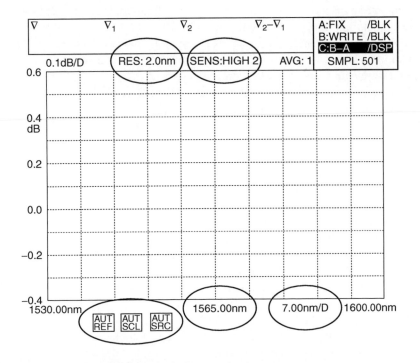

FIGURE 35.3 Test parameters displayed on the optical spectrum analyser.

Note: the keypad can also be used to enter the resolution. Type "20" and then press the nm/ENTER key.

6. Set the sensitivity required for the measurements as follows:

   Press the SENS soft key.

   Turn the rotary knob, until a sensitivity of HIGH1 is displayed.

   Note: The UP and DOWN arrows on the keypad can also be used to select the proper sensitivity.

To select the sampling size, press the SAMPLING soft key. Using the keypad, enter 501 and then press the nm/ENTER key.

7. Set the parameters of the vertical axis of the OSA graph as follows:

   Press the LEVEL key.

   Press the AUTO REF LEVEL soft key, until the AUT-REF icon becomes highlighted on the bottom of the OSA screen.

   Press the MORE 1/2 soft key to go to the next page of the LEVEL menu.

   Press the AUTO SUB SCALE soft key, until the icon AUT-SCL becomes highlighted on the bottom of the OSA screen.

   At the bottom of the display, both the AUT-REF and the AUT-SCL should be displayed. If it is not, press the soft keys AUTO REF LEVEL and AUTO SUB SCALE again.

8. Set up the horizontal axis of the graph to measure the proper wavelength range for the experiment as follows:

   Press the CENTER key.

   Press the CENTER soft key.

   Using the keypad, enter the wavelength 1545 nm, and then press the nm/ENTER key.

   Press the SPAN key, and then press the SPAN soft key.

   Using the keypad, enter the wavelength range (50 nm for the SMF DUT, 30 nm for the pass band filter, and 50 nm for the WDM DUT).

   Note: a reference reading must be taken any time the wavelength range, wavelength span, or centre wavelength is changed. The OSA only stores the data for the wavelengths selected and displayed on the screen. Any wavelength outside the range displayed on the screen was neither analysed nor stored in memory by the analyser.

Therefore, take a new set of reference readings every time the horizontal axis is changed, as follows:

   Press the nm/ENTER key.

   Press the PEAK SEARCH key.

   Press the MORE ½ soft key, to display the next page of options for the PEAK SEARCH menu.

   Press the AUTO SEARCH soft key. The AUT-SRC icon should appear on the bottom of the screen. If not, press the AUTO SEARCH soft key again.

   Press the SWEEP key, for measurement sweeping.

   Press the SINGLE soft key. This will initialize the new settings of the OSA for single sweep measurement.

   Note: The settings are displayed on the OSA screen.

   1. Take the reference reading, with the BBNS source connected to the OSA.

   2. Store the reference reading as follows:

      Press the TRACE key.

Activate Trace A, using the TRACE ABC soft key.

Press the WRITE A soft key.

Press the SWEEP key. Then press the REPEAT soft key, to get continuous measurement.

Verify that the source power is stable and does not change with every sweep. When t the source power is stable, press the SINGLE soft key.

When the single sweep is completed, a power distribution can be seen, which is similar to that shown in the BBNS user's manual.

Press the TRACE key again, and then press the FIX A soft key. Doing so will store the reference reading in memory.

Press the BLANK A soft key, to hide the reference stored.

Activate Trace B and press the WRITE B soft key. Press BLANK B.

Activate Trace C.

Press the soft key CALCULATE C.

Press the B–A (B/A) soft key.

Press the DISPLAY C soft key.

The following settings should now appear in the top right corner of the OSA screen: A:FIX/BLK, B:WRITE/BLK, C:B/A/DSP.

3. Press the SWEEP key and press the REPEAT soft key. The readings should be a flat line centered at 0 dB.

4. The peak insertion loss (due to noise in the signal or dirt on the connections and measurement error) will be indicated in the top left corner of the OSA display, under ▼PEAK. The wavelength at which this loss is located will also be displayed.

   Note: The reference reading is now complete. The OSA is ready to start testing the DUT.

5. Record the reference IL and wavelength. If the reference IL is greater than 0.2 dB, redo the reference set-up procedure, steps 2–10.

It is strongly recommended to print out all the graphs obtained during this experiment. Press the COPY button, and tear off the printed graph from the OSA printer.

To identify the graphs, write a descriptive text on the label field portion of the OSA screen, which will appear on the printout. To add/modify text on the label, press the DISPLAY key, then press the LABEL soft key. Using the rotary knob, the up/down and left/right arrows, and the keypad, write the desired text. When satisfied with the text, press the DONE soft key. Print the identified graph by pressing the COPY key. If there is not enough blank paper below the graph to tear the roll without loosing part of the graph, press the FEED key until there is enough clearance to tear off the chart paper.

35.10.7.1.1 Measuring the IL of a wavelength-independent DUT

The following is the testing procedure to measure the IL of a wavelength-independent DUT (such as a fibre optic cable, switch, circulator, and coupler).

1. Disconnect the reference fibre from the OSA and connect it to the input port of the DUT.
2. Connect the OSA to the DUT output port corresponding to the channel to be tested.
3. A line should appear on the OSA screen. The line should be mostly flat, since the DUT does not present any wavelength dependency. The lowest IL value will be indicated in the top left corner of the OSA screen, along with the corresponding wavelength.

Note: If the loss line shows ripples or waves instead of being mostly flat, these problems are probably caused by dirt on connections, damage in the fibre optic cables causing an etalon effect (two partially reflective surfaces parallel to each other in the optical path cause interference), or by a bad or damaged anti-reflecting coating of the internal components inside the DUT.

Using an OSA to measure the loss of a wavelength independent DUT does not add significant value to the test results that could be obtained using a regular and less expensive power meter.

35.10.7.1.2 Measuring the IL of a wavelength-dependent DUT

The following is the Testing procedure to measure the IL of a wavelength-dependent DUT.This test can use a passband filter, WDM device, or Fabry-Perot etalon as a DUT.

1. Disconnect the reference fibre from the OSA, and connect it to the input port of the DUT.
2. Connect the OSA to the DUT output port corresponding to the channel to be tested.
3. The optical spectrum for the DUT should appear on the OSA screen. The peak IL value will be indicated in the top left corner of the OSA screen, along with the corresponding wavelength. Record the peak IL value and associated wavelength. Figure 35.4 shows a chart printout from the OSA for the DUT.
4. Calculate the FWHM, or in other words the -3 dB bandwidth (-3 dB BW), as follows: Press the MARKER key.

   Set the horizontal line marker one at the top of the transmission peak.

   Place the horizontal line marker two to be 3 dB lower than the horizontal line marker one (the difference between the two markers will be indicated on the display).

   Place the vertical line markers three and four at the intersection of the line marker two and the trace.

   The difference between vertical line markers three and four is displayed on the OSA screen. This is the -3 dB BW, in nanometres.

   For the -20 dB BW, follow the same instructions but place the horizontal line marker two at 20 dB below the horizontal line marker one.

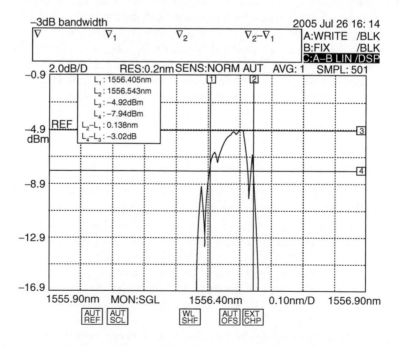

FIGURE 35.4 A chart printout from the optical spectrum analyser for the device under the test.

Note: It is evident that many valuable optical parameters can be determined from this test for wavelength-dependent DUT. The information given by the power loss versus wavelength graph gives more details about the behaviour of the device tested than what could be obtained from using only a power meter and a single-wavelength laser source. For example, from these graphs one can calculate the FSR of a filter device (distance, in nanometres, between consecutive transmission peaks), or calculate the flatness (difference in IL between the maximum and minimum transmission peaks over a specified wavelength range). In addition, the data illustrated on the OSA display panel can be saved and downloaded to a computer to be analysed in an Excel spreadsheet or similar application, and the graph obtained can be printed either via the OSA or a computer. Customers often like to have these graphs for the devices they purchase, because it gives them more detail about their new device than just data alone.

35.10.7.2 Testing Mechanical Properties of Fibre Optic Devices

The following is the procedure for testing mechanical properties of fibre optic devices

1. Connect the SMF cable with the FC/PC connector to the laser output port, and turn the laser on. The light source usually needs up to 30 min to warm-up and stabilize. In the meantime, gather all the material and equipment, and set up for the measurements.
2. If the DUTs do not have connectors on their input and/or output ports, a connection is needed to attach their input port to the light source and their output port to the power meter. Gather all the necessary tools and prepare the fibre tips for connections. Clean the tips of the fibres properly and set them aside.
3. If the DUTs have connectors on their input/output ports, clean the connectors properly. Find the appropriate mating sleeve to connect the DUT's input port to the laser source fibre.
4. Turn on the power meter; the initialization and warm-up process will take a few seconds.
5. For non-connectorized DUTs, connect the SMF cable with the FC/PC connector to the input port of the power meter. Connect the laser source fibre to the measurement fibre of the power meter. Take a reference reading.
6. For connectorized DUTs, connect the laser source fibre directly to the power meter input port. Take a reference reading.
7. Repeat the procedure explained in Section 35.14.1 to set up the OSA.

35.10.7.3 Testing A Fibre Optic Cable Using an Optical Spectrum Analyser

The following describes the process employed for fibre optic cable testing using an optical-spectrum analyser.

1. Set up the experimental apparatus, in the way illustrated in Figure 35.5.
2. Measure the light loss in the optical materials, using the light attenuation equation.
3. Prepare a fibre optic cable with a connector on one end. Make sure that the connector is the correct type, matching the input terminal connection of the OSA.
4. Prepare the other end of the cable, as explained in the fibre optic cable end preparation section of the fibre optic cables chapter. Cleave the end of the cable at a 90° angle. Clean the fibre end.
5. Mount the stripped end of the fibre into a copper fibre holder, with the tip of the fibre extending beyond the end of the copper fibre-holder. Now insert the copper fibre holder into the lens/fibre cable holder/positioner assembly.

FIGURE 35.5 Fibre optic cable testing apparatus set-up.

6. Using a HeNe laser source, project the light into the lens/fibre cable holder/positioner assembly, ensuring that the light is fully coupled into the fibre cable end.
7. Turn the laser on.
8. Turn off the laboratory lights before taking measurements.

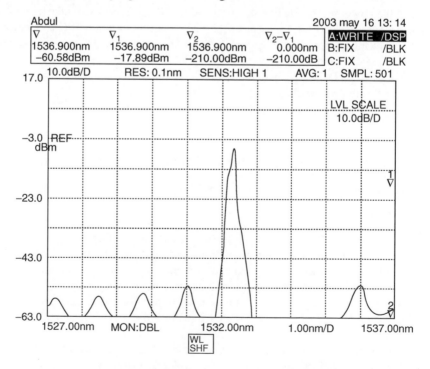

FIGURE 35.6 A sample of collected data printed on a chart from the optical spectrum analyser.

TABLE 35.1
Loss Versus Wavelength Measurements Using the OSA

Device Under Test DUT	Reference Max Loss		DUT Peak IL		- 3 dB BW	- 20 dB BW
	Wavelength (nm)	Loss (dB)	Wavelength (nm)	Loss (dB)	(nm)	(nm)
WDM device						
Fabry-Perot Etalon						
Pass band Filter						
Fiber Optic Cable						

9. Using the OSA, measure the laser power and wavelength of the laser source. Figure 35.6 shows a sample of data collected by the OSA printed on a paper chart. Ask the supervisor/instructor for the OSA operational instructions and referencing the source. Record the results for port 1 in Table 35.1.

10. Turn on the laboratory lights after completing the measurements.

35.10.8 DATA COLLECTION

35.10.8.1 Testing A Fibre Optic Device Using an Optical Spectrum Analyser

Fill out Table 35.1 with the test values of the DUT.

35.10.8.2 Testing Mechanical Properties of Fibre Optic Devices

Fill out Table 35.1 with the test values of the DUT.

35.10.8.3 Testing A Fibre Optic Cable Using an Optical Spectrum Analyser

Fill out Table 35.1 with the test values of the fibre optic cable.

35.10.9 CALCULATIONS AND ANALYSIS

35.10.9.1 Testing A Fibre Optic Device Using an Optical Spectrum Analyser

1. Print the graphs obtained and attach them to the report.
2. Record the maximum loss obtained during the reference measurements.
3. Present the peak IL and wavelength of the tested devices.
4. Calculate the BW as requested and record the results.

35.10.9.2 Testing Mechanical Properties of Fibre Optic Devices

1. Print the graphs obtained and attach them to the report.
2. Record the maximum loss obtained during the reference measurements.

3. Present the peak IL and wavelength of the tested devices.
4. Calculate the BW as requested and record the results.

35.10.9.3 Testing A Fibre Optic Cable Using an Optical Spectrum Analyser

1. Print the graph obtained and attach to report.
2. Record the maximum loss obtained during the reference measurements.
3. Present the peak IL and wavelength of the tested fibre optic cable.
4. Calculate the BW as requested and record the results.

35.10.10 Results and Discussion

35.10.10.1 Testing A Fibre Optic Device Using an Optical Spectrum Analyser

1. From the BBNS power distribution illustrated on the OSA screen (obtained before the reference measurements), discuss the importance of taking a reference reading prior to performing any optical measurements
2. Depending on the optical range available for the BBNS light source, noise in the reference readings is probably visible towards the edges of the graphs, especially for wider ranges. Compare the flatness of the reference line at the center and the edges of the wavelength range; explain any difference in flatness, called noise.
3. From the manipulations performed during this experiment, and knowing that power meters, OSAs, and other testing equipment are usually equipped with parallel, serial or GPIB ports, explain why, from an industrial point of view, some testing procedures should be automated instead of manual.

35.10.10.2 Testing Mechanical Testing Properties of Fibre Optic Devices

1. As explained in Section 35.17.1.
2. List all the items that should appear on a test report to be sent out to a customer for one particular device chosen among those tested during this lab. Justify the list of items (why they should appear on the report) and state the reason why some (if any) of the tested parameters should not appear on the customer's report.

35.10.10.3 Testing A Fibre Optic Cable Using an Optical Spectrum Analyser

1. As explained in Section 35.17.1.
2. List all the items that should appear on a test report to be sent out to a customer for the fibre optic cable tested. Justify the list of items (why they should appear on the report), and state the reason why some (if any) of the tested parameters should not appear on the customer's report.

35.10.11 CONCLUSION

Summarize the important observations and findings obtained in this lab experiment.

35.10.12 SUGGESTIONS FOR FUTURE LAB WORK

List any suggestions for improvements using different experimental equipment, procedures, and techniques for any future lab work. These suggestions should be theoretically justified and technically feasible.

35.11 LIST OF REFERENCES

List any references that were used in the report. Use one format in writing the references. Never mix reference formats in a report.

35.12 APPENDICES

List all of the materials and information that are too detailed to be included in the body of the report.

FURTHER READING

Camperi-Ginestet, C., Kim, Y. W., Wilkinson, S., Allen, M., and Jokerset, N. M., Micro-opto-mechanical devices and systems using epitaxial lift off, JPL, Proceedings of the Workshop on Microtechnologies and Application to Space Systems, pp. 305–316, (SEE N94-29767 08-31), Category Solid-State Physics, Georgia Inst. of Tech., Atlanta, U.S.A., June 1993.

Chee, J. K. and Liu, M., Polarization-dependent parametric and raman processes in a birefringent optical fiber, *IEEE J. Quantum Electron.*, 26, 541–549, 1990.

Dennis, Derickson, *Fiber Optic Test and Measurement*, Prentice Hall, Englewood Cliffs, NJ, 1998.

Duton, Harry J.R, *Understanding Optical Communications*, IBM, Prentice Hall, Inc, Englewood Cliffs, NJ, 1998.

Edmund Industrial Optics, *Optics and optical instruments catalog*, Edmund Industrial Optics, Barrington, NJ, 2004.

Gerd, Keiser, *Optical Fiber Communications*, 3rd ed., McGraw-Hill, UK, 2000.

Goff, David R., *Fiber Optic Reference Guide: A Practical Guide to the Technology*, 2nd ed., Butterworth-Heinemann, Sudbury, MA, 1999.

Golovchenko, E., Mamyshw, P. V., Pilipetskii, A. N., and Dianiv, E. M., Mutual influence of the parametric effects and stimulated raman scattering in optical fibers, *IEEE J. Quantum Electron.*, 26, 1815–1820, 1990.

Hagness, S. C., Rafizadeh, D., Ho, S. T., and Taflone, A., DTD Microcavity simulations: design and experimental realization of waveguide-coupled single-mode ring and whispering-gallery-mode disk resonators *IEEE J. Lightwave Technol.*, 15, 2157–2164, 1997.

Hibino, Kenichi, Error-compensating phase measuring algorithms in a fizeau Interferometer, *Opt. Rev.*, 6, 529–538, 1999.

Hibino, Yoshinori, Recent advances in high-density and large-scale AWG multi/demultiplexers with higher index-contrast silica-based PLCs, *IEEE J. Sel. Top. Quantum Electron.*, 8, 1090–1101, 2002.

Hine, T. J., Cook, M., and Rogers, G. T., An Illusion of relative motion dependent upon spatial frequency and orientation *Vis. Res.*, 35, 3093–3102, 1995.

Ho, M. C., Vesaka, K., Marhic, M., Akasaka, Y., and Kazovsky, L. G., 200-nm-bandwidth fiber optical amplifier combining parametric and raman gain *J. Lightwave Technol.*, 19, 977–981, 2001.

Horng, H. E., Chich, J. J., Chao, Y. H., Yang, S. Y., Hony, C. Y., and Yang, H. C., Designing Optical-fiber modulators by using,magnetic fluids *Opt. Lett.*, 30 (5), 543–545, 2005.

IGI Consulting, Inc., *Optical Amplifiers: Technology and Systems*, Global Information, Inc., Boston, MA, 1999.

Jackel, J., GoodMan, M. S., Bron, J. E., Tomlinson, W. B., Chang, G. K., Igbal, M. Z., and Song, G. H., Acousto-optic tunable filters (AOTF's) for multiwavelength optical cross-connects: crosstalk considerations *J. Lightwave Technol.*, 14, 1056–1066, 1996.

Javier, Alda, Laser and gaussian beam propagation and transformation, In *Encyclopedia of Optical Engineering*, Barry Johnson, R. et al., Eds., Marcel Dekker, New York, 2002.

JDS Uniphase Corporation, Opto-Mechanical Switches, SN Series, JDS Uniphase Corporation, San Jose, California, U.S.A., 2003.

Kao, Charles K., *Optical Fiber Systems: Technology, Design and Applications*, McGraw-Hill, New York, 1982.

Kuhn, K., *Laser Engineering*, Prentice Hall, Englewood Cliffs, NJ, 1998.

Laude, J. P., *DWDM Fundamental, Components, and Applications*, Artech House, Boston, 2002.

Li, X., Chem, J., Wu, G., and Ye, A., Digitally tunable optical filter based on DWDM thin film filters and semiconductor optical amplifiers *Opt. Express*, 13, 1346–1350, 2005.

Parry-Hill, Matthew J. and Michael W. Davidson, *Fiber optics testing*, National High Magnetic Field Laboratory, Florida State University, Tallashassee, Florida, U.S.A., September 2006.

Ralston, J. M. and Chang, R. K., Spontaneous-raman-scattering efficiency and stimulated scattering in silicon, *Phys. Rev.*, B2, 1858–1862, 1970.

Robillard, J. J. and Luna-Moreno, D., All-optical switching with fast response variable-index materials, *Opt. Eng.*, 42, 3575–3578, 2003.

Salah, B. E. A. and Teich, M. C., *Fundamentals of Photonics*, John Wiley and Sons, New York, 1991.

Sato, Y. and Aoyama, K., OTDR in optical transmission sdystems using Er-Doped fiber amplifiers containing optical circulators, *IEEE Photon. Technol. Lett.*, 3, 1001–1003, 1991.

Shen, Y. R. and Bloembergen, N., Theory of stimulated brillouin and Raman scattering, *Phys. Rev.*, A137, A1787–A1805, 1964.

Shen, L. P., Hvang, W.-P., Chum, G. K., and Jian, S. S., Design and optimization of photonic crystal fibers for broad-band dispersion compensation *IEEE Photon. Technol. Lett.*, 15, 540–542, 2003.

Sugimoto, N., Shintaku, T., Tate, A., Tervi, H., Shimoko200, M., Ishii, E., and Inone, Y., Waveguide polarization-independent optical circulator *IEEE Photon. Technol. Lett.*, 11, 355–357, 1999.

Yeh, C., *Handbook of fiber Optics: Theory and Applications*, Academic Press, San Diego, 1990.

Yeh, C., *Applied Photonics*, Academic Press, San Diego, 1994.

Zhang, Lin and Yang, Changxi, Polarization splitter based on photonic crystal fibers, *Opt. Express*, 11, 1015–1020, 2003.

Section VI

Safety

36 Photonics Laboratory Safety

36.1 INTRODUCTION

Our lives are filled with hazards created by electrical power supply, lasers, chemicals, and a diversity of equipment used in laboratories and classrooms. While studying science and engineering and performing experiments, students will learn to identify hazards and to protect themselves. Students will also learn to take care of their health and safety while working in laboratories. A safer and healthier learning and working environment should be created so that students have the opportunity to live safely and more healthily.

The following list of safety reminders is a brief compilation of generally accepted practices and should be adopted or modified to suit the unique aspects of each working environment, school policy, and local and/or set of Provincial and Federal codes. The intent of this chapter is to stimulate thinking about important safety considerations for students in laboratories.

36.2 ELECTRICAL SAFETY

The importance of electrical safety cannot be overstated. Electrical accidents can result in property damage, personal injury, and sometimes death. Ensuring electrical safety in laboratories and classrooms is important for students and staff. Students can learn to have a healthy respect for electricity and to spot potential electrical hazards anywhere. Respecting electricity does not mean that one should fear it; rather, one should just use it properly and wear personal protective equipment.

36.2.1 FUSES/CIRCUIT BREAKERS

The most common protection against property damage from circuit overloads (too much current) and overheating is the use of fuses and circuit breakers. All electrical circuits in laboratories are required to be protected by these means. When too much current flows in a circuit, the circuit

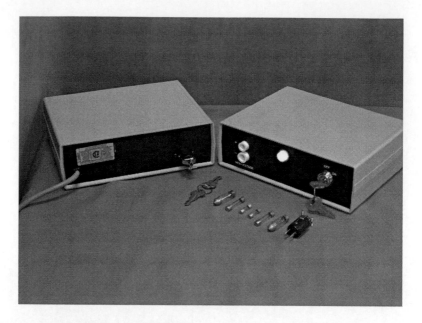

FIGURE 36.1 Types of fuses.

becomes hot and could melt the wire insulation, emit caustic fumes, and start a fire. An overload may also burn out and damage devices and instruments. Electronic equipment commonly has fuses to protect the components from overloads. A fuse is essentially a short strip of metal with a low melting point. When the current in a fused circuit exceeds the fuse rating, for example, 3, 5, 15, or 20 amps, the heat melts or vaporizes the fuse strip. The fuse blows, and the circuit is opened. Figure 36.1 shows types of fuses used in most laboratory electric instruments.

Fuses and circuit breakers should be the correct current rating for the circuit. If the correct rating is unknown, a certified electrician can identify and label it. A fuse should always be replaced with another of the same rating. Determine the reason why a fuse blew or a circuit breaker tripped, before replacing the fuse or resetting the breaker. Figure 36.1 also shows a few types of fuses and a power supply. The fuse will need to be plugged in on the back of the power supply. Plug in the electrical cord; turn on the key switch on the front panel; and turn the power supply on, as shown in Figure 36.1. After finishing with the power supply, remember to turn off the key and unplug the fuse.

A common problem is that the insulation may become worn on, for example, an extension cord, device wire, or instrument cord. If bare wires touch each other, or if a high-voltage or hot wire touches ground, this is called a short circuit, since the path of the circuit is effectively shortened. A low-resistance path to ground is created, causing a large current, which blows the protecting fuse.

Circuit breakers are more commonly used today instead of fuses in large equipment and houses, as shown in Figure 36.2. If the current in a circuit exceeds a certain value, the breaker is activated, and a magnetic relay (switch) breaks or opens the circuit. The circuit breaker switch can be reset or closed manually.

In either case, whether a circuit is opened when a fuse blows or when a circuit breaker trips, steps should be taken to remedy the cause. Remember, fuses and circuit breakers are safety devices. When fuses blow and open a circuit, they are indicating that the circuit is overloaded or shorted. Or they may be indicating the presence of another problem. In any case, a certified technician must investigate the source of the problem.

FIGURE 36.2 A circuit breaker panel.

36.2.2 Switches ON/OFF

Figure 36.3 shows samples of ON/OFF switches, which are used in computers, lighting systems, and instruments.

36.2.3 Plugs

Switches, fuses, and circuit breakers are always placed on the hot (high-voltage) side of the line, to interrupt power flow to the circuit element. Fuses and circuit breakers may not, however, always protect from electrical shock. To prevent shock, a grounding wire is used. The circuit is then completed (shorted) to ground, and the fuse in the circuit is blown. This is why many electrical

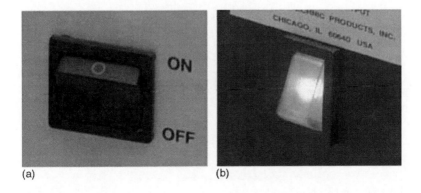

(a)   (b)

FIGURE 36.3 Examples of switches.

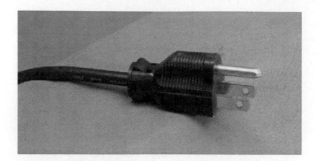

FIGURE 36.4 A three-prong plug.

tools and appliances have three-prong plugs, as shown in Figure 36.4. In the wall receptacle, this connection runs to ground.

When trying to plug in a two-prong plug that will not fit, do not use force. Instead, turn the plug over and try again. Figure 36.5 shows a two-prong plug. One of the prongs is bigger than the other, making the plug polarized. Polarizing in the electrical sense refers to a method of identification by which proper connections can be made. The original purpose of these types of plugs was to act as a safety feature. The small slit in the receptacle is the hot side, and the large slit is the neutral or ground side, if properly connected. The housing of an appliance could then be connected to the ground side all the time via a three-prong plug. A receptacle or appliance not wired (polarized) properly can be dangerous. The polarization is ensured with a dedicated third grounding wire as in a three-prong plug system, which is the accepted safety system. The original two-prong polarized plug system remains as a general backup safety system, provided it is wired properly.

Ensure the plug type fits the receptacle. Never remove the ground pin (the third prong) to make a three-prong plug fit into a two-conductor outlet; doing so could lead to an electrical shock. Never force a plug into an outlet if it does not fit. Plugs should fit securely into outlets. Avoid overloading electrical outlets with too many devices.

36.2.4 WALL OUTLETS

Figure 36.6 shows a wall outlet, which is used to connect computer and extension cords. Avoid using wall outlets with loose fitting plugs. They can overheat and lead to fire. Ask a certified technician to replace any missing or broken wall plates.

FIGURE 36.5 Two-prong plugs.

FIGURE 36.6 A wall outlet.

36.2.5 CORDS

Ensure the cords are in good condition. Check cords for cut, broken, or cracked insulation. Protect flexible cords and cables from physical damage. Ensure they are not placed in traffic areas. Cords should never be nailed or stapled to the wall, table, baseboard or to another object. Do not place cords under a device or computer; do not rest them under any object. Cords can create tripping hazards and may be damaged if walked upon. Allow slack in flexible cords to prevent tension on electrical terminals.

Check that extension power bars are not overloaded, as demonstrated in Figure 36.7. Figure 36.7(a) shows an overloaded extension power bar, while Figure 36.7(b) shows a bar not overloaded. Additionally, extension power bars should only be used on a temporary basis; they are not intended for use as permanent wiring. Ensure that the extension power bars have safety closures.

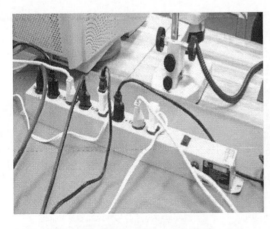
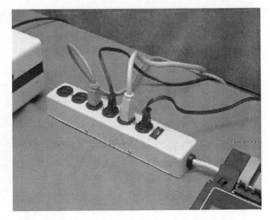

(a)   Extension power bar is overloaded.   (b)   Extension power bar is not overloaded.

FIGURE 36.7 An extension power bar.

FIGURE 36.8 (**See colour insert following page 512.**) Electrostatic discharge warning symbols and signs.

36.2.6 GROUND FAULT CIRCUIT INTERRUPTERS

Ground fault circuit interrupters (GFCIs) can help prevent electrocution. They should be used in any area where water and electricity may come into contact, especially near a sink or basin. Water and electricity do not mix; they create an electrical shock. When a GFCI senses current leakage in

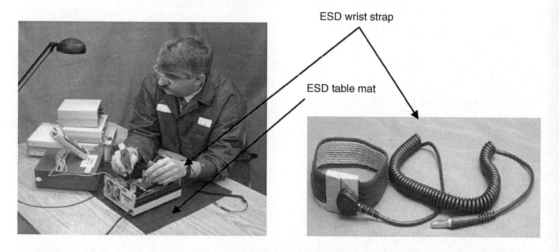

FIGURE 36.9 Working with ESD wrist strap and table mat.

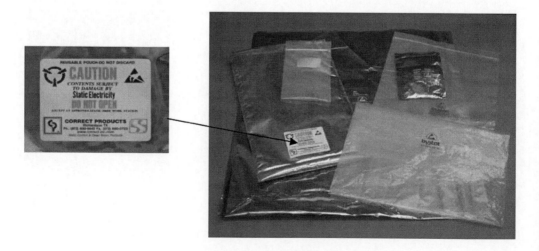

FIGURE 36.10 (**See colour insert following page 512.**) ESD bags.

an electrical circuit, it assumes that a ground fault has occurred. It then interrupts power quickly enough to help prevent serious injury due to electrical shock. GFCIs should be regularly tested according to the manufacturer's instructions to ensure they are working properly. Some benches are connected to true ground to be electrostatic discharge (ESD) compliant. This compliance is very important for devices and equipment that are very sensitive to ESD. Figure 36.8 shows ESD warning symbols and signs.

Figure 36.9 shows an ESD wrist strap and table mat used in handling an ESD sensitive device. The straps and mats should be connected to the true ground before handling a sensitive device. The strength of the charge on a human body is enough to destroy an ESD sensitive device. Each person should discharge his or her electrostatic charge before entering an environment sensitive to ESD. The discharge devices are usually located at the entrance of sensitive areas. An ESD heel strap is also available to wear when handling devices and walking in an environment sensitive to ESD.

Figure 36.10 shows ESD bags used to package devices sensitive to ESD. Available in various sizes, the bags have printed labels.

36.3 LIGHT SOURCES

The wattage rating should be checked for all bulbs in light fixtures, table lamps, and other light sources, to make sure they are the correct rating for the fixture. Bulbs must be replaced with another of the same wattage rating; bulbs' wattage rating must not be higher than recommended. If the correct wattage is unknown, check with the manufacturer of the fixture. Ensure that the bulbs are screwed in securely; loose bulbs may overheat. Different gas light sources (e.g., hydrogen, mercury, neon), as shown in Figure 36.11, are used in laboratories for light-loss measurements and for spectrometers and optical applications. These lamps operate at much higher temperatures than those of standard incandescent light bulbs. Never place a lamp where it could come in contact

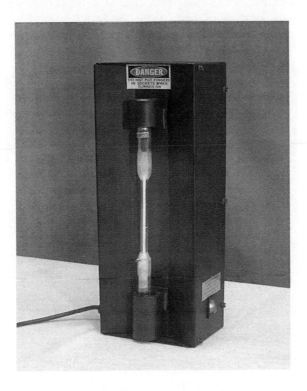

FIGURE 36.11 A mercury light source.

with any combustible materials or the skin. Be sure to turn the lamp off before leaving the laboratory for an extended period of time. Note that laser light sources have special provisions and, therefore, special precautions must be taken to operate them.

36.4 DEVICES AND EQUIPMENT

If a device or piece of equipment repeatedly blows a fuse or trips a circuit breaker, or if it has given you a shock, report the incident immediately to your supervisor/instructor. Unplug the device and remove it to have it repaired or replaced.

36.5 AUDIO–VISUAL AND COMPUTER PERIPHERALS

Audio–visual and computer equipment must be checked and kept in good working condition. Ask the technician to load the printer with paper and replace the toner. Report the faulty equipment to the technician for repair.

36.6 HANDLING OF FIBRE OPTIC CABLES

Fibre optic cables are made from a glass strand, covered with a polymer jacket. They are very thin and rigid, with sharp ends. Handle fibre optic cable with care during inspection, cleaning, and preparation of the fibre optic cable ends. Fibre optic cables should be cleaned using the cleanser recommended by the manufacturer. Follow the recommended procedure for each fibre optic cable type during cleaning, handling, assembling, packaging, and storage. When cleaving a fibre optic cable, the loose scrap material is hard to see and can be very dangerous. Dispose of loose scrap immediately in a properly designated container. Do not touch the end of a stripped fibre optic cable or a loose (scrap) piece of fibre. Fibre easily penetrates skin, and a fibre shard could break off. Do not rub your eyes when handling fibre optic cables; this would be extremely painful and requires immediate medical attention. Follow all safety procedures and regulations, and always wear the

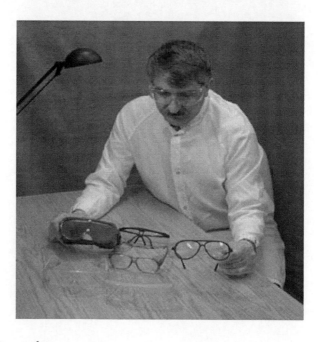

FIGURE 36.12 Safety goggles.

required personal protective safety equipment. Use safety goggles with side shields and wear protective rubber gloves or finger cots, when handling fibre optic cables. Figure 36.12 shows different types of safety goggles. Always treat fibre optic cables as a potential hazard. Never look directly at the fibre optic cable ends during fibre optic assembly and testing.

36.7 EPOXY ADHESIVES AND SEALANTS

Epoxy adhesives and sealants are essential components in the manufacturing of optical devices. There are different types and colours, depending upon the application. Epoxy adhesives come in several forms. One-part, two-part, and ultraviolet (UV) systems are the most common. A graded index (GRIN) lens can be glued to a beamsplitter with an epoxy. Sealant materials are used in the packaging of optical devices.

When using adhesives and sealant materials, be aware of their specifications. Specifications, applications, and handling procedures of these materials are found on the Material Safety Data Sheets (MSDS), which are available from the manufacturer or distributor. They may also be downloaded from a number of web sites. The adhesives and sealants are also very hazardous during storage, handling, and application. Prolonged or repeated exposure may cause eye or skin irritation. If contact does occur, wash the contact area immediately and seek medical help. Use safety goggles with side shields and wear protective rubber gloves or finger cots when handling adhesives and sealants. Follow all safety procedures and regulations, read the MSDS carefully, and wear the required personal protective safety equipment.

36.8 CLEANING OPTICAL COMPONENTS

Optical surfaces have to be clean and free of dust and other particles, which can range in size from tenths to hundreds of microns in diameter. Their comparative size means that they can cover a part of the optical surfaces, and thus degrade the reflection or transmission quality of the data transmission in telecommunication systems. There are many standard procedures for cleaning optical surfaces. Before starting any cleaning procedure, locate the following standard equipment:

1. Cleaning material (Denatured Ethanol)
2. Cotton swabs
3. Tissue
4. Safety goggles
5. Finger cots or rubber gloves
6. Compressed air
7. Disposal container
8. Microscope with a magnification range of about 50X
9. Infrared sensor card
10. Additional cleaning equipment:
    Ultrasonic bath
    Warm water and liquid soap
    Premoistened cleaning wipes
    Polymer film

Some optical components (e.g., lenses, mirrors, prisms, beamsplitters) have special coatings, such as antireflection coatings, that are sensitive to solvents, grease, liquid, and mechanical abrasion. Take extra care and choose appropriate cleaning liquid and swabs when cleaning optical components with these coatings. The following is the preferred cleaning procedure for optical components:

1. Wear rubber gloves and safety goggles.
2. Hold a lens or a mirror by the rim and a prism by the corners. Clean the optical component using a new dry or dampened swab with the recommended solvent. Rub the surfaces of the lens, using small circular movements, or one-directional movement on plane prism surfaces.
3. Blow away any remaining lint with compressed air. This step depends on the optical component size and surface conditions. Check the air quality from the compressor before using to clean optical components.

Some optical devices consisting of several optical components may not always be sealed completely. Therefore, use the recommended procedure to clean optical component surfaces without leaving any residue that could reduce the optical performance.

When cleaning any optical interface, disable all sources of power, such as the end of the ferrule on a fibre connector. Under no circumstances should you look into the end of an optical device in operation. Light from a laser device may not be visible, but it can seriously damage the human eye.

36.9 OPTIC/OPTICAL FIBRE DEVICES AND SYSTEMS

There has been a significant increase in the use of optic/optical fibre devices and systems. As optic/optical fibre devices become more common, it is important to understand the associated hazards. Optical devices typically use a laser as a light source. Not all lasers are created equal. They are classified based on their output wavelength and power. Since they operate over a wide range of wavelengths and power outputs, the hazards arising from their use vary substantially.

Lasers are classified into four classes. Laser sources conformant to Class 1 and Class 2 do not cause serous damage, but the use of eye protection should be taken into consideration. Class 3 and Class 4 lasers are powerful and can cause serious damage. Therefore, it is important to determine the class type of any optical equipment before working with it, assess the associated hazard, and comply with the safety requirements.

It is always a good practice to handle optical devices and measuring instruments with care. Normally, these devices and instruments are very expensive and sensitive, and they may present a potential hazard if not used properly. Follow the recommended procedures for each device or instrument to ensure proper handling during assembly, testing, packaging, and storage.

36.10 CLEANING CHEMICALS

Before the application of an epoxy or sealant, all surfaces should be treated using the recommended cleaning material. When using cleaning materials, be aware of appropriate precautions. Read all the information regarding cleaning materials in the MSDS. All types of cleaning materials are potentially hazardous; they may be flammable (even at low temperatures) and may pose other exposure risks. Use safety goggles with side shields and wear appropriate protective rubber gloves or finger cots. Follow all safety procedures and regulations. Use a ventilation hood when working with cleaning chemicals and epoxy adhesives, sealants, or any material producing fumes.

36.11 WARNING LABELS

There are various types of warning labels used in buildings, transportation, services, and industry to warn users about the level of danger ahead. Warning labels sometimes are called safety signs or safety messages. Safety signs clearly communicate by choosing the proper design and wording to suit safety needs. Standard signs, such as traffic warning signs and construction work labels, are available for general warnings.

Safety signs are divided into three general categories: danger, warning, and caution. They are also available in different sizes and colours, and with different graphics. Sometimes, a standard header can be used to create a new sign to suit a specific need. It is very important to use warning labels in laboratories to alert students to any source of danger. These dangers may come from devices, instruments, chemicals, lasers, sounds, vibrations, and biological hazards. Students should be introduced, in advance, to each source of danger in laboratories and be shown the required personal protective safety equipment. Everybody must remember to consider safety first.

36.12 LASER SAFETY

A laser beam is a parallel, narrow, coherent, and powerful light source. It is increasingly powerful when concentrated by a lens. It is a hazard to human eyes and skin even at very low power.

All lasers are classified based on their potential power. These classifications are from the American National Standards Institute (ANSI Standard Z136.1-1993) entitled American National Standard for Safe Use of Lasers, and Z136.3 (1996), American National Standard for Safe Use of Lasers in Health Care Facilities, the Canada Labor Code, and Occupational and Safety and Health Legislation (L-2-SOR/86-304).

Needing to be adhered to when using laser devices, these standards and codes are universally recognized as definitive documents for establishing an institution, such us a school, factory, or hospital. Their basic classification system has been adopted by every major national and international standards board, including the Center for Devices and Radiological Health (CDRH) in the U.S. Federal Laser Product Performance Standard, which governs the manufacture of lasers in the United States.

Lasers are typed into four clases, with some subclasses: Class 1, Class 2, Class 2a, Class 3a, Class 3b, and Class 4. Higher numbers reflect an increased potential to harm users. Figure 36.13 shows laser warning labels, which are required to identify hazard from laser light sources.

The following criteria are used to classify the hazard level of lasers:

1. Wavelength: If the laser is designed to emit multiple wavelengths, the classification is based on the most hazardous wavelength.
2. Continuous Wave: For continuous wave (CW) or repetitively pulsed lasers, the average power output (Watts) and limiting exposure time inherent in the design are considered.
3. Pulse: For pulsed lasers, the total energy per pulse (Joule), pulse duration, pulse repetition frequency, and emergent-beam radiant exposure are considered.

Details of the laser classifications are listed below:

Class 1 lasers are laser devices with very low output power (between 0.04 and 0.40 mW), and they operate in the lower part of the visible range ($450 \, nm < \lambda < 500 \, nm$). These lasers are generally considered to be safe when viewed indirectly. Some examples of Class 1 laser devices include CD players, scanners, laser pointers, and small measurement equipment. Figure 36.14(a) shows the human eye, while Figure 36.14(b) shows an eye cross-section. Laser light in the visible range entering the human eye is focused on the retina and causes damage. The most likely effect of intercepting a laser beam with the eye is a thermal burn, which destroys the retinal tissue. Never view any Class 1 laser beam directly.

Class 2 lasers are devices with low output power ($< 1 \, mW$ of visible CW), and operate in the visible range ($400 \, nm < \lambda < 700 \, nm$). This class of laser could cause eye damage, if the beam is directly viewed for a very short period of time (more than 0.25 s). Some examples of Class 2 lasers include classroom demonstration laser sources and laser-source devices for testing and telecommunications. Never view any Class 2 laser beam directly.

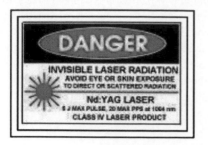

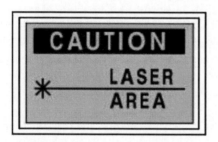

FIGURE 36.13 (See colour insert following page 512.) Laser warning labels.

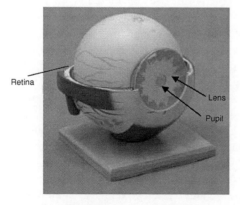

(a) Eye ball

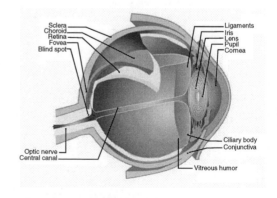

(b) Eye cross-section

FIGURE 36.14 The human eye.

Class 2a lasers are low-output power devices, which are considered to be visible-light lasers. This class of laser causes injury only when viewed directly for more than 0.25 second. This class must be designed so that intentional viewing of the laser beam is not anticipated. A supermarket bar-code scanner is a typical example of a Class 2a laser device. Never view any Class 2a laser beam directly.

Class 3 lasers are divided into two subgroups (Class 3a and Class 3b lasers).

Class 3a lasers are intermediate power devices; they are allowed to exceed the output power limit of Class 2 lasers by no more than a factor of five, or have visible light power less than 5 mW. They are considered CW lasers. Often they will have an expanded beam diameter so that no more than 1 mW can enter a fully dilated pupil, which is 7 mm in diameter. Some examples of Class 3a laser devices are laser scanners, laser printers, and laser-source devices for testing and telecommunications. Direct viewing of a laser in this class could be hazardous to the eyes. Never view any Class 3a laser beam directly. Although the beam wavelength may not be visible to the human eye, it can cause damage to the eye and skin. Laser safety goggles for appropriate wavelength are required when working with this class.

Class 3b lasers are intermediate power devices; they output between 5 and 500 mW of CW, or else pulsed 10 J/cm^2 power. They are considered to be CW lasers. Scattered energy (diffuse reflection) is not considered hazardous in most situations, unless the laser source is operating near its upper power limit and the diffuse target is viewed at close range. Some examples of Class 3b lasers are laser-source devices for testing. Never view any Class 3b laser beam directly or indirectly (by viewing any reflection from the surrounding surfaces). The laser beam wavelength may not be visible to the human eye, but it causes damage to the eye and skin immediately, with no time to react. Laser safety goggles for the appropriate wavelength are required when working with this class of laser.

Class 4 lasers are high-power devices; they output more than 500 mW of CW, or else pulsed 10 J/cm^2 power. They are considered to be very high-power lasers. Some applications of Class 4 laser devices include the following: surgery, drilling, cutting, welding, and micromachining. For the use of Class 4 lasers, all types of reflections (whether direct, specular, or diffuse) are extremely hazardous to the eyes and skin. Class 4 laser devices can also be a fire hazard. Much greater control is required to ensure the safe operation of this type of laser device. Never view any Class 4 laser beam directly or indirectly (any reflection by surrounding surfaces). Be cautious of this type of laser. The laser beam wavelength may not be visible to the human eye, but it can immediately cause damage to the eye and skin, with no time to react. Laser safety goggles for the appropriate wavelength are required when working with this class of laser.

Always follow all safety procedures and regulations, and wear the required the appropriate personal protective safety equipment when using lasers. Never look directly or indirectly at a laser beam. Each institute should create appropriate safety procedure to guide students and staff toward the creation of a safe working environment. Each laser laboratory has to be controlled by a designated instructor/professor certified in laser safety. All laser safety requirements should be implemented in a laser laboratory. It is recommended to have an introduction course and workshop in laser safety for each laser classification.

Knowing the classification of a particular device and comparing the information in Table 36.1 will usually eliminate the need to measure laser radiation or perform complex analyses of hazard potential.

36.13 LASER SAFETY TIPS

1. Do not enter the Nominal Hazard Zone (NHZ). This zone is established according to the procedures described in ANSI Z136.1-1993. Enter this area accompanied by a designated instructor/professor certified in laser safety. Do not put any body part or clothing in the way of a laser beam.

TABLE 36.1
Institutional Programme Requirements

Class	Power	Class Control Measures	Medical Surveillance	Safety & Training Programme
1	No more than 0.04–0.40 mW	Not applicable	Not applicable	Not required
2	Less than 1 mW of visible, continuous wave light	Applicable	Not applicable	Recommended
2a	Less than 1 mW of visible, continuous wave light	Applicable	Not applicable	Recommended
3a	From 1 to 5 mW of continuous wave light	Applicable	Not applicable	Required
3b	From 5 to 500 mW of continuous wave light	Applicable	Applicable	Required
4	More than 500 mW of continuous wave light	Applicable	Applicable	Required

2. Notice and comply with the signs and labels (shown in Figure 36.13) posted on laboratory door, devices, and equipment.
3. Wear the recommended eyewear and other protective equipment. Use laser safety goggles when you are in a laser laboratory, or in the vicinity of one, as shown in Figure 36.15.
4. Comply with the laser safety controls in the facility.
5. Attend laser safety training and workshops.
6. Update laser safety training and workshops, as needed.
7. While assembling and operating laser devices, it is important to remember that laser beams can cause severe eye damage. Keep your head well above the horizontal plane of the laser beams at all times. Use white index cards to locate beamspots along the various optical paths.
8. When moving optical components, mirrors, or metal tools through the laser beams, the beam may reflect laser light momentarily at your lab partner or you. If there is a possibility of an accidental reflection during a particular step in an operation, then temporarily block or attenuate the laser beam until all optical components are in their proper place.

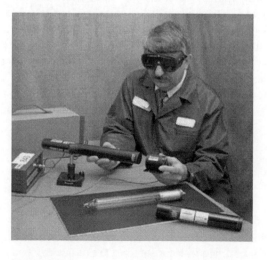

FIGURE 36.15 Wear laser safety goggles.

It is a good policy to be aware of any stray laser beam reflections, and to warn anybody of any danger. If you are unsure of how to proceed safely with a given step in the operation of the laser device, ask the professor/instructor for assistance.

36.14 INDOOR AIR QUALITY

Concerns with indoor air quality (IAQ) have increased since energy conservation measures were instituted in office buildings during the 1970s. These measures minimized the infiltration of outside air and contributed to the buildup of indoor air contaminants. IAQ generally refers to the quality of the air in a work environment. Other terms related to IAQ include indoor environmental quality (IEQ) and sick building syndrome. Complaints about IAQ range from simple complaints, such as the air smelling odd, to more complex situations, where the air quality causes illness and lost work time. It may not be easy to identify a single reason for IAQ complaints because of the number and variety of possible sources, causes, and varying individual sensitivities.

IAQ problems can be caused by ventilation system deficiencies, overcrowding, the presence of tobacco smoke, microbiological contamination, outside air pollutants, and off-gassing from materials in the building and mechanical equipment. Related problems may also include comfort problems caused by improper temperature and relative humidity conditions, poor lighting, and unacceptable noise levels, as well as adverse ergonomic conditions, and study-related psychosocial stressors. Typical symptoms may include headaches, unusual fatigue, itching or burning eyes, skin irritation, nasal congestion, dry or irritated throats, and nausea.

Ventilation is one of the most common engineering controls used to control emissions, exposures, and chemical hazards in the workplace. Other workplace environmental factors, including temperature, humidity, and odours, are also controlled with nonindustrial ventilation systems commonly known as heating, ventilating, and air-conditioning (HVAC) systems.

Management should have created guidelines for:

1. IAQ
2. Building air quality (BAQ)
3. Investigations, recommendations on sampling instrumentation and methods
4. Guidelines for management to prevent or alleviate
5. IAQ problems and take acute health effects of major indoor air contaminants.

Management should have an overview of:

1. Sources of indoor air pollution, and health problems
2. Ventilation, control, ventilation standards and building codes, and ventilation system problems
3. Solutions for air cleaners and resolving problems.

36.15 OTHER CONSIDERATIONS

These considerations apply to all students, staff, and management,

1. Laboratory injuries and illnesses are usually preventable by simply following safety precautions in school throughout the year.
2. Never overload circuits, power bars, or connectors.
3. Lead innovative and cooperative efforts to improve laboratory safety and health and the quality of student life.

FIGURE 36.16 A certified technical staff member.

4. Do not use or work with any device or equipment until it has been checked by qualified and authorized personnel in charge of the laboratory operation, as shown in Figure 36.16.
5. Everyone must wear personal protective equipment (e.g., safety goggles, protective gloves, ground connection, insulated tools) when working with electrical or laser equipment and chemicals.
6. Immediately report any damaged electrical or laser devices and equipment to the professor/instructor for immediate corrective action.
7. Staff should promote safety awareness among students.
8. Staff should teach safe work practices, at the beginning of each new laboratory session.
9. The NHZ should be established for each laser system.
10. Management should create and maintain a safe and healthy work, and study environment.
11. Management, staff, and students should understand the human and economic impact of poor safety and health in laboratories and classrooms.
12. Management should create a safety checklist, and maintenance and auditing programmes for each laboratory.
13. Eye protection should be worn at all times.
14. Eating, drinking, and smoking are not allowed in laboratories.
15. Unauthorized personnel should not be present in the laboratory or area, whether lasers are operating or not.
16. Laboratory coats must be worn when handling cleaning, corrosive, toxic, or flammable materials. Gloves should be worn when necessary, especially when handling corrosive and highly toxic materials.
17. Never work alone in a laboratory or workshop.
18. If a colleague is doing something dangerous, point the action out immediately and inform the supervisor.
19. Know where safety equipment (e.g., eyewash, shower, extinguishers, emergency exits, first aid kit) is located and how to use it.
20. Know where the MSDS and Workplace Hazardous Materials Information System (WHMIS) are located and how to use them.
21. Know where the emergency phones and alarms are located and how to use them.
22. Know how to clean up chemical spills using the appropriate agents.

23. Preplanned experiments and a properly organized work area can eliminate a lot of potential safety problems. Clean-up and decontamination must be routine parts of the experimental procedure for all students.
24. Wash your hands after handling chemicals and before leaving the laboratory.
25. Ensure the laboratory safety programme complements science.

FURTHER READING

Agrawal, G. P. and Dutta, N. K., *Semiconductor Lasers*, 2nd ed., Van Nostrand, New York, 1993.

Black, Eric, *An Introduction to Pound-Drever-Hall Laser Frequency Stabilization LIGO*, California Institute of Technology & Massachusetts Institute of Technology, California, Massachusetts, MA, U.S.A., 2000.

Canadian Health and Safety Legislation, Ecolog Canadian Health and Safety Legislation, Federal, Provincial, and Territorial Acts, Regulations, Guidelines, Codes, Objectives, Workers' Compensation, and WHMIS Legislation, 2000.

Charschan, S., *Lasers in Industry*, Van Nostrand, New York, 1972.

Cornsweet, T. N., *Visual Perception*, Academic Press, New York, 1970.

Davis, Christopher C., *Lasers and Electro-Optics, Fundamental and Engineering*, Cambridge University Press, New York, 1996.

Duarte, F. J. and Piper, J. A., Narrow-linewidth, high prf copper laser-pumped dye laser oscillators, *Appl. Opt.*, 23, 1391–1394, 1984.

Duarte, F. J., *Tunable Lasers Handbook*, Elsevier, Amsterdam, 1999.

Hood, D. C. and Finkelstein, M. A., Sensitivity to light handbook of perception and human performance, In *Sensory Processes and Perception*, Boff, K. R., Kaufman, L., and Thomas, J. P., Eds., Vol. 1, Wiley, Toronto, 1986.

Kuhn, K., *Laser Engineering*, Prentice Hall, Englewood Cliffs, NJ, 1998.

McComb, Gordon, *The laser Cookbook—88 Practical Projects*, McGraw-Hill, New York, 1988.

Nanni, C. A. and Alster, T. S., Laser-assisted hair removal: side effects of Q-switched Nd:YAG, long-pulsed ruby, and alexandrite lasers, *J. Am. Acad. Dermatology*, 2 (1), 165–171, 1999.

Nichols, Daniel R., *Physics for Technology with Applications in Industrial Control Electronics*, Prentice Hall, Englewood Cliffs, NJ, 2002.

Salah, B. E. A. and Teich, M. C., *Fundamentals of Photonics*, Wiley, New York, 1991.

SETON, Signs, labels, tags, and workplace safety, Catalog, 2006.

Tao, W. K. and Janis, R. R., *Mechanical and Electrical Systems in Buildings*, Prentice Hall, Englewood Cliffs, NJ, 2001.

Thompson, G. H. B., *Physics of Semiconductor Laser Device*, Wiley, Chichester, 1980.

Topping, A., Linge, C., Gault, D., Grobbelaar, A., and Sanders, R., A review of the ruby laser with reference to hair depilation, *Ann. Plast. Surg.*, 44, 668–674, 2000.

Venkat, Venkataramanan, *Introduction to Laser Safety*, Photonics Research, Ontario, Canada, 2002.

Yeh, Chai, *Handbook of Fiber Optics: Theory and Applications*, Academic Press, San Diego, 1990.

Section VII

Miscellaneous

Appendix A: Details of the Devices, Components, Tools, and Parts

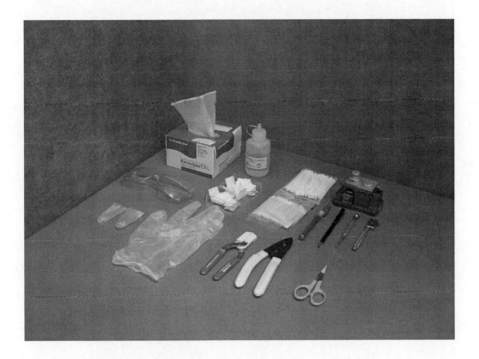

FIGURE A.1 Fibre cable end preparation kit.

FIGURE A.2 250 μm diameter/500 meters long fibre optic cable.

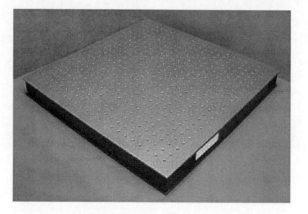

FIGURE A.3 2×2 ft breadboard.

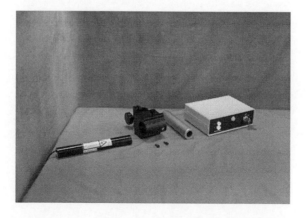

FIGURE A.4 HeNe laser source, laser power supply, and laser mount assembly.

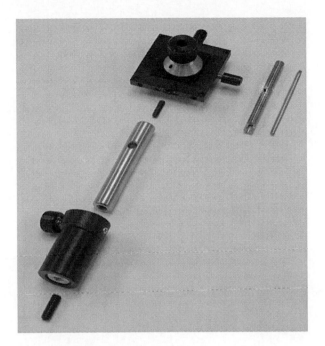

FIGURE A.5 Fibre optic cable holder/positioner assembly.

FIGURE A.6 Lens/fibre cable holder/positioner assembly.

FIGURE A.7 Lens and lens holder/positioner assembly.

FIGURE A.8 Fibre cable holders.

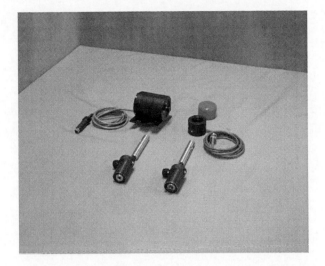

FIGURE A.9 Laser sensors.

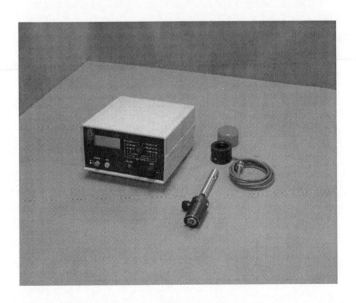

FIGURE A.10 Laser power meter with matching laser power detector.

FIGURE A.11 Laser power meter and laser power detectors.

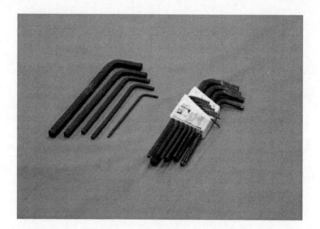

FIGURE A.12 Allen key set.

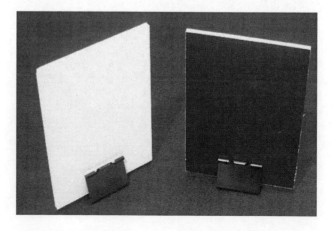

FIGURE A.13 Black/white card and cardholder.

FIGURE A.14 GRIN lens/fibre cable holder/positioners assembly.

FIGURE A.15 Rotation stage.

FIGURE A.16 Translation stage.

FIGURE A.17 XYZ translation stage.

FIGURE A.18 Multi-translation stage.

FIGURE A.19 Lab jack.

FIGURE A.20 HeNe laser clamp.

FIGURE A.21 Connector holder/positioner.

FIGURE A.22 Cube prism holder/positioner assembly.

FIGURE A.23 Convex lens and lens holder/positioner assembly.

FIGURE A.24 Prism and prism holder/positioner assembly.

(a)

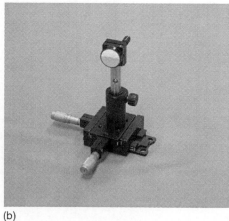

(b)

FIGURE A.25 (a) Mirror and mirror holder; (b) positioner assembly.

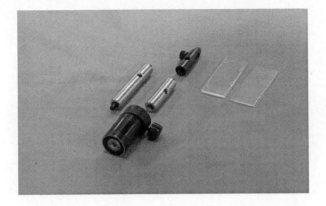

FIGURE A.26 Slide holder/positioner assembly.

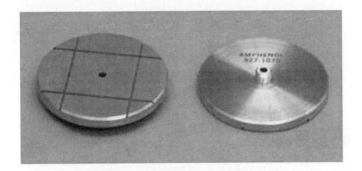

FIGURE A.27 Connector polishing disks (front and back sides).

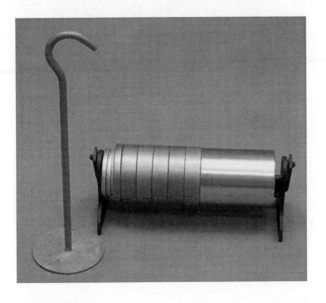

FIGURE A.28 Slotted gram weight sets.

FIGURE A.29 Pressure (squeeze) jig.

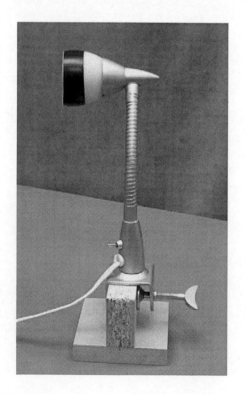

FIGURE A.30 Spot light source.

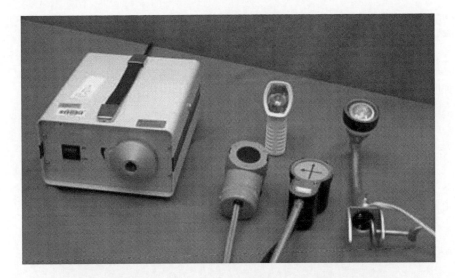

FIGURE A.31 Light sources.

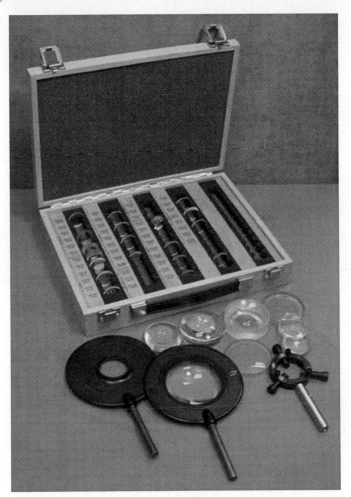

FIGURE A.32 Types of lenses.

FIGURE A.33 Types of prisms.

FIGURE A.34 Types of mirrors.

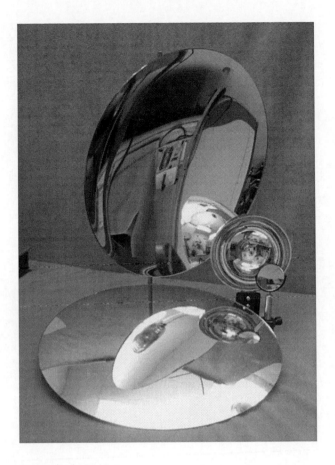

FIGURE A.35 Types of curved mirrors.

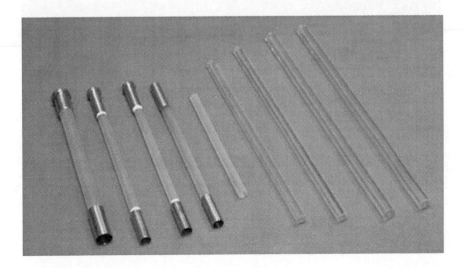

FIGURE A.36 Types of glass rods and tubes.

FIGURE A.37 Water tanks.

FIGURE A.38 Optical spectrum analyser.

Appendix B: Alignment Procedure of a Conventional Articulating Spectrometer

The following is the alignment procedure of a conventional articulating spectrometer using a prism to find the spectrum of a light source. The procedure is very sensitive and needs very fine alignment and focusing. Figure B.1 shows a conventional articulating spectrometer using a prism. The following steps walk you through the alignment of each part of the conventional articulating spectrometer. Figure B.2 shows the experiment set-up for spectrum test using the conventional articulating spectrometer.

APPENDIX B.1 TELESCOPE-COLLIMATOR COARSE ALIGNMENT

Orient the Collimator and the telescope so that they are collinear.
Tip: eye-ball the orientation first
Back-light the vertical slit in the Collimator
Tip: use the flashlight or table lamp
Adjust the slit width to be "narrow"
Align the vertical slit with the eyepiece cross-hair
Tip: adjust eyepiece in telescope to focus
Un-lock protractor ring and align with Vernier 0
Rotate the slit to a horizontal orientation

APPENDIX B.2 PRISM STAND ADJUSTMENT

Raise the prism stand to its maximum height
Orient the platform so that the prism holder is in the 10 O-clock position
Lower all stand adjustment screws so that the stand platform is flush

APPENDIX B.3 TELESCOPE ALIGNMENT

Place the mirror in the centre of the prism stand
Eyeball the placement to ensure that the mirror is facing the telescope
Adjust the eyepeice so that the eyepeice is ∼1 cm from the housing
Un-lock the Vernier dial
Carefully adjust the Vernier dial until the "green cross" is centred on the crosshair
Tip: use the platform adjustment screws to control the "green cross" vertically
Lock the Vernier dial

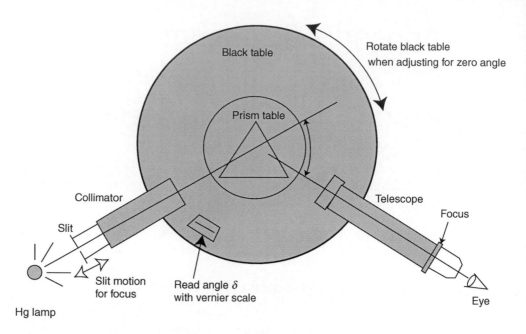

FIGURE B.1 A conventional articulating spectrometer.

APPENDIX B.4 COLLIMATOR-TELESCOPE FINE ALIGNMENT

Back-light the vertical slit in the collimator
Adjust the collimator vertically and horizontally to centre the slit on the crosshair
Tip: Use the adjustment screws on the collimator
Lock the protractor ring, the telescope and Vernier screws!

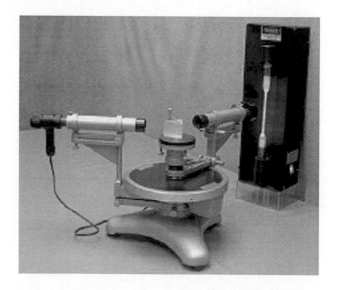

FIGURE B.2 Spectrum test using a conventional articulating spectrometer.

APPENDIX B.5 PRISM PLACEMENT

Place the prism on the centre of the stand
Tip: align the three corners of the prism with the scribed lines
Tip: ensure that the cloudy surface of the prism is facing the post holder
Lock the prism in position using the post holder
Place the mirror against the surface of the prism (facing the telescope)
Unlock the Vernier dial and adjust the mirror orientation
Tip: get the "green cross" re-centred
Lock the Vernier dial
Unlock the protractor ring
Align the protractor ring with the Vernier 0
Lock the protractor ring
This is the *Zero Reference Position* for the spectrometer

APPENDIX B.6 APEX ANGLE MEASUREMENT

The telescope should be aligned with the ejecting face of the prism
Move the mirror to the incident face of the prism
Tip: make sure that the mirror is flush
Unlock the telescope
Rotate the telescope to face the mirror
Tip: use the fine adjustment screw to precisely align the "green cross"
Record the displaced angle (this is the *Apex Angle* for the prism)
Tip: make 4 readings of the angle and determine the average

APPENDIX B.7 DEVIATION ANGLE MEASUREMENT

Return the spectrometer to the *Zero Reference Position*
Place a spectrum tube on top of the wooden box and in front of the slit
Tip: spectrum tubes get HOT…don't touch the tube!
The Zero Reference Position imposes an incident ray of 60 degrees
(a) Unlock the telescope and rotate to the spectrum lines
Align the crosshair with the left edge of the green spectrum line
Record the *incident angle* and the *deviation angle*
Return the telescope to the *Zero Reference Position*
Lock the telescope
Unlock the Vernier dial
Index the Vernier dial by rotating precisely 1 degree in the CCW direction
Hint: the new angle of incidence is now 59 degrees
Lock the Vernier Dial
Unlock the protractor ring and align with the Vernier 0
Lock the protractor ring
Go back to (a)
Repeat the above steps (from (a) on) until the incident angle is 40 degrees.

Appendix C: Lighting Lamps

APPENDIX C.1 INCANDESCENT LIGHT LAMPS

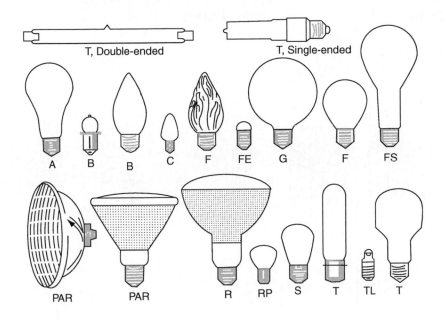

FIGURE C.1 Shape codes.

TABLE C.1
Shape Code and Application

	Shape Code	Application
A	Arbitrary (standard)	Universal use for home lighting
B	Bullet	Decorative
C	One shape	Used mostly for small appliances and indicator lamps
ER	Elliptical reflector	For substitution of incandescent R lamps
F	Flame	Decorative interior lighting
G	Globe	Ornamental lighting and some floodlights
P	Pear	Standard for street-railway and locomotive headlights
PAR	Parabolic aluminized	Used in spotlights and floodlights
S	Straight	Lower wattage lamps-sign and decorative
T	Tubular	Showcase and appliance lighting

APPENDIX C.1.1 LAMP DESIGNATION

For Example: 60A19, it means:
60: Wattage (60 W)
A: Bulb shape
19: Maximum bulb diameter, in eighths of an in.

APPENDIX C.2 TUNGSTEN HALOGEN LAMPS

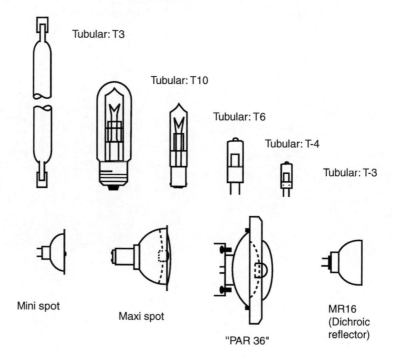

FIGURE C.2 Shape codes.

TABLE C.2
Shape Code and Type

Shape Code	Type
Tubular: T3	Line voltage tungsten halogen lamp-double ended
Tubular: T10	Line voltage tungsten halogen lamp-single ended
Tubular: T6	Line voltage tungsten halogen lamp-single ended
Tubular: T-4	Line voltage tungsten halogen lamp-without reflector
Tubular: T-3	Line voltage tungsten halogen lamp-without reflector
Maxi spot	Line voltage tungsten halogen lamp-with reflector
Mini spot	Line voltage tungsten halogen lamp-with reflector
PAR 36	Line voltage tungsten halogen lamp-PAR 36 reflector
MR 16	Line voltage tungsten halogen lamp-MR 16 reflector

APPENDIX C.3 FLUORESCENT LIGHT LAMPS

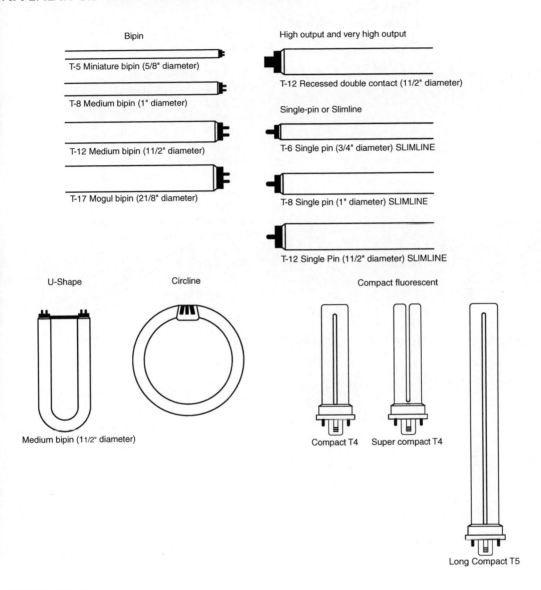

FIGURE C.3 Shape codes.

For Example: F40T12/CW, it means:
F: Fluorescent lamp
40: Wattage (40 W)
T: Tubular bulb shape
12: Maximum tube diameter — in eighths of an in. (12/8 = 1½ in.)
CW: Cool White Colour

TABLE C.3
Colour Code and Type

Colour Code	Type
C50	Chroma 50 (5000 K, CRI 90+)
C75	Chroma 75 (7500 K, CRI 90+)
CW	Cool white
CWX	Cool white deluxe
D	Daylight
LW	Lite white
LWX	Lite white deluxe
N	Normal
RW	Regal white
SP	Spectrum series
SPX	Spectrum series deluxe
WW	Warm white
WWX	Warm white deluxe
Deluxe	Means better CRI

APPENDIX C.4 MERCURY DISCHARGE LAMPS

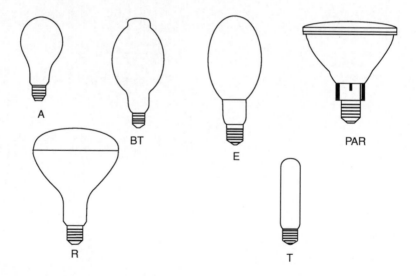

FIGURE C.4 Shape codes.

TABLE C.4
Shape Code and Type

Shape Code	Type
A	Arbitrary
BT	Bulged-tubular
E	Elliptical
PAR	Parabolic aluminized reflector
R	Reflector
T	Tubular

APPENDIX C.5 METAL HALIDE LAMPS

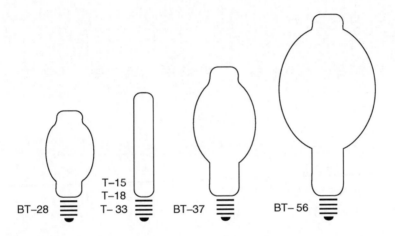

FIGURE C.5 Shape codes.

TABLE C.5
Shape Code and Type

Shape Code	Type
BT	Bulged-tubular
T	Tubular
Numbers	Indicate maximum diameter in eighths of an in.

APPENDIX C.6 HIGH-PRESSURE SODIUM LAMPS

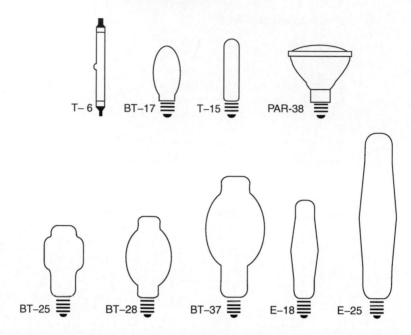

FIGURE C.6 Shape codes.

TABLE C.6
Shape Code and Type

Shape Code	Type
B	Bullet
BT	Bulged-tubular
E	Elliptical
PAR	Parabolic aluminized reflector
T	Tubular
Numbers	Indicate maximum diameter in eighths of an inch

APPENDIX C.7 COMPACT FLUORESCENT LAMPS

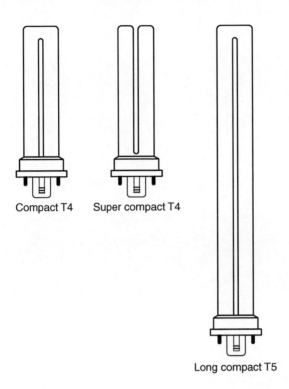

Compact T4   Super compact T4

Long compact T5

FIGURE C.7 Compact fluorescent lamps.

Appendix D: International System of Units (SI)

International System of Units (SI) (It is also called metric system). The modern form of the metric system, which has been developed by international standards. The SI is constructed from seven base units for independent physical quantities. The following tables showing these values are included below and are currently used worldwide (Table D.1 through Table D.6).

TABLE D.1
The Common Metric SI Prefixes

Multiplication Factor	Prefix Name	Prefix	Symbol
1 000 000 000 000 000 000 000 000	10^{24}	Yotta	Y
1 000 000 000 000 000 000 000	10^{21}	Zetta	Z
1 000 000 000 000 000 000	10^{18}	Exa	E
1 000 000 000 000 000	10^{15}	Peta	P
1 000 000 000 000	10^{12}	Tera	T
1 000 000 000	10^{9}	Giga	G
1 000 000	10^{6}	Mega	M
1 000	10^{3}	Kilo	k
100	10^{2}	Hecto	h
10	10^{1}	Deka	da
0.1	10^{-1}	Deci	d
0.01	10^{-2}	Centi	c
0.001	10^{-3}	Milli	m
0.000 001	10^{-6}	Micro	μ
0.000 000 001	10^{-9}	Nano	n
0.000 000 000 001	10^{-12}	Pico	p
0.000 000 000 000 001	10^{-15}	Femto	f
0.000 000 000 000 000 001	10^{-18}	Atto	a
0.000 000 000 000 000 000 001	10^{-21}	Zepto	z
0.000 000 000 000 000 000 000 001	10^{-24}	Yecto	y

TABLE D.2
Base Units

Quantity	Unit Name	Unit Symbol
Length	Metre	m
Mass	Kilogram	kg
Time	Second	s
Electric current	Ampere	A
Thermodynamic temperature	Kelvin	K
Amount of substance	Mole	mol
Luminous intensity	Candela	cd

TABLE D.3
SI Derived Units

Quantity	Unit Name	Unit Symbol	Expression in Terms of Other SI Units
Absorbed dose, specific energy imparted, kerma, absorbed dose index	Gray	Gy	J/kg
Activity (of a radionuclide)	Becquerel	Bq	1/s
Celsius temperature	Degree Celsius	°C	K
Dose equivalent	Sievert	Sv	J/kg
Electric capacitance	Farad	F	C/V
Electric charge, quantity of electricity	Coulomb	C	A s
Electric conductance	Siemens	S	A/V
Electric inductance	Henry	H	Wb/A
Electric potential, potential difference, electromotive force	Volt	V	W/A
Electric resistance	Ohm	Ω	V/A
Energy, work, quantity of heat	Joule	J	N m
Force	Newton	N	Kg m/s^2
Frequency (of a periodic phenomenon)	Hertz	Hz	1/s
Illuminance	Lux	Lx	lm/m^2
Luminous flux	Lumen	Lm	cd sr
Magnetic flux	Weber	Wb	V s
Magnetic flux density	Tesla	T	Wb/m^2
Plane angle	Radian	Rad	m/m
Power, radiant flux	Watt	W	J/s
Pressure, stress	Pascal	Pa	N/m^2
Solid angle	Steradian	Sr	m^2/m^2

Derived units are formed by combining base units and other derived units according to the algebraic relations linking the corresponding quantities. The symbols for derived units are obtained by means of the mathematical signs for multiplication, division, and use of exponents. Some derived SI units were given special names and symbols, as listed in this.

TABLE D.4
Conversion Factors from U.S. Customary Units to Metric Units

To Convert from	Multiply by	To Find
Inches	25.4	Millimetres
	2.54	Centimetres
Feet	30.48	Centimetres
Yards	0.91	Metres
Miles	1.61	Kilometres
Teaspoons	4.93	Millilitres
Tablespoons	14.79	Millilitres
Fluid ounces	29.57	Millilitres
Cups	0.24	Litres
Pints	0.47	Litres
Quarts	0.95	Litres
Gallons	3.79	Litres
Cubic feet	0.028	Cubic metres
Cubic yards	0.76	Cubic metres
Ounces	28.35	Grams
Pounds	0.45	Kilograms
Short tons (2000 lbs)	0.91	Metric tons
Square inches	6.45	Square centimetres
Square feet	0.09	Square metres
Square yards	0.84	Square metres
Square miles	2.6	Square kilometres
Acres	0.4	Hectares

TABLE D.5
Conversion Factors from Metric Units to U.S. Customary Units

To Convert from	Multiply by	To Find
Millimeters	0.04	Inches
Centimeters	0.39	Inches
Meters	3.28	Feet
	1.09	Yards
Kilometers	0.62	Miles
Milliliters	0.2	Teaspoons
Liters	0.06	Tablespoons
	0.03	Fluid ounces
	1.06	Quarts
	0.26	Gallons
	4.23	Cups
	2.12	Pints
Cubic meters	35.32	Cubic feet
	1.35	Cubic yards
Grams	0.035	Ounces

(continued)

Table D.5   *(Continued)*

To Convert from	Multiply by	To Find
Kilograms	2.21	Pounds
Metric ton (1000 kg)	1.1	Short ton
Square centimeters	0.16	Square inches
Square meters	1.2	Square yards
Square kilometers	0.39	Square miles
Hectares	2.47	Acres

Temperature conversion between Celsius and Fahrenheit $°C = (F - 32)/1.8$, $°F = (°C \times 1.8) + 32$

TABLE D.6
The Common Natural Temperatures

Condition	Fahrenheit (°)	Celsius (°)
Boiling point of water	212	100
A very hot day	104	40
Normal body temperature	98.6	37
A warm day	86	30
A mild day	68	20
A cool day	50	10
Freezing point of water	32	0
Lowest temperature Fahrenheit could obtain by mixing salt and ice	0	−17.8

Glossary

Aberration Distortion in an image produced by a lens or mirror caused by limitations inherent to some degree in all optical systems.

Absorption of Radiation The loss of light energy as it passes through a material. Loss is converted to other energy forms, which is usually heat (rise in temperature). The absorption process is dependent on the wavelength of the light and on the absorbing material.

Acceptance Angle The maximum angle over which the core of a fibre optic cable accepts incoming light. The angle is measured from the centreline of the core.

Active Medium Collection of atoms or molecules which can be stimulated to a population inversion, and emit electromagnetic radiation in a stimulated emission.

Adapter An adapter is a passive device used to connect two different connector types together.

Amplification The process in which the electromagnetic radiation inside the active medium within the laser optical cavity increase by the process of stimulated emission.

Amplitude The maximum value of a wave, measured from its equilibrium.

Angle of Incident (θ_i) The angle formed by an incident ray and the normal line to the optical surface at the point of incident.

Angle of Reflection (θ_{refl}) The angle formed by a reflected ray and the normal line to the optical surface at the point of reflection.

Angle of Refraction (θ_{refr}) The angle formed by a refracted ray and the normal line to the optical surface at the point of penetration. The ray is refracted (bent) while passing from one transparent medium to another having different refractive indices.

Angstrom A unit of measurement, equaling 10^{-10} m or 10^{-8} cm, usually used to express short wavelengths.

Aperture An adjustable opening in an instrument (like a camera) that controls the amount of light that can enter.

Attenuation The decrease in magnitude of the power of a signal in transmission media between points, normally measured in decibels (dB). Similarly, attenuation in a fibre optic cable is a measure of how of the signal injected into an optical fibre cable actually reaches the other end, usually expressed in decibels per kilometre (dB/km).

Backreflection Reflection of light in the direction opposite to that in which the light was originally propagating.

Bandwidth The measure of the information-carrying capacity of a fibre optic cable, normalized to a unit of MHz-km. This term is used to specify the capacity of a fibre cable.

Beam A bundle of light rays that are diverging, converging or parallel.

Beamsplitter An optical device that divides incident light into two components (magnetic and electric).

Beat-Length The length over which polarization rotates through 360° within an optical fibre and therefore a fundamental measure of the polarization maintaining ability of a polarization maintaining fibre.

bel (B) Unit of intensity of sound, named after Alexander Graham Bell. The threshold of hearing is 0 bel (10^{-12} W/m^2). The intensity is often measured in decibels (dB), which is a one-tenth of a bel.

Bend Radius   The amount that a fibre optic cable can bend before the risk of breakage or increase in attenuation. The minimum bend radius is dependent on the fibre optic cable diameter and type.

Bidirectional Device   An optical device which operates in both directions.

Binary Code   Code based on the binary number system (which uses a base of 2). In binary code any number can be expressed as a succession of ones and zeros. For example, the number 7 is 0111. These ones and zeros can then be interpreted and transmitted electronically as a series of "on" and "off" pulses, the basis for all computers and other digital equipment.

Birefringence   The fundamental principle by which polarization maintaining fibre works.

Birefringent   A birefringent material has distinct indices of refraction. The separation of light beam, as it passes through a calcite crystal object, into two diverging beams, commonly known as ordinary and extraordinary beams.

Bundle of Fibre Cables   A group of fibre cables packaged together in a unit.

Candle or Candela (cd)   A unit of luminous intensity.

Cascade   An arrangement of devices, each of which feeds into the next.

Chromatic Dispersion   A pulse-broadening and therefore bandwidth-limiting phenomenon which occurs because different wavelengths of light travel at different velocities.

Cladding   The layer of glass or other transparent material surrounding the light-carrying core of a fibre optic cable. The cladding has a lower refractive index than the core. The difference between the refractive indexes creates an interface that confines light propagation in the core.

Coating   One or more of thin layers of optical material applied to an optical surface to reduce reflection, create a mirror surface, absorb light or protect the surface.

Coherence   A property of electromagnetic waves which are in phase in both time and space. Coherent light has Monochromaticity and low beam divergence, and can be concentrated to high power densities. Coherence is needed for interference processes like holography.

Concave Grating Spectrometer   A spectrometer whose diffraction grating has a concave shape.

Concentricity   Core-cladding concentricity is the distance between the geometric centre of the core and the geometric centre of the cladding. Sometimes called eccentricity or concentricity error.

Concave Mirror   Mirror that curves inward like a "cave."

Collimate   To cause light rays to become parallel.

Connector   A device mounted on the end of a fibre optic cable that mates to a similar device to couple light into fibre cables. A connector joins a fibre cable end and a light source or detector in an optical device.

Continuous Wave (CW)   The output of a laser, which is operated, in a continuous rather than pulsed mode.

Control Area   It is an area in which the occupancy and activity of those present is subject to control and supervision for the purpose of protection from radiation, chemical, electrical, etc. hazards.

Conversion Efficiency   In an erbium doped fibre amplifier (EDFA), the ratio between the amplified signal output power and the power input from the pump laser.

Convex Lens   Curved outward. A lens with a surface shaped like the exterior surface of a sphere.

Convex Mirror   Mirror that curves outward. The virtual images formed is smaller and closer to the mirror than the object.

Core   The central part of a fibre optic cable, which conducts light. The core made from glass or plastic. It has a higher refractive index than the cladding.

Cornea   The transparent outer coat of the human eye, covering the iris and the crystalline lens. The cornea is the main refraction element of the eye.

Coupler   A device that connects two or more output ends, dividing one input between two or more outputs, or combining two or more inputs into one output. Couplers are used in telecommunication systems.

Coupling   Transfer of light into or out of a fibre optic cable. A coupling is used to launch light source into a fibre optic cable.

Cube Beamsplitter   Cube beamsplitters consist of matched pairs of right angle prisms cemented together along their hypotenuses.

Cut-Off Wavelength   The wavelength at which an optical fibre becomes single-moded. Below cut-off, the fibre will transmit more than one mode. Above cut-off the strength of the guidance is gradually reduced.

dB   Abbreviate for decibel. See bel.

Decibel (dB)   The standard unit used to express loss. Decibel is defined as ten times the base-ten logarithm of the ratio of the output signal to the input signal power.

Detector   A light-sensitive device that produces electrical signals when illuminated.

Diffraction Grating   A grooved optical element that has been deformed to reflect or transmit light of many colours. It acts like a prism to produce a spectrum.

Diffuse Reflection   The change of the spatial distribution of a beam of radiation when the beam is reflected in many directions by a rough surface or by a medium.

Diffusion   The flow of particles of a given species from high to low concentration regions by virtue of their random motions.

Dispersion   The separation of a light beams into its various wavelength components. All transparent materials have different indices of refraction for light of different wavelengths.

Distortion   An aberration in an optical element, which causes straight lines in the object, which are off the axis, to appear as, curved lines in the image.

EDFA   Erbium-doped fibre amplifier. A device incorporating erbium doped fibre to provide direct amplification of optical signals when pumped at either 980 or 1480 nm.

Electromagnetic Spectrum   See "Spectra". Also see graphic of spectrum.

Electron Volt (eV) Unit of energy   The amount of energy that the electron acquires while accelerating through a potential difference of 1 (V). 1 (eV) $= 1.6 \times 10^{-19}$ (J).

Excess Loss   Loss within a four directional coupler. It is defined as tcn times the base-ten logarithm of the ratio of the signal power between ports 2 and 3, and port 1.

Extinction Ration   In a polarization maintaining fibre, the ratio between the wanted and unwanted polarization states, expressed in decibels (dB). Highly dependent upon operating environment.

Ferrule   A part of a connector with a central hole, which contains and aligns a stripped fibre cable in a connector assembly.

Fibre Cable Bundle   A group of fibre cables packaged together in a unit.

Fibre Distributed Data Interface (FDDI)   A standard developed by ANSI for data transmissions through fibre optics systems. FDDI is capable of data transmissions of 100 Mb/s and above through a LAN incorporating 500 stations in a ring topology with a length of up to 100 km.

Fibre Grating   A selective reflector formed by inducing a periodic variation of refractive index within the core of an optical fibre.

Fibre Laser   A laser in which the gain-element is a length of rare-earth doped optical fibre.

Fibre Optic Cable   An optical cable used for light transmission in telecommunications. Fibre optic cables come in a great variety of configurations.

Focal Length (f)   The distance between the second principal plane or equivalent refracting plane of a lens and the lens focal point when the lens is imaging an object at infinity. In a positive lens, the focal length is measured on the side of the lens opposite to the object. In a negative lens, the focal length is measured on the same side as the object.

Focal Point   The point on the optical axis of a lens where light rays from a distant object point, will converge after being refracted by the lens.

Focus   The plane at which light rays from abject points form a sharp image after being refracted by a lens.

Fresnel Reflection   The reflection which occurs between parallel optical surfaces or at the interface where two materials having different refractive indices.

Fresnel Reflection Loss   Loss of signal power due to Fresnel reflection.

Fused Fibre   Two fibre cables are heated, placed under tension, and caused to create a taper coupler join.

Fusion Splice   A splice made by melting the ends of two fibre cables together by a spark shot so they form a permanent connection.

Gain   In an erbium doped fibre amplifier (EDFA), the ratio between the amplified signal output and the (un-amplified) signal input, expressed in dB.

Graded Index Fibre Cable   A fibre optic cable whose core has a nonuniform index of refraction. The core is comprised of concentric rings of glass whose refractive indices decrease radially from the centre of the core.

GRIN Lens   GRIN lenses an acronym for gradient index lenses. They have a cylindrical shape. One end is polished at an angle of 2, 6, 8, or 12°. The other end is polished at an angle of 2 or 90°.

Hertz (Hz)   The unit used to measure frequency. One Hertz equals one wave or cycle per second.

Homogeneous   A term used to describe any medium that is uniform in composition throughout.

Image   A likeness of an object formed by an optical element or system.

Image Distance   The distance between the equivalent refracting plane or second principal plane and the focal point measured on the optical axis.

Index of Refraction (*n*)   The ratio of the speed of light in a vacuum to the speed of light in a material.

Index-Matching Gel   A gel or fluid with a refractive index, which is matched to the refractive index of two fibre optic cores. It fills in the air gap between the fibre cable ends and reduces the Fresnel reflection, which occurs in the gap.

Intensity   The light energy per unit area.

Interferometer   The interferometer invented by the American physicist A. A. Michelson (1852–1931) is an ingenious device, which splits a light beam into two parts and then recombines them to form an interference pattern. The device can be used for obtaining accurate measurements of wavelength, precise length measurement, and to measure accuracy of an optical surface.

Jacket   A polymer (plastic, PVC, etc.) layer which covers the cladding/core layers of a fibre optic cable. The jacket has different colours and used as a mechanical protection layer.

Kevlar   A strong synthetic material used in strength members in a fibre optic cable. A pull wire can be fastened to Kevlar during fibre optic cable installation.

Laser   An acronym for "Light Amplification by the Stimulated Emission of Radiation." Lasers produce the coherent source of light for fibre optic telecommunication systems.

Laser Source   An instrument which produces monochromatic, coherent, collimated light.

Lens   One or more optical elements having flat or curved surfaces. If used to converge light rays, it is a positive lens; if used to diverge light rays, it is a negative lens. Usually made of optical glass, but may by molded from transparent plastic. Lenses are sometimes made from a natural or synthetic crystalline substance to transmit very short wavelengths (UV) or very long wavelengths (IR).

Light   The form of electromagnetic radiation with a wavelength range of ~400–700 nm. It generally travels in straight-line and exhibits the characteristics of both a wave and particle.

Lightguide   A fibre optic cable or fibre optic cable bundle.

Light Ray   The path of a single beam of light. In graphical ray tracing, a straight line represents the path along which the light travels.

Local Area Network   High-speed and high-capacity computer links used over relatively short distances (a few kilometres) to connect several buildings.

Loss (dB)   Attenuation of the power of a signal when it travels through an optical component. Normally measured in decibels (dB).

Lumen (lm)   A SI unit of luminous flux. One lumen is the luminous flux emitted per unit solid angle by a light source having an intensity of one candela (cd).

Luminous Intensity (I)   Luminous intensity measures the brightness of a light source. The unit of measure is the candle or candela (cd).

Lux (lx)   A unit of luminance equal to one lumen per square metre.

Mechanical Splice   A process whereby two optical fibres are joined together using mechanical means.

Metropolitan Area Networks (MAN)   An optical fibre or cable backbone interconnecting a number of LANs in a specific area.

Microbending   Tiny bends in a fibre optic cable, which allow light to leak-out the core and introduce loss.

Micrometre (μm)   One-millionth of a metre.

Microscope   An optical instrument used to inspect small objects, which provide power magnification.

Majority Carriers   Electrons in an n-type and holes in a p-type semiconductor.

Minority Carriers   Electrons in a p-type and holes in an n-type semiconductor.

Mirror   An optical element with a smooth, highly polished surface (plane or curve) for reflecting light. The reflecting surface is produced by a thin coating of gold, silver, or aluminium.

Mode   A term used to describe a light path(s) passing through a fibre optic cable as in single mode or multimode.

Mode Scramble   Mode scrambling is accomplished by bending a fibre optic cable in a corrugated. This causes light to leak out and for attenuation to increase.

Mode Scrambler   A device which bends a fibre optic cable to increase loss.

Monochromatic Light   Light is at one specific wavelength. The light out of a laser device is the monochromatic light.

Multimode Fibre   A fibre optic cable in which light travels in multiple modes.

Nanometre (nm)   One-billionth of a metre. The unit usually used in specifying the wavelength of light.

Noise   In an erbium doped fibre amplifier (EDFA), typically based on signal to noise ratio in a regime in which signal-spontaneous beat-noise and amplified signal shot-noise dominate.

Normal Line   A reference line constructed perpendicular to an optical surface.

Numerical Aperture (NA)   A measure of the divergence of the light emitted from the fibre, determined by the refractive index difference between the core and the cladding. The sine of the half the angle in which the core of a fibre optic cable can accept or transmit light.

Object   The source figure in by an optical system.

Object Distance   The distance from the first principal plane of a lens to the object.

Objective Lens   A lens that focuses light coming from an object. The objective lenses are used in microscopes, telescopes, etc.

Optical Axis   A straight line formed by joining the foci of lens.

Optical Coatings   Coatings specifically made for optical components (lenses, prisms, etc.) in light sensitive devices. There are many types of the coating materials. One coating helps to protect the optical components from scratches and wear. Some optical components are coated with antireflective (AR) layer(s) to reduce back reflection.

Optical Path   The sum of the optical distances along a specified light ray.

Optical Pumping   The excitation of the active medium in a laser by the application of light, rather than electrical discharge. Light can be from a conventional source like Xenon or Krypton lamp, or from another laser.

Optical Radiation   Ultraviolet, visible and infrared spectrum (0.35–1.4 μm) that falls in the region of transmittance of the human eye.

Optical Resonator   The mirrors (or reflectors) making up the laser cavity including the laser rod or tube. The mirrors reflect light back and forth to build up amplification.

Optical Surface   The reflecting or refracting surface of an optical element.

Phase   The position of a wave in its oscillation cycle.

Photon   A particle or packet of radiant electromagnetic energy representing a quanta of light.

Photonics   The field of science and engineering encompassing the physical phenomena and associated with the generation, transmission, manipulation, detection, and utilization of light.

Plastic Fibre Cable   A fibre optic cable having a plastic core and plastic cladding. Plastic fibre cables are typically used in applications where sensitivity and loss are not important.

Plenum   The air space between walls, under structural floors, and above drop ceilings, which can be used to route interconnection cabling in a building.

Polarization   Alignment of the electric and magnetic fields which comprise and electromagnetic wave. If all light waves from a source have the same alignment, then the light is said to be polarized.

Population Inversion   An excited state of matter, in which more atoms (or molecules) are in upper state than in a lower one. This is a required situation for a laser action.

Prism   An optical element, which is used to change the direction and orientation of a light beam. A prism has polished faces, which are used to transmit and reflect light.

Proximity Sensor   A device which senses distance from a reflecting surface.

Pulsed Laser   Laser, which delivers energy in the form of a single or sequence of laser pulses.

PVC(Polyvinyl chloride)   A material used in the manufacture of fibre optic cables jackets.

Quantum Efficiency   In the erbium doped fibre amplifier (EDFA), the actual conversion efficiency, expressed as a percentage of the maximum possible conversion efficiency (equal to the ratio of the pump and signal wavelengths).

Ray   Straight lines which represent the path of a light ray.

Rayleigh Scattering   The scattering of light, which results from small impurities in a material or composition.

Recombination   Recombination of an electron hole pair involves an electron in the conduction band (CB) falling in energy down into an empty state (hole) in the valence band (VB) to occupy it. The result is the annihilation of the EHP.

Recombination Current   It flows under forward bias to replenish the carriers recombining in the space charge (depletion) layer.

Rectangular Beamsplitters   Three prisms carefully cemented together along their hypotenuses. Polarization beamsplitters are used in optical devices where the output components are required to exit from the opposite side to the input signal. It also produces a lateral displacement between the two output components.

Reflection   The change in direction of a light ray when bounces off of a reflecting surface.

Refraction   The bending of a light ray as it passes from one transparent medium to another of different refractive index.

Refractive Index (n)   The ratio of the speed of light in a vacuum to the speed of light in a specific material.

Ribbon Fibre Cables   Cables in which many optical fibres are embedded in a plastic flat ribbon-like structure.

Right Angle Prism   A prism whose cross-section is a right angle triangle with two 45-degree interior angles. The prism faces, which are at right angles, are transmitting surfaces, while the hypotenuse face is a reflecting surface.

Saturation   Erbium doped fibre amplifier (EDFA), performance under conditions of total population inversion, which occur at high input powers.

Scattering   Loss of light due to the presence of atoms in a transparent material.

Sensor   A device which, responds to the presence of energy.

Sheath   An outer protective layer of a fibre optic cable. Fibre optic cables having an outer protective layer are suitable for indoor and outdoor cable installations.

Simplex Fibre Optic Cable   A term sometimes used to describe a single fibre optic cable.

Single Mode Fibre   A fibre optic cable in which the signal travels in one mode.

Snell's law   Describes the path that a light ray takes as it goes from one optical medium to another. It is also called Law of Refraction.

Secular Reflection   Several rays of a beam of light incident on a smooth, mirror-like, reflecting surface where the reflected rays are all parallel to each other.

Solar Flux   The amount of light from the sun.

Solid Angle   The ratio of the area on the surface of a sphere to the square of the radius of that sphere. It is expressed in steradians (sr).

Spectra, Spectrum   Spectra is the plural of spectrum, which is a series of energies (like light) arranged according to wavelength, or frequency. The electromagnetic spectrum is an array of radiation that is divided into a number of sub-portions, where the boundaries are only vaguely defined. They extend from the shortest cosmic rays, through gamma rays, x-rays, ultraviolet light, visible light, infrared radiation, microwave and all other wavelengths of radio energy.

Spectrograph   A spectroscope that measures wavelengths of light (spectra) and then displays the data as a graph. UVIS is an *imaging* spectrograph, which means it can also display the points of the graph as a picture (see Imaging Sepctrosopy).

Spectroscope   A machine (instrument) for producing and observing spectra.

Spectroscopic Measurements   The measurements taken by a spectrograph.

Spectrometer   A spectroscope equipped with the ability to measure wavelengths.

Speed of Light   In vacuum, approximately 3×10^8 m/s.

Splice   A method for joining the ends of two fibre optic cables. There are two primary methods for splicing fibre optic cables: fusion and mechanical.

Spontaneous Emission   Random emission of a photon by decay of an excited state to a lower level. Determined by the lifetime of the excited state.

Spot Size   A measure of the diameter of the beam of laser radiation.

Stimulated Emission   Coherent emission of radiation, stimulated by a photon absorbed by an atom (or molecule) in its excited state.

Strength Member   That part of a fibre optic cable composed of Kevlar armed yarn, steel strands, or fibreglass filaments. Strength members increase the tensile strength of a cable.

Switch   A device, which regulates or directs a signal in telecommunication systems.

Transverse Mode   The geometry of the power distribution in a cross section of a laser beam.

Transparent   The adjective used to describe a medium through which light can pass in a percentage.

Transverse Electro-Magnetic (TEM) Mode   Used to designate the shape of a cross section of a laser beam.

Total Internal Reflection   Total internal reflection of light occurs when light rays in a high-index medium exceed the critical angle (to the normal to a surface). This is the principal theory for explaining how light travels in the core of a fibre optic cable.

Ultraviolet Light(UV) (Extreme Ultraviolet and Far Ultraviolet)   A portion of the complete electromagnetic spectrum Ultraviolet is a portion of the spectrum that is a shorter wavelength than visible light; roughly, with a wavelength interval from 100 to 4000 Å. Ultraviolet radiation from the sun is responsible for many complex photochemical reactions like the formation of the ozone layer. Extreme and far ultraviolet wavelengths are different portions of the ultraviolet portion of the spectrum, with extreme being between 55.8 and 118 nm and far being between 110 and 190 nm.

V-Value   Also called normalized frequency. The fundamental relationship between numerical aperture, cut-off wavelength and core diameter.

Visible Light   Electromagnetic radiation, which is visible to the human eye. It has a wavelength range between 400 and 700 nm.

Wave   One complete cycle of a signal with a fixed period.

Waveguide   A structure, which guides an electromagnetic wave along its length. A fibre optic cable is an example of optical waveguide.

Wavelength (λ)   The period of a wave. Distance between successive crests, troughs, or identical parts of a wave.

Wavelength Division Multiplexing (**WDM**)   The process whereby multiple optical carriers of different wavelengths utilize the same optical fibre cable.

Index